APPLIED SURFACE THERMODYNAMICS

Second Edition

SURFACTANT SCIENCE SERIES

FOUNDING EDITOR

MARTIN J. SCHICK
1918–1998

SERIES EDITOR

ARTHUR T. HUBBARD
Santa Barbara Science Project
Santa Barbara, California

1. Nonionic Surfactants, *edited by Martin J. Schick* (see also Volumes 19, 23, and 60)
2. Solvent Properties of Surfactant Solutions, *edited by Kozo Shinoda* (see Volume 55)
3. Surfactant Biodegradation, *R. D. Swisher* (see Volume 18)
4. Cationic Surfactants, *edited by Eric Jungermann* (see also Volumes 34, 37, and 53)
5. Detergency: Theory and Test Methods (in three parts), *edited by W. G. Cutler and R. C. Davis* (see also Volume 20)
6. Emulsions and Emulsion Technology (in three parts), *edited by Kenneth J. Lissant*
7. Anionic Surfactants (in two parts), *edited by Warner M. Linfield* (see Volume 56)
8. Anionic Surfactants: Chemical Analysis, *edited by John Cross*
9. Stabilization of Colloidal Dispersions by Polymer Adsorption, *Tatsuo Sato and Richard Ruch*
10. Anionic Surfactants: Biochemistry, Toxicology, Dermatology, *edited by Christian Gloxhuber* (see Volume 43)
11. Anionic Surfactants: Physical Chemistry of Surfactant Action, *edited by E. H. Lucassen-Reynders*
12. Amphoteric Surfactants, *edited by B. R. Bluestein and Clifford L. Hilton* (see Volume 59)
13. Demulsification: Industrial Applications, *Kenneth J. Lissant*
14. Surfactants in Textile Processing, *Arved Datyner*
15. Electrical Phenomena at Interfaces: Fundamentals, Measurements, and Applications, *edited by Ayao Kitahara and Akira Watanabe*
16. Surfactants in Cosmetics, edited by Martin M. Rieger (see Volume 68)
17. Interfacial Phenomena: Equilibrium and Dynamic Effects, *Clarence A. Miller and P. Neogi*
18. Surfactant Biodegradation: Second Edition, Revised and Expanded, *R. D. Swisher*
19. Nonionic Surfactants: Chemical Analysis, *edited by John Cross*
20. Detergency: Theory and Technology, *edited by W. Gale Cutler and Erik Kissa*
21. Interfacial Phenomena in Apolar Media, *edited by Hans-Friedrich Eicke and Geoffrey D. Parfitt*

APPLIED SURFACE THERMODYNAMICS

Second Edition

Edited by

A. Wilhelm Neumann
University of Toronto
Toronto, Ontario, Canada

Robert David
University of Toronto
Toronto, Ontario, Canada

Yi Zuo
University of Hawaii at Manoa
Honolulu, Hawaii, U.S.A.

CRC Press
Taylor & Francis Group
Boca Raton London New York

CRC Press is an imprint of the
Taylor & Francis Group, an **Informa** business

CRC Press
Taylor & Francis Group
6000 Broken Sound Parkway NW, Suite 300
Boca Raton, FL 33487-2742

First issued in paperback 2017

ISBN-13: 978-0-8493-9687-8 (hbk)
ISBN-13: 978-1-138-11637-5 (pbk)

Library of Congress Cataloging-in-Publication Data

Applied surface thermodynamics. -- 2nd ed. / editors, A.W. Neumann, Robert David, Yi Zuo.
    p. cm. -- (Surfactant science series ; v. 151)
Includes bibliographical references and index.
ISBN 978-0-8493-9687-8 (hardcover : alk. paper)
    1. Surfaces (Physics) 2. Thermodynamics. 3. Surface chemistry. 4. Surface tension. 5. Contact angle. I. Neumann, A. W. (A. Wilhelm), 1933- II. David, Robert, 1977- III. Zuo, Yi, 1976- IV. Title. V. Series.

QC173.4.S94A67 2011
541'.33--dc22
                                                           2010028241

Visit the Taylor & Francis Web site at
http://www.taylorandfrancis.com

and the CRC Press Web site at
http://www.crcpress.com

To the memory of Ladislav Boruvka

Contents

Preface

This book deals with different aspects of surface science and its applications. While the chapters are freestanding, there is a logical progression: knowledge of previous chapters will enhance the understanding of subsequent ones.

This second edition is both an extension and reorganization of the material in the first edition. Major additions include the recent progress in axisymmetric drop shape analysis (ADSA, Chapter 3); a new chapter on image processing methods for drop shape analysis (Chapter 4); a new chapter on advanced applications and generalizations of ADSA (Chapter 5); recent studies of contact angle hysteresis (Chapter 7); a new chapter on contact angles on inert fluoropolymers (Chapter 8); and an updated presentation of line tension and the drop size dependence of contact angles (Chapter 13).

Philosophically, the book is firmly anchored in Gibbsian thermodynamic thinking. The first two chapters generalize Gibbs classical theory of capillarity including discussions of highly curved interfaces. The next three chapters discuss liquid–fluid interfacial tension and its measurement, using drop shape techniques. Chapters 6 through 9 are contact angle chapters, dealing with experimental procedures, thermodynamic models, and the interpretation of contact angles in terms of solid surface tension. Chapter 10 discusses theoretical approaches to determine solid surface tension, whereas Chapters 11 and 12 deal with interfacial tensions of particles and their manifestations. Finally, Chapter 13 contains material on drop size dependence of contact angles and line tension.

Chapter 1 presents a generalized theory of capillarity. The approach is entirely Gibbsian; however, it differs from the classical theory in that it is not restricted to moderately curved liquid–fluid interfaces. It also includes line phases in addition to surface and volume phases. While the mathematical complexities of the minimization of the resulting free energy function are considerable (but not presented here), we believe that this generalized theory is, conceptually, easier to grasp than the classical theory. The second half of this chapter presents simple derivations of the generalized Laplace equation of capillarity, as well as a verification of the fundamental equation for surfaces from hydrostatic considerations. The chapter concludes by demonstrating that the appropriate free energy function for capillary systems is the grand canonical potential, and that the Gibbs dividing surface can be shifted freely within the framework of the generalized theory, but not the classical theory.

Chapter 2 deals exclusively with axisymmetric liquid–fluid interfaces. While it is more restrictive geometrically than Chapter 1, it is more general in that it does, among other things, consider compressible as well as incompressible interfaces. Its importance may well lie in such insights as the interplay between line tension and compressibility in surfactant monolayers.

The next three chapters deal with the determination of liquid–fluid interfacial tension, primarily via drop shape methods.

Chapter 3 describes recent advances in a methodology called ADSA. The methodology is based on numerical integration of the Laplace equation of capillarity and computer-based matching of experimental drop profiles to theoretical ones, with the interfacial tension as one of the parameters in the optimization scheme. The methodology is applicable to sessile and pendant drops/bubbles alike. Illustrations of its applicability include, among others, pressure dependence of interfacial tension, ultra-low liquid–liquid interfacial tensions, and the use of ADSA as a film balance.

Chapter 4 delves further into image processing techniques, which are a critical component of drop shape methods for automated measurement of interfacial properties with high accuracy. Advanced methods are presented for overcoming noise encountered in images of biological samples, for example, lung surfactant preparations and bacterial cell lines.

Chapter 5 discusses three advanced applications of ADSA in depth, and presents an alternative drop shape algorithm originating in ADSA. The first application concerns the use of ADSA to study lung surfactant, a phospholipid–protein complex that lowers surface tension in the lung, allowing normal breathing. The second application is the development of ADSA as a fully functional miniature Langmuir film balance with the capacity of subphase replacement. In the third application, ADSA is extended to analyze the shapes of liquid drops subjected to external electric fields. Finally, an alternative algorithm called theoretical image fitting analysis (TIFA) is presented. TIFA measures interfacial properties from whole drop images without the use of edge detection to extract liquid–fluid profiles.

Chapters 6 through 10 discuss measurement of contact angles and their use for the determination of interfacial tensions involving a solid phase.

Chapter 6 offers a general guide to many of the techniques used to measure contact angles. It focuses on two versions of the drop shape method ADSA, namely ADSA-Profile and ADSA-Diameter. Also provided in this chapter are protocols for the careful preparation and handling of solid surfaces and liquids for conducting contact angle measurements.

Chapter 7 deals with conceptual aspects of contact angles on imperfect solid surfaces. The main topics include thermodynamic models of contact lines on chemically heterogeneous and rough solid surfaces. Experimental studies of contact angle hysteresis are presented, emphasizing its link not only to heterogeneity and roughness, but also to sorption. Contact angles on surfaces covered by a thin liquid film are considered as well.

Chapter 8 presents a detailed study of contact angles of various liquids on certain well-characterized polymer surfaces. Contact angles of different liquids on the same low-energy solid surface follow a pattern. Minor deviations from this pattern are interpreted in terms of specific solid–liquid interactions.

Chapter 9 presents two approaches for measuring solid surface tensions from contact angle data—surface tension components and an equation of state—with the focus on the latter. The formulation of an equation of state is described and independent experimental data using a variety of methods are used to examine each approach. The chapter concludes by describing a method for predicting solid surface tensions from molecular properties.

Chapter 10 discusses approaches for measuring solid surface tensions that rely on the shapes of solid–liquid solidification fronts and use the Gibbs–Thomson equation. Gradient theory and Lifshitz theory are used to estimate solid–liquid interfacial tensions. These results strongly favor the contact angle approaches of Chapter 9. A critique of the applicability of the Gibbs–Thomson equation to solidification fronts is presented.

Chapters 11 and 12 concern the thermodynamic study of particles at interfaces.

Chapter 11 deals with wetting phenomena of particles, starting with a discussion of approaches to determine contact angles and surface tensions of particles. It is concluded that direct approaches often do not yield useful information; therefore, indirect approaches, such as the sedimentation volume technique and the capillary penetration method, are considered in detail.

Chapter 12 deals exclusively with the behavior of small particles at solidification fronts. In the first instance the method is an excellent tool for testing surface energetic theories such as those considered in Chapter 9. In later sections the experiment is developed into a method to determine the surface tensions of the particles. Finally, experiments are described that allow determination of the actual force between polymeric particles and the solidification front, in which the microscopic observations of engulfment or rejection of a particle by the solidification front are interpreted using van der Waals interactions at the interface.

Chapter 13 revisits a topic of Chapters 1 and 2, line tension, and examines its possible link to the drop size dependence of contact angles. Theoretical approaches for estimating line tension are reviewed. The literature data for contact angles are then surveyed and their implications for the value of line tension discussed.

As a final note, we would like to acknowledge the extensive work of Zdenka Policova during the preparation of this book. Her contributions ranged from secretarial to organizational to providing cohesion between the authors. We would also like to thank Professor Daniel Kwok for his efforts in generating the outline of the book. In addition, we thank Tatjana Ljaskevic, Ting Zhou, and Regina Park for their help with formatting text and reproducing figures.

Editors

A. Wilhelm Neumann is a professor emeritus in the Department of Mechanical and Industrial Engineering at the University of Toronto. He received his PhD in chemical physics from the University of Mainz in 1962.

Robert David is a postdoctoral fellow in the Departments of Mechanical and Industrial Engineering and Cell and Systems Biology at the University of Toronto. He received his PhD in mechanical engineering from MIT in 2006.

Yi Zuo is an assistant professor in the Department of Mechanical Engineering at the University of Hawaii at Manoa. He received his PhD in mechanical and industrial engineering from the University of Toronto in 2006.

Contributors

Arash Bateni
Teradata Corporation
Toronto, Ontario, Canada

Ladislav Boruvka (Deceased)
Department of Mechanical and
    Industrial Engineering
University of Toronto
Toronto, Ontario, Canada

M. Guadalupe Cabezas
Department of Mechanical, Energy,
    and Materials Engineering
University of Extremadura
Badajoz, Spain

Miguel Cabrerizo-Vílchez
Applied Physics Department
University of Granada
Granada, Spain

Pu Chen
Department of Chemical Engineering
University of Waterloo
Waterloo, Ontario, Canada

Robert David
Department of Mechanical and
    Industrial Engineering
University of Toronto
Toronto, Ontario, Canada

John Gaydos
Department of Mechanical and
    Aerospace Engineering
Carleton University
Ottawa, Ontario, Canada

Mina Hoorfar
School of Engineering
University of British Columbia
    Okanagan
Kelowna, British Columbia, Canada

Daniel Kwok
Department of Mechanical and
    Manufacturing Engineering
University of Calgary
Calgary, Alberta, Canada

Dongqing Li
Department of Mechanical and
    Mechatronics Engineering
University of Waterloo
Waterloo, Ontario, Canada

Julia Maldonado-Valderrama
Institute of Food Research
Norwich Research Park
Norwich, United Kingdom

Elio Moy
Ashland-Tech Inc.
Scarborough, Ontario, Canada

A. Wilhelm Neumann
Department of Mechanical and
    Industrial Engineering
University of Toronto
Toronto, Ontario, Canada

Yehuda Rotenberg
Rotech Development Corporation
Integral Design Associates
Toronto, Ontario, Canada

Jan Spelt
Department of Mechanical and
    Industrial Engineering
University of Toronto
Toronto, Ontario, Canada

Hossein Tavana
Department of Biomedical
    Engineering
University of Michigan
Ann Arbor, Michigan

Junfeng Zhang
School of Engineering
Laurentian University
Sudbury, Ontario, Canada

Yi Zuo
Department of Mechanical
    Engineering
University of Hawaii at Manoa
Honolulu, Hawaii

1 Outline of the Generalized Theory of Capillarity

John Gaydos, Yehuda Rotenberg, Pu Chen,
Ladislav Boruvka, and A. Wilhelm Neumann

CONTENTS

1.1 OUTLINE OF THE GENERALIZED THEORY OF CAPILLARITY

1.1.1 INTRODUCTION

It is the purpose of this chapter to give an introduction to the macroscopic thermodynamics of interfaces, sometimes called the theory of capillarity. The entire foundation of classical thermodynamics in general and surface thermodynamics in particular was laid by Gibbs [1–3], who created a "pure statics of the effects of temperature and heat" [4]. The development given here is Gibbsian in the sense that it is based on his concept of the dividing surface; however, the treatment of curved surfaces and three-phase lines is more general [5] than Gibbs treatment and, we believe, easier to grasp than Gibbs moderate curvature approximation.

The chapter consists of two parts. The first part is a concise outline of the generalized theory of capillarity. The emphasis is on establishing the proper fundamental equation for curved interfaces and the derivation of the generalized mechanical equilibrium conditions. The second part discusses applications, implications, and corollaries of the theory presented in the first part.

1.1.2 THE FUNDAMENTAL EQUATION FOR BULK PHASES

The relation postulated by Josiah Willard Gibbs between the internal energy of a simple thermodynamic bulk system $U^{(V)}$ and the various modes of energy transfer between the system and the surroundings is called the fundamental equation; allowing for heat transfer, mechanical work, and chemical work, we may characterize the change in the internal energy $dU^{(V)}$ of the system as

$$dU^{(V)} = TdS^{(V)} - PdV + \sum_j \mu_j dM_j^{(V)}, \tag{1.1}$$

where T denotes the bulk temperature, $S^{(V)}$ is the total entropy, P is the pressure, V is the volume, μ_j is the chemical potential of component j, and $M_j^{(V)}$ represent the total mass of component j within the bulk system. Quite often the chemical potentials, μ_j, are expressed in terms of energy per unit mole, N_j; however, in this case we shall define the chemical potential as an energy per unit mass, M_j. These quantities, N_j and M_j, represent the total number of moles and total mass in the system of component j, respectively. The superscript (V) denotes the fact that all quantities in Equation 1.1 are volume or bulk phase quantities. Equation 1.1 also implies that there is a functional relationship between the internal energy and the other quantities given above, which may be written as

$$U^{(V)} = U^{(V)} \left[S^{(V)}, V, M_j^{(V)} \right]. \tag{1.2}$$

Thus, for a simple thermodynamic system the internal energy is completely specified by the independent extensive properties given in Equation 1.2. Associated intensive parameters (i.e., parameters independent of the size of the system), namely, the temperature T, the pressure P, and the chemical potentials μ_j for $j = 1,2,...,r$ are defined from the fundamental equation by differentiating Equation 1.2 with respect to $S^{(V)}$, V, and, $M_j^{(V)}$ to obtain the usual definitions of the intensive quantities for a homogeneous phase, or

$$T = \left(\frac{\partial U^{(V)}}{\partial S^{(V)}} \right)_{V, \{ M_j^{(V)} \}} \tag{1.3}$$

$$-P = \left(\frac{\partial U^{(V)}}{\partial V} \right)_{S^{(V)},\left\{ M_j^{(V)} \right\}}, \tag{1.4}$$

and

$$\mu_j = \left(\frac{\partial U^{(V)}}{\partial M_j^{(V)}} \right)_{S^{(V)},V,\left\{ M_{k \neq j}^{(V)} \right\}} \quad \text{for } j = 1,2,\dots,r \text{ chemical components,} \tag{1.5}$$

where the curly brackets { } indicate that all component masses are held constant. The other familiar forms of the fundamental equation are expressed either as a variational relation (e.g., see Equation 1.1), or as the integrated relation

$$U^{(V)} = TS^{(V)} - PV + \sum_j \mu_j M_j^{(V)}, \tag{1.6}$$

which is known as the Euler equation. Finally, an isolated, composite, simple thermodynamic system is governed by a minimum principle. At equilibrium, the internal energy of the composite system is less than it is for any other thermodynamic state having the same overall entropy, volume, and chemical components. The mathematical formulation of this principle leads to equality of temperature, pressure, and chemical potentials throughout the composite system at equilibrium. The reader requiring more background on these fundamental matters may wish to refer to Callen [6] or any one of a number of other excellent texts [7–20].

The above development assumed tacitly that the physical properties are homogeneous throughout the simple system. When a fluid phase is placed in an external force field, a redistribution of matter takes place and the bulk phase becomes heterogeneous, although it remains macroscopically continuous and isotropic [21,22]. Therefore, the thermodynamic parameters of a bulk phase inside a finite volume will generally depend on the external field and the shape of the volume. The fundamental Equation 1.2 for a bulk phase may then be considered either in terms of the extensive quantities inside an infinitesimal volume as done by Gibbs or, equivalently, in terms of volume densities of these parameters. The densities may be defined either as quantities per unit mass or as quantities per unit volume. Both definitions are used in thermodynamics; however, when physical properties like mass vary with location as would occur for a gas or fluid in a gravitational field our preference is to define the density as a quantity per unit volume. This latter approach is consistent with the definition taken in fluid mechanics [13,21–23]. It is this latter form of the fundamental equation that may be applied to both homogeneous and heterogeneous bulk phases to determine all equilibrium thermodynamic properties or tendencies of a bulk phase. Thus, the fundamental equation per unit volume is written as

$$u^{(v)} = u^{(v)} \left[s^{(v)}, \rho_j^{(v)} \right], \tag{1.7}$$

where $u^{(v)} = U^{(V)}/V$, $s^{(v)} = S^{(V)}/V$, $\rho_j^{(v)} = M_j^{(V)}/V$, and the specific volume superscript (v) is introduced at this point to indicate those quantities whose definition is dependent upon the existence of a well-defined volume element. The symbol $\rho_j^{(v)}$ denotes the r component densities present in the bulk phase. The intensive parameters (i.e., the temperature T, the pressure P, and the chemical potentials μ_j for $j = 1,2,\ldots,r$) are now defined from the fundamental Equation 1.7 as

$$T = \left(\frac{\partial u^{(v)}}{\partial s^{(v)}} \right)_{\{\rho_j^{(v)}\}} \tag{1.8}$$

$$\mu_j = \left(\frac{\partial u^{(v)}}{\partial \rho_j^{(v)}} \right)_{s^{(v)}, \{\rho_{k \neq j}^{(v)}\}} \quad \text{for } j = 1, 2, \ldots, r \text{ chemical components,} \tag{1.9}$$

and

$$P = Ts^{(v)} + \sum_j \mu_j M_j^{(v)} - u^{(v)}, \tag{1.10}$$

where Equation 1.10 is identical to the (integrated) Euler relation. The fundamental equation may be written in the differential form as

$$du^{(v)} = Tds^{(v)} + \sum_j \mu_j dM_j^{(v)}. \tag{1.11}$$

The extensive quantities for a finite volume V are calculated from their respective volume densities by volume integrals; that is

$$U^{(V)} = \iiint_V u^{(v)} dV, \tag{1.12}$$

$$S^{(V)} = \iiint_V s^{(v)} dV, \tag{1.13}$$

$$M_j^{(V)} = \iiint_V \rho_j^{(v)} dV \quad \text{for } j = 1, 2, \ldots, r. \tag{1.14}$$

In the absence of external force fields the densities $u^{(v)}$, $s^{(v)}$, and $\rho_j^{(v)}$ are constant at every location throughout the entire bulk phase at equilibrium [i.e., the bulk phase is homogeneous only when $\nabla \varphi(\vec{r}) = 0$ where $\varphi(\vec{r})$ represents the potential of the external field given as a function of position]. In this case, we have the simple identities

$$U^{(V)} = Vu^{(v)}, \, S^{(V)} = Vs^{(v)}, \, M_j^{(V)} = V\rho_j^{(v)} \quad \text{for } j = 1, 2,\ldots, r. \tag{1.15}$$

Otherwise, the total energy of a bulk phase consists of both the internal energy and the potential energy associated with the external field. If the volume density of the total energy is denoted as $e_t^{(v)}$ and the volume density of the total mass of the phase as $\rho_t^{(v)}$, then

$$e_t^{(v)} = u^{(v)} + \rho_t^{(v)}\varphi, \tag{1.16}$$

where

$$\rho_t^{(v)} = \sum_j \rho_j^{(v)}, \tag{1.17}$$

and the total quantities are determined from the integral relations given in Equations 1.12 through 1.14.

1.1.3 GENERALIZATION OF THE CLASSICAL THERMODYNAMICS OF SURFACES

In this section, we shall demonstrate how the Gibbsian formalism of bulk thermodynamics can be extended to the surface phase.

A molecule in the neighborhood of the interface between any two bulk phases will experience a different environment than if that same molecule were deep within the bulk of a similar bulk phase. Consequently, the density of the various components and their energy and entropy densities in the neighborhood of the interface will be different from the corresponding densities in the bulk phases. Furthermore, the time-averaged densities will most likely change continuously from one bulk phase to another and not in a discontinuous or stepwise fashion. Thus, for an equilibrium system that is subjected to external body and/or surface forces and that is sufficiently small (i.e., the boundaries are not infinitely removed) so that the interfacial regions need consideration one finds that $u^{(v)}$, $s^{(v)}$, and $\rho_j^{(v)}$ will be functions of position that will vary slowly through each bulk phase due to the influence of gravity or other external body forces and will, in general, vary rapidly across each interface. Figure 1.1 presents a simple model of the time-averaged density $\rho^{(v)}$ of a liquid in equilibrium with its saturated vapor across the liquid–vapor interface. In most cases of interest, the influence of the interface is limited and does not extend beyond several molecular diameters (about $10^{-9}–10^{-8}$ metres) into the bulk phases. Therefore, the bulk phases forming the interface may in most cases be considered as extending uniformly right up to the interface. The most noticeable exceptions occur with liquid–vapor interfaces near the critical point of the fluid. To reiterate, the interface formed between two bulk phases is in reality a thin region in which the physical

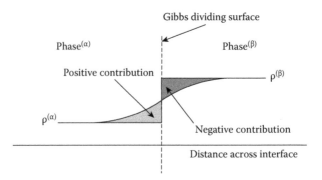

FIGURE 1.1 Schematic of the time-averaged variation of the density across a deformable liquid-fluid interface showing the placement of the dividing surface and the regions of the density profile that contribute to the final value for the surface density.

properties (like the density) vary rapidly and continuously from the bulk proper-ties of one phase to the bulk properties of the other phase. However, because this interface region is very thin, it may be considered to a first approximation, as done by Gibbs [1,2,24], as a mathematical, two-dimensional boundary between two bulk phases that extend uniformly right up to the mathematical *dividing* surface (shown as a dashed line in Figure 1.1). From the Gibbsian viewpoint, the dividing surface or surface of discontinuity is a mathematically constructed surface of only two dimen-sions that is "sensibly" placed within the thin interface region to separate the bulk regions that make contact when forming the interface. Bulk properties in each phase are assumed to persist uniformly right up to the dividing surface and the excess properties, formed as a result of this assumed model for the interface, are attributed to the dividing surface. Therefore, the dividing surface that is initially constructed as a geometrical surface of bulk separation may be transformed into a thermodynamic, autonomous system governed by a suitable fundamental equation for the interface, which is dependent only on excess or surface quantities [25]. Surface or *excess* quan-tities like the surface internal energy are defined as

$$U^{(A)} = U^{(V)}_{total} - U^{(\alpha)} - U^{(\beta)}, \tag{1.18}$$

where the quantities denoted by the superscripts (α) and (β) are those for the hypo-thetical bulk systems (α) and (β) that extend right up to the dividing surface. The variable $U^{(A)}$ represents the difference in energy between the real system, of total energy $U^{(V)}_{total}$, with an interface of nonzero volume and a hypothetical system of energy $U^{(\alpha)} + U^{(\beta)}$ in which the bulk phases (α) and (β) are completely uniform right up to the dividing surface. Thus, $U^{(A)}$ represents the excess energy arising from the Gibbsian model of an interface or equivalently it represents the total internal energy of the surface phase. In the same sense $S^{(A)}$ is the corresponding total surface entropy for an interface formed by bulk phases (α) and (β). Finally, $M^{(A)}_j$ is the total surface mass of the *j*th component. The superscript (A) is used to denote a surface or two-dimensional phase variable. It should also be noted that in the process of introducing a dividing surface to model the interface region one ensures that the total energy, total entropy, and total mass of the system do not change (i.e., they are all conserved).

The fundamental equation for a planar interface without external field influence is readily established by considering modes in which the internal energy $U^{(A)}$ of the interface can be changed. There is, just as in the case of the bulk phases, the possibility of transfer of mass and heat into and out of the interface. Instead of the volume, as in the case of the bulk phase, we are now concerned with the interfacial area, A. The work done in generating an interfacial area increment dA is γdA where γ is the interfacial tension. Overall we obtain

$$dU^{(A)} = TdS^{(A)} + \gamma dA + \sum_j \mu_j dM_j^{(A)}, \tag{1.19}$$

so that the fundamental equation for the total surface may be expressed as

$$U^{(A)} = U^{(A)}\left[S^{(A)}, A, M_j^{(A)}\right]. \tag{1.20}$$

The interfacial tension, defined as

$$\gamma = \left(\frac{\partial U^{(A)}}{\partial A}\right)_{S^{(A)}, \{M_j^{(A)}\}}, \tag{1.21}$$

is the two dimensional counterpart of the three dimensional bulk pressure P. Both quantities are isotropic (same magnitude in all directions), but the surface or interfacial tension is tensile rather than compressive like the pressure. The term γdA represents the work required to change the area of the planar dividing surface just as the term $-PdV$ represents the work required to increase the volume of the system. Therefore, Equation 1.19 holds for all positions and variations of a planar dividing surface.

In the presence of an external field we require that the arguments of the surface internal energy density $u^{(a)}$ be all surface densities of extensive surface; that is, of the dividing surface quantities. Ultimately, this means that we assume that it is physically meaningful to discuss surface densities defined at a point on the dividing surface. Thus, surface densities defined at a point in the dividing surface will be considered in exactly the same manner as volume densities defined at a point in the bulk fluid. We notice that, as far as the geometric variables are concerned, the fundamental equation for bulk phases is complete since a volume region has no extensive geometric variables (besides its volume), and hence no geometric point-variables upon which the volume densities $u^{(v)}$ could be assumed to depend. Likewise, we require that a fundamental equation for surfaces be complete as far as the geometric variables are concerned. Other than $u^{(a)}$ and $\rho_j^{(a)}$ where $j = 1, 2, \ldots, r$, the only other required variables are geometric ones; any other properties would have to be considered also in the fundamental equation for bulk phases and the resulting theory would be more general (e.g., electrocapillarity) than presently desired.

For a planar dividing surface we can see that a surface domain in two-dimensional space (analogous to a volume region in three-dimensional space) has no extensive geometric properties other than its surface area. Therefore, the complete fundamental equation for planar surfaces is identical to the one suggested by Gibbs more than a century ago, namely

$$u^{(a)} = u^{(a)} \left[s^{(a)}, \rho_j^{(a)} \right]. \tag{1.22}$$

The corresponding extensive or total quantities are defined in analogy with the bulk phase definitions provided by Equations 1.12 through 1.14; however, all integrations are carried out over the surface instead of the volume. Next, we consider the case of a nonplanar, curved surface.

For a nonplanar surface there are geometric quantities in addition to the surface area A that need to be considered. Physically, one needs to realize that work can be performed on the system by bending the interface (i.e., by changing its curvature). The best known quantities for describing the curvature of a surface at a point are the so called principal (orthogonal) curvatures $c_1 = 1/R_1$ and $c_2 = 1/R_2$, where R_1 and R_2 are the principal radii of curvature [26,27]. If one were to enter the analysis with the principal radii of curvature, the problem would quickly become intractable, because c_1 and c_2 are not differential invariants. Therefore, it is desirable to replace c_1 and c_2 by equivalent curvature related quantities that are invariant. With these considerations in mind, the simplest geometric parameters that possess the desired characteristics are the first (mean) curvature J and the second (Gaussian) curvature K defined by

$$J = c_1 + c_2 \quad \text{and} \quad K = c_1 c_2. \tag{1.23}$$

Using J and K as the two scalar differential invariants of the surface, permits one to write the generalized fundamental equation in the energetic density form as

$$u^{(a)} = u^{(a)} \left[s^{(a)}, \rho_j^{(a)}, J, K \right]. \tag{1.24}$$

In analogy with bulk phases we introduce, in addition to temperature and chemical potentials, three intensive parameters for the dividing surface as

$$C_J = \left(\frac{\partial u^{(a)}}{\partial J} \right)_{s^{(a)}, \left\{ \rho_j^{(a)} \right\}, K}, \tag{1.25}$$

$$C_K = \left(\frac{\partial u^{(a)}}{\partial K} \right)_{s^{(a)}, \left\{ \rho_j^{(a)} \right\}, J}, \tag{1.26}$$

and

$$\gamma = u^{(a)} - Ts^{(a)} - \sum_{j} \mu_j \rho_j^{(a)} - C_J J - C_K K, \tag{1.27}$$

where T is the surface temperature, μ_j is the surface chemical potential of component j, C_J, and C_K are certain mechanical or curvature potentials that may be called the first and second bending moments of the surface and γ (the surface analogy of $-P$) is the surface tension. The differential form of Equation 1.24 can be written as

$$du^{(a)} = Tds^{(a)} + \sum_{j} \mu_j d\rho_j^{(a)} + C_J dJ + C_K dK, \tag{1.28}$$

and the integrated (Euler relation) form is given by Equation 1.27. The corresponding surface density of the total energy of the dividing surface is given by

$$e_t^{(a)} = u^{(a)} + \rho^{(a)} \varphi, \tag{1.29}$$

where

$$\rho^{(a)} = \sum_{j} \rho_j^{(a)} \tag{1.30}$$

is the surface density of the total mass in the surface phase.

The analogous extensive parameters for a finite area A of a surface phase are calculated from their respective surface densities by surface integrals. The corresponding extensive (because of their explicit definition in terms of the surface area) curvature terms are given by

$$\mathsf{J} = \iint J dA, \tag{1.31}$$

and

$$\mathsf{K} = \iint K dA. \tag{1.32}$$

The quantities J and K although less often encountered than J and K, have been discussed previously in the differential geometry literature [27]. Specifically, we are assuming that the curvature densities indicate that the surface is of sufficient continuous and local uniformity that a continuum definition for the curvature is possible. This is analogous to assuming that the bulk system is sufficiently near a continuum that one may define a localized density $\rho^{(v)}$ everywhere within the bulk system. A discussion of the physical applicability of such a *point-thermodynamic* approximation may be found in Rowlinson [28].

For a homogeneous portion of a dividing surface, with surface area A and constant curvatures, the extensive parameters are expressed as

$$U^{(A)} = Au^{(a)}, \quad S^{(A)} = As^{(a)}, \quad M_j^{(A)} = A\rho_j^{(a)} \quad \text{for } j = 1, 2,..., r \tag{1.33}$$

and

$$J = JA, \quad K = KA. \tag{1.34}$$

The fundamental equation for a homogeneous dividing surface expressed in terms of the extensive quantities is then given by

$$U^{(A)} = U^{(A)}[S^{(A)}, A, M_1^{(A)}, M_2^{(A)},..., M_r^{(A)}, \mathrm{J}, \mathrm{K}], \tag{1.35}$$

and its differential and integrated forms by

$$dU^{(A)} = TdS^{(A)} + \gamma dA + \sum_j \mu_j dM_j^{(A)} + C_J d\mathrm{J} + C_K d\mathrm{K}, \tag{1.36}$$

and

$$U^{(A)} = TS^{(A)} + \gamma A + \sum_j \mu_j M_j^{(A)} + C_J \mathrm{J} + C_K \mathrm{K}. \tag{1.37}$$

Equations 1.35 through 1.37 are the surface analogs of the fundamental equations for bulk phases (i.e., Equations 1.2, 1.1, and 1.6, respectively).

1.1.4 EXTENSION TO THREE-PHASE LINEAR SYSTEMS

The extrapolation of Gibbsian thermodynamics to composite systems with linear phases was alluded to by Gibbs in a footnote when he stated [29]:

> We may here remark that a nearer approximation to the theory of equilibrium and stability might be obtained by taking special account, in our general equations, of the lines in which surfaces of discontinuity meet. These lines might be treated in a manner entirely analogous to that in which we have treated surfaces of discontinuity. We might recognize linear densities of energy, of entropy, and of the several substances which occur about the line, also a certain linear tension.

Following this line of thought we extrapolate the properties of the dividing surfaces by assuming that the fundamental equation for each surface holds at every point of the surface right up to the dividing line. Thus, the actual parameters of internal energy, entropy, and masses in the vicinity of the dividing line are represented partially by the extrapolated volume densities, partially by the extrapolated surface densities, and partially by whatever is remaining (i.e., the excess) that is attributed to the dividing line in the form of linear densities $u^{(l)}$, $s^{(l)}$, and $\rho_j^{(l)}$ for $j = 1, 2, \dots r$. In order to set up a general fundamental equation for dividing lines, additional geometric variables must be considered. These additional geometric variables must be:

1. Scalar geometric parameters
2. Linear densities of extensive properties just as $u^{(l)}$, $s^{(l)}$, and $\rho_j^{(l)}$
3. The lowest order scalar differential invariants of a line on a surface or of a line as an intersection of surfaces

There are two types of geometric quantities relevant for the description of three-phase lines: contact angles and linear curvature terms. There are, at any point on a three-phase contact line, at least three contact angles that can be defined. We denote these angles as

$$\theta_{jk} \qquad \text{where } (jk) = (12), (23), (31). \tag{1.38}$$

It is also well known from differential geometry that for a line in three-dimensional space, the curvature κ and the torsion τ, specified at each point on the line, are the only two geometric parameters that are needed to describe the line in space. Despite their close analogy with the surface parameters J and K, the curvature of the dividing line cannot be characterized exclusively by κ and τ because they do not possess a relation to the dividing surfaces as required by the third condition. Therefore, as thermodynamic subsystems, the dividing lines are nonautonomous to a greater extent than the dividing surfaces. A set of curvature terms that satisfies all three requirements is

$$\kappa_{nj}, \kappa_{gj}, \tau_{gj} \qquad \text{for } j = 1, 2, 3, \tag{1.39}$$

where κ_{nj}, κ_{gj}, and τ_{gj} are the normal curvature, the geodesic curvature, and the geodesic torsion of the dividing line at each point along the length of the line [26,27], relative to the jth dividing surface. This set of variables, listed in Equations 1.38 and 1.39, is highly symmetrical at the expense of not being entirely independent since the variables are related by the relations

$$\sum_{(jk)} \theta_{jk} = 2\pi, \tag{1.40}$$

or, for a three-phase system, by

$$\sum_{(jk)}^{(12),(23),(31)} \theta_{jk} = \theta_{12} + \theta_{23} + \theta_{31} = 2\pi, \tag{1.41}$$

and

$$\kappa_{nj}^2 + \kappa_{gj}^2 = \kappa^2, \qquad (1.42)$$

or, for a three-phase system, by

$$\kappa_{n1}^2 + \kappa_{g1}^2 = \kappa_{n2}^2 + \kappa_{g2}^2 = \kappa_{n3}^2 + \kappa_{g3}^2 = \kappa^2. \qquad (1.43)$$

Thus, by analogy with the fundamental equations for bulk and surface phases, the general fundamental equation for dividing lines in the density formalism is

$$u^{(l)} = u^{(l)}\left[s^{(l)}, \rho_j^{(l)}, \theta_{jk}, \kappa_{nj}, \kappa_{gj}, \tau_{gj}\right]. \qquad (1.44)$$

These parameters characterize the way in which the specific energy $u^{(l)}$ depends on the shape or curvature of the dividing line in three dimensions, as defined by the curvatures κ_{nj}, κ_{gj}, and torsion τ_{gj} of the dividing line at each point along the length of the line, and by the angles θ_{jk} that are formed by the jth and kth adjacent dividing surfaces at the point in which they intersect to form the dividing line. In analogy with Equation 1.27, which defines the surface tension γ, we are led to the definition of line tension σ as

$$\sigma = u^{(l)} - Ts^{(l)} - \sum_j \mu_j \rho_j^{(l)} - \sum_{(jk)} C_{\theta_{jk}} \theta_{jk} - \sum_j \left(C_{nj}\kappa_{nj} + C_{gj}\kappa_{gj} + C_{\tau j}\tau_{gj}\right), \qquad (1.45)$$

where $C_{\theta_{jk}}$, C_{nj}, C_{gj}, and $C_{\tau j}$ are linear mechanical potentials or bending moments of the equilibrium dividing line, entirely analogous to the bending moments introduced in Equations 1.25 and 1.26; and σ (the linear analog of the surface tension γ) is the line tension. The differential and integrated forms of the linear fundamental equation are expressed as

$$du^{(l)} = Tds^{(l)} + \sum_j \mu_j d\rho_j^{(l)} + \sum_{(jk)} C_{\theta_{jk}} d\theta_{jk} + \sum_j \left(C_{nj}d\kappa_{nj} + C_{gj}d\kappa_{gj} + C_{\tau j}d\tau_{gj}\right) \qquad (1.46)$$

and by Equation 1.45, which is the linear version of the Euler relation. The summations with indices j and (jk) are over all the dividing surfaces and over all pairs of adjacent dividing surfaces (nonrepeated summation) that are connected to a particular dividing line, respectively.

For a homogeneous segment of dividing line (with completely uniform linear densities) of length L, the fundamental equation in terms of extensive parameters can be expressed as

$$U^{(L)} = U^{(L)}\left[S^{(L)}, L, M_j^{(L)}, \Theta_{jk}, K_{nj}, K_{gj}, T_{gj},\right]. \qquad (1.47)$$

The quantities Θ_{jk}, K_{nj}, K_{gj}, and T_{gj} are introduced, in general, as differentially invariant, extensive parameters defined by

$$\Theta_{jk} = \int_L \theta_{jk} dL, \tag{1.48}$$

$$K_{nj} = \int_L \kappa_{nj} dL, \tag{1.49}$$

$$K_{gj} = \int_L \kappa_{gj} dL, \tag{1.50}$$

$$T_{gj} = \int_L \tau_{gj} dL. \tag{1.51}$$

The differential and integrated forms of $U^{(L)}$ may be written down in a similar manner to the expressions given previously for $u^{(l)}$ in the density formalism.

In summary, the fundamental equation for dividing lines, as given by Equation 1.44, defines the energy of the dividing line in terms of the "linear densities" to which Gibbs alluded and in terms of suitable differentially invariant, intensive parameters that uniquely define the curvature and the contact angle configuration of the line at every point in three dimensions.

In addition to the confluent zones represented by dividing surfaces (interfaces) and the ones represented by dividing lines, a fluid system may contain yet another dimensional class of confluent zone, namely that represented by dividing points. When several dividing lines intersect, the common point of intersection may be described in terms of excess properties of the point in analogy with the excess property descriptions of both dividing surfaces and lines [5].

1.1.5 MECHANICAL EQUILIBRIUM CONDITIONS

The descriptive formalism of equilibrium thermodynamics for fluid systems is based on the fundamental equations described above and a minimum principle. We shall consider a multicomponent, multiphase fluid system governed by the above fundamental equations (i.e., Equations 1.7, 1.24, and 1.44). Any particular configuration of the total system in which the thermodynamic parameters are distributed in compliance with the fundamental equations and also in compliance with the constraints on and within the system is called a possible state of the system. In our case, this means that we maintain the total entropy and the total mass of each component in the system as a constant. Therefore, the fundamental equations determine and describe the thermodynamic states in all parts of the fluid system, while the minimum principle is a necessary condition that allows determination of the equilibrium states from the multitude of thermodynamic states allowed by the governing fundamental equations. Mathematically, the thermal, chemical, and mechanical equilibrium conditions are obtained through application of the calculus of variations.

Gibbs applied the criterion necessary for equilibrium of a volume region to the internal portion of a fluid system with the condition of isolation imposed by enclosing the internal portion of the composite system with an imaginary envelope or bounding wall [30]. Following this approach, we may write the necessary condition for equilibrium of a composite system with volume, area, line, and point phases as

$$\delta(E_t)_{S_t^{(V)},V,\{M_{ij}^{(V)}\}} = 0, \tag{1.52}$$

where the total energy $E_t = U_t + \Omega_\phi$ represents the total internal energy and external field energies of the composite system. The expression for U_t is given by Equation 1.55 below, while the corresponding expression for Ω_ϕ is given by an identical Equation 1.55 if one replaces $u^{(v)}$ with $\rho^{(v)}\phi$, $u^{(a)}$ with $\rho^{(a)}\phi$, $u^{(l)}$ with $\rho^{(l)}\phi$, and $U^{(0)}$ with $M^{(0)}\phi$ (cf. the form of Equations 1.16 and 1.29). The three subsidiary conditions denoted by the subscripted quantities above are necessary if one requires that the calculus of variations problem remain equivalent to the problem stated by Gibbs for an isolated composite system. In other words, an isolated system does not permit the transfer of heat, mass, or work across its outer boundary. If these restrictions are imposed on our system and on the formulation of the calculus of variations problem that accompanies the system, then we must force all dissipation processes to vanish, restrict the total mass of each species in the system to remain fixed and require that all outer boundary variations that would perform work be zero. We impose the first condition that all dissipation processes vanish in the composite system by requiring that the total entropy remain fixed. Imposition of the second condition simply requires that the mass of each species remain constant. The final boundary condition, which requires that no virtual work be possible on the outer wall, requires that position variations

$$\delta r \big|_{\{A_{w,j}\}} = 0, \tag{1.53}$$

and normal displacement variations

$$\delta \hat{n} \big|_{\{L_{w,k}\}} = 0, \tag{1.54}$$

where $\{A_{w,j}\}$ denotes the union of all internal surfaces that would intersect the bounding wall during a variation and $\{L_{w,k}\}$ denotes the union of all internal contact lines that would intersect the bounding wall during a variation. The first condition fixes the "imaginary" bounding wall by imposing the condition that all internal surfaces remain unvaried along the bounding wall while the second condition fixes the unit normals to the dividing surfaces along all contact lines that contact the bounding wall.

The outer wall may have arbitrary shape, however, to insure that the total internal energy U_t is unambiguously determined it is necessary to place certain geometric constraints on the manner in which internal surfaces, lines, and points contact the outer wall. Specifically, it shall be required that no portion of a dividing surface, with the exception of its boundary lines or points (i.e., no amount of its area), lie on the outer wall. In addition, it shall also be required that no segment of a dividing line, with the exception of its end points (i.e., no amount of its length), may lie on the outer wall. Finally, it is necessary to require that a dividing point not be an outer wall point. If any of these conditions are violated, then one would obtain a constrained variation or the mechanical equilibrium conditions for the dividing surfaces, lines, or points would be connected to the geometric shape of the imaginary bounding surface of the composite, fluid system.

The total free energy is divided into parts assigned to the bulk, surface, line, and point regions of the composite system. If the total number of bulk phases, dividing surfaces, dividing lines and dividing points inside the composite system are denoted by the symbols V_{k1}, A_{k2}, L_{k3}, and P_{k4}, respectively, then it is possible to write the total internal energy of the system as

$$U_t = \sum_{k=1}^{V_{k1}} \iiint_{V_k} u^{(v)} dV + \sum_{k=1}^{A_{k2}} \iint_{A_k} u^{(a)} dA + \sum_{k=1}^{L_{k3}} \int_{L_k} u^{(l)} dL + \sum_{k=1}^{P_{k4}} (U^{(o)})_{P_k}, \qquad (1.55)$$

where V_k denotes a particular volume region with a particular specific internal energy and it is one out of a total of V_{k1} volume regions that contribute to the composite system. Likewise, A_k, L_k, and P_k denote particular dividing surfaces, lines, and points, respectively. The k subscripts on the symbols V_{k1}, A_{k2}, L_{k3}, and P_{k4} acquire values, in general, such that $k1 \neq k2 \neq k3 \neq k4$. However, these seemingly unrelated quantities are in fact connected by a topological or combinatorial quantity, denoted by the symbol χ, which is called the Euler characteristic [31–35]. For any compact surface in three-dimensional space, the Euler characteristic χ is related to the geometric genus of the surface g_s by the relation $\chi = 2(1 - g_s)$. Furthermore, if the surface can be segmented and represented by a large number of regions or patches, then the number of vertices P_{s4}, edges L_{s3}, and patches A_{s2} are related to the Euler characteristic by the expression $\chi = A_{s2} - L_{s3} + P_{s4}$. A surface that is representable in this fashion is known as a differential geometric surface.

Upon solution of the calculus of variations Problem 1.52, one finds that the condition of thermal equilibrium in isolation is

$$T = \bar{T}. \qquad (1.56)$$

Physically, this states that the equilibrium temperature $\bar{T}$ is the same in all bulk phases, dividing surfaces, and linear regions. Similarly, considering the chemical components to be independent (with no chemical reactions permitted), one finds that the conditions of chemical equilibrium for each component are

$$\mu_j + \phi = \bar{\mu}_j \quad \text{for } j = 1, 2, \ldots, r \text{ chemical components} \qquad (1.57)$$

throughout the system, where $\bar{\mu}_j$ are the equilibrium chemical potentials of the chemical constituents of the system at the reference surface, $\phi(r) = 0$.

In addition to the thermal and chemical equilibrium conditions that are given by Equations 1.56 and 1.57, there are two kinds of mechanical equilibrium conditions that also arise from the mathematical analysis. One condition is for liquid–fluid interfaces (i.e., dividing surfaces), while the other condition is for three-phase dividing lines. The simplest forms of these mechanical equilibrium conditions are explicitly derived in Chapter 2 for both incompressible and compressible axisymmetric capillary systems.

When the surface fundamental Equation 1.24 is used, it can be shown that the condition of mechanical equilibrium across each dividing surface is given by [5]

$$J\gamma + 2KC_J - \nabla_2^2 C_J - K\nabla_2^* \cdot (\nabla_2 C_K) + \rho^{(a)}\hat{n} \times \nabla\phi = P^{(\beta)} - P^{(\alpha)}, \qquad (1.58)$$

where $\hat{n}$ is the unit normal by which the dividing surface is oriented, $P^{(\beta)}$ is the pressure in the bulk phase for which $\hat{n}$ is directed outward, and ∇_2^2 and ∇_2^* are surface differential operators [27]. Relation 1.58 expresses the balance that exists in equilibrium between the internal surface forces and the forces external to the dividing surface, namely gravity and the pressure difference; it is the most general form of the Laplace equation of capillarity. When gravity or the surface mass density is negligible, the last term on the left-hand side of Equation 1.58, $\rho^{(a)}\hat{n} \times \nabla\phi$, can be dropped. Hence, if the surface mechanical potentials γ, C_J, and C_K are constant along the dividing surface the condition of mechanical equilibrium for each dividing surface reduces to

$$J\gamma + 2KC_J = P^{(\beta)} - P^{(\alpha)} = \Delta P. \qquad (1.59)$$

The condition of mechanical equilibrium for dividing lines (i.e., the most general case of the Neumann relation corresponding to Equation 1.58) is extremely complex and available elsewhere [5]. However, if the linear curvature potentials $C_{\theta jk}$, C_{nj}, C_{gj}, and $C_{\tau j}$ and the line tension σ are also constant along the dividing line, then the equilibrium condition reduces to

$$\left(\sigma + \sum_{(jk)} C_{\theta_{jk}}\theta_{jk}\right)\vec{\kappa} = \sum_j (\gamma^{(j)} + (C_{\theta_{jj1}} - C_{\theta_{jj2}})\kappa_{nj})\hat{m}_j, \qquad (1.60)$$

where the subscripts on the angle, denoted as $j1$ and $j2$, are arranged such that the order of the dividing surfaces clockwise about the dividing line is $j1$, j, and $j2$. The vectors $\vec{\kappa}$ and $\hat{m}$ are defined in terms of the surface normal $\hat{n}$ direction and the tangent normal to the dividing line $\hat{t}$ so that

$$\hat{m} = \hat{t} \times \hat{n} \quad \text{and} \quad \vec{\kappa} = \dot{t} = \ddot{r}, \qquad (1.61)$$

where the latter two quantities denote arc-length derivatives of the (nonnormal) tangent vector and the position vector, respectively.

Relation 1.60 represents the generalized Neumann triangle relation. It serves as the natural boundary condition for Equation 1.59 and balances the internal forces in a dividing line with the forces external to the line, namely gravity and surface forces, to determine the equilibrium shape of the dividing surfaces meeting at the dividing line. The conditions of mechanical equilibrium at dividing points may be found elsewhere and will not be discussed further [5].

In a manner similar to that used by Gibbs we shall apply the formalism just developed to a dividing line at a solid surface. We consider a rigid, insoluble solid with a smooth surface and assume that the two solid–fluid dividing surfaces and the dividing line are governed by fundamental equations of the same form as those for a fluid system. For convenience, we shall introduce the following changes in notation. The bulk phases, previously labeled by the subscripts j or k, will be denoted by the superscripts (s) for the solid phase, (l) for liquid, and (v) for vapor, gas, or second liquid, while the dividing surfaces, previously labeled by the subscripts j, will now be denoted by the double superscripts (sv) for solid–vapor surface, (lv) for liquid–vapor, and (sl) for solid–liquid. Thus, in place of Equation 1.60, the condition of mechanical equilibrium for a dividing line at a solid surface is given by [5]

$$\left(\sigma + \left(C_{\theta_l} - C_{\theta_v}\right)\theta_l\right)\kappa_{gs} + \left(\gamma^{(lv)} + \left(C_{\theta_l} - C_{\theta_v}\right)\left(\kappa_{ns}\cos\theta_l - \kappa_{gs}\cos\theta_l\right)\right)\cos\theta_l$$
$$= \gamma^{(sv)} - \gamma^{(sl)} + \left(C_{\theta_l} - C_{\theta_v}\right)\kappa_{ns}. \tag{1.62}$$

This relation is the generalized Young equation of capillarity for a solid–liquid–vapor interface; it serves, *inter alia*, as a boundary condition for the generalized Laplace equation of capillarity given by Equation 1.59.

A further restriction that brings the above results closer to the classical equations is obtained by neglecting all remaining curvature and contact angle potentials. The surface equilibrium condition, Equation 1.59, then reduces to

$$J^{(lv)}\gamma^{(lv)} = P^{(l)} - P^{(v)} = \Delta P, \tag{1.63}$$

which is identical to the classical Laplace equation of capillarity. Conditions 1.60 and 1.62 also simplify to

$$\sum_j \gamma^{(j)}\hat{m}_j - \sigma\vec{\kappa} = \vec{0}, \tag{1.64}$$

and

$$\sigma\kappa_{gs} + \gamma^{(lv)}\cos\theta_l = \gamma^{(sv)} - \gamma^{(sl)}. \tag{1.65}$$

The mechanical equilibrium conditions expressed in Equations 1.63 through 1.65 apply locally at every point on the two-phase dividing surface and the three-phase dividing line, respectively. At the lowest level of generality, where one ignores the effect of curvature and contact angle potentials on the condition(s) of mechanical

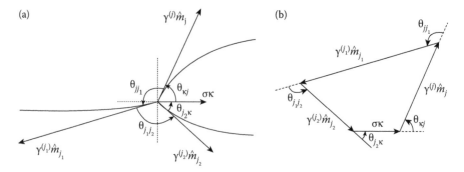

FIGURE 1.2 Schematic of (a) a side view of an axisymmetric liquid lens at the contact line, and (b) a quadrilateral composition of the three surface tensions, one line tension, and angles present at the contact line.

equilibrium, the classical Laplace equation of capillarity, Equation 1.63, as originally stated in 1805 is recovered; however, the classical Young equation of capillarity and the classical Neumann triangle relation must be corrected via additional terms that represent the influence of the line tension σ into the overall "force" balance at the three-phase contact line. Equation 1.64 is equivalent to two orthogonal scalar relations. One of these relations may be shown to be equivalent to the Young equation of capillarity (cf. derivation in Chapter 2 for axisymmetric capillary systems), while the second relation represents the mechanical equilibrium condition in a direction that is orthogonal to the first relation. In general, the curvature, $\vec{g}$, of the contact line is arbitrarily directed; however, for axisymmetric systems the curvature must lie perpendicular to the gravitational vector $\vec{g}$ and within the same plane as $\hat{m}_j$ for $j = 1$, 2, 3 as sketched in Figure 1.2.

If we employ the notation in Figure 1.2, we may write the vector expression given by Equation 1.64 as two scalar relations; one resulting from projecting the vectors into the horizontal plane and another from a corresponding projection into the vertical plane [36]. Both relations represent balances of surface and linear forces in their respective planes. In the horizontal direction one obtains a relation that is similar to the Young equation of capillarity and given by the expression

$$\gamma^{(j)} \cos\theta_{\kappa j} + \gamma^{(j_1)} \cos\theta_{j_1 \kappa} + \gamma^{(j_2)} \cos\theta_{j_2 \kappa} + \sigma\kappa = 0, \tag{1.66}$$

while the projection into the vertical plane yields the result

$$\gamma^{(j)} \cos\theta_{\kappa j} + \gamma^{(j_1)} \sin\theta_{j_1 \kappa} + \gamma^{(j_2)} \sin\theta_{j_2 \kappa} = 0, \tag{1.67}$$

where $\theta\vec{\kappa}_j$ is the angle of contact between the curvature vector $\vec{\kappa}$ of the contact line and the jth dividing surface, $\theta^{j_s}{}_k$ is the angle of contact between the curvature vector and the j_sth dividing surfaces (where $s = 1$ or $s = 2$) and $\theta_{j_1 j_2}$ is the angle between the j_1th dividing surface and the j_2th dividing surface; that is, the angle between the surface unit vectors $\hat{m}_{j_1}$ and $\hat{m}_{j_2}$ (cf. the definition of the "cusp angle" in Chapter 2). One should notice that with this particular orientation (i.e., horizontal and vertical) one obtains mechanical equilibrium equations at the line of contact that are identical to the classical expressions, save for the one additional term $\sigma\kappa$ that appears in Equation 1.66. If these two projection relations are squared and added, then one obtains a "cosine-rule" expression among the various forces at the contact line. The expression contains three angles of contact and is given by

$$(\gamma^{(j)})^2 + (\gamma^{(j_1)})^2 + (\gamma^{(j_2)})^2 + 2\gamma^{(j)}\gamma^{(j_1)}\cos\theta_{jj_1}$$
$$+ 2\gamma^{(j)}\gamma^{(j_2)}\cos\theta_{j_2j} + 2\gamma^{(j_1)}\gamma^{(j_2)}\cos\theta_{j_1j_2} = (\sigma\kappa)^2. \tag{1.68}$$

It is possible to eliminate two of these three angles in favor of the cusp angle $\theta_{j_1j_2}$ by performing a rotation of the horizontal and vertical axes. This result, which is demonstrated elsewhere [36], yields the interesting relation

$$\left(\gamma^{(j)}\right)^2 + \left(\sigma\kappa\right)^2 + 2\sigma\kappa\gamma^{(j)}\cos\theta_{\kappa j} = \left(\gamma^{(j_1)}\right)^2 + \left(\gamma^{(j_2)}\right)^2 + 2\gamma^{(j_1)}\gamma^{(j_2)}\cos\theta_{j_1j_2}, \tag{1.69}$$

which permits one to see that one recovers the classical "cosine-rule" Neumann triangle relation among the surface tensions when the magnitude of $\sigma\kappa$ is insignificant. In addition, the presence of the $\sigma\kappa$ term changes the Neumann triangle into a quadrilateral as illustrated in Figure 1.2b and if the quantity $\vec{\sigma\kappa}$ vanishes, then the quadrilateral transforms into a triangle.

1.1.6 FREE ENERGY VARIATION AND ALTERNATIVE CURVATURE MEASURES

It is possible to transform the constrained calculus of variations problem, stated by Equation 1.52, into an unconstrained (free energy) calculus of variations problem if we use the method of Lagrange multipliers to modify the complete energy integral E_t beforehand. This approach will also be discussed in more detail in the second section of this chapter. The modified energy integral (free of constraints) is given by the expression

$$\Omega_t = U_t - \lambda S_t - \sum_{j=1}^{r} \lambda_j M_{tj}, \tag{1.70}$$

where λ and λ_j are the Lagrange multipliers for the entropy constraint and the jth component mass constraint. Any variation of the total free energy Ω_t, together with the boundary conditions (Equations 1.53 and 1.54), can be handled as an unconstrained problem. The Lagrange multipliers can be evaluated from the boundary conditions. The final equilibrium conditions are obtained by eliminating the Lagrange multipliers using the constraint conditions that the total entropy and the total mass of each component must remain fixed. Two specific examples are presented in detail in Chapter 2. Accordingly, the variation of the total free energy can be written as

$$\delta\Omega_t = \sum_{k=1}^{V_{k1}} \delta\Omega_k^{(v)} + \sum_{k=1}^{A_{k2}} \delta\Omega_k^{(a)} + \sum_{k=1}^{L_{k3}} \delta\Omega_k^{(l)} + \sum_{k=1}^{P_{k4}} \delta\Omega_k^{(o)} = 0, \tag{1.71}$$

where

$$\Omega_k^{(v)} = \iiint_{V_k} \omega^{(v)} dV \quad \text{with} \quad \omega^{(v)} = u^{(v)} + \omega_\phi^{(v)} - \lambda s^{(v)} - \sum_{j=1}^{r} \lambda_j \rho_j^{(v)}, \tag{1.72}$$

$$\Omega_k^{(a)} = \iiint_{A_k} \omega^{(a)} dA \quad \text{with} \quad \omega^{(a)} = u^{(a)} + \omega_\phi^{(a)} - \lambda s^{(a)} - \sum_{j=1}^{r} \lambda_j \rho_j^{(a)}, \tag{1.73}$$

$$\Omega_k^{(l)} = \iiint_{L_k} \omega^{(l)} dL \quad \text{with} \quad \omega^{(l)} = u^{(l)} + \omega_\phi^{(l)} - \lambda s^{(l)} - \sum_{j=1}^{r} \lambda_j \rho_j^{(l)}, \tag{1.74}$$

and

$$\Omega_k^{(o)} = U^{(o)} + \Omega_\phi^{(o)} - \lambda S^{(o)} - \sum_{j=1}^{r} \lambda_j M_j^{(o)}. \tag{1.75}$$

The solution of the calculus of variations problem posed by Equation 1.71 will depend critically upon the choice of parameterization and upon the generality of the functional expression that is adopted for the specific free energies $\omega^{(v)}$, $\omega^{(a)}$, $\omega^{(l)}$ and point energies $\Omega^{(o)}$. We shall return to this problem in the first part of the next section.

The earliest attempts at solving this problem of determining the mechanical equilibrium conditions that would render the integrals stationary usually considered a capillary system as a composite system of at most three bulk phases with three surface phases and one contact line of mutual intersection. Any mobile interface that existed between adjacent deformable bulk phases was considered to possess an energy that was proportional to the surface area of the interface. In virtually all cases, this proportionality factor was treated as a constant or uniform tension on the surface. The only real exception to this state of affairs, until the studies of Buff and Saltsburg [37–42] and Hill [43], was the impressive fundamental capillarity work of Gibbs [44].

The mechanical equilibrium condition for the surface that arises from the solution to this calculus of variations problem simplifies approximately to a problem that renders the area of the interface a minimum, or

$$\delta \iint dA = 0. \tag{1.76}$$

When solved, this problem yields a minimal surface of negligible thickness and mass whose mean curvature $J = c_1 + c_2$ vanishes. If the surface bounds a bulk phase of fixed volume, a constraint must be added to the problem that leads to a surface of constant, but not vanishing, mean curvature. The unique properties of these surfaces with either fixed or zero mean curvature soon captivated the imagination and interest of many mathematicians. In both cases, the problem was restricted by fixing the position of the boundary so that no boundary conditions occur because the boundary was not free to vary. In addition, alternative surface integral expressions such as

$$\delta \iint J^2 dA = 0, \tag{1.77}$$

designed by Poisson in the nineteenth century to characterize the potential energy of a membrane started to appear [45]. Another example, in 1889, was provided by Casorati [46]

$$\delta \iint (J^2 - 2K) dA = 0, \tag{1.78}$$

where $K = c_1 c_2$ is the Gaussian curvature. It might be argued that, as was done by Nitsche [47], a more appropriate surface integral to investigate would be

$$\delta \iint \Phi(J, K) dA = 0, \tag{1.79}$$

where $\Phi(J, K)$ denotes a positive, symmetric but not necessarily homogeneous function of the mean and Gaussian curvatures. Simple polynomial examples are: $\Phi = a + bJ^2 - cK$, with both constants b and c much less than a [47] and $\Phi = b(J - J_0)^2 + cK$ [48]. If $\Phi = \psi(J) - cK$, then the Euler–Lagrange equation, which is a necessary condition for the variation of the surface integral to vanish, is given by [47]

$$\Delta_b \frac{\delta \Psi}{\delta J} + \left[(J^2 - K) \frac{\delta \Psi}{\delta J} - J\Psi \right] = 0, \tag{1.80}$$

where Δ_b denotes the Beltrami operator [49]. For the special case $\Phi = J^2$, the differential Equation 1.80 reduces to

$$\Delta_b J + \frac{J}{2}(J^2 - 4K) = 0 \tag{1.81}$$

and was derived by Schadow in 1922 [50]. Regardless what particular expression is adopted for the surface energy, the Euler–Lagrange equation arising from the calculus of variations problem Equation 1.79 is lengthy and involves the fourth-order derivatives of the position vector for the surface. Recent mathematical investigations have centered on the Expression 1.77 and its higher dimensional extensions. The case of surfaces with nonfixed or free boundaries "requires the discussion of appropriate boundary conditions and has not attracted much attention so far" [49,51]. Extensions and elucidations of Gibbs and Buff's efforts, which consider nonfixed boundary conditions, by Murphy [52], Melrose [53–55], Cahn and Hoffman [56,57], Helfrich [48], Boruvka and Neumann [5], Scriven et al. [58,59], Rowlinson and Widom [60,61], Alexander and Johnson [62,63], Shanahan and de Gennes [64–67], Markin et al. [68,69] and Kralchevsky et al. [70–73] have been primarily directed at the determination of the appropriate mechanical equilibrium conditions across a surface (i.e., Laplace's equation) and at a contact line boundary (i.e., either Young's equation or Neumann's triangle relation) for quite general differential geometric surfaces. However, a certain amount of contention among these investigators has occurred over the particular functional expression that one might expect for the free energies [74–76].

One result of importance in this connection is the Gauss–Bonnet theorem, which states for sufficiently smooth boundaries that

$$\iint K dA = 2\pi\chi = 4\pi(1 - g_s). \tag{1.82}$$

As a consequence of this result, we see that for any calculus of variations problem in which the genus g_s of the surface is fixed and the surface integral is homogeneous of first degree with Gaussian curvature K, that the Euler-Lagrange equation will be unaffected by the presence of the Gaussian curvature [77]. Therefore, the variation problems given by Expressions 1.77 and 1.78 are equivalent for closed surface systems.

1.2 APPLICATIONS, IMPLICATIONS, AND COROLLARIES

1.2.1 INTRODUCTION

The preceding part of this chapter contains an outline of a generalized theory of capillarity. While the generalized theory uses all of the available building blocks of thermodynamics, it does not explicitly attempt a comparison with the historical development of this field; specifically, it does not develop, nor compare with, Gibbs classical theory of capillarity. The most important differences between the classical and the generalized theory are as follows:

1. The classical theory invokes a "moderate curvature" approximation, whereas the generalized theory should be applicable for high curvature situations as well. This is best appreciated by considering mechanical equilibrium conditions (e.g., Equations 1.58 and 1.59), which, depending on the level of generalization, contain various curvature related terms.
2. The classical theory does not consider contact lines and hence did not introduce any thermodynamic definition for the line tension. The generalized theory introduces a line tension term into the Young equation for situations

of moderate curvature, implying that line tension is not necessarily a small quantity that might be safely omitted.

3. The classical theory introduces contact angles as a form of boundary condition that is separate from the thermodynamic formalism. This fact may well have enhanced general doubts with respect to the status and significance of contact angles. In contrast, the generalized theory introduces contact angles as thermodynamic quantities that are just as fundamental as, say, curvatures or interfacial area.

4. Positioning and shifting of dividing surfaces is a matter of practical and theoretical importance; however, the question of the legitimacy of such shifts has often been disregarded. As no a priori choice of a dividing surface (e.g., to coincide, as in Gibbs classical theory, with the position of the surface of tension) is necessary in the generalized theory we may expect greater flexibility from this theory than from the classical one with regards to this choice.

It is the purpose of the subsequent sections in this chapter to explore the differences between the two theories and to corroborate further the generalized theory of capillarity.

1.2.2 THE FREE ENERGY REPRESENTATION

A thermodynamic investigation into the equilibrium of any system begins with the selection of a suitable thermodynamic potential (i.e., a fundamental equation) and the appropriate equilibrium principle or condition. The fundamental equation describes the thermodynamic states in all parts of the fluid system, while the minimum principle determines only the equilibrium states possible from the multitude of thermodynamic states permitted by the fundamental equation. Various forms or representations of the minimum principle and the fundamental equation are possible. Conversion between the various expressions of the fundamental equation (i.e., the thermodynamic potentials) is performed by means of a mathematical technique known as a Legendre transformation [6,11,14,17,19]. Using this technique, parameters defining the fundamental equation may be replaced by their corresponding intensive quantities. Therefore, in essence, it becomes possible to reformulate the thermodynamic formalism so that parameters like the entropy, volume, or interfacial areas that are not easily manipulated experimentally, may be replaced by quantities like the temperature, pressure, and surface tension that are much easier to control. Below, we shall consider some of the alternative, Legendre transformed versions of the fundamental equations for capillary systems.

As noted by Callen [6], the energy formulation (i.e., internal energy plus gravitational) is not really suited for capillary systems because the representation does not take advantage of the thermal equilibrium present in the system; that is, temperature is constant throughout and known. The next thermodynamic potential one frequently finds considered is the Helmholtz function. In this representation the entropy as an independent variable is replaced by the temperature that is kept constant throughout the system. The Helmholtz function is "admirably" [78] suited to assure thermal equilibrium since the search for configurations that are at complete equilibrium is reduced to configurations that already are at thermal equilibrium. However, the equilibrium principle for the Helmholtz function still requires fixed component masses inside a fixed system volume that eliminates the possibility of considering open systems. If the Helmholtz function is used, then the desired constant pressure within each phase and the composition of the phase can only be obtained indirectly. The next thermodynamic potential, the Gibbs function, is rejected immediately for capillary systems because it requires that each pressure be controlled by a pressure reservoir [79]. This is impossible for a small bubble or drop phase surrounded by another larger fluid phase since it is obvious that the smaller phase does not have a pressure reservoir. At this point, the well-known thermodynamic potentials have been exhausted. Thus, to no surprise, it is the Helmholtz function that is usually selected when treating capillary systems.

Conceptually, the relevant Legendre transformations have not really been exhausted, because neither the Helmholtz nor the Gibbs potential considers the possibility of changes in mass or mole numbers, and hence the possibility of chemical equilibrium with one or more components, expressed by the equality of chemical potentials. Thus, the thermodynamic potential in which the independent variables "entropy" and "masses" of the individual chemical constituents are replaced, respectively, by the temperature and the chemical potentials is a suitable fundamental equation for investigating capillary systems. This thermodynamic potential, often called the *grand canonical potential* and denoted by the symbol Ω, does not seem to have been used much in the field of thermodynamics (e.g., Gibbs refers to it once without a name), although it is well known in statistical mechanics [80]. When it comes to capillary systems there are many instances of either the Helmholtz or the Gibbs functions being used in applications where the free energy or grand canonical potential would have been far more suitable and appropriate.

As a consequence, since the conditions of thermal and chemical equilibrium are the same throughout the system this presents the possibility of using the conditions of thermal and chemical equilibrium beforehand to reduce the minimum (internal energy) problem described above in Equation 1.52 to that formulated in Equation 1.71. Evidently, in the reduced minimum problem, the state of complete equilibrium is sought only among those thermodynamic states that already are in thermal and chemical equilibrium. Thus, using the equilibrium conditions that exist between the temperature and the chemical potentials throughout the system (i.e., Equations 1.56 and 1.57) we may write the grand canonical potential density for the bulk phase (cf. Equation 1.72) as

$$\omega^{(v)} = u^{(v)} - Ts^{(v)} - \sum_j \mu_j \rho_j^{(v)}, \tag{1.83}$$

where all the quantities are to be evaluated at the equilibrium temperature $T = \bar{T}$ and chemical potentials $\mu_j = \bar{\mu}_j - \phi$ for the chemical components labeled $j = 1, 2,$ ... , r. In essence, Equation 1.83 defines a Legendre transformation from the specific volume internal energy $u^{(v)}$ to the specific grand canonical potential

$$\omega^{(v)} = \omega^{(v)}(T, \mu_1, \mu_2, \ldots, \mu_r), \tag{1.84}$$

which is the specific free energy representation of the fundamental equation for bulk phases that are known to be in thermal and chemical equilibrium. Expression 1.84 simultaneously replaces the entropy density by the temperature and the mass densities by the chemical potentials as the independent parameters in the fundamental equation. The differential form of the fundamental equation is obtained by taking a total differential of Equation 1.83 and using Equation 1.11 for $du^{(v)}$ to obtain

$$d\omega^{(v)} = -s^{(v)}dT - \sum_j \rho_j^{(v)} d\mu_j. \tag{1.85}$$

A comparison with the Euler relation given by Equations 1.10 and 1.83 yields the simple connection that

$$\omega^{(v)} = -P, \tag{1.86}$$

which shows that the negative of the pressure in a bulk phase is the expression for the specific free energy. Alternatively, the quantity $\omega^{(v)} dV = -PdV$ can be interpreted as representing the work done on the bulk system when there is an associated volume change dV. The contribution of the bulk phases to the total free energy, Ω_j, is then written as

$$\Omega^{(V)} = \iiint\limits_V \omega^{(v)}(\overline{T}, \overline{\mu}_1 - \phi, \overline{\mu}_2 - \phi, \dots, \overline{\mu}_r - \phi)\, dV. \tag{1.87}$$

Despite its appearance, this expression is considerably reduced, in the sense that the independent functions of $\omega^{(v)}$ that remain in the integrand of Equation 1.87 are known, so that $\omega^{(v)}$ becomes a known function of position through the given external potential $\phi(\vec{r})$. However, to evaluate $\Omega^{(v)}$ one still needs to know the exact functional relation for the fundamental equation, $\omega^{(v)}(\vec{r})$.

The reduction of the dividing surface part of Ω_Σ can be carried out in complete analogy with that of the bulk phase. The conditions of thermal and chemical equilibrium permit one to use Equation 1.24 to write (cf. Equation 1.73) the specific free energy as

$$\omega^{(a)} = u^{(a)} - Ts^{(a)} - \sum_j \mu_j \rho_j^{(a)}, \tag{1.88}$$

which introduces the free energy representation of the fundamental equation for surfaces as

$$\omega^{(a)} = \omega^{(a)}\left[T, \mu_1, \mu_2, \dots, \mu_r, J, K\right]. \tag{1.89}$$

The differential form of Equation 1.89 is given by

$$d\omega^{(a)} = -s^{(a)}dT - \sum_j \rho_j^{(a)}d\mu_j + C_J dJ + C_K dK \qquad (1.90)$$

and, from Equations 1.88 and 1.27, the surface version of the Euler relation is given by [81,82]

$$\omega^{(a)} = \gamma + C_J J + C_K K, \qquad (1.91)$$

which defines the specific free energy of a dividing surface. Thus, only in the restrictive case of a flat interface, where both curvatures J and K are identically equal to zero, will the surface free energy $\omega^{(a)}$ be equal to the surface tension γ. The contribution of the dividing surfaces to the total free energy function Ω_Σ becomes

$$\Omega^{(A)} = \iint_A \omega^{(a)}\left[\overline{T}, \overline{\mu}_1 - \phi, \overline{\mu}_2 - \phi, \ldots, \overline{\mu}_r - \phi, J, K\right]dA, \qquad (1.92)$$

where the integrand becomes a known function of the potential $\phi(\vec{r})$ and the surface curvatures J and K on each dividing surface in the system. Once again, the functional expression for $\omega^{(a)}(\vec{r})$ remains unknown.

Reduction of the total free energy Ω_Σ into its separate geometric contributions, when the system also contains linear and point phases, follows directly and in an analogous manner to that of the bulk and surface phases discussed above. After a suitable reduction, the total free energy Ω_Σ remains a thermodynamic potential with the same extremum properties (yielding the same solution) as any other suitable thermodynamic potential. Mathematically, the difference between the total energy and the total free energy extremum formulations is that the constraints in the first definition (total entropy and masses remaining constant) are replaced by the subsidiary conditions $T = \overline{T}$ (a constant) and $\mu_j + \varphi = \overline{\mu}_j$ (a constant) in the second definition, such that both problems yield the same solution. The transformations between such conjugate extremum problems are known as *involutory* transformations [83]. Finally, the advantage of employing the free energy Ω_Σ is that there is a direct connection between the variation $\delta\Omega_\Sigma$ and the virtual work, as demonstrated later.

1.2.3 A SIMPLE DERIVATION OF THE GENERALIZED LAPLACE EQUATION

In this section we shall apply the free energy methodology, which decouples the thermal and chemical equilibrium conditions from the mechanical equilibrium conditions, to a particular class of two phase capillary system with the explicit purpose of arriving at the mechanical equilibrium condition for a dividing surface (i.e., the Laplace equation of capillarity) in a more direct manner. Only spherical and cylindrical interfacial geometries will be considered for illustration; however, the approach is completely general and applies to any geometry.

Physically, the particular class of two-phase capillary systems we shall consider is one in which one phase contains a small bubble or droplet of the second phase, either attached to a phase boundary or free. The effect of gravity is neglected in the latter case. The larger phase is kept at constant temperature, pressure, and composition either directly through contacts with appropriate reservoirs, or indirectly by using a rigid adiabatic enclosure that is sufficiently large that the appearance and growth of the second phase will not affect the first phase appreciably. This general prescription covers a rather large class of capillary systems, such as homogeneous and heterogeneous nucleation systems, capillary condensation systems, and some of the model systems for investigation of contact angle hysteresis, and so forth.

Regardless of the system considered, the free energy Ω_Σ can be written as

$$\Omega_\Sigma = \iint_A \omega^{(a)} dA - \iiint_V \Delta P dV, \tag{1.93}$$

where geometries A and V represent the bubble area and volume, $\omega^{(a)}$ is given by Equation 1.91, and ΔP is the pressure difference across the interface. If gravity is neglected, then all intensive parameters, including the pressure, remain constant in the system and Equation 1.93 simplifies to [81]

$$\Omega_\Sigma = \gamma A + C_J J + C_K K - \Delta PV. \tag{1.94}$$

A completely identical approach was used above when we restricted our attention to mechanical equilibrium conditions (i.e., Equations 1.59, 1.60, and 1.62) where all curvature and contact angle potentials, along with the surface tension, remained constant. Now, according to the free energy minimum principle, the equilibrium radius of either a spherical or cylindrical (as constrained between parallel plates) bubble is determined by the condition

$$\left(\frac{d\Omega_\Sigma}{dR}\right)_{T,\{\mu_j\}} = 0, \tag{1.95}$$

where the temperature and chemical potentials are held constant. Differentiating Equation 1.94, according to the constraints above, yields

$$\left(\frac{d\Omega_\Sigma}{dR}\right)_{T,\{\mu_j\}} = \gamma \frac{dA}{dR} + C_J \frac{dJ}{dR} + C_K \frac{dK}{dR} - \Delta P \frac{dV}{dR} = 0 \tag{1.96}$$

after employing the surface Gibbs–Duhem relations

$$d\gamma + J dC_J + K dC_K = 0 \quad dT = d\mu_j = 0. \tag{1.97}$$

Case 1: The Cylindrical Bubble

The relevant geometric parameters for a section of cylindrical bubble, constrained between parallel plates, with radius R and length L are

$$V = \pi R^2 L \qquad \text{(Volume)}$$

$$A = 2\pi RL \qquad \text{(Surface area)}$$

$$J = JA = 2\pi L \qquad \text{(Total Mean curvature)}$$

$$K = KA = 0 \qquad \text{(Total Gaussian curvature)}$$

Therefore, Equation 1.96 becomes

$$\left(\frac{d\Omega_\Sigma}{dR} \right)_{T,\{\mu_j\}} = 2\pi\gamma L - 2\pi\Delta PRL = 0, \tag{1.98}$$

which means that the surface mechanical equilibrium radius of a cylindrical bubble has to satisfy the simple well-known equation

$$\gamma \left(\frac{1}{R} \right) = \Delta P. \tag{1.99}$$

Case 2: Spherical Bubble

The relevant geometric parameters for a spherical bubble of radius R are

$$V = \frac{4}{3}\pi R^3 \qquad \text{(Volume)}$$

$$A = 4\pi R^2 \qquad \text{(Surface area)}$$

$$J = 8\pi R \qquad \text{(Total Mean curvature)}$$

$$K = 4\pi \qquad \text{(Total Gaussian curvature)}$$

Therefore, Equation 1.96 becomes

$$\left(\frac{d\Omega_\Sigma}{dR} \right)_{T,\{\mu_j\}} = 8\pi\gamma R + 8\pi C_J - 4\pi\Delta PR^2 = 0, \tag{1.100}$$

which means that the surface mechanical equilibrium radius of a spherical bubble has to satisfy

$$\gamma \left(\frac{2}{R} \right) + C_J \left(\frac{2}{R^2} \right) = \Delta P. \tag{1.101}$$

Obviously, both Equations 1.99 and 1.101 are just specialized forms of the generalized Laplace equation of capillarity given by Equation 1.59.

1.2.4   A DIRECT DERIVATION OF THE GENERALIZED LAPLACE EQUATION

Consider a simple, two-phase capillary system with constant surface tension and curvature potentials (i.e., mechanical potentials) along the entire interface. That is, we restrict our consideration to an interfacial system that is homogeneous over the dividing surface. The free energy minimum principle states that, for a system in equilibrium, the variation of the total free energy $d\Omega_\Sigma$ vanishes (i.e., $d\Omega_\Sigma = 0$). This total free energy Ω_Σ consists of two free energy terms for the two bulk phases and one free energy term for the interface between the bulk phases; that is,

$$\Omega_\Sigma = \Omega^{(V_1)} + \Omega^{(V_2)} + \Omega^{(A)}, \tag{1.102}$$

with the total volume of the system maintained constant; that is,

$$V_\Sigma = V^{(V_1)} + V^{(V_2)} \quad \text{(A constant value)}. \tag{1.103}$$

If the free energy, using Equations 1.83 to 1.87, for each homogeneous, bulk phase is written as

$$\Omega^{(V)} = U^{(V)} - TS^{(V)} - \sum_j \mu_j M_j^{(V)} = -PV, \tag{1.104}$$

and the corresponding expression for the surface phase, using Equations 1.88 and 1.91, is used to obtain

$$\Omega^{(A)} = U^{(A)} - TS^{(A)} - \sum_j \mu_j M_j^{(A)} = \gamma A + C_J \mathsf{J} + C_K \mathsf{K}, \tag{1.105}$$

one may obtain the complete variation, in extensive notation, from Equations 1.102 through 1.105 as

$$(d\Omega_\Sigma)_{T,\{u_j\}} = \gamma \delta A + C_J \delta \mathsf{J} + C_K \delta \mathsf{K} - \Delta P \delta V = 0. \tag{1.106}$$

We simplify all terms above by considering a small variation, δz, produced in a direction that is normal to the dividing surface defined by the explicit function $z = (x, y)$ [83]. As illustrated in Figure 1.3, a small rectangular portion of dividing surface, with side-lengths x and y and area $A = xy$, is considered for variation. The corresponding variations of all other variables can be obtained readily. First, the variation of the bulk phase volume can be written as

$$\delta V = A \delta z. \tag{1.107}$$

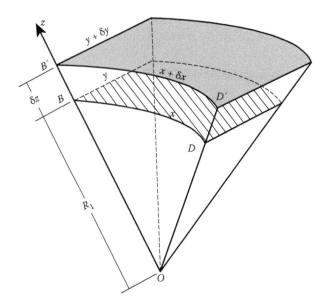

FIGURE 1.3 The variation of the surface area due to the variation δz.

At constant temperature and constant chemical potentials, the variation of the total free energy of the combined bulk phases is obtained from Equation 1.104 as

$$d\Omega^{(V)} = -P_1\delta V_1 - P_2\delta V_2 = -P_1 A\delta z + P_2 A\delta z = -\left(P_1 - P_2\right)A\delta z. \qquad (1.108)$$

The corresponding variation of the dividing surface area, δA, caused by the normal displacement δz can be written as

$$\delta A = (x+\delta x)(y+\delta y) - xy \approx x\delta y + y\delta x. \qquad (1.109)$$

In order to find these length variations δx and δy, the following equality can be obtained using the geometric similarity between triangles ΔOBD and $\Delta OB'D'$ as shown in Figure 1.3

$$\frac{x+\delta x}{x} = \frac{R_1 + \delta z}{R_1}, \qquad (1.110)$$

where R_1 is the radius of the curvature c_1 of the area along side x; that is, side BD. As a consequence, Equation 1.110 can be simplified to

$$\delta x = \frac{x}{R_1}\delta z = xc_1\delta z. \qquad (1.111)$$

Similarly, δy can be expressed as

$$\delta y = yc_2\delta z, \tag{1.112}$$

where c_2 is the curvature of the area along side y. Substituting these two results, Equations 1.111 and 1.112, into Equation 1.109 results in

$$\delta A = xyc_2\delta z + yxc_1\delta z = (c_1 + c_2)xy\delta z = JA\delta z, \tag{1.113}$$

which represents the change in area of the planar patch in terms of the mean curvature, the original area and the normal displacement.

The next term in Equation 1.106 that we consider involves the variation of the total mean curvature

$$\delta J = \delta(JA) = A\delta J + J\delta A, \tag{1.114}$$

where δJ can be evaluated as

$$\delta J = \delta(c_1 + c_2) = \delta\left(\frac{1}{R_1} + \frac{1}{R_2}\right) = -\left(\frac{1}{R_1^2} + \frac{1}{R_2^2}\right)\delta z = \left(c_1^2 + c_2^2\right)\delta z = \left(2K - J^2\right)\delta z. \tag{1.115}$$

Substitution of Equations 1.113 and 1.115 into Equation 1.114 results in

$$\delta J = 2KA\delta z. \tag{1.116}$$

Similarly, the third term in Equation 1.106, involving the variation of the total Gaussian curvature δK can be obtained as

$$\delta K = A\delta K + K\delta A = -KJA\delta z + KJA\delta z = 0. \tag{1.117}$$

After using Equations 1.113, 1.116, and 1.117 in Equation 1.106, the variation of the three free energy of the surface terms can be expressed as

$$\delta\Omega^{(A)} = \gamma\delta A + C_J\delta J + C_K\delta K = \gamma JA\delta z + 2C_J KA\delta z. \tag{1.118}$$

Finally, substitution of Equations 1.108 and 1.118 into Equation 1.106 yields an expression that is, subject to the constraints of this derivation, the Laplace equation of capillarity as written in Equation 1.59. Once again, by assuming as done here that the surface tension and the curvature potentials are constant a priori, one simplifies the mathematical complexity. Furthermore, these examples illustrate in a simple fashion and by using a number of common geometries that the free energy representation appropriate for a nonmoderately curved capillary system produces a modified form of the Laplace equation; that is, Equation 1.59. This version of the Laplace equation that acts as the mechanical equilibrium relation across the nonmoderately curved surface includes the bending moment C_J and the Gaussian curvature K in an additional term. Only if the product of these two quantities is small with respect to the surface tension term will this modified Laplace equation simplify to the well-known classical version.

1.2.5   HYDROSTATIC APPROACH TO CAPILLARITY

The fundamental equation for surfaces, Equation 1.24, satisfies completely all known thermodynamic requirements: The independent properties are all densities of extensive quantities and the curvatures are differential invariants. However, it is still a postulated equation and, as such, it remains in need of confirmation and physical interpretation. Furthermore, in retaining the second curvature term, we have ignored the question of whether this term is or is not negligible for all practical purposes [85,86].

It is at this point that the hydrostatic approach to capillarity makes its contribution [87]. Buff and Saltsburg [37-42] have shown that this nonthermodynamic approach to capillarity, based on the introduction of an interfacial stress tensor field, is capable of an independent confirmation of some of the results of the generalized thermodynamic theory.

In Buff's hydrostatic theory, the excess hydrostatic equation is integrated across the interface. This procedure leads to the Laplace equation as one of the equilibrium conditions. However, our motivation is not merely to re-derive an equilibrium condition, but to corroborate the form of the proper fundamental equation for curved interfaces. The necessary connection between the hydrostatic approach and thermodynamics can be obtained by extending the concept of virtual work. Since, in a hydrostatic analysis, thermal and chemical equilibrium are tacitly assumed, the principle of virtual work will be equivalent to the minimum principle for the free energy (i.e., the grand canonical potential), as developed above.

If one has a single-component, two-phase capillary system with a plane parallel interfacial zone, then all physical properties like the density ρ, the normal stress σ_N, and tangential stress σ_T (see Figure 1.4) will vary along a direction λ that is oriented in some fashion across the interfacial zone. In the presence of a gravitational field, these properties will change slightly with position; however, this effect will not be explicitly considered or illustrated here. The orientation of λ is arbitrary until one attempts to model the interfacial zone as a two-dimensional (dividing) surface; then the unit normal to the surface provides the orientation. Extrapolated and excess quantities with respect to the surface are defined in terms of both the position of the dividing surface and the orientation of the surface. Surface excess quantities like $u^{(a)}$, $s^{(a)}$, $\rho_j^{(a)}$, or $\sigma^{(a)}$ are assigned to a particular point on the surface and are obtained by integration through the interfacial zone in the direction of the surface's unit normal $\hat{n}$. For example, when the dividing surface is positioned at $\lambda = 0$ (see Figure 1.5) the excess tangential stress σ_{TE} is defined by

$$\sigma_{TE} = \sigma_T - (\sigma_T)_{\text{Extrapolated}}, \tag{1.119}$$

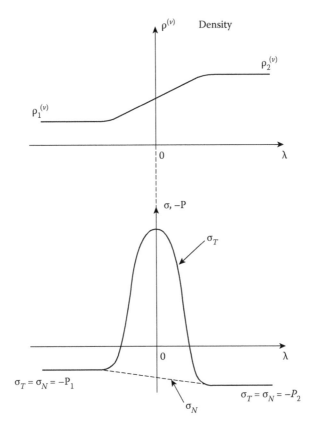

FIGURE 1.4 Schematic of the time-averaged variation of the density and the stress across a deformable liquid–vapor interface where λ is a coordinate directed normal to the interface, $\rho^{(v)}$ is the mass density, and σ_N and σ_T are the normal and tangential stresses, respectively.

so that

$$\sigma_{TE} = \sigma_T + P_{E-} \quad \text{for} \quad \lambda < 0, \tag{1.120}$$

and

$$\sigma_{TE} = \sigma_T + P_{E+} \quad \text{for} \quad \lambda > 0, \tag{1.121}$$

where Equation 1.119 is analogous in form to Equation 1.18 and the subscript $E\pm$ on the pressure denotes an extrapolated quantity. In general, one could define an interfacial excess stress tensor, s_E, as

$$\mathbf{s}_E = \mathbf{s} + P_E \mathbf{1}, \tag{1.122}$$

where s is the stress tensor, P_E is the extrapolated pressure, and $\mathbf{1}$ denotes a unit tensor. The positive sign on the right-hand side denotes the fact that work done by

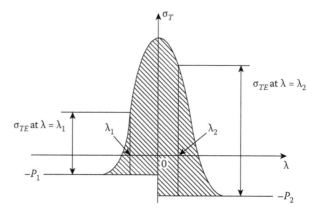

FIGURE 1.5 Schematic of the time-averaged variation of the stress across a deformable liquid–vapor interface, where λ is a coordinate directed normal to the interface, σ_T and σ_{TE} are the tangential and excess tangential stresses, and P_1 and P_2 are the pressures in the two adjacent bulk phases. The shaded portion corresponds to the magnitude of the surface tension, $\gamma = \sigma^{(a)}$, calculated from Equation 1.123 when the position of the dividing surface is selected as $\lambda = 0$. Alternative positions of the dividing surface are denoted by λ_1 and λ_2.

the stress on the surface is tensile, whereas the three-dimensional analog, $-PdV$, is compressive. As a consequence of Definition 1.119, the surface excess stress $\sigma^{(a)}$, or the surface tension γ is defined by [88]

$$\gamma = \sigma^{(a)} = \int \sigma_{TE} d\lambda, \tag{1.123}$$

where the integration is across the interfacial zone and the magnitude of $\sigma^{(a)}$ corresponds to the shaded region of Figure 1.5. In general, if either the position or the orientation of the surface changes, then $\sigma^{(a)}$ will also change. For example, if dividing surfaces are positioned at $\lambda = \lambda_1$ and $\lambda = \lambda_2$ as shown in Figure 1.6, then the σ_{TE} stress distributions will differ by the shaded area and the two surface excess stresses will be related by

$$\sigma_2^{(a)} = \sigma_1^{(a)} - \Delta P \Delta \lambda, \quad \text{when} \quad \Delta \lambda = \lambda_2 - \lambda_1, \tag{1.124}$$

and ΔP is the pressure difference across the interface. It is obvious from this relation that $\sigma_2^{(a)}$ will never equal $\sigma_1^{(a)}$ for arbitrary shifts $\Delta \lambda$ unless $\Delta P = 0$, and this pressure condition is satisfied only for plane parallel surfaces whose radii of curvature are infinite. Similar considerations also apply to the excess densities and the corresponding surface excess densities [89].

While external body forces as well as external surface forces may well be operative in general, we are concerned here only with virtual work arising from internal forces. This virtual work due to internal forces consists of two parts: that

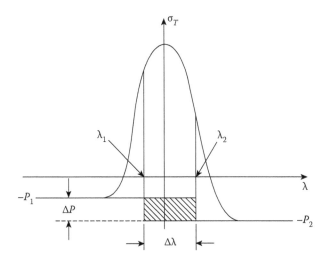

FIGURE 1.6 Schematic of the time-averaged variation of the stress across a deformable liquid–vapor interface, where λ is a coordinate directed normal to the interface, σ_T and σ_{TE} are the tangential and excess tangential stresses, and P_1 and P_2 are the pressures in the two adjacent bulk phases. The shaded rectangle corresponds to the amount by which the surface tension $\gamma = \sigma^{(a)}$ changes if the dividing surface is shifted from a position at $\lambda = \lambda_1$ to a position at $\lambda = \lambda_2$.

due to the extrapolated bulk pressure, P_E, and that due to the excess stress tensor s_E; that is,

$$\delta W_{iE} = -\iiint_V s_E : (\nabla \delta r) dV, \tag{1.125}$$

where $\nabla \delta r$ is the virtual strain tensor, V is the volume of the interfacial layer, and the colon between the symbols represents the double dot product of two tensors. The excess stress tensor, s_E, is zero outside the interfacial region. The reason for considering the internal work δW_{iE} is that it corresponds closely to the variation of the grand canonical potential as discussed above, which considers work done by interfacial tension and the two bending moments.

To make Equation 1.125 tractable, two approximations are needed. The first one requires a division of the excess stress tensor into tangential and normal components. The form of the corresponding excess stress tensor will be

$$s_E = \sigma_{TE} 1_2 + \sigma_{NE} \hat{n}\hat{n}, \tag{1.126}$$

where the first term on the right-hand side represents the two identical, isotropic, tangential stress components and the second term represents the normal stress component of the interface [87]. The second approximation or restriction limits the variation of the shape of the interface. The stress tensor field as expressed by

Equation 1.126 is exact only for a spherical interface, and in this case the normal and tangential excess stress components (i.e., σ_{TE} and σ_{NE}) will be some function of the interfacial thickness alone. Thus, even when the curvature of the surface is not uniform (constant) the implicit assumption invoked via this construction is that the constant density surfaces within the interface are taken as parallel and that the stress is transversely isotropic. To this end, consider a family of parallel surfaces inside the interfacial region. While the dividing surface can be deformed arbitrarily the surfaces must remain parallel and are not allowed to slip with respect to one another.

With these restrictions in place it may be shown mathematically [87] that the volume integral in Equation 1.125 may be replaced by an equivalent surface integral such that

$$\delta W_{iE} = \delta W_i^{(A)}, \tag{1.127}$$

where

$$\delta W_i^{(A)} = -\iint_A (X_0 \delta dA - X_1 \delta dJ + X_2 \delta dK). \tag{1.128}$$

The factors X_k in Equation 1.128 are given by the moment expressions

$$X_k = \int \lambda^k \sigma_{TE} d\lambda \quad \text{for} \quad k = 0,1,2, \tag{1.129}$$

where λ is a coordinate directed normal to the interface. The limits of integration over λ are implied by the vanishing of $\mathbf{s}_E$ outside the interfacial region.

The free energy representation provides a convenient link between the thermodynamic and the hydrostatic approaches to capillarity. Consider a single bulk phase inside a fixed volume, where both the volume and the external system boundary are fixed. In this case, the free energy equilibrium principle simply requires that the first variation of $\Omega^{(V)}$ vanishes. Thus, from Equations 1.85 and 1.87 and allowing for arbitrary variations of the position vector, $\mathbf{r}$, inside the volume V, we have

$$\delta\Omega^{(V)} = \iiint_V \rho^{(v)} \nabla\phi \cdot \delta\mathbf{r}dV - \iiint_V P(\nabla \cdot \delta\mathbf{r})dV = 0, \tag{1.130}$$

where the first term is the virtual work done by the external field (gravity), and the second term

$$\delta W_i^{(V)} = \iiint_V P(\nabla \cdot \delta\mathbf{r})dV = \iiint_V P\delta dV \tag{1.131}$$

is the virtual work done by internal forces. After employing Gauss's divergence theorem for the virtual work of internal forces, Equation 1.130 reduces to the well-known equation of hydrostatics

$$\nabla P = -\rho^{(v)} \nabla \phi. \tag{1.132}$$

For the more complicated case of two bulk phases, separated by an interface, it can be shown that the first variation of the total free energy of the system is

$$\delta \Omega_\Sigma = \delta \Omega^{(V)} + \delta \Omega^{(A)} = \iiint_V \rho^{(v)} \nabla \phi \cdot \delta \mathbf{r} dV - \iiint_V P \delta dV$$

$$+ \iint_A \rho^{(a)} \nabla \phi \cdot \delta \mathbf{r} dA - \iint_A \left(\gamma \delta dA + C_J \delta dJ + C_K \delta dK \right) = 0, \tag{1.133}$$

where the volume integral extends over the volume of both bulk phases; the third term represents the (negative) virtual work of gravity on the interface and the last term is the work done by internal forces on the interface. Further mathematical treatment of the above equation yields the generalized Laplace equation for the interface, along with the surface analog of the hydrostatic condition given by Equation 1.132 [84].

From the above general thermodynamic theory of capillarity in the free energy formalism, and specifically by Equation 1.133, it was shown that the following expression holds for the free energy and hence for the virtual work of internal forces in a dividing surface

$$\delta W_i^{(A)} = - \iint_A \left(\gamma \delta dA + C_J \delta dJ + C_K \delta dK \right), \tag{1.134}$$

where, as before, γ is the surface tension and C_J and C_K are the first and the second bending moments, respectively. Equation 1.134 is matched term by term to Equation 1.128 by setting

$$\gamma = X_0, C_J = X_1 \quad \text{and} \quad C_K = X_2. \tag{1.135}$$

Physically, this identification shows that the three thermodynamic quantities γ, C_J, and C_K correspond to the first three moments of the tangential excess stress component, σ_{TE}, about the dividing surface at $\lambda = 0$. Furthermore, the close correspondence between Equations 1.128 and 1.134 indicates that the hydrostatic approach to capillarity is equivalent and consistent with the mechanical portion of the general thermodynamic theory. Thus, the agreement between the two approaches suggests that the form of the fundamental equation for surfaces, Equation 1.24, with extensive geometric curvatures given by the total mean curvature, J, and the total

Gaussian curvature, K, is the proper expression required to generalize the theory of capillarity.

1.2.6 HYDROSTATIC DERIVATION OF THE GENERALIZED LAPLACE EQUATION

Having confirmed the form of the fundamental equation in the generalized theory of capillarity, the next logical step is to confirm the generalized Laplace equation by using a hydrostatic approach. An interfacial system may be treated as a continuum mechanical system; in other words, spatial smoothing is carried out for the variables under consideration limited to situations where the surface mechanical potentials are constant along the dividing surface [84]. Following Equation 1.132, the basic equation of hydrostatics is written as

$$\nabla \cdot \mathbf{s} = \rho^{(v)} \nabla \phi, \qquad (1.136)$$

where s has replaced $-P$ in Equation 1.132, to account for the fact that the interfacial stress tensor is anisotropic. However, inside a bulk phase and away from the influence of the interface, the stress tensor is isotropic and the hydrostatic equation may be written as

$$\nabla P = -\rho_o^{(v)} \nabla \phi, \qquad (1.137)$$

where the symbol $\rho_o^{(v)}$ is used to denote the volume density of mass well inside the bulk.

Subtracting Equation 1.137 from Equation 1.136 and using the definition of the interfacial excess stress tensor, Equation 1.122, one obtains

$$\nabla \cdot \mathbf{s}_E = \rho_E^{(v)} \nabla \phi, \qquad (1.138)$$

where $\rho_E^{(v)}$ is the excess density of mass assigned to the surface; that is, $\rho_E^{(v)} = \rho^{(v)} - \rho_o^{(v)}$.

In preparation for the following analysis, we outline mathematical identities for dividing surfaces that are parallel to one another. First, the volume element dV can be written in terms of the principal curvatures, c_1 and c_2, of the surface as [39–41,84]

$$dV = (1 + c_1\lambda)(1 + c_2\lambda)dAd\lambda, \qquad (1.139)$$

where λ denotes the distance along the normal to the surface placed at $\lambda = 0$ and dA denotes an area element of the dividing surface. As shown in Figure 1.7, the radius $r_1(\lambda)$ of the surface A', with principal curvature $c_1(\lambda)$, at a point on this surface can be expressed as

$$r_1(\lambda) = r_1 + \lambda, \qquad (1.140)$$

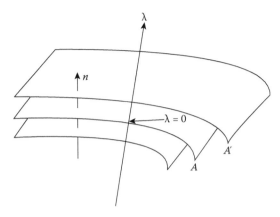

FIGURE 1.7 Coordinate system for parallel surfaces, where surface A' is displaced from the surface A by a normal displacement λ, where λ is a coordinate directed normal to the surfaces considered.

where r_1 is the radius of curvature of the surface A defined at the location $\lambda = 0$. Invoking the definition $r_1 = 1/c_1$ permits one to express Equation 1.140 as

$$c_1(\lambda) = \frac{c_1}{1 + \lambda c_1}. \tag{1.141}$$

Similarly, the relationship between the other principal curvature $c_2(\lambda)$ of surface A' and c_2 can be written as

$$c_2(\lambda) = \frac{c_2}{1 + \lambda c_2}. \tag{1.142}$$

Next, the divergence of the unit surface normal $\hat{\mathbf{n}}$ can be written as [84]

$$\nabla_2 \cdot \hat{\mathbf{n}} = c_1(\lambda) + c_2(\lambda) = J(\lambda), \tag{1.143}$$

where $J(\lambda)$ is the mean curvature of shifted surface A'.

It follows from Equation 1.138 that, using Equations 1.126 and 1.143, the normal component of the hydrostatic equilibrium condition (i.e., the condition corresponding to the Laplace equation), is given by [84]

$$(\nabla \times \mathbf{s}_E) \times \hat{\mathbf{n}} = \nabla \times (\mathbf{s}_E \times \hat{\mathbf{n}}) - \nabla \hat{\mathbf{n}} : \mathbf{s}_E$$

$$= \nabla \times \left(\sigma_{NE} \hat{\mathbf{n}} \right) - \left(c_1(\lambda) + c_2(\lambda) \right) \sigma_{TE} \tag{1.144}$$

$$= \rho_E^{(v)} \nabla \phi \times \hat{\mathbf{n}}$$

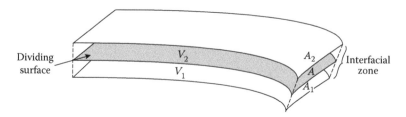

FIGURE 1.8 Schematic of the dividing surface A splitting the interfacial system into two subvolumes, V_1 and V_2, which are bounded by surfaces A and A_1, and by A and A_2.

The dividing surface A splits the whole system into two subvolumes, V_1 and V_2, which are bounded by surfaces A and A_1, and by surfaces A and A_2, respectively. Surfaces A_1 and A_2 denote the outer boundaries of the system (see Figure 1.8). Integration of Equation 1.144 over the regions V_1 and V_2, with the use of the divergence theorem, leads to [84]

$$\iint \sigma_{NE} dA - \iint \sigma_{NE} dA_1 = \iiint_{V_1} \left\{ (c_1(\lambda) + c_2(\lambda)) \sigma_{TE} + \rho_E^{(v)} \nabla \phi \times \hat{\mathbf{n}} \right\} dV_1, \quad (1.145)$$

and

$$-\iint \sigma_{NE} dA + \iint \sigma_{NE} dA_2 = \iiint_{V_2} \left\{ (c_1(\lambda) + c_2(\lambda)) \sigma_{TE} + \rho_E^{(v)} \nabla \phi \times \hat{\mathbf{n}} \right\} dV_2, \quad (1.146)$$

where σ_{TE} is expressed by Equations 1.120 and 1.121, and σ_{NE} can be obtained similarly. This interfacial excess stress σ_{NE} is zero inside the bulk and away from the influence of the interface so the integrations at the outer boundaries A_1 and A_2 in Equations 1.145 and 1.146 disappear. Summation of these two expressions results in the cancellation of the area integrals for σ_{TE} at the chosen dividing surface A. Substitution of Equations 1.120 and 1.121 into this summation, with the use of Equations 1.139, 1.141, and 1.142, leads to

$$\iint \left\{ (P_1 - P_2) - (c_1 + c_2) \int \sigma_{TE} d\lambda - 2c_1 c_2 \int \sigma_{TE} \lambda d\lambda \right.$$
$$\left. - \int \rho_E^{(v)} \nabla \phi \cdot \hat{\mathbf{n}} (1 + c_1 \lambda)(1 + c_2 \lambda) d\lambda \right\} dA = 0 \quad (1.147)$$

The assumption of parallel dividing surfaces and constant densities along each dividing surface insures that the area integration drops out. Hence, Equation 1.147 is simplified to

$$P_1 - P_2 = \gamma J + 2C_J K + \int \rho_E^{(v)} \nabla \phi \cdot \hat{\mathbf{n}} (1 + J\lambda + K\lambda^2) d\lambda. \quad (1.148)$$

where γ and C_J are defined in Equations 1.123, 1.129, and 1.135. When the external force is gravity

$$\nabla\phi = g\hat{\mathbf{k}}, \tag{1.149}$$

where g is the gravitational constant and $\hat{\mathbf{k}}$ is a unit vector directed opposite to the gravity force. Substitution of Equation 1.149 into Equation 1.148 results in

$$P_1 - P_2 = \gamma J + 2C_J K + \rho^{(a)} g\hat{\mathbf{k}} \cdot \hat{\mathbf{n}}, \tag{1.150}$$

where density $\rho^{(a)}$ is defined as

$$\rho^{(a)} = \int \rho_E^{(v)}(1 + c_1\lambda)(1 + c_2\lambda)\,d\lambda, \tag{1.151}$$

which is the surface density of mass in the surface phase. These results demonstrate that Equation 1.150 agrees with Equation 1.58 for the situation were the surface mechanical potentials are constant along the dividing surfaces; that is, the surface gradients of the curvature potentials are zero so both $\nabla_2^2 C_J$ and $K\nabla_2^* \cdot (\nabla_2 C_K)$ in Equation 1.58 vanish. Finally, if the effect of the surface mass density $\rho^{(a)}$ is negligible, then Equation 1.150 will reduce to Equation 1.59.

1.2.7 INVARIANCE OF THE FREE ENERGY AGAINST SHIFTS OF THE DIVIDING SURFACE

We have explained earlier that, in order to satisfy conservation requirements for the system, excess quantities defined in the Gibbsian model of capillarity, such as mass and energy, have to be attributed to the dividing surface. Obviously, shifting the dividing surface within the interfacial region will change the extensive properties of each of the homogeneous bulk phases. As a consequence, the corresponding surface excess quantities will normally also change. In certain circumstances, it is desirable to be able to shift the dividing surface. The motivation for shifting dividing surfaces often stems from the realization that various mathematical simplifications can be obtained by placing the dividing surface in such a position that certain excess properties vanish. A well-known example is the Gibbs adsorption equation, which relates the excess surface density per unit area of a solute to the isothermal derivative of the interfacial tension with respect to the activity of the solute [90]. In deriving the Gibbs adsorption equation, the dividing surface is positioned such that the excess surface density of the solvent is zero.

In the classical theory of capillarity as developed by Gibbs, shifting of the dividing surface is not possible, in general. This limitation arises because Gibbs dealt only with systems of moderate curvature and because he specifically chose the position of the dividing surface at the so-called *surface of tension* position [91]. It stands to reason that one cannot simply change one of the assumptions on which the theory rests without affecting the theory. Gibbs was very much aware of the constraints that he imposed on the formalism. Thus, when he considered, for instance, the surface tension γ, he was very careful to distinguish between its value at the surface of tension and its value at any other dividing surface location. As a tension, Gibbs noted, its position is at the surface that is called the surface of tension and nowhere else [92]. Clearly Gibbs had no intention of generalizing his analysis beyond systems with moderate curvatures [74]. However, other researchers investigating the shifting of dividing surfaces have apparently disregarded the limitations of Gibbs's classical theory [61,76,93,94]. It is the purpose of this section first to show that dividing surfaces cannot be shifted within the classical theory and then that such shifts are possible within the framework of the generalized theory of capillarity. The strategy used to demonstrate this is to calculate the free energy for two arbitrary positions of the dividing surface and then to impose the condition of invariance of the total free energy of the system against shifts of the dividing surface for both theoretical frameworks.

In the absence of gravity, a system consisting of a bulk fluid phase and an interface has a free energy given by Equation 1.94. In the classical theory of capillarity there is no explicit dependence of the free energy on curvature, thus the second and third terms in Equation 1.94 are not present in the Gibbsian formalism. Next, we consider a specific geometry (e.g., a spherical capillary system), and we let the total free energy of this capillary system with respect to one position of a dividing surface, say at $r = R$, be denoted by Ω_Σ and we let Ω_Σ^* be the value of the total free energy with respect to another varied position, say at $r = R^*$. The free energy of the system will only be conserved, without physically changing the size of the composite system, if the position of the dividing surface is changed such that the total free energy is invariant, or

$$\Omega_\Sigma\left(T,\mu_j,r=R\right)=\Omega_\Sigma^*\left(T,\mu_j,r=R^*\right). \tag{1.152}$$

For a classical or Gibbsian description of this spherical capillary system, the formalism requires that

$$\gamma A - V\Delta P = \gamma^* A^* - V^* \Delta P, \tag{1.153}$$

where A, V and γ are, respectively, the area, the volume, and the surface tension of a spherical drop whose surface radius occurs at $r = R$, and A^*, V^* and γ^* are, respectively, the area, the volume, and the surface tension of the drop in the same system but with a radius $r = R^*$. It should be noted that ΔP, the pressure difference across the interface, is the only property contained in the free energy expression that does not, and must not, change when the dividing surface is shifted. The equilibrium condition for the system is the Laplace equation

$$\gamma\left(\frac{2}{R}\right)=\Delta P \quad \text{for position} \quad r = R \tag{1.154}$$

and, after shifting the dividing surface

$$\gamma^* \left(\frac{2}{R^*} \right) = \Delta P \text{ for position } r = R^*. \tag{1.155}$$

Combining these three relations, substituting the proper expressions for the areas and volumes, and solving the system of equations for R^* and γ^* in terms of R and γ shows that the only solution is $R = R^*$ and $\gamma = \gamma^*$; that is, shifting the dividing surface is not possible.

On the other hand, shifting of the dividing surface in the generalized formalism is possible. This can be shown as follows: From Equation 1.94 we see that, in order to calculate the difference in the free energy Ω_Σ for the two dividing surfaces, it is necessary to be able to calculate the changes in the surface quantities γ, C_J, and C_K upon changing the position of the dividing surface. This becomes feasible through the use of Equations 1.128 through 1.131. It can be shown that, when shifting the dividing surface from one location, say position "1," to another location, say position "2," the values of the quantities γ, C_J, and C_K at the second position can be expressed as functions of the equivalent properties at the original position by [95]

$$\gamma^{(2)} = \gamma^{(1)} - \Delta\lambda\Delta P, \tag{1.156}$$

$$C_J^{(2)} = C_J^{(1)} + \gamma^{(1)}\Delta\lambda - \frac{1}{2}(\Delta\lambda)^2\Delta P, \tag{1.157}$$

and

$$C_K^{(2)} = C_K^{(1)} + 2C_J^{(1)}\Delta\lambda + \gamma^{(1)}(\Delta\lambda)^2 - \frac{1}{3}(\Delta\lambda)^3\Delta P, \tag{1.158}$$

where ΔP is the pressure difference across the interface and $\Delta\lambda = \lambda_2 - \lambda_1$ is the displacement of (i.e., distance between) the dividing surfaces. It is seen that a parallel shift of a dividing surface causes changes γ, C_J, and C_K that depend on their original values, on ΔP and on the amount of the shift, $\Delta\lambda$. The dependencies on the stress tensor component σ_{TE} in Equation 1.129 cancel out. From Equation 1.94 and Equations 1.156 through 1.158 the difference in the overall free energy of the system for the two positions of the dividing surface becomes

$$\Omega_\Sigma^{(2)} - \Omega_\Sigma^{(1)} = \gamma^{(1)} \{ A^{(2)} - A^{(1)} + \Delta\lambda J^{(2)} + (\Delta\lambda)^2 K^{(2)} \}$$

$$+ C_J^{(1)} \{ J^{(2)} - J^{(1)} + 2\Delta\lambda K^{(2)} \} + C_K^{(1)} \{ K^{(2)} - K^{(1)} \} \tag{1.159}$$

$$+ \Delta P \left\{ V^{(1)} - V^{(2)} - \Delta\lambda A^{(2)} - \frac{1}{2}(\Delta\lambda)^2 J^{(2)} + \frac{1}{3}(\Delta\lambda)^3 K^{(2)} \right\}.$$

At this point, further progress can be made by considering specific capillary geometries. The cases of cylindrical and spherical geometries are most easily dealt with.

Case 1: The Cylindrical Bubble

The relevant geometric parameters for a cylindrical bubble of radius R and length L are $V = \pi R^2 L$ (volume), $A = 2\pi RL$ (surface area), $J = JA = 2\pi L$ (first total curvature) and $K = KA = 0$ (second total curvature). Using all these relations with the appropriate subscripts in Equation 1.159 along with $\Delta\lambda = \lambda_2 - \lambda_1 = -R_2 - (-R_1) = R_1 - R_2$, it turns out that $\Omega^{(2)} - \Omega^{(1)} = 0$. Thus, the free energy for a cylindrical bubble is independent of the position of the dividing surface. It should be noted that in this case the second bending moment C_K is immaterial since it is multiplied by $K = 0$, but the first bending moment C_J is indispensable in making the free energy of the system independent of the position of the dividing surface.

Case 2: The Spherical Bubble

The relevant geometric parameters for a spherical bubble of radius R are the bubble's $V = 4/3\pi R^3$ (volume), $A = 4\pi R^2$ (area), $J = JA = 8\pi R$ (first total curvature), and $K = KA = 4\pi$ (second total curvature). Again, Equation 1.159 becomes $\Omega^{(2)} - \Omega^{(1)} = 0$, which means that the free energy for a spherical bubble is independent of the position of the dividing surface. However, in this case both curvature potentials, C_J and C_K, of the generalized thermodynamic theory are necessary to achieve this result.

We conclude that shifting of the dividing surface is possible within the framework of the generalized theory of capillarity without violating the conservation requirement for the total free energy of the system.

1.2.8   SUMMARY AND CONCLUSIONS

The generalized theory of capillarity outlined in the first section, while following Gibbs thinking very closely, is a more general theory than the classical theory. It has no limitations on curvature of the interface, and it considers the role of the contact line explicitly. The approach identifies contact angles as fundamental properties, just as, say, curvatures, surface area, surface entropy, or the contact line length. In spite of the considerable mathematical complexities, the theory is conceptually simpler than the classical theory. There is no moderate curvature assumption or explicit choice of a dividing surface. The choice of a dividing surface position remains open. The approach implements the thermodynamic method, as described by Callen [6], for capillary systems.

The grand canonical potential is identified as the appropriate specific free energy for a capillary system. It is shown that the difference between the surface tension and the free energy of a surface is contained in the curvature terms. For a flat surface, the specific free energy and the surface tension become identical.

One of the main features of the generalized theory is the introduction of two extensive curvature terms into the fundamental equation for surfaces. While these curvature terms also satisfy the necessary requirements of invariance, the fundamental equation remains, just like the fundamental equation for any nonsimple system, a postulate, and hence is in need of corroboration. To this end, a hydrostatic approach is presented through which an expression for the virtual work of the internal forces in the surface is developed. As the principle of virtual work is equivalent to the minimization of the free energy, a link with the thermodynamic theory is possible. It turns out that the first three moments of the virtual work expression correspond, in turn, to the surface tension and the two curvature potentials, reaffirming the correctness of the postulated fundamental equation for surfaces.

The derivation of the mechanical equilibrium conditions is mathematically so complex that corroboration through simpler types of analyses is desirable. Thus, two derivations of the generalized Laplace equation, where constancy of surface tension and curvature potentials is assumed a priori for curved interfaces, are given: one based on hydrostatics and the other on thermodynamics. These two approaches are independent but both lead to the generalized Laplace equation, Equation 1.59. In addition, a simple derivation of the generalized Laplace equation is given for both cylindrical and spherical interfaces.

The generalized theory of capillarity introduces into both the Young equation and the Neumann triangle relation a line tension term; even when at low or moderate curvatures, the classical Laplace equation is still applicable.

Finally, it is a convenience, if not a necessity, to be able to shift dividing surfaces. It is shown that the classical theory of capillarity does not allow shifting of dividing surfaces without a violation of the conservation of free energy. In the generalized theory, on the other hand, a shifting of dividing surfaces is possible.

REFERENCES

1. J. W. Gibbs. *The Scientific Papers of J. Willard Gibbs*. Vol. 1, 55–371. Dover, New York, 1961.
2. J. W. Gibbs. *Collected Works of J. W. Gibbs*. Vol. 1, Yale University Press, New Haven, CT, 1928.
3. A. I. Rusanov. "The Centennial of Gibbs' Theory of Capillarity." In *The Modern Theory of Capillarity*. Edited by F. C. Goodrich and A. I. Rusanov, 1–18. Akademie-Verlag, East Berlin, 1981.
4. C. Truesdell, ed. *Rational Thermodynamics*. 2nd ed., 20. Springer-Verlag, New York, 1984.
5. L. Boruvka and A. W. Neumann. *Journal of Chemical Physics* 66 (1977): 5464.
6. H. B. Callen. *Thermodynamics*. 1st ed. John Wiley & Sons, New York, 1960.
7. R. Defay and I. Prigogine. *Surface Tension and Adsorption*. Collaboration with A. Bellemans. Translated by D. H. Everett. Longmans-Green, London, 1966.
8. E. A. Guggenheim. *Thermodynamics*. 7th ed. North-Holland, Amsterdam, 1977.
9. R. Haase. *Thermodynamics of Irreversible Processes*. Dover, New York, 1990.
10. T. L. Hill. *Thermodynamics of Small Systems*. W.A. Benjamin, New York. Part I, 1963; Part II, 1964.
11. J. G. Kirkwood and I. Oppenheim. *Chemical Thermodynamics*. McGraw-Hill, New York, 1961.
12. A. G. McLellan. *The Classical Thermodynamics of Deformable Materials*. Cambridge University Press, New York, 1980.
13. I. Müller. *Thermodynamics*. Pitman Advanced Publishing Program, London, 1985.
14. A. Münster. *Classical Thermodynamics*. John Wiley, New York, 1970.
15. D. R. Owen. *A First Course in the Mathematical Foundations of Thermodynamics*. Springer- Verlag, New York, 1984.

16. I. Prigogine and R. Defay. *Chemical Thermodynamics*. Translated by D. H. Everett. Longmans-Green, London, 1954.
17. H. Reiss. *Methods of Thermodynamics*. Blaisdell, Toronto, 1965.
18. A. Sanfeld. *Thermodynamics of Charged and Polarized Layers*. John Wiley, New York, 1968.
19. L. Tisza. *Generalized Thermodynamics*. MIT Press, Cambridge, MA, 1966.
20. L. C. Woods. *The Thermodynamics of Fluid Systems*. Clarendon Press, Oxford, 1986 (Corrected version).
21. R. Aris. *Vectors, Tensors, and the Basic Equations of Fluid Mechanics*. Prentice-Hall, Englewood Cliffs, NJ, 1962.
22. L. D. Landau, E. M. Lifshitz, and L. P. Pitaevskii. *Electrodynamics of Continuous Media*. 2nd ed. Translated by J. B. Sykes, J. S. Bell, and M. J. Kearsley. Pergamon, New York, 1984.
23. P. G. Drazin and W. H. Reid. *Hydrodynamic Stability*. Cambridge University Press, New York, 1981.
24. F. G. Donnan, A. E. Haas, and P. M. M. Duhem. *A Commentary on the Scientific Writings of J. Willard Gibbs*. Vol. 1. Yale University Press, New Haven, CT, 1936.
25. R. Defay and I. Prigogine, *op. cit.*, ref. [7], p. xvi.
26. A. J. McConnell. *Applications of Tensor Analysis*. Dover, New York, 1931.
27. C. E. Weatherburn. *Differential Geometry of Three Dimensions*. 2 Vols. Cambridge University Press, London, 1930.
28. J. S. Rowlinson. *Chemical Society Reviews* 12 (1983): 251.
29. J. W. Gibbs, *op. cit.*, ref. [1], p. 288.
30. J. W. Gibbs, *op. cit.*, ref. [1], pp. 62, 278.
31. R. Courant. *What is Mathematics?* 258–64. Oxford University Press, Toronto, 1947.
32. A. Goetz. *Introduction to Differential Geometry*. 256–58. Addison-Wesley, Toronto, 1970.
33. M. M. Lipschutz. *Schaum's Outline Series: Theory and Problems of Differential Geometry*. 242–46. McGraw Hill, Toronto, 1969.
34. R. S. Millman and G. D. Parker. *Elements of Differential Geometry*. 188–91. Prentice-Hall, Englewood Cliffs, NJ, 1977.
35. R. Osserman. *A Survey of Minimal Surfaces*. 85–86. Dover, New York, 1986.
36. P. Chen, J. Gaydos, and A. W. Neumann. *Langmuir* 12 (1996): 5956.
37. F. P. Buff. *Journal of Chemical Physics* 19 (1951): 1591.
38. F. P. Buff. *Journal of Chemical Physics* 23 (1955): 419.
39. F. P. Buff. *Journal of Chemical Physics* 25 (1956): 146.
40. F. P. Buff and H. Saltsburg. *Journal of Chemical Physics* 26 (1957): 23.
41. F. P. Buff and H. Saltsburg. *Journal of Chemical Physics* 26 (1957): 1526.
42. F. P. Buff. "The Theory of Capillarity." In *Encyclopedia of Physics, Vol. X, Structure of Liquids*. Edited by S. Flügge, 281–304. Springer-Verlag, Berlin, 1960.
43. T. L. Hill. *Journal of Chemical Physics* 19 (1951): 1203.
44. J. W. Gibbs. *op. cit.*, ref. [1], pp. 219–331.
45. S. D. Poisson. *Mem. Cl. Sci. Mathem. Phys., Part 2*. 167–225. Institute de France, 1812.
46. F. Casorati. *Acta Mathematica* 14 (1890–1891): 95.
47. J. C. C. Nitsche. *Lectures on Minimal Surfaces*. Vol. 1, 23–24. Cambridge University Press, New York, 1989.
48. W. Helfrich. *Zeitschrift fur Naturforschung* 28C (1973): 693.
49. C. E. Weatherburn. *op. cit.*, ref. [27], p. 243.
50. G. Thomsen. *Über konforme Geometrie. I. Grundlagen der konformen Flächentheorie*. 31–56. Abh. Math. Sem., University of Hamburg 3, 1924.

51. U. Dierkes, S. Hildebrandt, and A. Küster. *Minimal Surfaces*. 2 Vols. Springer-Verlag, New York, 1992.
52. C. L. Murphy. PhD Thesis, *Thermodynamics of Low Tension and Highly Curved Interfaces*, University of Minnesota, Department of Chemical Engineering, University Microfilms, Ann Arbor, MI, 1966.
53. J. C. Melrose. *Industrial & Engineering Chemistry* 60, no. 3 (1968): 53.
54. J. C. Melrose. "Thermodynamic Aspects of Capillarity." In *Applied Thermodynamics*. Chairman K.-C. Chao, 249–66. American Chemical Society, Washington, DC, 1968.
55. J. C. Melrose. "Thermodynamics of Surface Phenomena." In *Proceedings of the International Conference on Thermodynamics*. Edited by P. T. Landsberg, 273–86. Cardiff, UK, Butterworths, London, 1970.
56. D. W. Hoffman and J. W. Cahn. *Surface Science* 31 (1972): 368.
57. J. W. Cahn and D. W. Hoffman. *Acta Metallurgica* 22 (1974): 1205.
58. L. E. Scriven. "Equilibrium Bicontinuous Structures." In *Surfactants in Solution*. Edited by K. L. Mittal. Vol. 2, 877–93. Plenum Press, New York, 1977.
59. R. E. Benner, Jr., L. E. Scriven, and H. T. Davis. "Structure and Stress in the Gas-Liquid-Solid Contact Region." In *Faraday Symposia of the Chemical Society, Vol. 16, Structure of the Interfacial Region*. Edited by D. H. Whiffen, 169–90. London, 1981.
60. J. S. Rowlinson and B. Widom. *Molecular Theory of Capillarity*. Clarendon Press, Oxford, 1982.
61. J. S. Rowlinson. *Journal of the Chemical Society Faraday Translation II* 79 (1983): 77.
62. J. I. D. Alexander and W. C. Johnson. *Journal of Applied Physics* 58 (1985): 816.
63. W. C. Johnson and J. I. D. Alexander. *Journal of Applied Physics* 59 (1986): 2735.
64. M. E. R. Shanahan and P. G. de Gennes. "Equilibrium of the Triple Line Solid/Liquid/Fluid of a Sessile Drop." In *Adhesion 11*. Edited by K. W. Allen, 71–81. Elsevier Applied Science, New York, 1987.
65. M. E. R. Shanahan. *Journal of Adhesion* 20 (1987): 261.
66. M. E. R. Shanahan. *Revue de Physique Appliquée* 23 (1988): 1031.
67. M. E. R. Shanahan. *Journal of Physics D: Applied Physics* 23 (1990): 321.
68. V. S. Markin, M. M. Kozlov, and S. L. Leikin. *Journal of the Chemical Society Faraday Translation II* 84 (1988): 1149.
69. M. M. Kozlov and V. S. Markin. *Journal of Colloid and Interface Science* 138 (1990): 332.
70. T. D. Gurkov and P. A. Kralchevsky. *Colloids and Surfaces* 47 (1990): 45.
71. I. B. Ivanov and P. A. Kralchevsky. "Mechanics and Thermodynamics of Curved Thin Films." In *Thin Liquid Films*. Edited by I. B. Ivanov, 49–129. Marcel Dekker, New York, 1988.
72. P. A. Kralchevsky. *Journal of Colloid and Interface Science* 137 (1990): 217.
73. P. A. Kralchevsky and T. D. Gurkov. *Colloids and Surfaces* 56 (1991): 101.
74. J. Gaydos, L. Boruvka, and A. W. Neumann. *Langmuir* 7 (1991): 1035.
75. V. S. Markin, M. M. Kozlov, and S. L. Leikin. *Colloid Journal of the USSR* 51 (1989): 768.
76. V. S. Markin and M. M. Kozlov. *Langmuir* 5 (1989): 1130.
77. M. M. Lipschutz. *op. cit.,* ref. [33], p. 246.
78. H. B. Callen. *Thermodynamics*. 2nd ed., 157. John Wiley, New York, 1985.
79. H. B. Callen. *op. cit.,* ref. [78], p. 115.
80. R. C. Tolman. *The Principles of Statistical Mechanics*. 511. Dover, New York, 1938.
81. L. Boruvka, Y. Rotenberg, and A. W. Neumann. *Langmuir* 1 (1985): 40.
82. L. Boruvka, Y. Rotenberg, and A. W. Neumann. *Journal of Physical Chemistry* 89 (1985): 2714.

83. R. Courant and D. Hilbert. *Methods of Mathematical Physics.* 2 Vols. Interscience Publications, New York, 1937.
84. P. Chen, S. S. Susnar, M. Pasandideh-Fard, J. Mostaghimi, and A. W. Neumann, *Advances in Colloid and Interface Science* 63 (1996): 179.
85. C. A. Miller. *Journal of Dispersion Science Technology* 66 (1985): 159.
86. C. A. Miller and P. Neogi. *Interfacial Phenomena: Equilibrium and Dynamic Effects.* Marcel Dekker, New York, 1985.
87. L. Boruvka, Y. Rotenberg, and A. W. Neumann. *Journal of Physical Chemistry* 90 (1986): 125.
88. G. Bakker. *Kapillarität und Oberflächenspannung, Wien-Handbuch der Exp. Phys. VI,* Leipzig, 1928.
89. J. Gaydos. *Langmuir* 10 (1994): 3365.
90. A. W. Adamson. *Physical Chemistry of Surfaces.* 5th ed. John Wiley, New York, 1990.
91. J. W. Gibbs. *op. cit.,* ref. [1], p. 227.
92. J. W. Gibbs. *op. cit.,* ref. [1], p. 234.
93. S. Ono and S. Kondo. "Molecular Theory of Surface Tension in Liquids." In *Encyclopedia of Physics, Vol. X, Structure of Liquids.* Edited by S. Flügge, 134–280. Springer-Verlag, Berlin, 1960.
94. G. Navascues. *Reports on Progress in Physics* 42 (1979): 1131.
95. Y. Rotenberg, L. Boruvka, and A. W. Neumann. *Langmuir* 2 (1986): 533.

2 Thermodynamics of Simple Axisymmetric Capillary Systems

John Gaydos and A. Wilhelm Neumann

CONTENTS

2.1 THE MECHANICS OF AXISYMMETRIC INCOMPRESSIBLE EQUILIBRIUM SYSTEMS

A generalized, thermodynamic approach to systems with nonignorable surface and line regions in Chapter 1 derived extended mechanical equilibrium relations for both the surface and the line boundary regions of uncharged capillary systems [1]. The conditions so obtained are generalizations of the Laplace equation of capillarity and the corresponding Young equation or Neumann triangle relation originally considered by Gibbs [2]. In the general formulation these relations, on the one hand, are quite complex because they include a detailed analysis of the higher-order curvature dependence of both the surface and the line boundaries. On the other hand, the relations are limited since they are restricted to one particular form of the free energy, namely the specific free energy expression $\omega^{(\alpha\beta)} = \gamma_\infty^{(\alpha\beta)} + C_J J + C_K K$, where $\gamma_\infty^{(\alpha\beta)}$ denotes the usual surface tension between the bulk phases (α) and (β) where these bulk phases can be solid, liquid, or vapor. The remaining two terms denote free energy contributions associated with the surface's mean curvature $J = c_1 + c_2$ and its Gaussian curvature $K = c_1 c_2$ where c_1 and c_2 are the principal normal curvatures of the surface [3]. The factors C_J and C_K denote the associated bending moments, respectively. Furthermore, these very general mechanical equilibrium conditions are so general that they are not of immediate necessity in many experimental studies that deal with either spherically symmetric or axisymmetric systems.

In this chapter an investigation of common axisymmetric systems (see Figures 2.1 and 2.4) that include contact lines illustrates the generalized theory and its use in many common experimental situations. Along with detailed calculations that show the manner by which the mechanical equilibrium conditions (i.e., the Laplace equation, Young equation, and Neumann triangle relation) are derived using calculus of variations, this section also provides a mathematical justification for the influence of surface-phase compressibility on the estimate of the line tension. In other words, the slope of a plot of $\cos\theta_1$ (the cosine of the contact angle) versus $1/R$ (the reciprocal of the contact line radius) does not yield the line tension directly [4] if the surface phase can be noticeably compressed upon going from one size of drop to another size where θ_1 denotes the usual contact angle and R is the radius of the drop's solid-liquid surface. It is also demonstrated for liquid–fluid lens systems that, without line tension, the underlying surface (that is, the curve that we shall denote by the function $\xi_1(r)$ in Figure 2.4), has a cusp at the location where all surfaces intersect to form the contact line. As a direct consequence of this cusp occurrence, the underlying surface may have a point of infinite curvature along the axisymmetric curve that forms the boundary between the lower liquid and the remainder of the surface system.

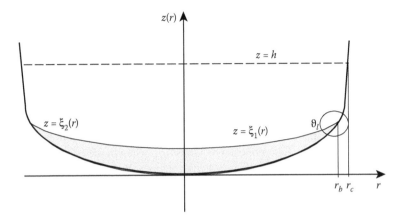

FIGURE 2.1 Schematic of an idealized solid–liquid–fluid capillary system with one deformable liquid–fluid interface and one contact line at radial position $r = r_b$.

Consequently, one must, even for this relatively common situation, either abandon the "moderate curvature approximation" of Gibbs (see discussion in Chapter 1) or introduce a boundary or line tension to prevent this cusp occurrence for moderately curved systems. Thus, consideration of line tension effects would seem unavoidable.

This chapter contains a formulation of what we consider to be the proper fundamental equations for bulk, surface, and linear regions of an axisymmetric capillary system. For these systems, governed by these fundamental equations, the explicit equilibrium conditions and the nature of the internal forces are derived by exact mathematical means using calculus of variations and the principle of energy minimization. This energy principle can be stated as follows: A possible state of an open system is a stable equilibrium state if and only if its total free energy is a local minimum on the set of all possible states of the system with the same thermal and chemical conditions.

We consider a typical axisymmetric solid surface bounding a liquid–fluid system to have a geometry as shown in Figure 2.1. In Figure 2.1, the boundary surfaces are illustrated by the curve $\xi_2(r)$, which is the axisymmetric curve that separates the lower solid phase from the liquid and fluid phases and the curve $\xi_1(r)$, which separates the liquid phase from the upper fluid phase. These curves are arbitrary functions that generate their axisymmetric surface by revolving their curve about the vertical axis of symmetry, denoted here as the $z(r)$ axis. The contact line is included directly into the deliberations by endowing the boundary or contact line with a specific linear free energy $\omega^{(l)} = \omega^{(l)}(T, \mu)$ that is equal to a constant line tension $\sigma_\infty^{(l)} = \sigma_\infty^{(l)}(T, \mu)$ in the first approximation. This notation is used to indicate that the linear free energy and the line tension are dependent on the thermal and chemical condition of the capillary system. When operative, the line tension, as shown by the contact line boundary condition or Young equation, will influence the magnitude of the contact angle. One conclusion of importance is that for a liquid in an axisymmetric cylinder or capillary tube, even a very large line tension value will have no effect on the contact angle that the liquid forms with the solid surface of the cylinder.

2.1.1 The Solid–Liquid–Fluid System

In the arrangement illustrated in Figure 2.1, a liquid is shown to be in equilibrium with another fluid (this situation also includes the case where the other fluid is the liquid's vapor) and it is in contact with a rigid solid wall [5]. The solid does not react chemically in any fashion with the liquid and the wetting liquid forms a unique contact angle on the wall. Whenever a liquid–fluid surface comes into contact with a solid phase there are a number of associated phenomena that contribute to the equilibrium shape of the interface. In this section, these phenomena are characterized in terms of the principle of virtual work, first proposed by Gauss [6], whereby all effects behave such that the energy of the total system from all contributions does not vary under arbitrary virtual displacements that are compatible with the system's constraints. The system's total energy will consist of contributions from the bulk, surface, and line regions of the composite system that are subject to a single constraint (as described below).

If a free surface separating two fluids (either both liquids or one liquid and one gas) is to exist and be in equilibrium, then the elements of a given fluid in the neighborhood of the interface must be more attracted to fluid elements of their own kind than to fluid elements of the other fluid near the interface. If this were not the case, then the interfacial configuration would soon vanish as the two fluids would mix and destroy the interface. The difference in the strength of attraction between the two fluids in the vicinity of the interface causes a reduction in the density of the fluid at that location since each distinct fluid phase attempts to pull itself away from the zone of contact. The energy associated with this pulling away of mass from the interfacial region must be proportional to the interfacial area so that

$$\Omega^{(\alpha\beta)} = \omega^{(\alpha\beta)} A^{(\alpha\beta)} \quad \text{where} \quad (\alpha\beta) = \{(lv),(lf)\}, \tag{2.1}$$

where $\Omega^{(\alpha\beta)}$ is the grand canonical potential of either a liquid–vapor (lv) or a liquid–fluid (lf) interface, $\omega^{(\alpha\beta)}$ is the specific (with respect to area) free energy, and $A^{(\alpha\beta)}$ denotes the boundary area separating the two bulk phases. In addition to the energy associated with this mobile interface, there is also an analogous surface energy corresponding to a liquid or fluid when it wets a solid, rigid boundary. In this case, the solid boundary or wall will not vanish (i.e., as in the case of liquids that mix to form solutions) if the net attraction of fluid elements in the neighborhood of the wall is toward the wall since the wall is rigid. What does occur is a translation of the liquid–fluid boundary parallel to the solid boundary. The energy corresponding to this effect can be written as

$$\Omega^{(s\alpha)} + \Omega^{(s\beta)} = \omega^{(s\alpha)} A^{(s\alpha)} + \omega^{(s\beta)} A^{(s\beta)}, \tag{2.2}$$

whereas $\omega^{(s\alpha)}$ and $\omega^{(s\beta)}$ are specific free energies for the solid–liquid and the solid–fluid surfaces, $A^{(s\alpha)}$ is the solid wall area wetted by the liquid that must include the three-phase contact line formed by the intersection of solid–liquid–fluid phases, and $A^{(s\beta)}$ is the solid wall area wetted by the other fluid.

For exactly the same reason that surface tension arises among two fluids that are more attracted to themselves than to each other, one realizes that an analogous effect may occur in the three-phase intersection zone that characterizes the meeting of solid–liquid–fluid phases. This contact line force is known as the line tension. It is likely to be very small in comparison with the other energy contributions in most systems and only subtle changes to quantities like the contact angle or line curvature may be expected. The energy associated with the line tension contribution [4,7] is given by the product of the specific, linear free energy of the contact line region $\omega^{(s\alpha\beta)}$ and the total length of the contact line $L^{(s\alpha\beta)}$, or

$$\Omega^{(s\alpha\beta)} = \omega^{(s\alpha\beta)} L^{(s\alpha\beta)}. \tag{2.3}$$

In most capillary systems, the dominant energy contribution to the system is from the intrinsic bulk free energy of each phase (i.e., $\omega^{(v)} = -P^{(v)}$, where $P^{(v)}$ is the pressure of the volume phase). Also present is a gravitational potential energy, denoted by φ per unit mass, which is a function of position within the system. The contribution of this external field to the energy is integrated over the entire volume of interest that includes both the liquid-fluid boundary and the three-phase contact line boundary.

Finally, if the amount of liquid present in the system is finite, then one must consider either the mass or the volume to be a constraint placed upon the system. For an incompressible system, a common method of handling the restriction is to introduce a Lagrangian multiplier, say λ, which multiplies the liquid volume $V^{(l)}$ to form a new energy term.

2.1.2 THE PRINCIPLE OF VIRTUAL WORK

According to the principle of virtual work, the total free energy, denoted herein by Ω_Σ, must vanish for any arbitrary variation that does not violate the constraints placed upon the system. For an incompressible system, the volume constraint is the only constraint that is present. One finds—in complete agreement with intuition— that one obtains a mechanical equilibrium condition across the liquid-fluid interface and another condition along the three-phase contact line where all three phases intersect to form the contact line. No condition exists across the solid boundary since it is considered to be rigid and not subject to any variation. The next sections develop axisymmetric free energy expressions for bulk, surface, and line regions of the system that sum to give this total free energy. For simplicity, we consider a single-component, incompressible, three-phase system as shown in Figure 2.1 consisting of a solid bowl enclosing the two-phase, single component.

2.1.3 FREE ENERGY EXPRESSIONS FOR THE BULK REGIONS

The free energy of each bulk phase is written as an integral over the volume of the phase in question. For the lower liquid phase illustrated in Figure 2.1 the volume integral is given by

$$\Omega^{(l)} = 2\pi \int_0^{r_b} |r| \int_{\xi_2}^{\xi_1} \omega^{(l)} dz dr, \tag{2.4}$$

where $\omega^{(l)}$ represents the specific free energy density of the liquid phase and the second integral implies that the volume considered is between the solid surface and the liquid–fluid surface. For an isotropic fluid phase, the specific free energy $\omega^{(l)}$ is equal to the negative of the local pressure (i.e., $\omega^{(l)} = -P^{(l)}$). For the upper fluid phase the corresponding term is

$$\Omega^{(v)} = 2\pi \int_0^{r_b} |r| \int_{x_i}^h \omega^{(v)} dz dr + 2\pi \int_{r_b}^{r_c} |r| \int_{\xi_2}^h \omega^{(v)} dz dr, \tag{2.5}$$

where $\omega^{(v)}$ represents the specific free energy density of the upper fluid phase and the second integral implies that the volume considered is between the liquid-fluid surface and the upper boundary of the system. If it is assumed that the intrinsic bulk free energy is completely independent of any contribution to the free energy from any external potential field, then it may be assumed that both $\omega^{(l)}$ and $\omega^{(v)}$ are independent of the vertical z-coordinate. Thus, after a single integration, the total bulk expressions are given as

$$\Omega^{(V)} = \Omega^{(l)} + \Omega^{(v)}$$

$$= 2\pi \left(\int_0^{r_c} \omega^{(v)} (h - \xi_2) r dr + \int_0^{r_b} (\omega^{(l)} - \omega^{(v)}) \xi_1 r dr - \int_0^{r_b} (\omega^{(l)} - \omega^{(v)}) \xi_2 r dr \right). \tag{2.6}$$

2.1.4 EXTERNAL BODY FORCES

The presence of an external gravitational potential or force field also influences the total expression for the system's free energy. This influence arises because the external field changes the local value of the stress tensor and any change in the stress tensor will mean that there will be an alteration in both the pressure (in the bulk regions) and the surface tension (in the surface regions). In the derivation to follow, the gravitational field strength g is assumed to be uniform over the extent of the system and to be directed opposite to the axis of symmetry (i.e., the z-axis as shown in Figure 2.1).

The free energy associated with the gravitational potential acting on the bulk regions of the system is given by two expressions. For the lower liquid the appropriate term is

$$\Omega_g^{(l)} = 2\pi \int_0^{r_b} |r| \int_{\xi_2}^{\xi_1} \rho^{(l)} gz dz dr, \tag{2.7}$$

where $\rho^{(l)}$ represents the density of the liquid phase and the second integral implies that the volume considered is between the solid surface and the liquid-fluid surface. For the upper fluid phase the corresponding term is

$$\Omega_g^{(v)} = 2\pi \int_0^{r_b} |r| \int_{\xi_1}^h \rho^{(v)} gz dz dr + 2\pi \int_{r_b}^{r_c} |r| \int_{\xi_2}^h \rho^{(v)} gz dz dr, \tag{2.8}$$

where $\rho^{(v)}$ represents the density of the upper fluid phase and the integrals imply that the volume considered is between the liquid-fluid surface and the upper boundary of the system. It is also possible to write free energy expressions for the influence of gravity on the density of both the surface phases and the linear phase that is associated with the contact line. Essentially, one needs only the geometric expressions for the surface area and the contact line length to be able to write the free energy expressions.

The free energy associated with the gravitational potential of the liquid–solid interface is given by

$$\Omega_g^{(sl)} = \iint_{A^{(sl)}} \rho^{(sl)} g \xi_2 dA = 2\pi \int_0^{r_b} \rho^{(sl)} g \xi_2(r)\left(1+[\xi_2'(r)]^2\right)^{1/2} r dr, \qquad (2.9)$$

where $\rho^{(sl)}$ is the surface density (defined per unit area) of the liquid-solid interface and $\xi_2'(r)$ denotes differentiation of the curve $\xi_2(r)$ with respect to the radius. The curve $\xi_2(r)$ is an arbitrary, axisymmetric function that describes the position of the liquid–solid interface (as shown in Figure 2.1). Analogous expressions arise for the fluid–liquid interface and the fluid–solid interface. They are given, respectively, by

$$\Omega_g^{(lv)} = \iint_{A^{(lv)}} \rho^{(lv)} g \xi_1 dA = 2\pi \int_0^{r_b} \rho^{(lv)} g \xi_1(r)\left(1+[\xi_1'(r)]^2\right)^{1/2} r dr, \qquad (2.10)$$

where $\xi_1(r)$ is the function for the position of the axisymmetric fluid-liquid interface and

$$\Omega_g^{(sv)} = \iint_{A^{(sv)}} \rho^{(sv)} g \xi_2 dA = 2\pi \int_{r_b}^{r_c} \rho^{(sv)} g \xi_2(r)\left(1+[\xi_2'(r)]^2\right)^{1/2} r dr, \qquad (2.11)$$

where $\rho^{(lv)}$ is the surface density of the fluid-liquid interface and $\rho^{(sv)}$ is the surface density of the fluid–solid interface. Finally, to complete the specification of the gravitational energy expressions for this system one can include a linear term to represent the free energy associated with the mass content of the three-phase contact line as

$$\Omega_g^{(slv)} = 2\pi r_b \rho^{(slv)} g \xi_1(r_b), \qquad (2.12)$$

where $2\pi r_b$ is the total circumference or length of the contact line and $\rho^{(slv)}$ represents the linear density (per unit length) of the contact line. The contact line free energy may also be expressed as an integral expression by writing

$$\Omega_g^{(slv)} = 2\pi \int_0^{r_b} \rho^{(slv)} g \xi_1 dr. \qquad (2.13)$$

Their total contribution to the final summed energy would, in most cases, be quite small and we shall not discuss their potential influence. However, in a casual sense it is possible to conclude that in situations where these effects are not ignorable one would find that the value of specific surface potential energy would depend upon the direction of the external gravitational field as a result of the dependence of $\Omega_g^{(\alpha\beta)}$ on the slope ξ_j' of the surface denoted by superscript $(\alpha\beta)$.

2.1.5 FREE ENERGY EXPRESSIONS FOR THE SURFACE AND LINE REGIONS

The free energy or grand canonical potential for each interface region is expressed in the following manner. For the liquid–solid interface one has

$$\Omega^{(sl)} = \iint_{A^{(sl)}} \omega^{(sl)} dA = 2\pi \int_0^{r_b} \omega^{(sl)} \left(1+\left[\xi_2'(r)\right]^2\right)^{1/2} r\,dr, \qquad (2.14)$$

where $\omega^{(sl)}$ denotes the specific surface grand canonical potential or specific surface free energy (defined per unit area) of the liquid–solid interface and $\xi_2'(r)$ denotes differentiation of the solid surface curve $\xi_2(r)$ with respect to the radius. For the fluid–liquid interface and the fluid–solid interface the corresponding expressions are given by

$$\Omega^{(lv)} = \iint_{A^{(lv)}} \omega^{(lv)} dA = 2\pi \int_0^{r_b} \omega^{(lv)} \left(1+\left[\xi_1'(r)\right]^2\right)^{1/2} r\,dr, \qquad (2.15)$$

and

$$\Omega^{(sv)} = \iint_{A^{(sv)}} \omega^{(sv)} dA = 2\pi \int_{r_b}^{r_c} \omega^{(sv)} \left(1+\left[\xi_2'(r)\right]^2\right)^{1/2} r\,dr. \qquad (2.16)$$

Finally, to complete the specification of the energy expressions for this system a linear term to represent the energy content of the three-phase contact line is included as

$$\Omega^{(slv)} = 2\pi r_b \omega^{(slv)}, \qquad (2.17)$$

where $2\pi r_b$ is the total circumference or length of the contact line and $\omega^{(slv)}$ represents the specific free energy (per unit length) of the contact line. The contact line free energy may also be expressed as an integral expression by writing

$$\Omega^{(slv)} = \frac{2\pi}{r_b} \int_0^{r_b} \omega^{(slv)} r_b dr, \qquad (2.18)$$

provided the specific linear free energy $\omega^{(slv)}$ is a constant that is independent of the radius.

For a system that is closed and isolated from the outside universe, both the heat content (entropy) and the mass of the system must remain constant. However, if both deformable phases are open and incompressible, then the mass constraint can be replaced by an equivalent volume constraint. The volume of each portion of the deformable system, sketched in Figure 2.1, is given by

$$V^{(l)} = 2\pi \int_0^{r_b} (\xi_1 - \xi_2) r\,dr, \tag{2.19}$$

and

$$V^{(v)} = 2\pi \int_0^{r_b} (h - \xi_1)\,r\,dr + 2\pi \int_{r_b}^{r_c} (h - \xi_2) r\,dr, \tag{2.20}$$

where $V^{(l)}$ is the volume of the lower liquid phase and $V^{(v)}$ is the volume of the upper fluid phase forming the capillary system of interest. Obviously, if the volume of one phase in the system is incompressible, then both the total volume and the volume of the other phase in the two phase fluid-liquid system must be constant. Mathematically, this means that the volume constraint is considered to consist of requiring that either volume $V^{(l)}$ or $V^{(v)}$ remain constant and variations on the total energy Ω_Σ to determine the mechanical equilibrium conditions are performed subject to this volume constraint.

2.1.6　THE VARIATIONAL PROBLEM

The mechanical equilibrium conditions arise directly from the conditions that are necessary to insure that the total free energy integral is stationary. The total free energy integral, ignoring both surface and line gravitational energies, is given by the energy summation

$$\Omega_\Sigma = \Omega^{(l)} + \Omega^{(v)} + \Omega_g^{(l)} + \Omega_g^{(v)} + \Omega^{(sl)} + \Omega^{(sv)} + \Omega^{(lv)} + \Omega^{(slv)} \tag{2.21}$$

or, more explicitly, as

$$\frac{\Omega_\Sigma}{2\pi} = \int_0^{r_b} f(r)r\,dr - \int_0^{r_b} \left\{ \frac{1}{2}\Delta\rho g \xi_2^2 - \frac{\omega^{(slv)}}{r_b} + \left(\omega^{(l)} - \omega^{(v)}\right)\xi_2 \right\} r\,dr$$
$$+ \int_0^{r_c} \left\{ \frac{1}{2}\rho^{(v)} g\left(h^2 - \xi_2^2\right) - \omega^{(sv)}\left(1 + [\xi_2']^2\right)^{1/2} + \omega^{(v)}\left(h - \xi_2\right) \right\} r\,dr, \tag{2.22}$$

where the density difference $\Delta\rho = \rho^{(l)} - \rho^{(v)}$ must be greater than zero for the system to be stable and where

$$f(r) = \omega^{(lv)}\left(1 + [\xi_1']^2\right)^{1/2}\left(\omega^{(sl)} - \omega^{(sv)}\right)\left(1 + [\xi_2']^2\right)^{1/2} + \frac{1}{2}\Delta\rho g \xi_1^2 + \left(\omega^{(l)} - \omega^{(v)}\right)\xi_1 \tag{2.23}$$

is a composite function that depends implicitly on the radius.

The calculus of variations problem for this situation is written in the form

$$\delta(\Omega_\Sigma + \lambda V^{(l)}) = 0, \qquad (2.24)$$

so that the volume constraint is included directly in the function that is subject to the variations. The solid is assumed to be rigid and undeformable so that no variation in the position of the curve $z = \xi_2(r)$ is possible; however, the boundary point $r = r_b$ is considered to slide along the solid and continuous variations in the position of the curve $z = \xi_1(r)$ are permissible. It is also assumed that the bulk, surface, and line phases or regions are perfectly homogeneous energetically, that the liquid and fluid phases are completely deformable, and that each phase is discrete so that mixing and adsorption effects are ignored. This free energy integral, with dependent variable $\xi_1(r)$, gives rise to a calculus of variations problem with one free end point at $r = r_b$. The Laplace equation of capillarity for the liquid-fluid interface arises from the Euler-Lagrange equation for the curve $\xi_1(r)$, subject to the subsidiary condition that the system is incompressible with constant volume. The contact line equilibrium condition, called the Young equation of capillarity, comes from the transversality condition that applies at the end point $r = r_b$.

It is possible to collect three functions of the form

$$f(r,z,z') = f(r,\xi_j,\xi_j') \quad \text{where } j = 1,2, \qquad (2.25)$$

which are given by the expressions

$$f_1(r,\xi_1,\xi_1') = \omega^{(lv)}\left(1+[\xi_1']^2\right)^{1/2} + \frac{1}{2}\Delta\rho g\xi_1^2 + \{\lambda + (\omega^{(l)} - \omega^{(v)})\}\xi_1, \qquad (2.26)$$

$$f_2(r,\xi_2,\xi_2') = \omega^{(sl)}\left(1+[\xi_2']^2\right)^{1/2} + \frac{1}{2}\rho^{(v)}g(h^2 - \xi_2^2) + \omega^{(v)}(h-\xi_2), \qquad (2.27)$$

and

$$f_3(r,\xi_2,\xi_2') = (\omega^{(sl)} - \omega^{(sv)})\left(1+[\xi_2']^2\right)^{1/2} - \frac{1}{2}\Delta\rho g\xi_2^2 + \frac{\omega^{(slv)}}{r_b} - \{\lambda + (\omega^{(l)} - \omega^{(v)})\}\xi_2. \qquad (2.28)$$

Using these definitions, the calculus of variations problem, defined by Equation 2.24, can be recast into the simple integral forms

$$\delta\left(\int_0^{r_b} f_1(r,\xi_1,\xi_1')rdr + \int_0^{r_c} f_2(r,\xi_2,\xi_2')rdr + \int_0^{r_b} f_3(r,\xi_2,\xi_2')rdr\right) = 0. \qquad (2.29)$$

The necessary conditions for three stationary integrals consist of three Euler-Lagrange equations for each of the surfaces under consideration and two transversality conditions for the behavior of the contact line [8–13]. The first integral, which corresponds to the liquid-fluid surface term, yields the Euler-Lagrange equation

$$\frac{\partial(f_1 r)}{\partial \xi_1} - \frac{d}{dr}\frac{\partial(f_1 r)}{\partial \xi_1'} = 0. \tag{2.30}$$

The lower end point of the integration range, at $r = 0$, is fixed by the symmetry of the system to vary in such a fashion that only its elevation changes. Therefore, the radial position of the lower end point remains fixed and its elevation (i.e., position along the z-axis) satisfies the natural boundary condition

$$\left(\frac{\partial(f_1 r)}{\partial \xi_1'}\right)_{r=0} = 0, \tag{2.31}$$

which may be evaluated to show that the slope $\xi_1'(r = 0)$ of the liquid-fluid surface is zero at the origin of the system. This conclusion is then used to determine the value of the arbitrary constant (i.e., the undetermined multiplier λ). The upper end point, at $r = r_b j \cdot$ is free to slide along the solid surface according to the transversality condition

$$(f_1 r) + (\xi_2' - \xi_1')\frac{\partial(f_1 r)}{\partial \xi_1'} = 0. \tag{2.32}$$

The second integral has no free end points (since the incompressible liquid assumption means that the system's total volume is fixed by the condition $r = r_c$ or $z = h$ being constant; see Figure 2.1) so there are no transversality conditions for this term. The corresponding Euler-Lagrange equation is given by

$$\frac{\partial(f_2 r)}{\partial \xi_2} - \frac{d}{dr}\frac{\partial(f_2 r)}{\partial \xi_2'} = 0. \tag{2.33}$$

Finally, for the third integral in Equation 2.29 the Euler-Lagrange equation is

$$\frac{\partial(f_3 r)}{\partial \xi_2} - \frac{d}{dr}\frac{\partial(f_3 r)}{\partial \xi_2'} = 0, \tag{2.34}$$

and the only transversality condition at $r = r_b$ is given by

$$(f_3 r) + (\xi_2' - \xi_2')\frac{\partial(f_3 r)}{\partial \xi_2'} = 0, \tag{2.35}$$

which shows explicitly that the last term vanishes. Physically, this result occurs because the solid surface was assumed to be rigid and unable to undergo shear.

2.1.7 THE MECHANICAL EQUILIBRIUM CONDITIONS

The mathematical manipulations required to solve Equations 2.30 through 2.35 are tedious but straightforward. The solutions for Equations 2.30 through 2.35 are given [14], respectively, by

$$\Delta \rho g r \xi_1 + \lambda r + \left(\omega^{(l)} - \omega^{(v)} \right) r - \omega^{(lv)} \left(1 + \left[\xi_1' \right]^2 \right)^{-3/2} \left(\xi_1' + \left[\xi_1' \right]^3 + r \xi_1'' \right) = 0, \qquad (2.36)$$

$$\xi_1'(r = 0) = 0, \qquad (2.37)$$

and

$$\left(\frac{1}{2} \Delta \rho g \left[\xi_1(r_b) \right]^2 + \lambda \xi_1(r_b) + \left(\omega^{(l)} - \omega^{(v)} \right) \xi_1(r_b) \right) \left(1 + \left[\xi_1'(r_b) \right]^2 \right)^{1/2}$$
$$+ \omega^{(lv)} \left(1 + \left[\xi_1'(r_b) \right] \right) \left[\xi_2'(r_b) \right] = 0, \qquad (2.38)$$

for the first integral. In Expression 2.38, we used $\xi_1(r = r_b) = \xi_2(r = r_b)$ to insure that the curve representing the liquid-fluid surface contacts or intersects the curve representing the solid-liquid surface at radius r_b that corresponds to the contact line. Equation 2.33 yields

$$\rho^{(v)} g r \xi_2 + \omega^{(sv)} \left(1 + \left[\xi_2' \right]^2 \right)^{-3/2} \left(\xi_2' + \left[\xi_2' \right]^3 + r \xi_2'' \right) = 0, \qquad (2.39)$$

for the Euler-Lagrange equation of the second integral in Equation 2.29. For the third integral in Equation 2.29, one obtains Equation 2.34 that becomes

$$\Delta \rho g r \xi_2 + \lambda r + \left(\omega^{(l)} - \omega^{(v)} \right) r + \left(\omega^{(sl)} - \omega^{(sv)} \right) \left(1 + \left[\xi_2' \right]^2 \right)^{-3/2} \left(\xi_2' + \left[\xi_2' \right]^3 + r \xi_2'' \right) = 0, \qquad (2.40)$$

and the transversality condition Equation 2.35 for the third integral gives

$$\left(\omega^{(sl)} - \omega^{(sv)} \right) \left(1 + \left[\xi_2'(r_b) \right]^2 \right)^{1/2} - \frac{1}{2} \Delta \rho g \xi_2^2(r_b)$$
$$+ \frac{\omega^{(slv)}}{r_b} - \lambda \xi_2(r_b) - \left(\omega^{(l)} - \omega^{(v)} \right) \xi_2(r_b) = 0. \qquad (2.41)$$

Once again, the equality $\xi_1(r = r_b) = \xi_2(r = r_b)$ was used in Relation 2.41 to insure contact between the two curves at the contact line.

Rearranging Equation 2.36 gives the surface mechanical equilibrium condition for the liquid–fluid interface as

$$\lambda = \frac{\omega^{(lv)}}{r}\left(1+[\xi_1']^2\right)^{-3/2}\left(\xi_1'\{1+[\xi_1']^2\}+r\xi_1''\right)-\Delta\rho g\xi_1-\left(\omega^{(l)}-\omega^{(v)}\right)$$

$$= \omega^{(lv)}\left(\frac{\xi_1'}{r\left(1+[\xi_1']^2\right)^{1/2}}+\frac{\xi_1''}{\left(1+[\xi_1']^2\right)^{3/2}}\right)-\Delta\rho g\xi_1-\left(\omega^{(l)}-\omega^{(v)}\right), \tag{2.42}$$

or

$$\omega^{(lv)}\left(\frac{1}{R_1^{(lv)}}+\frac{1}{R_2^{(lv)}}\right)-\Delta\rho g\xi_1 = \lambda+\left(\omega^{(l)}-\omega^{(v)}\right), \tag{2.43}$$

where $R_1^{(lv)}$ and $R_2^{(lv)}$ are the principal radii of curvature of the liquid–fluid interface [15–18].

The next step in the evaluation of Equation 2.43 is to determine an expression for the single Lagrange multiplier λ. At the $r=0$ plane of symmetry, where $z=\xi_1(r=0)$, the $\xi_1(r)$ curve has a local extremum so that its slope vanishes (i.e., $\xi_1'=0$ at $r=0$). This location provides a convenient and necessary boundary in which to evaluate the multiplier λ. A careful observation of Equation 2.42 reveals that all quantities can be evaluated at this location, but that the first term inside the brackets is indeterminate since it yields a value of 0/0 at the point $(0, \xi_1(r=0))$. However, by applying l'Hôpital's rule [19] to this term, one finds that a well-defined value is obtained at this point, which is given by

$$\frac{\xi_1'}{r\left(1+[\xi_1']^2\right)^{1/2}} = \frac{\xi_1''}{\left(1+[\xi_1']^2\right)^{1/2}+r\xi_1'\xi_1''\left(1+[\xi_1']^2\right)^{-1/2}} = \frac{\xi_1''(r=0)}{1}. \tag{2.44}$$

Substituting this result into Equation 2.42 and evaluating all remaining terms at the symmetry point $(0, \xi_1(r=0))$ where the result in Equation 2.37 applies yields

$$\lambda = 2\xi_1''(r=0)\omega^{(lv)}-\Delta\rho g\xi_1(r=0)-(\omega^{(l)}-\omega^{(v)}). \tag{2.45}$$

The second derivative of the function $\xi_1(r)$ may be replaced by its radius of curvature, $R^{(lv)}$, evaluated in the plane of Figure 2.1. The general expression for the radius of curvature of a planar arc is given by [15–18]

$$\frac{1}{R^{(lv)}} = \frac{\xi_1''}{\left(1+[\xi_1']^2\right)^{3/2}} \quad \text{where } \xi_1'' > 0. \tag{2.46}$$

When Relation 2.46 is evaluated at the symmetry point, one obtains the result that

$$\frac{1}{R_o^{(lv)}} = \xi_1''(r = 0), \tag{2.47}$$

where the subscript o has been added to indicate the location at which the radius of curvature is evaluated (i.e., the origin). Replacing the function ξ_1'' by its radius of curvature at the origin yields an expression for the constant λ of

$$\lambda = \frac{2\omega^{(lv)}}{R_o^{(lv)}} - \Delta\rho g \xi_1(r = 0) + \left(P_o^{(l)} - P_o^{(v)}\right), \tag{2.48}$$

where the pressure of the bulk phase, near to but not at the location of the interface, is derived from the specific grand canonical potentials $\omega^{(l)}$ and $\omega^{(v)}$. The final version of Equation 2.30 is obtained by replacing λ using the expression above so that

$$\omega^{(lv)}\left(\frac{1}{R_1^{(lv)}} + \frac{1}{R_2^{(lv)}}\right) - \Delta\rho g\left(\xi_1(r \neq 0) - \xi_1(r = 0)\right) = \frac{2\omega^{(lv)}}{R_o^{(lv)}}. \tag{2.49}$$

This equation can also be expressed in terms of the bulk liquid and the bulk fluid phase pressures, which are measured on opposite sides of the liquid-fluid surface but in close proximity to the surface. The pressure just below the surface in the liquid phase will be denoted by $P_o^{(l)}(r)$ to indicate that it is measured at a distance r from the axis of symmetry, that it is measured just below the surface in the liquid phase, and that it is a relative pressure with respect to the datum position $r = 0$ as indicated by the subscript O. In the upper fluid phase, the pressure is denoted by $P_o^{(v)}(r)$. Once again, the pressure is measured with respect to the datum at $r = 0$ and at a distance r from the axis of symmetry; however, in this case the measurement is made above the surface in the vapor phase. When these definitions of pressure are used, one obtains

$$\omega^{(lv)}\left(\frac{1}{R_1^{(lv)}} + \frac{1}{R_2^{(lv)}}\right) = \Delta P^{(lv)}(r), \tag{2.50}$$

where the pressure jump across the liquid-vapor surface is defined as

$$\Delta P_o^{(lv)}(r) = P_o^{(l)}(r) - P_o^{(v)}(r) = \frac{2\omega^{(lv)}}{R_o^{(lv)}} + \Delta\rho g\left(\xi_1(r \neq 0) - \xi_1(r = 0)\right), \tag{2.51}$$

and it is greater than zero since the liquid pressure is greater than the fluid (vapor) pressure.

Relation 2.50 represents the mechanical equilibrium condition for the liquid–fluid surface when the specific surface free energy $\omega^{(lv)}$ is fixed but not necessarily a uniform constant at all points on the surface. It is very similar to the common form of Laplace's equation of capillarity (compare this result to Equation 2.61 below); however, it is not identical because the traditional form of Laplace's equation of capillarity assumes that the specific free energy is $\omega^{(lv)} = \gamma_\infty^{(lv)}$ (where the symbol $\gamma_\infty^{(lv)}$ denotes the usual constant surface tension for a flat surface) at every point on the surface whereas Relation 2.50 requires only that the energy $\omega^{(lv)}$ have a fixed (but potentially stratified) value on the surface.

The boundary condition of the liquid–fluid interface at the solid wall, which is known as the Young equation of capillarity, is derived from the end point relation that is given by Equation 2.38. The constant multiplier λ in Equation 2.38 can be eliminated from the final form of the boundary condition by using Equation 2.41 since $\xi_1(r = r_b) = \xi_2(r = r_b)$ in both of these relations. This equality insures that contact between liquid-fluid surface and the solid is maintained at the contact line regardless of the variation considered. Equation 2.38 can be written in the form

$$\frac{1}{2}\Delta\rho g\xi_1^2 + \lambda\xi_1 + \omega^{(lv)}\left(1+[\xi_1']^2\right)^{1/2} + \omega^{(lv)}\xi_1'\left(\xi_2'-\xi_1'\right)\left(1+[\xi_1']^2\right)^{-1/2} = 0, \qquad (2.52)$$

where both $\xi_j(r)$ surface functions are evaluated at the contact line radius $r = r_b$. Using Equation 2.41, one has that

$$\lambda\xi_2(r_b) = \lambda\xi_1(r_b) = \left(\omega^{(sl)} - \omega^{(sv)}\right)\left(1+[\xi_2'(r_b)]^2\right)^{1/2} - \frac{1}{2}\Delta\rho g\xi_1^2(r_b) + \frac{\omega^{(slv)}}{r_b}. \qquad (2.53)$$

Substituting the quantity to the right of $\lambda\xi_1$ from Equation 2.53 into Equation 2.52 eliminates λ and yields

$$\frac{1}{2}\Delta\rho g\xi_1^2 + \left(\omega^{(sl)} - \omega^{(sv)}\right)\left(1+[\xi_2'(r_b)]^2\right)^{1/2} - \frac{1}{2}\Delta\rho g\xi_2^2 + \frac{\omega^{(slv)}}{r_b}$$
$$+ \omega^{(lv)}\left(1+[\xi_1'(r_b)]^2\right)^{1/2} + \omega^{(lv)}\xi_1'\left(\xi_2'-\xi_1'\right)\left(1+[\xi_1'(r_b)]^2\right)^{-1/2} = 0, \qquad (2.54)$$

whereupon the two terms involving the density difference $\Delta\rho$ also cancel out. Factoring Equation 2.54 further yields

$$\omega^{(lv)}\left(1+[\xi_1'(r_b)]^2\right)^{1/2}$$
$$\times\left\{1 + \frac{\omega^{(sl)}-\omega^{(sv)}}{\omega^{(lv)}}\frac{\left(1+[\xi_2']^2\right)^{1/2}}{(1+[\xi_1']^2)^{1/2}} - \frac{\xi_1'(\xi_1'-\xi_2')}{1+[\xi_1']^2} + \frac{1}{r_b}\frac{\omega^{(slv)}}{\omega^{(lv)}}\frac{1}{\left(1+[\xi_1']^2\right)^{1/2}}\right\} = 0. \qquad (2.55)$$

Multiplying the terms inside the {...} brackets by $1+[\xi_1']^2$ and simplifying yields

$$1+\xi_1'\xi_2'+\frac{\omega^{(sl)}-\omega^{(sv)}}{\omega^{(lv)}}\left(1+[\xi_1']^2\right)^{1/2}\left(1+[\xi_2']^2\right)^{1/2}+\frac{1}{r_b}\frac{\omega^{(slv)}}{\omega^{(lv)}}\left(1+[\xi_1']^2\right)^{1/2}=0, \quad (2.56)$$

which represents the Young equation of capillarity in terms of the $\xi_j(r)$ surface functions.

Equation 2.56 can be expressed in the more commonly quoted form by introducing the angle relations

$$\xi_1'=\tan\vartheta_1 \quad \text{and} \quad \xi_2'=\tan\vartheta_2 \quad (2.57)$$

between the slope or tangent of the $\xi_j(r)$ functions and the angles ϑ_j between the tangent and the horizontal (see Figure 2.2). With these definitions replacing the slopes $\xi_j'(r)$ in Equation 2.56, one has a further simplification to

$$\cos\left(\vartheta_1-\vartheta_2\right)+\frac{\omega^{(sl)}-\omega^{(sv)}}{\omega^{(lv)}}+\frac{1}{r_b}\frac{\omega^{(slv)}}{\omega^{(lv)}}\cos\vartheta_2=0. \quad (2.58)$$

A quick perusal of Figure 2.2 shows that the contact angle ϑ_l, the difference between the slope of the liquid–fluid interface, and the slope of the solid wall as measured through the liquid, can be defined by the relation

$$\vartheta_l=\vartheta_2-\vartheta_1. \quad (2.59)$$

Using this definition in Equation 2.58 enables one to obtain

$$\omega^{(sv)}-\omega^{(sl)}=\omega^{(lv)}\cos\vartheta_l+\omega^{(slv)}\frac{\cos\vartheta_2}{r_b}, \quad (2.60)$$

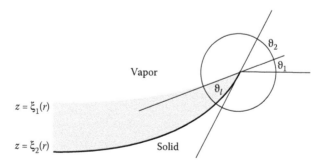

FIGURE 2.2 Schematic of the contact line region of an idealized solid–liquid–fluid capillary system showing the surface slope angles ϑ_j where $j=1,2$. The relation of slope to angle is provided by Equation 2.57 and the angle difference, $\vartheta_2-\vartheta_1$, represents the contact angle ϑ_l.

which is the Young equation of capillarity or the boundary condition for the liquid–fluid interface on the rigid solid at the position $r = r_b$. In contrast to Equation 2.54, this expression is, in terms of the specific free energies of the surfaces that intersect at the contact line, the specific free energy of the contact line and the contact angle of the liquid on the solid. Once again, we have assumed that the surface free energies are fixed (or stratified) in value on the surface and we have included the line tension effect explicitly with the last term in Equation 2.60.

2.1.8 THE MECHANICAL EQUILIBRIUM CONDITIONS FOR MODERATELY CURVED BOUNDARIES

A boundary in a capillary system is considered to be moderately curved if the local or specific free energy of the boundary (either surface or line) is a pure tension in the fashion suggested by Gibbs [2]. Thus, for a moderately curved surface and line region where the radius of curvature of the boundary is large one obtains the equalities

$$\omega^{(jk)} = \gamma_\infty^{(jk)} \quad \text{where} \quad (jk) = \{(sl),(sv),(lv)\} \quad \text{and} \quad \omega^{(slv)} = \sigma_\infty^{(slv)} \tag{2.61}$$

between the specific surface or line free energies and the usual tensions of the boundary regions. Employing these definitions, one may write the Laplace equation of capillarity (Equation 2.49) as

$$\gamma_\infty^{(lv)}\left(\frac{1}{R_1^{(lv)}} + \frac{1}{R_2^{(lv)}}\right) - \Delta\rho g \xi_1 = \lambda + \left(P^{(v)} - P^{(l)}\right), \tag{2.62}$$

or, after eliminating λ, as

$$\gamma_\infty^{(lv)}\left(\frac{1}{R_1^{(lv)}} + \frac{1}{R_2^{(lv)}}\right) = P_o^{(l)}(r=0) - P_o^{(v)}(r=0) + \Delta\rho g\left(\xi_1(r \neq 0) - \xi_1(r=0)\right), \tag{2.63}$$

and the Young equation of capillarity (Equation 2.60) as

$$\gamma_\infty^{(sv)} - \gamma_\infty^{(sl)} = \gamma_\infty^{(lv)}\cos\vartheta_l + \sigma_\infty^{(slv)}\frac{\cos\vartheta_2}{r_b}. \tag{2.64}$$

2.1.9 INTERPRETATION OF THE SURFACE BOUNDARY CONDITION

Strictly speaking, it has not been shown that the liquid–fluid interface is stable, since the Euler-Lagrange equation that is used here only yields a stationary value for the integral in question and not an extremum value. Mathematically, the conditions necessary for a stable stationary value are obtained from the second-order variation of the integral in question. This procedure is complex but tractable. However, on physical grounds, it should be obvious that the liquid-fluid interface will be stable if the density of the lower liquid phase is greater than the density of the upper fluid phase. If the local value of gravity is zero or sufficiently close to zero, the solution to Equation 2.63 is straightforward and requires the function $\xi_1(r)$ to represent a section of a circular arc. This result is physically realistic since the system will attempt to minimize its free energy by reducing its surface area to that of a sphere. From dimensional analysis, it is recognized that external body forces like gravity are characterized by length scales that are of the order (length)^3, whereas surface forces like surface tension are of characteristic length scale (length)^2 so that smaller capillary systems will be nearer to a spherical shape than larger systems [20]. Similarly, if the liquid and fluid phases have nearly identical densities, one would also expect that surface geometry could be well approximated by spherical shapes.

Finally, it should be noted that Equation 2.63 is not a new result, but was derived initially by Laplace [21] who showed that the pressure difference across a deformable surface is directly proportional to the surface's mean curvature J. However, Equation 2.63 only arises when the specific surface free energy equals the planar surface tension, or $\omega^{(lv)} = \gamma_\infty^{(lv)}$ at all points on the surface. When this condition is not satisfied one must use the more general Expression 2.50. This situation could occur, for example, if $\omega^{(jk)} = \gamma_\infty^{(jk)} + \mathbf{p} \cdot \mathbf{E}$ where $\mathbf{p}$ denotes the polarization and $\mathbf{E}$ the electric field at the surface. In 1830, Gauss proposed an alternative, conceptually superior method that has been adopted throughout this chapter, based on the principle of virtual work [22]. According to this approach, the energy of a mechanical system in equilibrium is invariant under arbitrary virtual displacements that are consistent with the constraints. For the axisymmetric system considered here, which is in thermal and chemical equilibrium *a priori*, we have considered arbitrary virtual displacements subject to a constant volume constraint that leaves the system's total free energy invariant. Similar approaches have been used by others to derive Expressions 2.63 and 2.64 [23–28].

2.1.10 Interpretation of the Contact Line Boundary Condition

If the surface of the solid wall is parallel to the z-axis, then $\vartheta_2 = \pi / 2$ and $\cos \vartheta_2 = 0$. In this case, the wall forms a solid of revolution (i.e., a capillary tube), whose major axis is the z-axis and whose radius is r_b. For any solid container that assumes this cylindrical shape, the classical Young equation, without the line tension term, is recovered as

$$\gamma_\infty^{(sv)} - \gamma_\infty^{(sl)} = \gamma_\infty^{(lv)} \cos \vartheta_l, \tag{2.65}$$

and this relation describes the behavior of the liquid's surface at the solid boundary. Thus, it may be seen from this result that one cannot measure the effect of line tension on the shape of the three-phase contact line with a cylindrical geometry. In contrast, the magnitude of the term containing the line tension is a maximum when the slope of the solid wall is zero; that is, when $\cos \vartheta_2 = 1$ and the solid wall is a horizontal plane so that the liquid phase achieves the shape of a sessile drop.

The Young equation, as stated in Equation 2.65, is often derived using a mechanistic approach that considers the surface tensions $\gamma_\infty^{(\alpha\beta)}$ to be two-dimensional stresses that can be resolved in a horizontal (i.e., in the tangential plane of the solid-liquid surface) direction. This approach was vehemently disfavored by Bikerman [29,30] who strongly advocated that Young's equation be rejected on the grounds that it failed to properly consider the vertical force balance in a direction perpendicular to the solid surface. According to Bikerman, there must be some deformation or strain in the solid to compensate for the force $\gamma_\infty^{(\alpha\beta)} \sin \vartheta_l$ that is directed away from the solid surface. However, our analysis of an undeformable solid surrounding a liquid-fluid system shows quite clearly that the Young equation by itself is the only boundary condition at the three-phase contact line. Furthermore, the uncompensated stress on the solid-liquid surface predominantly arises from the weight of the liquid and not from surface tension effects. This weight-related loading of the solid is approximately equal to

$$\Delta \rho g \left(\xi_1 - \xi_2 \right) \left(1 - \sin \vartheta_2 \right) \tag{2.66}$$

plus the effect of the pressure jump across the solid-liquid surface, see Equation 2.40. Finally, if the underlying substrate is deformable, then the Young Equation 2.64 must be replaced by two scalar relations [14].

2.1.11 NONMODERATELY CURVED BOUNDARIES

If it is assumed that the dividing surface is not in the surface of tension position of Gibbs [2,31,32] and that $C_1 + C_2 \neq 0$ (Gibbs's notation for bending moments is used) as a result, then on a shifted dividing surface, the mechanical behavior of the interface can still be represented by a uniform surface tension, $\gamma_\infty^{(\alpha\beta)}$. However, on the shifted dividing surface, the values of both the surface tension and the surface area would be different from the values for these quantities on the dividing surface in the surface of tension position [33]. This conclusion, however, violates physical reality. It is well known from mechanics that if there is a pure tension along one surface, then a shifted surface must be described by a tension and one or two distributed bending moments; otherwise, the system described by the shifted surface will not be statically equivalent to the original, unshifted surface [34].

When one uses the classical fundamental equation (see Chapter 1) given by

$$U^{(\alpha\beta)} = U^{(\alpha\beta)} \left[S^{(\alpha\beta)}, A, M_1^{(\alpha\beta)}, M_2^{(\alpha\beta)}, \dots, M_\eta^{(\alpha\beta)} \right], \tag{2.67}$$

one finds that the surface of tension position for the dividing surface is the only position that is in equilibrium. The discussion of this restriction may seem slightly excessive; however, one often encounters cases in the literature where the dividing surface is shifted to another position even when Equation 2.67 is taken as the expression for the fundamental equation and the classical Laplace equation, as given by Equation 2.63, is taken as the mechanical equilibrium condition across the surface. This restriction is not a severe drawback to the classical theory of capillarity since Gibbs envisaged the theory dealing primarily with surfaces that may be regarded as

nearly planar; that is, surfaces where the radii of curvature are very large in proportion to the thickness of the nonhomogeneous interface. However, as established by
Rotenberg, Boruvka, and Neumann [35], the position of the dividing surface in this
case is invariant if total energy is to be conserved.

What should be apparent from this argument is that any fundamental equation
that leads to a differential expression in the form

$$dU^{(\alpha\beta)} = TdS^{(\alpha\beta)} + \gamma^{(\alpha\beta)}dA + \sum_{j=1}^{\eta}\mu_j^{(\alpha\beta)}dM_j^{(\alpha\beta)} + C_1dc_1 + C_2dc_2 \qquad (2.68)$$

is limited. When one employs this equation to describe a system with a dividing
surface not in the surface of tension position, one finds that it is not even able to
reproduce the classical behavior of an interface, let alone any high-curvature refine-
ments [35]. Any comparison of the generalized theory of capillarity with the theory
of Gibbs based on an equation

$$U^{(\alpha\beta)} = U^{(\alpha\beta)}\left[S^{(\alpha\beta)}, A, M_j^{(\alpha\beta)}, c_1, c_2\right] \qquad (2.69)$$

should realize that this form of the fundamental equation is not suitable if one wishes
to describe surface systems with arbitrarily curved surfaces. We believe that Gibbs
was well aware of this deficiency but that he used Equation 2.68 only as a procedural
means to get to a formulation for a fundamental equation for moderate curvature
without explicit curvature terms. These difficulties with Gibbs's theory were recti-
fied in the generalized theory of capillarity [1,14,33,35–38], which is based on the
fundamental equation

$$U^{(\alpha\beta)} = U^{(\alpha\beta)}\left[S^{(\alpha\beta)}, A, M_j^{(\alpha\beta)}, J, K\right], \qquad (2.70)$$

where the total mean and Gaussian curvatures [15–18] are defined by the extensive,
rather than intensive, curvatures

$$J = \iint JdA \qquad (2.71)$$

and

$$K = \iint KdA \qquad (2.72)$$

with the mean and Gaussian curvatures defined by $J = c_1 + c_2$ and $K = c_1c_2$,
respectively.

The definition of the surface tension, which is completely analogous to the definition of the pressure, follows directly from the assumed form of Equation 2.70 as

$$\gamma^{(\alpha\beta)} = \left(\frac{\partial U^{(\alpha\beta)}}{\partial A}\right)_{S^{(\alpha\beta)}, M_j^{(\alpha\beta)}, J, K} \qquad (2.73)$$

or

$$\gamma^{(\alpha\beta)} = \left(\frac{\partial \Omega^{(\alpha\beta)}}{\partial A}\right)_{T,\mu_j,J,K} \tag{2.74}$$

after a suitable Legendre transformation from the total internal energy $U^{(\alpha\beta)}$ to the total grand canonical potential $\Omega^{(\alpha\beta)}$. If all surface properties are uniform or constant at each point on the surface, then one may replace the integral expressions in Equations 2.71 and 2.72 for J and K by JA and KA, respectively. These expressions for J and K are not as general as those considered in the generalized theory of Boruvka and Neumann [1], but are sufficiently general to illustrate the generalized theory for nonmoderately curved, axisymmetric surfaces [39,40]. Finally, with the moderate curvature restriction removed in the fashion suggested by the generalized theory of capillarity, one finds that the specific surface free energy has three terms

$$\omega^{(\alpha\beta)} = \gamma_\infty^{(\alpha\beta)} + C_J J + C_K K, \tag{2.75}$$

where C_J and C_K are mechanical curvature potentials. For our axisymmetric, liquid-vapor surface this surface free energy can be written as (see Equations 2.42 and 2.43)

$$\omega^{(lv)} = \gamma_\infty^{(lv)} + C_J^{(lv)} \frac{r\xi_1'' + \xi_1'(1+[\xi_1']^2)}{r(1+[\xi_1']^2)^{3/2}} + C_K^{(lv)} \frac{\xi_1'\xi_1''}{r(1+[\xi_1']^2)^2} \tag{2.76}$$

to show the explicit curvature dependence of the specific free energy; that is, $\omega^{(lv)}(r,\xi_1,\xi_1',\xi_1'')$. The importance of these conclusions cannot be overestimated: for highly curved surfaces, which are described by Equation 2.75 or Equation 2.76, one finds that the presence of these additional terms influences or modifies virtually all mechanical equilibrium conditions. The specific surface free energies, as defined in Equation 2.61, are no longer constant and, as a direct result, all expressions in the necessary stationary conditions become significantly more complicated (i.e., the Euler-Lagrange equations and the transversality conditions). Expressions for the free energy with alternative curvature measures are provided elsewhere [14,41].

The situation is somewhat simpler when the surface is axisymmetric; however, even in this case one can easily see from Equation 2.76 that the specific surface free energy $\omega^{(lv)}$ is not a constant but is a function of the variables r, $\xi_1(r)$, $\xi_1'(r)$ and $\xi_1''(r)$. Consequently, if one wishes to visualize the influence of the assumption that the specific surface free energy is not a constant but is dependent upon higher-order bending or curvature terms, then it will be necessary to generalize the previous analysis sufficiently that one may consider integrals up to and including the second derivative of the surface function $\xi_1(r)$. The details of this analysis are presented by Gaydos [14].

As described in Chapter 1, the additional quantities characterizing the contact line involve the collection of contact angles $\vartheta_{jj'}$ that represent the contact angles in the bulk phases between the jth and j'th dividing surfaces meeting at the contact line, the curvatures and the torsion of the contact line [1]. For the axisymmetric arrangement illustrated in Figure 2.1, the collection of angles can be represented by just three contact angles: ϑ_v for the vapor phase angle between the solid-vapor surface and the liquid-vapor surface measured in the usual counterclockwise fashion around the contact point, ϑ_l for the liquid phase contact angle between the liquid-vapor surface and the solid–liquid surface, and ϑ_s for the solid phase angle between the solid–liquid surface and the solid–vapor surface (see Figure 2.3). These angles are not independent but are constrained by the relation

$$\sum_{(jj')} \vartheta_{jj'} = \vartheta_v + \vartheta_l + \vartheta_s = 2\pi, \tag{2.77}$$

which indicates that any two of the three angles or a related quantity like the surface slopes ξ_1' and ξ_2' (see Equation 2.57) can be used as independent quantities for the contact line. In general, a contact line can curve in the plane of the jth surface with a geodesic curvature κ_{gj} and it can curve perpendicular to this surface in the normal plane with normal curvature κ_{nj} [1,17]. These curvatures are not independent but are related by the relation

$$\kappa_{gj}^2 + \kappa_{nj}^2 = \kappa^2, \tag{2.78}$$

which, in our axisymmetric case, has the expression

$$\kappa = \frac{1}{r_b}. \tag{2.79}$$

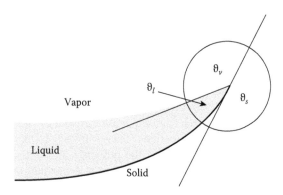

FIGURE 2.3 Schematic of the contact line region of an idealized solid–liquid–fluid capillary system showing the bulk phase angles: ϑ_s (within the solid phase), ϑ_l (the usual contact angle measured within the liquid phase), and ϑ_v (within the vapor phase).

This result would suggest that the radius r_b could be used in our axisymmetric situation as an independent variable to characterize the curvature of the contact line since only one of the three quantities κ_{gj}, κ_{nj} or κ is independent. The final quantity for consideration is the torsion or twist of the contact line. For the axisymmetric cases shown in Figures 2.1 to 2.3, the contact line is created by revolving a contact point around the system's vertical axis and this one-dimensional point cannot have a twist so the torsion quantity is not relevant here [7]. To summarize, the generalized theory of capillarity, described in Chapter 1, is highly symmetrical in its choice of contact line variables at the expense of not being entirely independent. An alternative view, with an explicit choice of variables, has also been considered for a three-phase system [42]. However, if we maintain the benefit of this symmetry we obtain the linear free energy for the contact line as

$$\omega^{(l)} = \sigma^{(l)} + \sum_{(jjl)} C_{\vartheta_{jjl}} \vartheta_{jjl} + \sum_{(j)} \left(C_{gj}\kappa_{gj} + C_{nj}\kappa_{nj} + C_{\tau j}\tau_j \right), \qquad (2.80)$$

which can be simplified for the axisymmetric case to

$$\omega^{(l)} = \sigma^{(l)} + C_{\vartheta_v} \vartheta_v + C_{\vartheta_l} \vartheta_l + C_{\vartheta_s} \vartheta_s + \sum_{(j=sl,sv,lv)} \left(C_{gj}\kappa_{gj} + C_{nj}\kappa_{nj} \right) \qquad (2.81)$$

subject to the constraints (Equations 2.77 and 2.78). If a particular choice of angles and curvatures is used, then it is possible to simplify Equation 2.81 further to

$$\omega^{(l)} = \sigma^{(l)} + C_{\vartheta_v} \vartheta_v + C_{\vartheta_l} \vartheta_l + C_\kappa \kappa \qquad (2.82)$$

without any subsidiary constraints. This situation may appear somewhat simpler; however, even in this case, one can easily see that the specific surface free energy $\omega^{(l)}$ is not a constant but is a complicated function of the surfaces that meet to form the contact line; that is,

$$\omega^{(l)} = \omega^{(l)} \left(r, \xi_1, \xi_2, \xi_1', \xi_2', \xi_1'', \xi_2'' \right). \qquad (2.83)$$

Consequently, if one wishes to visualize the influence of the assumption that the specific contact line free energy is not a constant but is dependent upon angles (or slopes), higher-order bending or curvature terms, then it will be necessary to generalize the previous analysis sufficiently that one may consider integrals up to and including the second derivative of the surface functions. The details of this analysis are presented by Gaydos [14].

2.2 THE NEUMANN TRIANGLE (QUADRILATERAL) RELATION

The main emphasis in this section will be upon the three-phase fluid contact line condition known as the Neumann triangle relation. The Neumann triangle relation [31,32] is the appropriate boundary condition when the three surfaces that intersect to form the contact line are all deformable (see location $r = ra$ in Figure 2.4). Some, estimates of the line tension have been performed using capillary systems that are arrayed in this fashion (i.e., the three surfaces forming the contact line are deformable). When one or two of the surfaces intersecting to form the contact line are rigid, then the previous Young equation is the appropriate boundary condition.

2.2.1 THE SOLID–LIQUID-LIQUID–FLUID SYSTEM

The axisymmetric arrangement considered in Figure 2.1 is easily adjusted to include another immiscible liquid in the composite system. The additional (small) quantity of liquid is required to have a density between that of the lower liquid and the upper fluid of Figure 2.1 and to float in a zone between these two bulk phases when in equilibrium. Both the upper fluid and the lower liquid are in contact with a rigid solid wall. Analogous considerations to those discussed above also apply to this situation. Once again, the state of the composite system is characterized in terms of the principle of virtual work, whereby the energy of the total system from all contributions is unvaried under arbitrary virtual displacements that are compatible with the system's constraints [8–13].

2.2.2 FREE ENERGY EXPRESSIONS FOR THE BULK REGIONS

For the system illustrated in Figure 2.4, the bulk phase free energy of the lowest liquid is given by the integral

$$\Omega^{(l)} = 2\pi \int_0^{r_b} |r| \int_{\xi_1}^{\xi_2} \omega^{(l)} dz dr. \qquad (2.84)$$

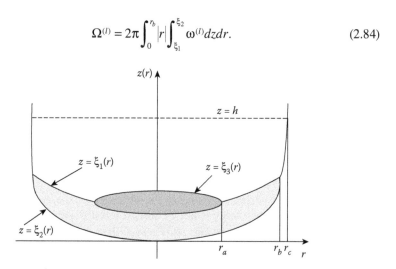

FIGURE 2.4 Schematic of an idealized solid–liquid-liquid–fluid capillary system with three deformable liquid–fluid or liquid–liquid surfaces and one deformable contact line at radial location $r = r_a$.

For the middle fluid phase, the expression is

$$\Omega^{(o)} = 2\pi \int_0^{r_a} |r| \int_{\xi_1}^{\xi_2} \omega^{(o)} dz dr, \tag{2.85}$$

while for the upper fluid phase, the term is

$$\Omega^{(v)} = 2\pi \left\{ \int_0^{r_a} |r| \int_{\xi_3}^{h} \omega^{(v)} dz dr + \int_{r_a}^{r_b} |r| \int_{\xi_1}^{h} \omega^{(v)} dz dr + \int_{r_b}^{r_c} |r| \int_{\xi_2}^{h} \omega^{(v)} dz dr \right\}, \tag{2.86}$$

where $\omega^{(k)}$ for $k = l, o, v$ represents the specific free energy density of the bulk phases. When the contribution of the intrinsic bulk free energy is independent of the contribution to the free energy from the external potential field, then the $\omega^{(k)}$; $k = l, o, v$ quantities are independent of the vertical z-coordinate and the three free energy integrals can be integrated with respect to the z-coordinate to yield

$$\Omega^{(l)} = 2\pi \int_0^{r_b} \omega^{(l)} (\xi_1 - \xi_2) r dr, \tag{2.87}$$

$$\Omega^{(o)} = 2\pi \int_0^{r_a} \omega^{(o)} (\xi_3 - \xi_1) r dr \tag{2.88}$$

and

$$\Omega^{(v)} = 2\pi \left\{ \int_0^{r_a} \omega^{(v)} (\xi_1 - \xi_3) r dr + \int_0^{r_b} \omega^{(v)} (\xi_2 - \xi_1) r dr + \int_0^{r_c} \omega^{(v)} (h - \xi_2) r dr \right\}. \tag{2.89}$$

2.2.3 External Body Forces

The presence of an external gravitational potential or force field also influences the total expression for the system's free energy. When the gravitational field strength g is uniform over the extent of the system, the expressions for the free energy or grand canonical potential for the various portions of the system are given as three volume integrals. For the lower liquid, one has the expression

$$\Omega_g^{(l)} = 2\pi \int_0^{r_b} |r| \int_{\xi_2}^{\xi_1} \rho^{(l)} g z dz dr, \tag{2.90}$$

where $\rho^{(l)}$ represents the density of the lower liquid phase. An analogous expression applies for the middle liquid phase, of density $\rho^{(o)}$,

$$\Omega_g^{(o)} = 2\pi \int_0^{r_a} |r| \int_{\xi_1}^{\xi_3} \rho^{(o)} g z dz dr, \tag{2.91}$$

while the upper fluid phase expression is given by

$$\Omega^{(v)} = 2\pi \left\{ \int_0^{r_a} |r| \int_{\xi_3}^h \rho^{(v)} gz dz dr + \int_{r_a}^{r_b} |r| \int_{\xi_1}^h \rho^{(v)} gz dz dr + \int_{r_b}^{r_c} |r| \int_{\xi_2}^h \rho^{(v)} gz dz dr \right\}, \qquad (2.92)$$

where $\rho^{(v)}$ represents the density of the upper fluid phase.

2.2.4 Free Energy Expressions for the Surface and Line Regions

The free energy or grand canonical potential for each interface region is expressed in the following manner. For the liquid–solid interface, one has

$$\Omega^{(sl)} = \iint_{A^{(sl)}} \omega^{(sl)} dA = 2\pi \int_0^{r_b} \omega^{(sl)} \left(1 + [\xi_2'(r)]^2\right)^{1/2} r dr, \qquad (2.93)$$

where the independent variable r represents the radius of the system, $\omega^{(sl)}$ is the specific surface grand canonical potential or specific surface free energy (defined per unit area) of the liquid–solid interface, and $\xi_2'(r)$ denotes differentiation of the curve $\xi_2(r)$ with respect to r. For the fluid–liquid interface and the fluid–solid interface, the corresponding expressions are given by

$$\Omega^{(lv)} = \iint_{A^{(lv)}} \omega^{(lv)} dA = 2\pi \int_{r_a}^{r_b} \omega^{(lv)} \left(1 + [\xi_1'(r)]^2\right)^{1/2} r dr \qquad (2.94)$$

and

$$\Omega^{(sv)} = \iint_{A^{(sv)}} \omega^{(sv)} dA = 2\pi \int_{r_b}^{r_c} \omega^{(sv)} \left(1 + [\xi_2'(r)]^2\right)^{1/2} r dr. \qquad (2.95)$$

These three expressions are essentially the same as the expressions given by Equations 2.14 through 2.16. For this particular lens arrangement, the new contributions to the surface energies arise from the interface between the middle liquid and the upper fluid and from the interface between the middle liquid and the lower liquid. The surface contributions are given, respectively, by

$$\Omega^{(lo)} = \iint_{A^{(lo)}} \omega^{(lo)} dA = 2\pi \int_0^{r_a} \omega^{(lo)} \left(1 + [\xi_1'(r)]^2\right)^{1/2} r dr \qquad (2.96)$$

and

$$\Omega^{(ov)} = \iint_{A^{(ov)}} \omega^{(ov)} dA = 2\pi \int_0^{r_a} \omega^{(ov)} \left(1 + [\xi_3'(r)]^2\right)^{1/2} r dr. \qquad (2.97)$$

Finally, to complete the specification of the energy expressions for this system, one must include two linear terms to represent the energy content of the three-phase contact lines at $r = r_b$ (i.e., the previous Expressions 2.12 and 2.13 apply) and at $r = r_a$. At the location $r = r_b$, the solid–liquid–fluid three-phase contact line energy is given by the expression

$$\Omega^{(slv)} = 2\pi r_b \omega^{(slv)} = \frac{2\pi}{r_b} \int_0^{r_b} r_b \omega^{(slv)} dr, \tag{2.98}$$

where $2\pi r_b$ is the total circumference or length of the contact line and $\omega^{(slv)}$ represents the specific free energy (per unit length) of the $r = r_b$ contact line. The corresponding energy expression for the liquid-liquid-fluid contact line at the $r = r_a$ location is given by

$$\Omega^{(lov)} = 2\pi r_a \omega^{(lov)} = \frac{2\pi}{r_a} \int_0^{r_a} r \omega^{(lov)} dr, \tag{2.99}$$

where $2\pi r_a$ is the total circumference or length of the liquid-liquid-fluid contact line and $\omega^{(lov)}$ represents the specific free energy (per unit length) of this line. Once again, both specific linear free energies, $\omega^{(slv)}$ and $\omega^{(lov)}$, must be independent of the radius r and constant. Furthermore, in equating the function and the line integral we must also require that r_a be constant under the integral sign. It is possible to drop this last requirement, as demonstrated by Gaydos [14].

2.2.5 CONSTRAINTS

If one assumes that all phases are incompressible, then the volumes of those bulk portions that compose the system must remain constant. Thus, for a completely incompressible system one can use a constant volume constraint as an equivalent replacement to the statement of mass conservation. The constant volume constraints (i.e., there are two constraints for the system illustrated in Figure 2.4) act as additional integral constraints or subsidiary conditions on the range of permissible variations. In addition to these two integral constraints, there is one nonholonomic constraint that restricts r_a to be less than r_b. This latter condition insures that the liquid-liquid-fluid contact line $L^{(lov)}$ exists, that it is unique, and that it does not coincide with the solid–liquid–fluid line $L^{(slv)}$. The volume of each portion of the system is given by

$$V^{(l)} = 2\pi \int_0^{r_b} (\xi_1 - \xi_2) r dr, \tag{2.100}$$

$$V^{(o)} = 2\pi \int_0^{r_a} (\xi_3 - \xi_1) r dr, \tag{2.101}$$

and

$$V^{(v)} = 2\pi \left\{ \int_0^{r_a} (h - \xi_3)r\,dr + \int_{r_a}^{r_b} (h - \xi_1)r\,dr + \int_{r_b}^{r_c} (h - \xi_2)r\,dr \right\}, \qquad (2.102)$$

where $V^{(l)}$ is the volume of the lower liquid phase, $V^{(o)}$ is the volume of the middle liquid phase, and $V^{(v)}$ is the volume of the upper fluid phase. Given that all phases are incompressible, one knows that by holding $V^{(l)}$ and $V^{(o)}$ fixed one will, by consequence of

$$V^{(l)} + V^{(o)} + V^{(v)} = \text{a constant}, \qquad (2.103)$$

also hold $V^{(v)}$ constant. Thus, the introduction of the two integral constraints (i.e., Equations 2.100 and 2.101) will necessitate the introduction of two Lagrange multipliers, say λ_1 and λ_2.

2.2.6 THE VARIATIONAL PROBLEM

The mechanical equilibrium conditions arise directly from the conditions that are necessary to insure that the total free energy integral Ω_Σ is stationary subject to its constraints. The total free energy integral is given by the summation of the contributions from all bulk, surface, and linear regions in the system and is given (for the system illustrated in Figure 2.4) by the expression

$$\Omega_\Sigma = \sum_{j=l,o,v} (\Omega^{(j)} + \Omega_g^{(j)}) + \Omega^{(sl)} + \Omega^{(sv)} + \Omega^{(lv)} + \Omega^{(lo)} + \Omega^{(ov)} + \Omega^{(slv)} + \Omega^{(lov)}, \qquad (2.104)$$

or, after some manipulation, by

$$\begin{aligned}
\frac{\Omega_\Sigma}{2\pi} = \int_0^{r_a} &\left\{ (\omega^{(o)} - \omega^{(v)})(\xi_3 - \xi_1) + \frac{1}{2} g \Delta \rho^{(ov)} (\xi_3^2 - \xi_1^2) \right. \\
&+ \left. (\omega^{(lo)} - \omega^{(lv)})(1 + [\xi_1'(r)]^2)^{1/2} + \omega^{(ov)}(1 + [\xi_3'(r)]^2)^{1/2} + \frac{\omega^{(lov)}}{r_a} \right\} r\,dr \\
+ \int_0^{r_b} &\left\{ (\omega^{(l)} - \omega^{(v)})(\xi_1 - \xi_2) + \frac{1}{2} g \Delta \rho^{(lv)} (\xi_1^2 - \xi_2^2) \right. \\
&+ \left. (\omega^{(sl)} - \omega^{(sv)})(1 + [\xi_2'(r)]^2)^{1/2} + \omega^{(lv)}(1 + [\xi_1'(r)]^2)^{1/2} + \frac{\omega^{(slv)}}{r_b} \right\} r\,dr \\
+ \int_0^{r_c} &\left\{ \omega^{(v)}(h - \xi_2) + \frac{1}{2} g \rho^{(v)}(h^2 - \xi_2^2) + \omega^{(sv)}(1 + [\xi_2'(r)]^2)^{1/2} \right\} r\,dr.
\end{aligned} \qquad (2.105)$$

Consequently, for this composite arrangement, the complete calculus of variation problem is written in the mathematical form

$$\delta\left(\Omega_{\Sigma} + \lambda_1 V^{(l)} + \lambda_2 V^{(o)}\right) = 0, \tag{2.106}$$

so that the two volume constraints are included directly in the function, which is subject to the variations. Each constraint insures that a particular volume remains constant. Variations are permissible in both the position of the curves $z = \xi_1(r)$ and $z = \xi_3(r)$ and in the position of the contact points $r = r_a$ and $r = r_b$. The solid is assumed to be rigid so that no variation in the position of the curve $z = \xi_2(r)$ is possible; however, the boundary point is considered to slide along the solid. Unlike the solid-liquid-fluid system considered above, the liquid lens system considered here yields three, rather than just one, Laplace equations. In essence, there is a Laplace equation for each deformable surface in the system. Along with the three Laplace equations and the previous contact line equilibrium condition; that is, the Young equation of capillarity that arises from the transversality condition at the point $r = r_b$, a new contact line condition—called the Neumann triangle (quadrilateral) relation—appears from the boundary conditions at the point $r = r_a$.

The complete calculus of variations problem is easier to handle if the total free energy integral is broken up into five terms such that each integral or functional has an integrand of the form

$$f\left(r, z(r), z'(r)\right) = f\left(r, \xi_j(r), \xi_j'(r)\right) \quad \text{where} \quad j = 1, 2, 3. \tag{2.107}$$

The integrands are given by the longish expressions

$$f_{1a}\left(r, \xi_1, \xi_1'\right) = \left(\omega^{(lo)} - \omega^{(lv)}\right)\left(1 + [\xi_1']^2\right)^{1/2} - \frac{1}{2}\Delta\rho^{(ov)}g\xi_1^2$$
$$+ \frac{\omega^{(lov)}}{r_a} - \left\{\lambda_2 + \left(\omega^{(o)} - \omega^{(v)}\right)\right\}\xi_1, \tag{2.108}$$

$$f_{3a}\left(r, \xi_3, \xi_3'\right) = \omega^{(ov)}\left(1 + [\xi_3'(r)]^2\right)^{1/2} + \frac{1}{2}\Delta\rho^{(ov)}g\xi_3^2 + \left(\lambda_2 + \omega^{(o)} - \omega^{(v)}\right)\xi_3, \tag{2.109}$$

$$f_{1b}\left(r, \xi_1, \xi_1'\right) = \omega^{(lv)}\left(1 + [\xi_1']^2\right)^{1/2} + \frac{1}{2}\Delta\rho^{(lv)}g\xi_1^2 + \left(\lambda_1 + \omega^{(l)} - \omega^{(v)}\right)\xi_1, \tag{2.110}$$

$$f_{2b}\left(r, \xi_2, \xi_2'\right) = \left(\omega^{(sl)} - \omega^{(sv)}\right)\left(1 + [\xi_2']^2\right)^{1/2} - \frac{1}{2}\Delta\rho^{(v)}g\xi_2^2 + \frac{\omega^{(slv)}}{r_b} - \left\{\lambda_1 + \left(\omega^{(l)} - \omega^{(v)}\right)\right\}\xi_2, \tag{2.111}$$

and

$$f_{2c}\left(r,\xi_2,\xi_2'\right)=\omega^{(sv)}\left(1+[\xi_2'(r)]^2\right)^{1/2}+\frac{1}{2}\rho^{(v)}g\left(h^2-\xi_2^2\right)+\omega^{(v)}\left(h-\xi_2\right). \quad (2.112)$$

Using these definitions, the calculus of variations problem, defined by Equation 2.106, can be recast into the form of the integral expression

$$\delta\left(\int_0^{r_a} f_{1a}\left(r,\xi_1,\xi_1'\right)rdr+\int_0^{r_a} f_{3a}\left(r,\xi_3,\xi_3'\right)rdr+\int_0^{r_b} f_{1b}\left(r,\xi_1,\xi_1'\right)rdr\right.$$

$$\left.\int_0^{r_b} f_{2b}\left(r,\xi_2,\xi_2'\right)rdr+\int_0^{r_c} f_{2c}\left(r,\xi_2,\xi_2'\right)rdr\right)=0. \quad (2.113)$$

Given that each integral above has the same functional dependence as the integrals in Equation 2.29 it stands to reason that the same necessary conditions for a stationary integral that apply to the integrals in Equation 2.29 also apply to each of the integrals present in Equation 2.113. These conditions consist of five Euler-Lagrange equations for each of the five surfaces present in the composite system and possibly two boundary conditions for each end point of the integral.

If the radial variation δr is not independent of the surface variation $\delta\xi_j$, then the two possible boundary conditions are combined into one transversality condition that describes the manner by which the boundary point slides or varies along a specific fixed curve (e.g., the solid wall curve defined by $z=\xi_2(r)$). If the radial variable is not permitted to vary (i.e., $\delta r=0$), then one boundary condition vanishes and from the other condition one obtains the natural boundary condition, as discussed above, in which the boundary point may only vary vertically at fixed r (e.g., this was the condition attached to the boundary point $r=0$ that yielded Equation 2.31). Finally, if a relationship does not exist between the independent and the dependent variables, then there are two boundary relations or conditions for each end point of the integral (i.e., each integral has a total of four boundary conditions two at each end).

Recognizing that the integrand $f_{1b}(r,\xi_1,\xi_1')$, in Equation 2.110 corresponds closely to the integrand $f_1(r,\xi_1,\xi_1')$ considered in Equation 2.26, permits one to write the appropriate Euler-Lagrange equation as

$$\frac{\partial\left(f_{1b}r\right)}{\partial\xi_1}-\frac{d}{dr}\frac{\partial\left(f_{1b}r\right)}{\partial\xi_1'}=0, \quad (2.114)$$

with natural boundary condition

$$\left(\frac{\partial\left(f_{1b}r\right)}{\partial\xi_1'}\right)_{r=0}=0, \quad (2.115)$$

and transversality condition

$$\left(f_{1b}r\right)+\left(\xi_2'-\xi_1'\right)\frac{\partial\left(f_{1b}r\right)}{\partial\xi_1'}=0. \tag{2.116}$$

Similarly, the integrand $f_{2b}\left(r,\xi_2,\xi_2'\right)$ corresponds closely to the integrand $f_3\left(r,\xi_2,\xi_2'\right)$ considered in Equation 2.28, so that the appropriate Euler-Lagrange equation is given by

$$\frac{\partial\left(f_{2b}r\right)}{\partial\xi_2}-\frac{d}{dr}\frac{\partial\left(f_{2b}r\right)}{\partial\xi_2'}=0, \tag{2.117}$$

with the transversality condition at $r=r_b$ as

$$\left(f_{2b}r\right)+\left(\xi_2'-\xi_2'\right)\frac{\partial\left(f_{2b}r\right)}{\partial\xi_2'}=0. \tag{2.118}$$

Finally, the integrand $f_{2c}\left(r,\xi_2,\xi_2'\right)$ corresponds closely to the integrand $f_2\left(r,\xi_2,\xi_2'\right)$ considered in Equation 2.27, with corresponding Euler-Lagrange equation

$$\frac{\partial\left(f_{2c}r\right)}{\partial\xi_2}-\frac{d}{dr}\frac{\partial\left(f_{2c}r\right)}{\partial\xi_2'}=0. \tag{2.119}$$

This integral has no free end points since the incompressible liquid assumption means that the system's total volume is fixed by the condition $r=r_c$ or $z=h$ being constant (see Figure 2.4) and thus no boundary condition of any kind exists for this term.

The two new integrands, not previously considered in the solid-liquid-fluid system, are given by the integrals whose integrands are $f_{1a}\left(r,\xi_1,\xi_1'\right)$ and $f_{3a}\left(r,\xi_3,\xi_3'\right)$. The integral with integrand $f_{1a}\left(r,\xi_1,\xi_1'\right)$ involves the behavior of the liquid-liquid surface (i.e., defined by the curve $z=\xi_1(r)$) below the lens from the origin to the contact line position at $r=r_a$ where the liquid-liquid surface intersects the liquid-fluid surface (i.e., defined by the curve $z=\xi_3(r)$). The integral with integrand $f_{3a}\left(r,\xi_3,\xi_3'\right)$ involves the behavior of the liquid-fluid surface (i.e., defined by the curve $z=\xi_3(r)$) that forms the upper surface boundary of the lens. The appropriate Euler-Lagrange equation and boundary conditions for the first integral that is defined in Equation 2.113 are

$$\frac{\partial\left(f_{1a}r\right)}{\partial\xi_1}-\frac{d}{dr}\frac{\partial\left(f_{1a}r\right)}{\partial\xi_1'}=0, \tag{2.120}$$

with natural boundary condition

$$\left(\frac{\partial(f_{1a}r)}{\partial\xi_1'}\right)_{r=0} = 0, \tag{2.121}$$

and boundary conditions

$$\left((f_{1a}r) - \xi_1'\frac{\partial(f_{1a}r)}{\partial\xi_1'}\right)_{r=r_a} = 0 \tag{2.122}$$

for the independent variation, and

$$\left(\frac{\partial(f_{1a}r)}{\partial\xi_1'}\right)_{r=r_a} = 0 \tag{2.123}$$

for the dependent variation. Similarly, the appropriate Euler-Lagrange equation and boundary conditions for the second integral that is defined in Equation 2.113 are

$$\frac{\partial(f_{3a}r)}{\partial\xi_3} - \frac{d}{dr}\frac{\partial(f_{3a}r)}{\partial\xi_3'} = 0, \tag{2.124}$$

with natural boundary condition

$$\left(\frac{\partial(f_{3a}r)}{\partial\xi_3'}\right)_{r=0} = 0, \tag{2.125}$$

and boundary conditions

$$\left((f_{3a}r) - \xi_3'\frac{\partial(f_{3a}r)}{\partial\xi_3'}\right)_{r=r_a} = 0 \tag{2.126}$$

for the independent variation, and

$$\left(\frac{\partial(f_{3a}r)}{\partial\xi_3'}\right)_{r=r_a} = 0 \tag{2.127}$$

for the dependent variation.

2.2.7 THE CLASSICAL MECHANICAL EQUILIBRIUM CONDITIONS

The mathematical manipulations required to solve Equations 2.120 through 2.127 are completely analogous to those performed previously and, for this reason, we shall restrict comments in this section to those conditions that are either new or slightly different from those discussed already. However, it would be remiss not to point out a few explicit connections between the relations above and the corresponding relations in this section.

The close connection between the two integrands $f_1(r,\xi_1,\xi_1')$ (see Equation 2.26), and $f_{1b}(r,\xi_1,\xi_1')$ means that the Euler-Lagrange Equation 2.30 and the boundary conditions Equations 2.26 and 2.27 also apply, without much change, to the integrand $f_{1b}(r,\xi_1,\xi_1')$. The essence of the change in description between the solid–liquid–vapor system considered above and the three-phase system considered in this section is the necessity of allowing for the possibility that a cusp might exist in the function $z = \xi_1(r)$ at the location $r = r_a$. Thus, it is more accurate to consider the function $\xi_1(r)$ as two piecewise continuous segments with one segment running from zero to r_a and the other segment from r_a to the solid–liquid–fluid contact line location at r_b. Similar statements, without the necessity of considering any cusps, apply for the pair of integrands $f_3(r,\xi_2,\xi_2')$ and $f_{2b}(r,\xi_2,\xi_2')$ and for the pair $f_2(r,\xi_2,\xi_2')$ and $f_{2c}(r,\xi_2,\xi_2')$. Consequently, it is not necessary to describe in detail, since this was already performed above, the manner by which the Laplace and Young equations of capillarity follow from these integrals. For the new integrands $f_{1a}(r,\xi_1,\xi_1')$ and $f_{3a}(r,\xi_3,\xi_3')$, the situation, with regards to the Laplace equation, is similar to the situation already encountered.

A comparison of the integrands involved shows that all of these expressions have the same form and will, as a consequence, result in a Laplace equation for each of the deformable surfaces involved in the composite system. This conclusion is to be expected and simply asserts the intuitive claim that an identical mechanical equilibrium relation should also exist at the liquid-fluid (*lo*) surface and at the liquid-vapor (*ov*) surface.

The most significant difference between the mechanical equilibrium conditions considered here and those considered above involves the variational end point conditions at the location $r = r_a$, where the three surfaces (*lv*), (*lo*), and (*ov*) intersect to form a contact line (denoted in superscript). The liquid-fluid-vapor (*lov*) boundary, unlike the solid-liquid-vapor (*slv*) contact line, is not restricted to vary along a fixed $\xi_2(r)$ curve, but is permitted to deform freely provided the three surfaces remain connected. Mathematically, the connectivity is effected by requiring the equality $\xi_1(r = r_a) = \xi_3(r = r_a)$. When both the specific surface free energies and the specific linear free energies are assumed constant, the end point conditions that result from the solution of Equations 2.122, 2.123, 2.126, and 2.127 (see [14] for details) yield the following relations. In the radial direction, variations in the horizontal $\hat{r}$ direction must satisfy the condition

$$\left(\omega^{(lo)} - \omega^{(lv)}\right)\left(1+[\xi_1']^2\right)^{-1/2} + \frac{\omega^{(lov)}}{r_a} + \omega^{(ov)}\left(1+[\xi_3']^2\right)^{-1/2} = 0, \qquad (2.128)$$

while in the vertical $\hat{z}$ direction the condition is given by

$$\left(\omega^{(lo)}-\omega^{(lv)}\right)\xi_1'\left(1+[\xi_1']^2\right)^{-1/2}+\omega^{(ov)}\xi_3'\left(1+[\xi_3']^2\right)^{-1/2}=0. \qquad (2.129)$$

Equations 2.128 and 2.129 may be expressed in a form that is similar to the Young equation of capillarity by introducing the relations

$$\xi_1'=\tan\vartheta_1^< \qquad \text{when } r=r_{a-0} \qquad (2.130)$$

and

$$\xi_1'=\tan\vartheta_1^> \qquad \text{when } r=r_{a+0} \qquad (2.131)$$

between the slope or tangent of the $\xi_1(r)$ liquid-fluid surface function and the angles $\vartheta_1^<$ and $\vartheta_1^>$. The symbol $\vartheta_1^<$ denotes the angle between the horizontal line, which passes through the contact line at $r=r_a$ and the tangent to the $\xi_1(r)$ curve that may be defined by approaching the $r=r_a$ point from a value of the radius that is less than r_a (see Figure 2.5). The symbol $\vartheta_1^>$ denotes the angle between the horizontal line that passes through the contact line at $r=r_a$ and the tangent for the $\xi_1(r)$ curve that may be defined by approaching the $r=r_a$ point from a value of the radius that is greater than r_a. Using these two symbols for the angle that the $\xi_1(r)$ curve forms with the horizontal through the point $r=r_a$ permits one to investigate situations where the surface defined by the curve $\xi_1(r)$ is not continuous, but has a cusp at an isolated number of axisymmetric gradient singularities (in the arrangement shown in Figure 2.5 there is only one gradient singularity at $r=r_a$). Therefore, by employing the definitions in

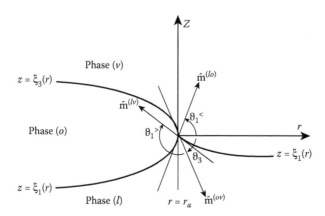

FIGURE 2.5 Magnified view of the region around the deformable contact line at $r=r_a$ showing the local (r, z) coordinate system that is fixed to the contact point and about which the rotation $\alpha_\theta=\vartheta_1^<$ occurs. Also shown are the orientations of the angles $\vartheta_1^<, \vartheta_1^>$ and ϑ_3 as well as the directions of the unit surface tangent vectors $\hat{\mathbf{m}}^{(lo)}, \hat{\mathbf{m}}^{(lv)}$ and $\hat{\mathbf{m}}^{(ov)}$.

Equations 2.130 and 2.131, it is possible to write the horizontal or radial variation condition (Equation 2.128), after simplifying, as

$$\omega^{(lo)}\cos\vartheta_1^< - \omega^{(lv)}\cos\vartheta_1^> + \frac{\omega^{(lov)}}{r_a} + \omega^{(ov)}\cos\vartheta_3 = 0, \tag{2.132}$$

where $\xi_3' = \tan\vartheta_3$. For this situation, the $\vartheta_1^<$ angle is associated with the liquid-fluid (lo) surface, which occurs (as shown in Figure 2.5) at a radial location $r \leq r_a$. Analogously, the $\vartheta_1^>$ angle is associated with free energy $\omega^{(lv)}$ and the liquid-vapor (lv) surface, which occurs at a radial location $r \geq r_a$. Likewise, we may use these angle definitions to write the vertical variation condition (Equation 2.129) as

$$\omega^{(lo)}\sin\vartheta_1^< - \omega^{(lv)}\sin\vartheta_1^> + \omega^{ov}\sin\vartheta_3 = 0. \tag{2.133}$$

Alternatively, one may consider these two relations as representing an in-plane (i.e., the plane formed by the radial $\hat{\mathbf{r}}$ vector and the vertical $\hat{\mathbf{z}}$ vector) force balance between the three surface energies and the single line energy that are involved in the formation of the liquid-fluid-vapor (lov) boundary. If one adopts this perception of these two in-plane orthogonal scalar boundary conditions, then one can express the in-plane force balance as a two-dimensional matrix expression. When this is done, one can write, in terms of the unit vectors, that

$$\mathbf{f}_{rz} = \{f_r \quad f_z\}\begin{Bmatrix} \hat{\mathbf{r}} \\ \hat{\mathbf{z}} \end{Bmatrix} = f_r\hat{\mathbf{r}} + f_z\hat{\mathbf{z}} \tag{2.134}$$

where the components of the force vectors are

$$f_r = \omega^{(lo)}\cos\vartheta_1^< - \omega^{(lv)}\cos\vartheta_1^> + \frac{\omega^{(lov)}}{r_a} + \omega^{(ov)}\cos\vartheta_3 \tag{2.135}$$

and

$$f_z = \omega^{(lo)}\sin\vartheta_1^< - \omega^{(lv)}\sin\vartheta_1^> + \omega^{(ov)}\sin\vartheta_3. \tag{2.136}$$

It is obvious from Equations 2.132 and 2.133 that these force components will vanish, as required, when the composite system is in static equilibrium.

Equations 2.128 and 2.129 can also be combined and written as an in-plane vector relation. In the $(\hat{\mathbf{r}}, \hat{\mathbf{z}})$ plane, the direction of each surface at the point where they intersect to form the liquid-fluid-vapor (lov) contact line is given by an outward unit normal, which will be defined by the vector $\hat{\mathbf{m}}^{(jk)}$ where $(jk) = (lv), (lo), (ov)$, see Figure 2.5. The curvature of the contact line is written in terms of a curvature vector $\kappa^{(lov)}$ that has two in-plane components

$$\kappa^{(lov)} = \left\{\frac{\hat{\mathbf{r}}}{r_a} \quad 0 \cdot \hat{\mathbf{z}}\right\}^{\mathrm{T}}, \tag{2.137}$$

where the superscript T on the vector indicates the transpose. The two scalar relations (Equations 2.135 and 2.136) combine with Equation 2.137 to give the equivalent vector relation

$$\omega^{(lo)}\hat{\mathbf{m}}^{(lo)} + \omega^{(lv)}\hat{\mathbf{m}}^{(lv)} + \omega^{(ov)}\hat{\mathbf{m}}^{(ov)} + \omega^{(lov)}\kappa^{(lov)} = 0. \tag{2.138}$$

If the scheme, analogous to that proposed in Equation 2.61, for replacing the specific surface free energies $\omega^{(jk)}$ in terms of their corresponding surface tensions is adopted and the linear term is dropped, one obtains the classical Neumann triangle relation from Equation 2.138.

2.2.8 AN ALTERNATIVE EXPRESSION FOR THE NEUMANN TRIANGLE RELATION

At that location in the composite system where the three deformable surfaces intersect to form a contact line one may select the origin of the horizontal and vertical (r, z) coordinate system so that it coincides with the contact line (see Figure 2.5). A positive rotation of the coordinate system about an axis that passes through the contact point and points outward from the plane of Figure 2.5 (i.e., the rotation turns in an anticlockwise sense) of magnitude α_ϑ will rotate the old coordinate axes (r, z) to the new coordinate axes $(\bar{r}, \bar{z})$. In this new coordinate system, the vector $\mathbf{f}_{rz}$ whose old components were f_r and f_z are now related to the new component values by the relations [43,44]

$$\left\{ \begin{matrix} \bar{f}_r \\ \bar{f}_z \end{matrix} \right\} = \begin{bmatrix} \cos\alpha_\vartheta & \sin\alpha_\vartheta \\ -\sin\alpha_\vartheta & \cos\alpha_\vartheta \end{bmatrix} \left\{ \begin{matrix} f_r \\ f_z \end{matrix} \right\}. \tag{2.139}$$

The Young equation, as expressed by Equation 2.64, for a moderately curved solid–liquid–fluid boundary may also be thought of as a horizontal force balance. In complete analogy with the Young equation one may express the boundary condition for the liquid-fluid-vapor (lov) boundary by rotating the force vector $\mathbf{f}_{rz}$ about the contact line point at $r = r_a$ by an amount $\alpha_\vartheta = \vartheta_1^<$ to yield a force vector $\bar{\mathbf{f}}_{rz}$ with new components. Substituting the expressions for f_r and f_z from Equations 2.135 and 2.136, respectively, into Equation 2.139 gives the new components as

$$\bar{f}_r = \omega^{(lo)} + \omega^{(ov)}\cos(\vartheta_1^< - \vartheta_3) - \omega^{(lv)}\cos(\vartheta_1^< - \vartheta_1^>) + \frac{\omega^{(lov)}}{r_a}\cos\vartheta_1^< \tag{2.140}$$

and

$$\bar{f}_z = \omega^{(lv)}\sin(\vartheta_1^< - \vartheta_1^>) - \omega^{(ov)}\sin(\vartheta_1^< - \vartheta_3) - \frac{\omega^{(lov)}}{r_a}\sin\vartheta_1^<. \tag{2.141}$$

It is possible to simplify these two expressions further by using the following two definitions (see Figure 2.6 for angular orientations). First, we note that

$$\vartheta_c + \vartheta_1^< - \vartheta_1^> = 0, \tag{2.142}$$

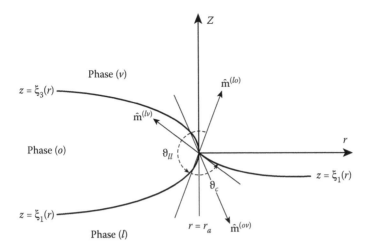

FIGURE 2.6 Magnified view of the region around the deformable contact line at $r = r_a$ showing the orientation of both the cusp angle ϑ_c and the liquid-lens contact angle ϑ_{ll}.

where all three angles are defined in an anticlockwise or positive sense. The cusp angle, ϑ_c, represents the angle between the two outward surface normals $\hat{\mathbf{m}}^{(lo)}$ and $\hat{\mathbf{m}}^{(lv)}$. Second,

$$\vartheta_{ll} = \vartheta_1^< - \vartheta_3,$$ (2.143)

where ϑ_{ll} denotes the liquid-lens contact angle that is defined in an anticlockwise sense so that it is also positive.

Using Equation 2.142 it is possible to eliminate the angle difference $(\vartheta_1^< - \vartheta_1^>)$ in favor of the cusp angle ϑ_c and from Equation 2.143 to eliminate angle ϑ_3 in favor of ϑ_{ll}. Therefore, with these substitutions, the force vector components in the rotated coordinate system become

$$\bar{f}_r = \omega^{(lo)} + \omega^{(ov)} \cos \vartheta_{ll} - \omega^{(lv)} \cos \vartheta_c + \frac{\omega^{(lov)}}{r_a} \cos \vartheta_1^<$$ (2.144)

and

$$\bar{f}_z = -\omega^{(lv)} \sin \vartheta_c - \omega^{(ov)} \sin \vartheta_{ll} - \frac{\omega^{(lov)}}{r_a} \sin \vartheta_1^<.$$ (2.145)

Once the components of the force vector are known, then static equilibrium requires that all forces vanish. This, in turn, means that Equations 2.144 and 2.145 are expressible as

$$\omega^{(lv)} \cos \vartheta_c - \omega^{(lo)} - \omega^{(ov)} \cos \vartheta_{ll} - \frac{\omega^{(lov)}}{r_a} \cos \vartheta_1^< = 0$$ (2.146)

and

$$\omega^{(lv)} \sin \vartheta_c + \omega^{(ov)} \sin \vartheta_{ll} + \frac{\omega^{(lov)}}{r_a} \sin \vartheta_{\overline{1}}^{<} = 0. \tag{2.147}$$

In contrast with Equations 2.128 and 2.129, these expressions are in terms of the specific free energies of the surfaces that intersect at the contact line, the specific free energy of the contact line, the liquid-lens contact angle ϑ_{ll}, and the cusp angle ϑ_c. For moderately curved surface and line regions, where the radius of curvature of the boundary is effectively quite large, one obtains the equalities

$$\omega^{(jk)} = \gamma_\infty^{(jk)} \text{ where } (jk) = \{(lv),(lo),(ov)\} \text{ and } \omega^{(lov)} = \sigma_\infty^{(lov)} \tag{2.148}$$

between the specific surface free energies and the tensions of the boundary regions. When these constant tensions are inserted into Relations 2.146 and 2.147 one obtains the mechanical equilibrium conditions

$$\gamma_\infty^{(lv)} \cos \vartheta_c - \gamma_\infty^{(lo)} - \gamma_\infty^{(ov)} \cos \vartheta_{ll} - \frac{\sigma_\infty^{(lov)}}{r_a} \cos \vartheta_{\overline{1}}^{<} = 0 \tag{2.149}$$

and

$$\gamma_\infty^{(lv)} \sin \vartheta_c + \gamma_\infty^{(ov)} \sin \vartheta_{ll} + \frac{\sigma_\infty^{(lov)}}{r_a} \sin \vartheta_{\overline{1}}^{<} = 0. \tag{2.150}$$

These conditions together, or an equivalent in-plane orthogonal scalar set, are known as the Neumann quadrilateral relations for the liquid-lens (*lov*) boundary or contact line [45]. In the absence of the line tension terms one recovers the classical Neumann triangle relations as expected. The scalar set of equations (i.e., represented by either Equations 2.149 and 2.150 or by Equations 2.132 and 2.133 with the $\omega^{(jk)}$ values replaced with surface tension $\gamma_\infty^{(jk)}$ values) or the corresponding in-plane vector relation (i.e., represented by either Equation 2.134 with components defined by Equations 2.135 and 2.136 or by Equation 2.139 with components defined by Equations 2.140 and 2.141) is the most general boundary condition possible when the specific surface free energies forming the boundary are assumed constant (e.g., expressed by Equation 2.148).

Equation 2.149, which represents the force balance in the direction of the tangent to the liquid-fluid (*lo*) surface at the point $r = r_{a-0}$, is similar to the Young equation of capillarity derived in Equation 2.64. The orthogonal force balance condition, Equation 2.150, has no counterpart in the solid–liquid–fluid system since the supporting substrate (i.e., the solid surface boundary defined by the curve $z = \xi_2(r)$) was assumed to be completely rigid and undeformable. When deformation is possible (e.g., the liquid-liquid surface defined by the curve $z = \xi_1(r)$ undergoes deformation by the fluid lens) an additional force balance condition is possible. This condition, given by Equation 2.150, states that a nonzero cusp angle, ϑ_c, is a real possibility when a liquid lens of nonzero lens angle, ϑ_{ll}, is present. According to the classical theory of capillarity, which excludes the possibility of line tension, a nonzero value of ϑ_{ll} means that the cusp angle ϑ_c must also be nonzero. Obviously, a nonzero cusp angle ϑ_c means that there is a discontinuity in the slope of the $z = \xi_l(r)$ curve that describes the liquid-fluid surface below the lens. The presence of the discontinuity implies that either the moderate curvature approximation for the liquid-fluid surface defined by $z = \xi_l(r)$ is inappropriate, or that the classical exclusion of the line tension is inappropriate. In Chapter 13, discussing line tension and a liquid in contact with a strip-wise wall, we shall see another example of this type of discontinuity [46–48]. However, it should be realized that no real system with a surface phase will behave as if its surface is exactly two-dimensional. For example, if the underlying substrate were solid, then a cusp-like solution at some location on the solid surface would not be realized since the liquid would either move along the solid's surface or the solid would fracture/deform at that contact line location before the curvature became too large.

2.3 THE MECHANICS OF AXISYMMETRIC COMPRESSIBLE EQUILIBRIUM SYSTEMS

In our previous derivations it was assumed that the capillary systems were incompressible so that the volumes of the respective bulk phases were constant. In many situations this incompressible assumption provides a satisfactory model of the true physical system, but in some arrangements the requirement of incompressibility is too restrictive. In particular, unsupported and, to some degree, supported films may exercise enormous changes in surface density or mass concentration in their transition from a gas-like structure to a liquid-like structure. In this section the mathematical procedure to model these systems as compressible phases will be developed so that the appropriate adjustments to the previous incompressible mechanical equilibrium conditions are possible.

2.3.1 FREE ENERGY EXPRESSIONS FOR THE BULK, SURFACE, AND LINE REGIONS

We shall begin with the case discussed above where a liquid, in equilibrium with another fluid, makes contact with a rigid solid wall (see Figure 2.1). The free energy of each bulk, surface, and linear phase is written as an integral over the volume of the phase in question. These expressions remain unchanged from those developed above. Similarly, the presence of an external gravitational potential influences the expression for the system's total free energy and is given by the expressions developed above.

2.3.2 COMPRESSIBLE SYSTEM CONSTRAINTS

Once again, it is required to determine what, if any, quantities are held constant in the system when the system is subjected to arbitrary, weak variations [12]. In all previous sections, it was assumed that all phases were incompressible so that we could assume that, in the variational problem, the volume of the system and of certain portions of the system would remain constant. For a compressible system, however, it is necessary to replace the constant volume constraint with an equivalent constant mass constraint.

The mass expressions for the two bulk phases in the system are

$$M_j^{(l)} = 2\pi \int_0^{r_b} |r| \int_{\xi_2}^{\xi_1} \rho_j^{(l)} dz\,dr = 2\pi \int_0^{r_b} \rho_j^{(l)} (\xi_1 - \xi_2) r\,dr \qquad (2.151)$$

and

$$M_j^{(v)} = 2\pi \int_0^{r_b} |r| \int_{\xi_1}^{h} \rho_j^{(v)} dz\,dr + 2r \int_{r_b}^{r_c} |r| \int_{\xi_2}^{h} \rho_j^{(v)} dz\,dr$$

$$= 2\pi \int_0^{r_b} \rho_j^{(v)} (h - \xi_1) r\,dr + 2\pi \int_{r_b}^{r_c} \rho_j^{(v)} (h - \xi_2) r\,dr, \qquad (2.152)$$

where $M_j^{(l)}$ represents the total mass in the lower liquid bulk phase of the jth chemical component and $M_j^{(v)}$ represents the total mass of the jth component in the upper fluid phase. When summed, the bulk contributions can be written in the form

$$M_j^{(l)} + M_j^{(v)} = 2\pi \left\{ \int_0^{r_c} \rho_j^{(l)} (h - \xi_2) r\,dr + \int_0^{r_b} \left(\rho_j^{(l)} - \rho_j^{(v)} \right) (\xi_1 - \xi_2) r\,dr \right\}. \qquad (2.153)$$

The total surface and line mass contributions are given by

$$M_j^{(sl)} = 2\pi \int_0^{r_b} \rho_j^{(sl)} \left(1 + [\xi_2'(r)]^2 \right)^{1/2} r\,dr, \qquad (2.154)$$

$$M_j^{(lv)} = 2\pi \int_0^{r_b} \rho_j^{(lv)} \left(1 + [\xi_1'(r)]^2 \right)^{1/2} r\,dr, \qquad (2.155)$$

$$M_j^{(sv)} = 2\pi \left\{ \int_0^{r_c} \rho_j^{(sv)} \left(1 + [\xi_2'(r)]^2 \right)^{1/2} r\,dr - \int_0^{r_b} \rho_j^{(sv)} \left(1 + [\xi_2'(r)]^2 \right)^{1/2} r\,dr \right\}, \qquad (2.156)$$

and

$$M_j^{(slv)} = 2\pi r_b \rho_j^{(slv)} \qquad (2.157)$$

where the bracketed superscripts denote the solid–liquid, liquid–vapor, and solid–vapor surface phases and the superscript (slv) denotes the solid–liquid–vapor contact line.

In many capillary systems it may seem irrefutable that the surface and line mass contributions are insignificant, however, for completeness, we shall define the total mass of the composite capillary system as the sum of the mass terms that are given by Equations 2.153 through 2.157, or

$$M_{\Sigma j} = M_j^{(l)} + M_j^{(v)} + M_j^{(sv)} + M_j^{(sl)} + M_j^{(lv)} + M_j^{(slv)}. \qquad (2.158)$$

As written, this expression is general. However, should a jth chemical component not be present in a particular phase, then the appropriate mass term would have a value of zero in that phase. Thus, for a closed, isolated composite system, the appropriate constraint on a compressible system is that the total mass $M_{\Sigma j}$ remain fixed for all potential variations of the system.

2.3.3 The Variational Problem

As with our previous derivations, the mechanical equilibrium conditions arise directly from the conditions that are necessary to insure that the total free energy integral, Equation 2.21, is stationary, subject to the mass constraint given by Equation 2.158. Mathematically, the calculus of variations problem for the compressible capillary system is written in the form

$$\delta\left(\Omega_\Sigma + \mu_j M_{\Sigma j} = 0\right) \qquad (2.159)$$

so that the mass constraint is included directly in the function that is subject to the variations. The solid is assumed to be rigid so that no variation in the position of the curve $z = \xi_2(r)$ is possible; however, the contact line boundary point at $r = r_b$ is considered to slide along the solid. Consequently, the above free energy integral with dependent variable $\xi_1(r)$ gives rise to a calculus of variations problem with one free end point at r_b. In addition, and unlike the previous situations considered, the upper boundary of the system, where $r = r_c$, is not fixed since preservation of mass is not equivalent to the preservation of the system's total volume at a fixed constant. The Laplace equation of capillarity for the liquid-fluid interface comes from the Euler-Lagrange equation for the curve $\xi_1(r)$, subject to the subsidiary condition that the system is compressible so that the total mass is constant. The contact line equilibrium condition, called the Young equation of capillarity, comes from the transversality condition that applies at the point $r = r_b$.

2.3.4 The Bulk Mass Only Solution

If we substitute the appropriate expressions into Equation 2.159 and assume that all surface densities and the line density vanish, then we may write the expression for $(\Omega_\Sigma + \mu_j M_{\Sigma j})$, after division by 2π, as

$$\int_0^{r_b}\left\{\omega^{(lv)}\left(1+\left[\xi_1'(r)\right]^2\right)^{1/2} + \left(\omega^{(sl)} - \omega^{(sv)}\right)\left(1+\left[\xi_2'(r)\right]^2\right)^{1/2} + \frac{1}{2}\Delta\rho g\xi_1^2\right.$$

$$+\left(\omega^{(l)} - \omega^{(v)}\right)\xi_1 + \mu_j\left(\rho_j^{(l)} - \rho_j^{(v)}\right)\xi_1\Big\}rdr + \int_0^{r_c}\left\{\frac{1}{2}\rho^{(v)}g\left(h^2 - \xi_2^2\right) + \omega^{(sv)}\left(1+\left[\xi_2'\right]^2\right)^{1/2}\right.$$

$$+\omega^{(v)}\left(h - \xi_2\right) + \mu_j\rho_j^{(v)}\left(h - \xi_2\right)\Big\}rdr - \int_0^{r_b}\left\{\frac{1}{2}\Delta\rho g\xi_2^2 - \frac{\omega^{(slv)}}{r_b} + \left(\omega^{(l)} - \omega^{(v)}\right)\xi_2\right.$$

$$+\mu_j\left(\rho_j^{(l)} - \rho_j^{(v)}\right)\xi_2\Big\}rdr,$$

$$(2.160)$$

where $\Delta\rho = \rho^{(l)} - \rho^{(v)} = \sum_{j=1}^{\eta}\rho_j^{(l)} - \sum_{j=1}^{\eta}\rho_j^{(v)}$ denotes the summed density difference and η denotes the number of individual chemical components. For each unique component there will be a Lagrange multiplier μ_j. Thus, the term $\mu_j(\rho_j^{(l)} - \rho_j^{(v)})$ in Equation 2.160 represents a summation over all chemical components and we have dropped the explicit summation sign for convenience.

We follow the methodology above and define three integrand functions to be of the form $f(r,z,z') = f(r,\xi_k,\xi_k')$ where $k = 1, 2$ so that we can write the three expressions

$$f_1 = f_1(r,\xi_1,\xi_1') = \omega^{(lv)}\left(1+[\xi_1']^2\right)^{1/2} + \frac{1}{2}\Delta\rho g\xi_1^2 + (\omega^{(l)} - \omega^{(v)})\xi_1 + \mu_j(\rho_j^{(l)} - \rho_j^{(v)})\xi_1,$$

(2.161)

$$f_2 = f_2(r,\xi_2,\xi_2') = \omega^{(sv)}\left(1+[\xi_2']^2\right)^{1/2} + \frac{1}{2}\rho^{(v)}g(h^2 - \xi_2^2) + (\omega^{(v)} + \mu_j\rho_j^{(v)})(h - \xi_2),$$

(2.162)

and

$$f_3 = f_3(r,\xi_2,\xi_2') = (\omega^{(sl)} - \omega^{(sv)})\left(1+[\xi_2']^2\right)^{1/2} - \frac{1}{2}\Delta\rho g\xi_2^2$$

$$+ \frac{\omega^{(slv)}}{r_b} - (\omega^{(l)} - \omega^{(v)})\xi_2 - \mu_j(\rho_j^{(l)} - \rho_j^{(v)})\xi_2.$$

(2.163)

Using these definitions, the calculus of variations problem, defined by Equation 2.159, can be recast into the form of the integral expression

$$\delta\left(\int_0^{r_b} f_1(r,\xi_1,\xi_1')r\,dr + \int_0^{r_c} f_2(r,\xi_2,\xi_2')r\,dr + \int_0^{r_b} f_3(r,\xi_3,\xi_3')\,r\,dr\right) = 0. \quad (2.164)$$

Once the variational problem is expressed in this form, it is easy to see that each of the above integrals has the same functional dependence as the integrals in Equation 2.29. Consequently, the same necessary conditions for a stationary integral, which applied to the integrals above in Equation 2.29, also apply to each of the integrals present in Equation 2.164. These conditions consist of three Euler-Lagrange equations for each of the surfaces under consideration, two transversality conditions for the behavior of the contact line and one mass-related condition for the outer boundary at $r = r_c$ that affects the second integral in Equation 2.164. The Euler-Lagrange equations and the boundary conditions for the problem posed by Equation 2.164 are identical with the expressions given by Equations 2.30 through 2.35. The only difference occurs because the upper $r = r_c$ limit on the second integral of Equation 2.164 is not fixed, but must be free to slide along the solid surface curve $\xi_2(r)$ if one wishes to maintain the total mass $M_{\Sigma j}$ of the composite system constant for arbitrary variations of the liquid-vapor or liquid-fluid surface.

2.3.5 MECHANICAL EQUILIBRIUM CONDITIONS FOR THE BULK MASS ONLY SOLUTION

Before proceeding with the derivation of the mechanical equilibrium conditions for a compressible capillary system it is necessary to make the additional assumption—in this section only—that the specific surface and line free energies are constants. The mathematical manipulations required to solve Equations 2.30 through 2.35 with Equations 2.161 through 2.163 are straightforward when we refer to the detailed calculations above. However, the presence of the Lagrange multipliers μ_j in the defining integrand $f_1(\ldots)$, rather than the single Lagrange multiplier λ, causes slight alterations to the expressions so that the final expressions for Equations 2.30 through 2.32 are

$$\omega^{(lv)}\left(1+[\xi_1']^2\right)^{-3/2}\left(\xi_1'+[\xi_1']^3+r\xi_1''\right)-(\omega^{(l)}-\omega^{(v)})r-\Delta\rho gr\xi_1-\mu_j\left(\rho_j^{(l)}-\rho_j^{(v)}\right)r=0,$$
$$(2.165)$$

$$\xi_1'(r=0)=0,\tag{2.166}$$

and

$$\left(\frac{1}{2}\Delta\rho g[\xi_1(r_b)]^2+\mu_j\left(\rho_j^{(l)}-\rho_j^{(v)}\right)\xi_1(r_b)+(\omega^{(l)}-\omega^{(v)})\xi_1(r_b)\right)\left(1+[\xi_1'(r_b)]^2\right)^{1/2}$$
$$+\omega^{(lv)}\left(1+[\xi_1'(r_b)]\right)\left[\xi_2'(r_b)\right]=0$$
$$(2.167)$$

for the first integral. Comparison of these three equations (i.e., Equation 2.165 through 2.167 with Equations 2.36 through 2.38) show that the essence of this change is the replacement of the Lagrange multiplier λ with the products $\mu_j(\rho_j^{(l)}-\rho_j^{(v)})$. In a similar manner, it can be shown that the change to the integrand $f_2(\ldots)$ results in a change to the surface expression so that Equation 2.33 yields

$$\omega^{(v)}r+\rho^{(v)}gr\xi_2+\mu_j\rho_j^{(v)}r+\omega^{(sv)}\left(1+[\xi_2']^2\right)^{-3/2}\left(\xi_2'+[\xi_2']^3+r\xi_2''\right)=0 \quad (2.168)$$

for the Euler-Lagrange equation for the second integral in Equation 2.164. For the third integral in Equation 2.164 one finds that Equation 2.34 becomes

$$\Delta\rho gr\xi_2+\mu_j\left(\rho_j^{(l)}-\rho_j^{(v)}\right)r+\left(\omega^{(l)}-\omega^{(v)}\right)r$$
$$+\left(\omega^{(sl)}-\omega^{(sv)}\right)\left(1+[\xi_2']^2\right)^{-3/2}\left(\xi_2'+[\xi_2']^3+r\xi_2''\right)=0$$
$$(2.169)$$

and the transversality condition, Equation 2.35, for the third integral gives

$$\frac{1}{2}\Delta\rho g\xi_2^2(r_b) - \frac{\omega^{(slv)}}{r_b} + \mu_j\left(\rho_j^{(l)} - \rho_j^{(v)}\right)\xi_2(r_b) + \left(\omega^{(l)} - \omega^{(v)}\right)\xi_2(r_b)$$

$$+ \left(\omega^{(sv)} - \omega^{(sl)}\right)\left(1 + \left[\xi_2'(r_b)\right]^2\right)^{1/2} = 0.$$

(2.170)

Equation 2.165 permits one to derive both an expression for the Lagrange multipliers μ_j and an expression for the surface mechanical equilibrium condition for the liquid-fluid interface as

$$\mu_j\left(\rho_j^{(l)} - \rho_j^{(v)}\right) = \frac{\omega^{(lv)}}{r}\left(1 + \left[\xi_1'\right]^2\right)^{-3/2}\left(\xi_1'\left\{1 + \left[\xi_1'\right]^2\right\} + r\xi_1''\right) - \Delta\rho g\xi_1 - \left(\omega^{(l)} - \omega^{(v)}\right)$$

$$= \omega^{(lv)}\left(\frac{\xi_1'}{r\left(1 + \left[\xi_1'\right]^2\right)^{1/2}} + \frac{\xi_1''}{r\left(1 + \left[\xi_1'\right]^2\right)^{3/2}}\right) - \Delta\rho g\xi_1 - \left(\omega^{(l)} - \omega^{(v)}\right)$$

(2.171)

or

$$\omega^{(lv)}\left(\frac{1}{R_1^{lv}} + \frac{1}{R_2^{(lv)}}\right) - \Delta\rho g\xi_1 = \mu_j\left(\rho_j^{(l)} - \rho_j^{(v)}\right) + \left(\omega^{(l)} - \omega^{(v)}\right),$$

(2.172)

where $R_1^{(lv)}$ and $R_2^{(lv)}$ denote the principal radii of curvature for the liquid-fluid interface. The Lagrange multipliers μ_j are evaluated at the symmetry point $[0, \xi_1(r = 0)]$ using the same procedure that was illustrated above for the single multiplier λ. After these steps are performed, the expressions for the constants μ_j is given by

$$\mu_j\left(\rho_j^{(l)} - \rho_j^{(v)}\right) = 2\omega^{(lv)}\xi_1''(r = 0) - \Delta\rho g\xi_1(r = 0) - \left(\omega^{(l)} - \omega^{(v)}\right)$$

(2.173)

or by

$$\mu_j\left(\rho_j^{(l)} - \rho_j^{(v)}\right) = \frac{2\omega^{(lv)}}{R_o^{(lv)}} - \Delta\rho g\xi_1(r = 0) + \left(P_o^{(l)} - P_o^{(v)}\right)$$

(2.174)

after the function $\xi_1''(r = 0)$ is replaced by its radius of curvature at the origin and the specific bulk free energies $\omega^{(l)}$ and $\omega^{(v)}$ are replaced by the pressure of the bulk phase, near to but not at the location of the interface. All quantities in Definitions 2.173 and 2.174, including the density difference $\Delta\rho$, which is a function of the radius, are evaluated at the origin. The final version of Equation 2.30 is obtained by replacing all $\mu_j(\rho_j^{(l)} - \rho_j^{(v)})$ terms, using the expression above (i.e., Equation 2.174), so that one obtains the previous expression for the Laplace equation (i.e., Equation 2.50). In the section above, the Lagrange multiplier λ was shown to be related to particular volumes in the composite system. In this section, the Lagrange multipliers μ_j, defined by either Equation 2.173 or Equation 2.174, have units of joules per kg and represent the chemical potentials of the jth chemical components. The three terms that occur in Equation 2.173 represent contributions to the chemical potentials from the bulk phases, the surface boundary, and the gravitational field strength. The key result from this derivation, for the situation in which all phases are compressible and all of the mass in the system resides only in the two bulk phases, is that the surface mechanical equilibrium condition is identical with the relation that was derived above, Equation 2.50, where we assumed that all phases were incompressible.

The boundary condition of the liquid-fluid interface at the solid wall (i.e., the Young equation of capillarity) is derived from the end point relation that is given by Equation 2.167. The constant multipliers μ_j and the terms in which they occur in Equation 2.167 can be eliminated from the final form of the boundary condition by using Equation 2.170 since $\xi_1(r = r_b) = \xi_2(r = r_b)$ in both relations, which apply only at the location of the contact line. Equation 2.167 can be written in the form

$$\frac{1}{2}\Delta\rho g\xi_1^2 + \left(\mu_j\left(\rho_j^{(l)} - \rho_j^{(v)}\right)\right) + \omega^{(l)} - \omega^{(v)}\right)\xi_1 + \omega^{(lv)}\left(1 + \left[\xi_1'\right]^2\right)^{1/2}$$

$$+ \omega^{(lv)}\xi_1'\left(\xi_2' - \xi_1'\right)\left(1 + \left[\xi_1'\right]^2\right)^{-1/2} = 0,$$

(2.175)

where both $\xi_k(r)$, $k = 1, 2$ surface functions are evaluated at the contact line radius $r = r_b$. From Equation 2.170, we have that

$$\mu_j\left(\rho_j^{(l)} - \rho_j^{(v)}\right)\xi_2(r_b) = \mu_j\left(\rho_j^{(l)} - \rho_j^{(v)}\right)\xi_1(r_b) = \frac{\omega^{(slv)}}{r_b} - \frac{1}{2}\Delta\rho g\xi_2^2(r_b)$$

$$+ \left(\omega^{(sl)} - \omega^{(sv)}\right)\left(1 + \left[\xi_2'(r_b)\right]^2\right)^{1/2} - \left(\omega^{(l)} - \omega^{(v)}\right)\xi_2(r_b).$$

(2.176)

Substituting the quantity to the right of $\mu_j(\rho_j^{(l)} - \rho_j^{(v)})\xi_1(r_b)$ from Equation 2.176 into Equation 2.175 eliminates all chemical potential terms and yields

$$\frac{1}{2}\Delta\rho g\left(\xi_1^2 - \xi_2^2\right) + \left(\omega^{(l)} - \omega^{(v)}\right)\left(\xi_1 - \xi_2\right) + \frac{\omega^{(slv)}}{r_b} + \omega^{(lv)}\left(1 + \left[\xi_1'\right]^2\right)^{1/2}$$

$$+ \omega^{(lv)}\xi_1'\left(\xi_2' - \xi_1'\right)\left(1 + \left[\xi_1'\right]^2\right)^{-1/2} + \left(\omega^{(sl)} - \omega^{(sv)}\right)\left(1 + \left[\xi_2'\right]^2\right)^{1/2} = 0,$$

(2.177)

which is identical with the Equation 2.54 obtained previously for the boundary condition at the liquid-fluid surface in contact with the solid wall. Upon simplifying this expression, we obtain the Young equation of capillarity in the form given by Equation

2.60. If the specific surface and line free energies are given in terms of their surface and line tensions, then the Laplace equation and the Young equation are expressible as Equations 2.63 and 2.64, respectively. It can also be shown that, although we shall not perform the calculation, the appropriate Neumann triangle (quadrilateral) relation for the bulk mass only situation is given by the usual scalar force balance relations (i.e., Equations 2.132 and 2.133 or Equations 2.149 and 2.150).

2.3.6 MECHANICAL EQUILIBRIUM CONDITIONS FOR THE CASE OF CONSTANT PHASE DENSITY

In the previous two sections, it was assumed for simplicity that the total mass in the composite system was distributed such that there was no excess mass (see Chapter 1 for a definition of excess mass) present at any of the interfaces or at the solid–liquid–fluid three-phase contact line. Therefore, in essence, it was assumed that the total mass $M_{\Sigma j}$, defined in Equation 2.158, was composed of only two contributions, namely, $M_j^{(l)}$ and $M_j^{(v)}$. When the composite system's mass is restricted to occur only within the bulk phases one recovers the incompressible forms of the Laplace equation and the Young equation of capillarity. However, in the section to follow we shall demonstrate that the form of the mechanical equilibrium conditions changes when nonignorable amounts of mass are permitted to exist within the surface and linear phases.

It is still possible to cast the variational problem in the form of three integral expressions subject to arbitrary, weak variations as written in Equation 2.164, however, we must modify the integrand functions $f_k(r,\xi_l,\xi_l')$ where $k = 1, 2, 3$ and $l = 1, 2$ that were given in Equations 2.161 through 2.163 to the expressions

$$
\begin{aligned}
f_1 = f_1(r,\xi_1,\xi_1') &= \left(\omega^{(lv)} + \rho^{(lv)} g\xi_1 + \mu_j \rho_j^{(lv)}\right)\left(1+[\xi_1']^2\right)^{1/2} + \frac{1}{2}\Delta\rho g\xi_1^2 \\
&+ \left(\mu_j\left(\rho_j^{(l)} - \rho_j^{(v)}\right) + \left(\omega^{(l)} - \omega^{(v)}\right)\right)\xi_1,
\end{aligned}
\tag{2.178}
$$

$$
\begin{aligned}
f_2 = f_2(r,\xi_2,\xi_2') &= \left(\omega^{(sv)} + \rho^{(sv)} g\xi_2 + \mu_j \rho_j^{(sv)}\right)\left(1+[\xi_2']^2\right)^{1/2} \\
&+ \frac{1}{2}\rho^{(v)} g\left(h^2 - \xi_2^2\right) + \left(\omega^{(v)} + \mu_j \rho_j^{(v)}\right)\left(h - \xi_2\right),
\end{aligned}
\tag{2.179}
$$

and

$$
\begin{aligned}
f_3 = f_3(r,\xi_2,\xi_2') &= \frac{1}{r_b}\left(\omega^{(slv)} + \rho^{(slv)} g\xi_2 + \mu_j \rho_j^{(slv)}\right) - \frac{1}{2}\Delta\rho g\xi_2^2 \\
&- \left(\mu_j\left(\rho_j^{(l)} - \rho_j^{(v)}\right) + \left(\omega^{(l)} - \omega^{(v)}\right)\right)\xi_2 \\
&+ \left[\left(\omega^{(sl)} + \rho^{(sl)} g\xi_2 + \mu_j \rho_j^{(sl)}\right) - \left(\omega^{(sv)} + \rho^{(sv)} g\xi_2 + \mu_j \rho_j^{(sv)}\right)\right]\left(1+[\xi_2']^2\right)^{1/2}.
\end{aligned}
\tag{2.180}
$$

If one compares Equations 2.161 and 2.178, 2.162 and 2.179 or 2.163 and 2.180, it will be readily apparent that the latter three relations are mathematically similar in form to the previous three relations, but with a different type of specific surface free energy quantity. Thus, in each of the three Integrands 2.161 through 2.163 we have effectively replaced the specific surface free energy $\omega^{(\alpha\beta)}$ with a three-term quantity $\omega^{(\alpha\beta)} + \rho^{(\alpha\beta)}g\xi_k + \mu_j\rho_j^{(\alpha\beta)}$ where $(\alpha\beta) = \{(sv),(sl),(lv)\}$. It is possible to take advantage of this structure by defining three specific surface free energies for these compressive systems as

$$\omega_c^{(sl)} = \omega^{(sl)} + \rho^{(sl)}g\xi_2 + \mu_j\rho_j^{(sl)}, \tag{2.181}$$

$$\omega_c^{(lv)} = \omega^{(lv)} + \rho^{(lv)}g\xi_1 + \mu_j\rho_j^{(lv)}, \tag{2.182}$$

$$\omega_c^{(sv)} = \omega^{(sv)} + \rho^{(sv)}g\xi_2 + \mu_j\rho_j^{(sv)}, \tag{2.183}$$

and the one compressive specific linear free energy as

$$\omega_c^{(slv)} = \omega^{(slv)} + \rho^{(slv)}g\xi_2 + \mu_j\rho_j^{(slv)}, \tag{2.184}$$

where the density factor in the gravitational term is a sum over all j chemical components (e.g., $\rho^{(sl)} = \Sigma_{j=1}^{\eta}\rho_j^{(sl)}$). It should he noted that even when the specific free energies and the phase densities are all constant, the presence of the surface functions ξ_k (where $k = 1, 2$) in all of these definitions means that these specific, compressive free energies are not constant. Consequently, if one wishes to replace the triplet of terms that occur in the Integrands 2.178 through 2.180 with these compressive free energies, then one must, when evaluating the Euler-Lagrange equation and the boundary conditions, remember that these energies are not constant.

For the purpose of illustration and comparison with the results above, only the first Euler-Lagrange Equation 2.30 with Integrand 2.178 and corresponding Definition 2.182 will be considered. The other two Euler-Lagrange equations are evaluated in a similar manner. For the case when the specific surface free energy is position dependent, one has from the Euler-Lagrange Equation 2.30

$$\Delta\rho g r\xi_1 + \left(\mu_j\left(\rho_j^{(l)} - \rho_j^{(v)}\right)\right) + \omega^{(l)} - \omega^{(v)}\right)r - \omega_c^{(lv)}\left(1 + [\xi_1']^2\right)^{-3/2}\left(\xi_1' + [\xi_1']^3 + r\xi_1''\right)$$

$$+ r\left(1 + [\xi_1']^2\right)^{-1/2}\frac{\partial\omega_c^{(lv)}}{\partial\xi_1} - r\xi_1'\left(1 + [\xi_1']^2\right)^{-1/2}\frac{\partial\omega_c^{(lv)}}{\partial r} = 0.$$

$$\tag{2.185}$$

This expression permits one to write the Lagrange multipliers as

$$\mu_j\left(\rho_j^{(l)}-\rho_j^{(v)}\right)+\omega^{(l)}-\omega^{(v)}=\omega_c^{(lv)}\left(\frac{\xi_1'}{r\left(1+\left[\xi_1'\right]^2\right)^{1/2}}+\frac{\xi_1''}{\left(1+\left[\xi_1'\right]^2\right)^{3/2}}\right)$$

$$-\Delta\rho g\xi_1-\left(1+\left[\xi_1'\right]^2\right)^{-1/2}\frac{\partial\omega_c^{(lv)}}{\partial\xi_1}+\xi_1'\left(1+\left[\xi_1'\right]^2\right)^{-1/2}\frac{\partial\omega_c^{(lv)}}{\partial r}.$$

$$(2.186)$$

where the first term in brackets is the mean curvature $J^{(lv)}$; see Equations 2.75 and 2.76.

At the axis of symmetry where $r=0$, the slope of the liquid-vapor surface vanishes; that is, $\xi_1'(r=0)=0$ as indicated by Equation 2.166, and the mean curvature simplifies to $J^{(lv)}=2\xi_1''$ so that Equation 2.186 also simplifies to

$$\mu_{oj}\left(\rho_{oj}^{(l)}-\rho_{oj}^{(v)}\right)+\omega_o^{(l)}-\omega_o^{(v)}=2\xi_1''(0)\omega_{oc}^{(lv)}-\Delta\rho_o g\xi_1(0)-\frac{\partial\omega_c^{(lv)}}{\partial\xi_1}(0),\qquad(2.187)$$

where the subscript o symbol is used to denote quantities evaluated at the axis of symmetry location and the density difference $\Delta\rho_o$ is the sum over j of all density differences $(\rho_{oj}^{(l)}-\rho_{oj}^{(v)})$ also evaluated at the axis of symmetry. If the expression for $\omega_c^{(lv)}$ from Equation 2.182 is substituted into the partial derivative in Equation 2.187 to obtain

$$\frac{\partial\omega_c^{(lv)}}{\partial\xi_1}(r=0)=\rho_o^{(lv)}g=\sum_{j=1}^{\eta}\rho_j^{(lv)}(r=0)g,\qquad(2.188)$$

then the chemical potentials can be written as

$$\mu_{oj}\left\{\left(\rho_{oj}^{(l)}-\rho_{oj}^{(v)}\right)-2\xi_1''(0)\rho_{oj}^{(lv)}\right\}=2\xi_1''(0)\omega_o^{(lv)}-\Delta\rho_o g\xi_1(0)$$

$$-\rho_{oj}^{(lv)}g\left(1-2\xi_1(0)\xi_1''(0)\right)-\left(\omega^{(l)}-\omega^{(v)}\right)$$

$$(2.189)$$

or

$$\mu_{oj}=\frac{\dfrac{2\xi_1''(0)\omega_o^{(lv)}}{\rho_{oj}^{(l)}-\rho_{oj}^{(v)}}-g\xi_1(0)\dfrac{\Delta\rho_o}{\rho_{oj}^{(l)}-\rho_{oj}^{(v)}}-g\left(1-2\xi_1(0)\xi_1''(0)\right)\dfrac{\rho_{oj}^{(lv)}}{\rho_{oj}^{(l)}-\rho_{oj}^{(v)}}-\dfrac{\omega_o^{(l)}-\omega_o^{(v)}}{\rho_{oj}^{(l)}-\rho_{oj}^{(v)}}}{1-2\xi_1''(0)\dfrac{\rho_{oj}^{(lv)}}{\rho_{oj}^{(l)}-\rho_{oj}^{(v)}}}.$$

$$(2.190)$$

Furthermore, if the component density $\rho_{oj}^{(lv)} \ll (\rho_{oj}^{(l)} - \rho_{oj}^{(v)})$, then we can approximate the denominator of Equation 2.190 as

$$\left(1 - 2\xi_1''(0)\frac{\rho_{oj}^{(lv)}}{\rho_{oj}^{(l)} - \rho_{oj}^{(v)}}\right)^{-1} \approx 1 + 2\xi_1''(0)\frac{\rho_{oj}^{(lv)}}{\rho_{oj}^{(l)} - \rho_{oj}^{(v)}} \tag{2.191}$$

so that the chemical potential can be written as

$$\mu_{oj} \approx \frac{2\xi_1''(0)\omega_o^{(lv)}}{\rho_{oj}^{(l)} - \rho_{oj}^{(v)}} - g\xi_1(0)\frac{\Delta\rho_o}{\rho_{oj}^{(l)} - \rho_{oj}^{(v)}} - \frac{\omega_o^{(l)} - \omega_o^{(v)}}{\rho_{oj}^{(l)} - \rho_{oj}^{(v)}}$$
$$+ 2\xi_1''(0)\frac{\rho_{oj}^{(lv)}}{\rho_{oj}^{(l)} - \rho_{oj}^{(v)}}\left(\frac{2\xi_1''(0)\omega_o^{(lv)}}{\rho_{oj}^{(l)} - \rho_{oj}^{(v)}} - g\xi_1(0)\frac{\Delta\rho_o}{\rho_{oj}^{(l)} - \rho_{oj}^{(v)}} - \frac{\omega_o^{(l)} - \omega_o^{(v)}}{\rho_{oj}^{(l)} - \rho_{oj}^{(v)}}\right). \tag{2.192}$$

Comparison of this expression with the chemical potential Expression 2.173 shows that the presence of a nonzero surface density $\rho_{oj}^{(lv)}$ slightly modifies the constant values of the chemical potentials μ_{oj} when it is assumed that most of the mass within the composite system is present in the bulk liquid phase as density $\rho_{oj}^{(l)}$ so that the inequality just above Equation 2.191 applies. However, if most of the mass is present within a surface film between two low density bulk phases, then one might consider treating both densities $\rho_{oj}^{(l)}$ and $\rho_{oj}^{(v)}$ as small with respect to $\rho_{oj}^{(lv)}$. Under this approximation, Equation 2.189 could be simplified to

$$\mu_{oj} \approx \frac{\omega_o^{(l)} - \omega_o^{(v)}}{2\xi_1''(0)\rho_{oj}^{(lv)}} - \frac{\omega_o^{(lv)}}{\rho_{oj}^{(lv)}} + \frac{g}{2}\frac{1 - 2\xi_1(0)\xi_1''(0)}{\xi_1''(0)} \tag{2.193}$$

or

$$\mu_{oj} \approx \frac{1}{2}\frac{R_o^{(lv)}}{\rho_{oj}^{(lv)}}\left(\omega_o^{(l)} - \omega_o^{(v)} + g\rho_{oj}^{(lv)} - \frac{2\omega_o^{(lv)}}{R_o^{(lv)}}\right). \tag{2.194}$$

If the definition of μ_{oj}, provided by Equation 2.189, is used to eliminate these quantities in favor of the variables on the right-hand side of Equation 2.189, then it is possible to substitute the result for μ_{oj} from Equation 2.189 into Equation 2.185 to obtain an expression that is the Laplace equation of capillarity for the arrangement where the liquid-vapor surface phase has nonvanishing surface density and the specific surface free energy $\omega^{(lv)}$ is a constant. The specific steps require one to write Equation 2.189 as

$$\mu_{oj}\left\{\left(\rho_{oj}^{(l)} - \rho_{oj}^{(v)}\right) - 2\xi_1''(0)\rho_{oj}^{(lv)}\right\} + \left(\omega_o^{(l)} - \omega_o^{(v)}\right) =$$
$$2\xi_1''(0)\omega_o^{(lv)} - \Delta\rho_o g\xi_1(0) - g\rho_{oj}^{(lv)}\left(1 - 2\xi_1(0)\xi_1''(0)\right), \tag{2.195}$$

and Equation 2.186 as

$$\mu_{oj}\left\{\left(\rho_{oj}^{(l)}-\rho_{oj}^{(v)}\right)-2\xi_1''(0)\rho_{oj}^{(lv)}\right\}+\left(\omega_o^{(l)}-\omega_o^{(v)}\right)=\omega_{oc}^{(lv)}J_o^{(lv)}-\Delta\rho_o g\xi_1(0)$$

$$-2\xi_1''(0)\mu_{oj}\rho_{oj}^{(lv)}-\left(1+[\xi_1']^2\right)^{-1/2}\frac{\partial\omega_c^{(lv)}}{\partial\xi_1}(0)+\xi_1'\left(1+[\xi_1']^2\right)^{-1/2}\frac{\partial\omega_c^{(lv)}}{\partial r}(0).$$

$$(2.196)$$

Equating the right-hand side of Equations 2.195 and 2.196 yields, after some rear-ranging, the expression

$$\omega_c^{(lv)}J^{(lv)}=2\xi_1''(0)\left(\omega_o^{(lv)}+\rho_o^{(lv)}g\xi_1(0)+\mu_{oj}\rho_{oj}^{(lv)}\right)+\left(\Delta\rho\xi_1-\Delta\rho_o\xi_1(0)\right)g$$

$$-\rho_o^{(lv)}g+\left(1+[\xi_1']^2\right)^{-1/2}\frac{\partial\omega_c^{(lv)}}{\partial\xi_1}-\xi_1'\left(1+[\xi_1']^2\right)^{-1/2}\frac{\partial\omega_c^{(lv)}}{\partial r}$$

$$(2.197)$$

for the Laplace equation. After substituting the result from Equation 2.188 for one partial derivative and a zero for the partial derivative with respect to r, one obtains the compressible form of the Laplace equation as

$$\omega_c^{(lv)}J^{(lv)}=2\xi_1''(0)\omega_{oc}^{(lv)}+\left(\Delta\rho(r\neq0)\xi_1(r\neq0)-\Delta\rho(r=0)\xi_1(r=0)\right)g$$

$$+\left(\rho^{(lv)}(r\neq0)\left(1+[\xi_1']^2\right)^{-1/2}-\rho^{(lv)}(r=0)\right)g,$$

$$(2.198)$$

where we have assumed that the densities $\rho_j^{(l)}$ and $\rho_j^{(v)}$ do not vary appreciably with position in the gravitational field. It is interesting to note that, in comparison with Equation 2.50, the apparent specific free energy is not equal to $\omega_c^{(lv)}=\gamma_\infty^{(lv)}$, but it is (approximately) a combination of the surface tension $\gamma_\infty^{(lv)}$, the gravity term $\rho^{(lv)}g\xi_1$, and the terms $\mu_j\rho_j^{(lv)}$ where the chemical potentials μ_j are constant as specified by either Equation 2.190 or its approximation, Equation 2.192.

Evaluation of the boundary condition at the solid–liquid–fluid contact line to determine the form of the Young equation of capillarity requires the evaluation of the transversality conditions Equations 2.32 and 2.35 at $r=r_b$. These transversality conditions involve a term of the form $\partial(f_k r)/\partial\xi_k'$ that must be evaluated because the surface energies $\omega_c^{(\alpha\beta)}$ are position dependent. Consequently, for the integrand defined by Equation 2.178 the transversality condition Equation 2.32 simplifies to

$$\left(\frac{1}{2}\Delta\rho g\xi_1^2+\left[\mu_j\left(\rho_j^{(l)}-\rho_j^{(v)}\right)+\omega^{(l)}-\omega^{(v)}\right]\xi_1\right)\left(1+[\xi_1']^2\right)^{1/2}+\omega_c^{(lv)}\left(1+\xi_1'\xi_2'\right)=0,$$

$$(2.199)$$

while the transversality condition Equation 2.35, for the integrand defined by Equation 2.180, simplifies to

$$\left(\omega_c^{(sl)} - \omega_c^{(sv)}\right)\left(1+\left[\xi_2'\right]^2\right)^{1/2} - \frac{1}{2}\Delta\rho g\xi_2^2 + \frac{\omega_c^{(slv)}}{r_b} - \left(\mu_j\left(\rho_j^{(l)} - \rho_j^{(v)}\right) + \omega^{(l)} - \omega^{(v)}\right)\xi_2 = 0.$$
(2.200)

At the contact point $r = r_b$, where all surfaces need to meet, one must have $\xi_1(r_b) = \xi_2(r_b)$; otherwise, the two surfaces described by these curves would not intersect. With this connectivity requirement, Equation 2.199 can be written as

$$\left(\mu_j\left(\rho_j^{(l)} - \rho_j^{(v)}\right) + \omega^{(l)} - \omega^{(v)}\right)\xi_1 = \omega_c^{(lv)}\left(1+\xi_1'\xi_2'\right)\left(1+\left[\xi_1'\right]^2\right)^{-1/2} - \frac{1}{2}\Delta\rho g\xi_1^2.$$
(2.201)

Substituting this expression for the last term in Equation 2.200 eliminates the quantities μ_j, $\omega^{(l)}$ and $\omega^{(v)}$, and yields the Young equation of capillarity expressed in terms of the surface functions $\xi_k(r)$ where $k = 1$ or 2 as

$$\left(\omega_c^{(sl)} - \omega_c^{(sv)}\right)\left(1+\left[\xi_2'\right]^2\right)^{1/2} + \frac{\omega_c^{(slv)}}{r_b} - \omega_c^{(lv)}\left(1+\xi_1'\xi_2'\right)\left(1+\left[\xi_1'\right]^2\right)^{-1/2} = 0$$
(2.202)

or

$$1+\xi_1'\xi_2' + \frac{\omega_c^{(sl)} - \omega_c^{(sv)}}{\omega_c^{(lv)}}\left(1+\left[\xi_1'\right]^2\right)^{1/2}\left(1+\left[\xi_2'\right]^2\right)^{1/2} + \frac{1}{r_b}\frac{\omega_c^{(slv)}}{\omega_c^{(lv)}}\left(1+\left[\xi_1'\right]^2\right)^{1/2} = 0.$$
(2.203)

Eliminating the surface derivatives ξ_k' in favor of the surface angles ϑ_k, following the procedure above, yields the Young equation of capillarity or the boundary condition of the liquid-fluid interface on the solid at the position $r = r_b$ as

$$\omega_c^{(sv)} - \omega_c^{(sl)} = \omega_c^{(lv)}\cos\vartheta_l + \omega_c^{(slv)}\frac{\cos\vartheta_2}{r_b}.$$
(2.204)

In contrast to Equation 2.203, this expression is in terms of the specific, compressible free energies of the surfaces that intersect at the contact line; the specific, compressible free energy of the contact line and the contact angle of the liquid on the solid. However, unlike the situation presented in Equation 2.60, the specific, compressible free energies $\omega_c^{(\alpha\beta)}$ are functions that depend upon the surface tension if $\omega^{(\alpha\beta)} = \gamma^{(\alpha\beta)}$, the surface excess densities $\rho_j^{(\alpha\beta)}$, and the orientation of the surface, through the surface function $\xi_k(r)$, with respect to the gravitational field.

For an axisymmetric liquid lens, the appropriate Neumann triangle (quadrilateral) relation for a compressible system is similar to the previous Expression 2.138, but with the specific surface free energies $\omega^{(\alpha\beta)}$ replaced by the specific, compressible surface free energies $\omega_c^{(\alpha\beta)}$ and the specific linear free energy $\omega^{(slv)}$ replaced by $\omega_c^{(slv)}$.

2.3.7 INTERPRETATION OF THE COMPRESSIBLE MECHANICAL EQUILIBRIUM CONDITIONS

Unlike the incompressible variational problem posed above, which considered arbitrary virtual displacements subject to a constant volume constraint, there are very few, if any, comparable calculations involving a constant mass constraint. However, from the calculations above, we can see that for a composite capillary system in which nearly all of the mass occurs in the bulk phases that the mechanical equilibrium conditions (i.e., the Laplace equation, Young equation, and Neumann triangle or quadrilateral relation) are identical to the expressions that occur when one assumes that the system is incompressible. On the other hand, when one does not restrict the system's mass to lie exclusively in the bulk phases, then the form of all mechanical equilibrium conditions changes to reflect this more complicated picture of the interfacial regions. Stated succinctly, it is as if the compressible capillary system obeys exactly the same mechanical equilibrium conditions but with a slightly different definition for the specific surface and line free energies. These definitions are given explicitly by Equations 2.181 through 2.184. In essence, the specific, compressible free energies $\omega_c^{(\alpha\beta)}$ are nonconstant functions that depend upon the surface tension if $\omega^{(\alpha\beta)} = \gamma^{(\alpha\beta)}$, surface excess densities $\rho_j^{(\alpha\beta)}$, and the orientation of the surface with respect to the gravitational field through the surface function $\xi_k(r)$. Consequently, the compressible version of the Laplace equation (e.g., Equation 2.198) is not identical to the classical, incompressible version of the Laplace equation (e.g., Equation 2.50) when the terms $\rho^{(\alpha\beta)}g\xi_k + \mu_j\rho_j^{(\alpha\beta)}$ are a significant fraction of the magnitude of $\omega_c^{(\alpha\beta)}$. From the boundary conditions (e.g., the Young Equation 2.204 or the Neumann triangle (quadrilateral) relation) similar conclusions also follow. The effect of this difference between $\omega^{(\alpha\beta)}$ and $\omega_c^{(\alpha\beta)}$ is most easily illustrated by a simple example.

If we consider the case of a sessile drop resting on a horizontal solid surface, then the solid surface's curve is $\xi_2(r) = 0$. Furthermore, if we assume that all solid surface excess densities; that is, $\rho_j^{(sl)}$ and $\rho_j^{(sv)}$ vanish and we also require that the specific surface free energies are constants; that is, $\omega^{(sl)} = \gamma_\infty^{(sl)}$ and $\omega^{(sv)} = \gamma_\infty^{(sv)}$, then we can write Equation 2.204 as

$$\gamma_\infty^{(sv)} - \gamma_\infty^{(sl)} = \omega_c^{(lv)}\cos\vartheta_l + \frac{\sigma_\infty^{(slv)}}{r_b}, \qquad (2.205)$$

where we have also assumed that, for simplicity, the linear excess densities vanish and the specific linear free energy equals $\sigma^{(slv)}$. Recognizing that the left-hand side of Equation 2.205 will still apply when the capillary system is very large; that is, as $r \to \infty$, we can write

$$\gamma_\infty^{(sv)} - \gamma_\infty^{(sl)} = \omega_{\infty c}^{(lv)}\cos\vartheta_{\infty l}. \qquad (2.206)$$

Combining Equations 2.205 and 2.206 to eliminate the solid surface tension difference we obtain the expression

$$\cos\vartheta_l = \frac{\omega_{\infty c}^{(lv)}}{\omega_c^{(lv)}}\cos\vartheta_{\infty l} - \frac{\sigma_{\infty}^{(slv)}}{\omega_c^{(lv)}}\frac{1}{r_b}, \tag{2.207}$$

which was used by Gaydos and Neumann [4] and others [49–51] to determine the magnitude and sign of the line tension $\sigma_{\infty}^{(slv)}$. However, in those situations in which the terms $\rho^{(\alpha\beta)}g\xi_1 + \mu_j\rho_j^{(\alpha\beta)}$ are a significant fraction of the magnitude of $\omega_c^{(\alpha\beta)}$ it must be expected that this procedure will lead, in general, to incorrect estimates of the angle $\vartheta_{\infty l}$ and the line tension. If we write the two specific surface free energies in Equation 2.207, using Equation 2.182, as

$$\omega_c^{(lv)}(\xi_1) = \left(\gamma_{\infty}^{(lv)} + \mu_j\rho_j^{(lv)}\right) + \rho^{(lv)}g\xi_1(r) \tag{2.208}$$

and

$$\omega_{\infty c}^{(lv)}(\xi_1) = \left(\gamma_{\infty}^{(lv)} + \mu_j\rho_j^{(lv)}\right) + \rho^{(lv)}g\xi_1(\infty), \tag{2.209}$$

then we may calculate the ratio of these two quantities, which appears in Equation 2.207, as

$$\frac{\omega_{\infty c}^{(lv)}}{\omega_c^{(lv)}} = \frac{1 + \dfrac{\rho^{(lv)}g\xi_1(\infty)}{\gamma_{\infty}^{(lv)} + \mu_j\rho_j^{(lv)}}}{1 + \dfrac{\rho^{(lv)}g\xi_1(r)}{\gamma_{\infty}^{(lv)} + \mu_j\rho_j^{(lv)}}}, \tag{2.210}$$

where $\rho^{(lv)} = \sum_{j=1}^{\eta}\rho_j^{(lv)}$ and $\mu_j\rho_j^{(lv)} = \sum_{j=1}^{\eta}\mu_j\rho_j^{(lv)}$. When the gravitational contribution is small, we can write $\rho^{(lv)}g\xi_1(r) << (\gamma_{\infty}^{(lv)} + \mu_j\rho_j^{(lv)})$, and approximate Equation 2.210 as

$$\frac{\omega_{\infty c}^{(lv)}}{\omega_c^{(lv)}} \approx \left(1 + \frac{\rho^{(lv)}g\xi_1(\infty)}{\gamma_{\infty}^{(lv)} + \mu_j\rho_j^{(lv)}}\right)\left(1 - \frac{\rho^{(lv)}g\xi_1(r)}{\gamma_{\infty}^{(lv)} + \mu_j\rho_j^{(lv)}}\right) \approx 1 - \left(\frac{\rho^{(lv)}g}{\gamma_{\infty}^{(lv)} + \mu_j\rho_j^{(lv)}}\right)(\xi_1(r) - \xi_1(\infty)), \tag{2.211}$$

which will be close to unity (under this assumption) but not exactly equal to one since the liquid-vapor surface density term slightly alters this ratio unless the solid surface is horizontal. Should the inequality above Equation 2.211 not apply, then one must retain the exact expression given by Equation 2.210. When Equation 2.210 applies, then setting the ratio of specific free energies multiplying $\cos\vartheta_{\infty l}$ in Equation 2.207 equal to one may cause a slight error. In a similar manner, it is possible to approximate the value of the ratio

$$\frac{\sigma_{\infty}^{(slv)}}{\omega_c^{(lv)}}\frac{1}{r_b} = \frac{\sigma_{\infty}^{(slv)}}{\gamma_{\infty}^{(lv)} + \rho^{(lv)}g\xi_1(r) + \mu_j\rho_j^{(lv)}}\frac{1}{r_b}, \tag{2.212}$$

which appears in Equation 2.207 as

$$\frac{\sigma_\infty^{(slv)}}{\omega_c^{(lv)}}\frac{1}{r_b} \approx \frac{\sigma_\infty^{(slv)}}{\gamma_\infty^{(lv)}r_b}\left(1 - \frac{\rho^{(lv)}g\xi_1(r) + \mu_j\rho_j^{(lv)}}{\gamma_\infty^{(lv)}}\right) \tag{2.213}$$

when it is possible to assume that $\gamma_\infty^{(lv)} \gg \left(\rho^{(lv)}g\xi_1(r) + \mu_j\rho_j^{(lv)}\right)$.

Equation 2.213, which is an approximation, shows that the estimate of the line tension will be slightly influenced by the excess surface densities $\rho_j^{(lv)}$. For solid–liquid–vapor systems like those studied by Gaydos and Neumann [4] one might be able to argue that most, if not all, excess surface density effects would be insignificant given the magnitude of reported line tensions on these solid surfaces [51]. Similar conclusions about the difficulty of decoupling the influence of line tension from the influence of excess surface density effects also apply for thin fluid films. In these thin fluid film systems, the fraction of mass in the surface and line regions is considerably larger and the measurements of the line tension are considerably smaller [52–61] than comparable estimates for sessile drop systems. Consequently, these thin film systems might present an experimental situation where it is not possible to obtain an unambiguous estimate of the magnitude of the line tension because of the masking effect of the excess surface density term.

2.4 CONCLUSIONS

A generalized, thermodynamic approach to axisymmetric systems with nonignorable surface and line regions has been considered to derive extended mechanical equilibrium relations for both the surface and the line boundary regions of uncharged capillary systems. The conditions so obtained are generalizations of the Laplace equation of capillarity and the corresponding Young equation or Neumann triangle relation. The key relations are the Laplace equation of capillarity expressed by Equation 2.50, the Young equation of capillarity given by Equation 2.60, and the generalized Neumann triangle relation, Equation 2.138, which is actually a quadrilateral relation. It is possible to express Equation 2.138 as two orthogonal scalar relations as given by Equation 2.146 and 2.147. These key relations provide a mechanical equilibrium description for any open axisymmetric capillary system. A limited description of nonmoderately curved, axisymmetric capillary systems was considered in Section 2.11. The selection of suitable variables for both the surface phase and the linear phase in an axisymmetric capillary system was considered. It was shown that surface free energies related to the mean and Gaussian curvature can be carried over directly from the generalized theory [1] but that linear free energies involving angles, line curvatures, and line torsions should be altered for axisymmetric system to account for the capillary system's symmetry and for those angles that are likely to be experimentally obtainable through measurement.

For a compressible, axisymmetric capillary system the forms of the Laplace, Young, and Neumann relations remain almost the same provided the specific surface free energies $\omega^{(\alpha\beta)}$ (where the two separating bulk phases are denoted by the Greek symbols α and β) are replaced by the triplet of terms given by the compressive surface free energy $\omega_c^{(\alpha\beta)}$ or

$$\omega^{(\alpha\beta)} + \rho^{(\alpha\beta)} g \xi_k(r) + \mu_j \rho_j^{(\alpha\beta)}, \tag{2.214}$$

where subscript j denotes the chemical species present at the surface and subscript k denotes the surface curve. Thus, Equation 2.198 provides the compressible version of the Laplace equation of capillarity, while Equation 2.204 gives the corresponding Young equation of capillarity. One consequence of this "replacement" is that the energy $\omega_c^{(\alpha\beta)}$ requires one to deal with a more complicated version of the Euler-Lagrange equation and associated boundary conditions since this energy depends explicitly on the surface functions $\xi_k(r)$. However, one benefit of treating a capillary system as compressible is that it permits one to estimate the ratios

$$\frac{\omega_{\infty c}^{(lv)}}{\omega_c^{(lv)}} \quad \text{and} \quad \frac{\sigma_\infty^{(slv)}}{\omega_c^{(lv)}} \frac{1}{r_b} \tag{2.215}$$

occurring in Equation 2.207 provided one can obtain values for the surface excess densities $\rho_j^{(\alpha\beta)}$ and associated chemical potentials μ_j. For axisymmetric liquid-fluid systems, such as a sessile drop on a horizontal solid surface, these ratios are most likely given by

$$\frac{\omega_{\infty c}^{(lv)}}{\omega_c^{(lv)}} \approx 1 \quad \text{and} \quad \frac{\sigma_\infty^{(slv)}}{\omega_c^{(lv)}} \frac{1}{r_b} \approx \frac{\sigma_\infty^{(slv)}}{\gamma_\infty^{(lv)} r_b}, \tag{2.216}$$

since $\xi_1(r) - \xi_1(\infty) \approx 0$ in Equation 2.211 and surface tension will likely dominate in Equation 2.213, whereas for thin liquid films it is not as obvious that one can decouple the influence of line tension from excess surface density effects.

REFERENCES

1. L. Boruvka and A. W. Neumann. *Journal of Chemical Physics* 66 (1977): 5464.
2. J. W. Gibbs. *The Scientific Papers of J. Willard Gibbs.* Vol. 1, 55–371. Dover, New York, 1961.
3. E. Kreyszig. *Differential Geometry.* 129. Dover, New York, 1991.
4. J. Gaydos and A. W. Neumann. *Journal of Colloid and Interface Science* 120 (1987): 76; see reference list therein.
5. R. Finn. *Equilibrium Capillary Surfaces.* Springer-Verlag, Berlin, 1986.
6. R. Finn. *op. cit.,* ref. [5], p. 4.
7. J. Gaydos, C. J. Budziak, D. Li, and A. W. Neumann. "Contact Angles on Imperfect Solid Surfaces." In *Surface Engineering: Current Trends and Future Prospects.* Edited by S. A. Meguid, 100–113. Elsevier Applied Science, New York, 1990.
8. O. Bolza. *Lectures on the Calculus of Variations.* 2nd ed. Chelsea, New York, 1960.
9. R. Courant and D. Hilbert. *Methods of Mathematical Physics.* 2 Vols. Wiley-Interscience, New York, 1989.

10. A. R. Forsyth. *Calculus of Variations.* Cambridge University Press, Cambridge, 1927.

11. C. Fox. *An Introduction to the Calculus of Variations.* Dover, New York, 1963.

12. I. M. Gel'fand and S. V. Fomin. *Calculus of Variations.* Translated by R. A. Silverman. Prentice- Hall, Englewood Cliffs, NJ, 1963.

13. R. Weinstock. *Calculus of Variations.* Dover, New York, 1974.

14. J. Gaydos. Ph.D. Thesis, *Implications of the Generalized Theory of Capillarity.* University of Toronto, Department of Mechanical Engineering, Toronto, 1992.

15. A. Goetz. *Introduction to Differential Geometry.* Addison-Wesley, Toronto, 1970.

16. J. J. Koenderink. *Solid Shape.* MIT Press, Cambridge, MA, 1990.

17. C. E. Weatherburn. *Differential Geometry of Three Dimensions.* 2 Vols. Cambridge University Press, London, 1930.

18. W. Kaplan. *Advanced Calculus.* 2nd ed. Addison-Wesley, Reading, MA, 1973.

19. R. Courant. *Differential and Integral Calculus.* 2 Vols., 2nd. ed. Wiley-Interscience, New York, 1988.

20. M. E. R. Shanahan. "A Variational Approach to Sessile Drops." in *Adhesion 6.* Edited by K. W. Allen, 75–86. Applied Science, London, 1982.

21. P. S. Laplace. *Traité de Mécanique Céleste.* Supplements au Livre X, 1805 and 1806 resp. in (Euvres Complete Vol. 4, Gauthier-Villars, Paris. See also the annotated English translation by N. Bowditch (1839); reprinted by Chelsea, New York, 1966.

22. C. F. Gauss. *Principia Generalia Theoriae Figurae Fluidorum.* Comment. Soc. Regiae Scient. Gottingensis *Rec.* 7,1830.

23. J. W. Strutt (Lord Rayleigh). "On the Theory of Surface Forces." *Philosophical Magazine (5th series)* 30 (1890): 285, 456.

24. R. E. Johnson, Jr. *Journal of Physical Chemistry* 63 (1959): 1655.

25. R. E. Collins and C. E. Cooke, Jr. *Transactions of the Faraday Society* (London) 55 (1959): 1602.

26. T. Li. *Journal of Chemical Physics* 36 (1962): 2369.

27. W. Fox. *Journal of Chemical Physics* 37 (1963): 2776.

28. M. Bracke, F. De Bisschop, and P. Joos. *Progress in Colloid and Polymer Science* 76 (1988): 251.

29. J. J. Bikerman. *Journal of Physical Chemistry* 63 (1959): 1658.

30. J. J. Bikerman. *Physical Surfaces.* Chapter 6. Academic, London, 1970.

31. R. Defay and I. Prigogine. *Surface Tension and Adsorption.* Collaborated with A. Bellemans. Translated by D. H. Everett, 9–10. Longmans Green, London, 1966.

32. J. S. Rowlinson and B. Widom. *Molecular Theory of Capillarity.* Clarendon, Oxford, 1982.

33. J. Gaydos, L. Boruvka, and A. W. Neumann. *Langmuir* 7 (1991): 1035.

34. Y. C. Fung. *Foundations of Solid Mechanics.* Prentice-Hall, Englewood Cliffs, NJ, 1965.

35. Y. Rotenberg, L. Boruvka, and A. W. Neumann. *Langmuir* 2 (1986): 533.

36. L. Boruvka, Y. Rotenberg, and A. W. Neumann. *Journal of Physical Chemistry* 89 (1985): 2714.

37. L. Boruvka, Y. Rotenberg, and A. W. Neumann. *Langmuir* 1 (1985): 40.

38. L. Boruvka, Y. Rotenberg, and A. W. Neumann. *Journal of Physical Chemistry* 90 (1986): 125.

39. P. Neogi and S. E. Friberg. *Journal of Colloid and Interface Science* 127 (1989): 492.

40. J. C. Eriksson and S. Ljunggren. *Journal of Colloid and Interface Science* 2 (1992): 575.

41. J. Gaydos. "The Laplace Equation of Capillarity." In *Drops and Bubbles in Interfacial Research.* Vol. 6. Edited by D. Möbius and R. Miller, 1–56. Elsevier, Amsterdam, 1998.

42. F. Xia, W. G. Gray, and P. Chen. *Journal of Colloid and Interface Science* 261 (2003): 464.
43. A. W. Joshi. *Matrices and Tensors in Physics.* 2nd ed., 146–48. John Wiley, Toronto, 1984.
44. L. Mirsky. *An Introduction to Linear Algebra.* 233–36. Dover, New York, 1982.
45. P. Chen, J. Gaydos, and A. W. Neumann. *Colloids and Surfaces A* 114 (1996): 5956.
46. L. Boruvka and A. W. Neumann. *Journal of Colloid and Interface Science* 65 (1978): 315.
47. J. Gaydos and A. W. Neumann. *Advanced Colloids and Surfaces* 49 (1994): 197.
48. M. Hoorfar, A. Amirfazli, J. A. Gaydos, and A. W. Neumann. *Advanced Colloids and Surfaces* 114/115 (2005): 103.
49. J. A. Wallace and S. Schürch. *Journal of Colloid and Interface Science* 124 (1988): 452.
50. P. A. Kralchevsky, A. D. Nikolov, and I. B. Ivanov. *Journal of Colloid and Interface Science* 112 (1986): 132.
51. A. Amirfazli and A. W. Neumann. *Advances in Colloid and Interface Science* 110 (2004): 121.
52. W. D. Harkins. *Journal of Chemical Physics* 5 (1937): 135.
53. N. V. Churaev, V. M. Starov, and B. V. Derjaguin. *Journal of Colloid and Interface Science* 89 (1982): 16.
54. J. A. de Feijter and A. Vrij. *Journal of Electroanalytical Chemistry* 37 (1972): 9.
55. J. A. de Feijter, I. B. Rijnbout, and A. Vrij. *Journal of Colloid and Interface Science* 64 (1978): 258.
56. T. Kolarov and Z. M. Zorin. *Colloid Journal of the USSR* 42 (1980): 899.
57. A. Scheludko, B. V. Toshev, and D. Platikanov. "On the Mechanics and Thermodynamics of Three Phase Contact Line Systems." In *The Modern Theory of Capillarity.* Edited by F. C. Goodrich and A. I. Rusanov, 163–82. Akademie-Verlag, East Berlin, 1981.
58. V. M. Starov and N. V. Churaev. *Colloid Journal of the USSR* 42 (1980): 585.
59. D. Platikanov, M. Nedyalkov, and A. Scheludko. *Journal of Colloid and Interface Science* 75 (1980): 612.
60. D. Platikanov and M. Nedyalkov. "Contact Angles and Line Tension at Microscopic Three Phase Contacts." In *Microscopic Aspects of Adhesion and Lubrication.* Edited by J. M. Georges, 97–106. Elsevier Science, Amsterdam, 1982.
61. I. B. Ivanov, P. A. Kralchevsky, and A. D. Nikolov. *Journal of Colloid and Interface Science* 112 (1986): 97.

3 Axisymmetric Drop Shape Analysis (ADSA)

Mina Hoorfar and A. Wilhelm Neumann

CONTENTS

3.1   INTRODUCTION

Numerous methodologies have been developed for the measurement of interfacial properties including contact angles and surface tensions. Contact angles are most commonly measured by aligning a tangent with the profile of a sessile drop at the point of contact with the solid surface. Liquid surface tension measurements commonly involve the determination of the height of a meniscus in a capillary or on a fiber or a plate. Some of the major methods including the Wilhelmy plate technique and Du Nouy ring method [1–5], the drop weight method [2–5], the oscillating jet method [3,4], the capillary wave method [3,4], and the spinning drop method [2–5] are briefly reviewed in this section. An overview of these techniques reveals that, in most instances, a balance must be struck between the simplicity, the accuracy, and the flexibility of the methodology. Alternative approaches to obtain the interfacial properties are drop shape methods developed to determine the liquid–vapor or liquid–liquid interfacial tensions and the contact angle from the shape of a sessile drop, pendant drop, or captive bubble. These methods are widely used due to their simplicity and accuracy. In this section, the drop shape methods are described in detail.

3.1.1   WILHELMY PLATE AND DU NOUY RING METHOD

The Wilhelmy plate technique is both accurate and relatively simple to use. The standard method of conducting an experiment with this technique is to raise the liquid sample until it just touches a thin platinum plate, which is suspended from an electrobalance. The measured downward force acting on the plate is related directly to the liquid surface tension [6]. The additional pulling force acting on the plate after touching the liquid sample, F is

$$F = p\gamma_{lv} \cos\theta - V\rho g, \qquad (3.1)$$

where p is the perimeter of the plate, V is the volume of displaced liquid by the submerged plate, ρ is the density of the liquid, and g is gravitational acceleration. In general, if only the measurement of surface tension is desired, then the plate is roughened to produce a zero contact angle (complete wetting). The contact angle may be determined, provided the surface tension is known.

This approach has a relatively high degree of accuracy (results are typically given with an error of approximately 0.2%), but there can be a number of experimental complications: vapor adsorption at parts of the hang-down mechanism, solute adsorption/precipitation on the plate or cylinder, and the possibility of swelling and absorption of the liquid by the solid are typical examples. A major disadvantage of the Wilhelmy plate technique is the requirement of a relatively large amount of liquid. The use of a large reservoir can also make it difficult to maintain a high degree of purity, which is of critical importance to all surface tension measurements since the introduction of impurities (even minute quantities) can dramatically affect the interfacial properties [7]. Finally, this method relies on perfect wetting (a zero contact angle) with the solid measuring probe, a condition that cannot always be ensured.

The commonly used Du Nouy ring method [3,5] is similar, in principle, to the Wilhelmy plate except that a circular loop of wire is used in place of a platinum plate. The method does not offer any advantage and is perhaps more awkward because of the requirement that the ring be kept horizontal (to within 1°) and the susceptibility of the ring to damage.

3.1.2 DROP WEIGHT METHOD

More than a century ago, a method based on the drop weight was proposed [2–5]. In this method, the weight of a drop falling from a capillary is measured. The weight of the drop falling off the capillary correlates with the interfacial tension through the following equation [8–10]

$$\gamma = \frac{V\Delta\rho\, g}{2\pi r f},\tag{3.2}$$

where V is the drop volume, r is the radius of the capillary, and f is an empirical factor tabulated as a function of r / R_c (R_c is a characteristic dimension defined as $V^{1/3}$ [9–11]).

The measurement of the interfacial tension with the drop weight technique is very simple but sensitive to vibrations [5]. More precisely, the vibrations of the apparatus can cause premature separation of the drop from the end of the capillary before the drop reaches the critical size. Also, the surface tension measurements of multicomponent solutions, in which adsorption occurs, may not reflect equilibrium saturation of the solutes at the interface [5]. There is another concern about the results obtained from the drop weight method when it is used especially for surface tension measurements of solutions. The reason lies in the fact that the method depends on an empirical factor. While one might argue that such a factor, even if in error, should still register small changes in surface tension, it is conceivable that the concomitant change in viscosity with solute concentration might affect the empirical factor.

3.1.3 OSCILLATING JET METHOD

The oscillating jet method is a dynamic method for surface tension measurement. The oscillating jet is generated by forcing the liquid through an elliptical orifice that produces a jet with properties of standing waves [3,4,12]. In the absence of viscosity and compressibility, the surface tension is related to measurable physical properties including the wavelength of the oscillations, liquid density, mean radius of the orifice (i.e., average of the minimum and maximum radii), and flow rate [4]. In this method, the wavelengths are measured by passing parallel light waves perpendicular to the jet stream [4]. Bohr [13] applied the method to a real liquid by taking into account the influence of the velocity profile of the jet on surface tension.

The oscillating jet method has been used to study the surface tension of surfactant solutions [3]. As the surface tension of the liquid jet changes, because of surfactant diffusion in the air/liquid interface, each succeeding wave will have longer wavelength that corresponds to a lower surface tension value.

The main attraction of this method is its capability to record a very early surface age [4]; that is, the length of time from surface formation to some specified time (usually until a measurement is taken). This method also gives reasonably accurate values of surface tension even if the jet velocity profile is not included in the calculations. A major problem associated with this method is the cost of equipment, which can be prohibitive [4]. The nature of the parameters necessitates measurement with sophisticated equipment, especially because of the high degree of accuracy required in measuring wavelengths.

3.1.4 CAPILLARY WAVE METHOD

If a deep body of liquid is perturbed by a vibrator, the surface of the liquid will oscillate where the wavelength of the surface waves depends on the liquid surface tension and gravity [3]. Such waves are called capillary-gravitational waves. Kelvin formulated the theory of capillary-gravitational waves [14]. His theory leads to the following relationship

$$v^2 = \frac{g\lambda}{2\pi} + \frac{2\pi\gamma}{\rho\lambda}, \tag{3.3}$$

where v is the velocity of propagation, λ is the wavelength, g is acceleration due to gravity, γ is surface tension, and ρ is the liquid density. It is clear in Equation 3.3 that the propagation velocity is determined by gravity for long waves and by surface tension for short waves (i.e., capillary waves) [4].

Experimentally, the waves are measured as standing waves, and the situation might be thought to be a static one. However, individual elements of the liquid in the surface region undergo a roughly circular motion, and the surface is alternately expanded and compressed [3]. As a result, damping occurs even with a pure liquid, and much more so with solutions or film-covered surfaces for which transient surface expansions and compressions may be accompanied by considerable local surface tension changes and by material transport between surface layers [15].

The theory of the capillary waves (Equation 3.3) is more complicated for viscous liquids, especially for surfactant solutions with viscoelastic surface properties [4]. Similar to the oscillating jet method, the theory of oscillations at a flat interface is based on the analysis of the Navier-Stokes hydrodynamic equations and the boundary conditions at an interface (see details [15–18]). It should be mentioned that capillary waves are spontaneously present due to small temperature and hence density fluctuations [3]. These minute waves (about 5 Å amplitude and 0.1 mm wavelength) can be detected by laser light-scattering techniques [19–22]. Again, similar to the oscillating jet method, the major problem associated with the capillary method is the cost of equipment. In general, this method is more complex than other methods in both theoretical and experimental aspects. Another disadvantage of this method can be the large amount of liquid required for each experiment.

3.1.5 SPINNING DROP METHOD

In this technique, a drop of liquid (or a bubble) is suspended in a denser liquid, and both the drop and the surrounding liquid are contained in a horizontal tube spun about its longitudinal axis [4]. As a result of spinning, gravity has little effect on the shape of the drop. At low rotational velocities (ω), the drop (bubble) has an ellipsoidal shape, but when ω is sufficiently large, it becomes cylindrical. Under the latter condition, the interfacial tension is calculated from the following equation [23]:

$$\gamma = \frac{1}{4} r^3 \Delta \rho \omega^2, \tag{3.4}$$

where r is the radius of the cylindrical drop, and $\Delta \rho$ is the density difference between the drop and the surrounding liquid.

The spinning drop method can be used to determine ultralow interfacial tensions down to 10^{-6} mJ/m^2 [5]. Another advantage of the spinning drop method is its applicability to the determination of surface tension of highly viscous liquids when many traditional methods are unsuitable. For instance, this method is appropriate for polymer melts with a viscosity of 300–500 Pa·s [4]. In these experiments, a solid polymer is initially placed in the tube that is heated to the melting temperature of the polymer while spinning in an oven with a control window [4]. However, the experiments show a smooth drop profile forms only in the case when at least one of the phases is of a rather high viscosity. For instance, the experiments are unsuccessful for bubbles in easily mobile liquids like water.

3.1.6 DROP SHAPE TECHNIQUES

Drop shape methods have been developed to determine the liquid–vapor or liquid–liquid interfacial tensions and the contact angle from the shape of a sessile drop, pendant drop, or captive bubble. In essence, the shape of a drop is determined by a combination of surface tension and gravity effects. Surface tension tends to make a drop spherical whereas gravity tends to elongate a pendant drop or flatten a sessile drop. When gravity and surface tension forces are comparable, then, in principle, one can determine the surface tension from an analysis of the shape of the drop.

The advantages of drop shape methods are numerous. In comparison with a method such as the Wilhelmy plate technique, only small amounts of the liquid are required. Drop shape methods are easy to handle. They can be used in many difficult experimental conditions such as studies of temperature or pressure dependence of liquid–fluid interfacial tensions. Also, they do not depend on adjustable parameters to determine interfacial tensions and contact angles. Drop shape methods have been applied to materials ranging from organic liquids to molten metals and from pure solvents to concentrated solutions. Also, since the profile of the drop may be recorded by digital images, it is possible to study interfacial tensions in dynamic systems, where the properties are time dependant.

Mathematically, the balance between surface tension and external forces, such as gravity, is reflected in the so-called Laplace equation of capillarity. The Laplace equation is the mechanical equilibrium condition for two homogeneous fluids separated by an interface [24,25]. It relates the pressure difference across a curved interface to the surface tension and the curvature of the interface:

$$\gamma\left(\frac{1}{R_1} + \frac{1}{R_2}\right) = \Delta P, \tag{3.5}$$

where R_1 and R_2 are the two principal radii of curvature, and ΔP is the pressure difference across the interface. In the absence of any external forces other than gravity, ΔP may be expressed as a linear function of the elevation:

$$\Delta P = \Delta P_0 + (\Delta \rho) g z, \tag{3.6}$$

where ΔP_0 is the pressure difference at a reference plane, and z is the vertical height of the drop measured from the reference plane. Thus, for a given value of γ, the shape of a drop may be determined from known physical parameters, such as density and gravity, and known geometrical quantities, such as the radius of curvature at the apex. The inverse (i.e., determination of the interfacial tension, γ, from the shape) is also possible, in principle, although this is a much more difficult task.

Mathematically, the integration of the Laplace Equation 3.5 is straightforward only for cylindrical menisci; that is, menisci for which one of the principal curvatures, $1/R$, is zero. For a general irregular meniscus, mathematical analysis would be more difficult. For the special case of axisymmetric drops, numerical procedures have been devised. Fortunately, axial symmetry is not a very significant restriction for most pendant drop and sessile drop systems.

The earliest efforts in the analysis of axisymmetric drops were those of Bashforth and Adams [26]. They generated sessile drop profiles for different values of surface tension and radius of curvature at the apex of the drop. This was long before digital computers appeared and their work required tremendous labor. Hence, the task of determining interfacial tension and contact angle from the actual profile became a matter of interpolation from their tables, which contained the solutions of the differential equations describing the profile. Blaisdell [27] and Tawde and Parvatikar [28] extended Bashforth and Adams' tables, and Fordham [29] and Mills [30] generated equivalent tables for pendant drops (see Paddy [31]).

Hartland and Hartley [32] collected numerous solutions for determining the interfacial tensions of axisymmetric fluid-liquid interfaces of different shapes. A FORTRAN computer program was used to integrate the appropriate form of the Laplace equation and the results were presented in tabulated form. However, the major source of errors in their methods stems from data acquisition. The description of the whole surface of the drop is reduced to the measurement of a few preselected critical points, which are compatible with the tables used. These points are critical since they correspond to special characteristics, such as inflection points on the interface, and they must be determined with high accuracy. Also, for the determination of the contact angle, the point of contact with the solid surface where the three phases meet must be established. However, these measurements are not easily obtained. Furthermore, the use of these tables is limited to drops of a certain size and shape range.

Malcolm and Paynter [33] proposed another method for the determination of contact angle and surface tension from sessile drop systems. However, as with some of the previous approaches, the data points are specific geometric points on the drop interface and the method is also limited to sessile drops with contact angles greater than 90°.

Maze and Burnet [34,35] developed a more satisfactory scheme for the determination of interfacial tensions from the shape of sessile drops. They developed a numerical algorithm consisting of a nonlinear regression procedure in which a calculated drop shape is fitted to a number of arbitrarily selected and measured coordinate points on the drop profile by varying two parameters until a best fit is obtained. In other words, the measured drop shape (one-half of the meridian section) is described by a set of coordinate points and no particular significance is assigned to any of the points. In order to start the calculation, reasonable estimates of the drop shape and size are required; otherwise the calculated curve will not converge to the measured one. The initial estimates are obtained, indirectly, using values from the tables of Bashforth and Adams [26]. Despite the progress in strategy, there are several shortcomings in this algorithm. For example, the error function (i.e., the difference between the theoretical drop profile and the experimental drop profile) is computed by summing the squares of the horizontal distance between the theoretical drop profile and the experimental profile. This computation is not particularly suitable for sessile drops whose shapes are strongly influenced by gravity. For instance, large drops of low surface tensions tend to flatten near the apex. Therefore, any data point near the apex may cause a large error, even if it lies very close to the best-fitting curve, and lead to considerable bias of the solution. Also, the identification of the apex of the drop is of paramount importance since it acts as the origin of the calculated curves. Huh and Reed [36] developed a similar approach, but they used a poor approximation of the normal distance to define the objective function [37] and the apex point must still be predetermined by the user. Also, their method is applicable only to sessile drops with contact angles greater than 90°.

Anastasiadis et al. [38] proposed another technique that couples digital image processing with robust shape comparison routines [39]. In general, it is recognized that "outliers," which are erroneous data points of an experimental drop profile, have a large impact on the results of the least squares method. Instead of comparing individual points on the two curves (as it is compared in the least squares method), the robust shape comparison routines compare vectors or line segments on the experimental profile with the corresponding vectors on the theoretical profile. Anastasiadis et al. claimed that their technique requires considerably less computer power and that it is intrinsically resistant to the presence of outliers. The major problem in their technique is that it requires the specification of a reference point on the profile to which the positions of all other points on the profile are related. The reference point can be either the drop apex or the "center" of the drop, defined as the intersection of the vertical axis of symmetry and the horizontal maximum diameter. Thus, the accuracy of the results depends not only on the accuracy of the drop profile coordinates but also on the accuracy of the determination of the reference point [40,41].

Rotenberg et al. [42,43] developed a more powerful technique, Axisymmetric Drop Shape Analysis-Profile (ADSA-P), which predates the work of Anastasiadis et al. The ADSA-P technique fits the measured profile (i.e., experimental curve) to a Laplacian curve. To evaluate the discrepancy between the theoretical Laplacian curve and the actual profile, an objective function is defined as the sum of the squares of the normal distances between the measured points and the calculated curve. This function is minimized by a nonlinear regression procedure, yielding the interfacial tension and the contact angle in the case of a sessile drop. The location of the apex of the drop is assumed to be unknown and the coordinates of the origin are considered as independent variables of the objective function. Thus, the drop shape can be measured from any convenient frame and any measured point on the surface is equally important. A specific value is not required for the surface tension, the radius of curvature at the apex, or the coordinates of the origin. The program requires as input several coordinate points along the drop profile, the value of the density difference across the interface, the magnitude of the local gravitational constant, and the distance between the base of the drop and the horizontal coordinate axis. An initial guess of the location of the apex and the radius of curvature at the apex are not required. The solution of the ADSA-P program yields not only the interfacial tension and contact angle, but also the volume, surface area, radius of curvature, and contact radius of the drop. Essentially, ADSA-P employs a numerical procedure that unifies the method for both the sessile drop and the pendant drop. No table is required, nor is there any drop size restriction on the applicability of the method.

The simplicity and accuracy of ADSA-P were further improved as Cheng [40,44] implemented image processing techniques to detect the edge of the drop automatically. He incorporated an automated edge detection technique into the ADSA-P program that considerably improved the accuracy of the results and the efficiency of the ADSA-P technique developed by Rotenberg. Cheng et al. [40,44,45] also evaluated the performance of ADSA-P for both pendant and sessile configurations using synthetic drops. The points at five different locations on the profile were individually perturbed to test the influence of each location on the results. It was found that data points near the neck of a pendant drop or near the liquid-solid interface for a sessile drop have more impact on the results than points from other locations. The numerical scheme of ADSA-P developed by Rotenberg was found to give very accurate results except for very large and flat sessile drops, where the program failed. In addition, it was difficult to achieve perfect alignment of the camera with a plumb line; there are errors associated with the coordinates of the plumb lines defined manually on the screen of the computer using a mouse.

Finally, del Río [41,46] developed a new version of ADSA-P that overcomes the deficiencies of the numerical schemes of the original algorithm (i.e., Rotenberg algorithm) using more efficient and accurate numerical methods. The new algorithm uses the curvature at the apex (rather than the radius of curvature), permits an additional optimization parameter (i.e., the vertical misalignment of the camera), and gives improved initial estimates of the apex location and shape. The new version is written in the C language (rather than FORTRAN); it is also superior to the original program in terms of computation time and range of applicability [41]. Del Río also developed a new program called Axisymmetric Liquid-Fluid Interface (ALFI) that performs in an opposite manner to ADSA-P. It generates theoretical Laplacian curves by integrating the Laplace equation for known values of surface tension and curvature at the apex; that is, it essentially automates the procedure of Bashforth and Adams.

ADSA-P has been used to study a wide variety of systems, ranging from biological to industrial [47]. Recently, it has been found that despite the general success of ADSA-P, inconsistent results may be obtained for drops close to spherical shape. Preliminary experiments have shown that Rotenberg and del Río ADSA-P fail when dealing with near spherical drop shapes, although the latter has a somewhat larger range of applicability. For a large drop with a "well-deformed" shape (i.e., a drop with inflection points in the neck area), both ADSA-P algorithms perform accurately. For example, Figure 3.1a presents an image of a large pendant drop of water formed at the end of a Teflon tube with an outer diameter of 3 mm. The surface tension value obtained from both algorithms of ADSA-P is 72.25 ± 0.01 mJ/m^2 in good agreement with the literature value of the surface tension of water (72.28 mJ/m^2 at 24°C). The surface tension value is the mean of the surface tension values of twenty drops of the same size. The error limit is calculated using the standard deviation of the surface tension values of the twenty drops at the 95% confidence level. As the drop is made smaller and hence close to spherical in shape, the results of both ADSA-P algorithms deviate from the correct value. For example, for the small drop of water shown in Figure 3.1b, the surface tension values obtained from Rotenberg ADSA-P and del Río ADSA-P are 83.02 ± 0.27 mJ/m^2 and 79.32 ± 0.19 mJ/m^2, respectively. The surface tension values and the error limits were obtained in a similar fashion as explained above. Since water is a pure liquid, it is expected that its surface tension remains constant regardless of the size of the drop. Therefore, one can conclude that the surface tension value obtained from ADSA-P for small drops cannot be true. It is noted that the error limits (i.e., ±0.27 or ±0.19 mJ/m^2) do not include the true value for the surface tension of water. Thus, it is concluded that ADSA-P (or any drop shape technique) works accurately only for "well-deformed" drops. It is apparent that a criterion is needed to decide whether a given drop is well-deformed. Simply making drops very large is not always possible or desirable. For instance, when ADSA-P is used as a film balance, very interesting patterns are observed at high compression of a surface film (i.e., necessarily for small and hence relatively spherical drops; see Section 3.3.7.4). To evaluate the quality of the surface tension measurements, a

(a)   (b)

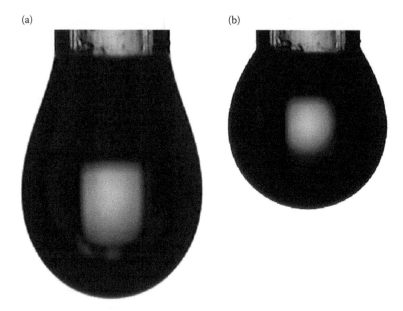

FIGURE 3.1   (a) An image of a large pendant drop of water formed at the end of a Teflon tube with an outer diameter of 3 mm. The drop is well-deformed with a maximum diameter of 4.7 mm. The surface tension value obtained from ADSA-P (i.e., the latest version) is 72.25 $\pm$ 0.01 (mJ/m^2), which agrees well with the literature value (i.e., 72.28 mJ/m^2 at 24°). (b) An image of a small drop of water formed at the end of a Teflon tube with an outer diameter of 3 mm. The drop is close to spherical shape with a maximum diameter of 4.5 mm. The calculated surface tension is 79.32 $\pm$ 0.19 (mJ/m^2). (Hoorfar, M. and Neumann, A. W., *Advances in Colloid and Interface Science,* 121, 25–49, 2006. With permission from Elsevier.)

quantitative criterion called "shape parameter" will be described in Section 3.3.6. In essence, the shape parameter determines the range of drop shapes in which ADSA-P succeeds or fails. Such a parameter was formulated using the fact that the curvature along the periphery of a spherical drop is constant whereas it changes markedly for a well-deformed drop.

For contact angle determinations, with most techniques it becomes increasingly difficult to make measurements for flat sessile drops with very low contact angles, say below 20° (Figures 3.2c and 3.2d). The accuracy of ADSA-P also decreases under these circumstances since it becomes more difficult to acquire accurate coordinate points along the edge of the drop profile. For these situations, it is more useful to view a drop from above and determine the contact angle from the contact diameter of the drop. Initially, Bikerman [48] proposed to calculate the contact angle from the contact diameter and volume of a sessile drop by neglecting the effects of gravity and assuming that the drops are sections of a sphere. Obviously, this simple approach is only applicable to small drops and/or to very large liquid surface tensions. A modified version of ADSA, called Axisymmetric Drop Shape Analysis-Contact Diameter (ADSA-CD), was developed by Rotenberg and later implemented by Skinner et al. [49]. This version does not ignore the effects of gravity. ADSA-CD requires the contact diameter, the volume and the liquid surface tension of the drop, the density

(a)

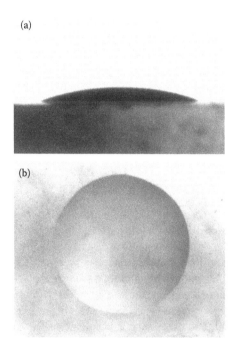

(b)

FIGURE 3.2 (a) side and (b) top view of a sessile drop with a small contact angle.

difference across the liquid–vapor interface, and the gravitational constant as inputs to calculate the contact angle by means of a numerical integration of the Laplace equation of capillarity (Equation 3.5).

It has been found that drop shape analysis utilizing a top view is quite useful for the somewhat irregular drops that often occur on rough and heterogeneous surfaces. In these cases, an average contact diameter leads to an average contact angle. The usefulness of ADSA-CD for averaging over irregularities in the three phase contact line proved to be such an asset that it became desirable to use it instead of ADSA-P for large contact angles as well. Unfortunately, for contact angles above 90°, the three phase line is not visible from above. For such cases, yet another version of ADSA has been developed by Moy et al. [50], called Axisymmetric Drop Shape Analysis-Maximum Diameter (ADSA-MD). The ADSA-MD technique is similar to ADSA-CD; however, it relies on the maximum equatorial diameter of a drop to calculate the contact angle. The ADSA-CD and ADSA-MD techniques have been unified into a single program called Axisymmetric Drop Shape Analysis-Diameter (ADSA-D). Chapter 6 provides a full description of ADSA-D.

This chapter provides an account of ADSA-P. First, the mathematical formulation of the Laplace equation of capillarity for axisymmetric fluid-liquid interfaces is introduced. Following this is a description of the numerical integration schemes, the formation of the objective function, the criterion for its minimization, and the ALFI program for the generation of Laplacian curves. The two ADSA-P algorithms (i.e., Rotenberg and del Río) are then compared for different theoretical drop shapes to evaluate the range of applicability of the ADSA-P algorithms. The experimental setup of ADSA-P for image acquisition and the automated image analysis process for detection of the experimental drop profile coordinates are illustrated. Also included are details of the shape parameter, a tool to control the quality of the surface tension measurements and to quantify the range of applicability of the ADSA-P algorithms. Finally, the applicability of ADSA-P is illustrated for the investigation of contact angles, pressure dependence of interfacial tensions, ultralow interfacial tensions, film balance measurements of insoluble monolayers, density measurement of polymer melts, and measurements of tissue surface tension.

3.2 LAPLACE EQUATION OF CAPILLARITY

As described above, the classical Laplace Equation 3.5 describes the mechanical equilibrium condition for two homogeneous fluids separated by an interface. Figure 3.3 illustrates a coordinate system for describing such a system. The two radii of curvature at any point of a curved surface can be obtained by erecting a normal to the surface at the point in question and then passing a plane through the surface that contains the normal. In general, the line of intersection between the plane and the surface is curved, thus generating the first radius of curvature. The second radius of curvature can be obtained by passing a second plane through the surface that is perpendicular to the first plane and also contains the normal. If the first plane is rotated through a full circle, the first radius of curvature will go through a minimum, and its value at this minimum is called the principal radius of curvature. The second principal radius of curvature is the corresponding radius in the second plane kept at right angles to the first. The pressure difference across the interface does not depend upon the manner in which R_1 and R_2 are chosen and it follows that the sum $(1 / R_1 + 1 / R_2)$ is independent of how the first plane is chosen. In the absence of external forces other than gravity, the pressure difference, ΔP, is a linear function of the elevation

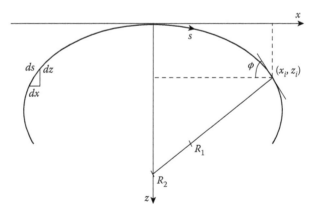

FIGURE 3.3 Definition of the coordinate system for two homogeneous fluids separated by an interface. At a point (x_i, z_i), the turning angle is ϕ. The arc length, s, is measured along the drop. R_1 and R_2 are the two principal radii of curvature; R_1 turns in the plane of the paper, and R_2 rotates in the plane perpendicular to the plane of the paper.

as described by Equation 3.6. Since the interface is assumed to be symmetric about the z-axis, the principal radius of curvature, R_1, is related to the arc length, s, and the angle of inclination of the interface to the horizontal, ϕ, by (see Figure 3.3)

$$\frac{1}{R_1} = \frac{d\phi}{ds}. \qquad (3.7)$$

The second radius of curvature is given by

$$\frac{1}{R_2} = \frac{\sin\phi}{x}. \qquad (3.8)$$

Due to the axial symmetry of the interface, the curvature at the apex is constant in all directions and the two principal radii of curvature are equal; that is,

$$\frac{1}{R_1} = \frac{1}{R_2} = \frac{1}{R_0} = b, \qquad (3.9)$$

where R_0 and b are the radius of curvature and the curvature at the origin, respectively. Then, from Equation 3.5, the pressure difference at the origin (i.e., at $s = 0$) can be expressed as

$$\Delta P_0 = 2b\gamma. \qquad (3.10)$$

Substituting Equations 3.7, 3.8, and 3.10 into Equation 3.5 and defining the capillary constant, c, yields

$$\frac{d\phi}{ds} = 2b + cz - \frac{\sin\phi}{x} \qquad (3.11)$$

$$c = \frac{(\Delta\rho)g}{\gamma}, \qquad (3.12)$$

where the capillary constant, c, has positive values for sessile drops and negative values for pendant drops.

Equation 3.11 together with the geometrical relations

$$\frac{dx}{ds} = \cos\phi \qquad (3.13)$$

$$\frac{dz}{ds} = \sin\phi \qquad (3.14)$$

forms a set of first-order differential equations for x, z, and ϕ as functions of the arc length, s, with the boundary conditions

$$x(0) = z(0) = \phi(0) = 0. \tag{3.15}$$

Also, at $s = 0$

$$\frac{d\phi}{ds} = b. \tag{3.16}$$

Therefore, the complete shape of the Laplacian axisymmetric fluid-liquid interface curve is obtained by simultaneous integration of the above set of equations (i.e., Equations 3.11, 3.13, and 3.14) for given values of b and c. However, there is no known general analytical solution for this system of equations, and a numerical integration scheme must be used to generate the Laplacian curves.

3.3 AXISYMMETRIC DROP SHAPE ANALYSIS: PROFILE (ADSA-P)

The flowchart presented in Figure 3.4 shows the general procedure of ADSA-P for the determination of the interfacial properties from the shapes of pendant or sessile drops. The drop profile coordinates (i.e., the experimental profile) are obtained from the image of the drop using an image analysis process. The experimental profile and physical properties are the input to numerical schemes that are used to fit a series of Laplacian curves with known surface tension values to the experimental profile. The best fit gives liquid-fluid interfacial tension, contact angle (in the case of sessile drops), drop volume, surface area, radius of curvature at the apex, and the radius of the contact circle (formed between the liquid and solid in sessile drop experiments).

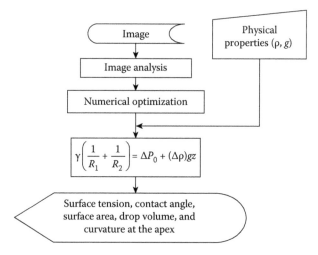

FIGURE 3.4 General procedure of Axisymmetric Drop Shape Analysis-Profile (ADSA-P). (Hoorfar, M. and Neumann, A. W., *Advances in Colloid and Interface Science*, 121, 25–49, 2006. With permission from Elsevier.)

3.3.1 Numerical Procedure

The numerical procedure of ADSA-P consists of three parts: (1) the integration of the Laplace equation for known values of b and c, (2) the formation of the objective function based on the error between the experimental profile and the theoretical curve obtained from the previous step, and (3) the minimization of the objective function (through an optimization procedure) to calculate the optimization parameters. One of the differences between the two ADSA-P algorithms (i.e., Rotenberg and del Río) is the number of optimization parameters utilized in the numerical procedure. The Rotenberg algorithm uses four optimization variables; that is, $1/b$, c/b^2, x_0, and z_0 (x_0 and z_0 are the coordinates of the apex). However, the del Río algorithm uses five optimization parameters; that is, b, c, x_0, z_0, and α (the angle of the vertical alignment). Although the misalignment of the camera is corrected through the image analysis process (see Section 3.3.5), in some experimental situations (e.g., image acquisition with very low magnification) it is difficult to achieve perfect vertical alignment. The del Río ADSA-P also resolved the convergence problem of the original algorithm for very flat sessile drops. As the sessile drop becomes very large and flat (i.e., sessile drops with ultralow surface tensions) the Rotenberg algorithm becomes unstable since the radius of curvature at the apex, R_0, becomes very large. The del Río algorithm overcomes this limitation by replacing the radius of curvature at the apex ($1/b$) with the curvature at the apex (b), which approaches zero for very flat sessile drops. Also, the del Río algorithm uses the capillary constant, $(\Delta \rho)g/\gamma$, instead of the Bond number, $(\Delta \rho)gR_0^2/\gamma$, which was used in the Rotenberg algorithm, to make all optimization parameters independent of each other. In addition to the optimization parameters, the two ADSA-P algorithms utilize different numerical methods for both the integration and the optimization procedures explained in the following sections.

3.3.1.1 Integration of the Laplace Equation

The integration of the Laplace equation is necessary not only to construct tables like those of Bashforth and Adams [26], and Hartland and Hartley [32], and to generate theoretical drop profiles or "synthetic" drops that can be used to evaluate the performance of ADSA-P [42,45], but it is also a very important part of the ADSA-P algorithm. There are several numerical methods available to integrate systems of differential equations for initial value problems. A review of some of the best numerical integration schemes is presented by Press and colleagues [51]. The ideal scheme would perform a given integration with a minimum required computational time and with the highest degree of numerical accuracy. The three most commonly used numerical methods for solving initial value problems for ordinary differential equations (ODEs) are: Runge–Kutta, Predictor-Corrector, and Richardson extrapolation. The method most widely used is the fourth-order Runge–Kutta procedure because of its accuracy. However, it is not usually the fastest. Predictor-Corrector methods, since they use past information, are somewhat more difficult to begin and they are useful only for smooth problems. Richardson extrapolation uses the powerful idea of extrapolating a computed result to the value that would have been obtained if the step size had been very much smaller than it actually is. In particular, extrapolation to zero step size is the desired goal. When combined with a particular way of taking individual steps (the modified midpoint method) and a particular kind of extrapolation (rational function extrapolation), Richardson extrapolation produces the Bulirsch-Stoer method. In recent years, Bulirsch-Stoer has become the method of choice in many applications. Each of the three types of methods can be organized to monitor internal consistency. This allows numerical errors, which are inevitably introduced into the solution, to be controlled by automatic adaptive changing of the fundamental step size. It is recommended that adaptive step size control always be implemented, but this can fail for rather complicated systems of ODEs.

After testing and comparing these methods, it was found that the Bulirsch-Stoer scheme executes much faster than Runge–Kutta (two to three times faster, depending on the system of equations), especially when the adaptive step size control is implemented [41]. The reason for such improved speed is that step sizes 16 to 32 times larger may be taken for the same degree of precision. The results obtained with the above integration schemes agree very well (to the numerical precision specified) with those published by Hartland and Hartley [32] and Couch [52].

The original version of ADSA-P (i.e., Rotenberg ADSA-P) employs an implicit second-order Euler method, whereas the latest version of ADSA-P (i.e., del Río ADSA-P) uses the Bulirsch-Stoer method, leaving Runge Kutta for those cases where the former fails. The Bulirsch-Stoer method is a more efficient numerical integrator than the Euler method as it includes higher-order error terms. Also, the use of adaptive step size in the del Río ADSA-P improves the convergence of the numerical integrator.

3.3.1.2 Error Estimation and Formation of the Objective Function

In essence, ADSA-P fits a Laplacian curve to the experimentally recorded profile. Thus, the first step in the analysis of a drop is the determination of the deviation of its profile from the shape dictated by the Laplace equation of capillarity, Equation 3.5. Experimental profile points U_i, $i = 1, 2,..., N$ that describe the meridian interface are compared with $u = u(s)$, a calculated Laplacian curve, by computing the normal distance, d_i, between U_i and u as illustrated in Figure 3.5. The U_i points are obtained from image analysis techniques (Section 3.3.5). The "error" for the ith point, e_i, can

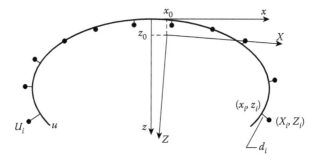

FIGURE 3.5 Comparison between experimental points and a Laplacian curve. (Hoorfar, M. and Neumann, A. W., *Advances in Colloid and Interface Science*, 121, 25–49, 2006. With permission from Elsevier.)

be computed as the square of the minimum distance, d_i, assuming that the coordinate systems for the experimental and the calculated profile coincide.

$$e_i = \frac{1}{2}d_i^2 = \frac{1}{2}\left[(x_i - X_i)^2 + (z_i - Z_i)^2\right],$$ (3.17)

where (X_i, Z_i) is an experimental point and (x_i, z_i) is the point on the Laplacian curve closest to it. However, the two coordinate systems do not coincide in general; thus, their offset and rotation angle must be considered. Then, Equation 3.17 can be written (dropping the subscript i) as

$$e = \frac{1}{2}(e_x^2 + e_z^2),$$ (3.18)

$$e_x = x - x_0 - X\cos\alpha + Z\sin\alpha,$$ (3.19)

$$e_z = z - z_0 - X\sin\alpha - Z\cos\alpha,$$ (3.20)

where (x_0, z_0) is the offset between the coordinate systems and α is the rotation angle.

However, to evaluate these equations for every experimental point, it is necessary to determine the closest point on the Laplacian curve. The distance between any experimental point and the computed Laplacian profile is a function of the arc length, s, and its minimum or normal distance corresponds to that for which

$$\frac{de}{ds} = f(s) = 0,$$ (3.21)

where derivatives of Equations 3.18 through 3.20 are taken for constant values of the parameters b, c, x_0, z_0, and α. Equation 3.21 must be solved numerically for s. The most efficient method to do so is the iterative Newton-Raphson method:

$$s^{i+1} = s^i - \frac{f(s^i)}{f'(s^i)},$$ (3.22)

where the second derivatives of Equations 3.18 through 3.20 with respect to s are required to calculate $f'(s^i)$. Newton–Raphson must be initialized with a first guess for s to ensure convergence to a minimum. Thus, a value of s—which is in the proximity of the experimental profile point U_i—is provided. The Laplace equation is then integrated to that estimated value of s, the derivatives of Equations 3.18 through 3.20 are evaluated, and a new value of s is obtained using Equation 3.22. Iteration is continued until convergence is achieved, with the final outcome being the identification of all (x_i, z_i) points.

To measure the agreement between the measured or experimental profile and an assumed or calculated Laplacian curve, a merit or objective function, E, is defined as the sum of the weighted individual errors

$$E = \sum_{i=1}^{N} w_i e_i, \tag{3.23}$$

where e_i can be evaluated from Equation 3.18 and w_i is a weighting factor to account for the influence of the location of the ith point on the fitted curve. Until more information is available on the appropriate form of the weight values, the w_i can be set equal to 1.0.

The value of the objective function, E, is a function of a set of parameters, $\mathbf{a}$, with elements a_k, $k = 1, ..., M$. The goal of this analysis is to calculate the values of a_k that minimize E; that is, to find the parameter set $\mathbf{a}$ that gives the best fit between the measured points and a Laplacian curve. The objective function E will assume a single absolute minimum value at one point in the M-dimensional space of E. In the latest version of ADSA-P, $\mathbf{a}$ is the vector of five parameters ($M = 5$) or any subset of it

$$\mathbf{a} = \begin{bmatrix} b & c & x_0 & z_0 & \alpha \end{bmatrix}^t, \tag{3.24}$$

where t signifies that Equation 3.24 is a transpose matrix.

As seen in Equations 3.18 through 3.20, e_i is a function of the parameter set $\mathbf{a}$ as well as of the position of the point (x_i, z_i) determined by its arc length, s_i. Moreover, s_i itself clearly depends upon the values of the parameters in $\mathbf{a}$. Thus, the objective function, E, can be expressed as

$$E(\mathbf{a}) = \sum_{i=1}^{N} w_i e_i \left[s_i(\mathbf{a}), \mathbf{a} \right]. \tag{3.25}$$

The necessary conditions for an extremum in the value of E are

$$\frac{\partial E}{\partial a_k} = \sum_{i=1}^{N} \frac{\partial e_i}{\partial a_k} = 0, \qquad k = 1, \cdots, M. \tag{3.26}$$

3.3.1.3 Optimization Procedure: Newton's Method

The extremum conditions of E, Equations 3.26, form a set of nonlinear algebraic equations in the variables a_k, $k = 1, ..., M$. In order to solve for these variables, an iterative solution is required. There exist several methods to solve these systems of equations [53,54]. In fact, there are no general or perfect methods for solving systems of nonlinear equations. Every method has advantages and disadvantages, and the choice of a particular method depends on the characteristics of the problem. The best known and most powerful one is Newton's method (also known as the Newton-Raphson method) for several variables. The iterative procedure of Newton's method can be expressed as

$$\mathbf{a}^{i+1} = \mathbf{a}^i - \Delta\mathbf{a}^i, \tag{3.27}$$

where $\mathbf{a}^i$ is the vector of unknown variables at the ith iteration step, and $\Delta\mathbf{a}^i$ is a correction vector resulting from the solution of the associated linear system

$$H(\mathbf{a}^i)\Delta\mathbf{a}^i = E(\mathbf{a}^i), \tag{3.28}$$

where $H(\mathbf{a}^i)$ is a Hessian matrix and the components of the vector $E(\mathbf{a}^i)$ are the first partial derivative terms obtained from Equations 3.26 and evaluated at the ith step. One important advantage of this algorithm is that the value of the objective function and its first and second partial derivatives are all evaluated with the same degree of accuracy since they can be evaluated analytically in terms of ordinary first-order differential expressions that can then be integrated numerically.

The Newton method is very easy to implement and its asymptotic convergence rate is quadratic. It has the disadvantage that partial derivatives are required, which are not easily evaluated in many cases, and the method is unpredictable if only a poor initial approximation to the solution is available. Several methods attempting to overcome these limitations have evolved from Newton's method. Amongst them, the method of incremental loading, steepest descent, and the Levenberg-Marquardt method are well known, together with several Newton-like methods. For instance, Rotenberg ADSA-P uses incremental loading to alleviate the local minima problem; however, this method reduces the convergence rate of the original Newton-Raphson method. The latest version of ADSA-P, on the other hand, uses the original Newton-Raphson method, but the user can adopt the Levenberg-Marquardt method when good initial values are unavailable. The Levenberg-Marquardt method needs only the first derivative and takes advantage of numerical properties particular to least square problems, which are known to be globally convergent, although at a much slower rate.

3.3.2 GENERATION OF LAPLACIAN CURVES USING ALFI

Prior to the use of ADSA-P in actual experiments, it is desirable if not necessary to evaluate its accuracy and applicability with synthetic drops obtained by a numerical integration of the Laplace Equation 3.5. Experimental drops suffer from errors associated with the identification of the true edge of the drop, and thus the coordinates of the profile. On the other hand, simulated drops eliminate these errors and, in fact, allow the effects of various experimental errors to be estimated by perturbation of drop profile points.

Del Río has developed a program called ALFI that not only generates Laplacian curves for given values of c, that is, $(\Delta\rho)g/\gamma$, and b (i.e., curvature at the apex), but also facilitates the simulation of experimental errors in the drop profile. Some of the features of the ALFI program are as follows:

- The coordinates of the origin of the drop, x_0 and z_0, are specified by the user. In the measurement of the interfacial properties using ADSA-P, the coordinates of the origin are obtained through the optimization process as they cannot be readily obtained from the experimental profile.
- The profile can be rotated through a specified angle, α, allowing for tests of the effect of the vertical misalignment of the camera.
- The drops can be scaled in both coordinates by the values X_s and Z_s to permit studies on the effect of errors of scaling and/or aspect ratio.
- Every point on the profile can be randomly perturbed in the normal direction with a specified maximum perturbation, δ_{max}.

To perturb the points of the Laplacian curve by a distance δ, and to translate and rotate the coordinate system, the following relations are used (as illustrated in Figure 3.6):

$$X = x + \delta \sin \phi, \tag{3.29}$$

$$Z = z - \delta \cos \phi, \tag{3.30}$$

$$X = \frac{1}{X_s}\left[(x - x_0)\cos\alpha + (z - z_0)\sin\alpha\right], \tag{3.31}$$

$$Z = \frac{1}{Z_s}\left[(z - z_0)\cos\alpha - (x - x_0)\sin\alpha\right], \tag{3.32}$$

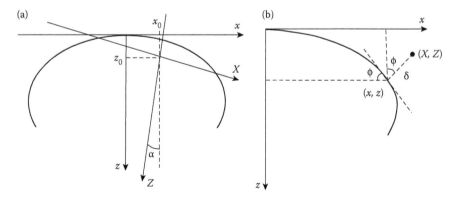

FIGURE 3.6   (a) Translation and rotation of the coordinate system; (b) normal perturbation of the points of a Laplacian curve by a distance δ.

where (x, z) are the coordinates of the original Laplacian point, (X, Z) are the coordinates of the new point, α is the angle of rotation of the coordinate system, and δ is the normal perturbation distance computed as a random number in the range $[-\delta_{max}, +\delta_{max}]$.

The ALFI program is written in C language using the Bulirsch-Stoer integration scheme and it performs much faster than the FORTRAN program written by Hartland and Hartley [32] using Runge-Kutta. The ALFI program can be run interactively or in batch mode using input files. Another useful feature is that comments are allowed in the input file, making it much easier to modify the file from one run to another. The output of the program can be directed to a file or to standard output, usually the console.

The program generates a number of points according to the step size, Δs, provided by the user without affecting the precision of the results. A precision of 10^{-10} is always used. The program integrates the Laplace equation until a specified value of s is reached or until the angle ϕ is greater than 180° for sessile drops or negative for pendant drops. The ALFI program can generate the whole drop profile or only half of it at the request of the user. Although the algorithm of ALFI is relatively simple, it has been found to be useful for general applications in the study of capillary phenomena.

3.3.3 COMPARISON OF TWO ADSA-P ALGORITHMS

In essence, the del Río algorithm was developed to overcome certain deficiencies of the original numerical method especially for flat sessile drop shapes, occurring for ultralow surface tensions [46]. As an illustration, Figure 3.7 shows an extremely flat synthetic drop generated using ALFI for a known surface tension value. The coordinates of this drop profile are given to the two ADSA-P algorithms and it is observed that Rotenberg ADSA-P fails to converge to the correct surface tension value, whereas del Río ADSA-P determines the correct surface tension value of 0.001 mJ/m² to the fourth decimal place.

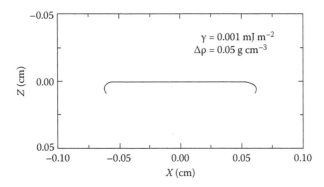

FIGURE 3.7 Analysis of an extremely flat synthetic drop ($R_0 = 1000$ cm, volume $= 0.1$ μl, equatorial diameter 1.2 mm) of known surface tension, $\gamma = 0.001$ mJ/m². The del Río algorithm gives the correct γ value, but the Rotenberg algorithm fails to converge to the correct surface tension value.

Despite the general success of the latest version of ADSA-P, inconsistent results can be obtained for nearly spherical drop shapes. In many studies, it is not possible to avoid drop shapes approaching spherical shape. For instance, in the study of film stability [55–61], interesting patterns are frequently observed at high compression; (i.e., necessarily for small and hence relatively spherical drops; see Section 3.3.7.4). Therefore, it is necessary to evaluate the performance of the ADSA-P algorithms for nearly spherical drop shapes. The strategy for such an evaluation is to use theoretical (i.e., ideal) drop coordinate points as input into ADSA-P instead of experimental points. Any deviation from the known surface tension, which is used as input to obtain the theoretical profile, is then obviously due to limitations of ADSA-P. Figure 3.8a shows a well-deformed theoretical drop shape where $b = 5.11$ (cm^{-1}) and $c = 13.62$ (cm^{-2}), which corresponds to a surface tension value of 72 mJ/m^2 provided that $\Delta \rho = 1$ (g/cm^3) and $g = 980.43$ (cm/s^2). For all intents and purposes, this profile is perfect and comparable to a hypothetical experimental profile obtained with eight significant figures. Surface tension values were obtained using the two ADSA-P algorithms for different cutoff levels. Experimentally, the cutoff level is selected as near to the holder as possible so that the largest portion of the experimental profile is given to ADSA-P. In this study, however, the cutoff level was lowered step by step toward the apex to scrutinize the performance of the two algorithms when only a portion of the drop profile is given to ADSA-P. At the "starting cutoff level" shown in Figure 3.8a, the two ADSA-P algorithms give the correct surface tension value to at least six significant places (i.e., well beyond any possible experimental accuracy). As the cutoff level is placed lower and lower, it becomes apparent that the del Río algorithm is more stable than the Rotenberg algorithm. Quite generally, the Rotenberg algorithm fails (i.e., $|\gamma_{ADSA} - \gamma_{true}| > 0.1$ mJ/m^2) in certain situations where the del Río algorithm continues to work properly. Finally, del Río ADSA-P also fails, although not as easily, when the drop shape near the apex becomes indistinguishable from spherical. Basically, at this portion of the drop profile near the apex, the Laplacian curves with different surface tension values are not distinguishable from each other. In this condition, any numerical scheme can become unstable and fail to converge to correct values. The above study was repeated for a more spherical theoretical drop where $b = 7.46$ (cm^{-1}) and $c = 13.62$ (cm^{-2}) (see Figure 3.8b). The results show that the del Río algorithm still performs better than the Rotenberg algorithm at the lower cutoff levels, but it requires a significantly larger portion of the profile near the apex (i.e., a much higher cutoff level) for a near spherical drop shape compared to a well-deformed drop shape (see Figure 3.8a and b).

To study the performance of the two algorithms closer to real experimental situations, the ideal drop profiles shown in Figure 3.8a and b were perturbed randomly by the equivalent of one pixel. Figure 3.9a and b show the perturbed well-deformed and close to spherical theoretical drop shapes, respectively. It is apparent that the two algorithms fail sooner for the perturbed profiles than for the ideal profiles.

The results of the above investigation show that the del Río algorithm also fails, although not as easily, as the given drop profile becomes indistinguishable from spherical. To illustrate further the source of the above limitation, the optimization scheme of the del Río algorithms was scrutinized using two different drop sizes of water; that is, a large (well-deformed) drop and a small (close to spherical in shape)

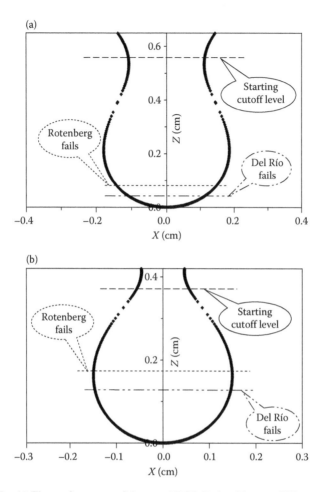

FIGURE 3.8   (a) The performance of the two ADSA-P algorithms at different cutoff levels for a well-deformed synthetic drop profile generated with eight significant figures for each coordinate point. The profile was generated for the values of $b = 5.11$ (cm^{-1}) and $c = 13.62$ (cm^{-2}). (b) The performance of the two ADSA-P algorithms at different cutoff levels for a synthetic drop profile that is close to spherical in shape. The profile was generated for the values of $b = 7.46$ (cm^{-1}) and $c = 13.62$ (cm^{-2}). (Hoorfar, M. and Neumann, A. W., *Advances in Colloid and Interface Science,* 121, 25–49, 2006. With permission from Elsevier.)

drop. The value of the objective function, E, that is, the sum of the normal distances between the points of the experimental profile and those of the theoretical curve, was recorded during the optimization process, which minimizes the objective function to find the best fitted Laplacian curve. First, the effect of the cutoff level (as shown in Figures 3.8a and 3.9a) on the performance of the optimization scheme is illustrated for the well-deformed drop of water at 24°C. Figure 3.10a shows the variations of E for different Laplacian curves with different values of γ as a function of b. Figure 3.10b shows the results of the same analysis at a lower cutoff level. The vertical axis E is presented in a logarithmic scale. It is clear that the global minimum

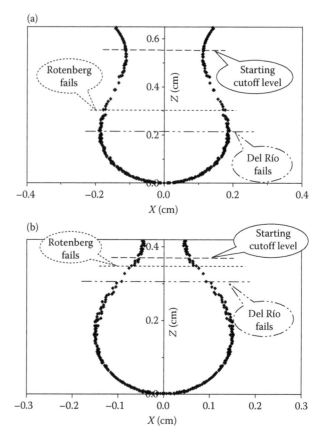

FIGURE 3.9   (a) The performance of the two ADSA-P algorithms at different cutoff levels for a well-deformed theoretical drop profile perturbed randomly by the equivalent of one pixel. (b) The performance of the two ADSA-P algorithms at different cutoff levels for a perturbed theoretical drop profile that is close to spherical in shape. (Hoorfar, M. and Neumann, A. W., *Advances in Colloid and Interface Science,* 121, 25–49, 2006. With permission from Elsevier.)

of the objective function is distinct when the cutoff level is near the capillary tube (see Figure 3.10a). This minimum value corresponds to a Laplacian curve generated for $b = 5.11$ (cm^{-1}) and $\gamma = 72.25$ mJ/m^2; that is, a value that is virtually identical with the known surface tension value of water (i.e., 72.28 mJ/m^2 at 24°C). Thus, ADSA-P performs accurately for a well-deformed drop that has inflection points (in the neck area). On the other hand, the objective function does not have a distinct minimum value when the cutoff level is placed lower and only the portion of the drop profile around the apex is given to ADSA-P (see Figure 3.10b). Specifically, the optimiza-tion process may choose any Laplacian curve with surface tension values within the range of 63.21 mJ/m^2 to 79.32 mJ/m^2. For the above cutoff level, the surface tension value obtained from ADSA-P is 63.21 mJ/m^2, which is obviously erroneous. This limitation is due to the fact that the Laplacian curves for different values of surface tension are not distinguishable near the apex. A similar effect has been observed for drops close to spherical in shape. For instance, Figure 3.11 shows the variations of

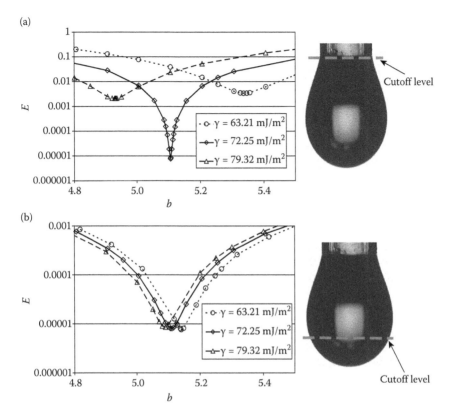

FIGURE 3.10   (a) The objective function for different Laplacian curves (with different values of b and γ) fitted to the experimental profile of a well-deformed drop of water; the cutoff level is near the capillary tube. (b) The objective function for different Laplacian curves fitted to the experimental profile of a well-deformed drop of water as the cutoff level is placed lower. (Hoorfar, M. and Neumann, A. W., *Advances in Colloid and Interface Science,* 121, 25–49, 2006. With permission from Elsevier.)

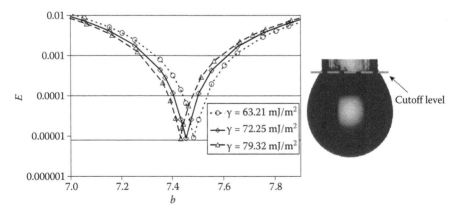

FIGURE 3.11   The objective function for different Laplacian curves fitted to the experimental profile of a small drop of water, which is close to spherical.

E as the optimization process minimizes the objective function for a small drop of water (i.e., near spherical shape). It is noted that there is no distinct minimum value for the objective function, which means that a wide range of theoretical Laplacian curves can satisfy the optimization process. Thus, the numerical scheme may easily become unstable and yield erroneous results. The surface tension value obtained from ADSA-P for this drop of water is 79.32 mJ/m^2, which differs significantly from the known surface tension value of water.

Further improvement of the numerical schemes of ADSA-P may have an incremental impact, but a big breakthrough is not expected. The del Río algorithm is mature, although limited by numerical truncation and accumulation of round-off errors that are the ultimate limitations of all numerical schemes. Theoretically, each drop shape corresponds to a certain surface tension value. However, for nearly spherical drop shapes, significantly different surface tension values correspond to only slightly different drop shapes. Thus, the numerical solver can easily become unstable and converge to a wrong value. In this situation, the output of the numerical scheme changes dramatically due to small variations of the input. This difficulty cannot be overcome within the framework of the numerical schemes. Thus, to increase the range of applicability of ADSA-P, it becomes necessary to improve the accuracy of the experimental profile coordinates by (1) improving the quality of the image and (2) using advanced image analysis techniques as explained in the following sections.

3.3.4   ADSA-P Setup

The image of a pendant or sessile drop is obtained using the experimental setup of ADSA-P shown in Figure 3.12. (See also Chapter 6 for the general experimental setup of ADSA-P.) A spot light source is used to illuminate the drop from behind. A

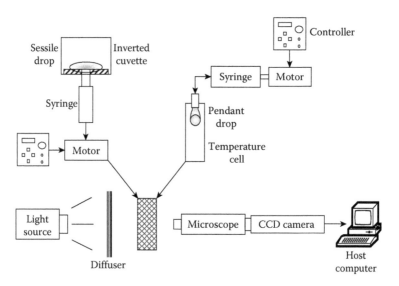

FIGURE 3.12   Schematic diagram of the experimental setup of ADSA-P for analysis of sessile and pendant drops.

heavily frosted diffuser is used in front of the light source to provide a uniformly lit background and to minimize heat transfer to the drop during image acquisition. In pendant drop experiments, the drop is formed inside a sealed quartz cuvette (Hellma Ltd) to minimize evaporation and to isolate the drop from vibration due to air currents. A Teflon stopper is used to seal the quartz glass cuvette that is located in a temperature/pressure cell. Usually, the pendant drop is formed at the end of either a Teflon capillary or a stainless steel holder (Section 3.3.6.2). The capillary (or holder) is connected to a 2.5 ml syringe (Gastight®, Hamilton Co.) by a Teflon tube with an outer diameter of 1.6 mm. The plunger of the syringe is connected to a stepper motor (Model 18705, Oriel Corp., USA), which is used to change the volume of the drop and facilitate the determination of surface tension in dynamic experiments. For a range in drop volume of 75 μl, the travel distance of the motor is 1500 steps. The available speed range of the motor is 0–500 steps per second.

In sessile drop experiments, the drop is formed on a solid surface (or on top of a stainless steel pedestal used in constrained sessile drop configurations explained in Section 3.3.6.5). The size of the drop is then increased by injecting the liquid from below the surface using the motorized syringe [62]. Such a mechanism leads to an increase in the drop volume and in the three-phase contact radius. The sessile drop arrangement is covered by an inverted cuvette to reduce contamination and evaporation. The entire experimental setup of ADSA-P is mounted on a vibration-free table (Model 78443-20, Technical Manufacturing Corp.).

A microscope and a CCD monochrome camera are used to obtain a magnified image of the drop. The CCD camera provides an analog video signal of the drop that is digitized using a frame grabber installed in a host computer. The digitized image consists of a fixed number of pixels that determines the resolution of the image. Each pixel specifies the intensity of light or gray level (in the black-and-white case) in a minute fractional area of the image. The gray level is registered using an 8-bit number, so it is defined in the 0–255 interval where 0 and 255 correspond to black and white, respectively. Thus, a digitized image is mathematically represented by an array of real numbers from 0 to 255. Once a digital image of the drop is generated and stored in memory, the image analysis and numerical schemes of the software detect the experimental profile and calculate the interfacial properties of the drop, respectively.

Figure 3.13 presents a typical output of ADSA-P obtained in a pendant drop experiment. The results were obtained using the del Río ADSA-P. This figure shows the output response to changes of volume and surface area of a pendant drop of cyclohexane. Cyclohexane was chosen because it is a cycloalkane and hence can be readily purified. The surrounding of the drop was maintained at 20°C. To obtain a wide range of drop sizes, the volume of the drop was changed continuously using a stepper motor. First, a large drop with a volume of 26 μl was formed at the end of a Teflon tube with outer and inner diameters of 3 mm and 2 mm, respectively. The volume of the drop was then decreased (at a rate of 3.2 μl/sec) until the drop was relatively small and close to spherical in shape. At this position the drop volume was 10 μl. The volume was then increased back to the initial size at the same rate. This procedure was repeated for 10 cycles. Since cyclohexane is a pure liquid (with purity greater than 99.9%), its surface tension is expected to remain constant regardless of

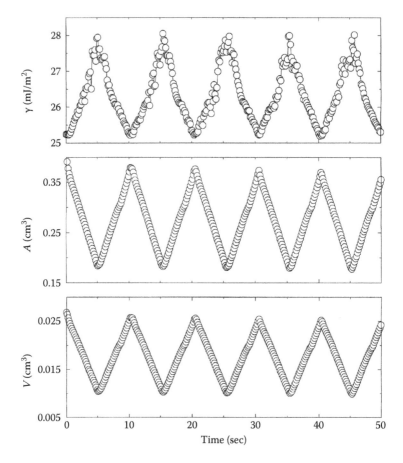

FIGURE 3.13   Surface tension responses to change of surface area of a pendant drop of cyclohexane in a dynamic cycling experiment (at 20°C).

the size of the drop. However, the results as determined by ADSA-P show that the surface tension value changes (in this case increases) as the surface area (or the volume) of the drop decreases. Clearly, such findings must be erroneous. One cycle of the above experiment is magnified and shown in Figure 3.14. For a large drop with a well-deformed shape (i.e., a drop with inflection points in the neck area), ADSA-P calculates the correct surface tension value; that is, the well-known value of the surface tension of cyclohexane (25.24 mJ/m^2 at 20°C) [63]. It is also noted that for a certain range of drop sizes, surface tension values are fairly constant within a range of ± 0.1 mJ/m^2. On the other hand, as the drop volume decreases, the drop shape becomes closer to spherical and the results start deviating from the correct value. This error is due to (1) the limitation of the numerical schemes (explained in Section 3.3.3) and (2) noise in the experimental profile. As discussed above, for nearly spherical drop shapes, significantly different surface tension values correspond to only slightly different drop shapes. Thus, even small errors in the experimental profile detected from the image of the drop can slightly shift the selected Laplacian curve,

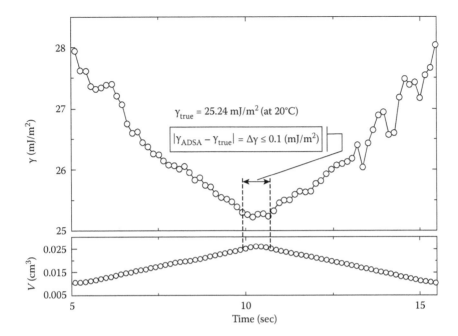

FIGURE 3.14 One cycle of a dynamic cycling experiment conducted for a pendant drop of cyclohexane (at 20°C).

which can cause the surface tension value to deviate dramatically from the true value. In essence, the accuracy of the experimental profile coordinates depends on the quality of the image and the performance of the image analysis technique; the latter is explained in Section 3.3.5.

The factors contributing to the quality of the image are focusing, the range of light wavelengths, quality of the lens, and camera resolution. Obviously, focusing has a major effect on the quality of the image. Cheng [40] has studied this effect and developed an automated focusing procedure that alleviates the problems associated with the manual focusing carried out by inspection. In this method, first, the image of the drop is acquired at a preliminary focus position obtained manually. Then, several images from different focus positions in the vicinity (i.e., some further and some closer) of the preliminary focus position are acquired. The Sobel gradient method [44,64–72] is applied to these images. The focus position of the image with the highest Sobel gradient at the edge is selected for focusing, since the highest gradient corresponds to the sharpest optical edge.

In the following sections, the effects of light source, microscope lens, and CCD camera on the quality of the image are described.

3.3.4.1 Light Source

White light sources are usually used to illuminate the pendant as well as the sessile drops. Composed of a wide range of wavelengths, white light undergoes chromatic aberration that can cause blurring of the image of the drop at the edge [73–78]. These

TABLE 3.1

Surface Tension Values of Large Drops (v = 26 μl) and Small Drops (v = 10 μl) of Cyclohexane Obtained from the Cycling Experiment Using Different Filters

White Light With	Wavelength Range (nm)	γ (mJ/m²) for Large Drop	γ (mJ/m²) for Small Drop
No filter	390–720	25.23 ± 0.02	27.99 ± 0.19
Light red[a]	600–700	25.23 ± 0.02	27.50 ± 0.17
Velvet green[a]	450–550	25.23 ± 0.02	27.29 ± 0.12
Mikkle blue[a]	400–500	25.24 ± 0.02	27.06 ± 0.08

Note: The literature value of the surface tension of cyclohexane is 25.24 mJ/m² at 20°C.

[a] These are commercial names for the filters purchased from LEE Filters Company (Toronto, Ontario).

chromatic effects can be reduced using light with a narrow range of wavelengths. Therefore, a band-pass filter (i.e., a filter that transmits wavelengths between the two cutoff wavelengths of the filter) can be used to pass only a narrow band of the visible wavelengths. The choice of an optical filter involves a tradeoff between the intensity and the bandwidth of light. In other words, the wider the bandwidth, the higher the intensity of light. An appropriate filter is expected to reduce the effect of chromatic aberration while maintaining sufficient intensity for the illumination.

The experiment with cyclohexane explained above was repeated using different optical filters with different ranges of wavelengths. The results of these experiments are summarized in Table 3.1. The results for the largest drops are given in the third column and those for the smallest drops in the fourth column. The values given are averages of the surface tension values in 10 cycles. The error limits were obtained at the 95% confidence level. It is apparent that for sufficiently large and hence well-deformed drops, ADSA-P produces correct surface tension values for all lighting conditions. On the other hand, for small drops, different lighting conditions yield different results that differ from the literature value.

The results also show that the use of any optical filter (red, green, or blue) reduces the discrepancy between the surface tension values obtained for large and small drops. This is presumably due to the fact that filters reduce chromatic effects and hence improve the quality of the image.

Finally, the results in the last row of Table 3.1 indicate that the use of the Mikkle blue filter (with a wavelength range of 400–500 nm) reduces the discrepancy between the surface tension values obtained for large and small drops more than the other filters examined here. Thus, blue filters are most effective, as expected from optics.

3.3.4.2 Microscope Lens

Geometrical distortion and spherical aberration are the most significant problems of a lens in an optical system regardless of the type of illumination used [75]. Spherical

TABLE 3.2

Surface Tension Values of Large Drops (v = 26 μl) and Small Drops (v = 10 μl) of Cyclohexane Obtained from the Cycling Experiment Using Two Different Choices of Aperture

Aperture	γ (mJ/m²) for Large Drop	γ (mJ/m²) for Small Drop
Partially open	25.31 ± 0.15	28.26 ± 0.26
Fully open	25.23 ± 0.02	27.99 ± 0.19

Note: The literature value of the surface tension of cyclohexane is 25.24 mJ/m² at 20°C.

aberration is caused by the failure of a lens to bring parallel rays of light into a single focus. Typically, the center and edges of a spherical lens have different focal points so that the images of objects, as seen through the whole surface of the lens, may be blurry.

A simple way to alleviate the effect of the spherical aberration is to step down the aperture and use only the center of the lens. However, this will increase the depth of field, which is not desired in this application because the image processing of ADSA-P requires a distinct image of the meridian plane. From this point of view, it is preferable to open the aperture as wide as possible, which will increase the spherical aberration. In general, a compromise between these conflicting requirements may have to be found. In the setup shown in Figure 3.12, an apochromatic lens (i.e., a series of lenses arranged in a row to reduce spherical aberration) is used, which minimizes the spherical aberration and hence allows the aperture to be fully open. Table 3.2 summarizes the surface tension values of cyclohexane measured with two choices of aperture (i.e., fully open and partially open) in the cycling experiment described before. In these experiments, a white light source was used. The results show that the discrepancy between the surface tension values obtained for large and small drops is smaller when the aperture is fully open.

In order to avoid the effect of geometrical distortion, a software module was incorporated into the original version of ADSA-P to eliminate the effect of the optical distortion of the lens using a calibration grid pattern engraved on an optical glass slide (see Section 3.3.5) [44].

3.3.4.3 Camera

Image resolution is expected to affect the results. Table 3.3 summarizes the surface tension values of cyclohexane measured in a cycling experiment with both high (1280 × 960) and low (640 × 480) image resolutions. For sufficiently large and hence well-deformed drops, ADSA-P produces correct surface tension values regardless of the image resolution and lighting condition. For small drops, on the other hand, the use of the higher image resolution (i.e., 1280 × 960 pixels) improves the results since the deviation between the surface tension values of large and small drops is reduced compared to the low image resolution. It should be noted that the use of a blue filter is a more effective way of improving the outcome than an increased image resolution.

TABLE 3.3

Surface Tension Values of Large Drops (v = 26 μl) and Small Drops (v = 10 μl) of Cyclohexane Obtained in a Cycling Experiment Using Two Different Image Resolutions and Two Lighting Conditions

Image Resolution (Pixel × Pixel)	Light	With Subpixel Resolution Algorithm	
		γ (mJ/m²) for Large Drop	γ (mJ/m²) for Small Drop
Low (640 × 480)	White	25.23 ± 0.02	27.99 ± 0.19
	Blue	25.24 ± 0.02	27.06 ± 0.08
High (1280 × 960)	White	25.23 ± 0.02	27. 81 ± 0.15
	Blue	25.24 ± 0.02	26.98 ± 0.07

Note: The literature value of the surface tension of cyclohexane is 25.24 mJ/m² at 20°C.

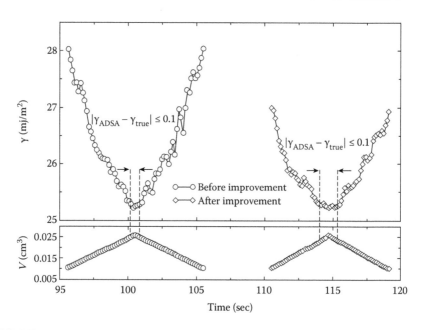

FIGURE 3.15 Experimental results before and after the improvements of the hardware.

The range of possible improvements by hardware modification is illustrated in Figure 3.15.

3.3.5 ADSA-P Image Processing

Once an image of a drop is obtained and stored in the computer memory, the ADSA-P program automatically finds the drop profile coordinates. Besides the quality of the image, the performance of the image analysis process has a significant effect on the accuracy of the experimental profile. Cheng [40,44] developed the original image analysis process of ADSA-P, which consists of four steps: (1) edge detection, (2) sub-pixel resolution, (3) correction of optical distortion, and (4) correction for the misalignment of the camera. First, the edge is found by applying the Sobel edge operator with a 3×3 convolution mask (see Chapter 4 for details) [40,64,65]. Basically, the image is divided into small areas that are 3×3 pixels in size. Then, a best least-squares plane is fitted through nine gray-level pixels of the 3×3 array, and the slope of this plane in the x and y directions is calculated. From these two directional gradients, the overall gradient of this plane is calculated. This procedure is then repeated for the whole digitized image such that each pixel point is the central point of the 3×3 array. In this fashion, a gray-level gradient for each pixel point of the whole image is determined, leading to a gradient image. The drop profile is established by searching for the pixels with maximum gradient along the drop profile. In other words, the drop profile is approximated by the pixel of the steepest gray-level gradient, moving from the outside of the drop image to the inside, through the boundary.

After the edge detection process, the error in each drop profile coordinate can be expected to be of order one pixel. For a sessile drop of 5 mm in contact diameter, this corresponds to an error of 20 μm. To further improve the precision of the detected edge, a subpixel resolution algorithm was implemented. This was achieved by fitting a gray-level profile perpendicular to the drop interface with the so-called natural cubic spline fit [40,79]. The end condition for the natural cubic spline fit is that the second derivative equals zero at each end, since the gray level profile across the drop interface should approach linearity at each end. For the digitized drop image, it is generally not possible to fit a natural cubic spline curve exactly perpendicular to the drop interface using the existing gray level values. This fitting can only be done in three directions: horizontal, vertical, and diagonal. Hence, the direction closest to the perpendicular is chosen from this set of three principal directions, and the natural cubic spline fit for the gray levels is calculated in this preferred direction for each drop profile point. Figure 3.16 shows a typical example of a gray level profile perpendicular to the drop interface fitted with a natural cubic spline curve. Point A is the pixel that is selected by the Sobel edge detection as the drop profile coordinate. Various choices can be made for the selection of more precise drop profile coordinates from the curve. There are two obvious choices: (1) the point that has the maximum slope on the fitted curve (i.e., point A), or (2) the point whose gray level is the midpoint of the high and low plateaus of the fitted curve. The first choice is found to be unsatisfactory in some cases because of high sensitivity to noise [40]. Therefore, the second is utilized (i.e., point B in Figure 3.16). Repeating the above procedure for all the pixels improves the accuracy of all the drop profile coordinates. It is possible to further improve the accuracy of each drop profile point using a lateral smoothing technique and hence further improve the precision of the results. After all the profile coordinates of a drop are determined to subpixel resolution, the drop profile is then divided into groups of five consecutive drop profile coordinates. For example, the data in the first group are $(x_1, y_1),...,(x_5, y_5)$ the data in the second group are $(x_2, y_2),...,(x_6, y_6)$ and the data in the ith group are $(x_i, y_i),...,(x_{i+4}, y_{i+4})$. The slope of each group of data is calculated by fitting a straight line between the first point and the last point of the group. The data are then rotated using a transformation matrix so

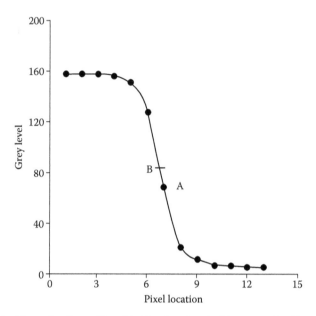

FIGURE 3.16   Example of a profile with 13 points fitted with a natural spline curve.

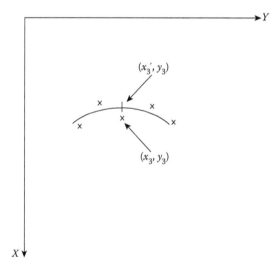

FIGURE 3.17   Example of the procedure in applying the smoothing technique for the first group of data after rotation. The point (x_3, y_3) is determined from natural spline fitting and the point (x'_3, y_3) is determined from the smoothing technique.

that the slope of the fitted straight line becomes zero. A second-order least squares polynomial, $x = ay^2 + by + c$, is then fitted to the group of rotated data points. The midpoint of the fitted polynomial is then calculated. Figure 3.17 shows an example of the procedure for the first group of data after rotation. As illustrated, to find the midpoint of the fitted polynomial for the first group of data, y_3 is substituted into the above least squares polynomial equation to evaluate the new value of x_3, say x_3'. The point (x_3, y_3) is then rotated back to the original slope and inputted as one of the drop profile coordinates. This smoothing technique is repeated for each group of data.

At this level, the coordinates of the experimental profile are presented in terms of pixels. However, the theoretical Laplacian curves (generated by integrating the Laplace equation of capillarity for known surface tension values) are obtained in centimeters. Therefore, a calibration grid is used to obtain the coordinates of the experimental profile in terms of centimeters. The grid is also used for correction of the optical distortion caused by the lens of the microscope. Cameras and microscopic lenses tend to produce slightly distorted images, and this distortion can cause errors in the final results, particularly in the interfacial tension [40]. To correct optical distortion, Cheng [40] used an approach similar to that of Green [66]. Consider the image of the grid (distorted because of the camera and microscope lens) and the original (corrected) grid. The grid consists of many small cells. Correction of optical distortion is accomplished by mapping each rectangular cell on the distorted image of the grid to the corresponding square cell on the corrected grid. Following this procedure, the optical distortion is corrected by mapping the image of the drop or its coordinates. A significant advantage of this method is that the aspect ratio of the digitizing board does not have to be known [40].

Finally, the misalignment of the camera with respect to the true vertical line as given by a plumb line is corrected by means of an appropriate transformation matrix [40]. At the beginning of the experiment, an image of a plumb line is acquired. The coordinates of the plumb line are detected and also corrected through the optical distortion correction process. Then, the angle of the plumb line with respect to a nominally vertical line is calculated. This angle represents the vertical misalignment of the camera, α. Although the misalignment of the camera is corrected through the above process, in some experimental situations it is difficult to achieve perfect vertical alignment. For instance, it is difficult to obtain the coordinates of the plumb line as well as the angle, α, between the plumb line and a true vertical line accurately when images are acquired with very low magnification. Therefore, in the numerical schemes of ADSA-P (see Section 3.3.1.3), α is also obtained through the optimization process.

In the original image analysis of ADSA-P developed by Cheng [40], the components of the image analysis process are mixed so that the study of the effect of each component as well as further improvement is difficult, if not impossible. Recently, the image analysis part has been redeveloped in a fully modular form [80,81] so that the significance of each module in the performance of ADSA-P can be evaluated separately. For instance, Table 3.4 shows the effect of each step of the image analysis part on the surface tension measurements obtained for large pendant drops of cyclohexane (after the improvement of the hardware explained in Section 3.3.4). Each column of the table contains: (1) the average of the surface tension values of the largest drops in 10 cycles, and (2) the range of error obtained based on the standard deviation of the surface tension values in 10 cycles at the 95% confidence level. It is clear that the surface tension value approaches the correct value (i.e., 25.24 mJ/m^2 at 20°C) and the error significantly decreases as each step is taken. Also, the effect of the optical distortion correction process is, at least in the case considered, more

TABLE 3.4

Effect of Each Step of the Image Analysis Part on the Surface Tension Measurements Obtained for Large Pendant Drops of Cyclohexane

Image Analysis Steps	Sobel Edge Detection	Cubic Spline Fit	Optical Distortion Correction	Camera Misalignment Correction
$\gamma \, (mJ/m^2)$	26.36 ± 0.11	26.30 ± 0.08	25.47 ± 0.04	25.24 ± 0.02

Note: The literature value of the surface tension of cyclohexane is 25.24 mJ/m^2 at 20°C.

significant than the other modules of the image analysis part. The above type of result has also been observed in other pendant drop experiments conducted with different liquids [82]. Obviously, the use of a microscope lens with higher quality will reduce the above effect.

Since the image analysis of ADSA-P is modular, it is possible to compare different edge detection techniques with the Sobel edge operator, implemented in the original image analysis process of ADSA-P. The Sobel operator was explained before. The other edge detection techniques considered here are Roberts, Prewitt, Laplacian of Gaussian (LoG), and Canny (see Chapter 4 for details of these different edge detectors). The Roberts operator [83] is the earliest operator developed to approximate the gradient in digital images. It uses a 2×2 convolution mask. The Prewitt operator [84] uses a 3×3 convolution mask (like the Sobel operator) and it performs better than the Sobel operator for a vertical edge. However, for a diagonal edge, the Sobel operator is superior [85]. The LoG [65] operator first smoothens the image by a convolution filter with a Laplacian of Gaussian mask. Then, it finds the edge as the location where the gradient of the gray level of the smoothened image is at the maximum. In other words, LoG locates the edge where the second derivative of the gray level is zero. Finally, the Canny operator [86] finds edges by searching for local maxima of the gradient of the digital image. The gradient is calculated using the derivative of a Gaussian filter. The method uses two thresholds to detect strong and weak edges, and includes the weak edges in the output only if they are connected to strong edges. Table 3.5 summarizes the surface tension values of large drops of cyclohexane obtained using different edge detection methods in two different conditions: (1) with drop profile corrections (i.e., cubic spline fitting, optical distortion correction, and misalignment correction of the camera), and (2) without drop profile corrections. It will become clear that the drop profile corrections have a considerable impact on the quality of the results no matter what edge detection method is used.

The effect of different edge detection techniques on ADSA-P output is shown in Table 3.6 for large and small drops. The results for cyclohexane indicate that the effect of different edge detection techniques is also significant for the small drops. The Canny method [86] may have particular merit, as the deviation between the surface tension values obtained for small and large drops is smaller than the values obtained using other edge detection techniques. The improvement of the results is possibly due to the fact that the Canny edge detection removes noise from the

TABLE 3.5

Surface Tension Values of Large Drops of Cyclohexane Obtained With and Without Drop Profile Corrections Using Different Edge Detection Techniques

	γ (mJ/m²) for Large Drop	
Edge Detection	With Drop Profile Corrections	Without Drop Profile Corrections
Sobel	25.24 ± 0.02	26.36 ± 0.08
Roberts	25.24 ± 0.02	26.38 ± 0.10
Prewitt	25.24 ± 0.02	26.37 ± 0.09
LOG	25.24 ± 0.02	26.41 ± 0.06
Canny	25.24 ± 0.02	26.26 ± 0.06

Note: The literature value of the surface tension of cyclohexane is 25.24 mJ/m² at 20°C.

TABLE 3.6

Surface Tension Values Obtained Using Different Edge Detection Techniques for Small and Large Drops of Cyclohexane Using all Drop Profile Corrections

Edge Detection	γ (mJ/m²) for Large Drop	γ (mJ/m²) for Small Drop
Sobel	25.24 ± 0.02	27.06 ± 0.08
Roberts	25.24 ± 0.02	27.18 ± 0.14
Prewitt	25.24 ± 0.02	27.11 ± 0.13
LOG	25.24 ± 0.02	26.91 ± 0.07
Canny	25.24 ± 0.02	26.79 ± 0.06

Note: The literature value of the surface tension of cyclohexane is 25.24 mJ/m² at 20°C.

image using a convolution Gaussian filter prior to the edge detection and that it uses two different threshold values for "strong" and "weak" edges. More detailed and advanced image analysis techniques for ADSA will be described in Chapter 4.

3.3.6 SHAPE PARAMETER

The results in Table 3.6 suggest that even after the improvement of the image quality (hardware) and use of more sophisticated edge detection techniques and efficient numerical schemes, there are still errors in surface tension measurement of ADSA-P for nearly spherical drops. In view of this limitation, it becomes necessary to identify the range of drop sizes for which surface tension values are obtained with a certain accuracy, say ± 0.1 mJ/m^2. For this purpose, a quantitative criterion called shape parameter is introduced. The shape parameter expresses quantitatively the difference in shape between a given experimental drop and a spherical shape. Such a parameter is formulated using the fact that the curvature along the periphery of a spherical drop is constant whereas it changes markedly for a well-deformed drop. Various definitions are possible. A possible definition for the shape parameter would be the difference between the drop volume and the volume of a sphere with radius R_0 (i.e., the radius of curvature at the apex of the drop). A seemingly simple way would be to use the volume of the drop obtained from ADSA-P and compare it to the volume of a sphere. However, it is not logical to use ADSA-P outputs to evaluate ADSA-P. In other words, the shape parameter should be obtained independently from ADSA-P. An alternative approach would be to calculate numerically the volume of the drop from the experimental profile of the drop and compare it to the volume of a sphere. A similar but less computationally involved approach is to calculate the difference between the projected area of the drop and an inscribed circle with radius R_0. This definition is preferred to formulate the shape parameter. Based on this definition, the shape parameter is zero for a completely spherical drop and larger than zero for a well-deformed drop. The above different definitions of the shape parameter were compared for different sizes and shapes of drops. The results show that the pattern of the shape parameter for different drop shapes is almost the same for all the above definitions and does not affect any conclusions one might want to draw.

To eliminate the effect of the image size or optical magnification of the drop, it is desired to define the shape parameter as a dimensionless parameter. One approach would be to normalize the shape parameter with the area of the inscribed circle. The difficulty with the above definition is that the shape parameter would not have an upper limit for the case of well-deformed drops. This difficulty is removed by normalizing the shape parameter with respect to the projected area of the drop instead of the circle. Thus, the final choice for the shape parameter is

$$P_s = \frac{\left| \int_0^{2\pi} \int_0^{r_\theta} r\,dr\,d\theta - \pi R_0^2 \right|}{\int_0^{2\pi} \int_0^{r_\theta} r\,dr\,d\theta}, \qquad (3.33)$$

where the numerator represents the absolute value of the difference between the projected area of the drop and the inscribed circle with radius R_0 (see the hatched area in Figure 3.18a); the denominator represents the projected area of the drop that is calculated numerically from the experimental profile obtained from the image of the drop. The radius R_0 is calculated by fitting tangent lines to the drop profile at any two points in the apex region, (X_1, Z_1) and (X_2, Z_2). These lines are described by the equations $Z = a_1 X + b_1$ and $Z = a_2 X + b_2$. The perpendicular lines to these tangent

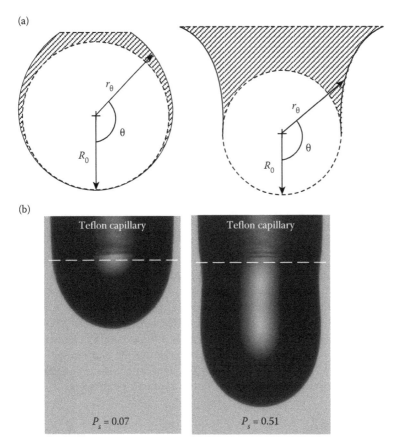

FIGURE 3.18 (a) The shape parameter P_s corresponds to the hatched area between the drop profile and an inscribed circle with a radius of R_0. (b) The shape parameter P_s for two drops of cyclohexane formed at the end of a Teflon capillary. (Hoorfar, M. and Neumann, A. W., *Advances in Colloid and Interface Science,* 121, 25–49, 2006. With permission from Elsevier.)

lines (i.e., the lines that are normal to the profile) intersect and form radii r_1 and r_2 equal to

$$r_1 = \frac{a_2(Z_2 - Z_1) + (X_1 - X_2)}{a_1 - a_2}\sqrt{a_1^2 + 1}, \qquad (3.34)$$

$$r_2 = \frac{a_1(Z_2 - Z_1) + (X_1 - X_2)}{a_1 - a_2}\sqrt{a_2^2 + 1}. \qquad (3.35)$$

If several pairs of points are selected such that they are close to the apex but far enough apart, their tangents (and normals) are not parallel. Thus, the radius of curvature at the apex can be averaged using Equations 3.34 and 3.35 for each pair of points. This approach gives better results than the commonly used method of fitting a circumference to some points in the apex region [41]. Also, in this way, the coordinates of the apex are not required. The value of the shape parameter for each drop profile is calculated using a program written using MATLAB® scripts [80].

The shape parameter obtained based on Equation 3.33 has two key features: (1) it shows the percentage of the projected drop area that deviates from the circle, and (2) it is bounded between zero and unity for pendant drops. This is normally true for sessile drops also, except for drops with very low surface tension where the drop can be quite flat even for large contact angles and the radius of curvature at the apex becomes quite large so that the tangent sphere will be quite large as well (see the results in Section 3.3.6.6). Figure 3.18b presents the shape parameter values (i.e., calculated based on Equation 3.33) of two different drop sizes of cyclohexane formed at the end of a Teflon capillary. It is noted that the shape parameter of the well-deformed drop (i.e., a large drop) is significantly larger than that of the near spherical drop (i.e., a small drop).

3.3.6.1   Critical Shape Parameter

The result of a pendant drop experiment is used to illustrate how the shape parameter can be used to quantify the range of applicability of ADSA-P. Figure 3.19 presents the surface tension measurements obtained for cyclohexane in a sequence of static experiments. In this type of experiment, the volume of the drop is increased/decreased slowly (i.e., at a rate of 0.1 μl/sec) in a finite number of intervals where the stepper motor is stopped for about one minute at the end of each interval to ensure that the drop reaches equilibrium. The experiment was conducted at 20°C. The drop was formed at the end of the Teflon capillary (with an outer diameter of

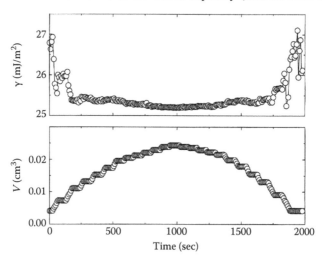

FIGURE 3.19   Experimental results of a sequence of static experiments conducted for a pendant drop of cyclohexane formed at the end of a Teflon capillary with an outer diameter of 3 mm. For a certain range of drop sizes shown in the surface tension graph, the difference between the surface tension values obtained from ADSA-P and the true surface tension of cyclohexane is less than ±0.1 (mJ/m²). (Hoorfar, M. and Neumann, A. W., *Advances in Colloid and Interface Science*, 121, 25–49, 2006. With permission from Elsevier.)

3 mm and an inner diameter of 2 mm). Again, cyclohexane was chosen because it is a cycloalkane and hence can be readily purified. Since cyclohexane is a pure liquid (with purity greater than 99.9%), its surface tension is expected to remain constant regardless of the size of the drop. However, the results of ADSA-P show that the surface tension value changes (in this case increases) as the volume (or the surface area) of the drop decreases. Obviously, the surface tension obtained for small drops must be erroneous.

It is clear that for a certain range of drop sizes (see Figure 3.19), the difference between the surface tension values obtained from ADSA-P and the literature value of the surface tension of cyclohexane (25.24 mJ/m² at 20°C) [63] is less than ±0.1 mJ/m². Outside of that range, the surface tension obtained from ADSA-P deviates considerably from the literature value. For each drop size in the above experiment, the value of the shape parameter was calculated based on Equation 3.33. Figure 3.20 shows the variation of the surface tension values versus the shape parameter calculated for each drop size. The vertical axis presents the relative error, ε_{rel}, which is defined as the difference between the surface tension value obtained from ADSA-P for each drop size and the true surface tension. The relative error has been normalized with respect to the true surface tension. The relative error is given by

$$\varepsilon_{rel} = \frac{\gamma_{ADSA} - \gamma_{true}}{\gamma_{true}} = \frac{\Delta\gamma}{\gamma_{true}}. \tag{3.36}$$

For the drops with larger values of the shape parameter, the error is relatively constant near zero; however, from a certain point on, it increases rapidly as the shape

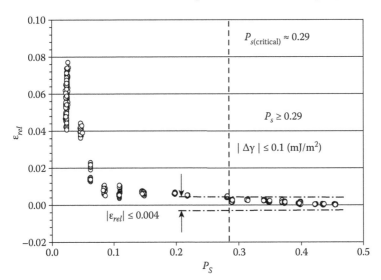

FIGURE 3.20 Relative error of surface tension as a function of the shape parameter for different drop sizes of a pendant drop of cyclohexane in a sequence of static experiments. The drop was formed at the end of a Teflon capillary with an outer diameter of 3 mm. (Hoorfar, M. and Neumann, A. W., *Advances in Colloid and Interface Science*, 121, 25–49, 2006. With permission from Elsevier.)

parameter decreases. The range of acceptable drop shapes depends on the desired accuracy. Here the desired accuracy is chosen as 0.1 mJ/m^2 (i.e., $|\Delta\gamma| \leq 0.1$), which corresponds to a relative error of 0.004 (i.e., $|\varepsilon_{rel}| \leq 0.004$) for the above experiment of cyclohexane. Based on the above value, the graph is divided into two parts. The cutoff line will be at the shape parameter of 0.29 for the above experiment. The threshold of 0.29 is referred to as the "critical shape parameter" (i.e., $P_{s\,(critical)} = 0.29$). A shape parameter above the critical value guarantees the error to be less than 0.1 mJ/m^2. The results in Figure 3.20 also show that the deviation of the surface tension values (obtained from ADSA-P) from the true value increases as the shape parameter decreases. This deviation can be positive (Figure 3.20) or negative (Figures 3.32 and 3.33). It will be shown in Section 3.3.6.6 that the sign of the error also depends on the numerical schemes.

Similar considerations will apply to other desired or needed accuracies. If a different accuracy is chosen, the value of the critical shape parameter as well as the range of applicability of ADSA-P will be different. For instance, for some purposes the accuracy needed may be as modest as ±1 mJ/m^2 (see Section 3.3.6.5). As a result, the critical value will be much smaller, and hence the range of the acceptable drop shapes will be significantly larger.

It is anticipated that the above critical shape parameter (i.e., $P_{s\,(critical)} = 0.29$) changes as certain experimental conditions change. The factors that may affect the critical shape parameter are: (1) the material, size, and shape of the holder used to form the drop; (2) liquid properties (such as density and surface tension); and (3) dynamic effects (such as momentum and viscous forces) that can influence the drop shape as the drop volume is changed rapidly (in dynamic experiments). The effects of these factors are scrutinized in the following sections.

3.3.6.2 Effect of the Material, Size, and Shape of the Holder on the Critical Shape Parameter

Different drop arrangements are shown in Figure 3.21. Figure 3.21a presents a pendant drop of a liquid formed at the end of a Teflon capillary. The second shape is a pendant drop formed at the end of a stainless steel holder (Figure 3.21b). In essence, the holder is an "inverted pedestal." The pedestal is used in a constrained sessile drop configuration shown in Figure 3.21c (see Chapter 5 for more details of the constrained sessile drop configuration). The edge of the holder is sharp with an angle of approximately 45° to prevent spreading of the liquid on the outer surface of the holder [87]. It is apparent that different types of holders produce different drop shapes. Thus, it is expected that the material and size of the holder affect the critical shape parameter and hence the range of applicability of ADSA-P. To elucidate this possibility, the experiment conducted with the Teflon capillary (Figures 3.19 through 3.20) was repeated using a stainless steel holder with the same diameter as the Teflon capillary (i.e., 3 mm). The chosen liquid was again cyclohexane. The drop is formed at the circular edge (i.e., the outer diameter) since the holder is hydrophilic. The experiment was conducted in a stepwise fashion. Figure 3.22 presents the relative error in surface tension as a function of the shape parameter. It is clear that the range of drop shapes yielding surface tensions with an accuracy of ±0.1 mJ/m^2 is larger in the case of the stainless steel holder than in the case of

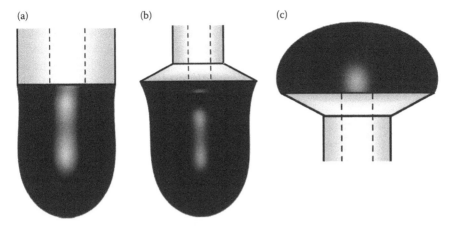

FIGURE 3.21 (a) Image of a pendant drop formed at the end of a Teflon capillary. (b) Image of a pendant drop formed at the end of a stainless steel holder. (c) Image of a constrained sessile drop formed on top of a stainless steel pedestal. (Hoorfar, M. and Neumann, A. W., *Advances in Colloid and Interface Science,* 121, 25–49, 2006. With permission from Elsevier.)

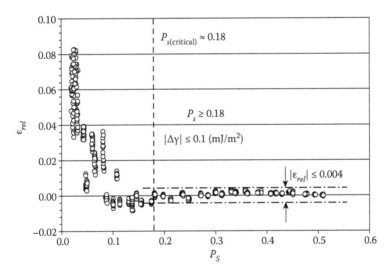

FIGURE 3.22 Relative error of surface tension as a function of the shape parameter for different drop sizes of a pendant drop of cyclohexane formed at the end of a stainless steel holder with an outer diameter of 3 mm and inner diameter of 0.5 mm. (Hoorfar, M. and Neumann, A. W., *Advances in Colloid and Interface Science,* 121, 25–49, 2006. With permission from Elsevier.)

the Teflon capillary (see Figure 3.20). For large values of the shape parameter, the error is relatively constant, near zero and bounded, verifying that ADSA-P performs accurately for well-deformed drops. On the other hand, for small values of the shape parameter, the relative error increases as the shape parameter decreases so that the accuracy of ADSA-P deteriorates although not as much as in the case of the Teflon capillary. Based on the desired accuracy (i.e., $|\Delta\gamma| \leq 0.1$ that corresponds to $|\varepsilon_{rel}| \leq 0.004$ for cyclohexane), the critical shape parameter was found as $P_{s\text{ (critical)}} = 0.18$. The comparison between Figures 3.20 and 3.22 shows that the critical shape parameter obtained for the case of the stainless steel holder (i.e., $P_{s\text{ (critical) steel}} = 0.18$) is significantly smaller than that obtained for the case of the Teflon tube (i.e., $P_{s\text{ (critical) Teflon}} = 0.29$). In other words, the more hydrophilic the material of the holder, the larger the range of applicability of ADSA-P. Thus, the stainless steel holder is recommended for surface tension measurements.

In addition to the material of the holder, its size could affect the critical shape parameter as it has been observed that different sizes of the holder produce different shapes. To illustrate the above effect, stainless steel holders with different sizes ranging from 1.5 mm to 6 mm in diameter were used. For each size of holder, the above experiment with cyclohexane was repeated and the values of the critical shape parameter were determined. Figure 3.23 shows the critical shape parameter as a function of the size of holder. It is clear that the larger the size of the holder, the smaller the critical shape parameter. In other words, larger holders provide larger ranges of drop shapes that are acceptable for surface tension measurements. Although holders with large diameters are generally recommended for surface tension measurements using the drop shape techniques, there is a limit. The diameter of the holder must be selected within a range where the effect of gravity and the effect of surface tension on the drop (formed at the end of the holder) are comparable. If the effect of gravity is significantly larger than the effect of surface tension, the drop will fall off the capillary (holder). On the other hand, the drop becomes close to spherical in shape if the effect of surface tension is significantly larger than that of gravity.

Another important inference from Figure 3.23 is that the critical shape parameter changes fairly linearly with the size of the holder. Therefore, it is possible to use interpolation to find the critical shape parameter for a given size of holder.

The results presented in this section suggest that the more hydrophilic and the larger the size of the holder, the larger the range of applicability. Thus, the use of the stainless steel holder with a large outer diameter is preferable. The circular edge of

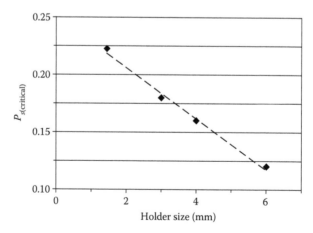

FIGURE 3.23 The critical shape parameter as a function of the outer diameter of the circular contact area of the holder for the pendant drop configuration. (Hoorfar, M. and Neumann, A. W., *Advances in Colloid and Interface Science,* 121, 25–49, 2006. With permission from Elsevier.)

the stainless steel holder should be a knife edge as shown in Figure 3.21b to prevent spreading of the liquid, especially with ultralow surface tension, over the exterior of the holder. However, the angle of the edge has no effect on the shape parameter since the drop is formed at the circular edge. This has been elucidated in the experiment conducted using a stainless steel capillary, which is essentially a holder with an edge angle of 90° [80,81]. The critical shape parameter was found as $P_{s\ (\mathrm{critical})} = 0.18$; that is, the same as that obtained for the case of the stainless steel holder with an edge angle of 45°. Thus, the edge has no effect on the critical shape parameter and only facilitates the experimental process by preventing the spreading of the liquid over the wall of the holder. Similar to the angle of the edge, the size of the inner diameter of the holder has no effect on the critical shape parameter since the holder is hydrophilic and the drop is attached (hinged) at the outer diameter. This was confirmed by an experiment using a holder with a larger inner diameter (i.e., 2 mm [80,81]). Nevertheless, the question of the inner diameter may arise in dynamic experiments where the momentum of the liquid that is pumped into or out of the drop depends on the size of the inner diameter.

3.3.6.3 Effect of Liquid Properties on the Critical Shape Parameter

The shape parameter and the critical shape parameter were scrutinized in the previous section for the effect of hydrophobicity and the shape of a capillary or other solid surface supporting the liquid drop. The shape parameter may very well also depend on liquid properties. This was elucidated experimentally using four liquids with different surface tensions and densities. Table 3.7 summarizes the results obtained for pendant drops of different liquids formed at the end of a stainless steel holder with an outer diameter of 3 mm. The results show that the critical shape parameter is constant from one liquid to another, and that surface tension values obtained from ADSA-P for large drops agree very well with those obtained from the literature [63]. Similar results were found in experiments using a Teflon capillary tube instead of a stainless steel holder [82]. Thus, it can be concluded that the liquid properties have no effect on the critical shape parameter. This finding can be explained by the fact that the shape parameter is a geometrical property and not a physical property of the drop. Thus, the critical shape parameter obtained for a certain material and size of

TABLE 3.7
The Values of the Critical Shape Parameter of Pendant Drops of Different Liquids Obtained With a Stainless Steel Holder of an Outer Diameter of 3 mm

Liquid	γ_{true} (mJ/m²)	γ_∞ (mJ/m²)	$P_{s\ (\mathrm{critical})}$
Cyclohexane	25.07 (at 21.5°)	25.06 ± 0.01	0.18
Hexadecane	27.47 (at 20°)	27.43 ± 0.02	0.17
Diethyl phthalate	36.54 (at 20°)	36.54 ± 0.01	0.18
Glycerol	63.35 (at 20°)	63.34 ± 0.01	0.19

holder can be used to identify the range of acceptable drop shapes in the experiments involving solutions or pure liquids with unknown surface tension values.

3.3.6.4   Impact of Dynamic Effects on the Critical Shape Parameter

The Laplace equation is an equilibrium condition, not necessarily applicable for flow situations. However, it is known that surface tension equilibrium is reached quickly, so that total quiescence of the liquid may not be strictly necessary. Of course, at sufficiently high flow rates of liquid into the drop, deviations from the Laplacian shape are to be expected. It will be shown below that ADSA-P can nevertheless be used up to remarkably large flow rates. Not unexpectedly, the effect of liquid flow expresses itself in an increase of the critical shape parameter. That is, larger deviations from spherical shape are necessary for the measurement to remain meaningful.

A systematic study of dynamic parameters in such a drop constellation might be difficult. In the experiments shown below, a constant volume flow rate into the drop will be used. Since the volume will change continuously, the constant volume flow rate cannot be expected to have a time invariant influence on the drop shape. In view of these complexities, the shape parameter provides a useful tool to establish confidence limits for the surface tension measurement.

Figure 3.24 presents the surface tension measurements obtained for cyclohexane in a dynamic cycling experiment. The drop was formed at the end of a Teflon capillary with an outer diameter of 3 mm. The size of the drop was changed at a rate of 3.2 μl/sec. The experiment was conducted at 20°C. For each size of the drop, the

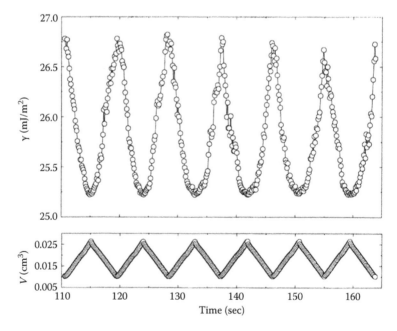

FIGURE 3.24   Experimental results of a pendant drop of cyclohexane in a dynamic cycling experiment. (Hoorfar, M. and Neumann, A. W., *Advances in Colloid and Interface Science,* 121, 25–49, 2006. With permission from Elsevier.)

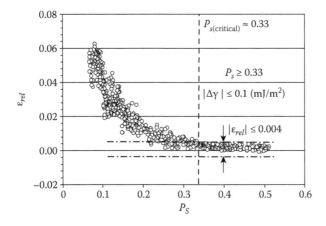

FIGURE 3.25 Relative error of surface tension as a function of the shape parameter for different drop sizes of cyclohexane in the dynamic cycling experiment. The drop was formed at the end of a Teflon capillary with an outer diameter of 3 mm.

shape parameter and relative error in surface tension measurement were calculated using Equations 3.33 and 3.36, respectively. Figure 3.25 shows the relative error as a function of the shape parameter for the dynamic experiment. Just as for static drops, for large values of the shape parameter the error is relatively constant and near zero, but it increases from a certain point on as the shape parameter decreases. The range of the acceptable drop shapes is obtained based on the desired accuracy. An error of ± 0.1 mJ/m^2 (i.e., $|\Delta\gamma| \leq 0.1$) corresponds to a relative error of 0.004 (i.e., $|\varepsilon_{rel}| \leq 0.004$) for cyclohexane. Based on the above value, the critical shape parameter for the above dynamic experiment was found as $P_{s\,(critical)} = 0.33$.

Comparison between Figures 3.20 and 3.25 shows that the critical shape parameter is smaller in the static experiment compared to the dynamic experiment. Since other experimental conditions (such as the material and size of the holder) are the same in the above two types of experiments, the difference in the values of the critical shape parameter is due to dynamic effects such as momentum of the liquid that is pumped into, or out of, the drop. Dynamic effects possibly deform the shape of the drop so that the underlying Laplace equation is no longer satisfied. ADSA-P responds by finding a value for the surface tension that minimizes the error function (Equation 3.23), assuming wrongly that the Laplace equation is satisfied. For a well-deformed drop shape, the relative errors in surface tension measurements in both types of experiment are small because (i) the role of dynamic effects is less pronounced for large drops as the amount of liquid that is pumped in and out of the drop represents a smaller volume fraction of the drop, and (ii) significantly different surface tension values correspond to significantly different drop shapes. Thus, the deformation of the drop due to dynamic effects is not significant enough to cause deviation of the surface tension values from the true value (the literature value). However, as the drop is made smaller, the role of dynamic effects is more pronounced. For nearly spherical drop shapes, significantly different surface tension values correspond to only slightly different drop shapes. Thus, even a small deformation can cause the results to deviate from the correct value. Therefore, in the dynamic experiment, not only is the critical shape parameter larger (i.e., $P_{s\,(critical)\,dynamic} = 0.33$ compared to $P_{s\,(critical)\,static} = 0.29$), but the error in the surface tension measurements also increases more significantly as the shape parameter decreases. For instance, for a drop shape with $P_s = 0.1$, the relative error obtained in the dynamic experiment is $|\varepsilon_{rel}| = 0.04$, which corresponds to an error of 1 mJ/m² in the surface tension measurements. However, for the drop shape with the same shape parameter (i.e., $P_s = 0.1$), the relative error is quite small (i.e., $|\varepsilon_{rel}| = 0.01$) in the static experiment.

3.3.6.5 Shape Parameter of Constrained Sessile Drops

A new sessile drop configuration "constrained sessile drop" [87–89] (see Figure 3.21c) has been developed in which the drop is formed on top of a machined and smooth circular stainless steel holder. The edge of the holder is sharp with an angle of approximately 45° to prevent spreading of the liquid on the outer surface of the holder. Since the holder is hydrophilic, the drop is attached (hinged) at the outer diameter, and hence remains axisymmetric even when increasing/decreasing the volume of the drop. Recently, the constrained sessile drop configuration has been extensively used to measure extremely low surface tensions of lung surfactant [87,88] (see Chapter 5), which cannot be measured in conventional pendant drop experiments due to film leakage [90]. The arrangement can also be used to measure surface tension and density of polymer melts simultaneously [89] (see Section 3.3.7.5). In view of the frequent use of the constrained sessile drop configuration in different experimental situations [87–89], the accuracy of the surface tension measurements of ADSA-P for this configuration has been studied. Several experiments were conducted using pure liquids again. One of the chosen liquids is diethyl phthalate with a surface tension of 36.54 mJ/m² (at 20°C) [63]. A pedestal with an outer diameter of 6 mm was used. Figure 3.26 shows the results obtained from ADSA-P. It is clear that for large and hence well-deformed drops, ADSA-P produces correct results. However, it appears that the range of the acceptable drop shapes for the above experiment is considerably smaller than that obtained from the pendant drop experiments conducted using the same size of the stainless steel holder. To quantify the above range, the shape parameter and the relative error were calculated for each drop size. Figure 3.27 shows the variation of the relative error over the shape parameter. It is noted that in general the values of the shape parameter of constrained sessile drops are significantly smaller than those calculated for pendant drops. In fact, for constrained sessile drops, the deviation of the drop shape from a spherical shape is quite small even for large drops. Based on the desired accuracy (i.e., $|\Delta\gamma| \leq 0.1$ that corresponds to $|\varepsilon_{rel}| \leq 0.003$ for diethyl phthalate), the critical shape parameter was identified as $P_{s\,(critical)} = 0.17$. It is also noted that the range of the drop shapes for which the shape parameter is larger than the critical value is quite small. The results for a different liquid, octamethylcyclotetrasiloxane (OMCTS) with a surface tension of 18.20 mJ/m² (at 24°C) [63], are shown in Figure 3.28. The comparison between Figures 3.27 and 3.28 shows that the range of drop sizes for which surface tensions are accurate is larger for the liquid with lower surface tension. This was expected, because the lower the surface tension of the liquid, the larger the range of the well-deformed drops. It is also noted that the critical shape parameter of OMCTS is identical to that of diethyl phthalate.

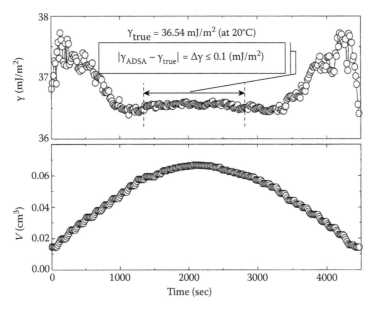

FIGURE 3.26 Experimental results for a constrained sessile drop of diethyl phthalate formed on top of a pedestal with an outer diameter of 6 mm. (Hoorfar, M. and Neumann, A. W., *Advances in Colloid and Interface Science,* 121, 25–49, 2006. With permission from Elsevier.)

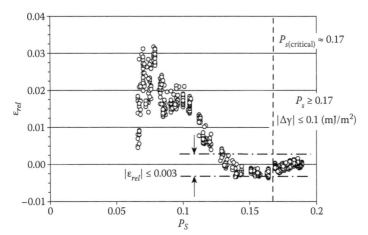

FIGURE 3.27 Relative error of surface tension as a function of the shape parameter for different drop sizes of a constrained sessile drop of diethyl phthalate formed on top of a pedestal with an outer diameter of 6 mm. (Hoorfar, M. and Neumann, A. W., *Advances in Colloid and Interface Science,* 121, 25–49, 2006. With permission from Elsevier.)

Thus, for this configuration, the critical shape parameter appears to be independent of the type of liquid.

The critical shape parameter is expected to depend on the diameter of the pedestal. This is illustrated for two additional diameters (i.e., 3.8 mm and 2.7 mm). Figure 3.29 shows that the critical shape parameter increases fairly linearly with the

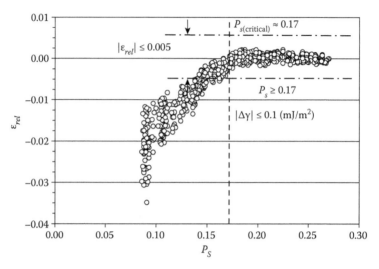

FIGURE 3.28 Relative error of surface tension as a function of the shape parameter for different drop sizes of a constrained sessile drop of OMCTS formed on top of a holder with an outer diameter of 6 mm. (Hoorfar, M. and Neumann, A. W., *Advances in Colloid and Interface Science*, 121, 25–49, 2006. With permission from Elsevier.)

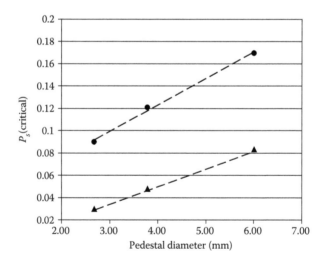

FIGURE 3.29 The critical shape parameter as a function of the outer diameter of the circular contact area of the pedestal for the constrained sessile drop configuration, for two different required accuracies: circles, 0.1 mJ/m²; triangles, 1 mJ/m².

diameter of the pedestal. Thus, it is possible to use interpolation and some extrapolation to find the critical shape parameter for any given diameter of the pedestal. Also, unlike the pendant drop configuration (see Figure 3.23), the critical shape parameter decreases with decreasing size of pedestal. The difference may be related to the effect of gravity on the shapes of pendant and constrained sessile drops. In the case of pendant drops, gravity elongates the drop whereas surface tension tends to make it more spherical. Obviously, larger holders allow for larger drops for which the effect of gravity, and hence the deformation of the drop, is more pronounced. On the other hand, for constrained sessile drops, the drop is pushed down on the pedestal by gravity. Thus, the deformation of the drop is more pronounced as the top portion (around the apex) becomes relatively flat. In this situation, the sides of the drop are pushed outward since the drop is hinged at the edge of the pedestal. It is expected that this effect is more pronounced for small sizes of the pedestal.

As an illustration, the constrained sessile drop configuration has been extensively used to study the behavior of a therapeutic lung surfactant (Bovine Lipid Extract Surfactant, BLES; see Chapter 5 for details). The adsorption and stability of the surfactant film for different concentrations of BLES have been studied using the surface tension results obtained from ADSA-P [87,88]. In such studies, involving changing drop size and changing surface tension, it is desirable to be able to evaluate the accuracy of the surface tension measurements prior to the interpretation of the results. Figure 3.30 presents the ADSA-P output obtained in a dynamic cycling experiment

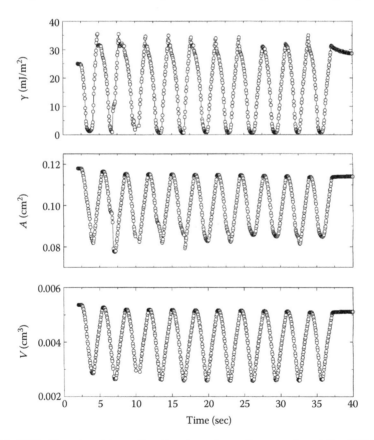

FIGURE 3.30 Experimental results of a dynamic cycling experiment for BLES with concentration of 2 mg/ml using the constrained sessile drop configuration. The drop was formed on top of a pedestal with an outer diameter of 2.7 mm. (Hoorfar, M. and Neumann, A. W., *Advances in Colloid and Interface Science,* 121, 25–49, 2006. With permission from Elsevier.)

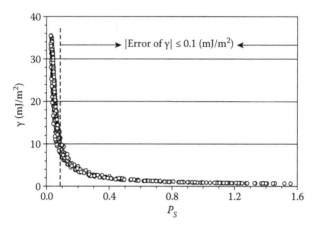

FIGURE 3.31 Surface tension as a function of the shape parameter for the dynamic cycling experiment of BLES with a concentration of 2 mg/ml. The accuracy of the surface tension values of the drop shapes for which the values of the shape parameter are larger than 0.09 is 0.1 mJ/m². (Hoorfar, M. and Neumann, A. W., *Advances in Colloid and Interface Science,* 121, 25–49, 2006. With permission from Elsevier.)

conducted for BLES with a concentration of 2 mg/ml. The drop was formed on top of a stainless steel pedestal with an outer diameter of 2.7 mm. The surfactant adsorbs at the surface of the drop and a film is formed. In each cycle, the film is compressed/ expanded by decreasing/increasing the drop surface area by changing the drop volume. The duration of each cycle is three seconds, which simulates the function of the lung in normal breathing. During the compression of the film, the surface tension changes from 35 mJ/m² to 0.6 mJ/m². For each size of the drop, the shape parameter was calculated in order to evaluate the accuracy of the results. Figure 3.31 presents the surface tension value as a function of the shape parameter for each size of the drop in the above experiment. It is observed that the value of the shape parameter increases as the surface tension decreases (or as the drop surface area is compressed). For low surface tension values, the shape parameter becomes quite large (even larger than 1) since the drop becomes quite flat. On the other hand, the shape parameter becomes quite small for higher surface tension values (e.g., 35 mJ/m²). For the 2.7 mm pedestal, the critical shape parameter is 0.09 (see Figure 3.29) regardless of the type of liquid or surface tension. Thus, drop shapes for which the values of the shape parameter are larger than 0.09 guarantee the error to be less than 0.1 mJ/m². The acceptable range of drop shapes corresponds to the surface tension values smaller than 10 mJ/m² (see Figure 3.31). It is noted that the larger surface tension values (i.e., larger than 10 mJ/m²), which were obtained for drop shapes outside the acceptable range, may not be reliable based on an accuracy of ±0.1 mJ/m². However, for the purposes of most lung surfactant studies, the desired accuracy can be relaxed to 1 mJ/m² for higher surface tension values. If a different accuracy is chosen, the value of the critical shape parameter will be different. More precisely, the larger the limit of the desired accuracy, the larger the range of applicability of ADSA-P or the smaller the value of the critical shape parameter. Figure 3.29 presents the values of the critical shape parameter obtained for different sizes of pedestal as the desired accuracy is extended to 1 mJ/m². For a 2.7 mm pedestal, the critical shape parameter was found as $P_{s\,\text{(critical)}}\big|_{\pm 1\,(\text{mJ/m}^2)} = 0.03$ (see Figure 3.29). Based on this value of the critical shape parameter, larger surface tension values (i.e., larger than 10 mJ/m²) obtained in the lung surfactant experiment are acceptable, within the accuracy of ±1 mJ/m².

3.3.6.6 Evaluation of the Numerical Schemes of ADSA-P Using Shape Parameter

The shape parameter is a useful criterion that can be used not only to evaluate the accuracy of the surface tension measurements, but also to study the effect of each part of ADSA-P (e.g., hardware, image analysis, and numerical scheme) on the range of applicability of the method. Specifically, the shape parameter has been used to evaluate the performance of two different numerical schemes of ADSA-P (i.e., Rotenberg [42] and del Río [41]) in a typical pendant drop experiment. Figure 3.32 presents the relative error of surface tension values calculated using the two ADSA-P algorithms for different drop sizes of a pendant drop of cyclohexane formed on a 4 mm outer diameter stainless steel holder. Based on the desired accuracy of ±0.1 mJ/m², the value of the critical shape parameter has been found for the results of each algorithm. The range of applicability of the more recent algorithm (i.e., del Río) is larger than the original algorithm (i.e., Rotenberg) as the critical shape parameter obtained for the more recent ADSA-P version (i.e., $P_{s\,\text{(critical)\,del Río}} = 0.16$) is smaller than that obtained for the original version (i.e., $P_{s\,\text{(critical)\,Rotenberg}} = 0.22$). Nevertheless, the error in the surface tension measurements always increases as the values of the shape parameter become smaller than the critical value. Interestingly, for the same image of the drop, the sign of the relative error obtained from the two algorithms is different.

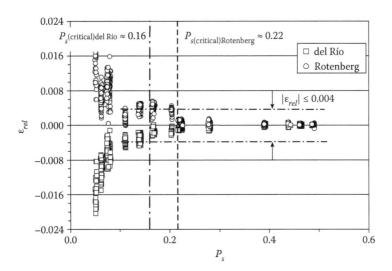

FIGURE 3.32 Range of applicability of the two ADSA-P algorithms in an experiment conducted with cyclohexane using a stainless steel holder with an outer diameter of 4 mm. (Hoorfar, M. and Neumann, A. W., *Advances in Colloid and Interface Science,* 121, 25–49, 2006. With permission from Elsevier.)

Similar observations were reported by other researchers [91–93] using numerical optimizations for the measurement of interfacial tensions from drop shape techniques (i.e., pendant and sessile drops). In their study, the surface tension values of large drops are calculated accurately; but, when the drop volume decreases, the surface tension value decreases. It is apparent from Figure 3.32 that such trends in the surface tension measurement do not necessarily represent a drop size dependence of surface tensions, but more likely an artifact of the measurement. Attempts were made to correlate the applicability of drop shape techniques to a more physical property, specifically the Bond number

$$B = \frac{\rho g R_0^2}{\gamma},\qquad(3.37)$$

where R_0 is the radius of curvature at the apex and γ is the literature value of surface tension.

Figure 3.33 shows that there is no clear pattern for the accuracy of the surface tension measurement and the Bond number. More precisely, a critical Bond number, in which the error in the surface tension measurement is less than ±0.1 (mJ/m²), apparently depends in part on liquid properties other than surface tension. On the other hand, a critical shape parameter that guarantees an error of less than ±0.1 (mJ/m²) is unique and independent of the type of liquid (see Table 3.7). These results are not surprising as the accuracy of the surface tension measurement using drop shape techniques (such as ADSA-P) depends on the performance of the numerical scheme and the accuracy of the detected experimental profile. Pragmatically, the

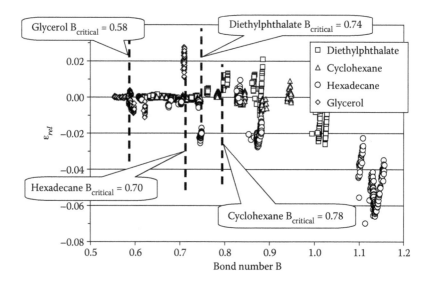

FIGURE 3.33 Relative error of surface tension as a function of the Bond number for different liquids.

performance of a drop shape technique cannot be evaluated by a single physical quantity.

3.3.7   APPLICATION OF **ADSA-P**

ADSA-P has been used to study a wide variety of systems, ranging from biological to industrial [47]. A few examples of such applications will be discussed below to illustrate the potential scope of ADSA-P.

3.3.7.1   Contact Angle Measurement

Contact angles provide insight into the interfacial tension of the solid phase through the use of Young's equation given as

$$\gamma_{sv} - \gamma_{sl} = \gamma_{lv} \cos\theta, \tag{3.38}$$

where the subscripts lv, sv, and sl in Equation 3.38 refer to liquid–vapor, solid–vapor, and solid–liquid interfaces, respectively, and θ is the contact angle formed between the liquid and the solid surface. The accurate measurement of contact angle is an avenue for the determination of solid interfacial tensions. The measurement of meaningful contact angles is not straightforward [6]. The issue is complicated by difficulties in preparing a proper solid surface, which must be smooth, homogeneous, and chemically inert (to the liquid used in the measurement). Satisfaction of these criteria is not a trivial task. The traditional means of contact angle measurement is a goniometer technique that depends on establishing a tangent to the drop at the three phase line. The procedure can lead to significant error. On the other hand, ADSA-P calculates the contact angle by integrating Equation 3.11 for the surface tension value and the radius of curvature obtained from the best fitted Laplacian curve. The technique is automated, which allows the acquisition and analysis of large amounts of data, making large scale dynamic studies feasible. As an illustration, Table 3.8 summarizes the ADSA-P surface tensions of 16 liquids and their calculated contact angles on FC-721 fluorocarbon compound, dip coated onto smooth sheets of mica. The reproducibility of the surface tension and contact angle values was generally better than 0.1 mJ/m² and 0.2°, respectively. More examples of using ADSA-P for measuring contact angle can be found in Chapter 6.

The high accuracy of ADSA-P also allows the study of the drop size dependence of contact angles, under certain conditions (see Chapter 13).

3.3.7.2   Pressure Dependence of Interfacial Tensions

The measurement of the pressure dependence of interfacial tension requires a very accurate method due to the fact that its value is quite small (typically of the order of 10^{-11} m). A large number of techniques can be used to measure interfacial tension, although drop shape methods are best suited for measuring γ under high pressure. Thus, ADSA-P with its high degree of accuracy, as illustrated above, is an appropriate method for such investigations and can be used to detect changes in interfacial tension over relatively small pressure ranges.

TABLE 3.8

Surface Tension Values, γ, of 16 Liquids (Measured by the Pendant Drop Method), and their Contact Angle Values, θ, on an FC-721 Coated Mica Surface, Measured by ADSA-P

Liquid	γ (mJ/m²)	± 95% Confidence Limits (mJ/m²)	Number of Drops	θ (deg.)	± 95% Confidence Limits (deg.)	Number of Drops
Decane	23.43	0.02	10	65.97	0.24	6
Dodecane	25.44	0.02	10	69.82	0.25	9
Tetradecane	26.55	0.05	10	73.31	0.14	6
Hexadecane	27.76	0.04	12	75.32	0.27	7
trans-Decalin	29.50	0.06	13	76.71	0.18	9
cis-Decalin	31.65	0.05	13	79.87	0.18	9
Ethyl cinnamate	38.37	0.03	10	88.20	0.14	10
Dibenzylamine	40.63	0.09	10	92.06	0.13	11
Dimethylsulfoxide (DMSO)	43.58	0.08	13	94.47	0.29	9
1-Bromonaphthalane	44.01	0.06	16	95.29	0.23	8
Diethylene glycol	45.04	0.07	16	96.84	0.17	7
Ethylene glycol	47.99	0.02	17	99.03	0.23	8
Thiodiglycol	54.13	0.11	10	103.73	0.24	7
Formamide	57.49	0.08	10	107.32	0.11	6
Glycerol	63.11	0.06	12	111.38	0.21	7
Water	72.75	0.06	10	119.05	0.17	6

For liquid–liquid pendant-drop experiments, a pressure/temperature (*P/T*) cell was utilized (Figure 3.34). The cell consisted of a 316 stainless steel cylinder with 1 inch thick optical glass windows fitted at each end. All lines and connections were 316 stainless steel. The cell was rated at 350 bars (5000 psi; 1 bar = 10^5 Pa) and 200°C. An Eldex HPLC pump was used to pressurize the system by pumping water into the *P/T* cell. Since the cell is sealed completely, it is possible to measure the interfacial tension at relatively high pressure.

The pressure and time dependences of the interfacial tension measured by ADSA-P at 21.5°C are presented in Figure 3.35 [94]. A pendant drop of *n*-decane was formed at the tip of a stainless steel needle immersed in water. Initially, the system was pressurized, then a new drop was formed, and a time-dependent study was undertaken using ADSA-P to acquire and analyze the drop images. The procedure was then repeated at a higher pressure. A time dependence for γ is observed at each pressure, *P*. This is probably due to surface active impurities that migrate to the liquid-liquid interface and decrease the interfacial tension. The isochronic γ-*P* plot reveals a linear relationship between γ and *P* in the pressure range studied. The $[\partial\gamma / \partial P]_T$ is found to be of the order of 10^{-11} m. Nevertheless, the observed pressure dependence is found to be significant at the 99% confidence level. This attests to the suitability of the apparatus and the ADSA-P methodology for the measurement

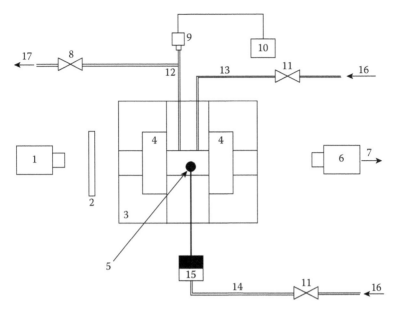

FIGURE 3.34   Schematic of the pressure/temperature (P/T) cell. 1: Light source, 2: diffuser, 3: P/T cell, 4: optically flat glass windows, 5: pendant drop, 6: microscope and CCD camera, 7: to digitizer, 8: discharge and relief valves, 9: pressure transducer, 10: pressure transducer indicator, 11: isolation valves, 12: tee, 13: bulk fluid line, 14: drop fluid line, 15: intermediate cell, 16: from the HPLC pump, and 17: to the discharge container.

of interfacial tensions as a function of pressure; contact angles may be studied in a similar manner. Moreover, the P/T cell readily facilitates the study of the temperature dependence of γ.

The accurate measurement of interfacial tension under high pressure is important in oil recovery [40,95]. At the beginning of the tertiary phase of oil recovery, approximately 70% of the petroleum is trapped in an oil reservoir by the capillary forces; that is, the interfacial tension of the oil-water system is quite large in proportion to the viscous forces so that pumping water into the reservoir will not liberate the oil. Thus, it is necessary to lower the interfacial tension by adding surfactants or raising the temperature of the reservoir [95]. However, if the interfacial tension is lowered too much, an emulsion could form that makes subsequent separation of the hydrocarbon from the water an expensive process. Clearly, an optimal use of surfactant is required and since oil recovery must be conducted under pressure, it is desirable to test various prospective surfactants in the laboratory to observe their effect on interfacial tension. ADSA-P has been extensively used to study the pressure or temperature dependence of interfacial tension for different liquid-liquid systems [95,96].

3.3.7.3 Ultralow Liquid–Liquid Interfacial Tensions

The ADSA-P methodology is particularly suited to measure very low interfacial tensions. For example, ADSA-P measurements on sessile drops of a mixture of

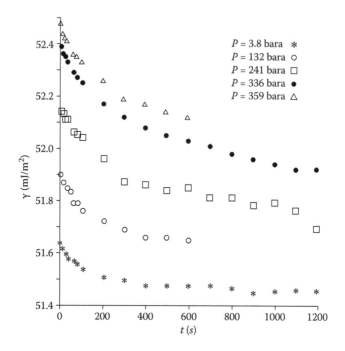

FIGURE 3.35 Measurement of the pressure dependence of the interfacial tension, γ, of a drop of *n*-decane immersed in water at constant temperature (21.5°C). The time dependence of γ may be attributed to system impurities.

$$\gamma = 5.57 \times 10^{-3} \text{ mJ/m}^{-2}$$

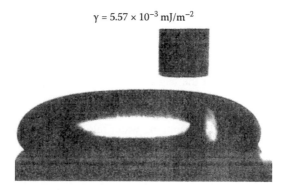

FIGURE 3.36 Sessile drops of various mixtures of dibutyl/dioctyl phthalate (1:1) and cholesterol in an aqueous solution of 0.005 M sodium dodecyl sulfate (SDS). The diameter of the pipet is 0.304 mm; γ is the computed interfacial tension.

dibutyl/dioctyl phthalate (1:1) and cholesterol immersed in an aqueous solution of 0.005 M sodium dodecyl sulfate (SDS) result in interfacial tensions in the range of $10^{-3}–10^{-4}$ mJ/m^2 [97]. A sessile drop of the organic liquid in the aqueous phase formed on an FC-721-coated glass surface is shown in Figure 3.36. The diameter of the pipette is 0.304 mm and provides a scale for the size of the drops. The drop is part of a series of 10 drops evaluated. The average value of the interfacial tension for the

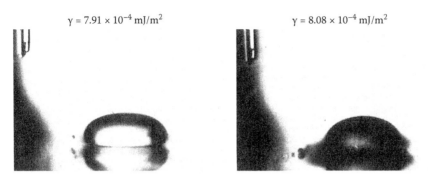

$\gamma = 7.91 \times 10^{-4}\ mJ/m^2$ $\gamma = 8.08 \times 10^{-4}\ mJ/m^2$

FIGURE 3.37 Two sessile drops of the SDS/cholesterol system. In this case, the concentration of cholesterol in the dibutyl/dioctyl phthalate mixture is 0.031 M. The diameter of the pipet is 0.040 mm; γ is the computed interfacial tension.

whole series was $(5.45 \pm 0.17) \times 10^{-3}\ mJ/m^2$, where the error represents the 95% confidence limits. Two drops of even lower interfacial tension are shown in Figure 3.37. This system is similar to Figure 3.36 except that the cholesterol concentration was increased. The pipette tip is 0.04 mm in diameter and the diameter of the droplet was approximately 0.3 mm. In this case, interfacial tensions of $7.91 \times 10^{-4}\ mJ/m^2$ and $8.08 \times 10^{-4}\ mJ/m^2$ were obtained.

Another example of a system of low interfacial tension is a series of phase-separated aqueous polymers [98]. Dextran (MW: 2×10^6) and polyethylene glycol (PEG; MW: 2×10^4) dissolved in 0.9% NaCl were used to make 11 two-phase systems of concentrations between 3 and 13.6 wt% at 22°C. Droplets of the denser dextran-rich phase (volume $c.$ 0.1–5 μl) were formed on a clean glass plate immersed in the lighter PEG-rich bulk phase and photographed. Figure 3.38 shows a graph of the measured interfacial tensions by ADSA-P for the system plotted as a function of concentration. The above investigations illustrate the applicability of ADSA-P to systems with a broad range of interfacial tensions.

3.3.7.4 ADSA-P as a Film Balance

The alteration of the drop volume in a controlled manner combined with the monitoring of the interfacial tension and area changes make ADSA-P suitable for film-balance measurements [99]. The possibility of using surface tension measurements to obtain the surface pressure depends on the well-known relation

$$\pi = \gamma_0 - \gamma, \tag{3.39}$$

where π is the surface pressure, γ_0 is the surface tension of the pure liquid, and γ is the surface tension of the liquid covered with the monolayer. In a film balance, the monolayer film is expanded and compressed by a floating barrier separating the pure liquid from the liquid covered with the monolayer. The corresponding compression and expansion of the film can also be performed by decreasing and increasing the volume of a pendant drop. The experimental scheme is as follows. Initially, a few

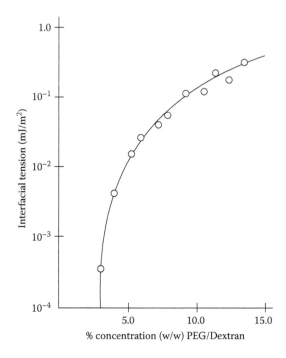

FIGURE 3.38 Functional relationship between polymer concentration and interfacial tension.

pictures of a pendant drop of the pure liquid are taken to determine γ. The desired amount of the insoluble surfactant is weighed and dissolved in a solvent (such as heptane), and a known amount of the surfactant solution is deposited onto the surface of the drop. Upon evaporation of the solvent, the drop carries an insoluble mono-layer. A sequence of images of the drop profile is acquired while the drop volume is decreased continuously until the drop becomes very small. Subsequently, the drop volume is increased to the original value. To ensure reproducibility of the results, the same cycle of compression and expansion is repeated. The measured surface ten-sions and surface areas can be transformed into the corresponding surface pressure as a function of the area per molecule by using Equation 3.39 and the known amount of insoluble surfactant on the drop surface. A typical result for a film of purified octadecanol on water with alteration of the surface area at the rate of 7.2 Å²/molecule -minute is illustrated in Figure 3.39. It is apparent that the two runs are quite simi-lar, illustrating the reproducibility of the results. Moreover, these measurements are in close agreement with film-balance results of the same sample of octadecanol. However, ADSA-P offers several distinct advantages over conventional film-balance methodology for determination of surface pressures. First, only small quantities of liquid and spreadable material are required. Second, both liquid-vapor and liquid-liquid interfacial tensions can be studied. Third, environment control (contamina-tion, temperature, and pressure) is a relatively straightforward matter. The latest development of ADSA-P as a film balance can be found in Chapter 5.

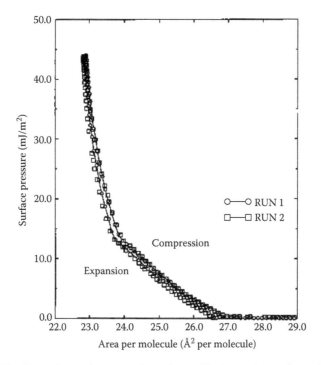

FIGURE 3.39 Expansion and compression of a purified monolayer of octadecanol on a pendant drop of water. Measurement of the liquid surface tension and drop surface area result in surface pressure measurements, which closely resemble film-balance measurements of the same system.

3.3.7.5 Simultaneous Determination of Surface Tension and Density of Polymer Melts

The surface tension of polymer melts is an important thermodynamic parameter that plays a key role in many processes such as wetting, coating, polymer blending, and the reinforcement of polymers with fibers [100,101]. However, the high viscosity and the limited thermal stability of polymer melts as well as the high temperatures cause difficulties in the determination of the interfacial properties of polymer melts.

Considerable efforts have been made to modify the Wilhelmy plate technique and drop shape methods for the surface tension measurement of polymer melts [102–107]. For the Wilhelmy technique, thin fibers are used as solid probes instead of the thin platinum plate or wire. In this way, no correction for buoyancy has to be made and hence the knowledge of the polymer melt density at elevated temperature is not necessary [105]. A drawback of the Wilhelmy plate technique is that the surface tension is not measured directly. Specifically, since the measured quantity equals the so-called wetting tension, $\gamma \cos \theta$, complete wetting of the fiber through the polymer melt (contact angle $\theta = 0°$) is required to obtain the surface tension γ [104,105].

Drop shape methods have the advantage over the Wilhelmy technique that the surface tension is obtained directly from the shape of the pendant or sessile drops so that complete wetting is not crucial. Also, drop shape techniques can be used to study both liquid-vapor and liquid-liquid interfacial tensions of polymer melts. However, since gravity is involved, the density of the polymers at elevated temperature is required. Traditionally, the density of polymer melts has been obtained separately using time-consuming methods of dilatometry [108,109]. Recently, the ADSA-P algorithm has been modified [101] in a fashion that allows simultaneous measurement of the surface tension and density of a polymer melt within a single experiment. In the modified version of ADSA-P, the density is calculated from the mass and volume of the polymer melt drop. The mass of the polymer granulate is measured using a highly accurate microbalance (with an accuracy of ±0.002 mg) [100,101]. Through the minimization of the objective function, ADSA-P calculates b and c (i.e., $\Delta \rho g / \gamma$) as well as the drop surface area and volume. The value of surface tension is obtained based on c and the calculated density.

For the measurement of surface tension and density of polymer melts, a closed high-temperature chamber whose temperature can be precisely controlled is required. Also, a sessile drop configuration is more desirable since weighing is difficult for pendant drops because some of the polymer material is inside the capillary. However, for the accurate determination of the volume from the drop profile, the formation of highly axisymmetric sessile drops is required. This condition can readily be satisfied using a pedestal in the constrained sessile drop configuration. The details of the experimental apparatus and procedure have been explained by Wulf and colleagues [100]. As an illustration, Figure 3.40 presents the density of a polystyrene (PS-38) melt at elevated temperatures [100].

During several experiments with different polymer melts [100,101], the density values obtained from ADSA-P have been compared to the results obtained

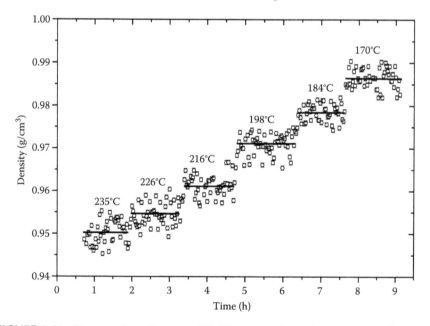

FIGURE 3.40 Density of a polystyrene (PS-38) melt at elevated temperatures. The bold lines are the mean densities at a specific temperature.

independently from the *PVT* method, which defines a temperature dependant density based on the density at the room temperature [110]. The results show that ADSA-P and *PVT* density values are comparable. However, unlike the *PVT* measurement, ADSA-P density measurement does not require a reference density [100,101].

The density measurement using ADSA-P is not limited to polymer melts. ADSA-P can also be used to measure the density of low molecular weight liquids at different temperatures.

3.3.7.6 Tissue Surface Tension

During the development of the embryo, tissues move and change shape due to motion of their constituent cells. Such intercellular migration also takes place during wound healing and tumor metastasis. The integrity of the tissue is meanwhile maintained by adhesion between cells, due to chemical bonds between molecules such as cadherin that protrude from the cell membranes.

Due to cell adhesiveness and motility, when an aggregate (cluster) of cells from a single tissue is removed from the embryo and kept alive in aqueous solution, it will slowly round up into a ball, maximizing mutual cell contact. In this and other respects, cell aggregates behave as if they possess surface tension, with each cell analogous to a liquid molecule [111]. Such tissue surface tensions can in fact be quantified by compressing rounded aggregates of cells between plates and measuring the resisting force [112].

In order to use a drop shape method to measure tissue surface tension, the shapes of cell aggregates must be deformed from their spherical equilibrium by a gravitational or equivalent force. Normal gravity may be insufficient because of the combination of the small sizes of aggregates, and the small density difference between the aggregates and the surrounding medium. The required deforming force can instead be provided by centrifugation [113,114].

Centrifuged aggregates adopt shapes resembling sessile drops (Figure 3.41). With a separate measurement of the density difference between cells and medium, the tissue surface tension can be determined by ADSA-P. Individual cells within

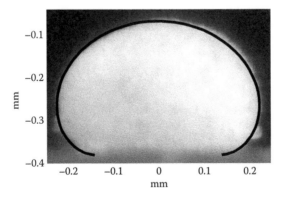

FIGURE 3.41 Example of a raw image of an aggregate of ectodermal cells from *X. laevis* with the Laplacian fit by ADSA superimposed. The measured aggregate/medium surface tension was 3.23 mJ/m^2.

aggregates are often delineated in images, making extraction of the profile edge somewhat tricky. For best results, manual elimination of edge noise may be needed. Unlike those of liquid drops, the profiles of aggregates are also bumpy due to the relatively small number of cells; this adds noise to the ADSA objective function and careful initial value finding is required in order to guarantee convergence [115]. Surface tensions of ectodermal cells from the embryo of the frog *Xenopus laevis* have been measured in this fashion [115], as well as with an earlier iteration of the method called ADSA-IP [114].

REFERENCES

1. L. Wilhelmy. *Annals of Physics* 119 (1863): 177.
2. S. Hartland. *Surface and Interfacial Tension: Measurement, Theory, and Applications.* Marcel Dekker, New York, 2004.
3. A. W. Adamson. *Physical Chemistry of Surfaces.* 5th ed. John Wiley & Sons, Inc., New York, 1990.
4. A. I. Rusanov and V. A. Prokhorov. *Interfacial Tensiometry.* Vol. 3. Edited by D. Möbius and R. Miller. Elsevier, Amsterdam, 1996.
5. J. Drelich, Ch. Fang, and C. L. White. "Measurement of Interfacial Tension in Fluid-Fluid Systems," In *Encyclopedia of Surface and Colloid Science.* Edited by P. Somasundaran and A. Hubbard. Marcel Dekker, Inc., New York, 2002.
6. A. W. Neumann and R. J. Good, "Techniques of Measuring Contact Angles." In *Experimental Methods in Surface and Colloid Science.* Vol. 11. Edited by R. J. Good and R. R. Stromberg. Plenum Press, New York, 1979.
7. D. N. Furlong and S. Hartland. *Journal of Colloid and Interface Science* 71 (1979): 301.
8. T. Tate. *Philosophical Magazine* 27 (1864): 176.
9. E. A. Boucher and M. J. B. Evans. *Proceedings of the Royal Society of London* A346 (1975): 349.
10. E. A. Boucher and H. J. Kent. *Journal of Colloid and Interface Science* 67 (1978): 10.
11. W. D. Harkins and F. E. Brown. *Journal of the American Chemical Society* 41 (1919): 499.
12. S. S. Dukhin, G. Kretzschmar, and R. Miller. *Dynamics of Adsorption at Liquid Interfaces: Theory, Experiment, Application.* Elsevier, Amsterdam, 1995.
13. N. Bohr. *Philosophical Transactions of the Royal Society Series A* 209 (1909): 281.
14. Lord Kelvin (W. Thomson). *Philosophical Magazine* 42 (1871): 368.
15. R. S. Hansen and J. Ahmad. *Progress in Surface and Membrane Science.* Vol. 4. Academic Press, New York, 1971.
16. E. H. Lucassen-Reynders and J. Lucassen. *Advances in Colloid and Interface Science* 2 (1969): 347.
17. J. A. Mann. In *Surface and Colloid Science.* Vol. 13. Edited by E. Matijević. Plenum Press, New York, 1984.
18. B. A. Noskov and N. N. Kochurova. In *Problems of Thermodynamics of Hetergeneous Systems and Theory of Surface Phenomena.* Vol. 7. Edited by A. V. Storonkin and V. T. Zhorov. Leningrad State University, Leningrad, 1985.
19. H. Löfgren, R. D. Newman, L. E. Scriven, and H. T. Davis. *Journal of Colloid and Interface Science* 98 (1984): 175.
20. S. Hård and R. D. Newman. *Journal of Colloid and Interface Science* 115 (1987): 73.
21. M. Sano, M. Kawaguchi, Y.-L. Chen, R. J. Skarlupka, T. Chang, G. Zographi, and H. Yu, *Review of Scientific Instruments* 57 (1986): 1158.

22. J. C. Earnshaw and R. C. McGivern. *Journal of Physics D: Applied Physics* 20 (1987): 82.
23. B. Vonnegut. *Review of Scientific Instruments* 13 (1942): 6.
24. T. Young. *Philosophical Transactions of the Royal Society of London* 95 (1805): 65.
25. P. S. Laplace. *Traité de Mécanique Céleste*. Supplement to Book 10. Gauthier-Villars, Paris, 1805.
26. F. Bashforth and J. C. Adams. *An Attempt to Test the Theory of Capillary Action*. Cambridge, London, 1892.
27. B. E. Blaisdell. *Journal of Mathematics and Physics* 19 (1940): 186.
28. N. R. Tawde and K. G. Parvatikar. *Indian Journal of Physics* 32 (1958): 174.
29. S. Fordham. *Proceedings of the Royal Society of London.* 194A (1948): 1.
30. O. S. Mills. *British Journal of Applied Physics* 4 (1953): 24.
31. J. F. Paddy. "Surface Tension. Part II. The Measurement of Surface Tension." In *Surface and Colloid Science*. Vol. 1. Edited by E. Matijević. Wiley, New York, 1968.
32. S. Hartland and R. W. Hartley. *Axisymmetric Fluid-Liquid Interfaces*. Elsevier, Amsterdam, 1976.
33. J. D. Malcolm and H. M. Paynter. *Journal of Colloid and Interface Science* 82 (1981): 269.
34. C. Maze and G. Burnet. *Surface Science* 13 (1969): 451.
35. C. Maze and G. Burnet. *Surface Science* 24 (1971): 335.
36. C. Huh and R. L. Reed. *Journal of Colloid and Interface Science* 91 (1983): 472.
37. J. F. Boyce, S. Schürch, Y. Rotenberg, and A. W. Neumann. *Colloids and Surfaces.* 9 (1984): 307.
38. S. H. Anastasiadis, J. K. Chen, J. T. Koberstein, A. P. Siegel, J. E. Sohn, and J. A. Emerson. *Journal of Colloid and Interface Science* 119 (1986): 55.
39. R. V. Hoggs. *An Introduction to Robust Estimation*. Academic Press, New York, 1979.
40. P. Cheng. *Automation of Axisymmetric Drop Shape Analysis Using Digital Image Processing*. PhD Thesis, University of Toronto, Toronto, 1990.
41. O. I. del Río. *On the Generalization of Axisymmetric Drop Shape Analysis*. MASc Thesis, University of Toronto, Toronto, 1993.
42. Y. Rotenberg. *The Determination of the Shape of Non-Axisymmetric Drops and the Calculation of Surface Tension, Contact Angle, Surface Area and Volume of Axisymmetric Drops*. PhD Thesis, University of Toronto, Toronto, 1983.
43. Y. Rotenberg, L. Boruvka, and A. W. Neumann. *Journal of Colloid and Interface Science* 93 (1983): 169.
44. P. Cheng, D. Li, L. Boruvka, Y. Rotenberg, and A. W. Neumann. *Colloids and Surfaces.* 43 (1990): 151.
45. P. Cheng and A. W. Neumann. *Colloids and Surfaces.* 62 (1992): 297.
46. O. I. del Río and A. W. Neumann. *Journal of Colloid and Interface Science* 196 (1997): 136.
47. P. Chen, D. Y. Kwok, R. M. Prokop, O. I. del Río, S. S. Susnar, and A. W. Neumann. "Axisymmetric Drop Shape Analysis (ADSA) and its Applications." In *Studies in Interface Science, Vol. 6, Drops and Bubbles in Interfacial Research*. Edited by D. Möbius and R. Miller, 61. Elsevier, Amsterdam, 1998.
48. J. J. Bikerman. *Industrial and Engineering Chemistry, Analytical Edition* 13 (1941): 443.
49. F. K. Skinner, Y. Rotenberg, and A. W. Neumann. *Journal of Colloid and Interface Science* 130 (1989): 25.
50. E. Moy, P. Cheng, Z. Policova, S. Treppo, D. Kwok, D. R. Mack, P. M. Sherman, and A. W. Neumann. *Colloids and Surfaces* 58 (1991): 215.
51. W. H. Press, B. P. Flannery, S. A. Teukolsky, and W. T. Vetterling. *Numerical Recipes the Art of Scientific Computing*. Springer Verlag, New York, 1986.

52. O. O. Couch. *Computer Image Measurement of Axisymmetric Fluid/Liquid Interfaces.* MASc Thesis. University of Toronto, 1992.
53. J. M. Ortega and W. C. Rheinboldt. *Iterative Solutions of Nonlinear Equations in Several Variables.* Academic Press, New York, 1970.
54. P. Rabinowitz, ed. *Numerical Methods for Nonlinear Algebraic Equations.* Gordon and Breach, London, 1970.
55. R. M. Prokop and A. W. Neumann. *Current Opinion on Colloid and Interface Science* 1 (1996): 677.
56. Y. Y. Zuo, M. Ding, A. Bateni, M. Hoorfar, and A. W. Neumann. *Colloids and Surfaces A: Physicochemical Engineering Aspects,* 250 (2004): 233.
57. Y. Y. Zuo, M. Ding, D. Li, and A. W. Neumann. *Biochimica et Biophysica Acta* 1675 (2004): 12.
58. R. M. Prokop, A. Jyoti, P. Cox, H. Frndova, Z. Policova, and A. W. Neumann. *Colloids and Surfaces B: Biointerfaces* 13 (1999): 117.
59. R. M. Prokop, P. Chen, A. Garg, and A. W. Neumann. *Colloids and Surfaces B: Biointerfaces.* Vol. 13, 59. Elsevier, Amsterdam, 1999.
60. J. J. Lu, J. Distefano, K. Philips, P. Chen, and A. W. Neumann. *Respiration Physiology* 115 (1999): 55.
61. J. J. Lu, L. M. Y. Yu, W. W. Y. Cheung, Z. Policova, D. Li, M. L. Hair, and A. W. Neumann. *Colloids and Surfaces B: Biointerfaces* 29 (2003): 119.
62. D. Li and A. W. Neumann. *Journal of Colloid and Interface Science* 148 (1992): 190.
63. J. J. Jasper. *Journal of Physical Chemistry Reference Data* 1, no. 4, 1972.
64. R. O. Duda and P. E. Hart. *Pattern Classification and Scene Analysis.* Wiley, New York, 1973.
65. M. Seul, L. O'Gorman, and M. J. Sammon. *Practical Algorithms for Image Analysis Description, Examples, and Codes.* Cambridge University Press, London, 2000.
66. W. B. Green. *Digital Image Processing.* Van Nostrand Reinhold, New York, 1983.
67. A. Rosenfeld and A. C. Kak. *Digital Image Processing.* Academic Press, New York, 1982.
68. K. C. Castleman. *Digital Image Processing.* Prentice-Hall, Englewood Cliffs, NJ, 1979.
69. W. K. Pratt. *Digital Image Processing.* Wiley-Interscience, New York, 1978.
70. M. P. Ekstrom. *Digital Image Processing Techniques.* Academic Press, Orlando, Florida, 1984.
71. D. H. Ballard and C. M. Brown. *Computer Vision.* Englewood Cliffs, Prentice-Hall, NJ, 1982.
72. L. S. Davis. *Computer Graphics and Image Processing* 4 (1975): 248.
73. E. Hecht. *Optics.* 3rd ed. Addison-Wesley Publishing Company, Reading, MA, 1987
74. W. Smith. *Optical Engineering: The Design of Optical Systems.* McGraw-Hill, New York, 2000.
75. H. Rutten and M. van Venrooij. *Telescope Optics: Evaluation and Design.* Chapter 4. Willmann-Bell, Inc., Richmond, VA, 1988.
76. K. H. Guo, T. Uemura, and W. Yang. *Applied Optics.* 24 (1985): 2655.
77. L. P. Thomas, R. Gratton, B. M. Marino, and J. A. Diez. *Applied Optics.* 34 (1995): 5840.
78. C. K. Chan, N. Y. Liang, and W. C. Liu. *Reviews of Scientific Instruments* 64 (1993): 632.
79. C. F. Gerald. *Applied Numerical Analysis.* 2nd ed. Addison-Wesley, Reading, MA, 1980.
80. M. Hoorfar. *Development of a Third Generation of Axisymmetric Drop Shape Analysis (ADSA).* PhD Thesis, University of Toronto, Toronto, 2006.
81. M. Hoorfar and A. W. Neumann. *Advances in Colloid and Interface Science* 121 (2006): 25–49.

82. M. Hoorfar, M. A. Kurz, and A. W. Neumann. *Colloids and Surfaces A: Physicochemical Engineering Aspects.* 260 (2005): 277–85.
83. L. G. Roberts. "Machine Perception of Three-Dimensional Solids." In *Optical and Electro-Optical Information Processing.* Edited by J. T. Tippett, D. A. Berkowitz, L. C. Clapp, C. J. Koester, and A. Vanderburgh, Jr. MIT Press, Cambridge, MA, 1965.
84. J. M. S. Prewitt. *Picture Processing and Psychopictorics.* Edited by B. S. Lipkin and A. Rosenfeld. Academic Press, New York, 1970.
85. I. E. Abdou and W. K. Pratt. *Proceedings of IEEE.* 57 (1979): 753.
86. J. Canny. *IEEE Transactions on Pattern Analysis and Machine Intelligence* 8, no. 6 (1986): 679.
87. L. M. Y. Yu, J. J. Lu, Y. W. Chan, A. Ng, L. Zhang, M. Hoorfar, Z. Policova, K. Grundke, and A. W. Neumann. *Journal of Applied Physiology* 97 (2004): 704–15.
88. L. M. Y. Yu. *Nonionic Polymers Enhance the Surface Tension of a Bovine Lipid Extract Surfactant.* MASc Thesis, University of Toronto, Toronto, 2004.
89. M. Wulf, S. Michel, K. Grundke, O. I. del Río, D. Y. Kwok, and A. W. Neumann. *Journal of Colloid and Interface Science* 210 (1999): 172.
90. J. J. Lu. *Surface Tension of Pulmonary Surfactant Systems.* PhD Thesis, University of Toronto, Toronto, 2004.
91. S. Y. Lin, L. J. Chen, J. W. Xyu, and W. J. Wang. *Langmuir* 11 (1995): 4159.
92. S. Y. Lin, W. J. Wang, L. W. Lin, and L. J. Chen. *Colloids and Surfaces A: Physicochemical Engineering Aspects* 114 (1996): 31.
93. A. T. Morita, D. J. Carastan, and N. R. Demarquette. *Colloid and Polymer Science* 280 (2002): 857.
94. S. S. Susnar, H. A. Hamza, and A. W. Neumann. *Colloids and Surfaces A: Physicochemical and Engineering Aspects* 89 (1994): 169.
95. S. S. Susnar. *Development of a Facility for Axisymmetric Drop Shape Analysis (ADSA).* PhD Thesis, University of Toronto, Toronto, 1999.
96. S. S. Susnar, C. J. Budziak, H. A. Hamza, and A. W. Neumann. *International Journal of Thermophysics* 13 (1992): 443.
97. Y. Rotenberg, S. Schürch, J. F. Boyce, and A. W. Neumann. In *Proceedings of the 4th International Symposium on Surfactants in Solutions.* Vol. 3. Edited by K. L. Mittal and B. Lindman, 2113. Plenum Press, New York, 1984.
98. J. F. Boyce, S. Schürch, Y. Rotenberg, and A. W. Neumann. *Colloids and Surfaces* 9 (1984): 307.
99. D. Y. Kwok, D. Vollhardt, R. Miller, D. Li, and A. W. Neumann. *Colloids and Surfaces A: Physicochemical and Engineering Aspects* 88 (1994): 51.
100. M. Wulf, S. Michel, W. Jenschke, P. Uhlmann, and K. Grundke. *PCCP, Physical Chemistry Chemical Phys*ics 1 (1999): 3899.
101. M. Wulf, S. Michel, K. Grundke, O. I del Río, D. Y. Kwok, and A. W. Neumann. *Journal of Colloid and Interface Science* 210 (1999): 172.
102. B. Song and J. Springer. *Journal of Colloid and Interface Sci*ence 184 (1996): 64.
103. D. Y. Kwok, L. K. Cheung, C. B. Park, and A. W. Neumann. *Polymer Engineering Sci*ence 38 (1998): 757.
104. B. B. Sauer and N. V. Dipaolo. *Journal of Colloid and Interface Sci*ence 144 (1991): 527.
105. K. Grundke, P. Uhlmann, T. Gietzelt, B. Redlich, and H.-J. Jacobasch. *Colloids and Surfaces A: Physicochemical Engineering Aspects* 116 (1996): 93.
106. C. J. Carriere, A. Cohen, and C. B. Arends. *Journal of Rheology* 33 (1989): 681.
107. G. Gianotta, M. Morra, E. Occhiello, F. Garbassi, L. Nicolais, and A. Damore. *Journal of Colloid and Interface Science* 148 (1992): 571.
108. P. Zoller. *Encyclopedia of Polymer Science and Engineering.* 2nd ed., Vol. 5. Edited by H. Mark, N. Bikales, and C. Overberger. John Wiley & Sons, New York, 1986.

109. P. Zoller. *Polymer Handbook.* 3rd ed. Edited by J. Brandrup and E. H. Immergut. John Wiley & Sons, New York, 1989.

110. J. Pionteck, S. Richter, S. Zschoche, K. Sahre, and K.-F. Arndt. *Acta Polym*erica 49 (1998): 192.

111. M. S. Steinberg. *Current Opinion in Genetics & Development* 17 (2007): 281.

112. R. A. Foty, G. Forgacs, C. M. Pfleger, and M. S. Steinberg. *Physical Review Letters* 72 (1994): 2298.

113. H. M. Phillips and M. S. Steinberg. *Proceedings of the National Academy of Sciences* 64 (1969): 121.

114. A. Kalantarian, H. Ninomiya, S. M. I. Saad, R. David, R. Winklbauer, and A. W. Neumann. *Biophysical Journal* 96 (2009): 1606.

115. R. David, H. Ninomiya, R. Winklbauer, and A. W. Neumann. *Colloids and Surfaces B* 72 (2009): 236.

4 Image Analysis for Axisymmetric Drop Shape Analysis

Yi Zuo and A. Wilhelm Neumann

CONTENTS

4.1   INTRODUCTION

As introduced in Chapter 3, Axisymmetric Drop Shape Analysis (ADSA) is a surface tension and contact angle measurement methodology based on drop shape analysis. ADSA determines the surface tension and contact angle from the experimental profile of drops or bubbles by means of computational parameter optimization. The drop/bubble profile can be obtained from a digital image either by manual digitization or by automatic image analysis. However, manual digitization suffers from several serious limitations: it is very time-consuming, the accuracy of the digitization is low, and much depends on the operator skill. Moreover, if photographic negatives or prints are involved in the manual digitization, there are storing problems as the prints may shrink, warp, or the contrast may fade after a period of time.

In recent years, due to the rapid development of computer vision and pattern recognition, digital image analysis has become a powerful tool for the drop shape methods to facilitate fully automatic measurements of surface tension and contact angle. In contrast to manual digitization, the use of image analysis has a number of advantages: the measurements can be carried out with minimum human intervention and are therefore less dependent on the skill of the experimenter. With automatic image analysis, it is feasible to study the dynamic properties of surfaces (e.g., dynamic surface tension, contact angle hysteresis, and rate-dependent contact angle). In these studies, usually a large amount of time-dependent images needs to be processed, which makes automation a necessity. Moreover, the accuracy of measurements relying on automatic image analysis is expected to be higher than for manual digitization due to the removal of human subjectivity.

This chapter reviews the image analysis techniques commonly used in drop shape methods for determining surface tension and contact angle. The focus is on the step-by-step development of two practical image analysis schemes, which are used in conjunction with the ADSA-P (Profile) and ADSA-D (Diameter) algorithms introduced in Chapter 3, for automatic measurements of surface tension and contact angle, respectively.

4.2   FUNDAMENTALS OF IMAGE ANALYSIS

A variety of image analysis techniques, such as image enhancement and image segmentation (including both edge detection and region detection), have been increasingly applied to a broad range of scientific and industrial applications involving surfaces, such as the study of cell adhesion, biomembranes, and surface energetics [1,2]. Specifically, in the determination of surface tension and contact angle using drop shape analysis, the essential operation of image analysis is to detect the drop/bubble profile; that is, the shape of two-phase interfaces (e.g., air–liquid, liquid–liquid interfaces). The most commonly used image analysis methods for edge detection are thresholding and derivative edge operators [2].

4.2.1   THRESHOLDING

Thresholding is a simple, noncontextual segmentation technique [2]. A thresholded image $g(x, y)$ is defined as

$$g(x,y) = \begin{cases} 1 & \text{if } f(x,y) > T \\ 0 & \text{if } f(x,y) \leq T \end{cases}, \tag{4.1}$$

where $f(x, y)$ is the gray level of point (x, y) in the original grayscale image (containing 256 gray levels for an 8-bit image); T is the global threshold value, which is a user-specified parameter determined prior to the segmentation. Pixels labeled 1 correspond to the objects, whereas pixels labeled 0 correspond to the background. After thresholding, a grayscale image is converted into a binary image composed of only black and white pixels that represent the foreground and background, respectively.

Due to its simplicity, global thresholding has been broadly used in drop shape methods for segmenting the drop/bubble profile [3–6]. The success of thresholding is crucially dependent on the selection of an appropriate threshold value to separate the object from the background. However, this is usually not an easy task. As demonstrated in Figure 4.1a, the gray level of a step edge varies continuously to form a slope instead of a sharp step. Hence, it is difficult to accurately determine the gray level that represents the edge. An intensity histogram is usually established a priori to provide a reference for determining the threshold value. However, only a histogram with a distinctive bimodal shape is capable of providing an appropriate threshold. This is usually the case for a clean and high contrast image with approximately equal areas of foreground and background. Most experimental drop/bubble images have fuzzy edges for which the thresholding method may not function well [7]. It has been found that the accuracy of surface tension measurement using thresholding as the edge detection method is generally one order of magnitude less than that using derivative methods [7,8]. Such a reduced accuracy is not acceptable for precise surface tension measurement and it may even fail for more sophisticated measurements involving the determination of drop/bubble geometry [9]. The effect of thresholding on the accuracy of measuring surface tension and drop/bubble geometry has been discussed in detail elsewhere [7,10].

4.2.2 DERIVATIVE EDGE OPERATORS

The main characteristic of an edge in a digital image is the discontinuity in intensity, usually corresponding to a sharp change of some physical properties, such as reflectivity and density, across an interface. Hence, the most popular approach to develop an edge operator is based on derivative algorithms [2,11]. Both the first- and second-order derivatives have been used. Figure 4.1 illustrates how these two derivatives are related to an edge. The first-order derivatives are usually determined by taking the intensity gradient. A local minimum or maximum of the gradient indicates an edge. Some pioneer gradient edge operators are the Roberts operator, the Prewitt operator, and the Sobel operator. The second-order derivatives are usually implemented by taking the Laplacian, in which an edge is located by finding the zero-crossings of the Laplacian. Laplacian of Gaussian (LoG) is the most popular operator in this category.

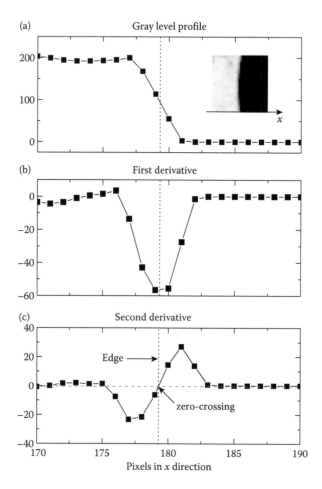

FIGURE 4.1 Illustration of the derivative methods for detecting a real step edge. (a) The gray level profile of the step edge, image of the step edge is shown in the insert; (b) the first derivative of the edge profile, the edge is indicated by the local minimum; and (c) the second derivative of the edge profile, the edge is detected by finding the zero-crossing.

The Sobel edge detector has been used in ADSA to analyze images of pendant/sessile drops (see also Chapter 3) [12]. The Sobel edge detector is one of the earliest gradient operators with small convolution masks: the gradient of intensity at each pixel is evaluated using its neighbors within a square region of 3×3 pixels. The convolution masks of the Sobel operator are shown in Equation 4.2.

$$S_x = \begin{bmatrix} -1 & -2 & -1 \\ 0 & 0 & 0 \\ 1 & 2 & 1 \end{bmatrix}, S_y = \begin{bmatrix} -1 & 0 & 1 \\ -2 & 0 & 2 \\ -1 & 0 & 1 \end{bmatrix}. \tag{4.2}$$

4.2.3   ADVANCED EDGE DETECTORS ROBUST AGAINST NOISE

Similar to thresholding, the efficiency of derivative edge detectors can be significantly reduced by noise. Noise in image analysis is an unexplained variation in intensity values [2]. It can be introduced into an image during the acquisition process from various sources. Some sources of noise are the uncertainty due to electronic devices (e.g., uncertainties of the sensors, fluctuation in the light intensity, salt-and-pepper noise in signal transmission), blur due to drop evaporation and condensation, and ambiguity due to poor focus. In certain studies such as a liquid–liquid system or a captive/pendant bubble surrounded by a turbid liquid, extensive noise can be introduced due to the existence of impurities and other insoluble substances. The study of lung surfactants (an aqueous suspension of insoluble phospholipid vesicles/aggregates, detailed in Chapter 5) using the captive bubble method is one such example.

Figures 4.2a through c show images of a pendant drop, a sessile drop, and a captive bubble in a lung surfactant suspension. It is noted that the images of the pendant/sessile drops exhibit a distinctive edge; that is, a dark object (the drop) against a bright background (the air). Thanks to the sharp edge of the pendant/sessile drop, the Sobel edge detector is usually adequate to extract an undisturbed edge (see Figures 4.2d and e for the Sobel detected edges from the drop images). In contrast, the captive bubble image shows extensive noise that prevents the Sobel detector from extracting a smooth edge (Figure 4.2f).

Most of the traditional edge detectors, such as Sobel and LoG, are not robust against noise due to the relatively small size of the convolution masks. Promoted

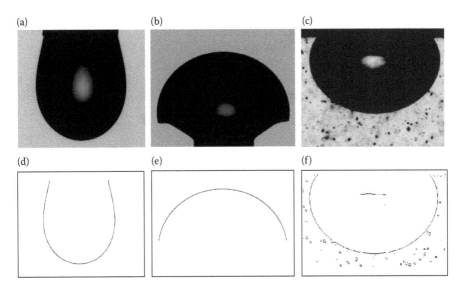

FIGURE 4.2   Typical images of three commonly used drop/bubble configurations for surface tension measurement: (a) pendant drop, (b) sessile drop, and (c) captive bubble. The Sobel detected edges of these three configurations are shown in (d) through (f), respectively.

by the rapid development of image analysis, recent edge detectors are increasingly strong in noise reduction. These algorithms are based on optimal filters [13,14], fuzzy techniques [15], neural networks [16], discrete singular convolution algorithms [17], and entropic methods [18,19]. All these advanced edge operators claim some level of noise-resistance and have the potential to be used in determining surface properties from noisy images. In the rest of the chapter, advanced edge detectors will be used with ADSA for automatic measurements of surface tension and contact angle.

4.3 IMAGE ANALYSIS FOR SURFACE TENSION MEASUREMENT USING ADSA-P

As illustrated in Figure 4.2, despite the success in processing images of pendant/ sessile drops, the Sobel edge detector is incapable of analyzing the captive bubble image because it is susceptible to noise [7]. In this section, a sophisticated image analysis scheme robust against noise will be developed for processing images of captive bubbles in turbid lung surfactant suspensions. This new image analysis scheme consists of four main parts:

(i) Edge detection: extraction of drop/bubble profile from the raw image
(ii) Edge smoothing: removal of outliers/noise from the extracted edge
(iii) Edge restoration: correction of optical distortion
(iv) Edge selection: selection of edge points for surface tension calculation using ADSA-P

Reliability of the scheme will be illustrated by analyzing images with different optical conditions, including images that are highly noisy and/or lacking in contrast. Accuracy of the scheme will be validated by measuring the surface tension of pure water. Automation of the scheme will be demonstrated by analyzing a sequence of images with only one set of user-specified parameters. Finally, a recent development in noise reduction using region detection will be introduced.

4.3.1 DEVELOPMENT OF THE IMAGE ANALYSIS SCHEME

4.3.1.1 Edge Detection

As introduced above, many advanced edge detectors have been developed. Among these methods, the Canny edge detector [13] was selected to be incorporated with ADSA due to its superior performance in eliminating fine noise [7]. The Canny is a sophisticatedly defined gradient operator and has been used as a standard method for edge detection. The Canny strategy satisfies three optimal criteria for performance evaluation: (i) good detection: a detected edge should have a high probability of matching the actual edge and a low probability of reporting a false edge; (ii) good localization: the distance between the detected edge and the actual edge should be as small as possible; and (iii) single response to one edge: a single edge should not cause a multiple-edge response. With these three criteria, Canny introduced an optimal filter, which can be efficiently approximated by the first derivative of a Gaussian function defined as

$$G(x) = \frac{1}{\sqrt{2\pi}\sigma_G} e^{-\frac{x^2}{2\sigma_G^2}}, \, G(y) = \frac{1}{\sqrt{2\pi}\sigma_G} e^{-\frac{y^2}{2\sigma_G^2}}, \tag{4.3}$$

where σ_G is the standard deviation of the Gaussian. Convolving an image with this filter smoothes ("blurs") the image, with the degree of blurring being determined by the value of σ_G. The smoothing step is followed by an optimal localization strategy which consists of a combination of nonmaximum suppression and hysteresis thresholding.

In the nonmaximum suppression, a pixel with a gradient magnitude that is not a local maximum is removed and the edge is thinned down to only one pixel width. If the gradient magnitude at a pixel is larger than that at its two neighbors in the gradient direction, the pixel is marked as an edge. Otherwise, the pixel is marked as background. In the hysteresis thresholding, two thresholds are applied to ensure an accurate and continuous edge. If the gradient magnitude at a pixel is above the high threshold (T_h), the pixel is marked as a definite edge. On the other hand, any pixels having gradient magnitudes less than the low threshold (T_l) are marked not to be edge pixels. Any pixels adjacent to the edge pixels and having gradient magnitudes greater than T_l are also selected to constitute the edge. The procedures of the Canny edge detector on analyzing a noisy captive bubble image are illustrated in Figure 4.3.

A straightforward procedure has been developed to determine the user-specified parameters for the Canny. These parameters are σ_G used in the Gaussian filter and T_h and T_l used in the hysteresis thresholding. The σ_G controls the amount of smoothing and T_h and T_l determine the continuity of the detected edge. Figure 4.4 shows the recommended procedures to set up the optimal input parameters: (i) an initial value of σ_G is assigned as 1.0 for a clean image and 2.0 for a noisy image; (ii) T_h is selected as 90%. That is, the gradient magnitude represented by T_h should be greater than that of 90% of the total pixels. The T_l is set as half of T_h; (iii) the Canny edge detection is performed; (iv) the quality of the detected edge is evaluated. If significant noise still remains in the extracted edge, σ_G is increased by 0.2; otherwise, σ_G is decreased by 0.2. Then, step three is repeated until a smooth edge is obtained using the smallest σ_G; and (v) after step three, if significant discontinuity exists in the extracted edge, T_h is decreased by 10%. Then, step three is repeated until there is no significant break in the extracted edge. The resultant σ_G, T_h, and T_l will be the optimal input parameters.

It should be noted that the Canny edge detector is not sensitive to the selection of the user-specified parameters. As will be shown later, the Canny would output accurate and consistent results with a wide range of input parameters. In most cases, the three input parameters can be further reduced down to only one (i.e., σ_G). It has been found that T_h equal to the value of 90% and T_l equal to one half of T_h (i.e., 45%) are adequate for most images.

4.3.1.2 Edge Smoothing

After edge detection, noise due to insufficient suppression in the Canny may still exist. As illustrated in Figure 4.5, there are usually two types of noise in a captive

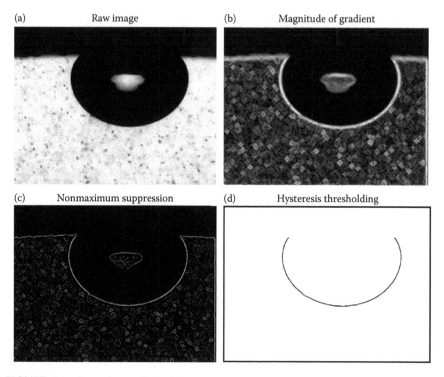

FIGURE 4.3 Procedures of the Canny edge detection: (a) the raw image; (b) the image of the magnitude of the gradient after smoothing using the first derivative of the Gaussian; (c) the image with thinned edge after nonmaximum suppression; and (d) the binary image after hysteresis thresholding.

bubble image depending on the relative distance to the main bubble profile: isolated noise (i.e., the noise far away from the main bubble profile) and adhering noise (i.e., the noise close to the main bubble profile). For best performance, separate edge smoothing techniques are necessary to remove these two types of noise. In most cases, the isolated noise can be simply removed by measuring edge cohesion: the binary image after edge detection is raster scanned from left to right and top to bottom. Any assumed edge pixel away from the main profile by 50 pixels is eliminated as isolated noise. Although applicable to most cases, the cohesion method may fail due to the difficulty of identifying the main profile a priori. If a noise pixel were erroneously picked as part of the main profile in the first place, all the subsequent cohesion measurement and noise removal would be wrong. To solve this problem, a region detection method for removing the isolated noise has been developed. Details of this new development are presented later in this chapter.

Adhering noise is more difficult to remove because of its adjacency to the main drop/bubble profile. A novel technique, called Axisymmetric Liquid Fluid Interfaces–Smoothing (ALFI-S), has been developed for removing adhering noise from a drop/bubble profile. Detailed in Chapter 3, ALFI is a numerical method of generating theoretical drop/bubble profiles governed by the Laplace equation of capillarity. In ALFI-S, the Canny-detected experimental bubble profile is fitted to the best matched

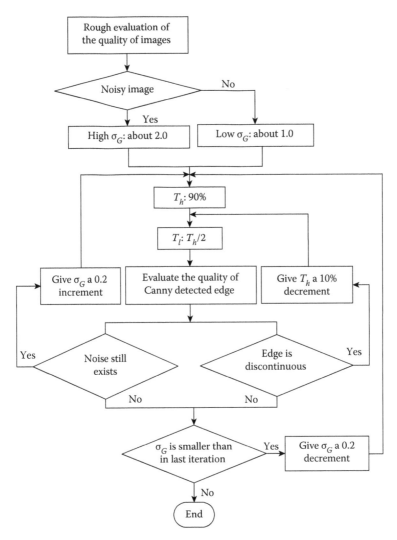

FIGURE 4.4   Flow chart to set up optimal parameters for the Canny edge detector.

theoretical profiles generated by ALFI. Then, a standard deviation (σ) is evaluated from the normal distance between the experimental profile and the theoretical profile. A 3σ criterion is used to eliminate outliers; that is, any assumed edge point that deviates from the best matched theoretical profile by more than three times σ is eliminated as noise. ALFI-S is run iteratively until no more noise is found.

4.3.1.3   Edge Restoration

All images inevitably suffer, more or less, from optical distortion due to the image acquisition hardware (microscope, camera, and digital video processor). This distortion can cause major error in the surface tension measurement and hence needs to be corrected [12]. To do so, an image of a calibration grid pattern (square pattern with 0.25 mm

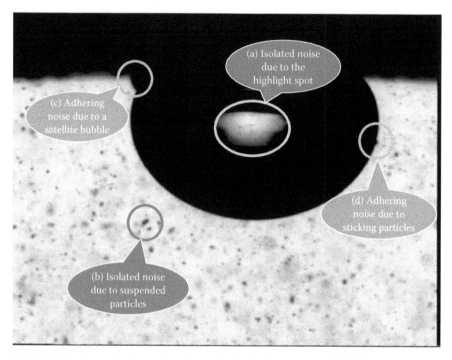

FIGURE 4.5 A sample image showing typical noise in a captive bubble image. There are four types of noise: (a) isolated noise due to the central light reflection zone; (b) isolated noise due to dark particles (surfactant aggregates in this case); (c) adhering noise due to satellite bubbles (minor bubbles formed during experiment); and (d) adhering noise due to dark particles.

spacing, Graticules Ltd., Tonbridge Kent, UK) on an optical glass slide is taken at the same position where the drop/bubble images are acquired. A mapping function based on the comparison between the distorted grid image and the original grid pattern (without optical distortion) is built. Subsequently, this mapping function is applied to all the drop/bubble images. The accuracy of this correction is ±1 pixel [12]. Details about the distortion correction algorithm can be found in Chapter 3 and elsewhere [12].

4.3.1.4   Edge Selection

There are two commonly used ways to select edge points for ADSA calculation. One apparent way is to use all the edge points on a bubble profile. An alternative is to use only a small fraction of randomly selected points along the profile and to repeat the selection and calculation several times (i.e., conduct multiple calculations on the same drop/bubble profile). The latter has been recommended in the past as it was thought to help average out the random errors associated with the edge points [12,20,21]. Moreover, by conducting multiple calculations on the same profile, this strategy is able to yield the 95% confidence intervals for all the ADSA results. This strategy also consumes less computer time due to the reduced total number of edge points to be processed. Cheng and Neumann [20] found that good ADSA results can be obtained by only calculating 20 randomly selected points 10 times (i.e., in total using 200 points).

Increasing the randomly selected points to 50 would significantly increase the accuracy [22]. Therefore, here we will use the calculation scheme of randomly selecting 50 points 10 times and compare it with the calculation using the whole profile.

4.3.2 EVALUATION OF THE IMAGE ANALYSIS SCHEME

4.3.2.1 Dependence of the User-Specified Parameters

Despite the fact that the performance of an edge detector is inevitably dependent on the relevant user-specified parameters, these parameters should not significantly affect the final results. That is, a desired image analysis scheme should function properly with the least user-interference to avoid subjectivity. We have examined the dependence of ADSA results on the primary user-specified parameter of the Canny edge detector (i.e., the σ_G of the Gaussian filter). The noisy captive bubble image shown in Figure 4.2c was analyzed using a wide range of σ_G. The results are shown in Table 4.1. One additional significant digit is deliberately kept in the mean values to demonstrate the repeatability. It is clear that over a large range of σ_G, from 1.0 to 4.0, the ADSA results (i.e., surface tension, bubble area, volume, and curvature at the bubble apex) are very consistent and

TABLE 4.1

ADSA Results as a Function of the Standard Deviation (σ_G) of the Gaussian Filter Used in the Canny Edge Detector

σ_G	Surface Tension (mJ/m²)	Area (cm²)	Volume (cm³)	Curvature at Apex (cm⁻¹)
1.0	23.65	0.3612	0.01906	4.271
1.2	23.68	0.3613	0.01906	4.272
1.4	23.74	0.3616	0.01908	4.274
1.6	23.74	0.3616	0.01908	4.274
1.8	23.74	0.3616	0.01908	4.274
2.0	23.72	0.3615	0.01908	4.273
2.2	23.70	0.3615	0.01907	4.272
2.4	23.70	0.3615	0.01907	4.272
2.6	23.70	0.3615	0.01907	4.272
2.8	23.64	0.3614	0.01906	4.270
3.0	23.63	0.3615	0.01906	4.269
3.2	23.64	0.3615	0.01906	4.269
3.4	23.63	0.3615	0.01906	4.269
3.6	23.64	0.3614	0.01906	4.270
3.8	23.64	0.3614	0.01906	4.270
4.0	23.64	0.3614	0.01906	4.270
Mean	$23.677 \pm 6.69 \times 10^{-4}$	$0.36146 \pm 1.71 \times 10^{-6}$	$0.019067 \pm 1.37 \times 10^{-7}$	$4.2713 \pm 2.85 \times 10^{-5}$

Note: The calculations are based on the captive bubble image shown in Figure 4.2c using the whole profile without correction of optical distortion; the mean values are shown with 95% confidence intervals.

there is no apparent trend or dependence of the results on the σ_G. The errors associated with the mean values are small. The results suggest that the Canny edge detector is not sensitive to the selection of the user-specified parameter.

4.3.2.2 Analysis of Sample Images

As shown in Figure 4.6, six captive bubble images are selected to represent a wide range of noise and contrast. A detailed description of the captive bubble method can be found in Chapter 5. Each captive bubble rests against the ceiling of a chamber and is surrounded by a liquid. These liquids are: (a) distilled water; (b) 0.5 mg/mL bovine lipid extract surfactant (BLES); (c) 0.5 mg/mL BLES + 30 mg/mL polyethylene glycol (PEG); (d) 0.5 mg/mL endogenous bovine surfactant; (e) 1.0 mg/mL BLES + 50 mg/mL PEG; (f) 0.8 mg/mL BLES + 27 mg/mL PEG. Here, (a) and (b) represent

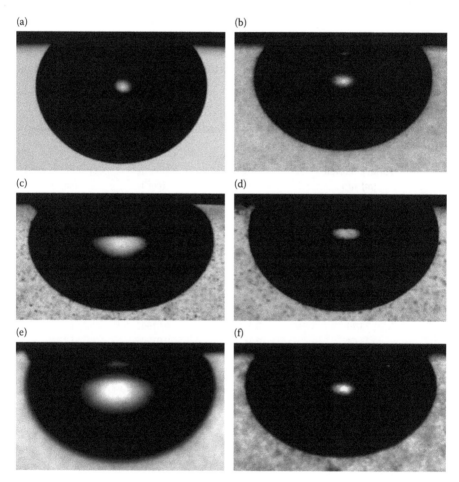

FIGURE 4.6 Six sample images of captive bubbles in different liquids. They are: (a) distilled water; (b) 0.5 mg/mL BLES; (c) 0.5 mg/mL BLES + 30 mg/ml PEG; (d) 0.5 mg/mL endogenous bovine surfactant; (e) 1 mg/mL BLES + 50 mg/ml PEG; (f) 0.8 mg/mL BLES + 27 mg/ml PEG.

clean images free of noise; (c) and (d) represent images with extensive noise; and (e) and (f) are examples of fuzzy images.

The extracted edges right after performing the Canny edge detection are shown in Figure 4.7. For each image, an optimal set of input parameters was developed based on the procedures shown in Figure 4.4. The σ_G used in the image analysis is summarized in Table 4.2. It is noted that even though some noise points still exist, the Canny edge detector successfully extracts all edges. For the clean images shown in Figure 4.6a and b, the extracted edges are very smooth. For the noisy and low contrast images, satisfactory edges are also obtained. The incomplete edge shown in Figure 4.7d is due to the nonuniform intensity distribution in the raw image. However, in spite of the deficiency on the left side, the right side of the edge is completely preserved. Since the bubble profile is assumed to be axisymmetric, one side of the bubble profile is adequate for the ADSA calculation. Figure 4.8 shows the smoothed edges after removing isolated noise using the cohesion method and removing adhering noise using ALFI-S. It is noted that ALFI-S is able to remove the fine noise in the

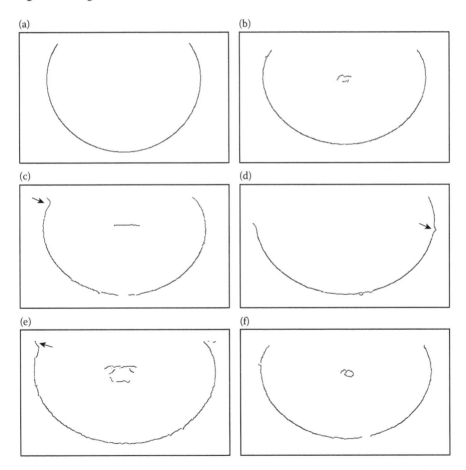

FIGURE 4.7 Extracted edges from the images shown in Figure 4.6 right after the Canny edge detection. The arrows point at fine adhering noise.

TABLE 4.2

Surface Tension (mJ/m²) Measured for the Images Shown in Figure 4.6

Images	$\sigma_G{}^a$	WP − DC[b]	WP + DC[c]	50 × 10 − DC[d]	50 × 10 + DC[e]
a	0.4	72.78	72.89	72.26 ± 1.11	73.17 ± 0.91
b	1.4	28.63	29.23	28.58 ± 0.42	29.62 ± 0.32
c	3.6	25.73	26.55	25.55 ± 0.18	26.34 ± 0.28
d	3.0	28.17	27.97	27.86 ± 0.24	27.88 ± 0.31
e	1.4	23.65	23.56	23.56 ± 0.21	23.75 ± 0.24
f	3.2	24.53	25.03	24.36 ± 0.34	24.91 ± 0.29

[a] The optimal standard deviation of the Gaussian filter used in the Canny edge detector, which is determined based on the procedure shown in Figure 4.4.
[b] Calculation using the whole profile (WP) without optical distortion correction (DC).
[c] Calculation using the whole profile with optical distortion correction.
[d] Calculation using randomly selected 50 points 10 times without optical distortion correction.
[e] Calculation using randomly selected 50 points 10 times with optical distortion correction.

edges, for example, the small bumps indicated by the arrows in Figure 4.7c, d, and e. After ALFI-S, all the edges are quite smooth.

The surface tensions measured from the images shown in Figure 4.6 are summarized in Table 4.2. The calculation for each edge is conducted in four different ways: calculations using the whole profile, with/without optical distortion correction (WP ± DC); and calculations using 50 randomly selected edge points 10 times with/without optical distortion correction (50 × 10 ± DC). The surface tensions calculated using randomly selected points are shown with 95% confidence intervals.

4.3.2.3 Experimental Validation

Previous studies have found that the accuracy-limiting step in ADSA is not the mathematical scheme but the image analysis [20]. Therefore, increasing the quality of the detected edge is expected to significantly enhance the accuracy of the surface tension measurement. To validate the accuracy, ADSA with the new image analysis scheme was used to measure the surface tension of distilled water at 20°C. Twenty images of a static captive bubble were taken within 10 seconds at a rate of two images per second. The first of these 20 images is shown in Figure 4.6a. Table 4.3 lists the surface tensions for these 20 images, calculated using the four different ways as described above (i.e., WP ± DC and 50 × 10 ± DC).

Mean values of the surface tensions calculated by the four different ways are very close. However, it is noted that the 95% confidence intervals associated with the mean values using randomly selected points are about three to four times greater than those obtained using the whole profile. The use of the whole profile produces the highest accuracy. It was not recommended in previous work mainly because it requires more computation time; however, this issue has become relatively trivial in view of the increasing availability of high power computers. Therefore, a calculation using the whole profile is recommended for accurate surface tension measurement.

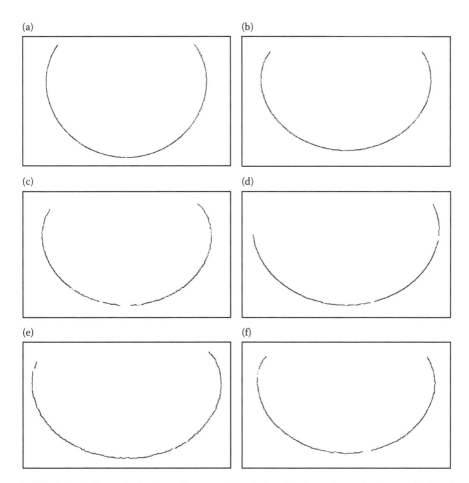

FIGURE 4.8 Smoothed edges after removing isolated noise using cohesion method and adhering noise using ALFI-S.

Another inference from Table 4.3 is that the effect of optical distortion on surface tension seems to be much less in this case than previously reported [12]. The previous study was based on the pendant drop configuration [12]. Possibly the geometrical arrangement of a captive bubble is not as sensitive to optical distortion as a pendant drop. Literature values for the surface tension of water at 20°C are given as 72.75 mJ/m^2 in [23] and 72.88 mJ/m^2 in [24]. As it stands, it is not possible to conclude if the distortion correction improves the results when considering the whole profile, as one of the two mean values agrees closely with one literature value and the other one with the second literature value. However, when randomly selected points (50 × 10) are used for calculation, a notable improvement in the surface tension (~0.1 mJ/m^2) is observed with the correction of optical distortion, as the mean value obtained without distortion correction is clearly too high. Be that as it may, it appears that ADSA with the new image analysis scheme can produce very accurate surface tension results.

TABLE 4.3

Surface Tension (mJ/m²) Measured for Distilled Water at 20°C

Bubble No.	WP – DC[a]	WP + DC[b]	50 × 10 – DC[c]	50 × 10 + DC[d]
1	72.78	72.89	72.26 ± 1.11	73.17 ± 0.91
2	72.80	72.91	73.75 ± 1.60	73.28 ± 0.81
3	72.75	72.86	72.05 ± 1.29	72.19 ± 1.42
4	72.92	73.04	73.38 ± 1.13	73.74 ± 1.42
5	72.85	72.97	73.41 ± 1.19	72.69 ± 1.11
6	72.91	73.02	73.67 ± 0.65	73.28 ± 0.77
7	72.53	72.63	71.20 ± 0.71	72.75 ± 1.67
8	72.53	72.64	72.09 ± 1.16	73.22 ± 1.28
9	72.76	72.85	72.37 ± 0.76	73.54 ± 1.49
10	72.87	72.99	73.14 ± 1.53	73.19 ± 0.76
11	72.36	72.46	72.68 ± 1.54	72.59 ± 0.64
12	72.52	72.63	72.01 ± 1.08	72.79 ± 1.05
13	72.69	72.79	73.42 ± 1.30	72.58 ± 0.99
14	72.64	72.75	74.42 ± 1.32	72.95 ± 1.72
15	72.85	72.99	73.38 ± 1.44	74.02 ± 0.97
16	72.58	72.69	73.95 ± 1.19	72.16 ± 1.58
17	73.07	73.13	73.06 ± 1.20	72.09 ± 1.01
18	72.87	72.97	73.82 ± 1.38	72.32 ± 1.15
19	72.88	72.99	73.55 ± 1.34	72.45 ± 0.92
20	72.94	72.94	73.05 ± 1.32	72.94 ± 0.85
Mean	72.755 ± 0.003	72.857 ± 0.002	72.984 ± 0.010	72.897 ± 0.008

[a] Calculation using the whole profile (WP) without optical distortion correction (DC).
[b] Calculation using the whole profile with optical distortion correction.
[c] Calculation using randomly selected 50 points 10 times without optical distortion correction.
[d] Calculation using randomly selected 50 points 10 times with optical distortion correction.

4.3.2.4 Automatic Analysis of Multiple Images

Automation of the above image analysis scheme is illustrated by analyzing a sequence of images corresponding to the aging of a lung surfactant film, using only one set of input parameters without further human intervention. (Details of the lung surfactant experiments can be found in Chapter 5.) To demonstrate the importance of the new image analysis scheme, the experimental results are compared to those calculated using ADSA in conjunction with thresholding [8].

Figure 4.9a and b shows the dynamic surface tensions measured from the thresholding version and the Canny version of ADSA, respectively. As seen from Figure 4.9a, the thresholding version results in considerable scatter, especially between 40 and 300 seconds. This scatter is due to the sudden appearance of a satellite bubble near the ceiling at 40.48 seconds (see the inserts in Figure 4.9b for the satellite bubble). The rapid formation of small satellite bubbles is a troublesome problem in captive bubble experiments and it usually occurs during bubble injection and dynamic

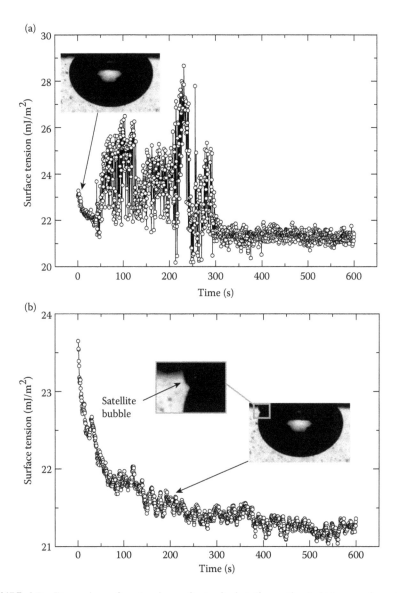

FIGURE 4.9 Dynamic surface tensions of a typical static captive bubble experiment (an oxygen bubble in a lung surfactant suspension) analyzed using (a) the thresholding version of ADSA; and (b) the Canny version of ADSA.

cycling [25]. From the point of view of image analysis, the satellite bubble constitutes major adhering noise. More importantly, due to its critical location (i.e., near the three-phase contact line), the satellite bubble has the most deleterious effect on the surface tension measurement [10].

Figure 4.9b shows the dynamic surface tensions measured from the Canny version of ADSA. The results obtained with this new image analysis scheme show a fairly smooth decrease of surface tension due to surfactant adsorption, as expected.

The superior performance of the new image analysis scheme is due to both the Canny edge detector and ALFI-S. First, the Canny edge detector is robust against noise. Second, ALFI-S is very powerful in removing the adhering noise and highly automatic owing to the iteration algorithm and the adaptive noise detection and removal (i.e., the 3σ criterion). Figure 4.10 shows the iteration information of ALFI-S associated with the analysis shown in Figure 4.9b. It is noted that the number of iterations and rejected points change automatically, depending on the quality of the images. For a less noisy image as shown in the insert in Figure 4.9a, the number of iterations is 5 and the number of rejected points is 24. In contrast, for an image with a satellite bubble as shown in the insert in Figure 4.9b, seven iterations are executed and 44 noise points are deleted automatically. After the automatic noise reduction, the number of remaining points on the smoothed edges is relatively constant.

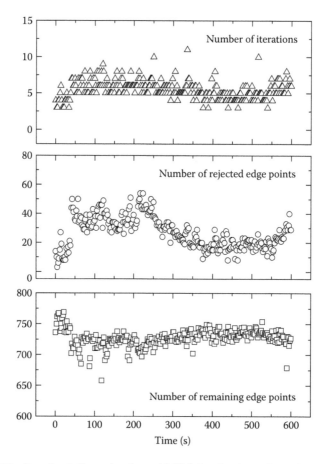

FIGURE 4.10 Iteration information from ALFI-S for the dynamic surface tension measurements shown in Figure 4.9b: number of iterations, number of rejected edge points, and number of remaining edge points.

4.3.3   FURTHER DEVELOPMENT IN NOISE REDUCTION

As alluded to above, the cohesion method to remove isolated noise may encounter difficulties in differentiating the bubble profile a priori through computer vision. In case isolated noise is erroneously picked as part of the bubble profile, the real edge will be in danger of being removed subsequently. This may cause a major error in ADSA calculations.

This difficulty has provoked new thinking on removing isolated noise in the pre-edge detection stage (i.e., the raw image). As shown in Figure 4.5, a typical captive bubble image features a bubble located in the center of the image, resting against the ceiling and surrounded by an aqueous suspension. The bubble accounts for a considerable portion of the image and shows remarkable uniformity of intensity lower than that of the background. The dominant area and the prominent contrast of the bubble against the background allow for easy localization of the bubble in the raw image and thus permit an alternative way of detecting and removing the isolated noise before edge detection. A component labeling based region detection technique has been developed for this purpose [26].

Component labeling refers to the process of detecting connected objects in a digital image [27]. A connected component in a digital image refers to a set of pixels in which each pixel is connected to all others [27]. In a 2-D image, the connectivity can be defined by 4-way or 8-way adjacency. The former only considers the nondiagonal neighbors, while the latter considers all eight possible neighbors of a pixel [27]. Finding connected components in a binary image is one of the most fundamental operations in computer vision and pattern recognition. Its applications cover a broad range of scientific and industrial fields, such as medical image processing, remote sensing, volume visualization, and character recognition [28]. After component labeling, a binary image is converted into a symbolic image in which each connected component is assigned a unique label [28].

In some sense, component labeling is a region-based binary image segmentation technique [29]. During component labeling, each image pixel is examined in the context of its neighbors and a region is grown by addition of new pixels if they are connected. Compared with contour-based segmentation methods (in which a region is identified by first determining its boundary pixels), the region-based methods are relatively insensitive to shape degradation and noise since these methods rely on the entire set of the interior region pixels [29]. Hence, it is possible and desirable to use component labeling for the purpose of noise (especially isolated noise) reduction.

The component labeling based noise reduction consists of three steps. First, the original grayscale image is converted into a binary image (black-and-white image) using Otsu's thresholding. In the second step, a modified two-pass sequential labeling process is performed on the binary image. In the first pass, the image is raster scanned and connected pixels are temporarily labeled to be one component. In the second pass, regions connected to each other but with different labels are merged into one component and all components are re-labeled correspondingly. After this step, information on the number of components, area (i.e., number of pixels), location (i.e., coordinates of pixels) and color (i.e., black or white, indicating foreground or background pixels) of each component is recorded. Also, the detected components are ranked in descending order based on their areas in the second pass of the component labeling. The components with the first and the second largest areas usually represent the background and the primary foreground object. The other components with much smaller areas represent the noise. After localizing the main object and the noise (i.e., knowing the coordinates of these components), in the third step, the noise can be safely removed. To do so, in the original grayscale image the intensities of the pixels in the noise components are replaced by the average intensity of their background neighbors, to provide a smoother transition from the regions of the original isolated noise to the background. This action is similar to the application of low-pass filters to the background of the image. The flowchart of the entire image analysis scheme used in ADSA, including the component labeling based noise reduction module, is shown in Figure 4.11.

It is noteworthy that although thresholding is used as an intermediate step in component labeling, it does not decrease the accuracy of the subsequent surface tension

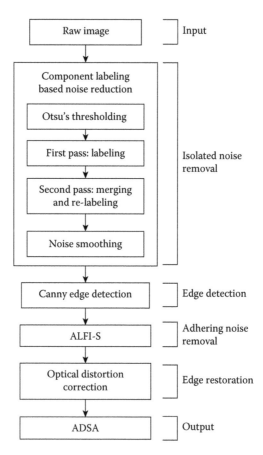

FIGURE 4.11 Flowchart of the innovative image analysis scheme used in ADSA-P. The new image analysis scheme consists of pre-edge detection isolated noise reduction using component labeling, Canny edge detector, adhering noise reduction using ALFI-S, and correction of optical distortion.

measurement. This is due to the fact that thresholding is used here only to facilitate the subsequent component labeling procedure rather than to segment the drop/bubble profile. The labeling procedure only collects information on the localization of the main object and the isolated noise in an image. Therefore, the grayscale image after component labeling still keeps the original drop/bubble contour. The detection of the drop/bubble profile still relies on the subsequent Canny edge detector.

The implementation of the component labeling method for smoothing a noisy captive bubble image is illustrated in Figure 4.12. First, the original grayscale image (Figure 4.12a, resolution 480×640) is converted into a black-and-white image (Figure 4.12b) using Otsu's thresholding. Next, the binary image is scanned for component labeling. Figure 4.12c shows the grayscale image after component labeling. In total, 89 components have been detected. The background accounts for the biggest component that consists of 177,324 pixels. The complex of the bubble and the ceiling accounts for the second biggest component, containing 126,658 pixels. The next is the reflection zone in the bubble center, containing 2118 pixels. The biggest isolated noise component contains 68 pixels and the smallest consist of only 1 pixel. Although too small to be caught by the eye, the component labeling method has successfully detected this 1-pixel noise, which proves that the method is highly sensitive and reliable.

After all the components have been labeled, the noise components are smoothed by replacing the original intensities of the noise pixels with the average of their neighbors'. In this way, the isolated noise "islands" in the background "ocean" are filled. The resultant image (Figure 4.12d) shows much smoother intensity transitions in the original noise regions as seen by comparison between Figure 4.12a and d. This less noisy image facilitates the subsequent edge detection. Figure 4.12e shows the Canny detected bubble profile, which is very smooth. Figure 4.12f shows the final smoothed edge after ALFI-S. It is noted that the small, almost imperceptible bump on the Canny detected edge (i.e., the adhering noise) is removed. The surface tension value calculated from this smoothed edge is 23.67 mJ/m^2, which is consistent with data reported before [10].

4.4   IMAGE ANALYSIS FOR CONTACT ANGLE MEASUREMENT USING ADSA-D

As discussed in Chapter 3, contact angles can be measured from the meridian profile of a sessile drop using ADSA-P. However, there are several limitations of ADSA-P for contact angle measurement: first, in cases where flat sessile drops (contact angle below 20°) are encountered, it becomes increasingly difficult to acquire accurate coordinate points along the drop profile; second, the roughness and/or heterogeneity inherent in nonideal substrates, such as biological surfaces, will cause deviations from axial symmetry and/or irregular three-phase contact lines in sessile drops. These limitations substantially reduce the accuracy of any method relying on the side view of a sessile drop for determining contact angles.

To overcome these shortcomings, an alternative drop shape method, ADSA-D (Diameter), can be used to measure contact angles using sessile drop images taken

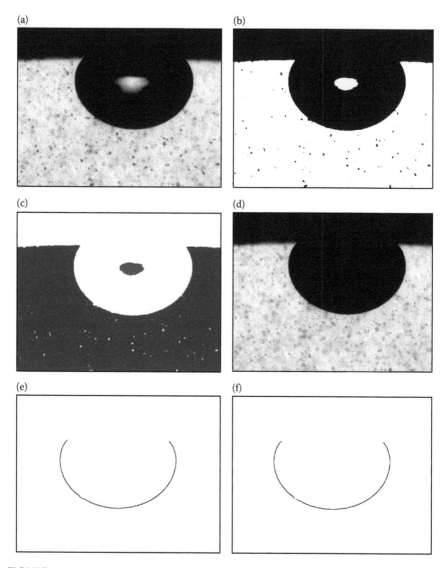

FIGURE 4.12 Implementation of the image analysis in ADSA for a noisy captive bubble image. (a) Original grayscale image; (b) binary image after Otsu's thresholding; (c) grayscale image showing 89 detected isolated components (each component is represented by one gray level) after component labeling; (d) grayscale image after removing isolated noise; (e) Canny detected edge (with pre-edge detection removal of the isolated noise); and (f) ALFI-S smoothed edge (removing adhering noise).

from the top view. In ADSA-D, the contact angle is determined by minimizing the difference between the actual drop diameter and the theoretical diameter obtained by solving the Laplace equation of capillarity. By viewing the drop from above, contact diameter (for contact angle less than 90°) or equatorial diameter (for contact angle larger than 90°) can be measured without a decrease in measurement accuracy due to flatness. Moreover, for noncircular sessile drops or those with irregular three-phase contact lines, an equivalent contact or equatorial diameter can be used to find the average contact angle. For some anisotropic surfaces, for example, wood or stone, the drop contour can be fitted to an ellipse through which the contact angles in different directions of the surfaces can be examined [30]. Detailed description of the ADSA-D algorithm can be found in Chapter 6.

In addition to the drop diameter, other input required in ADSA-D is the volume of the drop and the surface tension of the liquid. The volume of the sessile drop can be obtained precisely by means of the microsyringe used to form the drop. The surface tension of the liquid is either known or can be measured readily by other means (e.g., pendant drop experiments). Hence the only experimental parameter to be determined is the drop diameter. It can be determined by manual image digitization [31,32]: first, the perimeter of the drop is manually acquired by marking 8 to 10 points along the boundary; second, coordinates of this set of points are fitted to a circle using least squares fitting. Finally, the average drop diameter is calculated from the fitted circle.

Undoubtedly, manual digitization is time-consuming and can be subjective. Therefore, an automatic image analysis scheme was developed to replace the manual scheme [33]. As depicted in Figure 4.13, this image analysis scheme consists of three main parts:

(i) Noise reduction: reduction of the background noise using a combination of grayscale morphological filters and median filters
(ii) Edge detection: segmentation of the drop perimeter using binary morphological filters
(iii) Area detection: determination of the total area of the drop, as viewed from the top, using region growing. The area is then used to calculate the equivalent drop diameter, which will be used for the contact angle measurement using ADSA-D.

The robust design of the image analysis scheme enables it to accurately process many different types of drop images. As an example, the new image analysis scheme is illustrated by analyzing an image of a drop of water on a biological surface, *helicobacter mustelae*, as shown in Figure 4.14a. Accuracy of the image analysis scheme is tested by comparing with the contact angles measured from carefully performed manual digitization.

4.4.1 DEVELOPMENT OF THE IMAGE ANALYSIS SCHEME

4.4.1.1 Noise Reduction

The noise reduction consists of filtering on both the grayscale images and the intensity histograms. Grayscale morphological filters are first applied to the images. These filters are advantageous as they conserve the edges in the images while significantly reducing noise. The basic morphological filter operations are erosion and dilation. In an erosion operation, the center pixel of an $n \times n$ pixel neighborhood is

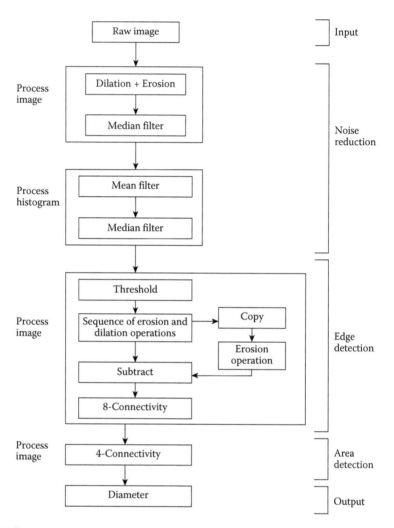

FIGURE 4.13 Flowchart of the image analysis scheme used in ADSA-D.

assigned the minimum grayscale value found in that neighborhood. A new image is produced by convolving the neighborhood with the entire image. A dilation operation is similar to the erosion operation, except the maximum grayscale in the neighborhood is assigned instead. These operations are applied in different sequences to produce the morphological filters. Several different neighborhood sizes and operation sequences were tested. Experiments showed that optimal noise reduction and edge preservation were achieved when an erosion operation followed by a dilation operation was applied to the image using a 9×9 pixel neighborhood. Further details describing the filter size and sequence selection process can be found in [2]. These morphological filters are particularly effective for images of drops on biological surfaces (e.g., bacterial layers; see Figure 4.14a), as these filters discard strands of bacteria touching the drop profile so that they are not considered to be part of the drop periphery.

As a result of applying these morphological filters, however, a degree of image pixelization (i.e., blurring due to reduced resolution) may be introduced to the images. Further noise reduction to eliminate the pixelization is achieved in a second step through the implementation of a 5×5 pixel median filter. The median filter involves a single operation similar to the erosion, but only the median grayscale in the neighborhood is assigned to the center pixel. The result of noise reduction on Figure 4.14a, by applying the morphological filters and the median filter, is shown in Figure 4.14b.

After the noise reduction, the grayscale image is converted into a binary image using thresholding. The threshold value used to segment the drop from the background is determined from a histogram smoothed by a 20 point neighborhood mean filter followed by a nine point neighborhood median filter. Details of the smoothing of the histogram can be found elsewhere [2]. The resultant binary image after histogram smoothing and thresholding is shown in Figure 4.14c.

4.4.1.2 Edge Detection

Before the edge detection, additional noise reduction is applied to the binary image using binary morphological filters. The operation of the binary morphological filters is similar to the grayscale morphological filters described above. It operates on images with only two rather than 256 gray levels. The image is filtered using the following sequence of operations: dilation, two erosions, and a final dilation. Neighborhood sizes used for all erosion and dilation operations are 3×3 pixels. To extract the edge, the image is first copied and then an erosion operation is applied to it. The transformed copy of the image is subsequently subtracted from the original image, resulting in a new image that only contains the edge, as shown in Figure 4.14d.

Aside from the edge of the drop, there may be other edges present in the image, for example, the substrate boundary, the reflection of the light from the drop center, or large strands of bacteria, as seen in Figure 4.14d. To establish the drop edge, the user is required to supply the coordinates of a point inside the drop. This point will be projected in the y direction of the image plane, until it reaches the drop edge. Once the edge is reached, the drop profile will be detected by measuring eight-connectivity. The resulting image that contains only the edge of the drop is shown in Figure 4.14e.

4.4.1.3 Area Detection

In order to calculate the equivalent drop diameter, it is necessary to find the drop area as viewed from the top (i.e., the summation of the pixels contained within the extracted edge). The true area can then be calculated by multiplying the area, in pixels, by an appropriate scale factor obtained from the calibration. The number of pixels inside the extracted edge is obtained using a region growing algorithm: beginning with a single pixel inside the edge contour, a region is formed and grown by

(a)   (b)

(c)   (d)

(e)   (f)

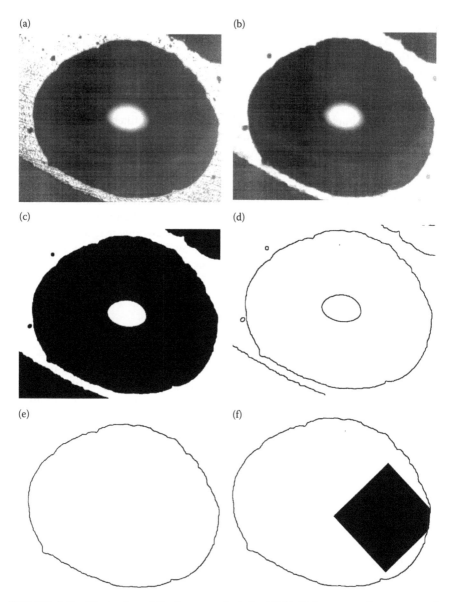

FIGURE 4.14 Illustration of the image analysis in ADSA-D for automatic contact angle measurement. (a) A typical image of a water droplet on a biological surface, *helicobacter mustelae*; (b) the image after noise reduction using a combination of a morphological filter (erosion and dilation operations) and a 5 × 5 median filter; (c) the binary image after histogram filtering and thresholding; and (d) the extracted edges. Aside from the drop edge there are other edges in this figure including substrate edges and the edge that results from the reflection of the light source on the drop; (e) the drop edge isolated from all other edges present in (d); and (f) region growing is used to compute the area within the edge contour in pixel units. The black region is the result after 20,000 iterations and contains 3,000 pixels.

iteratively adding four-connected pixels until reaching the edge boundary. During this process the number of pixels is counted. Figure 4.14f shows the grown region after 20,000 iterations. Once the area (A) is determined, the equivalent drop diameter (d) can be calculated as $d = 2\sqrt{A/\pi}$.

4.4.2 EVALUATION OF THE IMAGE ANALYSIS SCHEME

The accuracy of the automatic image analysis scheme was tested by comparing it to the contact angle results obtained from careful manual image processing. It should be noted that in both the manual and automatic image analysis schemes, the same core algorithms of ADSA-D were used for computing the contact angles. Hence, the differences in the contact angle measurements depend only on the diameters determined using the two different image analysis schemes.

Table 4.4 shows the contact angles measured using ADSA-D with the two image analysis schemes for water drops on a bacteria surface (i.e., *helicobacter mustelae*, see Figure 4.14a). It is found that the average contact angle difference between the two schemes is only 0.14°. The student *t*-test indicates no significant difference between the results obtained using these two methods. However, the automatic version is deemed to be more accurate as it calculates the real area within the extracted perimeter rather than the area within an estimated perimeter calculated by fitting a circle to the points on the profile. Thus, the automatic image analysis scheme should at least have an accuracy of

TABLE 4.4

Contact Angles of Water Drops on a *Helicobacter Mustelae* Surface, Measured Using ADSA-D with Manual Image Digitization and Automatic Image Analysis, Respectively

Drop No.	ADSA-D Calculated Contact Angles (°)		Difference (°)
	Manual Version	Automatic Version	
1	12.94	13.21	0.27
2	13.86	13.92	0.08
3	12.99	13.25	0.26
4	12.51	12.65	0.14
5	12.99	13.06	0.07
6	8.40	8.56	0.16
7	12.00	12.08	0.08
8	12.22	12.37	0.15
9	12.96	13.12	0.16
10	13.11	13.22	0.11
11	12.27	12.38	0.11
Mean ± SD	12.39 ± 1.42	12.53 ± 1.42	

Note: A typical image of such a sessile drop is shown in Figure 4.14a.

±0.5° (which is that of the manual scheme). More tests of the automatic image analysis scheme using different surfaces can be found elsewhere [33].

4.5 CONCLUDING REMARKS

Computer-based calculation of surface tension and contact angle from the shape of drops/bubbles is a well-established technique. ADSA, for example, determines surface tension and contact angle by numerical integration of the Laplace equation followed by a nonlinear least squares optimization (detailed in Chapter 3). These computational procedures are automatic and accurate. The only obstacle remaining for fully automatic measurements is the extraction of drop/bubble profiles from the digital images.

Image analysis provides a powerful tool for the automatic detection of drop/bubble profiles and hence contributes toward the fully automatic measurement of surface tension and contact angle. A key development in the image analysis scheme for real experimental images is noise reduction, as in the cases of surface tension measurement of lung surfactant using a captive bubble and of contact angle measurement on nonideal surfaces.

The two image analysis schemes developed in this chapter are both robust against noise. It should be noted that these image analysis schemes are independent of ADSA. They can be used as a standard software package in combination with any other surface tension and contact angle measurement algorithms, for example, surface tension measurement based on the evaluation of drop height to diameter ratios [21,34]. The individual image analysis techniques, such as ALFI-S and component labeling, are also free-standing. ALFI-S can be used as a standard filter for removing adhering noise from any Laplacian-type profile. The component labeling method developed in the surface tension measurement scheme would also be very suitable for contact angle measurement using ADSA-D. In such a context, the actual contour of a drop is not of importance but rather the area or the equivalent diameter of the drop is needed.

REFERENCES

1. A. W. Adamson. *Physical Chemistry of Surfaces.* Wiley, New York, 1990.
2. R. C. Gonzalez and R. E. Woods. *Digital Image Processing.* Prentice Hall, New York, 2002.
3. S. H. Anastasiadis, J. K. Chen, J. T. Koberstein, A. F. Siegel, J. E. Sohn, and J. A. Emerson. *Journal of Colloid and Interface Science* 119 (1987): 55.
4. H. H. J. Girault, D. J. Schiffrin, and B. D. V. Smith. *Journal of Colloid and Interface Science* 101 (1984): 257.
5. S. Y. Lin, K. Mckeigue, and C. Maldarelli. *AIChE Journal* 36 (1990): 1785.
6. J. Satherley, H. H. J. Girault, and D. J. Schiffrin. *Journal of Colloid and Interface Science* 136 (1990): 574.
7. Y. Y. Zuo, M. Ding, A. Bateni, M. Hoorfar, and A. W. Neumann. *Colloids and Surfaces A-Physicochemical and Engineering Aspects* 250 (2004): 233.
8. R. M. Prokop, A. Jyoti, M. Eslamian, A. Garg, M. Mihaila, O. I. del Rio, S. S. Susnar, Z. Policova, and A. W. Neumann. *Colloids and Surfaces A-Physicochemical and Engineering Aspects* 131 (1998): 231.

9. Y. Y. Zuo, D. Q. Li, E. Acosta, P. N. Cox, and A. W. Neumann. *Langmuir* 21 (2005): 5446.
10. Y. Y. Zuo, M. Ding, D. Li, and A. W. Neumann. *Biochimica Et Biophysica Acta-General Subjects* 1675 (2004): 12.
11. W. K. Pratt. *Digital Image Processing: PIKS Inside.* Wiley, New York, 2001.
12. P. Cheng, D. Li, L. Boruvka, Y. Rotenberg, and A. W. Neumann. *Colloids and Surfaces* 43 (1990): 151.
13. J. Canny. *IEEE Transactions on Pattern Analysis and Machine Intelligence* 8 (1986): 679.
14. J. Shen and S. Castan. *Cvgip-Graphical Models and Image Processing* 54 (1992): 112.
15. T. Kimura, A. Taguchi, and Y. Murata. *Electronics and Communications in Japan Part III-Fundamental Electronic Science* 83 (2000): 61.
16. H. S. Wong, T. Caelli, and L. Guan. *Pattern Recognition* 33 (2000): 427.
17. Z. J. Hou and G. W. Wei. *Pattern Recognition* 35 (2002): 1559.
18. C. Atae-Allah, M. Cabrerizo-Vilchez, J. F. Gomez-Lopera, J. A. Holgado-Terriza, R. Roman-Roldan, and P. L. Luque-Escamilla. *Measurement Science & Technology* 12 (2001): 288.
19. J. A. Holgado-Terriza, J. F. Gomez-Lopera, P. L. Luque-Escamilla, C. Atae-Allah, and M. A. Cabrerizo-Vilchez. *Colloids and Surfaces A-Physicochemical and Engineering Aspects* 156 (1999): 579.
20. P. Cheng and A. W. Neumann. *Colloids and Surfaces* 62 (1992): 297.
21. R. M. Prokop, O. I. del Rio, N. Niyakan, and A. W. Neumann. *Canadian Journal of Chemical Engineering* 74 (1996): 534.
22. O. I. del Rio and A. W. Neumann. *Journal of Colloid and Interface Science* 196 (1997): 136.
23. N. B. Vargaftik, B. N. Volkov, and L. D. Voljak. *Journal of Physical and Chemical Reference Data* 12 (1983): 817.
24. J. J. Jasper. *Journal of Physical Chemistry Reference Data* 1 (1972): 841.
25. G. Putz, J. Goerke, S. Schurch, and J. A. Clements. *Journal of Applied Physiology* 76 (1994): 1417.
26. Y. Y. Zuo, C. Do, and A. W. Neumann. *Colloids and Surfaces A: Physicochemical and Engineering Aspects* 299 (2007): 109.
27. A. K. Jain. *Fundamentals of Digital Image Processing.* Prentice-Hall, New York, 1989.
28. M. Seul, L. O'Gorman, and M. J. Sammom. *Practical Algorithms for Image Analysis, Description, Examples and Code.* Cambridge University Press, New York, 1999.
29. S. E. Umbaugh. *Computer Imaging: Digital Image Analysis and Processing.* CRC Press, Boca Raton, 2005.
30. M. A. Rodriguez-Valverde, M. A. Cabrerizo-Vilchez, P. Rosales-Lopez, A. Paez-Duenas, and R. Hidalgo-Alvarez. *Colloids and Surfaces A-Physicochemical and Engineering Aspects* 206 (2002): 485.
31. F. K. Skinner, Y. Rotenberg, and A. W. Neumann. *Journal of Colloid and Interface Science* 130 (1989): 25.
32. E. Moy, P. Cheng, Z. Policova, S. Treppo, D. Kwok, D. R. Mack, P. M. Sherman, and A. W. Neumann. *Colloids and Surfaces* 58 (1991): 215.
33. J. M. Alvarez, A. Amirfazli, and A. W. Neumann. *Colloids and Surfaces A-Physicochemical and Engineering Aspects* 156 (1999): 163.
34. J. D. Malcolm and C. D. Elliott. *Canadian Journal of Chemical Engineering* 58 (1980): 151.

5 Generalization and Advanced Application of Axisymmetric Drop Shape Analysis

Yi Zuo, Julia Maldonado-Valderrama, Miguel Cabrerizo-Vílchez, Arash Bateni, M. Guadalupe Cabezas, Robert David, and A. Wilhelm Neumann

CONTENTS

5.1 INTRODUCTION

In Chapters 3 and 4, the fundamental algorithms of Axisymmetric Drop Shape Analysis (ADSA) have been developed. Several classical applications of ADSA in determining surface tensions and contact angles have been discussed. In this chapter, more advanced applications and up-to-date generalization of ADSA will be introduced. These advanced applications are: (1) the use of ADSA to study interfacial and gas transfer properties of lung surfactant films, an example of biomedical application of ADSA (Section 5.2); and (2) the development of ADSA as a miniaturized Langmuir-type film balance, in which advanced experimental techniques are combined with ADSA to realize system miniaturization and integrity (Section 5.3). The generalizations involve: (1) the use of ADSA to study drop deformation in an electric field and its influence on surface tension (Section 5.4); and (2) the reformulation of ADSA to study external menisci, such as liquid bridges and liquid lenses (Section 5.5). Given the fact that ADSA has been extensively used all over the world in a variety of scientific and industrial fields, the applications and generalizations presented here are not complete or superior to those unmentioned. Instead, the intention is to show the versatility of ADSA and its extraordinary potential in both scientific and industrial applications.

5.2 ADSA FOR LUNG SURFACTANT STUDIES

5.2.1 INTRODUCTION

Lung surfactant is a complicated mixture of approximately 90% lipids and 10% proteins [1]. It forms a thin film at the air–water interface of alveoli and plays a crucial role in maintaining the normal respiratory mechanics by reducing the alveolar surface tension to near-zero values. Deficiency or dysfunction of lung surfactant causes respiratory distress syndrome (RDS), one of the major contributors to neonatal morbidity and mortality in industrialized countries [2]. Exogenous surfactant replacement therapy has been used as a standard therapeutic intervention for patients with RDS [3]. Different formulations, either synthetic or natural surfactants extracted from mammalian lungs, have been developed as surfactant substitutes. Owing to the surfactant therapy, the mortality rate of premature infants in the United States dropped by 24% between 1989 and 1990 and continued decreasing thereafter [4,5].

The clinical application of lung surfactant necessitates the in vitro assessment of its biophysical properties; that is, the properties related to highly dynamic and very low surface tensions. A variety of in vitro techniques, such as the Langmuir-Wilhelmy balance (LWB) [6], pulsating bubble surfactometer (PBS) [7], and captive bubble surfactometer (CBS) [8], have been developed for measuring surface tension of lung surfactant. Detailed discussion of these methods can be found elsewhere [9].

ADSA has been found to be particularly suitable for lung surfactant studies due to a number of facts: (1) the amount of liquid sample required in a drop shape technique such as ADSA is very small (as little as a few microliters), which minimizes the cost of the experimental materials. (2) ADSA is capable of simultaneously measuring surface tension and surface area, thus allowing for recording surface tension–area isotherms. This feature makes ADSA a microfilm balance [10], an intriguing alternative to the traditional Langmuir-type film balance. (3) ADSA allows measurement of dynamic surface tension. Therefore, it is possible to investigate the highly dynamic properties of lung surfactant, for example, rapid film formation and dynamic cycling at the physiologically relevant rate. (4) ADSA is capable of measuring very low surface tensions (less than 1 mJ/m^2) occurring in lung surfactant systems. (5) ADSA is highly automated and hence operation is less dependent on the skill of the experimenter (detailed in Chapter 4).

5.2.2 EXPERIMENTAL SETUP

Figure 5.1 shows the general experimental setup of ADSA. It consists of six fundamental subsystems: the drop/bubble configurations, the lighting system, the image acquisition system, the environmental control system, the liquid flow control system, and the antivibration system. Among these components, the drop/bubble configurations play a central role as they determine the applicability of each specific experimental setup. The three configurations used in lung surfactant studies will be discussed in the next section. The other five peripheral subsystems are described as follows.

The lighting system is composed of a light source (Newport Corp, Fountain Valley, CA) and a diffuser made of frosted glass, which is used to provide uniform incident light. If rigorous lighting conditions are required, monochromatic filters can be used to provide monochromatic illumination instead of white light [11].

The image acquisition system comprises a microscope (Apozoom, Leitz Wetzlar, Germany), a CCD camera (Cohu Corp), a digital video processor (Parallax Graphics, CA) and a computer. The microscope is equipped with a polarizing filter

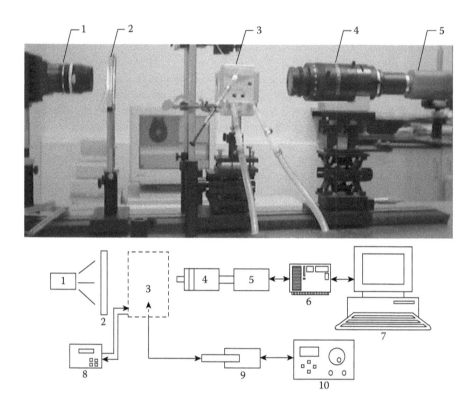

FIGURE 5.1 Picture and schematic diagram of the ADSA experimental setup. 1. light source; 2. diffuser; 3. thermostatic drop/bubble chamber; 4. microscope; 5. CCD camera; 6. digital video processor; 7. workstation; 8. water bath; 9. motorized syringe; 10. motor controller.

that reduces the glare and enhances the contrast of the image. The digital video processor performs both frame grabbing and image digitizing. Image acquisition can be performed at a speed of up to 30 images per second. Each image is digitized to a matrix of 640×480 pixels with 256 gray levels for each pixel, where 0 represents black and 255 represents white. The acquired images are stored in the computer for further analysis by the image processing program (detailed in Chapter 4).

The key environmental parameters to be controlled are temperature and humidity. To mimic the physiological conditions, the atmosphere surrounding the drop/bubble needs to be maintained at 37°C and saturated with water vapor. Different thermostatic chambers have been developed for different drop/bubble configurations. The temperature of these chambers is thermostatically maintained at 37 ± 0.2°C by a water bath (Neslab Instruments Inc, Portsmouth, NH).

The control of liquid flow is necessary for drop formation, in which a drop is grown, and for the subsequent dynamic cycling, in which a drop/bubble is compressed and expanded periodically. For a drop arrangement, the flow control is performed by directly adding or withdrawing liquid into or out of the drop by means of a motor-driven syringe (2.5 mL, Gastight, Hamilton Corp, Reno, NY). The rate and fashion of the motor movement (i.e., the liquid flow), is precisely controlled by a programmable motor controller (Oriel Instruments, Stratford, CT). For a bubble arrangement, a motor-driven syringe (5 mL, Gastight) is used to manipulate liquid into or out of the bubble chamber, thereby increasing or decreasing the pressure of the liquid subphase.

The entire experimental setup, except the computer, is mounted on a vibration-free table (Technical Manufacturing Corp, Peabody, MA).

5.2.3 DIFFERENT DROP/BUBBLE CONFIGURATIONS

The selection of an appropriate drop/bubble configuration depends on the purpose of the measurement and the desired accuracy. Three configurations have been used in conjunction with ADSA for lung surfactant studies: pendant drop (PD), captive bubble (CB), and constrained sessile drop (CSD). The main applications of these three configurations and their relative merits and limitations are summarized in Table 5.1.

5.2.3.1 Pendant Drop

The PD is an early developed drop configuration used for surface tension measurement. A drop is suspended at the end of a capillary made of Teflon or quartz (see Figure 5.2a). The other end of the capillary is connected to the liquid flow control system. The capillary commonly used has an inner diameter of 1.0 mm and an outer diameter of 3.0 mm. Volume of the drop varies from 10 to 20 μL, corresponding to a variation in the maximum diameter from 3.0 to 3.3 mm. The vertical alignment of the capillary is maintained by a metal guide tube, which is mounted onto

---

TABLE 5.1

Summary of Typical Applications and Relative Merits and Limitations of the Three Drop/Bubble Configurations in Conjunction with ADSA for Lung Surfactant Studies

Configurations	Main Applications	Advantages	Disadvantages
Pendant drop (PD)	Adsorption	• Easy to operate and clean • High accuracy (± 0.01 mJ/m^2)	• Film leakage at low surface tensions
Captive bubble (CB)	Film compressibility and stability	• Leakage proof	• Difficult to operate and clean • Limitation on the maximum concentration (1–2 mg/mL) • Uncontrolled humidity
Constrained sessile drop (CSD)	High surfactant concentrations and very low surface tensions	• Easy to operate and clean • Leakage proof • No concentration limitation • Full environmental control	• No apparent fundamental limitations

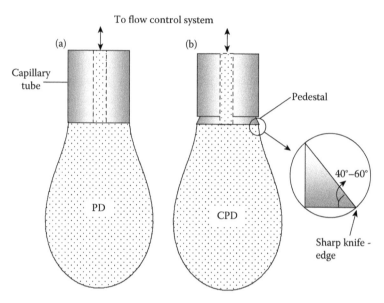

FIGURE 5.2 Schematics of pendant drops (PDs). (a) A PD formed at a capillary, and (b) a constrained PD (CPD) formed at a pedestal. The pedestal features a sharp knife-edge to prevent film leakage.

a three-way micromanipulator (Leica, Germany). The drop is enclosed in a quartz glass cuvette (Hellma), which is thermostatically controlled by a stainless steel temperature cell (Ramé-Hart). A reservoir of distilled water is placed in the cuvette well before starting the experiment to ensure a vapor-saturated atmosphere. A Teflon stopper is used to seal the cuvette to prevent evaporation and contamination from the outer environment.

In addition to the apparent advantages of simplicity and flexibility, the PD method has high accuracy (i.e., ± 0.01 mJ/m^2) [12]. However, the conventional PD arrangement suffers from the problem of film leakage [13]. Film leakage occurs for surface thermodynamic reasons: a film formed at an air–water interface spreads onto a solid support that contacts the film if the surface tension of the film is lower than that of the solid surface. For Teflon, this limiting surface tension is near 18 mJ/m^2. Due to the loss of film material from the air–water interface, the surface tension—area isotherms and the surface rheological properties measured using the PD could be erroneous.

To avoid film leakage, a special pedestal has been developed recently. As shown in Figure 5.2b, this pedestal is made of stainless steel and features a sharp knife-edge. Benefits from this pedestal are twofold: first, the hydrophilicity of the pedestal allows the formation of well-deformed drops that is favorable for accurate surface tension measurement [14]. Second, the sharp knife-edge is able to prevent the lung surfactant film from spreading over the pedestal at low surface tension (i.e., it prevents film leakage). This modified PD configuration is termed "constrained" pendant drop (CPD) due to the fact that the drop is confined by the sharp knife-edge of the pedestal. With the CPD, a surface tension as low as 1 mJ/m^2 can be recorded.

ADSA-PD is well suited for the study of lung surfactant adsorption as the limitation of film leakage is relatively unimportant in such a case. The equilibrium surface tension of lung surfactant films is approximately 22–25 mJ/m^2, which is well above the threshold value at which leakage may occur. ADSA-PD has been used to study the dependence of adsorption rate on the bulk concentration of lung surfactant [15] and the influence of nonionic polymers on the adsorption kinetics of lung surfactant [16]. With the addition of a subphase exchange system, ADSA-PD has been further developed to be a miniaturized Langmuir film balance that is capable of studying film penetration. Details of this development can be found in Section 5.3.

5.2.3.2 Captive Bubble

Figure 5.3 shows a schematic of a CB with adsorbed lung surfactant film. In a CB arrangement, a bubble with a volume of approximately 20 μL (~3 mm in diameter) is injected by a microsyringe into a chamber filled with a lung surfactant suspension. After injection, the bubble immediately rests against the ceiling and the shape of the bubble is controlled by the surface tension. A CB chamber currently used consists of three metal plates made of stainless steel and two viewing windows [17]. Before each experiment, the chamber is assembled by sandwiching the two windows within the metal plates. The middle plate provides a spacer of ~1 mL to hold the surfactant sample. The top of the reservoir is slightly concave to confine the bubble. Due to the hydrophilicity of the stainless steel ceiling, the air bubble is separated from the ceiling by a thin wetting film (estimated to be 100–500 nm thick [18]), thus eliminating film leakage. The temperature and gauge pressure in the chamber are continuously monitored by an ultrafine thermocouple (Omega Eng Inc, Laval, Quebec, Canada) and a pressure transducer (Validyne Eng Corp, Northridge, CA), respectively. A universal data acquisition card (Validyne) installed in a computer is used to simultaneously process both the temperature and pressure signals.

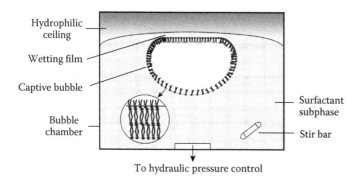

FIGURE 5.3 Schematic of the experimental setup of a captive bubble (CB) with adsorbed lung surfactant multilayers.

Precise humidity control in a CB is difficult; however, it has been long assumed that a CB would provide full humidification although ambient air is routinely used for bubble formation. Accumulating evidence, however, suggests otherwise: an air bubble may not be instantly saturated with water vapor, presumably due to a transport barrier on water evaporation posed by the rapidly adsorbed insoluble surfactant film [19]. Another limitation of the CB method is that the maximum surfactant concentration that can be tested is restricted to 1–2 mg/mL [20], far less than the physiological concentration. This constraint arises from optical limitations since lung surfactant suspensions become murky and eventually opaque at higher concentrations. This shortcoming can be removed by a recent development of a spreading technique in the CB [21].

Owing to its leakage-proof characteristics, ADSA-CB is suited for the study of film compressibility and stability, especially at low surface tensions. ADSA-CB has been used to investigate the stability of lung surfactant [22] and polymer enhanced lung surfactant films [23,24], and phase separation and transition of lung surfactant films [25]. Pison and his colleagues [26,27] used ADSA-CB to study surface dilatational properties, such as surface viscosity and elasticity, of dipalmitoyl phosphatidylcholine (DPPC) and DPPC/protein films. By spreading DPPC inside a bubble, ADSA-CB was also used to study the interaction between the monolayer and the evaporated spreading solvents [28]. In combination with gas chromatography, ADSA-CB has been used to study the dissolution characteristics of anesthetic vapors and gases [29].

5.2.3.3 Constrained Sessile Drop

The CSD is a novel drop configuration for surface tension measurement [30]. As shown in Figure 5.4, a sessile drop is sitting on a pedestal (similar to the pedestal used for the CPD but upside-down), which employs a horizontal sharp knife-edge to prevent film leakage. The pedestal is machined from stainless steel with a diameter of 2.5–4 mm. The angle between the horizontal and the lateral surfaces of the pedestal is in the range of 45°–60°. The pedestal has a central hole of 0.5 mm in diameter, through which the drop is connected to the liquid flow control system by a Teflon capillary. A sessile drop with a volume of 4~8 µL (dependent on the size of the pedestal) is formed by a motor-driven syringe (2.5 mL, Gastight). The time of forming the drop is less than 0.5 s, precisely controlled by a programmable motor controller (Oriel). This rapid drop formation ensures that the subsequent

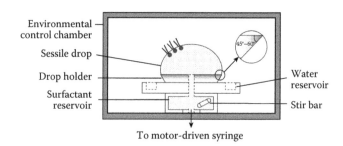

FIGURE 5.4 Schematic of the experimental setup of a constrained sessile drop (CSD) with spread lung surfactant monolayer.

adsorption of surfactant molecules occurs at a fresh, clean air–water interface. The film formation in a CSD can be also completed by spreading.

The CSD is enclosed in an environmental control chamber ($80 \times 35 \times 40$ mm^3) made of stainless steel. The chamber ensures sophisticated control of relative humidity (RH) and gas compositions, thus permitting the study of environmental effects, such as humidity and different gas compositions, such as carbon dioxide, on the surface activity of lung surfactant films.

So far, no apparent limitation of the CSD has been found. It eliminates both the problems of film leakage, as in the PD, and of concentration restriction, as in the CB. In addition, compared to the CB arrangement, the CSD is much simpler and easier to operate and clean, and it requires a much smaller amount of test liquid, typically 1% of that used in a CB experiment.

ADSA-CSD is very suitable for the measurement of very low surface tension of lung surfactant films formed at physiologically relevant surfactant concentrations. Preliminary tests have shown good agreement between the measurements with CSD and CB [30]. ADSA-CSD has been used in the study of the effect of humidity on the film stability [31], and the study of polymeric additives to lung surfactant [32].

5.2.4 TYPICAL APPLICATIONS

At least three biophysical properties of lung surfactant are essential to the normal respiratory physiology, especially in the neonatal period [33]. They are: (1) rapid film formation (i.e., within seconds) via adsorption from the alveolar hypophase; (2) low film compressibility (i.e., < 0.01 (mJ/m^2)$^{-1}$) associated with very low surface tension (i.e., near-zero values) during lung deflation; and (3) effective replenishment of the lung surfactant film during lung inflation. These biophysical properties can be evaluated using ADSA with different drop/bubble configurations. Some fundamental applications will be addressed here to show the applicability of ADSA in the study of lung surfactant. These applications are the study of: (1) the effect of bulk concentration on the adsorption kinetics [15], (2) the effect of compression ratio on film stability and compressibility [9,22], and (3) the very low surface tension at high surfactant concentrations [30]. Finally, a novel application of using ADSA to study interfacial gas transfer will be presented [18,34]. Details of the physiological and biophysical aspects of lung surfactant can be found elsewhere [1,2].

The lung surfactant used here is called bovine lipid extract surfactant (BLES Biochemicals Inc., London, ON, Canada). BLES is a clinically used lung surfactant and is commercially available. It is prepared from bovine natural lung surfactant obtained by bronchopulmonary lavage with organic extraction. BLES contains about 98% phospholipids and 2% proteins. BLES was stored frozen in sterilized vials with an initial concentration of 27 mg/mL. It was diluted to the desired concentration using 0.6% saline with 1.5 mM CaCl$_2$ on the day of experiment.

5.2.4.1 Study of Adsorption Kinetics Using a Pendant Drop

Figure 5.5 shows four adsorption curves of BLES at a concentration of 0.1 mg/mL. The measurements were conducted using ADSA-PD at 37°C. Time zero refers to

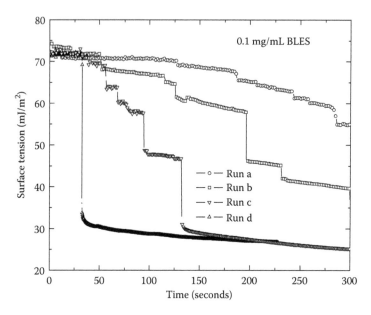

FIGURE 5.5 Four individual adsorption curves of 0.1 mg/mL BLES, measured using ADSA-PD at 37°C.

the end of drop formation at which a surface tension close to the value of a clean air–water interface (~70 mJ/m²) is recorded. After that, the surface tension decreases as a result of surfactant adsorption. It is found that the adsorption curves are not smooth; instead, a number of random, stepwise jumps occurring within a very short period (< 0.2 s) are observed. The magnitude of these jumps can be large (~35 mJ/m²) or moderate (1–5 mJ/m²). These sudden decreases in surface tension are referred to as adsorption clicks [35]. These clicks significantly enhance the adsorption toward reaching the equilibrium surface tension of 22–25 mJ/m². As seen from Figure 5.5, in run (a) a surface tension of only 55 mJ/m² is reached after 300 s. However, run (d) goes well below 30 mJ/m² in the first 100 s, due to a large click at about 30 s.

Figure 5.6 shows four adsorption curves of BLES at 10 mg/mL. It is found that these curves are very different from those at 0.1 mg/mL. First, the adsorption clicks are absent. As a result, the adsorption curves measured from individual runs are very consistent. Second, the surface tensions at time zero are already less than 28 mJ/m², indicating very rapid adsorption that occurs during drop formation. After time zero, the surface tension decreases rapidly only in the first 50 s, and levels off at the equilibrium value.

Adsorption at a series of BLES concentrations in the range of 0.1–10 mg/mL was also tested [15]. It was found that the adsorption clicks are not significant except for concentrations lower than 1 mg/mL. Table 5.2 collects the averaged surface tension values after 2, 20, and 300 s of the adsorption for BLES concentrations from 1 to 10 mg/mL. It can be seen that increasing BLES concentration from 1 to 10 mg/mL does not significantly enhance adsorption in a PD.

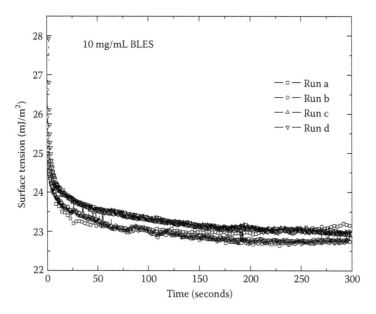

FIGURE 5.6 Four individual adsorption curves of 10 mg/mL BLES, measured using ADSA-PD at 37°C.

TABLE 5.2
Surface Tensions after 2 s, 20 s, and 300 s
of the Adsorption, for BLES Concentrations
in the Range of 1–10 mg/mL

Concentration	Surface Tension (mJ/m²)		
(mg/mL)	2 s	20 s	300 s
1	26.0 ± 0.1	24.7 ± 0.1	24.0 ± 0.2
2	25.6 ± 0.1	24.7 ± 0.1	23.9 ± 0.1
3	26.0 ± 0.1	24.5 ± 0.1	23.9 ± 0.1
6	25.8 ± 0.2	24.3 ± 0.3	23.6 ± 0.2
8	25.1 ± 0.1	24.0 ± 0.1	23.4 ± 0.1
10	24.9 ± 0.1	23.9 ± 0.1	22.9 ± 0.1

Note: Each value shows as an average of four individual runs with 95% confidence intervals.

It is concluded that the adsorption kinetics of lung surfactant depend strongly on the bulk concentration of the phospholipids. At a concentration as low as 0.1 mg/mL, the adsorption kinetics is controlled by the adsorption clicks, which may reflect the quick movement of large flakes of aggregated surfactant molecules into the air–liquid interface. The addition of these massive aggregates dramatically increases the surface concentration of the surfactant, thereby abruptly decreasing the surface tension. The fact that both the magnitude and the occurrence of the adsorption clicks are unpredictable agrees with this hypothesis. Increasing surface concentration up to 1 mg/mL significantly improves the adsorption kinetics. The air–water interface is quickly saturated with the surfactant molecules, thus preventing adsorption clicks. Further increasing the BLES concentration from 1 mg/mL to 10 mg/mL has no pronounced improvement on the in vitro adsorption kinetics. It should be noted that, in addition to the surfactant concentration, the adsorption kinetics of lung surfactant studied in a drop/bubble configuration should also be dependent on the area-to-volume ratio of this configuration. Hence direct comparison of these in vitro data established by a PD to surfactant adsorption in the lungs is difficult. However, taking into account the high surfactant concentration in the lungs (i.e., >3 mg/mL [36]), the formation of surfactant film in vivo due to adsorption is expected to be completed within only a few seconds, at the most.

5.2.4.2   Study of Film Stability and Compressibility Using a Captive Bubble

The stability and compressibility of surfactant films were studied using ADSA-CB. In these experiments, the surfactant films were continuously compressed and expanded at a rate of 5 s per cycle to simulate breathing. Figure 5.7 shows typical surface tension–relative area isotherms of a BLES film upon normal compression; that is, compression within the normal physiological range (no more than 30% area reduction [37]). Only the first five cycles are shown since the cycles afterward are essentially identical. It is found that except the first cycle, the compression and expansion

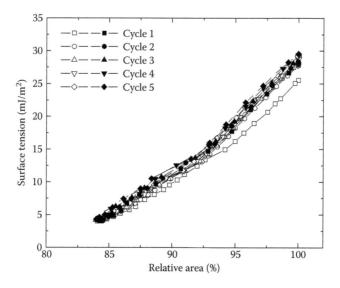

FIGURE 5.7   Surface tension–area isotherms of reversible cycling, showing overlapped compression (open symbols) and expansion (solid symbols) portions in each cycle. The BLES film was compressed at a normal compression ratio (~17%) using ADSA-CB at 37°C.

portions of these isotherms coincide completely; i.e., there is no surface tension–area hysteresis. A cycle with such a feature is termed a "reversible cycle," which indicates that no film collapse occurs [22]. The minimum surface tension reached by the ~15% film compression is less than 5 mJ/m^2 and the maximum surface tension at the end of expansion is no more than 30 mJ/m^2. The film compressibility, $dlnA/d\gamma$, calculated at 15 mJ/m^2 is only 0.0065 (mJ/m^2)$^{-1}$, which is close to the compressibility of pure DPPC films [8].

Figure 5.8 shows the surface tension–area isotherms of a BLES film upon overcompression (i.e., ~60% compression used here). These isotherms show completely different patterns from those at low compression ratios. First, significant hysteresis loops appear. A cycle with this feature is termed an "irreversible cycle" [22]. Second, except for the first cycle, a compression shoulder appears at a surface tension of 20–25 mJ/m^2. Third, two plateaus appear at the ends of compression and expansion, in which the surface tension only slightly varies even though the bubble area changes significantly. The compression plateau occurs at a surface tension near or below 1 mJ/m^2, which is an indication of film collapse [38]. The expansion plateau occurs at the surface tension of 30–35 mJ/m^2, in which the effect of film dilation is balanced by replenishment of the lung surfactant molecules from the subphase or from the multilayer structures associated with the interfacial monolayer.

From these in vitro tests, it is concluded that the natural surfactant film has a low compressibility; that is, it reaches low surface tension effectively upon only moderate area reduction. Overcompressing the films decreases the film stability (i.e., inducing film collapse). Film collapse causes pronounced surface tension hysteresis between

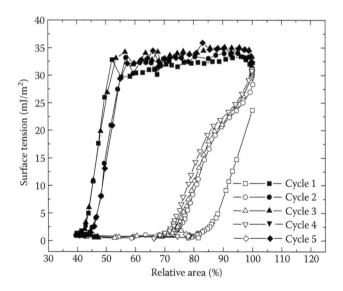

FIGURE 5.8 Surface tension–area isotherms of irreversible cycling, showing pronounced hysteresis loops of compression (open symbols) and expansion (solid symbols) portions in each cycle. The BLES film was overcompressed (~62%) using ADSA-CB at 37°C.

compression and expansion. This is unfavorable for the lungs since it causes a loss of
mechanical work; but it may be of interest in connection with mechanical ventilation
in clinical settings.

5.2.4.3 Study of High Surfactant Concentration Using a Constrained Sessile Drop

Figure 5.9a shows the surface tension-area-volume isotherms of a BLES film
recorded using ADSA-CSD. This experiment was conducted at room temperature.
The surfactant film was adsorbed from 5 mg/mL BLES and the film was cycled at
a relatively slow rate of 20 s per cycle. It is noted that near-zero surface tensions are
readily achieved.

Figure 5.9b is the enlargement of the rectangular region in Figure 5.9a, showing
the surface tensions during the first compression. The compression curve clearly
shows patterns of film collapse as indicated by the three surface tension jumps.
These jumps indicate the film instability upon overcompression. However, before
the first jump occurs, a surface tension as low as 0.23 ± 0.01 mJ/m^2 is recorded.
Reaching such a low surface tension clearly shows that there is no film leakage in
the CSD.

To simulate the physiological conditions, BLES at 5 mg/mL was also studied at
37°C. The BLES films were cycled at a rate of 3 s per cycle and with a compression
ratio of approximately 18%. Typical results are shown in Figure 5.10. It is noted that
a minimum surface tension of approximately 1 mJ/m^2 can be readily obtained in
the first cycle. Such a study that closely mimics the physiological conditions is not
readily feasible with other in vitro methods.

5.2.4.4 ADSA Studies Beyond Surface Tension: Gas Transfer Through Interfacial Films

ADSA has also been extended to study transport phenomena. In conjunction with a
CB, it was used as a miniaturized film balance to study the physicochemical effect
of lung surfactant films on interfacial oxygen transfer [18,34]. The interfacial gas
transfer, quantified by a steady-state mass transfer coefficient, was determined
by analyzing shrinkage of a pure oxygen bubble due to gas diffusion across the
surfactant film adsorbed on the bubble surface. Meanwhile, the surface tension was
continuously measured by analyzing the shape of the bubble. In addition to simultaneously acquiring surface tension and interfacial gas transfer, ADSA-CB offers
other advantages over the traditional Langmuir balance. First, only a small amount
of liquid sample is required; this allows the study of adsorbed lung surfactant films
at a reasonable cost. Second, no equilibrium in gas transfer is required; hence, the
measurements can be completed within a relatively short period (e.g., five minutes).
Third and most important, surface tension well below the equilibrium value can be
easily obtained and maintained for a prolonged period, owing to the leakage-proof
capacity of the CB technique [8]. Consequently, the effect of compressed lung surfactant films, at a low surface tension range, on interfacial oxygen transfer can be
readily studied.

The principle of the procedures is as follows: ADSA outputs surface tension
(γ), bubble area (A), volume (V), height (H_B), and curvature ($1/R_0$) at the bubble

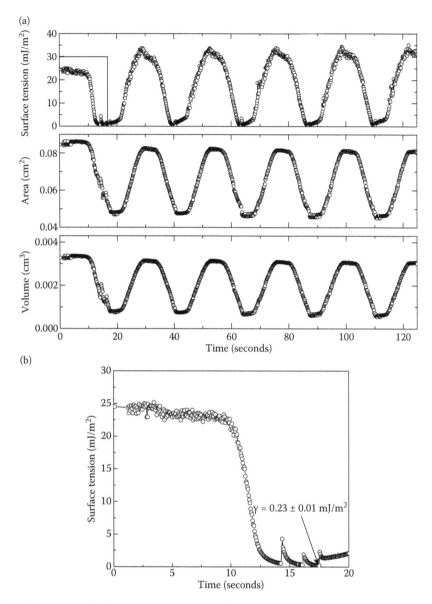

FIGURE 5.9 (a) Surface tension-area-volume isotherms for dynamic cycling of 5 mg/mL BLES; (b) enlargement of the rectangular region shown in (a). The dynamic cycling was conducted at a low rate of 20 s per cycle using ADSA-CSD at 23°C.

apex, as a function of time (t). Hence, the rate of bubble shrinkage due to gas diffusion can be estimated directly as $\Delta V/\Delta t$. To calculate the rate of mass transfer (M), the ideal gas law can be used to estimate the gas density, which is a function of the gas partial pressure in the bubble (P^g) at each time. Even though only pure oxygen was used to form the bubble, the bubble would be a mixture of oxygen

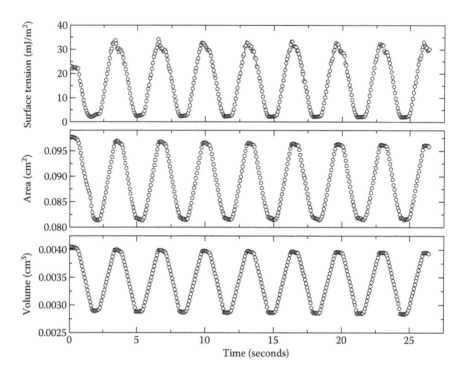

FIGURE 5.10 Surface tension-area-volume isotherms for dynamic cycling of 5 mg/mL BLES. The dynamic cycling was conducted at a high rate of 3 s per cycle using ADSA-CSD at 37°C.

and water vapor. P^g at time i can be evaluated from Dalton's law and the Laplace equation,

$$P_i^g = P_i^l + \Delta P_i - P^v = P_i^l + \left(\frac{2\gamma_i}{R_{0i}} + \frac{H_{Bi}}{2} \Delta \rho g \right) - P^v, \tag{5.1}$$

where P^l is the pressure of the liquid subphase, which was continuously measured by a pressure transducer (Validyne); ΔP is the average Laplace pressure in the bubble, which is equal to the sum of the Laplace pressure at the bubble apex and the averaged hydrostatic pressure increasing with bubble height (H_B). P^v is the partial pressure of water vapor, which is assumed to be constant and equal to the saturation pressure at the experimental temperature.

Using the mass transfer data calculated above, an average mass transfer coefficient (k_L) can be developed from a simple steady-state gas transfer model as follows,

$$k_L = \frac{M}{A_m (C_A^* - C_A^0)} = \frac{1}{N} \sum_{i=1}^{N} \frac{\Delta \left(\dfrac{P_i^g V_i}{\bar{R} T} \right)}{A_{li} \Delta t_i \left(\dfrac{P_i^g}{H} - C_A^0 \right)}, \tag{5.2}$$

where A_m is the mass transfer area, which changes with time due to bubble shrinkage and surface tension variation in the presence of surfactant. Here, the real-time lateral area (A_l) instead of the entire bubble area (A) was used. This is due to the extremely small thickness of the wetting film (as sketched in Figure 5.3) in contact with the bubble. This thin aqueous layer is most likely to be saturated with oxygen within seconds (i.e., right after bubble formation), thus preventing further gas transfer through this region [18]. C_A^* and C_A^0 are the gas concentrations at the gas–liquid interface and in the bulk liquid phase, respectively. After formation of the bubble (i.e., time zero), C_A^* can be assumed to be constant and equal to the saturation concentration in equilibrium with the gas in the bubble. In other words, C_A^* can be correlated to the gas partial pressure (P^g) in the bubble by Henry's law. H is Henry's law constant of oxygen in water at 37°C. C_A^0 is the average concentration of the dissolved gas in the bulk liquid, which is assumed to be unchanged throughout the observation. This assumption is valid only when the liquid phase is much larger than the gas phase and the amount of gas transfer is small (i.e., both the time and area of contact should be limited). To ensure that this assumption is satisfied, each gas transfer experiment was restricted to five minutes and the liquid subphase was replaced after each experimental run. k_L is averaged throughout the entire observation to rule out any spatial and temporal fluctuation. N is the number of discrete ADSA measurements, which depends on the rate of image acquisition. Hence, all the parameters used in Equations 5.1 and 5.2 are either output by ADSA (e.g., γ, V, A_l, $1/R_0$, H_B) or controlled during the experiment (e.g., T, P^l, C_A^0). More detailed validation and correction of this gas transfer model according to the experimental setup and protocol can be found elsewhere [18].

Figure 5.11 shows the surface tension of water and 0.5 mg/mL BLES as a function of subphase gauge pressure (i.e., different film compressions), where pressure equal to zero refers to the case of no compression. The measured surface tension of water is close to the literature value of 70 mJ/m² and relatively unchanged within the range of pressure variation. This suggests that the surface tension of water is relatively independent of the system pressure, which is in line with previous studies [39]. In contrast, the surface tension of BLES decreases with increasing subphase pressure. This is due to the lateral compression of the lung surfactant film adsorbed on the bubble surface. With increasing subphase pressure, the film is compressed to an increasing extent, thus increasing the surface density of the lung surfactant film and decreasing surface tension. The surface tension reduction in BLES is illustrated by bubble flattening shown in Figure 5.11. It should be noted that at moderate and high subphase pressures (i.e., 6, 8, and 10 psi), surface tensions less than 10 mJ/m² are recorded. Such low surface tensions are difficult, if not impossible, to reach and maintain in a conventional Langmuir trough method, due to film leakage.

Figure 5.12 shows the mass transfer coefficients in water and in 0.5 mg/mL BLES as a function of the subphase pressure. At low and moderate system pressures (i.e., 0–6 psi), k_L in water and in BLES change relatively in parallel as the system pressure increases. However, at the high pressure range (i.e., 6–10 psi), k_L in water and in BLES significantly diverges from each other: k_L in BLES decreases substantially as the system pressure increases from 6 to 10 psi; in contrast, k_L in water is relatively unchanged in the same range of pressure increase.

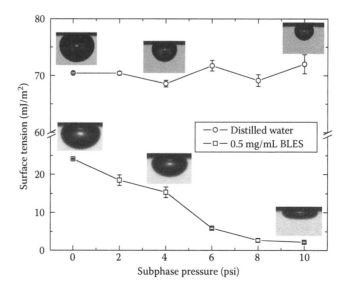

FIGURE 5.11 Surface tensions of water and 0.5 mg/mL BLES as a function of the subphase gauge pressure, respectively, measured using ADSA-CB at 37°C. The error bar shows the standard deviation of each measurement. Inserts are images of captive bubbles in water and in BLES at the corresponding subphase pressure.

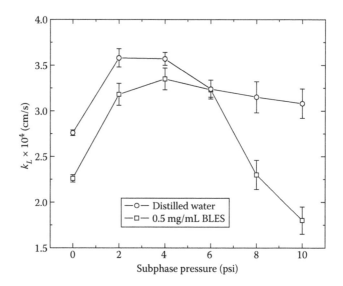

FIGURE 5.12 The mass transfer coefficients (k_L) in water and in 0.5 mg/mL BLES as a function of the subphase gauge pressure, respectively. The error bar shows the standard error of the mean of each measurement.

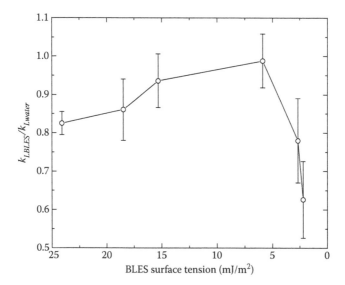

FIGURE 5.13 The normalized mass transfer coefficient in 0.5 mg/mL BLES to that in water (k_{LBLES}/k_{Lwater}) as a function of BLES surface tension.

Figure 5.13 shows the normalized k_L of oxygen in BLES to k_L in water at the same controlled subphase pressure, as a function of surface tension. Despite the relatively large errors associated with these normalized results after error propagation, it is evident that within the limitation of the maximum errors, decreasing the surface tension of a BLES film from the equilibrium value of ~24 mJ/m² to a low value of ~2 mJ/m² causes an approximately 25% decrease in oxygen transfer, indicating a significant surface resistance to oxygen transfer due to highly compressed lung surfactant films.

Direct measurement at the alveolar surface has established that the surface tension in the lungs likely varies from a value of no more than 30 mJ/m² near the total lung capacity (TLC) [40,41] to as low as 1 mJ/m² when deflating to the functional residual capacity (FRC) [42]. The experimental data, hence, cover the physiologically relevant surface tension range of normal tidal breathing. The results indicate a possible role of lung surfactant films as a one-way gate for oxygen transfer; that is, in facilitating oxygen transfer during inspiration due to the increased surface tension and hindering oxygen transfer during expiration due to the decreased surface tension. This hypothesis makes intuitive sense in that oxygen transfer from the inspired air (with a high oxygen partial pressure of approximately 159 mmHg) to the capillary blood occurs mainly during inspiration; after oxygenation of the capillary blood, the alveolar gas (with a lower oxygen partial pressure of approximately 105 mmHg) is exhaled out of the lungs. The lung surfactant film may play a role in preserving oxygen in the capillary blood during expiration.

5.3 ADSA AS A MINIATURIZED LANGMUIR FILM BALANCE

5.3.1 INTRODUCTION

In this section, an ADSA-based Langmuir film balance will be described, which consists of a pendant drop, a rapid-subphase-exchange technique and a fuzzy logic control algorithm. This new film balance allows the performance of noninvasive kinetic studies of adsorption/desorption and penetration and reaction of surface layers. Therefore, it arises as a versatile accessory to the PD technique and offers innumerable applications.

The studies of film penetration/desorption are of key importance in surface science [43]. Interactions between different adsorbed species or reversibility of protein adsorption are two examples of interest that still remain unclear in the literature. Unfortunately, there is a scarcity of experimental techniques suitable for studies of penetration/desorption from interfacial layers. The classical way was to obtain a surface layer in a Langmuir trough and then exchange the subphase by injection with a syringe underneath the surface layer [44,45]. The PD technique has advantages in these kinds of experiments due to its smaller dimensions requiring considerably smaller quantities of material and permits a more stringent control of the environmental conditions. However, a number of experimental complications had to be overcome. Cabrerizo et al. designed an innovative experimental accessory to the PD that enables the performance of reliable penetration/desorption experiments by using a coaxial double capillary system that enables a subphase exchange with no disturbance of the interface [46]. Since then, due to the interest of these kinds of studies, this technique has been recently applied and modified. For instance, Miller et al. employed a similar device [47] and Svitova et al. proposed a convection cell in which the liquid in a cuvette is exchanged at a known flow rate [48].

As a consequence, the subphase exchange now offers a wide range of possibilities and can be applied in the study of very different systems. In this section, first the experimental setup is briefly described and subsequently three different applications are explained in detail: a new methodology to obtain adsorbed protein monolayers, a study of the surface interaction of proteins and surfactants, and a direct observation of interfacial enzymatic hydrolysis. Such diverse applications point out the versatility of the technique.

5.3.2 EXPERIMENTAL SETUP

The experimental setup is a constant surface pressure penetration Langmuir balance based on ADSA, which is described in detail elsewhere [49]. In this device, the normal capillary tip has been substituted by an arrangement of two coaxial capillaries, each one being connected to one of the channels of a Hamilton Microlab 500 microinjector. These can operate independently, permitting to vary the interfacial area by changing the drop volume and to exchange the drop content by through-flow [50]. A schematic diagram of the arrangement of the coaxial capillaries connected independently to the two syringes of the microinjector is shown in Figure 5.14.

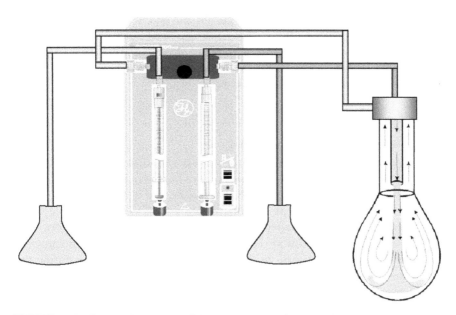

FIGURE 5.14 Schematic diagram of the arrangement of the coaxial capillaries connected independently to the two syringes of the microinjector.

The whole setup, including the image capturing, the microinjector, the ADSA algorithm, and the fuzzy pressure control is managed by a Windows® integrated program (DINATEN). The program fits experimental drop profiles, extracted from digital drop micrographs, to the Laplace equation of capillarity by using ADSA, and provides as outputs the drop volume V, the interfacial tension γ, and the interfacial area A. The interfacial/surface pressure values are obtained from the relationship $\pi = \gamma_0 - \gamma$, where π is the interfacial/surface pressure, γ_0 is the interfacial/surface tension of pure liquid–fluid interface, and γ is the interfacial/surface tension of the liquid covered with the surface active material.

Pressure and area are controlled by varying the drop volume with a modulated fuzzy logic PID algorithm (Proportional, Integral, and Derivative control). The π-A isotherms recorded in the monolayer studies are generated by changing the drop volume in a controlled manner and simultaneously measuring surface tension and surface area. To perform studies at liquid–liquid interfaces, a glass cuvette containing a drop of the denser liquid is filled with the less dense liquid.

The double capillary offers a wide range of possibilities toward the interfacial characterization of a system. The exchange is done by simultaneously extracting bulk liquid of the drop with one of the capillaries and injecting the new solution with the other capillary, illustrated in Figure 5.14. Cabrerizo-Vilchez et al. showed that this rapid subphase exchange does not disrupt monolayers at the interface and that it is complete if at least 250% of the drop volume is pumped through the drop [49]. This accessory allows multiple experiments such as a study of the possible desorption of adsorbed material [51] or the penetration of soluble surfactants into a previously adsorbed interfacial layer [52]. Moreover, it enables the formation of protein monolayers at liquid–fluid interfaces [51,53], and the study of the penetration of insoluble monolayers by some reactant dissolved in the subphase [54,55]. Finally, the combination of the subphase exchange technique with the constant pressure accessory permits the study of adsorption/penetration/reaction kinetics. These applications will be discussed in detail below.

5.3.3 Typical Applications

5.3.3.1 Adsorbed Protein Monolayers

Proteins constitute a significant group of natural emulsifiers and many of their functional properties are derived from the structure that they adopt at interfaces. In this sense, the monolayer technique arises as a useful tool in the study of protein interfacial conformation in the literature [43]. Monolayer studies have been frequently performed in conventional surface balances by applying Trurnit's method for spreading proteins at the air–water interface [56]. Most proteins are soluble in water and the Trurnit method solves this difficulty by denaturing the proteins prior to deposition onto the surface. However, because of the experimental difficulties added by the presence of a second liquid at the interface, there is a lack of experimental studies dealing with protein monolayers at the oil–water interface. Thus, the denaturation process of proteins at liquid interfaces remains fairly unclear in the literature. Likewise, monolayer studies of globular proteins, which are very resistant to denaturation, are also scarce in the literature. The subphase exchange technique provides a new methodology of studying protein monolayers, which overcomes the abovementioned difficulties with globular proteins and liquid–liquid interfaces. In order to test this novel methodology by comparison with literature results, a model protein, β-casein, was studied. β-casein is a well-known protein, present in dairy products, which has a molecular weight of 24 kda and a random coil structure in solution [44,57]. The oil chosen is a model alkane, tetradecane, which was purified prior to use in order to remove surface active contaminants.

The experimental procedure is schematized in Figure 5.15. A protein solution drop is formed at the tip of the capillary and kept at constant surface area while the protein adsorbs freely at the interface forming a layer. Once the desired surface pressure is attained, the bulk solution in the drop is substituted by the aqueous subphase. This is done by extracting simultaneously the bulk solution through the outer capillary and injecting through the inner one the same amount of aqueous subphase at the same flow rate. Details of the experimental procedure at the air–water interface can be found elsewhere [51] as well as for the oil–water interface [53]. After exchange, the behavior at constant surface area is analyzed and the effect of the interfacial pressure of the adsorbed β-casein film can be tested by changing the film pressure at which the subphase exchange is performed.

This technique proposed for obtaining protein films by adsorption from bulk solution has clear advantages over the conventional spreading methods; much less perturbation of the interface and minus diffusion into the bulk, absence of spreading solvents, no losses of protein remaining in the glass rod, and no external contamination of the interface. Furthermore, it generates a reliable interfacial structure

(a)   (b)

(c)

FIGURE 5.15 Visualization of the stages of the formation of an adsorbed protein mono-layer with methylene blue: (a) the protein adsorbs onto the drop surface, (b) the subphase is depleted of protein, and (c) an adsorbed protein monolayer remains at the interface.

exclusively attained by interfacial unfolding and promises to be very useful toward the better understanding of very complex systems practically inaccessible before.

Figure 5.16 shows the time evolution of the interfacial tension of a solution droplet of β-casein before and after the subphase exchange at three different interfacial pressures at the (a) air–water and (b) the tetradecane–water interface. Note that a refer-ence curve without exchange is included in both cases. After the subphase exchange at a fixed interfacial pressure, the behavior of the adsorbed protein layer was moni-tored at a constant surface area.

It can be seen in Figure 5.16 that in all cases, the values of the interfacial tension immediately before and after exchange remain unchanged, indicating that the protein film endures the process at all pressures. Moreover, once the subphase is depleted of protein, the interfacial tension remains essentially constant at both interfaces. This feature suggests not only that the protein is well attached to the interface, but also that the adsorption of β-casein onto the air–water and the tetradecane–water inter-face is probably accompanied by a conformational change. Moreover, the stability of

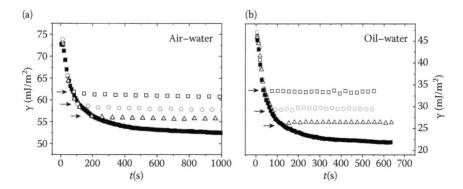

FIGURE 5.16 Reversibility of protein adsorption (β-casein 1.25×10^{-3} mol/m³) at (a) the air-water, and (b) the tetradecane–water interfaces. The arrows indicate the moment at which the subphase exchange was performed: no exchange (closed squares), exchange at $\pi = 5$ mJ/m² (open squares), exchange at $\pi = 10$ mJ/m² (open circles), and exchange at $\pi = 15$ mJ/m² (open triangles).

the interfacial tension after the exchange suggests that such conformational changes should occur upon contact with the interface and provide a new structural configuration to the protein that seems to be irreversibly anchored at both the air–water and the tetradecane–water interface. Hypothetically, an increase in interfacial tension would possibly indicate desorption of molecules from the adsorbed layer.

The stability of the adsorbed layers as deduced from Figure 5.16 suggests that they are suitable for compression-expansion isotherms. These are easily obtained by injecting and extracting clean buffer solution at a certain compression-expansion speed. The compression rate is a critical parameter in the acquisition of reliable isotherms. The films should be compressed slowly enough so that the π-A isotherms obtained represent the equilibrium isotherm and only when π(A) is a single-valued function for all compression-expansion cycles, the layer is considered stable and the data are used for the isotherms.

Figure 5.17 shows the π-A isotherms obtained for adsorbed β-casein layers at (a) the air–water and (b) the tetradecane–water interface. In order to discriminate whether we have an adsorbed monolayer at the interface or not, the experimental isotherms of spread β-casein monolayers at the same two interfaces have been included in each of the figures. The spread monolayer's isotherms have been taken from previous work and were obtained by means of an adaptation of Trurnit's method to the requirements of the PD technique [51,53]. At interfaces, the adsorbed and spread monolayer isotherms are represented versus the area of the drop in Figure 5.17 (upper axis), allowing a direct comparison of the respective isotherms. Furthermore, the spread protein monolayers are also represented versus the specific area of the protein, knowing the amount of protein spread onto the surface (lower axis). This provides structural information of the molecules at the two interfaces. The great similarity of the curves for the two methods is noteworthy.

Regarding the conformation adopted by β-casein at the air–water and the oil–water interface, two features arise from the comparison of Figure 5.17a and b: hysteresis

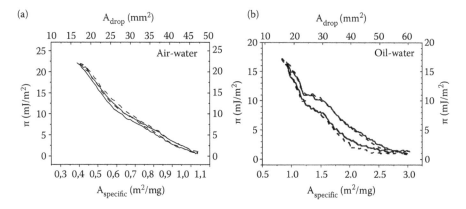

FIGURE 5.17 β-casein isotherms at (a) the air–water, and (b) the tetradecane–water interfaces. The solid lines represent spread protein monolayers (Trurnit's method) and the dashed lines represent adsorbed protein monolayers (subphase exchange).

phenomena and the larger surface area per molecule of β-casein at the oil–water interface. The former might be related to the enhanced solubility of the more hydrophobic amino acid residues of the protein in the oil. This interaction between protein and tetradecane seems to result in a partially irreversible unfolded configuration that is not completely recovered upon expansion of the interface. Accordingly, a further unfolding of the polypeptide chain might be expected, resulting in a higher effective area per molecule at the oil–water interface than found at the air–water interface. Details on the different conformations of β-casein monolayers at the air–water and oil–water interfaces can be found elsewhere [58].

5.3.3.2 Surface Interaction of Proteins and Surfactants

The interfacial behavior of mixtures of proteins and surfactants is a question of increasing interest because of important technological applications. The structural differences between both types of molecules can result in important differences in their interfacial structure as the behavior of mixed systems can be very complex. Despite its enhanced interest, the behavior of mixed protein/surfactant systems still presents unsolved questions. In this sense, the subphase exchange is a powerful tool in the study of the interaction of two different types of soluble surfactants at the surface. The use of this technique, combined with the usual adsorption studies, facilitates the interpretation of the surface pressure isotherms providing innovative structural information about the mixed surface layer. As an illustration, a protein and a soluble surfactant are studied. The protein is β-casein from bovine milk and the surfactant is Tween 20. The latter is a nonionic surfactant of a molecular weight of 1,228 Da. The detailed study can be found elsewhere [52].

Surface characterization of the mixed system is performed in two steps. On the one hand, the surface behavior of the mixed system is evaluated in terms of the competitive adsorption of the mixture onto the surface as compared to that of the individual components. To this end, the surface pressure isotherms of the protein, the surfactant and the mixture of both are recorded. On the other hand, the sequential adsorption is studied by evaluating the effect of surfactant added in the bulk using subphase exchange on a previously adsorbed protein layer at the surface. It will be seen that the combination of both strategies yields crucial information about the behavior of the system that is not readily available by other methods.

Figure 5.18a shows the surface pressure isotherms obtained for β-casein, Tween 20 and the mixed system under the same experimental conditions. The mixtures are formed with a constant concentration of β-casein of $5 \times 10^{-3} \mathrm{mol/m^3}$ and different concentrations of Tween 20 between 10^{-5} mol/m^3 and 1 mol/m^3. For the surface pressure isotherms of pure Tween 20 and the mixed system, the x-axis corresponds to the surfactant concentration. For the surface pressure isotherm of pure β-casein, the x-axis corresponds to the protein concentration. The surface pressure isotherms suggest that the mixed system shows two different patterns at the surface. At low surfactant concentrations, the system behaves like a pure protein solution, whereas at higher surfactant concentration the film shows properties similar to those of the pure surfactant solution.

In detail, at low surfactant concentration, the surface pressure remains practically constant and coincides with that of the protein alone at the concentration used in the mixture. Thus, it may be assumed that the final adsorbed layer is composed basically of protein and that the presence of surfactant does not significantly affect the surface tension. At a well-defined surfactant concentration, 10^{-2} mol/m^3, the behavior of the mixed system changes abruptly. From this concentration on, the surfactant seems to control the adsorption process, and the isotherm recorded for the mixed system practically coincides with that of the surfactant alone. The surface pressure increases very steeply in a narrow range of surfactant concentration and the saturation also occurs at the critical micelle concentration (cmc) of Tween 20 (2×10^{-2} mol/m^3).

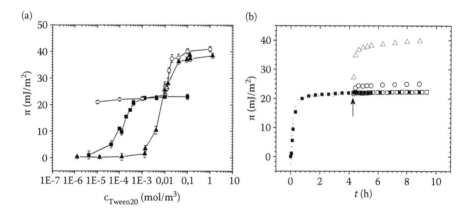

FIGURE 5.18 (a) Surface pressure isotherms of β-casein (solid squares), Tween 20 (solid triangles), and the mixture of 5×10^{-3} mol/m^3 β-casein and Tween 20 (hollow circles) at different concentrations; (b) adsorption kinetics of 5×10^{-3} mol/m^3 β-casein (solid squares). The arrow indicates the beginning of subphase exchange by buffer (hollow squares), by 10^{-3} mol/m^3 Tween 20 (hollow circles) and by 10^{-1} mol/m^3 Tween 20 (hollow triangles).

These experimental results suggest that the surface layer at this high concentration of Tween 20 is composed essentially of surfactant. Apparently, the surfactant might have forced the protein molecules out of the surface layer.

In order to elucidate the phenomena occurring in the surface layer and the possible interaction taking place between the adsorbed protein and the surfactant at the surface, sequential adsorption of the two components onto the surface has been studied by means of the subphase exchange technique. First, the β-casein is allowed to adsorb freely from the pure protein solution onto the surface until a saturated surface layer is achieved; that is, the change in the surface pressure is negligible. Second, the subphase is depleted of protein by exchanging the bulk solution with a buffer (0.05M Tris buffer of pH 7.4). As illustrated in the first application, the adsorbed β-casein layer remains intact after subphase exchange. Under these conditions, the evolution of the surface pressure, after adding Tween 20 to the bulk solution underneath the protein layer, provides direct information of the effect of the surfactant on such a surface protein layer.

Figure 5.18b shows the time evolution of the surface pressure of a solution of β-casein and the effect of exchanging the protein subphase by a surfactant solution at two different bulk concentrations, one below (10^{-3} mol/m^3) and one above (10^{-1} mol/m^3) the cmc of Tween 20. It is found that the surface pressure increases after the exchange with both surfactant concentrations. The increase in surface pressure obtained for the two concentrations clearly indicates that the Tween 20 is able to penetrate the preformed layer of β-casein. Furthermore, in view of the adsorption results of the mixed system (Figure 5.18a), this penetration into the surface layer apparently produces a complete displacement of the protein by the surfactant at the high surfactant concentration. This effect has also been reported by other authors [59]. Accordingly, the surface exchange technique allows an in situ observation of the displacement of the protein by the surfactant at high surfactant concentrations, and provides a satisfactory explanation of the behavior of the mixed system at the surface as shown in Figure 5.18a.

5.3.3.3   Interfacial Hydrolysis

A further application of the subphase exchange technique is to the study of the penetration of insoluble monolayers by soluble surfactants. Here, the exchange is applied to the study of the surface interaction between a lipase (porcine pancreatic phospholipase, PLA$_2$) and a phospholipid monolayer at the air–water interface. Lipases are very important compounds in nature since they are responsible for fat digestion. The digestion of lipids is based on the hydrolysis of dietary glycerides. Pancreatic lipase is able to induce a very fast hydrolysis reaction. This enzyme catalyzes the intraduodenal conversion of long chain triglycerides into the more polar β-monoglycerides and free fatty acids [60]. This process can be reproduced in the PD by looking into the effect of PLA$_2$ on a monolayer of phospholipids as displayed in Figure 5.19. The products of the hydrolysis are more soluble in the aqueous phase and, hence, the addition of PLA$_2$ may result in diminishing the interfacial concentration of phospholipids due to solubilization, thus reducing the surface pressure of the system. Therefore, the interest of this application of the subphase exchange is twofold since it also enables the in situ observation of the interfacial catalysis. Accordingly, in order

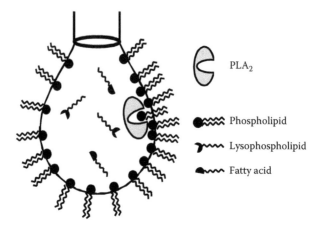

FIGURE 5.19 Schematic of the interfacial hydrolysis of a phospholipid monolayer by PLA$_2$.

to study the surface interaction of these two components and, more importantly, to try to obtain quantitative information of the hydrolysis phenomenon, the following experimental procedure can be used.

First, a spread monolayer of dimyristoyl phosphatidylcholine (DMPC) is formed on a drop of Tris buffer as described in detail elsewhere [54]. Once the spreading solvent has evaporated, the drop is enclosed in the cuvette and is grown to a larger size (typically 30×10^{-9} m^3, i.e., an area of 0.5×10^{-4} m^2), providing sufficient surface area for the monolayer to be in the gas-analogous state ($\pi < 1$ mJ/m^2). Then, the microinjector slowly decreases the drop volume, compressing the monolayer until the desired compression state is reached ($\pi > 5$ mJ/m^2). Then, the film pressure is kept constant by the controller for approximately one minute, allowing the film to equilibrate. Next, the subphase under the monolayer is exchanged with the same buffer containing PLA$_2$ and the pressure control is then activated. The setup maintains the surface pressure constant during the period of enzymatic hydrolysis. The hydrolysis products of DMPC are lyso-MPC and myristic acid, which are more soluble in the aqueous subphase than DMPC. Hence, after the hydrolysis the resultant molecules can be expelled from the monolayer and solubilized. As a result, due to the decrease of interfacial concentration, the surface area of the drop decreases to compensate for the decrease in surface concentration. This decrease continues until the drop is too small to provide reliable surface tension values. This process is repeated for various fixed interfacial pressures and the rate of area decrease depends on the decrease of interfacial concentration. The compression rate hence provides important information about the interfacial hydrolysis occurring in the system.

Figure 5.20a shows the time evolution of the relative interfacial area of a monolayer of DMPC after the subphase exchange with PLA$_2$ (3 g/L or $\sim$0.2 mol/m^3) occurring for each of the fixed surface pressures as explained above. The rate of change of the surface area can be interpreted as a measure of the activity of the enzyme as a function of the concentration of DMPC in the monolayer. In particular, PLA$_2$ seems to be very active for intermediate amounts of phospholipids at the interface

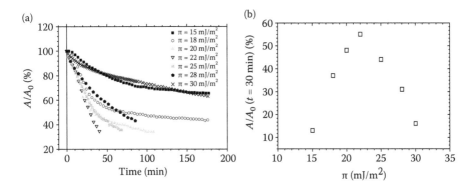

FIGURE 5.20 Enzymatic activity of PLA$_2$. (a) Time evolution of the relative interfacial area of a DMPC monolayer after the subphase exchange with PLA$_2$ (3 g/L), at fixed interfacial pressures; (b) reduced relative interfacial area of a DMPC monolayer 30 minutes after the subphase exchange with PLA$_2$ (3 g/L).

(i.e., intermediate values of the surface pressure), whereas the interfacial hydrolysis is substantially reduced as the concentration either increases or decreases. This complex behavior is shown more clearly in Figure 5.20b, in which the reduction of the relative interfacial area of the DMPC monolayer after 30 minutes is plotted for each of the interfacial pressures studied. The activity of the enzyme shows a very clear maximum at intermediate interfacial concentration of phospholipids. The actual concentration can be estimated from the isotherm of DMPC with no PLA$_2$ recorded in the same experimental conditions [54]. Moreover, this experimental methodology provides a clear visualization of the activity of the enzyme.

The subphase exchange applied in this manner to the study of interfacial catalysis offers many advantages compared to penetration studies in conventional film balances since a unique interface is combined with the fact that in a few seconds the whole subphase is replaced, avoiding further enzyme transport over macroscopic distances and providing a homogeneous enzyme subsurface concentration. Undoubtedly, this experimental procedure provides key and reliable information about the activity of the enzyme toward the better understanding of lipids digestion.

5.4 ADSA FOR ELECTRIC FIELDS (ADSA-EF)

5.4.1 INTRODUCTION

Understanding the influence of an electric field on surface properties of liquids is important from both fundamental and practical standpoints. Charged or electrified drops play a key role in various applications, ranging from microfluidic devices to agricultural treatments. Nevertheless, the effects of the electric field on surface properties of drops are not understood yet, mainly due to the lack of reliable tools and methodologies to measure such effects.

Generally, an electric field may have two kinds of effects on liquid drops or bubbles. First is that the shape of the drop is changed in the electric field. As shown in

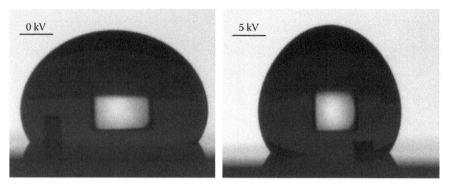

FIGURE 5.21 The effect of an electrostatic field on the shape of sessile drops of water on Teflon coated silicon wafers. The figure shows that the shape of the drop is changed significantly when a 5 kV electric potential is applied to the capacitor with 6 mm distance between the two plates. Note that the two drops are not of the same volume.

Figure 5.21, this is a pronounced effect and can be easily observed in an experiment. Second is the possible effect of the electric field on the surface tension of liquids. This is a relatively subtle effect and is more difficult to detect or measure experimentally. Any study on the effect of electric fields on the surface tension of liquids requires an accurate methodology and experimental procedure.

As described in Chapter 3, ADSA is a powerful methodology for surface tension measurements. However, the standard version of ADSA relies on the assumption that gravity is the only external force deforming the shape of a drop. This assumption is not valid in electric fields (see Figure 5.21), which limits the applicability of ADSA in the presence of such fields. This section presents a generalization of the ADSA methodology to account for both gravity and electric field as operative external forces. The new methodology is called Axisymmetric Drop Shape Analysis for Electric Fields (ADSA-EF).

In essence, the equilibrium shape of a drop in an electric field is determined by balancing the surface tension and the external forces, such as gravity and electric field. Surface tension tends to make a drop spherical, gravity tends to flatten a sessile drop and elongate a PD, and the electric field typically elongates a drop along the direction of the field. The mechanical equilibrium between the surface tension, the electric field and gravity can be described mathematically by the augmented Laplace equation of capillarity [61,62]

$$\gamma\left(\frac{1}{R_1} + \frac{1}{R_2}\right) = \Delta P_0 + (\Delta \rho)gz + \Delta P_e, \tag{5.3}$$

where γ is the surface tension, R_1 and R_2 are the two principal radii of curvature, ΔP_0 is the pressure difference across the interface at the reference (i.e., the apex of the drop), $\Delta \rho$ is the density difference across the interface, g is the gravitational acceleration, z is the vertical distance of any point on the drop surface from the reference, and

ΔP_e is the electrical pressure (i.e., the jump in the normal component of the Maxwell stress tensor across the interface).

The value of ΔP_e at each point of the drop surface depends on the intensity of the electric field at that point, which is unknown for most practical systems. Therefore, Equation 5.3 needs to be solved in conjunction with the Laplace equation for the electric field, which governs the distribution of the electrostatic potential [63–65]

$$\nabla^2 U = 0. \tag{5.4}$$

The values of the electric field along the drop surface can then be calculated as the gradient of the electric potential, U.

The principle behind the ADSA-EF algorithm is similar to that of ADSA. ADSA-EF generates numerical drop profiles as a function of surface tension at a given electric field. Then it calculates the actual value of the surface tension of a real drop by matching the numerical profiles with the shape of the experimental drop, taking the surface tension as an adjustable parameter. The distribution of the electric field along the drop surface is an input for the above calculations, which is not known a priori. No analytical approach is known for solving either the drop shape or electric field distribution for general conditions. Thus, numerical schemes have been developed for this purpose.

Despite the conceptual similarities between ADSA and ADSA-EF, the algorithm and the implementation of ADSA-EF are more complex. For instance, typical versions of ADSA deal with the gravitational force, which is known and constant over the drop surface; ADSA-EF, in addition to gravity, deals with the electric field force. This force is unknown and its magnitude and direction are variable from point to point over the drop surface. Calculation of the distribution of the electric field force at the drop surface is sophisticated and requires precise modeling of the experimental system.

Numerical calculation of drop shapes in the electric field is another challenge in the development of ADSA-EF. Calculation of drop shapes in gravity for given surface tension is fairly straightforward, while drop shape calculation in the presence of an electric field, even when the distribution of the operative forces is known, is not trivial, and by itself is the subject of many studies.

The overall algorithm of ADSA-EF and its main modules are shown in Figure 5.22. A drop is formed in the electric field and experimental images are acquired from its equilibrium shape. The drop profile is extracted from the image using an edge detector. The extracted profile along with the physical geometry of the system and the magnitude of the applied voltage is fed to the electric-field module. This module numerically solves Equation 5.4 to calculate the distribution of the electric field along the drop surface. The electric field distribution and the optimization parameters (e.g., surface tension and apex curvature) are then used as input for the drop-shape module that generates numerical drop profiles by solving Equation 5.3. The optimization scheme of ADSA-EF adjusts the optimization parameters to match the numerical profile with the experimental shape of the drop.

This part of the chapter focuses on the new components of ADSA-EF that do not exist in the other versions of ADSA. The electric-field and the drop-shape

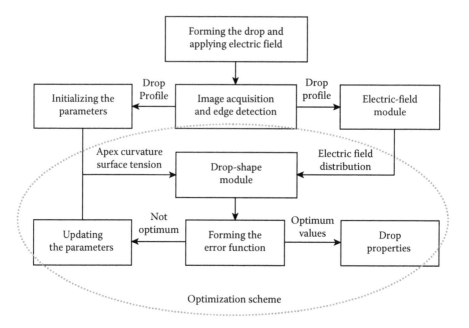

FIGURE 5.22 The algorithm and the structure of the ADSA-EF methodology. The two key modules; that is, the electric field and the drop shape modules, calculate the distribution of the electric field and simulate the shape of the drop, respectively. The optimization scheme (showed by dotted ellipse) calculates the optimum values of the surface tension and the apex curvature, by finding the best match between the numerical and the experimental drop profiles.

modules are the key components developed for ADSA-EF (see Figure 5.22). The former calculates the electric field distribution; the latter generates numerical drop shapes in an electric field, which is significantly different from that of the standard ADSA. The optimization scheme, too, was substantially improved for ADSA-EF. Such improvements were needed to ensure the convergence of the scheme to the global optimum (i.e., convergence of the numerical profile to the experimental drop shape). Furthermore, particular considerations were needed with respect to conducting experiments in the electric field. A new experimental configuration was designed for this purpose. These four components of ADSA-EF are described in detail below.

5.4.2 Constrained Sessile Drop Configuration for Electric Fields

Surface tension measurement using drop shape methods relies on careful experimental procedure and image acquisition from the equilibrium shape of the drop. Conducting such experiments in the presence of an electric field requires certain considerations that are described here. In particular, the development of a CSD configuration for electric fields is described.

As mentioned earlier, the effect of the electric field on the surface tension is fairly small. Measuring such an effect is a goal of ADSA-EF. This requires measurements at a wide range of electric fields to ensure that a detected effect is real and is not a consequence of experimental error. Pendant drop configuration would be an ideal experimental system for this purpose. However, early investigations showed that PDs are not stable at high electric fields, since the electric field detaches the drop from the holder. Consequently, early versions of ADSA-EF were limited to the sessile drop configuration. Typical images of sessile drops of water in an electric field are shown in Figure 5.21. A schematic of the experimental configuration for a sessile drop system is shown in Figure 5.23a. In this configuration the electric field is applied using a parallel plate capacitor.

Generally, it was found that sessile drop experiments are sufficient to show the overall effect of an electric field on surface tension, but they failed to reveal more

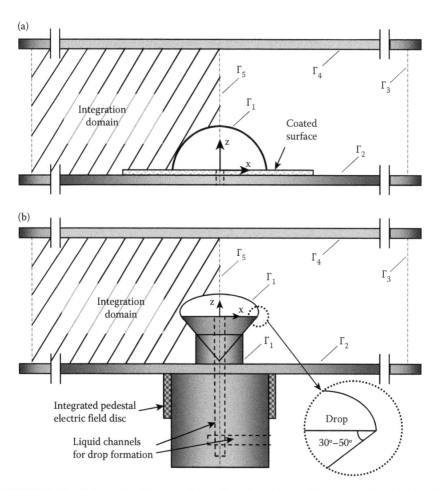

FIGURE 5.23 Schematic of the experimental configuration and the integration domain of ADSA-EF for (a) sessile and (b) constrained sessile drop systems. The pedestal and the lower disc of the capacitor were manufactured as an integrated part. The sharp edge of the pedestal prevents liquid leakage. The symbols Γ_i show the boundaries of the integration domain.

sophisticated patterns. More specifically, surface tension measurements using a sessile drop configuration may suffer from the following shortcomings:

1. Irregularities of the solid surface (e.g., roughness and heterogeneity) may affect the shape of the drop, and hence reduce the accuracy of surface tension results. A drop contact area may not be quite circular, and hence the drop shape violates the axisymmetric assumption of the methodology.
2. The procedure of sessile drop experiments is laborious. It requires careful preparation of a solid sample (i.e., a Teflon-coated silicon wafer) for each run of the experiment.
3. It is cumbersome to model the effect of the Teflon-coated silicon wafer on the distribution of the electric field. The coated wafer sits on the lower plate of the capacitor just beneath the drop (see Figure 5.23a), and hence it affects the distribution of the electric field on the drop surface. Modeling of such effects is not possible due to variable thickness of the coating, limited knowledge about the electrical properties of Teflon, and the complexity of numerical models dealing with nonconducting materials.

To overcome the above limitations, a CSD setup has been developed for ADSA-EF. As described in Section 5.2.3.3, the CSD is a sessile drop formed on a pedestal that employs a sharp knife-edge to prevent drop spreading at low surface tensions (see Figure 5.23b for a schematic of the CSD) [30,66,67]. Figures 5.24a and 5.24b show the experimental images of a pedestal and a CSD, respectively. Details of the design and manufacturing of the pedestal for the electric field can be found in [68].

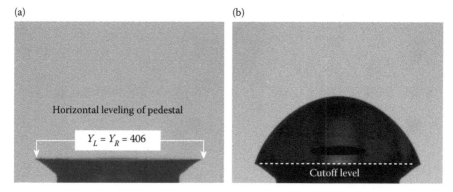

(a)   (b)

Horizontal leveling of pedestal

$Y_L = Y_R = 406$

Cutoff level

FIGURE 5.24 (a) Image of the pedestal before forming a drop, used to level the pedestal and to determine the cutoff level in pixels (i.e., Y_L and Y_R); and (b) the position of the pedestal and the camera remain unchanged during an experiment, hence the cutoff level is calculated once and is used to process images of several runs.

Development of the CSD for electric fields enhances experimental work in the electric field on several fronts:

1. The pedestal enforces a circular base for the drop, which in turn guarantees an axisymmetric shape. Consequently, this setup has the advantages of both PD configuration (i.e., accurate surface tension measurement) and sessile drop configuration (i.e., stability in a strong electric field).

2. The preparation of the pedestal for an experiment only involves sonication with alcohol and water, which is far easier than the coating process required for Teflon surfaces. Moreover, unlike coated surfaces, the pedestal is not disposable and does not interact with the liquids under consideration.

3. The influence of the pedestal on the distribution of the electric field can be modeled without difficulty, since (a) unlike Teflon coated surfaces, the geometry of the pedestal is precisely known; and (b) the pedestal is made of a stainless steel, hence it can be dealt with as a perfect conductor. That is, the surface of the pedestal forms an equipotential area, which simplifies the formulation of the problem (see the next section for more details).

4. Unlike typical sessile drop experiments, advance of the drop front is not necessary for CSDs. Experiments on solid surfaces often involve advancing of the three-phase line to avoid complications due to contact angle hysteresis. This is not required for experiments on a pedestal. Among advantages of a static (not advancing) experiment are: (a) higher reproducibility and better confidence limits of the results—this is essential for sensitive experiments in the electric field, (b) easier focusing of the camera and better image acquisition, and, most importantly, (c) better control of the stability of the drop. The stability of a drop in the electric field significantly depends on its volume (the larger the drop, the less stable it is). Hence, conducting an experiment with an advancing three-phase line (i.e., increasing the drop volume during the experiment) is not readily possible in a high electric field where the drop is close to its stability limit.

5. Horizontal leveling of the pedestal is straightforward. A common difficulty for sessile drop experiments is to ensure that the surface of the Teflon coated wafer is indeed horizontal; that is, the right and left contact points of the drop are at the same horizontal level in the digital image. This can be easily accomplished for a pedestal configuration using an image of the pedestal before forming the drop (see Figure 5.24a) and employing simple image processing techniques.

6. Detection of the cut-off levels (contact point of the drop) is simple for the new configuration. The position of the camera/pedestal system is kept unchanged during a round of experiment that may consist of several runs; hence, the cutoff levels need to be calculated only once (see Figure 5.24b). Inaccurate detection of cutoff levels is a major source of error in sessile drop methods.

Employing the above experimental configuration is fast and simple in the electric field and it leads to accurate and reliable measurements. The distortion of the electric

field caused by the pedestal-drop system should be calculated next. This calculation requires precise modeling of the experimental system (including the capacitor,
pedestal, and drop) and is performed in the electric field module of ADSA-EF. This
module is described next. Further details regarding the experimental procedure in
the electric field can be found in Section 5.4.6.

5.4.3 Electric Field Module

The electric field module of ADSA-EF calculates the distribution of the electrostatic
field along the surface of the drop by solving Equation 5.4 [61]. The electric field
values are then substituted in Equation 5.3 to calculate the shape of a drop in the
electric field [61]. It is known that the accuracy of the calculated drop shapes, as
well as the ultimate accuracy of ADSA-EF, significantly depends on the quality of
the numerical techniques used for the electric field calculation [68,69]. Suitability of
different numerical approaches for this purpose is examined, and the features of the
numerical model particularly developed for ADSA-EF are described here.

5.4.3.1 Mathematical Formulation

The mathematical formulation of the electrostatic field problem is derived by express-
ing Equation 5.4 as a family of two-dimensional differential equations [70,71].
Figure 5.23a and b show the integration domain and the boundaries for a case where
a conducting drop is formed on a Teflon-coated silicon wafer or a pedestal in a parallel plate capacitor. In this case, no electric field exists within the conducting drop or
the pedestal, so the integration domain is limited to the area outside the drop. Taking
advantage of the axisymmetric nature of the problem, the governing equation of this
problem in a cylindrical coordinate system can be expressed as

$$\frac{\partial^2 U}{\partial x^2} + \frac{\partial^2 U}{\partial y^2} + \frac{1}{x} \cdot \frac{\partial U}{\partial x} = f(x,y) \qquad \text{over } \Omega \qquad (5.5)$$

subject to the following boundary conditions:

$$U = 0 \qquad \text{on } \Gamma_1 \text{ and } \Gamma_2,$$
$$U = 1 \qquad \text{on } \Gamma_4,$$
$$\left(\frac{\partial U}{\partial x} \times \vec{i} + \frac{\partial U}{\partial y} \times \vec{j}\right) \times \vec{n} = 0 \qquad \text{on } \Gamma_5,$$
$$\left(\frac{\partial U}{\partial x} \times \vec{i} + \frac{\partial U}{\partial y} \times \vec{j}\right) \times \vec{n} = g(x,y) \qquad \text{on } \Gamma_3,$$

Ω is the integration domain, x is the radial (horizontal) coordinate, z is the vertical
coordinate (axis of symmetry), Γ_1 is the drop and pedestal boundary (grounded), Γ_2

is the lower plate of the capacitor (grounded), Γ_3 is a boundary at which the gradient of the electric potential is known; that is, $g(x,y)$, Γ_4 is the upper plate of the capacitor (hot plate), Γ_5 is the axis of symmetry at which the radial (horizontal) component of the electric field is zero, and $f(x,y)$ is the distribution of the possible charge density within the integration domain.

The above formulation is valid for various geometries and charge distributions. For a simple case where the drop is formed in a parallel disc capacitor, the boundary Γ_3 is defined far from the drop so that the electric field is uniform along Γ_3 and $g(x,y) = 0$. Moreover, for most practical cases, no free charge density exists within the capacitor, hence $f(x,y) = 0$. All the variables are cast into a dimensionless form so that the magnitude of the applied potential (i.e., the potential of the hot plate) and the distance between the capacitor plates are unity.

5.4.3.2 Numerical Scheme

The choice of the numerical method for solving the above system of partial differential equations is critical for accuracy and efficiency of ADSA-EF. In principle, this system can be solved using various numerical methods; however, the choice of the optimum method depends on the particular application. The feasibility of two different approaches (i.e., finite difference and finite element methods) for the electric field module of ADSA-EF is examined here.

As the first alternative, a second-order finite difference scheme with a Cartesian grid has been developed. This method employs a difference formula to calculate the magnitude of the electric potential at the nodal (mesh) points. The nodal points are distributed evenly over the integration domain. The difference formula is not applicable to the nodal points that do not coincide with the boundary (e.g., the points next to the curved drop boundary). Linear interpolation techniques were employed to approximate the electric potentials at these points [72,73].

The following third-order difference formula [73–75] was used to calculate the component of the electrostatic field along the radial coordinate (x direction) from the values of the electric potential at the nodal points

$$\vec{\nabla} U_x = \frac{2U_{i+4} - 9U_{i+3} + 18U_{i+2} - 11U_{i+1}}{6h}, \tag{5.6}$$

where h is the distance between the nodes and i is an index referring to the nodal points in the radial direction. A similar formula was used for the calculation of the component of the electrostatic field along the axis of symmetry (i.e., z direction).

The sensitivity of the numerical results to the number of mesh points is used to determine the optimum mesh size. The value of the electric field at the drop apex is considered for this purpose. Table 5.3 shows the numerical values of the dimensionless electric field at the apex of a hemispherical drop for different numbers of mesh points. The results converge to the true electric field, which is equal to three (see below), with increasing number of mesh points.

The accuracy of the finite difference results is further improved by calculating the order of the discretization error [73]. Let E_i be the calculated electric field at the drop apex for mesh length h_i. Taking the leading term in the discretization error proportional to h^p, the value of p can be estimated from [73]

$$2^p = \frac{E_2 - E_1}{E_3 - E_2},\tag{5.7}$$

where $h_1 = 2 \times h_2 = 4 \times h_3$. Using the values of Table 5.3 for 250, 500, and 1000 mesh points as input results in a value of $p = 0.938$ for the order of the discretization error. Then the magnitude of the electric field at any node can be calculated with higher accuracy as [73]

$$E = \frac{h_2^p \times E_1 - h_1^p \times E_2}{h_2^p - h_1^p}.\tag{5.8}$$

For instance, by substituting the results of Table 5.3 for 100 and 150 mesh points in Equation 5.8 the electric field at the apex is calculated as $E = 3.012$, significantly more accurate than the original values shown in Table 5.3. This analysis was conducted for different numbers of mesh points. It was found that by solving the problem using 200 and 400 mesh points and taking advantage of Equation 5.8, the electric field can be calculated fairly efficiently and with a maximum error of less than 1%.

TABLE 5.3

The Calculated (Numerical) Electric Field at the Apex of a Hemispherical Drop

Number of Mesh Points	Numerical Electric Field at the Apex
20	2.287
50	2.628
100	2.801
150	2.868
200	2.903
250	2.924
300	2.939
400	2.958
500	2.969
800	2.986
1000	2.992

Note: The first column corresponds to the number of mesh points in the radial direction (i.e. along the capacitor plate). Electric field values are dimensionless (see Figure 5.25). The results converge to the true value of the electric field, which is equal to three (see Figure 5.26), as the number of mesh points increase.

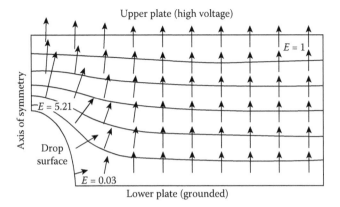

FIGURE 5.25 The distribution of the electrostatic field along with the equipotential lines calculated over the integration domain. The length of the arrows signifies the magnitude of the electric field. Dimensionless variables were used for the calculation, so that the electric field is one ($E = 1$) far from the drop, where the electric field is uniform. The maximum electric field (the highest density of equipotential lines) occurs at the apex of the drop. The electric field then decreases continually from the apex to the contact point.

Figure 5.25 illustrates the calculated distribution of the electrostatic field (arrows) and the equipotential lines. The dimensionless magnitude of the electric field is unity far from the drop where its distribution is uniform. The figure shows that the electric field is maximum at the apex, and it is almost zero at the contact point of the drop. This was anticipated since the drop is conducting and there is no charge density at the contact point. Furthermore, it should be noted that the plotted equipotential curves are dimensionless; that is, they are independent of the magnitude of the applied potential and only depend on the shape and geometry of the drop-capacitor system. (More numerical results regarding the effect of the drop geometry on the electric field are shown in Figure 5.28.)

In order to validate the formulation and the implementation of the numerical scheme, the results were compared with the analytical solutions that exist for the simple case of hemispherical drops. For this case, the method of images [76] can be used to calculate the distribution of the electric field at the drop surface as [77]:

$$E = 3\cos(\theta), \quad 0 \le \theta \le \frac{\pi}{2}, \quad (5.9)$$

where θ is the polar angle measured from the drop apex. Figure 5.26 shows the numerical and analytical electric fields calculated along the drop surface.

Further verification of finite difference results revealed a fluctuation in electric field values unless a very fine mesh (e.g., 1000×500 nodes) was used. This error is believed to be due to the fact that the nodal points at which the electric potential is calculated do not coincide with the curved boundary of the drop, causing an error in the numerical calculations [72–75].

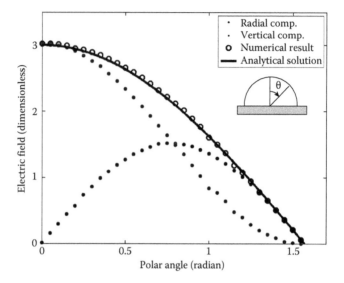

FIGURE 5.26 Comparison of the analytical solution (solid line) with the numerically calculated electric field at several points along the drop surface (circles). The good agreement validates the numerical scheme. The illustrated radial (arch-shaped) and vertical (S-shaped) components of the electric field were used in the calculation of the numerical electric field. The polar angle was measured from the apex to the contact point.

A Galerkin finite element scheme was developed as a second alternative for the electric field module and to remedy the above shortcoming [68,69]. This scheme takes advantage of various advanced capabilities of the finite element approach that lead to higher accuracy and performance. Examples of the features of this scheme are:

1. The mesh can be generated so that element vertices (i.e., the nodal points) coincide with the drop boundary.
2. Smaller elements can be used for the area close to the drop surface, where (a) the maximum gradient of the electric potential exists, and (b) the highest accuracy is required. Similarly, large elements can be used far from the drop where the electric field is almost uniform to gain computational efficiency.
3. The new scheme can account for various geometries, which allows study of different experimental configurations. For instance, it can be applied to pedestals with any given size and geometry, as well as to typical sessile drop configurations.
4. The scheme can be easily modified in the future to account for nonconducting liquids. For that case, the electric field needs to be calculated both inside and outside the drop. Hence, a discontinuity (caused by the drop surface) will exist in the integration domain. Such a problem can be easily formulated using a finite element method.

Several components of the Galerkin finite element scheme that were specifically developed for ADSA-EF (i.e., the techniques used for modeling the geometry, discretizing the domain, and calculating the electric field) are described in the rest of this section. Other technical details regarding the numerical scheme (e.g., deriving the weak formulation of the problem, deriving the characteristic matrices and vectors, and the elemental construction) are not described here.

5.4.3.3 Modeling the Geometry

The geometry of the experimental system (i.e., the capacitor, pedestal, and drop) should be precisely modeled as it defines the boundaries of the integration domain. An important advantage of the finite element scheme developed for ADSA-EF is its versatility to model various configurations. In particular, the scheme is applicable to both pedestal and sessile drop configurations (see Figure 5.27). The geometry of a given capacitor/pedestal is defined simply by providing the dimensionless coordinates of the corner points. However, defining the drop boundary is more complex. ADSA-EF provides several alternative approaches for defining the drop boundary. The choice of the approach depends on the particular application.

In the first approach, the boundary can be generated analytically as a section of an ellipse. This option is particularly useful to test and evaluate the scheme and to study theoretically the effect of the drop geometry on the distribution of the electric field [61]. Various hypothetical drop shapes can be generated by defining the horizontal and vertical diameters of the ellipse. Figure 5.28 shows the effect of drop geometry on the distribution of the electric field on the drop surface. This approach is applicable to both sessile drop and pedestal configurations. When a pedestal is used, the horizontal diameter is equal to the diameter of the surface of the pedestal, assuming

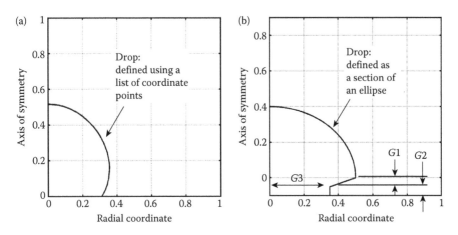

FIGURE 5.27 The geometry of the electric field problem defined for (a) an experimental sessile drop, and (b) a hypothetical constrained sessile drop defined using a portion of an ellipse with the vertical and horizontal diameters of 0.8 and 1.0 (dimensionless). The shape of any given pedestal is determined using three geometries, $G1$ to $G3$. The origin of the coordinate system is defined at the base of the drop.

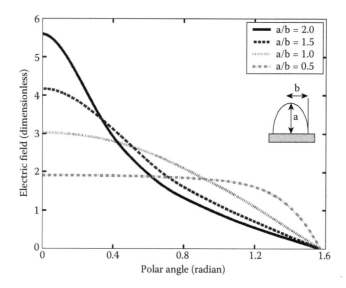

FIGURE 5.28 The effect of the drop geometry on the distribution of the electric field. Hypothetical drop shapes were generated analytically as portions of an ellipse.

that the drop completely covers the pedestal. As an illustration, Figure 5.27b shows the domain of the problem for a hypothetical drop formed on a pedestal.

As a second alternative, the drop boundary can be defined using a numerical drop profile, often generated by solving Equation 5.3 [61]. This option is required when no experimental drop shape is available, and particularly to predict the shape of a drop in the electric field using a successive approximation approach; that is, ALFI-EF (details of ALFI-EF can be found in [61,68]). In this case, the boundary is not described by a *continuous* analytical function; it is rather defined by a list of coordinate points connected by linear interpolation, a more complex algorithm. The ultimate accuracy of the scheme will depend on the number of coordinate points used to define the drop profile. For this reason, a fixed and small step size is preferred while Equation 5.3 is solved to generate the drop profile. (Employing an adaptive step size is not recommended for this purpose although it would allow larger steps.) Generally, it was found that defining a drop profile by 1000 coordinate points results in the same accuracy as a drop defined by an analytical function (see Figure 5.29).

A third option for defining the drop geometry is to use an experimental image of the drop in conjunction with edge detection. This option is useful when ADSA-EF is employed to measure surface tension by analyzing experimental drop images. Similar to the second option, the drop is defined by a list of coordinate points. However, it was found that the pixel coordinates obtained from the edge detection are not sufficient to provide enough accuracy, since (a) the number of pixel points (the image resolution) is inadequate, and (b) a profile extracted by an edge detection is not smooth enough for our purpose (see Figure 5.30). It was found that the jagged shape of an experimental profile significantly affects the calculated electric field, as the electric field tends to be higher at a sharp edge (see Figure 5.31). Interestingly, this effect becomes

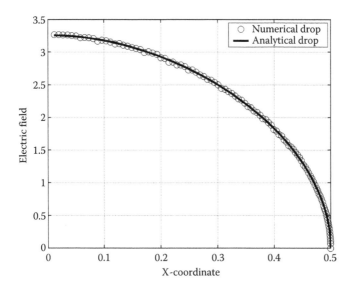

FIGURE 5.29   The electric field distribution (dimensionless) along the surface of a hemi-spherical sessile drop with diameter of 0.5, where the drop profile is defined analytically (solid line) and numerically using a list of 1000 coordinate points (circles). The good agreement suggests that the number of points in the numerical approach is sufficient to accurately define the boundary.

more significant as the number of nodal points on the drop surface increases (i.e., the mesh is refined) since it is more likely that a fine mesh detects the sharp edges on the profile. Consequently, experimental drop profiles are smoothed using a cubic spline method (see Figure 5.30). Then, a sufficient number of coordinate points is chosen from the smoothed curve and used to define the drop geometry.

Figure 5.27 shows the geometry of the numerical model defined for (a) an experimental sessile drop (i.e., the third alternative above), and (b) a hypothetical CSD configuration (i.e., the first alternative above). The shape of any given pedestal is determined using three parameters, $G1$ to $G3$. In this figure, $G1 = 0.05$, $G2 = 0.05$ and $G3 = 0.35$, and the horizontal and vertical diameters of the drop are equal to 0.50 and 0.40, respectively. The geometry is defined in a dimensionless form such that the distance between the two plates of the capacitor is unity. The origin of the coordinate system is always defined at the base of the drop. It was found that the electric field distribution is uniform beyond the radial coordinate of $x = 1$; therefore, for better computational efficiency, the integration domain was limited to a 1×1 square (i.e., the boundary Γ_3 is defined at $x = 1$; see Figure 5.23 for details).

5.4.3.4   Discretization of the Domain

After defining the boundaries of the model, the integration domain is discretized by a triangular mesh generator [78–80]. The quality of the mesh is controlled by defining the maximum area and the minimum angle constraints for the elements. The number of nodal points on the drop boundary is also prescribed to control the accuracy of the electric field calculated on the drop surface. The optimum combination

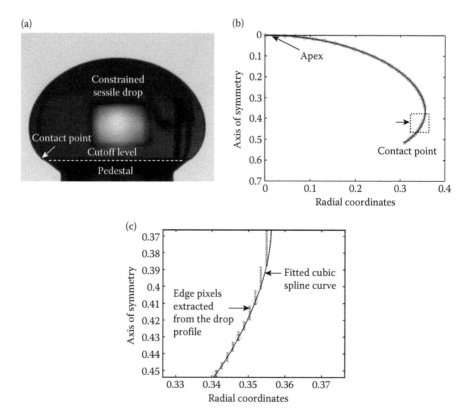

FIGURE 5.30 (a) An experimental image of a constrained sessile drop of water; (b) the experimental profile of the drop, extracted by the Canny edge detector, along with the fitted cubic spline curve; and (c) the magnified view of the section shown by the dotted square. The cubic spline curve was calculated using the coordinates of the 30 pixels highlighted by circles.

of these three parameters is determined by evaluating the accuracy of the resulting electric field versus the computational time.

The discretization of the domain leads to a system of linear equations. Two different numerical methods were used to solve the resulting system of equations: the direct Gauss-Jordan elimination [70] and the iterative conjugate gradient method [71]. Both methods were validated by comparing the results. However, conjugate gradient method was employed in the final version due to several advantages: (a) it is known to be efficient for finite element matrices; (b) it will converge to the final (exact) numerical solution using a finite number of iterations; and (c) it can be developed in a form that does not require assembly of the matrices. Hence, it is more efficient than other techniques regarding both time and memory usage [68].

5.4.3.5 Electric Field Calculation

Finally, the electric field distribution can be calculated from the values of the electric potential calculated at the nodal points. Gradient calculations are fairly

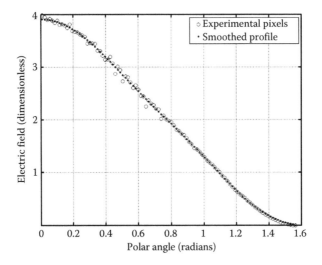

FIGURE 5.31   The electric field distribution calculated along the surface of the sessile drop shown in Figure 5.30. Employing experimental pixels to define the drop boundary resulted in a fluctuation in the electric field distribution (circles). This problem was solved by smoothing the drop profile using a cubic spline fit (see Figure 5.30).

straightforward for a finite element scheme compared to finite difference methods that require additional approximations [61]. In the first-order finite element scheme developed here, the electric potential is approximated linearly over each element:

$$u^k = u_1 \xi_1 + u_2 \xi_2 + u_3 \xi_3, \tag{5.10}$$

where u^k is the approximated (linear) electric potential over a given element k, the u_i's are the calculated nodal values, and ξ_i is the natural coordinate system for element k. Consequently, the slope (gradient) of the electric potential is simply calculated over each element as:

$$\begin{cases} \dfrac{\partial u^k}{\partial x} = u_1 \times \dfrac{\partial \xi_1}{\partial x} + u_2 \times \dfrac{\partial \xi_2}{\partial x} + u_3 \times \dfrac{\partial \xi_3}{\partial x} \\[3mm] \dfrac{\partial u^k}{\partial y} = u_1 \times \dfrac{\partial \xi_1}{\partial y} + u_2 \times \dfrac{\partial \xi_2}{\partial y} + u_3 \times \dfrac{\partial \xi_3}{\partial y} \end{cases}. \tag{5.11a}$$

This equation, along with the following relations, is employed to calculate the horizontal and vertical components of the electric field over a given element [71,81–84]:

$$\left. \frac{\partial \xi_i}{\partial x} \right|_{\Omega^k} = \frac{1}{2 \times area^k} \left(y_{[i+1]}^k - y_{[i+2]}^k \right), \tag{5.11b}$$

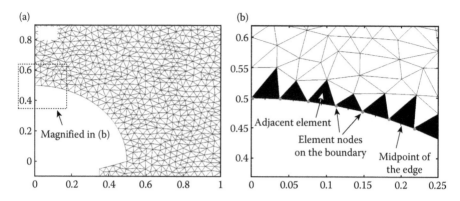

FIGURE 5.32 (a) A typical triangular discretization of the domain, and (b) a magnified view of the mesh close to the drop apex. The electric field is calculated over the elements adjacent to the drop boundary; that is, elements having two nodes on the boundary (shaded). The calculated value of the electric field over each element was assigned to the midpoint of the element edge bordering the drop surface.

$$\left.\frac{\partial \xi_i}{\partial y}\right|_{\Omega^k} = \frac{1}{2 \times \text{area}^k}\left(x_{[i+2]}^k - x_{[i+1]}^k\right),\tag{5.11c}$$

$$\text{area}^k = \frac{1}{2}\begin{vmatrix} 0 & 0 & 1 \\ x_{[i+1]}^k - x_{[i]}^k & y_{[i+1]}^k - y_{[i]}^k & 0 \\ x_{[i+2]}^k - x_{[i]}^k & y_{[i+2]}^k - y_{[i]}^k & 0 \end{vmatrix},\tag{5.11d}$$

where $x_{[i]}$ and $y_{[i]}$ are the coordinates of the vertex i ($i = 1$, 2, or 3) of the element and area^k is the area of the element k.

In order to calculate the distribution of the electric field along the surface of the drop, all the elements adjacent to the drop boundary (i.e., having two nodes on the boundary) were selected (see Figure 5.32). The calculated value of the electric field at each element was assigned to the center point of the element edge bordering the drop surface (see Figure 5.32). Figures 5.29 and 5.30 show the graph of calculated electric field as a function of the radial coordinate and the polar angle, respectively. In these graphs, each data point represents the electric field over an element adjacent to the drop boundary.

5.4.3.6   Evaluation and Tuning of the Electric Field Module

The accuracy of the scheme is evaluated by comparing the numerical electric field with the analytical solution that exists only for the apex (see Figure 5.26). This evaluation is particularly useful to optimize the discretization parameters.

The quality of the mesh generated in this module is controlled using three parameters: (a) the minimum angle of the elements, A, (b) the maximum area of the elements, S, and (c) the number of nodes on the drop boundary, N. Elements with equal angles, i.e., 60°, produce the best accuracy. However, it is difficult (often impossible)

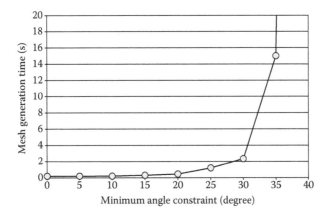

FIGURE 5.33 Time required for generating a mesh with parameters of $S = 0.001$ and $N = 100$ for a typical domain as a function of the minimum angle constraint, A. The graph suggests that a minimum angle of 30° is feasible with respect to the computational time. Calculations were performed using an Intel Pentium 1.6 GHz processor.

to discretize a domain using such triangles. Figure 5.33 shows that a minimum angle of 30° for the element is feasible with respect to the computational time.

The other two criteria both affect the size of the elements; hence, they must be evaluated simultaneously. The number of nodes on the drop boundary, N, determines the size of elements close to the drop, as well as the number of data points for the distribution of the electric field. The maximum area constraint, A, determines the general size of elements over the domain. Comprehensive analysis was performed to evaluate the effect of these parameters on the generated mesh and the calculated electric fields (details of this analysis can be found in [68]). Figure 5.34 shows that prescribing 100 nodes on the drop boundary using a maximum element size of 0.001 (dimensionless) results in a numerical error of the order of 0.01%. Such an error will have a negligible influence on the ADSA-EF results, and hence is accepted for this study. This error is one tenth of the error of the finite difference scheme described before (see Table 5.3).

Normally, the tolerance of the iterative solver needs to be evaluated and adjusted. The solver implemented in this module (i.e., conjugate gradient), converges to the exact numerical solution using a finite number of iterations (less than 500 iterations for a typical mesh). Therefore, no analysis was required to determine the tolerance for the solver.

5.4.4 DROP-SHAPE MODULE

The drop-shape module numerically integrates the Laplace equation of capillarity (i.e., Equation 5.3), to simulate the shape of conducting drops when both gravity and electric field are present. The values of the surface tension, drop apex curvature, and the electric field distribution are needed as input to this module; they are calculated by the optimization scheme and the electric field module (see Figure 5.22).

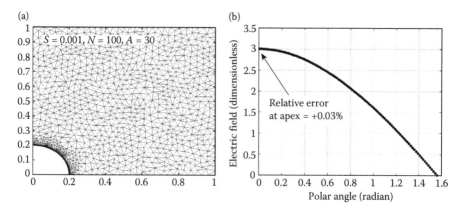

FIGURE 5.34 Evaluation of the electric field distribution along the surface of a hemispherical drop with a radius of 0.2 (dimensionless). An analytical solution exists for this problem indicating that the electric field is equal to 3 (dimensionless) at the apex. The electric field can be calculated accurately by prescribing 100 nodes along the drop boundary (i.e., $N = 100$), and by restricting the maximum size of the elements to 0.001 of the area of the integration domain (i.e., $S = 0.001$).

The calculated numerical drop profiles will then be fitted to the experimental ones through an optimization process (see Figure 5.22).

The electrical pressure on the right-hand side of Equation 5.3, ΔP_e, depends on the magnitude and direction of the electric field, as well as the permittivity of the fluids [77,85,86]

$$\Delta P_e = \frac{1}{2}[\varepsilon^{(a)}E_n^{(a)2} - \varepsilon^{(b)}E_n^{(b)2} + (\varepsilon^{(b)} - \varepsilon^{(a)})E_t^2], \qquad (5.12)$$

where E_n and E_t are the normal and the tangential components of the electric field at the drop surface, ε is the permittivity of the fluid, and superscripts a and b refer to the surrounding fluid and the drop liquid, respectively. When the drop is a conducting liquid there is no electric field inside the drop and the electric field is normal to the drop surface. Thus, the governing equation (i.e., Equation 5.3), for conducting drops can be simplified as

$$\gamma\left(\frac{1}{R_1} + \frac{1}{R_2}\right) = \Delta P_0 + (\Delta\rho) \times g \times z + \frac{1}{2}\varepsilon^{(a)} \times E_n^{(a)2}. \qquad (5.13)$$

The two-dimensional differential form of Equation 5.13 can be obtained when the principal radii of curvature (i.e., R_1 and R_2) are replaced with the corresponding differential terms for the axisymmetric shape (See Chapter 3)

$$\frac{1}{R_1} = \frac{d\phi}{ds}, \qquad (5.14a)$$

$$\frac{1}{R_2} = \frac{\sin\phi}{x},$$
(5.14b)

where ϕ is the angle of inclination of the interface to the horizontal and s is the arc length from the apex (see Figure 5.35).

Due to the symmetric nature of the problem, the curvature at the apex is constant in all directions; that is, the two principal radii of curvature are equal:

$$\frac{1}{R_1} = \frac{1}{R_2} = \frac{1}{R_0} = b \qquad \text{at the apex } (s = 0),$$
(5.15)

where R_0 is the radius of curvature and b is the curvature, both at the apex. Moreover, by defining the apex as origin (i.e., $z = 0$ at the apex), the gravitational term vanishes and Equation 5.13 for this point reduces to

$$2b\gamma = \Delta P_0 + \frac{1}{2}\varepsilon^{(a)} \times E_n^{(a)2} \qquad \text{at the apex } (s = 0).$$
(5.16)

The second term on the right-hand side of Equation 5.16 is a known constant and its value can be calculated using the electric field module. Therefore, the pressure difference at the reference (apex of the drop) can be expressed as

$$\Delta P_0 = \gamma\left(2b - \frac{K}{\gamma}\right) \qquad \text{at the apex } (s = 0),$$
(5.17)

where K is the known electric-field term in Equation 5.16. Substituting Equations 5.14 and 5.17 into Equation 5.13 yields

$$\frac{d\phi}{ds} = 2b - \frac{K}{\gamma} + \frac{1}{2\gamma}\varepsilon^{(a)} \times E_n^{(a)2} - \frac{\sin\phi}{x}.$$
(5.18)

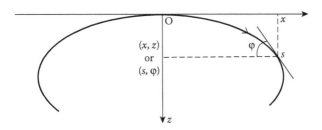

FIGURE 5.35 The coordinate system used for the integration of the drop shapes in the electric field (i.e., the drop-shape module).

Equation 5.18 together with the geometric relations

$$\frac{dx}{ds} = \cos\phi \qquad\qquad (5.19)$$

$$\frac{dz}{ds} = \sin\phi \qquad\qquad (5.20)$$

forms a set of first-order differential equations in terms of φ, x, and z as functions of the arc length, s. The boundary conditions for the drop-shape problem can be defined as (see Figure 5.35):

$$\phi(0) = x(0) = z(0) = 0 \qquad \text{at the apex } (s = 0), \qquad (5.21)$$

$$\frac{d\phi}{ds} = b \qquad \text{at the apex } (s = 0), \qquad (5.22a)$$

$$\frac{dx}{ds} = 1 \qquad \text{at the apex } (s = 0), \qquad (5.22b)$$

$$\frac{dz}{ds} = 0 \qquad \text{at the apex } (s = 0). \qquad (5.22c)$$

A fourth-order Runge–Kutta [70] scheme with an adaptive stepsize control was implemented to solve the above system of differential equations (i.e., Equations 5.18 through 5.22). The stability of the scheme was tested by generating numerical drop profiles for a wide range of input parameters (within the stability limit of the drop); that is, surface tension, apex curvature, and magnitude of the electric field. No limitation was observed with respect to the computational effort or stability of the program.

5.4.5 Development of an Automated Optimization Scheme

As described earlier, ADSA-EF calculates surface tension of liquids by fitting numerical drop profiles to the experimental profile of the drop. This requires calculation and minimization of an error function that quantifies the difference between the two profiles. In principle, this process is similar to that of the standard versions of ADSA. However, it was found that the optimization methods employed in current versions of ADSA fail to converge to the minimum error function when the electric field is high (i.e., >3 kV applied electric potential). Consequently, an improved optimization scheme was developed.

The deviation of the numerical curve from the experimental profile is quantified using a least square error function ε (γ, b, x_a, y_a) (See Chapter 3). The value of this function depends on surface tension, γ, and apex curvature, b, which determine the shape of the numerical profile, as well as on the coordinates of the apex, (x_a, y_a), which determine the position of the numerical curve with respect to the experimental profile. Therefore, minimization of the error function involves a four-parameter optimization.

The established Gauss-Newton method was employed to minimize the error function [87]. The method is known to be robust and reliable; however, as all multi-variable optimization methods, it essentially relies on good estimates (initial values) of the optimization parameters. Cabezas et al. [88,89] have shown that the key for calculation of sufficiently accurate initial values is to estimate the parameters in a particular order; that is, the coordinates of the apex, the curvature at the apex, and, finally, the surface tension.

In the new optimization module, the coordinates of the apex are estimated using a simple image processing procedure (see Figure 5.36). First, the experimental drop profile is extracted using an edge detection method. Next, the position of the drop axis is determined as follows:

1. an arbitrary horizontal row, j, of the digital image is chosen;
2. the two pixels of the drop profile corresponding to the row j of the image are identified; and
3. the center point of the two pixels is calculated.

The above steps are repeated for a number of different rows of the image and the position of the axis is calculated as the average of the x-coordinate of the center points. During this calculation, anomalous values (that may be caused by experimental noise) are rejected through a statistical process. Finally, the coordinate of the apex is calculated by intersecting the drop axis with the experimental drop profile. The last step requires interpolation of the drop profile, since the pixels of the profile are integers but the position of the drop axis is a real number (see Figure 5.36).

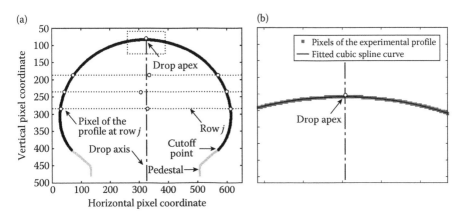

FIGURE 5.36 Estimating the position of the apex for the drop shown in Figure 5.30a: (a) the extracted experimental profile, several arbitrary horizontal lines, and the calculated axis of the drop; and (b) the magnified view of the apex area, a fitted cubic spline curve, and the intersection of the axis with the profile (i.e., the drop apex).

The curvature of the apex is estimated in the next step. It was shown [88,89] that numerical curves, which are calculated using a constant apex curvature and different surface tensions, coincide closely near the apex (i.e., over a vertical distance of about 1/6 of the drop length). This suggests that the shape of numerical (Laplacian) profiles close to the apex is a function of apex curvature only and is almost independent of surface tension. Taking advantage of this finding, the apex curvature is estimated as follows:

1. the coordinates of the apex are estimated as explained above;
2. numerical (Laplacian) drop profiles are generated using the estimated coordinates of the apex and an arbitrary value of surface tension; and
3. 1/6 of the experimental drop profile close to the apex is fitted to the numerical curve by adjusting the apex curvature as the only optimization parameter. The best fit represents the estimate of the apex curvature.

The value of the surface tension is estimated next by analyzing the whole drop profile. A Laplacian curve is fitted to the experimental profile using the estimated coordinates and curvature of the apex, and employing a single-parameter optimization. Figure 5.37 shows the resulting numerical curve after each stage of the estimating process.

Finally, the Gauss-Newton optimization method is employed to fit a Laplacian curve into the experimental profile through a four-parameter optimization, starting from the numerical curve obtained by the above estimating process. It was found that the estimated parameters are sufficiently accurate to guarantee the convergence of the algorithm to the "global" minimum. Furthermore, the above optimization

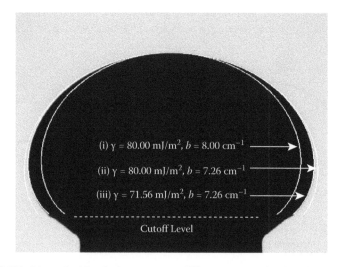

FIGURE 5.37 Numerical Laplacian curves at different stages of estimating the optimization parameters: (i) after the position of the apex is estimated but using initial (arbitrary) values of apex curvature and surface tension, (ii) after apex curvature, b, is estimated, and (iii) after the surface tension, γ, is estimated.

procedure is robust enough to converge to the optimum parameters regardless of the arbitrary initial values.

5.4.6 Experiments and Results

This section presents the results of an experimental study to investigate the effect of electric fields on surface tension of conducting liquids. This study became possible only after the development of advanced versions of ADSA-EF, as described above. The measurements are focused on three conducting liquids: distilled water (pH = 5.3), formamide, and propylene carbonate (with electrical conductivities of 4×10^{-8} ohm^{-1}cm^{-1}, 4×10^{-6} ohm^{-1}cm^{-1}, and 3×10^{-6} ohm^{-1}cm^{-1}, respectively, at room temperature) [90].

These liquids are considered as conductors in this study. Notz and Basaran [91,92] evaluated the validity of this assumption for distilled water. They showed that the conductivity of water is sufficiently high for similar drop experiments when the liquid flow rate in the drop is small. The other two liquids under study have higher electrical conductivities than distilled water, so it is believed that they also satisfy the underlying assumption of ADSA-EF.

A range of alcohols, from pentanol to undecanol, were also investigated in this study. The electrical conductivity of alcohols is of the order of 10^{-11} ohm^{-1}cm^{-1} at room temperature [90], which is less than that of the above three liquids. Therefore, ADSA-EF may not be strictly applicable to alcohols and the reported results should be interpreted with circumspection. Such results may, however, be valuable for means of comparison with other studies.

5.4.6.1 Experimental Procedure

The experimental procedure used for this study is fairly simple and straightforward. No sophisticated solid surface preparation is required. Before each experiment, the pedestal is sonicated in alcohol (three times, 15 minutes each), then in distilled water (15 minutes), and is dried under a heat lamp. Details of the optics and setups used for ADSA experiments can be found in Section 5.2.2 [93,94].

Each round of experiments consists of several runs carried out at different voltages (normally from 0 kV to 7 kV). At each run, a CSD is formed, the camera is focused, the electric field is applied, and image acquisition is started after five seconds. Experiments with suspended glass beads were conducted previously to ensure that no induced liquid flow exists in the drop as a result of the electric field [95]. Existence of such flow could affect the equilibrium shape of the drop.

Each run of the experiment lasts for only 10 seconds after the beginning of image acquisition. The liquids under investigation (i.e., water, formamide, and propylene carbonate) are polar with high surface tension; hence, such runs are preferred to reduce the chance of contamination of the drop during the experiment. In total, 11 images (at the rate of one image per second) were acquired during each run, providing 11 surface tension measurements at any given voltage. The pattern and scatter of the results were scrutinized and anomalous runs/observations were rejected or repeated.

It was found that the volume of the drop can significantly affect the measurements. Small drops are stable over a wider range of applied electric field; however, they are not well-deformed by gravity, and hence cannot produce accurate results at low voltages. Large drops, on the other hand, are less stable in the electric field and will spark or splash at a high voltage. Their shape, however, is sensitive to both gravity and electric field; hence, they can result in more accurate surface tension measurements. Another criterion is that the drop should be large enough to completely cover the surface of the pedestal (diameter of 6 mm), as assumed in the numerical model. Our experience showed that conducting drops with a volume within the range of 0.08–0.12 cm^3 are stable up to an electric potential of 6 kV and can produce reliable surface tension measurements throughout this electric field range.

All the results reported here were generated using drops of about 0.1 cm^3, a parallel plate capacitor with a distance of 12.5 mm between the plates, and a pedestal with integrated electric field disc with a diameter of 38 mm (see Figure 5.23). All experiments were conducted at room temperature (i.e., 24.5 ± 0.5°), and a RH of 50 ± 10%.

5.4.6.2 Experimental Results

Table 5.4 shows the measured surface tension of CSD of water in the electric field along with the calculated mean, standard error, and 95% confidence interval for each voltage. The mean values show an increase with increasing electric potential, indicating that the surface tension of water is affected by the electric field. Table 5.5 shows the summary of results (i.e., mean and 95% confidence interval) for the three conducting liquids under consideration (i.e., water, propylene carbonate, and formamide). Similarly, Table 5.6 summarizes the surface tension measurements for alcohols. Both Tables 5.5 and 5.6 show an increase in the surface tension of liquids due to the applied electric field.

The fact that surface tension depends on the electric field suggests that a surface tension gradient may exist along the drop surface, as the magnitude of the electric field varies along the surface (see Figure 5.26). Measuring or modeling such a gradient is beyond the scope of this study; hence, the reported results should be considered as the average (or effective) surface tension of the drop.

Table 5.7 compares the surface tension of alcohols obtained in this study (using ADSA-EF) with the corresponding results obtained in a different study, which was based on the interpretation of contact angle measurement in electric fields [95]. The contact angle measurements were performed using an Automated Polynomial Fitting (APF) method (see Chapter 6 for details of the APF method) [96] and interpreted in terms of surface tension using the equation of state for interfacial tensions (see Chapter 9 for the equation of state approach).

Both methodologies show the same trend and order of magnitude of the change in surface tension with the electric field. However, ADSA-EF results generally show a higher increase in the surface tension than the APF method. Such a difference between the two methodologies indicates that these studies are still in an early stage. Different experimental setups were used for the two studies: a stainless steel pedestal was used for the ADSA-EF experiments while a Teflon coated substrate was used for the APF measurements. These two setups generate different magnitudes of the electric field at the drop surface for a given applied voltage, and thus have different effects on the surface tension. Moreover, it should be noted that both

TABLE 5.4

The Surface Tension of Constrained Sessile Drops of Distilled Water in an Electric Field, Measured by ADSA-EF

Applied Voltage (V)	Surface Tension Measurements (mJ/m²)											Mean	Standard Error	Confidence Limit
0	71.73	71.99	73.42	72.29	72.37	72.50	72.60	72.79	72.98	73.07	73.18	72.63	0.52	0.31
1000	72.45	73.29	72.67	72.16	72.17	73.25	71.82	73.20	72.72	72.28	71.82	72.53	0.54	0.32
2000	72.81	73.75	73.44	72.21	72.68	73.83	73.27	73.56	73.39	73.62	73.49	73.28	0.50	0.30
3000	73.05	73.49	72.73	73.58	72.99	73.10	72.26	72.79	72.62	71.90	73.50	72.91	0.53	0.31
4000	73.60	74.75	74.94	74.45	74.15	72.77	74.92	73.04	72.99	74.22	73.10	73.90	0.83	0.49
5000	75.09	75.83	74.59	73.67	75.04	74.92	75.19	72.72	73.37	73.53	74.66	74.42	0.95	0.56
6000	74.85	75.24	75.44	74.85	76.92	76.12	75.12	74.38	75.29	75.88	75.33	75.40	0.69	0.41
7000	76.64	77.60	79.22	78.63	78.11	75.87	77.67	78.10	75.98	77.98	75.82	77.42	1.17	0.69

Note: The measurements were obtained from different images acquired at the frequency of 1 image/second. All measurements were conducted at room temperature (24.5 ± 0.5°C). The literature value of water surface tension is 72.13 mJ/m² in the absence of an electric field (at 24.5°C).

TABLE 5.5

The Surface Tension and 95% Confidence Interval of Constrained Sessile Drops of Distilled Water, Propylene Carbonate and Formamide in an Electric Field, Measured by ADSA-EF

Applied Voltage (V)	Average Surface Tension (mJ/m^2)		
	Water	Propylene Carbonate	Formamide
0	72.63 ± 0.31	41.39 ± 0.16	58.30 ± 0.22
1000	72.53 ± 0.32	41.25 ± 0.11	58.66 ± 0.17
2000	73.28 ± 0.30	41.22 ± 0.16	58.54 ± 0.22
3000	72.91 ± 0.31	41.47 ± 0.12	58.91 ± 0.28
4000	73.90 ± 0.49	41.60 ± 0.17	58.98 ± 0.19
5000	74.42 ± 0.56	41.93 ± 0.24	59.50 ± 0.47
6000	75.40 ± 0.41	41.87 ± 0.28	59.63 ± 0.50
7000	77.42 ± 0.69	42.27 ± 0.29	60.12 ± 0.57

Note: Each data point represents the average of a run of experiments. All measurements were conducted at room temperature (24.5 ± 0.5°C). The literature values for surface tension of propylene carbonate and formamide are 40.9 mJ/m^2 and 57.82 mJ/m^2, respectively, in the absence of an electric field (at 24.5°C).

ADSA-EF and APF results reported in Table 5.7 are approximate: ADSA-EF is not strictly applicable to alcohols (with intermediate conductivity) and APF results are indirect; that is, an interpretation of contact angle measurements using thermodynamic relations.

The level of agreement found in Table 5.7 is adequate to confirm that contact angle measurements can be used to deduce information about surface tension of liquids in general, and in the electric field in particular. Table 5.7 also compares the measurements of ADSA-EF at 0 kV with the corresponding results obtained from PD experiments. The PD experiments were conducted using current versions of ADSA designed for gravity and are believed to be accurate. The comparison generally shows a difference of less than 1%, verifying the accuracy of ADSA-EF at 0 kV.

Figure 5.38 graphically illustrates the mean surface tension of water in the electric field. A simple analysis of regression was carried out by fitting first- and second-order polynomials to the data points and calculating the correlation coefficients, R^2. A higher correlation coefficient was obtained for the second-order curves, suggesting that the influence of the electric field on the surface tension of conducting liquids is proportional to the square of the applied electric potential (or electric field). Similar results were obtained for all liquids under investigation. This interesting finding agrees with theory, since the electric field pressure at the drop surface (i.e., the jump in the normal component of the Maxwell stress tensor across the interface) is also proportional to the square of the electric field.

TABLE 5.6

The Surface Tension of Constrained Sessile Drops of Alcohols in an Electric Field, Measured by ADSA-EF

Applied Voltage (V)	Surface Tension Measurements (mJ/m^2)						
	Pentanol	Hexanol	Heptanol	Octanol	Nonanol	Decanol	Undecanol
0	26.13 ± 0.12	26.15 ± 0.69	26.86 ± 0.13	27.49 ± 0.02	27.87 ± 0.02	28.55 ± 1.04	29.28 ± 0.09
1000	26.15 ± 0.10	26.18 ± 0.54	26.91 ± 0.15	27.64 ± 0.05	27.79 ± 0.04	29.54 ± 0.55	29.26 ± 0.09
2000	26.36 ± 0.12	25.92 ± 0.63	27.31 ± 0.12	27.74 ± 0.14	27.91 ± 0.06	29.16 ± 0.64	29.36 ± 0.09
3000	26.34 ± 0.17	26.25 ± 0.37	27.07 ± 0.13	27.94 ± 0.20	28.06 ± 0.09	29.17 ± 0.48	29.36 ± 0.13
4000	26.45 ± 0.20	26.18 ± 0.45	27.37 ± 0.11	28.00 ± 0.30	28.25 ± 0.10	29.90 ± 0.58	29.50 ± 0.16
5000	26.77 ± 0.23	26.74 ± 0.53	27.48 ± 0.15	28.19 ± 5.03	28.46 ± 0.16	29.31 ± 0.62	29.88 ± 0.20
6000	27.06 ± 0.23	26.59 ± 0.58	27.98 ± 0.25	28.82 ± 0.24	28.74 ± 5.12	30.48 ± 0.64	30.35 ± 0.24
7000	27.67 ± 0.29		28.74 ± 0.26	30.04 ± 0.46	29.90 ± 0.23	30.96 ± 0.63	

Note: All the measurements were conducted at room temperature (24.5 ± 0.5°C). See Table 5.7 for the surface tension measurements using a pendant drop configuration at 0 kV.

TABLE 5.7
Comparison of ADSA-EF and APF to Determine the Effect of the Electric Field on Surface Tension of Alcohols

Methodology	Pendant Drop Vs. ADSA-EF			ADSA-EF		APF	
	γ_{lv}^{o} (mJ/	γ_{lv}^{o} (mJ/	Deviation at	γ_{lv}^{e}	Change in γ_{lv}	γ_{lv}^{e}	Change in γ_{lv}
Liquid	m²)	m²)	0 kV (%)	(mJ/m²)	(%)	(mJ/m²)	(%)
Pentanol	26.01	26.13	0.46	26.77	2.92	—	—
Hexanol	26.05	26.15	0.38	26.74	2.65	26.28	0.88
Heptanol	26.85	26.86	0.04	27.48	2.35	27.12	1.01
Octanol	27.50	27.49	−0.04	28.82	4.80	27.83	1.20
Nonanol	27.55	27.87	1.16	28.46	3.30	27.93	1.38
Decanol	28.29	28.55	0.92	29.31	3.61	28.76	1.66
Undecanol	28.88	29.28	1.39	29.88	3.46	—	—
Dodecanol	29.41	—	—	—	—	30.00	2.01

Note: The first row shows the methodologies used. The second column from left shows the surface tension of alcohols at 0 kV measured using a pendant drop configuration. The third and fourth columns show the measured surface tension at 0 kV using ADSA-EF and the percentage deviation from the corresponding pendant drop measurements. The fifth and sixth columns show the measured surface tensions at 5 kV using ADSA-EF and its percentage increase as a result of the applied voltage. Columns seven and eight show the calculated surface tensions at 4.5 kV using APF and the percentage increase as a result of the applied voltage. All the measurements were conducted at room temperature (24.5 ± 0.5°C).

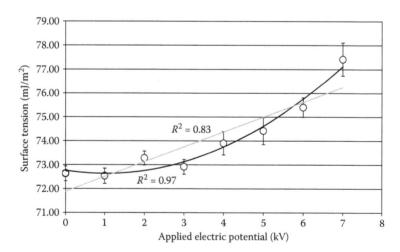

FIGURE 5.38   The surface tension of constrained sessile drops of water in the electric field, measured by ADSA-EF. A straight line and a second-order polynomial are fitted to the data points. The correlation coefficients, R^2, suggest that the influence of the electric field on surface tension is of second order. The minimum observed at 1 kV is unlikely to be physically real.

An alternative interpretation, assuming a threshold level for the electric field below which the liquid surface tensions are unaffected, cannot be excluded. This threshold could be about 1 or 2 kV for the water surface tensions shown in Figure 5.38. In this scenario, the data points beyond the threshold level fall roughly on a straight line. Furthermore, it is unlikely that the minimum observed at 1 kV (see Figure 5.38) has physical significance. It is believed that such minima are only due to experimental fluctuations.

Table 5.8 shows the least square error for the surface tension measurements of water in the electric field (corresponding to the results shown in Table 5.4). The error represents the deviation of the fitted Laplacian curve from the experimental drop profile, and hence the reliability of the ADSA-EF measurements. Figure 5.39 plots both the mean value of the least square errors (shown in Table 5.8) and the standard errors (calculated in Table 5.4) versus the applied electric potential. The graph shows that both errors slightly increase with increasing electric field. Then, there is a jump in the least square error at 7 kV.

The graph suggests that the surface tension measurements up to 6 kV are at the same level of reliability as those in the absence of an electric field (corresponding to the same order of magnitude of the least square error) but, the measurements at 7 kV should be interpreted with circumspection. However, it should be noted that the measured value at 7 kV falls on a smooth curve with the rest of the observations (See Figure 5.38). A possible explanation for a higher error at 7 kV is that the drops at such voltage are close to their stability limit, so their shape may deviate from the Laplace equation, which is valid only for stable and equilibrium conditions. This error analysis was repeated for formamide and propylene carbonate, resulting in a similar pattern for both liquids.

The operative mechanism for the observed increase in the surface tensions in the electric field is not understood yet. However, it is believed that rearrangement of liquid molecules in an electric field alters the intermolecular interactions and force balance, and hence changes the surface tension. For instance, it is known that the molecules and free charges in a conducting liquid react to the external electric field, as they rearrange to produce an internal field exactly equal but in the reverse direction to the external one [97]. As a result, no electric field can exist within a conducting drop. Similarly, an external electric field aligns polar (dipole) molecules of the liquid in the direction of the field [97]. The magnitude of this alignment depends on the dielectric constant of the liquid, as well as the temperature. The relation between the molecular alignment and its effect on the surface tension is expected to depend on the size or chain length of the molecules.

Alternatively, the observed increase in surface tensions could be a result of a similar decrease in the liquid density in the electric field. Such an effect (i.e., electrostriction) was reported before in the literature [98,99]. Further investigation is required to shed light on this issue. If the density change were found to be an operative mechanism, ADSA-EF would need to be modified to use the mass of a liquid drop, instead of its density, as input.

TABLE 5.8

Least Square Error (i.e., Deviation of the Fitted Laplacian Curve from the Experimental Profile) for the Surface Tension Measurement of Water in an Electric Field

Applied Voltage (V)	Least Square Error (Dimensionless)											Mean
0	0.0047	0.0043	0.0036	0.0039	0.0036	0.0036	0.0035	0.0034	0.0035	0.0035	0.0035	**0.0037**
1000	0.0064	0.0097	0.0076	0.0053	0.0043	0.0039	0.0039	0.0039	0.0039	0.0037	0.0036	**0.0051**
2000	0.0039	0.0039	0.0113	0.0042	0.0054	0.0063	0.0062	0.0071	0.0130	0.0128	0.0127	**0.0079**
3000	0.0089	0.0088	0.0038	0.0068	0.0069	0.0059	0.0060	0.0053	0.0053	0.0047	0.0042	**0.0061**
4000	0.0103	0.0111	0.0058	0.0101	0.0090	0.0090	0.0087	0.0078	0.0077	0.0067	0.0059	**0.0084**
5000	0.0058	0.0076	0.0053	0.0099	0.0102	0.0088	0.0088	0.0079	0.0068	0.0059	0.0067	**0.0076**
6000	0.0071	0.0077	0.0035	0.0078	0.0061	0.0048	0.0043	0.0053	0.0048	0.0042	0.0037	**0.0054**
7000	0.0344	0.0347	0.0426	0.0353	0.0354	0.0365	0.0374	0.0380	0.0378	0.0397	0.0414	**0.0376**

Note: The errors are dimensionless and correspond to the results of Table 5.4.

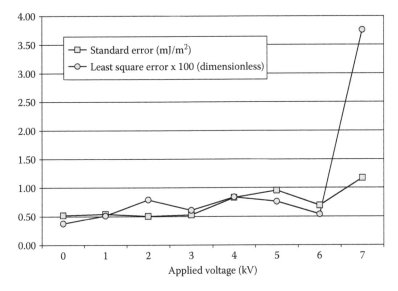

FIGURE 5.39 The least square errors (shown in Table 5.8) and the standard errors (calculated in Table 5.4) versus the applied electric potential. Overall, both errors increase slightly with increasing the voltage, except for a significant jump in the least square error at 7 kV.

5.5 AN ALTERNATIVE TO ADSA: THEORETICAL IMAGE FITTING ANALYSIS (TIFA)

5.5.1 INTRODUCTION

As detailed in Chapter 3, ADSA determines surface tension and contact angle by comparing the experimental shape of a drop or a bubble with the theoretical prediction obtained by solving the Laplace equation. An image of a drop/bubble configuration is acquired in an experiment, and the profile (i.e., the liquid–fluid interface) is extracted using edge detection. Then, an error function that measures the deviation of the theoretical curve from the experimental profile is defined. The value of the surface tension used to generate the theoretical curve, together with other parameters, is adjusted to minimize this error function (i.e., to find the best fit between the theoretical profile and the experimental one). The measured value of the surface tension is that corresponding to the best fit.

As discussed in Chapter 4, the accuracy of ADSA depends crucially on the quality of the drop/bubble profile extracted by edge detection. Current image analysis schemes of ADSA employ advanced edge detection methods, such as the Canny edge detector, together with other correction procedures, such as distortion correction and cubic spline interpolation (see Chapters 3 and 4 for details). Nevertheless, edge detectors may fail when the acquisition of sharp and clear images of the drop/bubble is not possible due to experimental or optical limitations. One example is CB experiments for lung surfactant research, where the images are fuzzy or noisy due to opacity of the surfactant suspension (see Chapter 4).

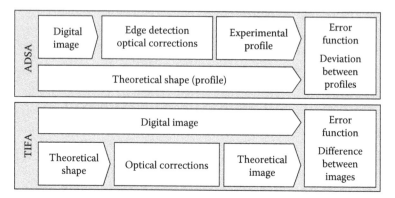

FIGURE 5.40 Comparison of the ADSA and TIFA procedures.

Two strategies have been developed for analyzing noisy images. One is the advanced image analysis scheme (i.e., edge detection and smoothing) developed in Chapter 4. The other one is called Theoretical Image Fitting Analysis (TIFA), to be discussed in detail in what follows.

In TIFA, entire theoretical and experimental images are compared, eliminating the need for an independent edge detection technique. The quantity to be minimized is then not the deviation (i.e., distance) between experimental and theoretical profiles, but the pixel-by-pixel intensity difference between experimental and theoretical images. The parameters determining the theoretical image (e.g., the surface tension) are adjusted to minimize the error; that is, to find the theoretical image that best fits the experimental one. The measured value of the surface tension (or other surface properties) provided by TIFA is that corresponding to the best fit.

More precisely, the gradient of the experimental image, rather than the raw image, is employed in TIFA in order to minimize the effect of contrast and lighting conditions, which may vary between experiments. A comparison between the ADSA and TIFA procedures is shown in Figure 5.40. One drawback of TIFA relative to ADSA is that, in its current form, the running time of TIFA is considerably longer (a few minutes vs. ~2 s per image with a 2 GHz CPU computer).

Originally, TIFA was developed for the analysis of images of drops and bubbles in a version called TIFA-Pendant Drops (TIFA-PD) [88,89,100]. The configurations analyzed by TIFA-PD utilize the presence of the drop/bubble apex. The latest version, TIFA-Axisymmetric Interfaces (TIFA-AI), is applicable to practically any axisymmetric fluid configuration, with or without apex [100,101]. In Section 5.5.2, the principles of TIFA are presented, and the following sections describe how TIFA is applied to different fluid configurations, such as drops and bubbles, configurations without apex, and liquid lenses.

5.5.2 FORMULATION OF THE OBJECTIVE FUNCTION

In essence, TIFA builds a theoretical image of the entire fluid configuration and compares it with the experimental configuration.

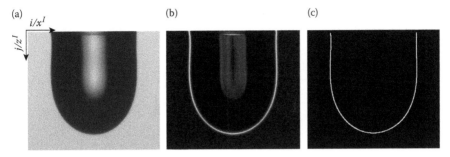

FIGURE 5.41   (a) Experimental image of a drop of cyclohexane in air, and coordinate system associated with the image; (b) corresponding gradient image after applying the 3×3 Sobel operator; and (c) theoretical gradient image corresponding to a calculated theoretical profile.

Figure 5.41a shows a typical image of an axisymmetric PD obtained in an experiment. Mathematically, a digital image is characterized by the gray level function $I(i, j)$ that takes a value between 0 (black) and 255 (white) for each pixel. The pixels (i, j) are numbered starting from the origin of the coordinate system (x^I, z^I) at the top left corner of the image. In TIFA, the gradient $G(i, j)$ of the experimental image $I(i, j)$ is calculated using the 3×3 Sobel operator (See Chapter 4 for details). Figure 5.41b shows the gradient image for the experimental image shown in Figure 5.41a. The value of the gradient function $G(i, j)$ is high (light pixel) near the contour line of the interface (where the gray level $I(i, j)$ changes rapidly) and low (dark pixel) away from the interface. In a sharp image, moving perpendicularly across the interface, the peak in the gradient may be several pixels wide; in a blurry image, it is even wider (see also Chapter 4). Thus, while the edge detection in ADSA selects just one of these pixels to represent the interface location, all the gradient information is preserved in TIFA (see the discussion below, as well as Figure 5.44, for more details).

The theoretical counterpart of $G(i, j)$ is a gradient image generated from the theoretical profile of the interface (numerically calculated by integrating the Laplace equation of capillarity). Since the theoretical profile and its gradient are continuous, and the experimental gradient image $G(i, j)$ is a pixelated (digital) image, a transformation must take place to allow the difference between them to be calculated. The details of this transformation will be described below; for the moment, we will assume that we have a pixelated theoretical gradient image $GT(i, j)$ that has the value 255 (white pixel) along the theoretical interface and zero (black pixel) everywhere else (see Figure 5.41c). In other words, $GT(i, j)$ is what $G(i, j)$ would be if the experimental profile lay exactly along the theoretical profile, and if the experimental image $I(i, j)$ was perfectly sharp. We may write

$$GT(i,j) = \begin{cases} 255 & \text{for} \quad (i,j) \text{ in the set } \{(i_p, j_p)\} \\ 0 & \text{otherwise} \end{cases} \qquad (5.23)$$

where $\{(i_p, j_p)\}$ are the pixels that constitute the theoretical profile. The set $\{(i_p, j_p)\}$ depends on certain unknown properties, such as the surface tension. Therefore, the value of those properties can be calculated by fitting the theoretical gradient image, $GT(i, j)$, to the experimental gradient image, $G(i, j)$. An error function is defined that describes the difference between the two images:

$$\varepsilon = \sum_{(i,j)} \left[G(i,j) - GT(i,j) \right]^2. \tag{5.24}$$

Substituting Equation 5.23 into Equation 5.24 yields

$$\varepsilon = K - 255 \times \sum_{(i_p, j_p)} \left[2 \times G(i,j) - 255 \right], \tag{5.25}$$

where K is a constant for each experimental image that does not play a role in the minimization of the error function. Note that the summation in Equation 5.25 is limited to the pixels $\{(i_p, j_p)\}$ corresponding to the theoretical interface. TIFA searches for the value of the unknown properties, such as the surface tension, leading to the set of pixels $\{(i_p, j_p)\}$ corresponding to the minimum error, Equation 5.25. In this sense, TIFA works similarly to any gradient based edge detection technique, searching for the pixels in the experimental image with maximum gradient. However, the pixels are not considered individually in TIFA, but as a set $\{(i_p, j_p)\}$, which corresponds to a Laplacian profile.

We now return to the subject of how the experimental gradient image is compared with the theoretical gradient image. As explained in Chapter 3, optical devices (e.g., microscope lenses) often produce distortion in the experimental images. These effects should be considered when building the theoretical image. Distortion of the images is discovered and corrected by analyzing the image of a calibration grid. Figure 5.42a shows an assumed ideal calibration grid and the coordinate system (x^G, z^G) associated with it. Note that symbols (i, j) are used for discrete (pixel) coordinate systems, while symbols (x, y), with various superscripts,

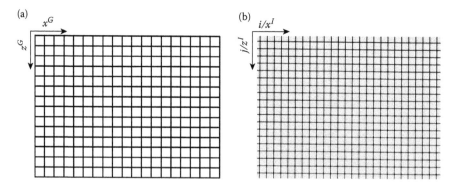

FIGURE 5.42 (a) Sketch of an ideal calibration grid and coordinate system (x^G, z^G) associated with the grid; and (b) image of the calibration grid acquired in an experiment, and coordinate system (x^I, z^I) associated with the image.

are used for continuous (analytical) coordinate systems. Figure 5.42b shows the (actual) distorted image of the grid and the corresponding coordinate system (x^I, z^I). In TIFA, the same calibration procedure as in ADSA is used to map any point of the grid to the corresponding point in the image. Two functions are constructed with the calibration results: the Optical Correction function, $f_{OC}(x^I, z^I)$, which calculates the coordinates (x^G, z^G) of any point in the grid coordinate system as a function of the corresponding coordinates (x^I, z^I) on the image; and its inverse the Optical Effects function, $f_{OE}(x^G, z^G)$, which calculates the coordinates of any point (x^I, z^I) in the image coordinate system as a function of the corresponding coordinates (x^G, z^G) on the grid.

The theoretical profile of the interface is given by the Laplace equation. The coordinate system (x^L, z^L) used to solve the Laplace equation is usually shifted and rotated with respect to the grid coordinate system (see Figure 5.43, for example). The position of the Laplace coordinate system in the grid coordinate system is defined by the position (x_0^G, z_0^G) of its origin (e.g., the apex for a drop), whereas its orientation is given by the angle α between the z^G axis and the direction of gravity (i.e., the z^L axis). The coordinates of the theoretical profile in the grid $\{(x_p{}^G, z_p{}^G)\}$ are therefore given by the relations:

$$x_p^G = x_0^G + x_p^L \times \cos\alpha - z_p^L \times \sin\alpha$$

$$z_p^G = z_0^G - x_p^L \times \sin\alpha - z_p^L \times \cos\alpha,$$

$$(5.26)$$

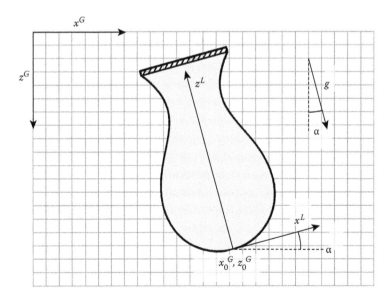

FIGURE 5.43 The Laplace coordinate system used to calculate the drop profile is shifted and rotated with respect to the grid coordinate system. The position is defined by the coordinates (x_0^G, z_0^G) of the apex of the drop and the orientation by the angle α.

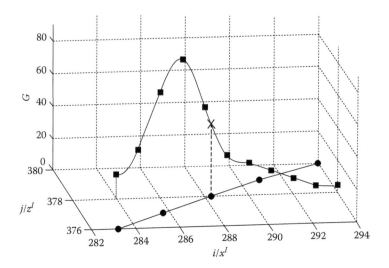

FIGURE 5.44 The line in the x-z plane is the continuous theoretical profile in the image coordinates. The dots are the points chosen on this line for each integer value of $z^l = j$. The squares represent the gray levels of the experimental gradient image G in a horizontal (x-direction) 11 pixel neighborhood of the particular profile point for the row $j = 378$. The line drawn through the squares is a cubic spline interpolation. The cross is therefore the interpolated experimental gradient at the profile point for the row $j = 378$. Similar 11-point splines could be drawn at each other integer value of $z^l = j$, but are omitted for clarity. If they were drawn, each would produce a corresponding heavy dotted line like the one shown; TIFA maximizes the sum of squares of the lengths of these lines.

where $\{(x_p^L, z_p^L)\}$ are the coordinates of the interface calculated by numerically solving the Laplace equation. Finally, the image coordinates of the theoretical profile $\{(x_p^I, z_p^I)\}$ can be calculated from the grid coordinates $\{(x_p^G, z_p^G)\}$ via

$$\{(x_p^I, z_p^I)\} = f_{OE}\left[\{(x_p^G, z_p^G)\}\right]. \tag{5.27}$$

Therefore, the shape of the theoretical profile in the image depends on the position and orientation of the Laplacian profile in the grid, characterized by the parameters $\{x_0^I, z_0^I, \alpha\}$. These parameters are initially unknown, so they must be calculated in the minimization of the error function, Equation 5.25.

In this way, the TIFA program transforms the continuous theoretical profile of the interface given by the Laplace equation into a continuous profile in the image coordinates. The continuous profile must then be discretized into a set of points $\{(i_p, j_p)\}$ for comparison with the experimental gradient image; different alternatives for accomplishing this were analyzed by Cabezas et al. [88]. It was found that the most advantageous method was to discretize the profile in only the z dimension, and to interpolate the experimental image in the x dimension. Thus, the discrete profile is formed by the points $x_p^I(z^l)$ of the continuous Laplacian profile for which the vertical coordinate takes an integer value $z^l = j$ (shown as dots in Figure 5.44). Then, the experimental image gradient is interpolated in the horizontal (x) direction to obtain the value $G(x_p^I, j)$, shown as a cross in Figure 5.44. A cubic spline over an 11-pixel neighborhood of the profile point is used for the interpolation. The values $G(x_p^I, j)$ obtained from the interpolation in each image row (each value of j) are then used in the error Equation 5.25.

In the current version of TIFA, the Nelder-Mead simplex technique is employed as the minimization method for all fluid configurations. As in most multidimensional minimization processes, this technique requires good initial values of the unknown parameters in order to converge to the global minimum. Different procedures for estimating initial values of the unknown parameters are used in different applications of TIFA.

In the following sections, TIFA is applied to various fluid configurations. For each configuration, the unknown parameters are identified, and an estimation procedure is developed for finding the initial values to be input into the final, Nelder-Mead minimization. Then, results calculated by TIFA are compared with those calculated by ADSA from the same images.

5.5.3 TIFA FOR DROPS AND BUBBLES

The theoretical profile of the drop is generated numerically by solving the Laplace equation of capillarity (see Axisymmetric Liquid Fluid Interfaces, ALFI, in Chapter 3). The shape of the drop is calculated as a function of a set of four parameters: the surface tension γ, the curvature at the apex b, the density difference $\Delta\rho$, and the gravity g. The last two are usually considered as known, leaving only $\{\gamma, b\}$ to be calculated in the optimization procedure. As explained above (Figure 5.43), the parameters determining the position and orientation of the drop profile in the image $\{x_0^I, z_0^I, \alpha\}$ are also unknown. Therefore, as in ADSA, the error function $\varepsilon(\gamma, b, x_0^I, z_0^I, \alpha)$ depends on five unknown parameters that must be adjusted to obtain the theoretical image that best fits the experimental one (i.e., the minimum error). Good estimates of these parameters need to be used as initial values in the minimization of the error.

A special estimation procedure has been designed for TIFA-PD. The order in which the parameters are estimated was found to be crucial. For the best results, the orientation of the drop and the position of the apex are estimated first, then the curvature at the apex, and finally the surface tension.

The orientation of the drop profile in the image (Figure 5.43) is defined by the angle α formed by gravity g and the vertical axis of the grid (z^G). The direction of gravity is estimated by analyzing a plumb-line image acquired in the experiment. Then, the position of the drop in the image is defined by the position of its apex (x_0^I, z_0^I). To estimate this, pairs of points on the left and right profiles are detected. The horizontal position of the apex is estimated as the mean value of those of the detected points. The initial value for the vertical coordinate of the apex is obtained by detecting the point where the axis intersects the profile of the drop (details on this procedure can be found elsewhere [88]).

The parameters $\{b, \gamma\}$ that define the shape of the drop are estimated next. The estimates of $\{\alpha, x_0^I, z_0^I\}$ obtained previously are used in this calculation. The shape

TABLE 5.9

Comparison of Surface Tension Values (mJ/m²), with 95% Confidence Intervals, Obtained by TIFA-PD and ADSA for Identical Images of PDs of Various Liquids

Liquid	TIFA-PD	ADSA
Cyclohexane	25.26 ± 0.02	25.24 ± 0.02
Hexadecane	26.92 ± 0.01	26.89 ± 0.01
Water	71.39 ± 0.03	71.37 ± 0.03
Glycerine	59.84 ± 0.01	59.77 ± 0.01

Note: Note that due to experimental error, the results may not correspond to literature values.

of the drop close to the apex is mainly determined by the curvature b. For that reason, this parameter can be estimated by using only a portion of the image near the apex. In particular, the error function Equation 5.25 is used such that the summation is applied only to pixels (i_p, j_p) close to the apex (1/8 of the length of the whole drop). This error function, $\varepsilon_{1/8}$, essentially depends only on b, being almost independent of γ. The apex curvature b is estimated by finding the minimum of $\varepsilon_{1/8}$ for an arbitrary value of γ. It has been proved that the final measured value of the surface tension does not depend on the arbitrary value of γ used in this step [89].

Once a good estimate for the curvature has been calculated, the shape of the whole drop mainly depends on the value of the surface tension. Therefore, the initial value of the surface tension is found by fitting the whole theoretical image to the experimental one while holding $\{\alpha, x_0^I, z_0^I, b\}$ fixed at their estimated values. Thus, the error function in this step depends only on the surface tension. A single-parameter minimization is used to estimate the value of the surface tension.

In the final step, the estimates of $\{\gamma, b, x_0^I, z_0^I, \alpha\}$ thus obtained are used as initial values in the full Nelder-Mead multivariate optimization.

The accuracy of TIFA-PD can be assessed by direct comparison with output for the same images from ADSA. Results from such a comparison are presented in Table 5.9, where excellent agreement between the two methods is seen. In Figure 5.45, TIFA-PD results are compared with those of ADSA for images from a CB experiment. Again, good agreement is found between the two methods.

5.5.4 TIFA FOR AXISYMMETRIC INTERFACES WITHOUT APEX

In some common experimental geometries, the liquid–fluid interface does not have an apex. Note that this is distinct from a situation in which an apex exists but is not

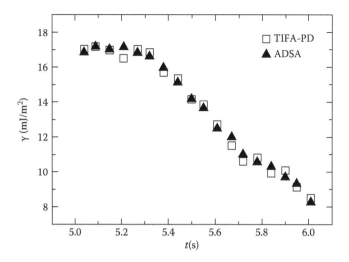

FIGURE 5.45 Surface tension values calculated by TIFA-PD (squares) and ADSA (triangles) for a series of images obtained as a function of time t using the captive bubble configuration. The liquid was a mixture of 1 mg/mL bovine lipid extract surfactant (BLES) and 30 mg/mL polyethylene glycol (PEG).

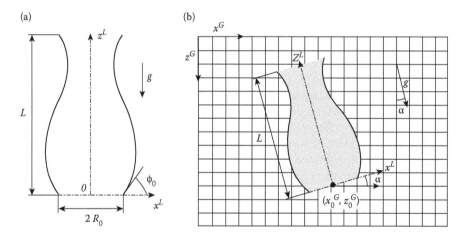

FIGURE 5.46 (a) Coordinate system (x^L, z^L) used in the numerical integration of the Laplace equation for axisymmetric liquid–fluid interfaces without apex; (b) position and orientation of the interface coordinate system in the grid.

visible in the image of the interface due to for example, an obstruction. Figure 5.46a shows an example of a geometry with no apex—the liquid bridge (another example, the liquid lens, will be discussed in the next section). In the absence of the apex, the origin of the coordinate system is located at one end of the interface profile. The Laplace equation shows that the shape of the interface depends on the set of parameters $\{\gamma, b_0, \Delta\rho, g\}$, where b_0 is the local mean curvature of the interface at the reference level $z^L = 0$. The shape of the interface is calculated by considering the boundary conditions at the reference level:

$$x^L(0) = R_0, \ z^L(0) = 0, \ \phi(0) = \phi_0, \tag{5.28}$$

where R_0 is the radius of the interface at the reference level and ϕ_0 the inclination of the interface at the same point [100,101]. The interface profile is calculated as a function of the set of six parameters $\{\gamma, b_0, \Delta\rho, g, R_0, \phi_0\}$, of which $\Delta\rho$ and g are known.

The position and orientation of the axisymmetric interface in the image are given by the set of parameters $\{x_0^I, z_0^I, \alpha\}$. It has been found, however, that the value of the coordinate z_0^I can be fixed at its initial estimate (detailed below) without loss of accuracy. So, TIFA-AI adjusts the values of six parameters $\{\gamma, b_0, R_0, \phi_0, x_0^I, \alpha\}$ to obtain the theoretical image that best fits the experimental one. Good initial estimates of these parameters need to be used in the minimization procedure to converge to the global minimum. The procedure used by TIFA-AI estimates the orientation of the interface and the position and radius at the reference level first, then the inclination and curvature at the reference level, and finally the surface tension.

As in TIFA-PD, the orientation of the interface α, given by the direction of gravity, is obtained by analyzing the plumb-line image acquired in the experiment. Analysis of the image of the axisymmetric interface allows four other parameters to be estimated as follows. A set of 15 points of the interface profile close to the reference level is detected at each side. The horizontal coordinate of the origin x_0^I is estimated as the mean value of those of the detected points. Then a parabola is fitted to the profile points, and the estimates of R_0, b_0, and ϕ_0 are calculated as those of the parabola at the reference level.

However, it was found that better estimates of the curvature and the inclination at the reference level are necessary for the multivariate minimization. The shape of the interface close to the reference level is mainly determined by the parameters b_0 and ϕ_0. Therefore, the values of these parameters can be estimated from a section of the interface close to the reference level. TIFA-AI calculates a second estimate of these parameters by fitting a theoretical image to the area close to the reference level in the experimental image. In particular, the error function Equation 5.25 is used such that the summation is applied only to pixels (i_p, j_p) close to the reference level (1/5 of the length of the interface). This error function, $\varepsilon_{1/5}$, essentially depends on just b_0 and ϕ_0, being almost independent of γ. TIFA-AI estimates the curvature and inclination at the reference level by finding the minimum of $\varepsilon_{1/5}$ for an arbitrary value of γ using as initial values of b_0 and ϕ_0 those calculated from the parabola. It has been proved that the arbitrary value used for γ does not affect the final measured value of the surface tension [101].

Once good estimates of the rest of the parameters have been calculated, the shape of the whole interface mainly depends on the value of the surface tension. TIFA-AI estimates the value of the surface tension by fitting the whole theoretical image to the experimental one. The estimated values of the orientation of the interface and of its position, radius, curvature, and inclination at the reference level are kept fixed at this stage so that the error function depends only on the surface tension. A

TABLE 5.10

Surface Tensions (mJ/m²), with 95% Confidence Intervals, Obtained from Liquid Bridge Images Processed by TIFA-AI, and from Pendant Drop Images Processed by TIFA-AI, TIFA-PD, and ADSA

Configuration		Liquid Bridge	Pendant Drop		
Liquid—Experiment		TIFA-AI	TIFA-AI	TIFA-PD	ADSA
Hexadecane	A	27.07 ± 0.02	26.93 ± 0.02	26.92 ± 0.01	26.89 ± 0.01
	B	27.10 ± 0.03	27.05 ± 0.02	27.03 ± 0.01	27.02 ± 0.01
	C	27.13 ± 0.01	27.02 ± 0.02	27.04 ± 0.01	27.00 ± 0.01
Water	A	71.95 ± 0.11	71.36 ± 0.05	71.39 ± 0.03	71.37 ± 0.03
	B	71.84 ± 0.04	71.85 ± 0.09	71.93 ± 0.08	72.12 ± 0.09
	C	71.93 ± 0.03	71.94 ± 0.07	72.12 ± 0.02	72.21 ± 0.04

Note: A series of 10 images of each configuration is obtained in each experiment. The results presented here are the averages of the measurements from the 10 images.

single-parameter minimization is used to estimate the value of the surface tension. Finally, the estimates of $\{\gamma, b_0, R_0, \phi_0, x_0{}', \alpha\}$ thus obtained are used as initial values in the full Nelder-Mead multivariate optimization.

Table 5.10 presents the results of TIFA-AI, TIFA-PD, and ADSA for identical images of PDs of hexadecane and water. (The reference level for TIFA-AI was set slightly away from the apex so that TIFA-AI ignores a small region of the image near the apex.) As well, results are shown for liquid bridges of the same liquids as analyzed by TIFA-AI. Good agreement is seen, with differences of about 1% and below.

Figure 5.47 displays results for a sequence of 16 images from an experiment with o-xylene in the sessile drop configuration. The drop was growing during the experiment to allow measurement of advancing contact angles. Experimental images were acquired approximately every 1.5 seconds and were processed using both ADSA and TIFA-AI. The results show good agreement between the two methods. However, the scatter of the measurements (i.e., the 95% confidence interval) is slightly higher for the TIFA-AI method, possibly because TIFA-AI uses less experimental information as it neglects a section of the drop near the apex.

5.5.5 TIFA FOR LIQUID LENSES

In line tension research (see Chapter 13), the liquid lens is a geometry that allows the tension of a liquid-liquid-fluid contact line to be measured. This configuration avoids the use of a solid surface and the accompanying complications of roughness and/or chemical heterogeneity. In order to hold the lens in place during the experiment, the lens may be suspended from a needle. However, the needle prevents formation of the lens apex, meaning that methods such as ADSA and TIFA-PD cannot be applied to analyze its shape. Thus, contact angles for suspended liquid lenses have

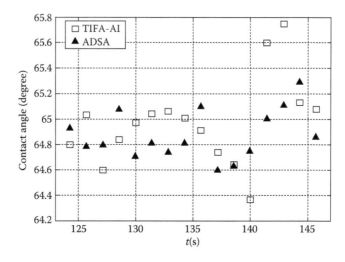

FIGURE 5.47 Contact angle of sessile drop of o-xylene on Teflon as a function of time t, calculated using ADSA and TIFA-AI for a sequence of images from a dynamic experiment. The diameter of the drop is about 1 cm.

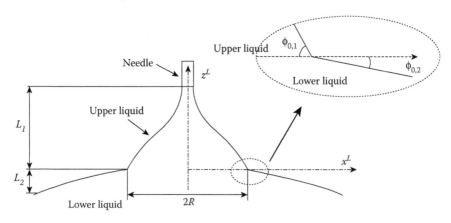

FIGURE 5.48 Liquid-lens profile and the associated coordinate system (x^L, z^L).

been previously measured only by fitting polynomials to the liquid surfaces in the vicinity of the contact line (see Chapter 6). The contact angles are subsequently used to measure deviation from the Neumann triangle relation, which, if found, may be attributed to line tension.

Liquid lenses are formed by two axisymmetric interfaces (see Figure 5.48), the shapes of which are each given by the Laplace equation. As explained above, the shape of any axisymmetric interface without apex depends on the set of parameters $\{\gamma_0, b_0, \Delta\rho, g, R_0, \phi_0\}$. For liquid lens problems, gravity g, the upper and lower interfaces' density differences, $\Delta\rho_1$ and $\Delta\rho_2$, and the interfacial tensions γ_1 and γ_2 are usually known. Fixing the reference level at the liquid-liquid contact line, the radius of both (upper and lower) interfaces at that level is equal to that of the three-phase line, R. Then, the shape of the profile of the whole configuration depends on the set of unknown parameters $\{R, b_{0,1}, \phi_{0,1}, b_{0,2}, \phi_{0,2}\}$. The parameters $\{x_0^l, z_0^l, \alpha\}$ determine the orientation and position of the fluid configuration in the image. Therefore, a set of eight parameters, $\{R, b_{0,1}, \phi_{0,1}, b_{0,2}, \phi_{0,2}, x_0^l, z_0^l, \alpha\}$, must be adjusted to find the theoretical image that best fits the experimental one. The minimization method requires good initial values of the parameters to converge to the global minimum.

The estimation procedure developed for this application of TIFA first estimates the orientation of the interface α from the plumb-line image, and then calculates the rest of the parameters by analyzing the liquid-lens image. A set of 10 profile points are detected for both interfaces (upper and lower) close to the contact line. A second-order polynomial is fitted to the set of points of the upper profile and a circle is fitted to the set of points of the lower one. The intersection of these curves corresponds to a diameter of the three-phase line, and is used to calculate the estimates of $\{R, x_0^l, z_0^l\}$. The curvatures and inclinations of the interfaces are calculated as those of the polynomial and the circle at the intersection point.

To assess accuracy, TIFA results for liquid lenses are compared to results obtained by APF in Figure 5.49 (see Chapter 6 for details of the APF method). While there is some discrepancy, since APF does not account for optical distortion, fits each side of the image separately, and only considers a small region of the image near the contact line, the TIFA results are expected to be more accurate.

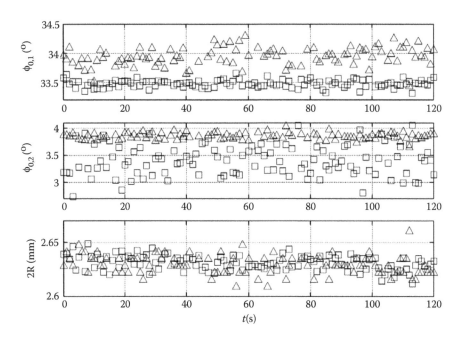

FIGURE 5.49 Results for TIFA (squares) and APF (triangles) for a liquid lens of dodecane on water, as a function of time t. Top panel: dodecane contact angle; middle panel: water contact angle; bottom panel: contact line diameter.

REFERENCES

1. F. Possmayer. In *Fetal and Neonatal Physiology*. Edited by R. A. Polin, W. W. Fox, and S. H. Abman. Vol. 2, 1014. W. B. Saunders Company, Philadelphia, PA, 1014.
2. R. H. Notter. *Lung Surfactants. Basic Science and Clinical Applications*. Marcel Dekker, Inc., New York, 2000.
3. B. Robertson and H. L. Halliday. *Biochimica et Biophysica Acta* 1408 (1998): 346.
4. B. Guyer, M. A. Freedman, D. M. Strobino, and E. J. Sondik. *Pediatrics* 106 (2000): 1307.
5. J. F. Lewis and R. A. Veldhuizen. *Annual Reviews of Physiology* 65 (2003): 613.
6. J. Clements. *Proceedings of the Society for Experimental Biology and Medicine* 95 (1957): 170.
7. G. Enhorning. *Journal of Applied Physiology* 43 (1977): 198.
8. S. Schurch, H. Bachofen, J. Goerke, and F. Possmayer. *Journal of Applied Physiology* 67 (1989): 2389.
9. Y. Y. Zuo, R. A. Veldhuizen, A. W. Neumann, N. O. Petersen, and F. Possmayer. *Biochimica et Biophysica Acta* 1778 (2008): 1947.
10. D. Y. Kwok, D. Vollhardt, R. Miller, D. Li, and A. W. Neumann. *Colloids and Surfaces A-Physicochemical and Engineering Aspects* 88 (1994): 51.
11. M. Hoorfar and A. W. Neumann. *Journal of Adhesion* 80 (2004): 727.
12. D. Li, P. Cheng, and A. W. Neumann. *Advances in Colloid and Interface Science* 39 (1992): 347.
13. R. M. Prokop and A. W. Neumann. *Current Opinion in Colloid and Interface Science* 1 (1996): 677.
14. M. Hoorfar, M. A. Kurz, and A. W. Neumann. *Colloids and Surfaces A-Physicochemical and Engineering Aspects* 260 (2005): 277.
15. J. J. Lu, L. M. Y. Yu, W. W. Y. Cheung, Z. Policova, D. Li, M. L. Hair, and A. W. Neumann. *Colloids and Surfaces B-Biointerfaces* 29 (2003): 119.
16. L. M. Y. Yu, J. J. Lu, I. W. Y. Chiu, K. S. Leung, Y. W. W. Chan, L. Zhang, Z. Policova, M. L. Hair, and A. W. Neumann. *Colloids and Surfaces B-Biointerfaces* 36 (2004): 167.
17. R. M. Prokop, A. Jyoti, M. Eslamian, A. Garg, M. Mihaila, O. I. del Rio, S. S. Susnar, Z. Policova, and A. W. Neumann. *Colloids and Surfaces A-Physicochemical and Engineering Aspects* 131 (1998): 231.
18. Y. Y. Zuo, D. Q. Li, E. Acosta, P. N. Cox, and A. W. Neumann. *Langmuir* 21 (2005): 5446.
19. Y. Y. Zuo, E. Acosta, Z. Policova, P. N. Cox, M. L. Hair, and A. W. Neumann. *Biochimica et Biophysica Acta-Biomembranes* 1758 (2006): 1609.
20. S. Schurch, H. Bachofen, and F. Possmayer. *Comparative Biochemistry and Physiology A-Molecular and Integrative Physiology* 129 (2001): 195.
21. J. R. Codd, S. Schurch, C. B. Daniels, and S. Orgeig. *Biochimica et Biophysica Acta* 1580 (2002): 57.
22. J. Y. Lu, J. Distefano, K. Philips, P. Chen, and A. W. Neumann. *Respiration Physiology* 115 (1999): 55.
23. J. J. Lu, W. W. Y. Cheung, L. M. Y. Yu, Z. Policova, D. Li, M. L. Hair, and A. W. Neumann. *Respiratory Physiology & Neurobiology* 130 (2002): 169.
24. J. J. Lu, L. M. Y. Yu, W. W. Y. Cheung, I. A. Goldthorpe, Y. Y. Zuo, Z. Policova, P. N. Cox, and A. W. Neumann. *Colloids and Surfaces B-Biointerfaces* 41 (2005): 145.
25. J. M. Crane, G. Putz, and S. B. Hall. *Biophysical Journal* 77 (1999): 3134.
26. N. Wustneck, R. Wustneck, V. B. Fainerman, R. Miller, and U. Pison. *Colloids and Surfaces B-Biointerfaces* 21 (2001): 191.
27. N. Wustneck, R. Wustneck, and U. Pison. *Langmuir* 19 (2003): 7521.

28. R. Wustneck, N. Wustneck, D. Vollhardt, R. Miller, and U. Pison. *Materials Science & Engineering C-Biomimetic and Supramolecular Systems* 8 (1999): 57.
29. N. Wustneck, R. Wustneck, U. Pison, and H. Mohwald. *Langmuir* 23 (2007): 1815.
30. L. M. Y. Yu, J. J. Lu, Y. W. Chan, A. Ng, L. Zhang, M. Hoorfar, Z. Policova, K. Grundke, and A. W. Neumann. *Journal of Applied Physiology* 97 (2004): 704.
31. E. J. Acosta, R. Gitiafroz, Y. Y. Zuo, Z. Policova, P. N. Cox, M. L. Hair, and A. W. Neumann. *Respiratory Physiology & Neurobiology* 155 (2007): 255.
32. Y. Y. Zuo, H. Alolabi, A. Shafiei, N. X. Kang, Z. Policova, P. N. Cox, E. Acosta, M. L. Hair, and A. W. Neumann. *Pediatric Research* 60 (2006): 125.
33. J. Goerke and J. A. Clements. In *Handbook of Physiology, Section 3: The Respiratory System*. Edited by A. P. Fishman, Vol. III, Part 1, 247. American Physiological Society, Bethesda, MD, 1986.
34. Y. Y. Zuo, E. Acosta, P. N. Cox, D. Li, and A. W. Neumann. *Langmuir* 23 (2007): 1339.
35. S. Schurch, H. Bachofen, J. Goerke, and F. Green. *Biochimica et Biophysica Acta* 1103 (1992): 127.
36. T. Kobayashi, A. Shido, K. Nitta, S. Inui, M. Ganzuka, and B. Robertson. *Respiration Physiology* 80 (1990): 181.
37. H. Bachofen, S. Schurch, M. Urbinelli, and E. R. Weibel. *Journal of Applied Physiology* 62 (1987): 1878.
38. S. Schurch, D. Schurch, T. Curstedt, and B. Robertson. *Journal of Applied Physiology* 77 (1994): 974.
39. S. S. Susnar, H. A. Hamza, and A. W. Neumann. *Colloids and Surfaces A-Physicochemical and Engineering Aspects* 89 (1994): 169.
40. S. Schurch, J. Goerke, and J. A. Clements. *Proceedings of the National Academy of Sciences* 73 (1976): 4698.
41. S. Schurch, J. Goerke, and J. A. Clements. *Proceedings of the National Academy of Sciences* 75 (1978): 3417.
42. S. Schurch. *Respiration Physiology* 48 (1982): 339.
43. F. MacRitchie. *Proteins at Liquid Interfaces*. Edited by D Moebius and R. Miller, 149. Elsevier, Amsterdam, 1998.
44. D. E. Graham and M. C. J. Phillips. *Colloid and Interface Science* 70 (1979): 427.
45. D. Vollhardt. *Materials Science Forum* 25 (2003): 541.
46. M. A. Cabrerizo-Vílchez. *Colloids and Surfaces A*, 156 (1999): 509.
47. R. Miller, D. O. Grigoriev, J. Krägel, A. V. Makievski, J. Maldonado-Valderrama, M. Leser, M. Michel, and V. B. Fainerman. *Food Hydrocolloids* 19 (2005): 479.
48. T. F. Svitova, M. J. Wetherbee, and C. J. J. Radke. *Colloid and Interface Science* 261 (2003): 170.
49. M. A. Cabrerizo-Vilchez, H. A. Wege, J. Holgado-Terriza, and A. W. Neumann. *Review of Scientific Instruments* 70 (1990): 2438.
50. Oficina Espanola De Patentes Y Marcas. Capilares coaxiales y procedimiento de intercambio para balanza de superficies de penetracion (Coaxial capillaries and exchange procedure for a penetration pendant drop film balance). Patent No. ES 2 153 296.
51. J. Maldonado-Valderrama, H. A. Wege, M. A. Rodríguez-Valverde, M. J. Gálvez-Ruiz, and M. A. Cabrerizo-Vílchez. *Langmuir* 19 (2003): 8436.
52. J. Maldonado-Valderrama, A. Martín-Molina, A. Martín-Rodríguez, M. A. Cabrerizo-Vílchez, M. J. Gálvez-Ruiz, and D. J. Langevin. *Physical Chemistry C* 111 (2007): 2715.
53. J. Maldonado-Valderrama, M. J. Gálvez-Ruiz, A. Martín-Rodriguez, and M. A. Cabrerizo-Vílchez. *Langmuir* 20 (2004): 6093.
54. H. A. Wege, J. Holgado-Terriza, M. J. Gálvez-Ruiz, and M. A. Cabrerizo-Vílchez. *Colloids and Surfaces B* 12 (1999): 339.
55. H. A. Wege, J. A. Holgado-Terriza, and M. A. Cabrerizo-Vílchez. *Journal of Colloid and Interface Science* 249 (2002): 263.

56. H. J. Trurnit. *Journal of Colloid and Interface Science* 15 (1960): 1.
57. B. C. Tripp, J. J. Magda, and J. D. Andrade. *Journal of Colloid and Interface Science* 173 (1995): 16.
58. J. Maldonado-Valderrama, M. J. Gálvez-Ruiz, A. Martín-Rodriguez, and M. A. Cabrerizo-Vílchez. *Colloids and Surfaces A* 270 (2005): 323.
59. P. Wilde, A. R. Mackie, F. Husband, P. Gunning, and V. Morris. *Advances in Colloid and Interface Science* 108 (2004): 63.
60. Z. S. Derewenda. *Advances in Protein Chemistry* 45 (1994): 1.
61. A. Bateni, S. S. Susnar, A. Amirfazli, and A. W. Neumann. *Langmuir* 20 (2004): 7589.
62. K. Adamiak. *Journal of Electrostatics* 51–52 (2001): 578.
63. E. Borzabadi and A. G. Bailey. *Journal of Electrostatics* 5 (1978): 369.
64. H. J. Cho, I. S. Kang, Y. C. Kweon, and M. H. Kim. *International Journal of Multiphase Flow* 22 (1996): 909.
65. O. A. Basaran and L. E. Scriven. *Journal of Colloid and Interface Science* 140 (1990): 10.
66. M. Wulf, S. Michel, K. Grundke, O. I. del Río, D. Y. Kwok, and A. W. Neumann. *Journal of Colloid and Interface Science* 210 (1999): 172.
67. L. M. Y. Yu. Master Thesis. Department of Mechanical and Industrial Engineering. University of Toronto, 2003.
68. A. Bateni. PhD Thesis, Department of Mechanical and Industrial Engineering. University of Toronto, 2005.
69. A. Bateni, A. Amirfazli, and A. W. Neumann. *Colloids and Surfaces A* 289 (2006): 25.
70. W. H. Press, B. P. Flannery, S. A. Teukolsky, and W. T. Vetterling. *Numerical Recipes The Art of Scientific Computing*. 710. Springer-Verlag, New York, 1986.
71. C. Johnson. *Numerical Solution of Partial Differential Equations by the Finite Element Method*. Cambridge University Press, Cambridge, 1987.
72. L. Fox. *Numerical Solutions of Ordinary and Partial Differential Equations*. Pergamon Press, New York, 1962.
73. G. D. Smith. *Numerical Solutions of Partial Differential Equations: Finite Difference Method*. 3rd ed. Oxford University Press, New York, 1985.
74. W. F. Ames. *Numerical Methods for Partial Differential Equations*. Academic Press Inc., London, 1977.
75. R. Li, Z. Chen, and W. Wu. *Generalized Difference Methods for Differential Equations*. Marcel Dekker, Inc., New York, 2000.
76. J. Jeans. *Mathematical Theory of Electricity and Magnetism*. 185. Cambridge University Press, Cambridge, UK, 1960.
77. O. A. Basaran and L. E. Scriven. *Journal of Colloid and Interface Science* 140 (1990): 10.
78. "Triangle: Engineering a 2D Quality Mesh Generator and Delaunay Triangulator." In *Applied Computational Geometry: Towards Geometric Engineering*. Edited by M. C. Lin and D. Manocha, Vol. 1148, 203. Lecture Notes in Computer Science. Springer-Verlag, Berlin, May 1996 (from the First ACM Workshop on Applied Computational Geometry).
79. J. Shewchuk. *Computational Geometry: Theory and Applications* 22 (2002): 21.
80. J. Shewchuk. *Delaunay Refinement Mesh Generation*. PhD thesis, Technical Report CMU-CS-97-137, School of Computer Science, Carnegie Mellon University, Pittsburgh, PA, May 18, 1997.
81. J. Donea and A. Huerta. *Finite Element Methods for Flow Problems*. John Wiley & Sons Ltd., West Sussex, England, 2003.
82. L. J. Segerlind. *Applied Finite Element Analysis*. John Wiley & Sons Inc., 1984.
83. M. J. Fagan. *Finite Element Analysis, Theory and Practice*. Longman Group, UK, 1992.

84. R.Wait and A. R. Mitchell. *Finite Element Analysis and Applications.* John Wiley & Sons Ltd., UK, 1985.
85. F. K. Wohlhuter and O. A. Basaran. *Journal of Fluid Mechanics* 235 (1992): 481.
86. M. T. Harris and O. A. Basaran. *Journal of Colloid and Interface Science* 170 (1995): 308.
87. J. E. Dennis. "Nonlinear Least-Squares." In *State of the Art in Numerical Analysis.* Edited by D. Jacobs, 269. Academic Press, New York, 1977.
88. M. G. Cabezas, A. Bateni, J. M. Montanero, and A. W. Neumann. *Applied Surface Science* 238 (2004): 480.
89. M. G. Cabezas, A. Bateni, J. M. Montanero, and A. W. Neumann. *Colloids and Surfaces A-Physicochemical Engineering Aspects* 255 (2005): 193.
90. J. R. Partington. "The Electrical Conductivity of Non-Metallic Liquids." In *International Critical Tables of Numerical Data, Physics, Chemistry and Technology.* Edited by E. W. Washburn, 1st electronic ed., 142. Knovel, New York, 2003.
91. P. K. Notz and O. A. Basaran. *Journal of Colloid and Interface Science* 213 (1999): 218.
92. J. R. Melcher and G. I. Taylor. *Annual Reviews of Fluid Mechanics* 1 (1969): 111.
93. P. Cheng, D. Li, L. Boruvka, Y. Rotenberg, and A. W. Neumann. *Colloids and Surfaces* 43 (1990): 151.
94. J. M. Alvarez, A. Amirfazli, and A. W. Neumann. *Colloids and Surfaces A* 156 (1999): 163.
95. A. Bateni, S. Laughton, H. Tavana, S. S. Susnar, A. Amirfazli, and A. W. Neumann. *Journal of Colloid and Interface Science* 283 (2005): 215.
96. A. Bateni, S. S. Susnar, A. Amirfazli, and A. W. Neumann. *Colloids and Surfaces A* 219 (2003): 215.
97. P. M. Fishbane, S. G. Gasiorowicz, and S. T. Thornton. *Physics for Scientists and Engineers.* 3rd ed. Pearson Education, Upper Saddle River, NJ, 2005.
98. J. C. Rasaiah, D. J. Isbister, and G. Stell. *Journal of Chemical Physics* 75 (1981): 4707.
99. Y. M. Shkel and D. J. Klingenberg. *Journal of Applied Physics* 80 (1996): 4566.
100. M. G. Cabezas. PhD Thesis. *Experimental Study of the Statics of Drops and Liquid Bridges. Surface Tension Measurement.* Universidad de Extremadura, Spain, 2005.
101. M. G. Cabezas, A. Bateni, J. M. Montanero, and A. W. Neumann. *Langmuir* 22 (2006): 10053.

6 Contact Angle Measurements: General Procedures and Approaches

Hossein Tavana

CONTENTS

6.1   INTRODUCTION

Contact angles have been a subject of interest in pure and applied sciences. Technologically, contact angles are utilized as a means of characterizing the wettability of a wide range of materials in various industries including aviation, automobiles, oil and gas, printing, and pharmaceuticals as well as wetting properties of biological surfaces such as cells, tissues, and lipids. From a purely scientific viewpoint, contact angles provide a unique means to evaluate solid surface tensions and line tension. The broad applicability of contact angles has generated great interest in developing measurement techniques. The measurement of contact angles with an acceptable accuracy and reproducibility is essential to many areas of applied surface thermodynamics. When first encountered, the measurement of contact angles appears to be quite straightforward. This apparent simplicity is, however, very misleading, and experience shows that the acquisition of thermodynamically significant contact angles requires rigorous experimental designs and reliable analysis tools.

The importance of establishing a proper advancing contact angle cannot be overemphasized since, on properly prepared solid surfaces, this is the only contact angle that is unique and of thermodynamic significance. The thermodynamic status of contact angles and advancing and receding angles will be discussed in detail in Chapter 7. In the following sections, available conventional and new methods of contact angle measurements will be overviewed and their utility will be discussed with regard to various types of solid surfaces on which contact angles are measured. It is emphasized upfront that of the many techniques developed for the measurement of contact angles [1], only a few are widely used today: axisymmetric drop shape analysis (ADSA), capillary rise at a vertical plate, and the goniometer telescope. These techniques will be presented in greater detail. Application of ADSA for the measurement of different modes of contact angles including static, advancing, and receding angles, contact angles on nonideal surfaces, contact angles on wetting surfaces, and contact angles on superhydrophobic surfaces will all be discussed. Temperature dependence of contact angles will be the next subject of discussion as an example of broad applicability of the capillary rise technique. Finally, techniques of preparation of nonbiological and biological solid surfaces will be presented.

6.2   MEASUREMENT OF CONTACT ANGLES: CONVENTIONAL TECHNIQUES

A widely used technique for contact angle measurement involves simply aligning a tangent with the sessile drop profile at the point of contact with the solid surface. This is most frequently done directly using a telescope equipped with a goniometer eyepiece. The telescope is tilted down slightly (1°–2°) out of the horizontal. On smooth reflecting surfaces, a portion of the profile is reflected by the surface and a cusp is created at the point of contact with the solid. Observations are facilitated by a brightly lit, diffuse background against which the drop appears as a black silhouette. The tangent is aligned to the profile at the contact point and the contact angle is measured. This is usually done at relatively high magnifications (up to 50×) that allow the detailed examination of the intersection of the drop profile and the solid surface and yield more accurate measurements compared to lower magnifications. To establish an advancing contact angle, it is best to slowly grow the sessile drop to a diameter of approximately 5 mm using a micrometer syringe and a narrow gauge, stainless steel or Teflon needle. The needle must not be removed from the drop as this may cause vibrations that can decrease the advancing contact angle to a value corresponding to a metastable state. Contact angles should be measured on both sides of the drop profile. Repeated addition of small amounts of liquid to the drop and advancing it over fresh areas of the solid yields a large number of contact angle data whose average is representative of a relatively large area of the surface. The goniometer technique is easy to implement and straightforward to use, nevertheless the results are subjective and dependent on the experience of the operator. With some training, contact angles with a best accuracy of ±2° may be obtained [1]. It is noted that an alternative strategy developed by Zisman and his coworkers may be used to form a sessile drop on a surface [2]. A fine platinum wire was dipped in the liquid and then was gently flicked to create a pendant drop hanging from the tip of the wire. The drop was slowly brought into contact with a solid surface, making it flow from the wire and form a sessile drop. This technique is easy to use; nevertheless, the kinetic energy associated with the flowing liquid and the detachment from the wire can vibrate the drop and result in a lower metastable contact angle.

Phillips and Riddiford analyzed photographs of sessile drop profiles with a "tangentometer" instrument, which was simply a plane mirror mounted at right angles to a straight edge and positioned so that it was normal to the photograph at the drop tip [3]. The device was rotated until the profile formed a smooth, continuous curve with its mirror image, causing the straight edge to be tangent at that point. This technique is subjective because identification of the point at which the profile edge and the reflected image merge to form a smooth curve relies on user expertise [4]. Langmuir and Schaeffer used the specular reflection from a drop surface to measure the contact angle [5]. A light source was pivoted about the three-phase line. The angle at which reflection from the drop surface just disappeared was simply identified as the contact angle.

The sessile drop approach has also proved useful for measuring contact angles on compressed pharmaceutical powder or granules [6–12]. However these contact angles should be interpreted with great caution as the roughness and porosity of the surface and possible solubility of the powder in the probe liquid may result in misleading data [13–15]. Alternatively, a so-called h-ε method may be used to obtain contact angles on compressed powders [16,17]. The sample is saturated with the probe liquid and a liquid drop is placed onto the surface. The contact angle is calculated from the height of the drop, cake porosity, and the liquid density and surface tension.

Contact angles can also be measured using the captive bubble method [18]. A bubble is formed beneath the solid surface while immersed in the liquid of interest. To establish the equivalent of an advancing contact angle, air is withdrawn from the bubble using a micrometer syringe. As with sessile drops, the needle should remain in the bubble throughout the experiment to ensure that the drop profile is not disturbed. The needle also serves to keep the bubble from drifting over the surface if the latter is not perfectly flat and horizontal. On smooth and homogeneous polymer surfaces, the captive bubble technique yields contact angles comparable to those from sessile drops.

Adam and Jessop developed the tilting plate method for contact angle measurements [19]. A flat solid surface is attached at one end to a beam and immersed in the probe liquid. It is then tilted until the meniscus becomes horizontal on one side of the plate. The angle the plate makes with the liquid surface is the contact angle. The major drawbacks of this method are the difficulty in establishing an advancing contact angle and the need for a relatively large plate and a large volume of high-purity liquid. Nevertheless, the method has been used for the measurement of contact angles as small as $10°$ [20].

Two rather specialized methods were also developed to measure contact angles less than approximately $60°$. Interference microscopy makes use of fringe patterns reflected from the drop surfaces to calculate the contact angle [21]. Fisher obtained comparable results for contact angles less than $30°$ by simultaneously measuring the mass of the drop and the radius of the three-phase line [22]. The contact angle was then derived from a semiempirical relationship involving these two quantities.

For a probe liquid with known surface tension, the Wilhelmy balance can be used to calculate contact angles from the measured force of the liquid on a plate of interest using the following relation [1]:

$$f = p\gamma_{lv}\cos\theta - V\Delta\rho g, \qquad (6.1)$$

where p is the perimeter of the plate, γ_{lv} is the liquid surface tension, θ is the contact angle, V is the volume of displaced liquid, $\Delta\rho$ is the difference in density between the liquid and air (or the second liquid), and g is the acceleration of gravity. If the perimeter of the plate is accurately known, the plate is of uniform composition and roughness along the entire perimeter, and absorption of the liquid by the solid is insignificant, the Wilhelmy balance can provide advancing and receding contact angles, free of operator subjectivity. The technique is the best means of measuring contact angles on individual fibers of known diameter. Alternatively, the goniometer and the capillary depression techniques can be employed to measure contact angles on individual fibers [23,24]. It is noted that the apparent contact angle of a sessile drop on a mat of fibers or on a woven fabric should not be interpreted as the contact angle that the liquid would make on an individual fiber of the same material.

The positions of the fibers affect the local geometry of the three-phase line and the apparent contact angle. For this reason, only qualitative tests of fabric wettability are available [25].

The Wilhelmy balance can also be used to measure the contact angle of a liquid in a capillary tube with identical inner and outer composition [1]. For tubes of a small enough diameter, the meniscus may be considered to be spherical and the capillary rise, h, is given by

$$h = \frac{2\gamma_{lv}\cos\theta}{rg\Delta\rho},$$
(6.2)

where r is the capillary radius. In many cases, both h and r can be determined optically.

The majority of contact angle studies in the literature rely on techniques involving sessile drops. However, these techniques pose certain limitations that should be taken into account, especially for the purpose of interpretation of contact angles in terms of surface energetics. An inherent limitation with sessile drop methods is that the camera is focused on the largest meridian section, and hence reflects only the contact angle at the point in which the meridian plane intersects the three-phase line. The presence of surface heterogeneity and/or roughness could well cause variations in the contact angle along the three-phase line. To avoid this problem, contact angles may be inferred from the contact diameter of the sessile drop (see Section 6.3.3). Nevertheless for drops up to approximately 1 cm in diameter, contact angles show a systematic change with the drop size, possibly due to line tension effects (see Chapter 13). An alternate popular technique that does not involve such complexities is the capillary rise at a vertical plate. A solid surface is aligned vertically and brought into contact with the probe liquid, which rises to a certain height, h, on the surface (Figure 6.1). Assuming that the vertical plate is infinitely wide, the Laplace equation integrates to [1]

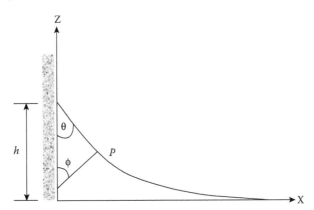

FIGURE 6.1 Schematic of capillary rise at a vertical plate, where ϕ is the angle between the vertical axis and the normal at a point on the liquid–vapor surface, P; θ is the contact angle; and h is the capillary rise at a vertical plate.

$$\sin \theta = 1 - \frac{\Delta \rho g h^2}{2\gamma_{lv}}. \tag{6.3}$$

If three parameters; that is, the density difference between liquid and vapor ($\Delta\rho$), the gravitational acceleration (g), and the liquid surface tension (γ_{lv}), are known, then the contact angle (θ) can be obtained from a measurement of the capillary rise (h) from Equation 6.3. Thus, the task of contact angle measurement is reduced to measuring a length, which can be performed optically with a very high degree of accuracy by means of a cathetometer. Since the contact line is observed directly, any irregularities due to imperfections of the solid surface can be detected and handled, through averaging.

The capillary rise technique can facilitate dynamic contact angle measurements too. A vertical plate is immersed into or withdrawn from the liquid at essentially stationary three-phase line and advancing and receding contact angles are measured, respectively [26–28]. For example, Sedev et al. employed an automated version of the capillary rise to study advancing and receding dynamic contact angles of n-octane on dry, pre-wetted, and soaked fluoropolymer FC-722 (3M Inc.) surfaces [27]. Such measurements provided preliminary evidence for changes in the surface properties of a polymer due to contact with liquids as inert as n-alkanes.

The technique of capillary rise can also be used to determine γ_{lv} and θ separately. For example when contact angles and surface tensions are time-dependent due to change in the temperature or adsorption at the interfaces, both of these parameters can be obtained through combining the capillary rise and the Wilhelmy plate techniques. Thus, measurements of contact angles and surface tensions are reduced to the measurements of the capillary rise, h, and the change in weight of the vertical plate as a function of time [1,29].

Crystalline and amorphous solid surfaces may undergo conformational and molecular rearrangement due to changes in temperature. Neumann et al. utilized the capillary rise technique, which is amenable to contact angle measurements at different temperatures, to study allotropic phase transitions in polytetrafluoroethylene (PTFE), n-hexatriacontane, cholesteryl acetate, and chlorinated rubber [30,31]. For example with the PTFE surfaces, the samples were heated to 330°C and then cooled down to room temperature at a rate of 36°C/h [30]. The capillary rise values and the corresponding contact angles, θ, and solid surface tensions, γ_{sv}, are shown in Figure 6.2. The data reveal a glass transition at −10°C, as well as two crystalline transitions near 17°C and 25°C. Changing the cooling rate of the solid surfaces shifted both the glass transition and the crystalline transitions states. Similar solid-solid phase transitions were observed with other solid surfaces using the capillary rise technique [31].

6.3 MEASUREMENT OF CONTACT ANGLES: NEW TECHNIQUES

Conventional techniques for the measurement of contact angles have greatly contributed to the progress in the field and much of the current understanding of contact angle phenomena is due to a wealth of information generated using these techniques.

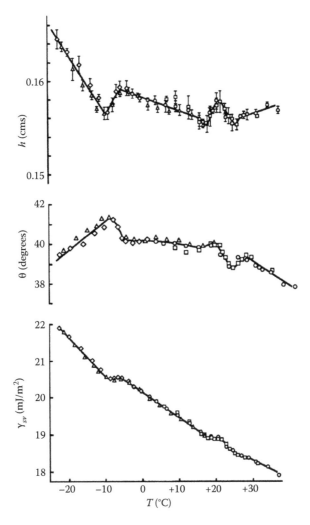

FIGURE 6.2   The temperature dependence of capillary rise, h, contact angles, θ, and solid surface tension, γ_{sv}, of PTFE. The peaks near $-10°C$, $17°C$, and $25°C$ are due to the glass transition and the two well-known allotropic transitions of the crystalline phase. Different symbols represent different runs. (Reprinted from Neumann, A. W. and Tanner, W., *Journal of Colloid and Interface Science*, 34, 1, 1970. With permission from Academic press.)

During the past two decades, significant advancements were made in the design and manufacture of precision hardware tools as well as in computational sciences. Incorporation of new hardware and software capabilities in the area of contact angle research has led to the development of experimental strategies and numerical schemes that enhance the precision of measurements. Today, methodologies such as ADSA [32,33] and theoretical image fitting analysis-axisymmetric interfaces (TIFA-AI) [34,35] enable contact angle measurements with a reproducibility of ±0.2°.

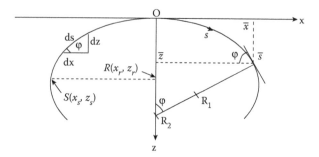

FIGURE 6.3 $R(x_r, z_r)$ is a user-defined reference point inside the drop where ADSA starts the search for a point $S(x_s, z_s)$ with the highest gradient magnitude along $z_s = z_r$ while $x < x_r$. The schematic also represents the definition of the coordinate system for two homogeneous fluids separated by an interface. At a point (x_i, z_i), the turning angle is φ. The arc length, s, is measured along the drop. R_1 and R_2 are the two principal radii of curvature; R_1 turns in the plane of the paper, and R_2 rotates in the plane perpendicular to the plane of the paper.

6.3.1 AXISYMMETRIC DROP SHAPE ANALYSIS-PROFILE (ADSA-P)

Determination of liquid surface tension and contact angle in ADSA is based on drop shape. Details have been presented in Chapter 3; nevertheless, the procedure is briefly discussed below for the purpose of contact angle determination from sessile drops in ADSA.

After an edge operator (e.g., Sobel) is applied to the image of an experimental drop, differences in the gray level of the pixel points in the vicinity of the drop edge are used to calculate a gray level gradient for each pixel and to construct a gradient image. To find drop profile coordinates from the gradient image to pixel resolution, a reference point inside the drop is determined by the user, for example, point $R(x_r, z_r)$ in Figure 6.3. Starting from this point, the program searches for a point $S(x_s, z_s)$ with the highest gradient magnitude along $z_s = z_r$ while $x < x_r$. Once point S is identified, the rest of the drop profile coordinates are determined in the clockwise and counter-clockwise directions from point S using a compass directional search method. In the case of sessile drops, the search must stop at the points of contact of the liquid–vapor interface with the solid surface.

Identification of the contact points primarily relies on comparing the x coordinates of successive pixel points along the drop profile. The program prompts the user to input whether the contact angle is less than, more than, or about 90°. For the case of a contact angle less than 90°, pixel points in the clockwise direction whose x coordinates are either equal to or greater than that of the preceding pixel point are identified as drop profile coordinates. If a pixel point $(i + 1)$ is reached whose x coordinate (x_{i+1}) is less than that of the preceding point (x_i), the search stops and the pixel point with the largest x coordinate (x_i) is selected as the right contact point of the sessile drop with the solid (Figure 6.4). A similar algorithm is applied in the counterclockwise direction to find the left contact point with the criterion being that the x coordinate of a pixel point (x_{i+1}) must not be greater than that of the preceding one (x_i); otherwise, the preceding pixel i is taken as the left contact point.

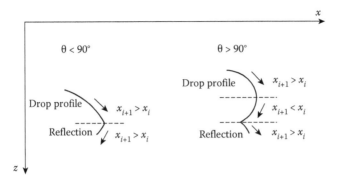

FIGURE 6.4   Criteria for the identification of contact points of the liquid–vapor interface with the solid surface for the cases of contact angle smaller and greater than 90°.

A similar strategy is used to find the contact points of sessile drops with a contact angle of greater than 90°. The right contact point is reached when the x coordinates of two consecutive pixel points along the extracted profile in the clockwise direction shift from $x_{i+1} < x_i$ to $x_{i+1} > x_i$ (Figure 6.4). And in the counterclockwise direction, the condition for finding the left contact point is change in the x coordinates of two consecutive pixel points from $x_{i+1} > x_i$ to $x_{i+1} < x_i$.

It often happens that two or three pixels with the same x coordinates exist at the contact point region of the drop profile. The strategy to resolve this issue in the current version of ADSA-P is to choose the upper pixel of two pixel points with a similar x coordinate and the middle pixel of three pixels points with the same x coordinate as the contact point. The rationale behind this strategy is as follows. Detection of contact points of the liquid–vapor interface with the solid surface is experimentally facilitated by slightly tilting the camera such that the edges of the drop close to the solid–liquid–vapor interface are reflected on the shiny surface of the solid. Applying an edge operator to the experimental image of a sessile drop results in a profile that consists of the actual drop profile and the reflection of parts of it close to the two contact points on the solid surface. When two pixel points with a similar x coordinate are obtained, it is assumed that the upper pixel belongs to the actual drop profile and the lower one to its reflection on the solid surface. In the case of three pixels of similar x coordinates, the middle pixel is taken as an average and is assumed as the contact point of the drop profile with the solid surface.

It is noted that there might be more than three pixels with similar x coordinates for sessile drops whose contact angles are close to 90°. The above strategy may introduce significant errors in such cases. However, an option has been incorporated in the ADSA-P program that allows the user to manually determine two points on the left and right sides of the drop, each being located on the same plane as the contact points are. The search for the contact point stops when a pixel point is reached along the extracted drop profile, whose z coordinate is similar to that of the manually determined point. Experience shows that, with some practice, using magnified images of the drop and selecting these two points from a region in close vicinity of the contact points produces consistent contact angle results.

This procedure of contact point selection is followed by correction of optical distortion, further refining of the drop profile to subpixel resolution through a cubic spline fit technique (cf. Chapter 3), rotation of the profile according to the vertical alignment of the camera, and in the case of a sessile drop, rotation to the horizontal based on the slope between a line connecting left and right contact points. At this stage, the following set of first-order differential Equation 6.4a through f is simultaneously integrated:

$$\frac{d\varphi}{ds} = 2b + cz - \frac{\sin\varphi}{x}, \tag{6.4a}$$

$$\frac{dx}{ds} = \cos\varphi, \tag{6.4b}$$

$$\frac{dz}{ds} = \sin\varphi, \tag{6.4c}$$

$$\frac{dV}{ds} = \pi x^2 \sin\varphi, \tag{6.4d}$$

$$\frac{dA}{ds} = 2\pi x, \tag{6.4e}$$

$$x(0) = z(0) = \varphi(0) = V(0) = A(0) = 0, \tag{6.4f}$$

where b is the curvature at the apex, $c = (\Delta\rho g)/\gamma$ is the capillary constant, and φ is the tangential angle that, for sessile drops, becomes the contact angle at the three-phase contact line. Although the surface area A and the volume V are not required to define the Laplacian profile, their integration along with the above set of equations does not require significant extra computational time. Details of the integration procedure are given in the numerical scheme of ADSA in Chapter 3. Once b and c are known, other parameters are readily calculated. As shown in Figure 6.3, φ is the turning angle measured between the tangent to the interface at (x, z) and the plane of reference. The contact angle is determined as the value of φ at the contact point by integrating the differential Equation 6.4 with the known coefficients.

Since determination of contact angles in ADSA-P is indirect and subsequent to computation of surface tension of the liquid, the accuracy of contact angle data very much depends on the value of surface tension. In Chapter 3, the accuracy of surface tension values from the ADSA-P algorithm has been discussed in detail. The influence of various parameters such as b, c, the number of points selected from the drop profile, and drop profile acquisition errors on the calculated surface tension from ADSA-P was investigated and compared to the surface tension of a computer-generated theoretical profile with a known surface tension. In all cases, ADSA-P surface tension reproduces the theoretical value with a precision of 10^{-5} mJ/m^2 (convergence criteria in the program). It is expected that the numerical integration of Equation 6.4 with known coefficients for the determination of contact angles will not generate any significant numerical error in the calculated results, as long as contact points are selected correctly.

A new variation of ADSA; that is, ADSA-NA (ADSA-No Apex), has been developed to broaden its application for drop configurations without an apex [36]. ADSA-NA is useful for configurations such as liquid bridges and floating lenses where a capillary protrudes into the liquid lens to provide mechanical stability to it. A methodology different from ADSA was also developed to determine surface tension and contact angles by analyzing the shape of axisymmetric liquid–fluid interfaces without use of apex coordinates and edge detection procedures; it is called Theoretical Image Fitting Analysis-Axisymmetric Interfaces (TIFA-AI) [34] (see Chapter 5). A recent study shows that contact angles from these two methods agree within 0.1° [36]. The agreement of results from these two methods with contact angle data from ADSA methodology using apex coordinates is approximately 0.2°–0.3°.

6.3.2 APPLICATIONS OF ADSA-P FOR CONTACT ANGLE MEASUREMENT

The applications of ADSA-P for the measurement of liquid–fluid interfacial tensions in various surface phenomena were presented in Chapter 3. A second major area of application for ADSA-P is the measurement of contact angles of different solid–liquid–fluid systems. Apart from its greater accuracy over conventional methods, the ADSA-P methodology facilitates measuring different modes of contact angles including static (θ_{stat}), advancing (θ_a), and receding (θ_r) contact angles over a wide range and in a highly reproducible manner. This enables studying various contact angle phenomena such as contact angle hysteresis, time dependence of contact angles, drop size dependence of contact angles, rate dependence of contact angles, and contact angles on superhydrophobic surfaces. These topics will be discussed in the light of contact angle measurements in ADSA-P next.

6.3.2.1 Static Contact Angles (θ_{stat})

Li and Neumann used a motor-driven syringe mechanism to measure static contact angles [37]. First a measurement stage was leveled using a sensitive bubble level. A test surface containing a hole of approximately 1 mm in diameter at the center was placed on the stage such that the needle of the syringe passed through the hole slightly above the test surface. Teflon tape was wrapped around the neck of the needle to seal the hole and prevent leakage of the test liquid during measurements. An initial small drop was formed by carefully depositing liquid on the surface to cover the needle tip. Then liquid was pumped slowly into the drop from below the surface until the radius of the three-phase contact line reached about 4 mm. The motor was stopped and the sessile drop was allowed to relax for about 30 seconds to reach equilibrium. Then three images of the drop were taken successively at time intervals of 30 seconds. More liquid was then injected into the drop until the contact line reached another desired size, and the above procedure was repeated. Finally, the images were processed and contact angles were reported as the average of measurements on several surfaces. It is noted that supplying liquid from below the surface, rather than depositing from above, was done to ensure that measured contact angles were proper "advancing contact angles." In contrast, if one deposits a drop from

TABLE 6.1

Average Static Contact Angles of Different Liquids Measured on PET Surfaces and the Corresponding Standard Deviation and 95% Error Limits

Liquid	γ_{lv} (mJ/m^2)	θ_{mean} (°)	Std. Dev. (°)	±95% Error Limits (°)	No. of Drops
Diethylene glycol	45.04	41.19	0.20	0.09	8
Ethylene glycol	47.99	47.52	0.20	0.10	7
Thiodiglycol	54.13	55.57	0.34	0.17	6
Formamide	57.49	61.50	0.37	0.18	6
Glycerol	63.11	68.10	0.27	0.13	6
Water	72.75	79.09	0.12	0.08	6

Note: Reprinted from Li, D. and Neumann, A. W., *Journal of Colloid and Interface Science*, 148, 190, 1992. With permission from Academic press.

the top, inevitable vibrations of the drop may lead to a contact angle intermediate between advancing and receding contact angles.

This strategy was employed to measure contact angles of different liquids on three different solid surfaces (FC-721 fluoropolymer, Teflon fluorinated ethylene propylene (FEP), and polyethylene terephthalate (PET) prepared by dip coating or heat pressing techniques [37]. Table 6.1 shows the results on PET surfaces. The accuracy of the contact angles is better than 0.2°. These carefully measured contact angles were used to modify the formulation of the equation of state for interfacial tensions, which is used to determine the surface tension of the solid substrates from contact angle and liquid surface tension (see Chapters 8 and 9). It is noted that existing static contact angles in the literature should be treated with caution when the goal is determination of solid surface tensions from contact angles. This is due to measurement errors inherent to goniometer and similar direct methods.

The static contact angle measurement strategy of Duncan et al. facilitated the investigation of the drop-size dependence of contact angles [38]. Figure 6.5 shows contact angles of *n*-dodecane on FC-721 surfaces from two sets of experiments started with an initial droplet diameter of about 1 mm and ended when the contact radius of the sessile drop was approximately 5 mm, as limited by the field of the camera lens. Each point in the plot represents the average calculated from the contact angle results of three successively recorded images of one sessile drop. The figure shows that the contact angles decrease by approximately 3° as the contact radius increases from approximately 1–5 mm. For large drops, the contact angle size dependence vanishes. This drop-size dependence of contact angles was interpreted as a line tension effect implying a value of the line tension in the range of 10^{-6} J/m (see Chapter 13).

6.3.2.2 Dynamic Advancing (θ_a) and Receding (θ_r) Contact Angles

The experimental approach of Kwok et al. was utilized to measure dynamic advancing (θ_a) and receding (θ_r) contact angles [39]. A motor-driven syringe was used to pump liquid steadily into the sessile drop from below the surface. A quartz cuvette

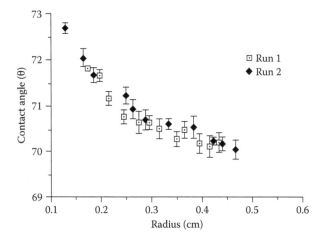

FIGURE 6.5 Typical results for the drop-size dependence of contact angles measured by ADSA-P. Contact angles decrease by approximately 3° as the contact radius increases from 1 to 5 mm. (Reprinted from D. Li et al., *Colloids and Surfaces*, 43, 195, 1990. With permission from Elsevier.)

TABLE 6.2

Advancing and Receding Contact Angles of *n*-Decane and *n*-Tridecane on Teflon AF 1600 Films at Different Rates of Motion of the Three-Phase Measured with ADSA-P

	n-Decane				*n*-Tridecane		
Rate (mm/ min)	θ_a (°)	Rate (mm/ min)	θ_r (°)	Rate (mm/ min)	θ_a (°)	Rate (mm/ min)	θ_a (°)
0.48	59.35 ± 0.24	0.69	53.33 ± 0.24	0.42	65.32 ± 0.24	0.50	59.30 ± 0.25
0.53	59.33 ± 0.32	0.72	53.30 ± 0.26	0.51	65.34 ± 0.24	0.60	59.48 ± 0.21
0.69	59.27 ± 0.19	0.84	53.32 ± 0.21	0.67	65.33 ± 0.29	0.74	59.49 ± 0.18
0.76	59.17 ± 0.25	1.01	53.36 ± 0.24	0.91	65.21 ± 0.22	0.84	59.24 ± 0.23
0.91	59.40 ± 0.26	1.12	53.33 ± 0.18	1.18	65.04 ± 0.27	1.13	59.21 ± 0.19
Mean	59.30 ± 0.06		53.33 ± 0.06		65.25 ± 0.11		59.34 ± 0.10

($5 \times 5 \times 5$ cm³) was used to isolate the drop from its surroundings, although it has been found that there is virtually no difference between measured contact angles with or without a cuvette. Dynamic advancing and receding contact angle measurements were performed, respectively, by continuously pushing or pulling the plunger of the motorized syringe. This resulted in an increase or decrease in the drop volume and advancing or retreating of the three-phase contact line. A schematic of this mechanism is shown in Figure 6.6. Measurements are usually performed at low rates of motion of the three-phase line in the range 0.1–1 mm/min. Table 6.2 shows a typical result of advancing and receding contact angle measurement with ADSA-P for

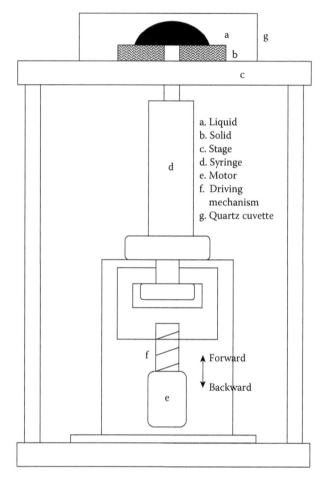

FIGURE 6.6 Schematic of a motorized syringe mechanism for dynamic contact angle measurements. (Reprinted from D.Y. Kwok et al., *Advances in Colloid and Interface Science*, 81, 167, 1999. With permission from Elsevier.)

n-decane and *n*-tridecane on Teflon AF 1600 surfaces. The data show the reproducibility of contact angle measurements from one experiment to another and extremely small confidence intervals. Indeed, for a given set of solid–liquid systems, similar contact angles were measured by different operators at different times [40]. This indicates the consistency and robustness of the ADSA-P methodology for dynamic contact angle measurements. Details about the reproducibility of ADSA-P contact angles are presented in Chapter 8.

It is important to note from Table 6.2 that contact angles obtained at this range of three-phase line velocity are rate independent. A simple test by Kwok et al. confirmed that for well-prepared smooth surfaces, low-rate advancing contact angles are essentially identical to static contact angles [39]. The test was performed as follows: A drop of *cis*-decalin was first selected to advance at a rate of 0.41 mm/min from approximately 0.36–0.41 cm while a sequence of images was recorded. The volume of the drop correspondingly increased from 0.05–0.07 cm³. The motor was then stopped and a sequence of drop images was acquired at $R = 0.41$ cm. The contact angles were found to be independent of slow rates of advancing (see Figure 8.2), suggesting that low-rate dynamic contact angles θ_{dyn} are identical to properly measured static contact angles θ_{stat}. This result also reconfirmed the validity of the experimental protocol established by Li and Neumann for static contact angle measurements [37]. Similar conclusions were reached by static and dynamic contact angle measurements for tetradecane/FC-721 and dodecane/Teflon (FEP) systems using the capillary rise technique [26,27].

The dynamic contact angle measurement capability of the ADSA-P methodology is a major advantage over conventional techniques for detecting various experimental complexities that may arise during contact angle measurements. Two such cases where measured contact angles vary with solid–liquid contact time are discussed below.

6.3.2.3 Time-Dependent Contact Angles

Extensive studies have shown that not all liquids yield essentially constant advancing contact angles [41–43]. Figure 6.7 shows the results for formamide on a poly(propene-*alt*-N-(n-propyl)maleimide) copolymer. It can be seen that as drop volume increases initially, the contact angle increases from 60° to 63° at essentially constant three-phase line radius. As the drop volume continues to increase, θ suddenly

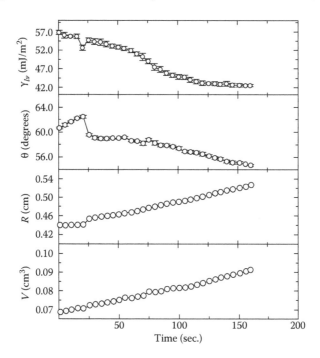

FIGURE 6.7 Low-rate dynamic contact angles of formamide on a poly(propene-*alt*-N-(n-propyl) maleimide) copolymer. (Reprinted from Kwok, D. Y., Gietzelt, T., Grundke, K, Jacobasch, H.-J., and Neumann, A. W., *Langmuir*, 13, 2880, 1997. With permission from American Chemical Society.)

decreases to 60° and the three-phase contact line starts to move. As R increases further, the contact angle decreases slowly from 60° to 54°. The surface tension time plot indicates that the surface tension of formamide decreases with time. This suggests that dissolution of the copolymer occurs, causing γ_{lv} and θ to change from those of pure formamide. Thus, the resulting contact angles from this and similar solid–liquid systems cannot be meaningful. It would be virtually impossible for the goniometer technique to detect such complexities.

It is occasionally found that during contact angle measurements with certain solid–liquid systems, the drop front does not move smoothly on the solid surface, but rather shows a jerky motion [42–45]. As liquid is pumped into the drop and the volume increases, the drop front remains hinged on the surface (i.e., the contact angle increases at a constant contact radius). Then the three-phase line slips suddenly on the solid surface as more liquid is supplied. This is accompanied by an abrupt decrease in the contact angles and by a sudden increase in the contact radius. By supplying more liquid, the three-phase line again sticks to the solid surface at a new location, and the radius remains constant. This pattern is repeated as the measurement continues (cf. Chapter 7 for details). The thermodynamic significance of the corresponding contact angles is not well understood and it is difficult to decide unambiguously whether or not Young's equation is applicable. Therefore, these contact angles should be disregarded for interpretation in terms of surface energetics. While pronounced cases of stick-slip behavior can be observed by the goniometer, it is almost impossible to record an entire stick-slip period manually.

6.3.2.4 Rate Dependence of Contact Angles

It has been reported in the literature that contact angles may be dependent on the rate of motion of the three-phase contact line [46–53], especially for viscous liquids [50]. The dynamic contact angle measurement capability of ADSA-P has allowed studying this phenomenon using a large number of solid–liquid systems over a wide range of three-phase line velocity (i.e., ~0.2–10 mm/min). It was found that on smooth and homogeneous solid surfaces, both advancing and receding contact angles of liquids with a dynamic viscosity of well below 10 cP are independent of the rate of motion [54]. This is in agreement with the findings from other studies of contact angles of low viscosity liquids on different fluorinated solid surfaces using the sessile drop [55] and [26–28] the capillary rise techniques. However, as the viscosity of probe liquids increased to greater than ~10 cP, viscous forces affected the contact angles and increasing the rate of motion of the three-phase contact line caused advancing angles to increase and receding angles to decrease [54]. Thus, the velocity dependence of contact angles appeared to be an issue only for fairly viscous liquids (see Chapter 7 for details).

6.3.2.5 Small and Extremely Large Contact Angles

ADSA-P can also accommodate measurements of a wide range of contact angles with ease and convenience. Contact angles as low as 20°–25° have been measured with a reasonable accuracy. For example, cis-decalin yielded a contact angle of 28.81° on poly(propene-alt-N-(n-hexyl)maleimide) copolymer with an error of ±0.67° [41,44]. It is noted that a separate algorithm (i.e., ADSA-D) has been developed for more accurate measurements of small contact angles as discussed below. ADSA-P has also been applied to measure very large contact angles such as those of water on superhydrophobic surfaces of *n*-hexatriacontane [56]. An average contact angle of 170.90° was reported for this system with an error of only ±0.90° (cf. Chapter 7).

6.3.3 ADSA-D FOR MEASUREMENT OF SMALL CONTACT ANGLES AND CONTACT ANGLES ON NONIDEAL SURFACES

In practice, many solid–liquid pairs produce low contact angle values. For example, saline solutions on many biological surfaces typically produce contact angles less than 20° [57]. As the contact angle decreases, the accuracy of direct methods, such as a goniometer, is adversely affected. The resulting decrease in accuracy is due primarily to the difficulty in determining the location of the three-phase contact line (Figure 6.8). The same problem affects computational methods including ADSA-P, which rely on the profile of a drop to determine the contact angle. A second phase of ADSA methodology, ADSA-D (diameter), has been developed exclusively for contact angle studies in response to such experimental difficulties.

The ADSA-D program circumvents this problem by utilizing a top view of the drop. Essentially, the contact angle is computed numerically through minimizing the difference between the volume of the drop, as predicted by the Laplace equation of capillarity, and the experimentally measured volume. ADSA-D consists of

(a)   (c)

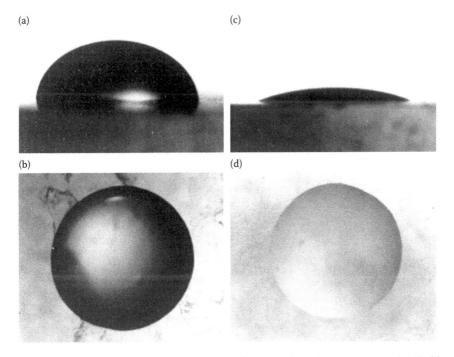

(b)   (d)

FIGURE 6.8 (a) Side and (b) top view of a sessile drop with a large contact angle; (c) side and (d) top view of a sessile drop with a small contact angle. (Reprinted from Skinner, F. K., Rotenberg, Y., and Newmann, A. W., *Journal of Colloid and Interface Science*, 130, 25, 1989. With permission from Academic press.)

two modules, ADSA-CD (contact diameter) for contact angles < 90° and ADSA-MD (maximum diameter) for contact angles > 90°. The contact diameter (or maximum diameter for contact angles greater than 90°), the volume and the surface tension of the liquid under consideration, the density difference across the liquid–fluid interface, and the gravitational constant are used as input to calculate the contact angle of the drop. The flatness of the drop does not affect the accuracy with which the contact diameter can be measured.

ADSA-CD was originally developed to measure the contact angles of very flat drops (drops with very low contact angles). Additionally, it has been a very useful approach for the measurement of contact angles of drops on nonideal surfaces, which are usually rough and heterogeneous. It is practically impossible to form an axisymmetric drop on such surfaces. The ADSA-CD routine circumvents this problem and utilizes an average contact diameter to determine the contact angle. This average contact diameter is obtained by a least-squares fit of a circle to the experimentally measured points along the three-phase line of the drop (Figure 6.9), and is used as the input to the ADSA-CD routine to determine the contact angle. Details of the digital image processing procedure for ADSA-D are presented in Chapter 4.

The generation of a nonsymmetrical drop on rough and heterogeneous surfaces is not limited to cases where the contact angle is less than 90°. For cases where the contact angle is greater than 90°, the three-phase line is obscured by the free surface of the drop in a top view. In such cases, the ADSA-MD routine utilizes the maximum or equatorial diameter instead of the contact diameter. The equatorial diameter can be obtained with relative ease by viewing the drop perpendicular to its apex. Other required inputs are similar to those for ADSA-CD and include the drop volume, the liquid–fluid interfacial tension, the liquid–fluid density difference, and the gravitational acceleration. ADSA-MD retains the same advantages that ADSA-CD has over conventional goniometric and drop shape methods for systems with nonideal substrates.

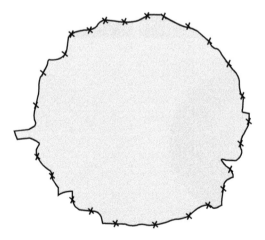

FIGURE 6.9 Schematic of the determination of the perimeter of a sessile drop on a video screen using a cursor controlled by a mouse to select perimeter points. (Reprinted from Duncan-Hewitt, W. C., Policova, Z., Cheng, P., Vargha-Butler, E. I., and Neumann, A. W., *Colloids and Surfaces*, 42, 391, 1989. With permission from Elsevier.)

6.3.3.1 Numerical Procedure

The ADSA-D program utilizes integration of the Laplace equation of capillarity, which may be expressed in differential form as Equation 6.4a, along with the geometric relations Equation 6.4b through e and the boundary conditions Equation 6.4f to determine the contact angle. For a sessile drop, the angle φ at the point of intersection of the solid surface with the fluid interface defines the contact angle, θ (see Figure 6.3). For this system with known surface tension, γ, density difference, $\Delta\rho$, and gravitational constant, g (i.e., the capillary constant, c, is an input parameter), the coordinates of the system may be expressed as

$$x = x(s, R_0), \quad z = z(s, R_0), \quad \varphi = \varphi(s, R_0),$$

where s is the arc length variable and R_0, the radius of curvature at the apex, is a geometrical parameter of the system. Thus, if the value of R_0 and the volume of the drop are assumed to be known, the above set of differential equations can be integrated to determine the contact angle. It is seen that the numerical procedure for ADSA-D is simpler than for ADSA-P.

To solve for the proper value of the parameter R_0, the volume of the drop is utilized, which is calculated as

$$V = \int \pi x^2 \sin\varphi ds. \tag{6.5}$$

With a known value of the surface tension of the drop, and experimentally measured values of the volume and the contact diameter, the volume can form the basis for the establishment of an objective function (which is to be minimized) for the determination of R_0, and thus the contact angle. The objective function, $\varepsilon(R_0)$, may be expressed as

$$\varepsilon(R_0) = \int_0^{S_c(R_0)} \pi x^2 \sin\varphi ds - VOL, \tag{6.6}$$

where $S_c(R_0)$ is a boundary point that depends on the value of R_0, and is equal to the value of the arc length at the contact point; that is, the value of s for which $x(s, R_0) - x_c = 0$, and VOL is the experimentally determined volume of the sessile drop. Minimization of this function leads to the determination of R_0, and hence to the determination of a Laplacian drop (via the above set of differential equations) with the given values of x_c, VOL, and c. The value of φ at S_c on the Laplacian curve is the contact angle, θ, of the experimental drop. The solution for R_0 is not available in an analytical form, and thus is sought by numerical means. The value of R_0 is calculated using Newton's method:

$$R_0^{(k+1)} = R_0^{(k)} - \frac{\varepsilon\left(R_0^{(k)}\right)}{\varepsilon'\left(R_0^{(k)}\right)}, \tag{6.7}$$

where $\varepsilon'(R_0)$ is calculated using the Leibniz rule. The value of R_0 in Equation 6.7 is updated until $\varepsilon(R_0) = 0$. This condition terminates the iteration and yields the value of R_0.

For the case of contact angles less than 90°, using the above set of differential equations with known R_0 and c, contact angle, θ, is calculated as the value of φ at the contact point $x = x_c$ (the ADSA-CD program). For drops with contact angles greater than 90°, the contact diameter of the drop, x_c, cannot be determined simply by viewing the drop from above and the maximum diameter is used instead. As an initial guess of the parameter R_0, it is assumed that the sessile drop has the shape of a hemisphere with diameter equal to the maximum diameter of the drop, d_{max}, so that $R_0 = d_{max}/2$. Once the parameter R_0 for the given sessile drop is known, the contact angle can be obtained by solving for the value of $S_c(R_0)$. However, since in this case the value of x_c is unknown, the value of $S_c(R_0)$ is obtained numerically by integrating Equation 6.5 until the condition

$$VOL - \int_0^{S_c(R_0)} \pi x^2 \sin\varphi \, ds = 0 \qquad (6.8)$$

is met. Once Equation 6.8 is satisfied, the values of $x(S_c)$ and $\varphi(S_c)$ correspond to the contact diameter and contact angle, respectively.

For a sessile drop with a given volume, VOL, and contact diameter, x_c, three configurations are possible depending on the solid–liquid system: the contact angle can be less than, equal to, or greater than 90°. Let the 90° drop with contact radius, x_c, have radius of curvature at the apex (or the origin of the coordinate system) $R_0^{90°}$. If the contact angle of a drop of diameter x_c is not 90°, it is either less than 90° ("wetting drop") or greater than 90° ("nonwetting drop"). The radius of curvature at the apex, R_0, for these two configurations is such that $R_0 > R_0^{90°}$; that is, these drops have a lower curvature than the 90° drop. Correspondingly, the volume of the wetting drop is less than that of the 90° drop and the volume of the nonwetting drop is greater than that of the 90° drop. To initialize ADSA-D, it is necessary to determine, for given values of VOL and x_c, which of the above is the case so that the ADSA-CD or ADSA-MD approach is utilized accordingly. On occasion, the user can give this information as input. But in many cases, especially for contact angles near 90°, it is difficult to make a sound judgment. The approach implemented in the program is as follows: (i) If the user knows whether the drop is wetting or nonwetting, solve the respective problem and exit; otherwise, (ii) assume the contact angle is greater than 90°, solve the corresponding problem for the given x_c and compute volume $VOL_{90°}$ for $\theta = 90°$ by numerically integrating the set of differential Equations 6.4a through f. If $VOL_{90°} \leq VOL$, the initial assumption was correct, compute θ, and exit; otherwise, (iii) solve the problem for contact angle less than 90°.

The ADSA-CD algorithm readily lends itself to averaging procedures and can be used to determine the average contact angle of irregular drops on rough and heterogeneous surfaces. These advantages also apply to measurements of large contact angles on hydrophobic surfaces using ADSA-MD. Both programs have simpler numerical analysis than ADSA-P since the liquid surface tension is used as an input.

6.3.3.2 ADSA-D Setup

A schematic of the experimental setup for ADSA-D is shown in Figure 6.10. Before any measurements are taken, the measurement stage is leveled by adjusting the set

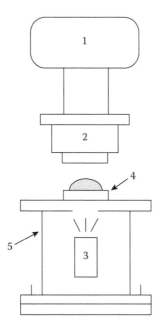

FIGURE 6.10   Schematic of the ADSA-D setup. (1) Camera, (2) stereomicroscope, (3) light source, (4) solid surface, and (5) leveling table. (Reprinted from Skinner, F. K., Rotenberg, Y., and Newmann, A. W., *Journal of Colloid and Interface Science*, 130, 25, 1989. With permission from Academic press.)

screws of the platform. A freshly prepared solid surface is then placed on the stage. Using a Gilmont micrometer syringe fitted with an angled needle, a drop is placed on the surface. The syringe-needle assembly is mounted on a stage with a micro-manipulator. This allows the pendent drop formed at the tip of the needle to be lowered slowly, minimizing vibrations. Images are acquired immediately after drop deposition. Drop images are acquired from above using a vertically mounted Wild Heerbrugg M7S Zoom stereomicroscope fitted with a Cohu 4800 CCD camera and a coaxial incident illuminator (Volpi Intralux 5000 fiber optic system). The images are stored in a computer for later analysis. The measurement location on the platform can be shifted to facilitate formation of another drop on the same surface.

6.3.3.3   Experimental Evaluation of ADSA-D Accuracy

ADSA-D has been evaluated for accuracy using experimental drops both for contact angles less and greater than 90°. Comparison of the output of ADSA-D with the results from ADSA-P provides a convenient means for the evaluation of the accuracy of the ADSA-D methodologies for contact angle measurements. Table 6.3 illustrates a direct comparison of the results obtained by ADSA-P and ADSA-D from the same drop for the case of contact angles greater than 90° [58]. Sessile drops of water were formed on smooth FC-721 coated mica sheets. Images of the profile and the top view of the drop were acquired simultaneously and used in subsequent analyses. The contact angles computed from ADSA-D were within 0.4° of the values computed from ADSA-P.

TABLE 6.3

Contact Angle Results from ADSA-P and ADSA-MD for the Same Sessile Drops of Water on FC-721 Coated Mica

Drop Number	Volume (mL)	Maximum Diameter (cm)	Contact Angle ± CL (°) ADSA-P	ADSA-MD
1	0.0892	0.6728	117.00 ± 0.13	117.34
2	0.0894	0.6735	117.20 ± 0.13	117.63

Note: Measurements were performed at 23°C. Errors are 95% confidence limits (CL). (Reprinted with permission from Moy, E., Cheng, P., Policova, Z., Treppo, S., Kwok, D., Mack, D. R., Sherman, P. M., and Neumann, A. W., *Colloids and Surfaces*, 58, 215, 1991. With permission from Elsevier.)

TABLE 6.4

Comparison of Contact Angles Measured by ADSA-CD, ADSA-P, and the Goniometer Techniques

Substrate[a]/Liquid[b]	Drop Number	ADSA-CD	ADSA-P	Goniometer
VM/EG	1	84.1 ± 0.4	83.2 ± 0.6	84
	2	84.9 ± 0.2	84.7 ± 0.5	83
	3	85.2 ± 0.3	85.2 ± 0.5	84
SM/EG	1	84.3 ± 0.3	83.6 ± 0.4	83
	2	84.3 ± 0.3	84.3 ± 0.6	84
VM/UN	1	19.5 ± 0.2	18.5 ± 0.6	18
	2	18.8 ± 0.2	20.9 ± 0.6	18
SM/UN	1	22.7 ± 0.2	21.8 ± 0.7	19
	2	21.2 ± 0.2	22.2 ± 0.6	20
	3	21.5 ± 0.2	22.2 ± 0.5	20

Note: The errors for ADSA measurements are the 95% confidence limits. The error of the direct measurements is estimated to be 3°. The contact diameters for ADSA-CD were determined using a digitization tablet. (Reprinted from Skinner, F. K., Rotenberg, Y., and Newmann, A. W., *Journal of Colloid and Interface Science*, 130, 25, 1989. With permission from Academic press.)

[a] The substrates were siliconized glass prepared by either a soaking method (SM) or a vapor deposition method (VM).

[b] The liquids were ethylene glycol (EG) or undecane (UN).

Table 6.4 compares ADSA-D and ADSA-P results for contact angles smaller than 90° obtained with ethylene glycol or undecane on siliconized glass substrates [59]. The solid surfaces were prepared by either the vapor deposition method (VM) or the soaking method (SM). The errors given for the ADSA methodologies are the 95% confidence limits. The results from the two ADSA techniques are in close agreement, confirming the accuracy of the ADSA-D program. For low contact angles (*n*-undecane), ADSA-D exhibits a higher degree of precision than ADSA-P. This is not unexpected due to the difficulties of measuring small contact angles from the profile of a drop as discussed above. The contact angle results for these solid–liquid systems using a direct goniometric measurement have an error of 3°. Both ADSA techniques provide greater accuracy and reproducibility than the direct method.

6.3.3.4   Applications of ADSA-D

The ADSA-D methodology is particularly useful for, although not limited to, measurement of contact angles on rough, heterogeneous, and hydrophilic surfaces. Biological materials are a prime example of such surfaces. The surface tension of biological particles, such as bacterial cells, plays an important role in processes such as cell adhesion and phagocytosis. Obviously, layers of cells are necessarily rough, and absorb water and other liquids so that the drop sinks into the layer of cells. In addition, they are typically hydrophilic, producing small time-dependent contact angles that make the identification of the contact point of the drop with the surface exceedingly difficult. ADSA-D circumvents these problems.

Figure 6.11 illustrates the typical shape of drops of double-distilled water formed on a layer of *Thiobacillus ferrooxidans* cells. For these drops, points at the drop perimeter were selected to estimate the average drop diameter. Complications due to small perturbations, possibly a result of "sticking" of the three-phase line at some inhomogeneity, or "fingers," which could form in particularly rough regions due to wicking, are avoided. Figure 6.9 shows a schematic of selected contact points. The results of such measurements on three different species of bacteria (*T. thiooxidans, Staphylococcus epidermidis,* and two strains of *T. ferrooxidans*) are illustrated in Figure 6.12 [60]. The bacterial layers are deposited on cellulosic membrane filters by suction. The membrane filter and the deposited layer of cells are then rapidly transferred to the surface of a freshly prepared 2% solidified agar plate. The agar plate acts as a water reservoir and decreases the rate at which the cells dry, thus maintaining them in their fully hydrated state. It is apparent that the measured contact angles remain in a fairly narrow range as the cell layers dry slowly over approximately

(a)   (b)

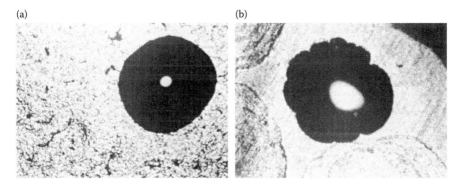

FIGURE 6.11   Images of sessile drops of water on a layer *of Thiobacillus ferrooxidans* cells. The contact angles calculated using ADSA-CD are (a) 12.7° and (b) 11.3°. (Reprinted from Duncan-Hewitt, W. C., Policova, Z., Cheng, P., Vargha-Butler, E. I., and Neumann, A. W., *Colloids and Surfaces*, 42, 391, 1989. With permission from Elsevier.)

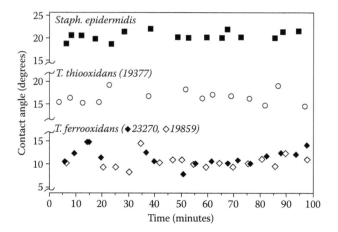

FIGURE 6.12 The contact angles of sessile drops of water on layers of different bacterial cells measured using ADSA-CD. The horizontal axis represents that time after formation of the layer at which the drops were deposited. The mean contact angles (±95% confidence limits) are: (■) 20.6 ± 0.9° (*Staphylococcus epidermidis*); (○) 16.9 ± 0.9° (*Thiobacillus thiooxidans*); (◆) 11.7 ± 1.0° (*T. ferrooxidans* strain 23270); and (◊) 10.5 ± 0.9° (*T. ferrooxidans* strain 19859). (Reprinted from Duncan-Hewitt, W. C., Policova, Z., Cheng, P., Vargha-Butler, E. I., and Neumann, A. W., *Colloids and Surfaces*, 42, 391, 1989. With permission from Elsevier.)

two hours. The average contact angles of 16 drops for *Staph. epidermidis* and *T. thiooxidans* are found to be 20.6 ± 0.9° and 16.9 ± 0.9°, respectively, where the errors represent the 95% confidence limits. Note that a relatively high degree of accuracy is attained from a relatively small number of measurements. The two strains of *T. ferrooxidans* do not exhibit a statistically significant difference in their contact angles.

Similar measurements were performed on *Helicobacter pylori* to characterize its surface hydrophobicity [61]. The adhesion of these bacteria to the gastric epithelium has been associated with both active gastritis and duodenal ulcer diseases in children and adults. The ADSA-D methodology may also be used to quantitate the surface hydrophobicity of different regions of the intestinal tract to which bacteria adhere during initial stages of infection. Figure 6.13 illustrates the variation of water contact angles measured on various intestinal segments of male New Zealand White rabbits. The data reveal that the small-intestinal segments are significantly less hydrophobic than the segments of the large bowel. These tissues are highly nonideal surfaces for contact angle measurements. The surface roughness leads to the formation of somewhat asymmetrical drops. The use of conventional methodologies, such as direct goniometric measurements, would be extremely difficult and prone to error in such cases. However, ADSA-D allows for an estimation of the contact or maximum diameter, thus providing an average contact angle for such drops.

6.3.4 AUTOMATED POLYNOMIAL FITTING (APF) METHODOLOGY

The calculation of surface tension and contact angle in ADSA-P relies on major assumptions that the drop is axisymmetric and Laplacian and gravity is the only

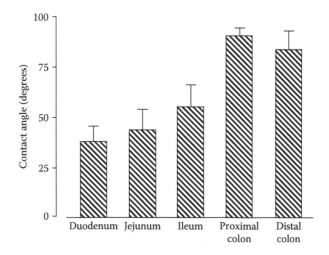

FIGURE 6.13 Contact angles of water formed on intestinal segments from normal adult rabbits measured using ADSA-CD. Results are expressed as average ±1 standard deviation. There are no differences in contact angles between regions of the small intestine (duodenum, jejunum, and ileum; ANOVA, $p > .05$) or large intestine (proximal colon and distal colon; ANOVA, $p > .05$). All small intestinal segments have lower contact angles compared with both segments of large bowel (ANOVA, $p < .05$).

operative external field. In the presence of an electric field, the Laplace equation contains an additional term and the ordinary ADSA approach is not applicable. An appropriate algorithm for such situations is termed ADSA-EF (electric field; see Chapter 5) [62]. In such situations and the more common ones where sessile drops are not axisymmetric, direct methods such as the goniometer can be used, but the reproducibility of the contact angle is subject to the experience of the operator and the best accuracy of the results is usually ±2°–3°.

To address the need for a robust method for calculating contact angles of non-axisymmetric sessile drops, Bateni et al. developed an automated polynomial fitting (APF) technique [63] that built upon an earlier version by del Rio et al. [64]. The new version of APF consisted of an image processing module and a curve fitting module. Contact angle experiments were performed at low rates of advancing of the three-phase contact line and high magnification images (35×) of the contact area of the sessile drop were acquired. In a highly magnified image, each pixel represents a smaller physical area when compared to lower magnifications and therefore, a smaller error will be caused by the digital resolution of the image. Also, magnified profiles have smaller curvature and can be better described by low order polynomials that are less sensitive to experimental noise. The performance of four different edge operators (Sobel, Canny, Prewitt, and Laplacian of Gaussian or LOG), were compared for extracting the drop profile from digital images and it was found that the LOG method generates the smoothest profiles with the least amount of background noise when its sensitivity threshold and filtering parameters are tuned. Figure 6.14 shows a high magnification image of the right contact area of a water sessile drop on

(a)   (b)

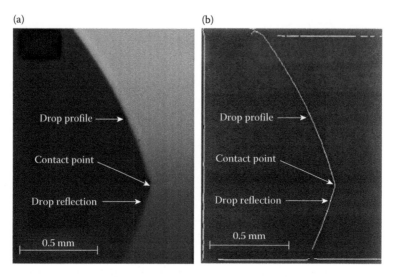

FIGURE 6.14 (a) A high magnification image of the right contact area of a water sessile drop on a polymethylmethacrylate (PMMA) surface, and (b) the corresponding extracted drop profile using the LOG method. (Reprinted with permission from Bateni, A., Susnar, S. S., Amirfazli, A., and Neumann, A. W., *Colloids and Surfaces* A, 219, 215, 2003. With permission from Elsevier.)

a polymethylmethacrylate (PMMA) surface and the corresponding extracted drop profile using the LOG method.

In the next step, the extracted drop profile was sent to the curve fitting module of APF. The left and right contact points of the drop with the surface were defined as the pixels with the lowest and the highest value of the x coordinate for contact angles smaller than 90°. Then, a series of coordinate points, P, close to the contact point were selected and a polynomial of the order O, which is expressed as $Y_{polynomial} = \sum_{i=0}^{O} a_i x^i$, was fitted to them. The contact angle was computed from the first derivative of the polynomial at the point of contact with the solid surface. Optimization of P and O was performed simultaneously because of strong correlation between them (higher order polynomials require more pixel coordinates in the fitting process). Analysis was carried out for $P = 10$–260 pixels with increments of 10 pixels and the order of polynomial $O = 1$–6. It was found that the largest range of stability of contact angles corresponds to 120–140 pixels and a third-order polynomial. Below and above this range of pixels, contact angles showed significant variations. Analysis of the standard error resulted in a minimum over a narrow range of 120–140 pixels for different orders of polynomial, including the third-order one. This confirmed that the third-order polynomial with 120–140 pixels for curve fitting gives optimum results. Measured contact angles were consistent and showed small scatter around the mean value (Figure 6.15). The relatively small 95% confidence limit values of the contact angles from different solid–liquid systems measured by APF are shown in Table 6.5.

Nevertheless, there is still a discrepancy of up to 1.5° between APF contact angles and those measured with ADSA-P. The difference is believed to be mainly due to

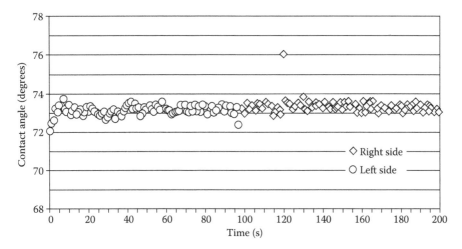

FIGURE 6.15 Advancing contact angles of water/PMMA system calculated by the APF program using a third-order polynomial and 130 pixel. (Reprinted with permission from Bateni, A., Susnar, S. S., Amirfazli, A., and Neumann, A. W., *Colloids and Surfaces* A, 219, 215, 2003. With permission from Elsevier.)

TABLE 6.5
Summary of Contact Angle Measurements by the
APF and the ADSA Schemes

	Contact Angle (°)	
Solid/Liquid System	APF	ADSA-P
Teflon AF 1600/Decane	56.92 ± 0.48	58.19 ± 0.45
Teflon AF 1600/Decanol	73.29 ± 0.49	73.83 ± 0.38
DF55/Formamide	69.62 ± 0.68	68.45 ± 0.73
DF13/Formamide	88.13 ± 0.84	86.47 ± 0.96

Note: Each contact angle represents the average of several runs and
the errors are the standard deviation. DF55 and DF13 are fluorinated polymers. Reprinted with permission from Bateni, A.,
Susnar, S. S., Amirfazli, A., and Neumann, A. W., *Colloids and
Surfaces* A, 219, 215, 2003. With permission from Elsevier.

---

different strategies these methods employ to determine contact angles. In ADSA-P the entire profile of the drop is analyzed and assuming the drop to be axisymmetric and Laplacian, a theoretical Laplacian curve is fitted to the drop profile. On the other hand in APF, only a small section of the drop profile close to the contact point is considered for the polynomial fitting. One consideration is that a polynomial cannot be a perfect fit to a Laplacian profile and only a Laplacian curve can properly describe it. Additionally, minor irregularities of the contact line due to surface inhomogeneity may affect the drop profile in the meridian plane close to the contact line, whereas the drop profile far from the contact line will reflect average properties of the solid surface. The local nature of the APF contact angle measurement renders it more sensitive to such irregularities.

6.4 TEMPERATURE DEPENDENCE OF CONTACT ANGLES

As discussed above in Section 6.2, there are certain experimental conditions where available sessile drop techniques for contact angle measurements are limited. One such difficulty is encountered in the measurement of the temperature dependence of contact angles, which is often not very pronounced and requires an experimental setup more complicated than those of sessile drop-based approaches. This problem was resolved by using the technique of capillary rise at a vertical plate and designing an experimental setup that accommodated measurements of temperature dependence of contact angles [31].

The setup consisted of various elements including the solid surface, a glass cell of relatively large surface area containing a test liquid, a liquid dispensing mechanism to compensate for the evaporation of liquid from the cell, and appropriate heating and cooling mechanisms. The assembled unit was installed inside a doubly enclosed chamber and several temperature control units helped monitor the temperature to maintain it uniform. To measure advancing angles, the vertically mounted solid surface was dipped into the bath of the test liquid maintained at a constant temperature. Dipping the plate was performed at a constant rate, typically in the range 10^{-3} mm/sec, and optical measurement of the capillary rise, h, was carried out by means of a cathetometer or a measuring microscope, with a precision of 2×10^{-4} cm. During measurements, the three-phase line remained essentially fixed with respect to the telescope of the cathetometer, while the immersion of the plate increased steadily. Receding angles were measured in a similar manner by withdrawing the plate from the liquid.

Neumann developed this setup to measure the contact angles of distilled water and long chain n-alkanes (n-decane, n-dodecane, n-tetradecane, and n-hexadecane) on siliconized glass at different temperatures [31]. It is noted that no contact angle hysteresis was observed on siliconized glass with several test liquids (water, four n-alkanes, and glycerol), presumably due to the smoothness and homogeneity of surfaces and the inertness of the surfaces with respect to the test liquids. Water contact angles showed an overall increase of about $2°$ by increasing the temperature from ~5 to ~50°C. For n-alkanes, contact angles showed a linear decrease with increase in temperature from ~20 to ~70°C. A larger decrease in contact angles was observed for liquids with shorter hydrocarbon chains. In all cases, contact angles were measured with very good reproducibility. Calculation of solid surface tension values from the contact angles of n-alkanes using an equation of state for interfacial tensions (see Chapters 8 and 9) indicated significant adsorption of hydrocarbon chains on the solid surface. This suggested that the contact angles were not measured on a surface consisting exclusively of methyl groups, but rather one that consisted of a considerable amount of CH_2 groups, suggesting adsorption of alkane molecules on the solid surface.

6.5   SOLID SURFACE PREPARATION TECHNIQUES

Contact angles can be used to characterize the fundamental surface properties of a solid material and to study the effective properties of the material in its natural or as-manufactured state. The acquisition of thermodynamically significant contact angle data for fundamental studies is largely dependent on the quality of the substrate surface. The effects of roughness and heterogeneity can easily overshadow the role of interfacial energetics. Therefore, in fundamental scientific studies it is important to make solid surfaces of high quality to ensure that the contact angles manifest the interactions between the solid and the liquid as given by Young's equation.

Historically, contact angle hysteresis has been proposed as a measure of non-ideality of rough and chemically heterogeneous surfaces. Recent studies have suggested that in addition to roughness and heterogeneity, other mechanisms can also cause contact angle hysteresis (see Chapters 7 and 8). If hysteresis is indeed due to heterogeneity consisting of high- and low-energy surface patches, then according to a simple model (see Chapter 7), both advancing and receding contact angles should be "Young contact angles" in the sense that they may be used in conjunction with Young's equation [31,65]. In this case, the advancing contact angle represents the equilibrium contact angle on an ideal surface composed entirely of the low-energy component, and the receding angle similarly corresponds to the higher-energy patches. Between these two angles are a range of metastable contact angles that are meaningless in terms of Young's equation. Since most real surfaces may be heterogeneous on a microscopic scale (e.g., due to presence of impurities), contact angles must always be measured in the advancing mode if the goal is to infer the solid surface tension. Based on analytical models, Neumann suggested that the scale of surface heterogeneities must be held below approximately 0.1 μm in order to eliminate hysteresis [31].

If surface roughness is the primary cause of hysteresis, the advancing contact angle is not a Young contact angle because it is influenced by the surface geometry as well as interfacial energetics. In practice, very smooth surfaces are required in order to eliminate all roughness effects. For example, on layers of organic pigment, contact angle hysteresis due to roughness was absent if roughness was less than 0.1 μm [66].

From a practical standpoint, it may not be necessary to completely eliminate all evidence of roughness and heterogeneity. If the average contact angle and liquid surface tension measurements for a variety of liquids all give the same solid surface tension via the equation of state (Chapters 8 and 9), then the surface is of sufficient quality. The required level of accuracy, whether contact angles are used for the interpretation of solid surface tensions or just as a measure of surface wettability in industrial settings, will often determine the acceptability of solid surface quality. The rest of this chapter discusses techniques for the preparation of solid surfaces. The methods are grouped according to the form of the solid material; that is, plate (sheet or film), fiber, powder, and so on.

6.5.1 Nonbiological Materials

A variety of techniques can be used to prepare solid surfaces for contact angle experiments, but regardless, morphological and energetic properties of a surface should be independent of the preparation technique. While there might be slight submicron morphological differences in well-prepared surfaces from different methods, they usually are not expected to alter the energetics of the surface. It is advisable that when a surface is prepared using a new approach, its morphological and energetic properties are examined, and if possible, compared with available data on the same type of surface prepared previously using other techniques. Surface morphology can be examined using different microscopic techniques such as atomic force microscope (AFM) and scanning electron microscope (SEM), whereas contact angle measurements provide information about energetics of the surface. Contact angles can also reveal indirect information about quality of solid surfaces. This can be achieved by measurements at different locations on the solid surface using accurate methods such as the capillary rise technique or ADSA. For example, with the former method, two types of investigations can be carried out: one along the vertical axis of the solid and the other on the horizontal axis. For the first test, the solid surface (plate) is immersed at a constant rate and the capillary rise, h, readings are taken at certain time intervals. For the horizontal investigation, the surface is held stationary. An image of the contact line is taken midway, and then the microscope is translated to the right and to the left, on the same vertical plane, by 10 steps of, say, 100 μm intervals. Following every step, an image is taken.

As an illustration, five surfaces—FC-721 dip coated on mica and on glass, heat-pressed FEP against mica and against quartz glass, and siliconized glass—were compared with respect to their surface quality. Measurements of contact angle as a function of the position along the five different surfaces at arbitrarily chosen constant immersion rates are shown in Figure 6.16, with 95% confidence limits [37]. The result for FC-721 coated on mica and immersed in 1-octanol at a constant immersion speed of 0.49 mm/min is shown in Figure 6.16a. There is less than 0.1° variation in θ over a total length of 14 mm. For comparison, Figure 6.16d shows the dodecane contact angle for FC-721 coated on a glass slide, at a constant immersion speed of 0.31 mm/min over a 40 mm length. This measurement indicates that there is very little variation in θ from point to point. The quality of this surface, reflected by the constant contact angle, is thus very similar to the FC-721 coated on mica.

A similar comparison is shown in Figure 6.16b and e for the heat-pressed FEP against mica and against quartz glass, respectively. Figure 6.16e reveals a significant difference in θ from point to point, suggesting that the surface is not uniformly smooth and/or homogeneous. Improvements in fabrication procedure and pressing against mica instead of quartz glass resulted in contact angles with very little variation for dodecane (Figure 6.16b). In the case of siliconized glass, nearly constant contact angles are measured for water on the surface over a total length of 14 mm (Figure 6.16c).

The lateral variation over the solid surface is illustrated in Figure 6.17 for four solid surfaces: FC-721, heat-pressed FEP against mica and against quartz glass, and heat-pressed polyethylene (PE) against quartz glass. The left Y-axis represents the calculated contact angle and the right Y-axis the corresponding capillary rise, h, on

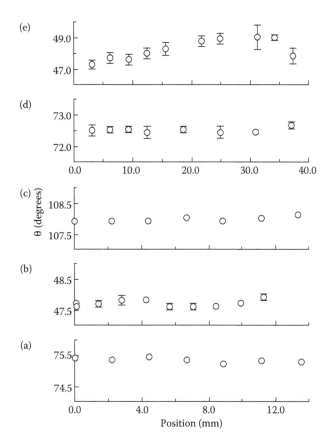

FIGURE 6.16 Contact angles as a function of position of (a) 1-octanol at a constant immersion speed of 0.49 mm/min on FC-721 coated on mica, (b) n-dodecane at a constant immersion speed of 0.31 mm/min on FEP heat pressed against mica, (c) water at a constant immersion speed of 0.49 mm/min on siliconized glass, (d) n-dodecane at a constant immersion speed of 0.31 mm/min on FC-721 coated on glass slide, and (e) n-dodecane at a constant immersion speed of 0.31 mm/min on FEP heat pressed against quartz glass. Note that parts (a), (b), and (c) have a different position scale than do (d) and (e). (Reprinted with permission from Kwok, D. Y., Budziak, C. J., and Neumann, A. W., *Journal of Colloid and Interface Science*, 173, 143–50, 1995. With permission from Academic press.)

the vertical plate. Figure 6.17a shows the contact angle and capillary rise as functions of lateral displacement for 1-octanol on FC-721 coated mica. The three-phase contact line on this surface was observed to be nearly straight and horizontal. Thus, the contact angle was found to be nearly independent of the lateral location on the surface. Figure 6.17b and c show similar plots for dodecane on FEP heat-pressed against quartz glass and against mica, respectively. As can be seen in Figure 6.17b, there is considerable scatter, which is indicative of a relatively poor surface quality. Heat pressing against mica instead of quartz glass yielded a straight contact line and nearly no variation in the contact angle along the surface (Figure 6.17c). In the case

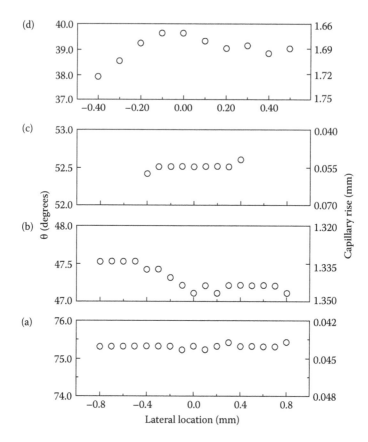

FIGURE 6.17 Capillary rise, h, and the corresponding contact angle, θ, as a function of lateral location for (a) 1-octanol on FC-721-coated mica, (b) n-dodecane on FEP heat pressed against quartz glass, (c) n-dodecane on FEP heat-pressed against mica, and (d) dimethylformamide on PE (polyethylene) heat pressed against quartz glass. Note that the lateral location scale of part (d) differs from that of (a) through (c). (Reprinted with permission from Kwok, D. Y., Budziak, C. J., and Neumann, A. W., *Journal of Colloid and Interface Science*, 173, 143–50, 1995. With permission from Academic press.)

of the PE surface and dimethylformamide shown in Figure 6.17d, a relatively large variation in contact angle of almost 2° is observed, over a very short distance.

These experiments show that contact angles can be employed to qualitatively evaluate solid surfaces. It is apparent that local roughness and/or heterogeneity of the solid surface can cause variations of the contact angle between different locations on the solid surface. It is noted that such variations cannot be detected by conventional contact angle measurement techniques, such as the goniometer technique, with an accuracy of ±2°.

The following subsections deal with preparation of organic and inorganic materials of a nonbiological nature. Section 6.5.2 covers the preparation of biological surfaces, such as proteins and cells, for contact angle measurement.

6.5.1.1   Heat Pressing

This method involves pressing a thermoplastic polymer (particles, film, or block) such as FEP, polyethylene, polypropylene, or polysulfone between two very clean, smooth surfaces at an elevated temperature [1,4]. Typically, glass plates are used as the platens, which are squeezed together with clamps in an oven or between the heated steel surfaces of a hydraulic heat press. The pressure and temperature employed during the heat-pressing process must be sufficient to cause the polymer to conform to the glass, but not so high as to cause any changes in the polymer chemistry. In practice to produce smooth surfaces, a thermoplastic must be heated to (or near) the glass transition temperature and then be pressed. The required temperature and pressure must be established empirically for each system. Care should be taken to avoid the fracture of polymer surfaces that may adhere to the platens. It has been found that immersing the polymer and platens together in warm water for 24 hours will facilitate the separation without damaging the surface.

6.5.1.2   Solvent Casting

When the solid material can be dissolved in a volatile solvent, solvent casting becomes an option for the preparation of smooth substrates. The method has been applied to a wide variety of materials including polymers [67], pharmaceuticals [68], and bitumen [69].

Although many variations of the basic procedure have been used, the following is a typical method. A solution of the substrate material with concentrations of the order of 1% is prepared with a volatile solvent. Glass microscope slides, cleaned with chromic acid, are mounted on a specially designed flat-bed centrifuge rotor head, which holds the slides in a horizontal position during centrifugation. A 500 μl drop of the solution is deposited on each 75×25 mm slide. The centrifuge is then operated at moderate speed and under slightly evacuated conditions (created by connecting the centrifuge to a vacuum pump or a water aspirator) until all the solvent has evaporated, leaving a thin, smooth film on the glass slide. The substrate selected for the film deposition is arbitrary as long as it has a smooth surface, can easily be cleaned, and is wettable by the solution.

On very hydrophobic substrates such as siliconized glass, thin free-standing films of solid solutes can be obtained by this technique. The solvent-casting method is also recommended for the preparation of hydrogel layers [70] as described in Section 6.5.2.

6.5.1.3   Dip Coating

Substrates may be dip coated in solutions or melts of the solid of interest. The technique has been applied to the fabrication of smooth surfaces of various materials including perfluorinated acids on platinum foil [71], fluorosurfactants on glass or mica [37,67,72], fluoropolymers on silicon, and to the production of elastomeric films of butyl rubber [73]. The morphology and the thickness of the resulting film depend on the concentration of the solution and on the speed of the substrate immersion into and withdrawal from the liquid. Generally, low solution concentrations and slow speeds (<1 mm/sec) result in a smooth surface. The immersion and withdrawal speed is best controlled by attaching the substrate to a variable speed electric motor.

6.5.1.4   Langmuir–Blodgett Film Deposition

Langmuir–Blodgett film deposition is basically a variation of the dip coating process [74–76]. The technique is used to create monomolecular layers of amphiphilic molecules on high-energy substrates such as quartz, glass, mica, and metals. Such long-chain organic molecules have a polar head group, which is well distinguished from the nonpolar hydrocarbon chain.

Typically, these films are deposited using the following procedure. A flat, shallow container, such as a Langmuir–Adam surface balance, is filled with water (or other suitable liquids) and the substrate to be coated is immersed in it. Then a solution of the amphiphilic material, in a solvent insoluble in water, is deposited dropwise onto the water, thereby forming an oriented monomolecular surface film upon evaporation of the solvent. This film can then be compacted by reducing the surface area until the surface pressure reaches a maximum. The solid substrate is then withdrawn from the water through this insoluble surface film at a very low speed (approximately 2 mm/min) while the surface pressure is kept constant. This creates an oriented monolayer of the amphiphilic material on the substrate.

The nature of the deposited monolayer depends on the interaction of the polar head group and the substrate surface. For example, if a glass microscope slide is raised up through a barium stearate monolayer spread on distilled water, the molecules in the film will be oriented with the hydrocarbon chains pointing outward, and hence the adsorbed film is hydrophobic [77]. When a previously coated plate is dipped back into the surfactant-coated water, a second oriented layer will be deposited and the coated surface again becomes hydrophilic as the head groups point outward. It should be noted that Langmuir–Blodgett films are not very stable, making them difficult to use for many contact angle studies. If the test liquid alters the deposited film, successive advancing contact angle measurements at the same location will generate inconsistent results.

6.5.1.5   Self-Assembled Monolayers

Monolayers of certain organic molecules can be produced on metal oxide or gold surfaces by means of self-assembly [78,79]. Alkanethiols on gold have proved to be a particularly useful system because the strong specific interaction between the sulfur atom and the gold surface results in the formation of relatively robust self-assembled monolayers upon immersion in dilute solutions. Moreover, by varying the tail group of the thiol molecules, a wide range of solid surface tensions can be achieved; water contact angles from 118° to less than 10° have been reported on treated gold surfaces [78,79].

Self-assembled monolayers of hydrolyzed octadecyltriethoxysilane (OTE) on cleaved mica have been found to be quite stable and well suited to surface investigations requiring exceptional smoothness and uniformity [80]. The coating procedure is relatively straightforward and involves the immersion of freshly cleaved mica in a solution of OTE and HCl in tetrahydrofuran and cyclohexane.

6.5.1.6   Vapor and Molecular Deposition Techniques

Vapor deposition has been found to yield very smooth surfaces for low molecular weight nonpolar and polar materials such as n-hexatriacontane and cholesteryl acetate [1,31], organic pigments [66], and some silicone compounds [59]. In all cases, the deposited material must not decompose at the elevated temperatures required for evaporation.

In general, vapor deposition is carried out in a vacuum chamber. The clean, smooth substrate is placed horizontally above the shuttered aperture of a Knudsen furnace containing the coating material in a ceramic dish. The temperature of the furnace is then raised and held constant until equilibrium is established, at which point the shutter is opened and a stream of the evaporating coating material rises to impinge and condense on the substrate. The smoothness of the resulting surface is dependent on the rate of evaporation and on the temperature of the substrate. The latter may be controlled separately with heating elements in the fixture used to hold the substrate above the Knudsen furnace.

Rather than depositing films by evaporation, radio frequency (RF) sputtering has been used to form PTFE films on quartz and stainless steel [81]. In this case, the cathode in a high vacuum chamber is bulk PTFE positioned before the target substrate along the discharge axis. The resulting sputtered films are between 0.05 and 0.06 μm thick, and have properties (e.g., fluorine to carbon ratio, cross linking, branching) that can be varied over a wide range.

Glow discharge polymerization is a third film deposition technique [82–84]. Plasma is generated, for example, by an inductively coupled (electrodeless) RF coil surrounding a glass vacuum chamber containing a clean glass substrate. Plasma of a gas such as argon is generated at a very low pressure and then a gaseous monomer is introduced into the chamber where it fragments. The molecular segments deposit on the substrate and the chamber walls where they crosslink and form a continuous film. The properties and uniformity of the resulting film are a complex function of many parameters such as the type of monomer and inert gas, pressure, flow rates, electrical power input, and chamber and substrate geometry.

6.5.1.7 Siliconization

There are a wide variety of silicone-based compounds that adsorb strongly to clean glass, producing high quality hydrophobic surfaces. Such surfaces have been manufactured by exposing glass to silicone oil at elevated temperature [66], or to silane solutions [72] or vapor [59]. For the latter, a typical procedure is to place clean glass slides in a vacuum desiccator containing a small amount of, for example, dimethyldichlorosilane. The desiccator is then evacuated and left at room temperature for 48 hours. After removal, the slides are thoroughly washed twice in toluene, followed by a final rinse in acetone and air drying.

Alternatively, clean glass surfaces can be placed in a solution of 20vol% dimethyldichlorosilane in highly pure n-hexane [66]. After 1 hour, the glass is removed and baked for a further hour at 100°C, followed by two rinses in toluene and one in acetone before drying. The resulting surfaces have an advancing water contact angle of 105° with very little hysteresis.

6.5.1.8 Surface Polishing

For certain materials, such as rock and coal, polishing is the typical method for the preparation of surfaces for contact angle measurements [85,86]. Polymers have also been prepared using this approach [87]; however, it is important to note that polished surfaces are mechanically deformed and may contain embedded contaminants from the polishing agents. The details of the procedure depend on the material being polished. For example, some polymer surfaces and natural crystals of fluoro- and hydroxyapatite might be polished on a cloth-covered rotating wheel, first with diamond paste (1 μm particle size) and then with a slurry of aluminum oxide (0.05 μm). The polished surfaces then require careful cleaning in an ultrasonic bath of distilled water in order to eliminate traces of the polishing media. For rocks and coal, small pieces are produced from the bulk sample and then wet-ground with a sequence of silicon carbide papers (e.g., numbers 220, 460, and 600) resulting in progressively smoother surfaces. Finer surfaces may then be obtained by polishing, as above, on a cloth-covered wheel saturated with a slurry of water and grit. To minimize the risk of surface contamination, commercial polishing compounds should be avoided, if possible, since they may contain unknown chemicals. Unfortunately, when the specimen is heterogeneous like coal, polishing exposes a new surface that may be significantly different from that of the natural surface.

Metals can also be electropolished to produce a smooth surface that is free of the deformed layer obtained by mechanical polishing [77]. The metal of interest is set up as the anode in a conducting liquid, undergoing a controlled corrosion reaction.

6.5.1.9 Preparation of Powders for Contact Angle Measurements

The process used for the preparation of powders depends on the method selected to measure contact angles. Powdered samples are used either in their original particulate form or they are compressed into a cake (tablet). The former implies that the powder will be analyzed using the packed-bed penetration methods of Washburn [6] or Bartell and Osterhof [7], the sedimentation volume technique (Chapter 11), or the solidification front technique (Chapter 12). In all of these cases, careful cleaning and degassing of the powders is essential to eliminate possible impurities and air bubbles that may adhere to the surface of particles. The solvents used to clean the powder should be chemically inert and should not swell the particles. Multiple washings in an ultrasonic bath are recommended. Alternatively, crystalline powders can be purified by recrystallization provided that the correct form of the crystal results. Polymorphic crystal forms can be a function of the solvent used and can have significantly different surface properties. The contact angle measurement techniques that use the actual powder rather than a compressed tablet usually perform best when the particle size distribution is as narrow as possible, achieved perhaps by sieving.

Contact angles measured directly on tablets of compressed powder are also improved if the particle size range is narrow. The surface structure (porosity and roughness) is more uniform, leading to less variability in the contact angles. Prior to the compression, therefore, the powder should be sized, recrystallized or otherwise cleaned by sonication in an appropriate solvent, and thoroughly dried in air or vacuum. The powder should be kept in a desiccator at constant temperature and relative humidity before tablet preparation. Tablets are made by compressing a known mass of powder (usually between 0.05 and 0.50 g depending on density, size, and shape) into a cake by using a hydraulic press and a highly polished die and punch. The die should be carefully cleaned with acetone and distilled water, and dried prior to tablet formation. A pressure of between 400 and 600 MPa applied for 2–5 minutes is found to be sufficient for most low molecular weight organic powders [8,10,12,68]. The exact pressure and time should be determined empirically in order to obtain the most reproducible contact angles. The use of a highly polished die provides a macroscopically smooth tablet surface; however, the surface is actually porous and, as noted above, may be altered by plastic deformation during compression [13].

Alternate methods of powder preparation for contact angle measurement include melting and subsequent dip coating on another substrate, solvent casting if the material is sensitive to heat, and heat pressing if the powder is a thermoplastic.

6.5.2 Biological Materials

This section discusses the preparation of various biological surfaces, specifically teeth and skin, cell and protein layers, liposomes, and hydrogels, for contact angle measurements.

6.5.2.1 Teeth and Skin

The wettability of tooth enamel and human skin is of interest in the study of pharmaceuticals, cosmetics, soaps, and other cleansing agents. The wettability of bovine incisors can be studied by first grinding flat the labial surface of the teeth and cutting it into small pieces a few square millimeters in area. Subsequently, the enamel slabs are polished with a slurry of aluminum oxide (0.05 μm) in distilled water. The slabs are then ultrasonically cleaned in distilled water and dried in an incubator overnight at 25°C. The polishing and cleaning procedure can be repeated on the same specimens several times without affecting the results [88–90].

Contact angles on the labial surfaces of the human upper central incisors can also be measured in situ. The preparation of the measurement site consists of not eating, drinking, or smoking after brushing the teeth prior to the measurement. The subjects are seated in a dental chair so that the labial surfaces are horizontal. The tooth surface is dried for 60 seconds and then an advancing sessile drop is established and photographed [91].

Wettability studies of human skin in situ utilize the dorsal areas of the ring and forefingers because the skin surface between the joints is relatively smooth and convenient for contact angle measurements [92]. Contact angles of sessile drops are studied with a goniometer equipped with a specially designed finger holder [93]. Measurements can also be done on the underside of the forearm.

Experiments may be performed in vitro with cleaned, excised skin (usually from the breast), which is washed with soap, ethanol, and distilled water, and then dried in hot air. Untreated human skin is very hydrophobic, having a water contact angle between 87° [94] and 100° [93]. Contact angles have been measured on rabbit intestines that were washed with sterile saline at 4°C [95]. Depending on the location of the sample and the age and condition of the rabbit, water contact angles varied between 38 and 93°.

Due to the presence of surface roughness, heterogeneity, and impurities on skin and teeth, it is necessary to average large numbers of contact angle measurements to obtain reliable data.

6.5.2.2 Bacteria, Cells, Proteins, and Liposomes

Contact angles have been used in a wide variety of studies of the wettability of bacteria [57,88,96–98], mammalian cells [96,97,99–101], plant cells [102–105], liposomes [11,12,106], and proteins [57,96,100,107,108]. Layers of these substances that are usually prepared by membrane filtration are hydrophilic because of their hydrated state. Due to different physiological constraints such as osmotic pressure, ionic strength, and chemical compatibility, a saline solution of 0.9 wt% (0.15 M) purified sodium chloride in deionized distilled water is usually recommended as the contact angle measuring liquid. When a drop of saline is placed on the surface of a highly hydrated biological layer, a zero contact angle is observed initially. When the excess surface water evaporates, the contact angles of subsequently placed drops become finite, progressively increasing with time to finally reach a plateau value. The contact angle corresponding to the plateau is taken as the characteristic contact angle.

(a) *Bacteria.* After being harvested from the culture broth, the bacteria are usually washed several times with physiological saline to remove contaminants. A layer of bacteria is formed on a membrane filter (typically cellulose acetate with a pore size < 0.45 μm) by depositing a small amount of the washed suspension on a filter in a glass holder and applying suction. The layer of bacteria should be washed at least three times with distilled water or saline. Suction is stopped after the final rinse, when most of the washing liquid has been removed. The bacterial layer should, however, remain slightly moist. This is facilitated by placing the filter membrane on a freshly prepared 2% solidified agar plate (see below) that acts as a water reservoir and decreases the rate at which the bacterial layer dries. This ensures that contact angles of water or saline drops can be measured while the bacteria are still in their fully hydrated state. As mentioned above, air drying is usually necessary to reach the optimal water content of the bacterial layer and obtain contact angle measurements independent of time. This condition might be judged by the disappearance of surface glossiness and the appearance of a more matted surface [96].

It should be noted that some filter membranes may contain surfactants that aid in wetting and facilitate the filtration process. Surfactants can contaminate the sessile drop liquid and affect the contact angle. The presence of surfactants in the membrane can be determined by soaking the membrane in ultra pure distilled water for at least 1 hour, then measuring the surface tension of the soaking water. The filter must be washed copiously with distilled water prior to use if the surface tension is too low [98].

(b) *Human and Animal Cells.* Layers of human and animal cells can be produced using membrane filtration as described above [96,100]. Care should be taken not to produce more than a monolayer coverage because of the excessive roughness that can complicate contact angle measurements. This is facilitated by controlling the number of cells per unit volume of the suspension applied to the membrane filter.

(c) *Proteins.* Protein layers have been prepared for contact angle measurements using in vitro and in vivo techniques. The former has been done in two ways: (i) A 1–2 wt% protein solution can be filtered through a cellulose acetate membrane without stirring the solution. As with cell layers, contact angles can be measured on the coated filter membrane after a period of air drying. (ii) Protein solution can be deposited directly onto a filter membrane with a pipette and allowed to partially dry. This procedure requires less protein and is somewhat faster. In both cases, it should be recognized that excessive drying may result in protein denaturation, giving rise to increased contact angles [96].

Contact angles have been measured on salivary protein layers formed in vivo. Clean and siliconized discs of glass and germanium were placed in the mouth for varying lengths of time prior to removal, rinsing, and measurement [107].

(d) *Plant Cells.* Contact angles have been used to model spontaneous immobilization of plant cells to different bioreactor substrates [102–105]. Cells of *Catharanthus roseus* were harvested from a culture, diluted in distilled water, and filtered through a series of nylon meshes (500, 350, and 210 μm) under gentle vacuum. The resulting suspension was then centrifuged and resuspended in distilled water three times. The final suspension consisted of more than 97% small aggregates (2–5 cells) with low levels of extracellular polysaccharides and proteins in the suspending liquid. Contact angles were then measured on a 1–2 mm layer produced by the membrane filtration of a suspension having a packed cell volume of 1% [102].

(e) *Liposomes.* Hydrated phospholipid vesicles, or liposomes, are extremely hydrophilic agents and useful as means of delivering encapsulated drugs to targeted tissues of the body. Steiner and Adam measured contact angles on phospholipid films deposited onto a number of substrates: glass microscope slides, glass siliconized with dimethyldichlorosilane, and cell culture substrates of both polystyrene and PMMA [11]. It was observed that liposome films, which were formed on very hydrophilic (glass) and very hydrophobic (siliconized glass) surfaces, became partly resuspended in the sessile drops of water, thereby invalidating the contact angle measurements. On the plastic cell culture surfaces, when the liposome deposition reached approximately 10 times monolayer coverage, the contact angle reached a minimum value of 15–20°, which corresponds to a very hydrophilic surface [11].

Liposome layers have also been produced using membrane filtration [12,106]. A few milliliters of a 5 wt% aqueous dispersion of liposomes was filtered through a cellulose membrane filter with a pore size of 0.45 μm. A thin liposome layer was formed by a slight vacuum suction for a few hours and was placed on 2% solidified agar to prevent dehydration.

6.5.2.3 Hydrogels

The aqueous nature of gels makes them an important component of biomedical applications involving cells, tissue, and blood [109]. Water contact angle measurements on hydrogels are complicated due to the flexibility of the interface that can undergo appreciable deformation. The procedures used to prepare hydrogel surfaces are different from those used with cells, and vary depending on the particular type of gel. The following examples illustrate the techniques used with several commonly used hydrogels.

For natural macromolecules such as gelatin, agar, or agarose, a 1–2 wt% solution is prepared by mixing the substance with cold distilled water or saline solution. The solution is then heated to 75°C and stirred continuously for approximately 20 minutes until the agar is completely dissolved. Excessive air bubbles in the gel are removed by applying a moderate vacuum. The solution is then poured carefully onto a chromic acid-cleaned glass surface (alternatively, on cotton gauze glued to the glass slide to retain the gel) to form a 2 mm thick layer, which is cooled to room temperature in a covered Petri dish. Using this solution, the gel formation is expected within 15 minutes. After gelation, the Petri dish is uncovered and placed in an oven at a low temperature for 30 minutes to allow excess water to evaporate from the gel surface. This surface is designated as the "air-exposed" surface. "Glass-exposed" surfaces can be prepared by pouring the agar solution into the 1.5 mm gap created by placing gauze between two glass plates. After cooling, one of the glass plates is separated from the agar. Similarly, "silane-exposed" gel surfaces are produced using dichlorodimethylsilane coated glass plates [101].

The preparation of synthetic polyacrylamide gels involves the polymerization of purified acrylamide monomers with the monomer cross linking agent bis (N,N'-methylene bis-acrylamide) using a free radical-generating catalyst system [100]. To produce gels with different surface properties, the concentration of the total gel composition and the percentage of the cross-linking agent can be varied. Prior to the polymerization reaction, the mixture should be subjected to vacuum suction in order to degas the solution as the reaction is strongly retarded by oxygen. After degassing, the solution with all components is poured into small Petri dishes where the gel is formed. The dishes must be filled completely. The gel is then soaked in distilled water and/or saline (to which 0.02% sodium azide is added as a bacteriostatic agent) for 4 days, changing the soaking liquid every day. This is done to completely remove unreacted monomers and to ensure that the gel is equilibrated with water or saline solution. After removing excess surface water in a drying oven, the contact angle of distilled water or saline (depending on the liquid used for equilibration) and the liquid content of the gel sample may be determined. A fresh preparation of gel should be used for each experiment.

Solvent casting has also been used to prepare polymeric hydrogels [70]. Andrade et al. prepared different PMMAs using dilute solutions of various methacrylate ester monomers containing the cross-linking agent hexamethylene diisocyanate. The solutions were cast onto aminopropylsilane-treated glass slides and resulted in the bonding of the polymer to the substrate as the polymer itself cross linked. The slides with the polymer film were then cured at 60°C for 24 hours, and vacuum dried at 60°C for another 24 hours to remove residual solvent. The thin gel coatings on the slides were then washed three times in distilled water, and immersed in water to allow them to equilibrate prior to contact angle measurement using the captive bubble technique.

It is noted that biological surfaces such as those of cells, tissues, and gels of extracellular matrices must be kept in an aqueous environment to remain viable. As a result, the surfaces will be either wetting or yield very small contact angles. The ADSA-D technique is a particularly useful approach for the measurement of contact angles on such surfaces with reasonable accuracy and reproducibility.

6.5.3 Cleaning and Handling Solid Surfaces

Surface cleaning is a critical process for ensuring good contact angle measurements as well as for the production of high quality coatings in industrial settings. There are different procedures to clean solid or semisolid surfaces. For glass or quartz substrates, the surfaces can first be washed with high-purity organic solvents, then soaked in chromic acid, and thoroughly rinsed with warm distilled water to remove acid residues. Cleaning glass or other substrates with a RF glow discharge in a vacuum chamber or in an ultrasonic bath are additional means of removing organic contaminants. High energy surfaces such as glass, mica, and metals can also be heated in a vacuum oven to over 200°C in order to remove adsorbed contaminants.

Proper handling and storage of clean solid surfaces is also crucial. Surfaces should never be touched with bare fingers or even with rubber gloves, which might themselves be covered with organic contaminants. Chromic acid-cleaned forceps with Teflon tips are recommended for handling samples by the edges, away from the areas that will be used for contact angle measurements. During handling and measurement, care should be taken not to breathe on the surface nor expose it to contaminated laboratory atmosphere. Contact angles should be measured in a clean environment or a chamber, away from motors, pumps, and sources of organic chemicals. High energy surfaces such as glass, metals, quartz, and mica should be stored under a high-purity organic solvent, inert gas, or high vacuum. Hydrogel surfaces may be stored under water or saline solution. It is always desirable to eliminate concern about storage contamination by measuring contact angles immediately after the surfaces are prepared. In some cases, such as heat pressing and hydrogel casting between glass plates, most of the preparation can be done in advance, with only the final separation and exposure of the surface performed immediately prior to the contact angle measurement.

REFERENCES

1. A. W. Neumann and R. J. Good. In *Surface and Colloid Science-Experimental Methods.* Edited by R. J. Good and R. R. Stromberg, Vol. 11. Plenum Press, New York, 1979.
2. W. A. Zisman. In "Contact Angle, Wettability and Adhesion." In *Advances in Chemistry Series, No. 43.* Edited by R. F. Gould, 2. American Chemical Society, Washington, DC, 1964.
3. M. C. Phillips and A. C. Riddiford. *Journal of Colloid and Interface Science* 41 (1972): 77.
4. J. K. Spelt, Y. Rotenberg, D. R. Absolom, and A. W. Neumann. *Colloids and Surfaces* 24 (1987): 127.
5. I. Langmuir and V. J. Schaeffer. *Journal of the American Chemical Society* 59 (1937): 2400.
6. B. W. Washburn. *Physical Review* 17 (1921): 273.

7. F. B. Bartell and P. M. Osterhof. *Industrial & Engineering Chemistry* 19 (1927): 1277.
8. G. Zografi and S. S. Tam. *Journal of Pharmaceutical Sciences* 65 (1976): 1145.
9. C. F. Lerk, M. Lagas, J. P. Boelstra, and P. Broersma. *Journal of Pharmaceutical Sciences* 66 (1977): 1480.
10. W. C. Liao and J. L. Zatz. *Journal of Pharmaceutical Sciences* 68 (1979): 488.
11. V. Steiner and G. Adam. *Cell Biophysics* 6 (1984): 279.
12. B. I. Vargha-Butler, S. J. Sveinsson, and Z. Policova. *Colloids and Surfaces* 58 (1991): 271.
13. G. Buckton and J. M. Newton. *Powder Technology* 46 (1986): 201.
14. D. T. Hansford, D. J. W. Grant, and J. M. Newton. *Powder Technology* 26 (1980): 119.
15. G. Buckton and J. M. Newton. *Journal of Pharmacy and Pharmacology* 37 (1985): 605.
16. N. W. F. Kossen and P. M. Heertjes. *Chemical Engineering Science* 20 (1965): 593.
17. P. M. Heertjes and N. W. F. Kossen. *Powder Technology* 1 (1967): 33.
18. J. B. Jones and A. W. Adamson. *Journal of Physical Chemistry* 72 (1968): 646.
19. N. K. Adam and G. Jessop. *Journal of the Chemical Society* 1925 (1925): 1863.
20. A. L. Spreece, C. P. Rutkowski, and G. L. Gaines. *Review of Scientific Instruments* 28 (1957): 636.
21. G. W. Longman and R. P. Palmer. *Journal of Colloid and Interface Science* 29:185 (1967).
22. L. R. Fisher. *Journal of Colloid and Interface Science* 72 (1977): 200.
23. A. M. Schwartz and F. W. Minor. *Journal of Colloid and Science* 14 (1959): 572.
24. A. M. Schwartz and C. A. Rader. *Proceedings of the IVth International Congress on Surface Activity.* Vol. 2, 383. Gordon & Breach, New York, 1964.
25. E. Weber. *Habilitationsschrift.* University of Stuttgart, 1968.
26. C. J. Budziak. PhD Thesis. *Thermodynamic Status of Static and Dynamic Contact Angles.* University of Toronto, 1992.
27. R. V. Sedev, C. J. Budziak, J. G. Petrov, and A. W. Neumann. *Journal of Colloid and Interface Science* 159 (1993): 392.
28. D. Y. Kwok, C. J. Budziak, and A. W. Neumann. *Journal of Colloid and Interface Science* 173 (1995): 143–50.
29. J. Kloubek and A. W. Neumann. *Tenside* 6 (1969): 4.
30. A. W. Neumann and W. Tanner. *Journal of Colloid and Interface Science* 34 (1970): 1.
31. A. W. Neumann. *Advances in Colloid and Interface Science* 4 (1974): 105.
32. Y. Rotenberg, L. Boruvka, and A. W. Neumann. *Journal of Colloid and Interface Science* 93 (1983): 169.
33. O. I. del Río. M.A.Sc. Thesis. *On the Generalization of Axisymmetric Drop Shape Analysis.* University of Toronto, 1993.
34. M. G. Cabezas, A. Bateni, J. M. Montanero, and A. W. Neumann. *Langmuir* 22 (2006): 10053.
35. R. David, M. K. Park, A. Kalantarian, and A. W. Neumann. *Colloid and Polymer Science* 287 (2009): 1167.
36. A. Kalantarian, R. David, and A. W. Neumann. *Langmuir,* 25 (2009): 14146.
37. D. Li and A. W. Neumann. *Journal of Colloid and Interface Science* 148 (1992): 190.
38. D. Duncan, D. Li, J. Gaydos, and A. W. Neumann. *Journal of Colloid and Interface Science* 169 (1995): 256.
39. D. Y. Kwok, R. Lin, M. Mui, and A. W. Neumann. *Colloids and Surfaces A: Physicochemical Engineering Aspects* 116 (1996): 63.
40. H. Tavana, C. N. C. Lam, K. Grundke, P. Friedel, D. Y. Kwok, M. L. Hair, and A. W. Neumann. *Journal of Colloid and Interface Science* 279 (2004): 493.
41. D. Y. Kwok, T. Gietzelt, K. Grundke, H.-J. Jacobasch, and A. W. Neumann. *Langmuir* 13 (1997): 2880.

42. H. Tavana, F. Simon, K. Grundke, D. Y. Kwok, M. L. Hair, and A. W. Neumann. *Journal of Colloid and Interface Science* 291 (2005): 497.
43. H. Tavana, D. Appelhans, R.-C. Zhuang, S. Zschoche, K. Grundke, M. L. Hair, and A. W. Neumann. *Colloid and Polymer Science* 284 (2006): 497.
44. D. Y. Kwok, C. N. C. Lam, A. Li, A. Leung, and A. W. Neumann. *Langmuir* 14 (1998): 2221.
45. H. Tavana, G. Yang, C. M. Yip, D. Appelhans, S. Zschoche, K. Grundke, M. L. Hair, and A. W. Neumann. *Langmuir* 17 (2006): 628.
46. G. E. P. Elliott and A. C. Riddiford. *Journal of Colloid and Interface Science* 23 (1967): 389.
47. A. C. Lowe and A. C. Riddiford. *Chemical Communications* 106 (1970): 387.
48. M. C. Philips and A. C. Riddiford. *Journal of Colloid and Interface Science* 41 (1972): 77.
49. R. E. Johnson, Jr., R. H. Dettre, and D. A. Brandreth. *Journal of Colloid and Interface Science* 62 (1977): 205.
50. T. D. Blake. In *Wettability.* Edited by J. C. Berg, 251. Marcel Dekker, New York, 1993.
51. J. B. Cain, D. W. Francis, R. D. Venter, and A. W. Neumann. *Journal of Colloid and Interface Science* 94 (1983): 123.
52. E. B. Dussan, V. *Annual Review of Fluid Mechanics* 1 (1979): 371.
53. K. Stoev, E. Ramé, and S. Garoff, *Physics of Fluids* 11 (1999): 3209.
54. H. Tavana and A. W. Neumann. *Colloids and Surfaces A: Physicochemical Engineering Aspects* 282–283 (2006): 256.
55. C. N. C. Lam, R. Wu, D. Li, M. L. Hair, and A. W. Neumann. *Advances in Colloid and Interface Science* 96 (2002): 169.
56. H. Tavana, A. Amirfazli, and A. W. Neumann. *Langmuir* 22 (2006): 5556.
57. A. W. Neumann, D. R. Absolom, D. W. Francis, S. N. Omenyi, J. K. Spelt, Z. Policova, C. Thomson, W. Zingg, and C. J. van Oss. *Annals of the New York Academy of Sciences* 416 (1983): 276.
58. E. Moy, P. Cheng, Z. Policova, S. Treppo, D. Kwok, D. R. Mack, P. M. Sherman, and A. W. Neumann. *Colloids and Surfaces* 58 (1991): 215.
59. F. K. Skinner, Y. Rotenberg, and A. W. Neumann. *Journal of Colloid and Interface Science* 130 (1989): 25.
60. W. C. Duncan-Hewitt, Z. Policova, P. Cheng, E. I. Vargha-Butler, and A. W. Neumann. *Colloids and Surfaces* 42 (1989): 391.
61. J. I. Smith, B. Drumm, A. W. Neumann, Z. Policova, and P. M. Sherman. *Infection and Immunity* 58 (1990): 3056.
62. A. Bateni, S. Laughton, H. Tavana, S. S. Susnar, A. Amirfazli, and A. W. Neumann. *Journal of Colloid and Interface Science* 283 (2005): 215.
63. A. Bateni, S. S. Susnar, A. Amirfazli, and A. W. Neumann. *Colloids and Surfaces A: Physicochemical Engineering Aspects* 219 (2003): 215.
64. O. I. del Río, D. Y. Kwok, R. Wu, J. M. Alvarez, and A. W. Neumann. *Colloids and Surfaces A: Physicochemical Engineering Aspects* 143 (1998): 197.
65. D. Li and A. W. Neumann. *Colloid and Polymer Science* 270 (1992): 498.
66. A. W. Neumann, D. Renzow, H. Reumuth, and I. E. Richter. *Fortschr. Kolloide u. Polymere* 55 (1971): 49.
67. B. B. Davidson and G. Lei. *Journal of Polymer Science* 9 (1971): 569.
68. T. R. Krishman, I. Abraham, and E. I. Vargha-Butler. *International Journal of Pharmaceutics* 80 (1992): 277.
69. E. I. Vargha-Butler, T. K. Zubovits, C. J. Budziak, and A. W. Neumann. *Energy & Fuels* 2 (1988): 653.
70. J. D. Andrade, R. N. King, D. B. Gregonis, and D. L. Coleman. *Journal of Polymer Science* 66 (1979): 313.

71. E. F. Hare, E. G. Shafrin, and W. A. Zisman. *Journal of Physical Chemistry* 58 (1953): 236.

72. J. K. Spelt. PhD Thesis. *Solid Surface Tension: The Equation of State Approach and the Theory of Surface Tension Components.* University of Toronto, 1985.

73. C. J. Budziak, E. I. Vargha-Butler, and A. W. Neumann. *Journal of Applied Polymer Science* 42 (1991): 1959.

74. D. R. Absolom, A. W. Neumann, W. Zingg, and C. J. van Oss. *Transactions of the American Society for Artificial Internal Organs* 25 (1979): 152.

75. K. B. Blodgett. *Journal of the American Chemical Society* 57 (1935): 1007.

76. I. Langmuir. *Journal of the Franklin Institute* 218 (1934): 143.

77. A. W. Adamson. *Physical Chemistry of Surfaces.* 5th ed. John Wiley, New York, 1990.

78. C. D. Bain, E. B. Troughton, Y. Tao, J. Eval, G. M. Whitesides, and R. O. Nuzzo. *Journal of the American Chemical Society* 111 (1989): 321.

79. C. D. Bain and G. M. Whitesides. *Angewandte Chemie-International Edition in English* 28 (1989): 506.

80. C. R. Kessel and S. Granick. *Langmuir* 7 (1991): 532.

81. J. J. Pireaux, J. P. Delrue, A. Hecq, and J. P. Dauchot. In *Physico-Chemical Aspects of Polymer Surfaces.* Edited by K. L. Mittal, Vol. 1, 53–81. Plenum Press, New York, 1983.

82. A. Dilks and B. Kay. In *Plasma Polymerization.* Edited by M. Shen and A. T. Bell. A.C.S. Symposium Series, No. 108, American Chemical Society, Washington DC, 1979.

83. A. Dilks and E. Kay. *Macromolecular* 14 (1981): 855.

84. H. Yasuda. *Journal of Polymer Science: Macromolecular Reviews* 16 (1981): 199.

85. E. I. Vargha-Butler, M. Kashi, H. A. Hamza, and A. W. Neumann. *Coal Preparation* 3 (1986): 53.

86. E. I. Vargha-Butler, D. R. Absolom, A. W. Neumann, and H. A. Hamza. In *Interfacial Phenomena in Coal Technology.* Edited by O. D. Botsaris and Y. M. Glazman, 33–84. Marcel Dekker, New York, 1989.

87. H. J. Busseher, A. W. J. van Pelt, H. P. De Jong, and J. Arends. *Journal of Colloid and Interface Science* 95 (1983): 23.

88. H. P. De Jong, A. W. J. van Pelt, and J. Arends. *Journal of Dental Research* 61 (1982): 11.

89. B. W. Darvell, M. D. Murray, and N. H. Ladizesky. *Journal of Dentistry* 15 (1987): 82.

90. J. W. van de Rijke, H. J. Busscher, J. J. Ten Bosch, and J. F. Perdok. *Journal of Dental Research* 68 (1989): 1205.

91. J. F. Perdok, A. H. Weerkamp, L. J. van Dijk, J. Arends, and H. J. Busscher. *Journal of Dental Research* 67 (1988): 701.

92. M. E. Ginn, C. M. Noyes, and B. Jungermann. *Journal of Colloid and Interface Science* 26 (1968): 146.

93. A. Rosenberg, R. Williams, and G. Cohen. *Journal of Pharmaceutical Sciences* 62 (1973): 920.

94. H. Schott. *Journal of Pharmaceutical Sciences* 60 (1971): 1893.

95. D. R. Mack, A. W. Neumann, Z. Policova, and P. M. Sherman. *American Journal of Physiology* 262 (1992): G171.

96. C. J. van Oss, C. F. Gillman, and A. W. Neumann. *Phagocytic Engulfment and Cell Adhesiveness as Cellular Surface Phenomena.* 39, 52, 53. Marcel Dekker, New York, 1975.

97. D. R. Absolom, W. Zingg, and A. W. Neumann. *Journal of Colloid and Interface Science* 112 (1986): 599.

98. W. C. Duncan-Hewitt, Z. Policova, P. Cheng, E. I. Vargha-Butler, and A. W. Neumann. *Colloids and Surfaces* 42 (1989): 391.

99. R. P. Smith, D. R. Absolom, J. K. Spelt, and A. W. Neumann. *Journal of Colloid and Interface Science* 110 (1986): 521.
100. D. R. Absolom, M. H. Foo, W. Zingg, and A. W. Neumann. In *Polymers as Biomaterials.* Edited by W. S. Shalaby, A. S. Hoffman, B. D. Ratner, and T. A. Horbett, 149–65. Plenum Press, New York, 1984.
101. C. J. van Oss, W. Zingg, O. S. Hum, and A. W. Neumann. *Thrombosis Research* 11 (1977): 183.
102. P. J. Facchini, A. W. Neumann, and F. DiCosmo. *Applied Microbiology and Biotechnology* 29 (1988): 346.
103. P. J. Facchini, A. W. Neumann, and F. DiCosmo. *Biotechnology and Applied Biochemistry* 11 (1989): 74.
104. P. J. Facchini, F. DiCosmo, L. G. Radvanyi, and Y. Giguere. *Biotechnology and Bioengineering* 32 (1988): 935.
105. P. J. Facchini, F. DiCosmo, L. G. Radvanyi, and A. W. Neumann. In *Methods in Molecular Biology.* Edited by J. W. Pollard and J. M. Walker, Vol. 6, 513–25. Humana Press, Totowa, NJ, 1990.
106. E. I. Vargha-Butler, M. Foldvari, and M. Mezei. *Colloids and Surfaces* 42 (1989): 375.
107. R. B. Baier and P. O. Glantz. *Acta Odontologica Scandinavica* 36 (1978): 289.
108. C. J. van Oss, D. R. Absolom, A. W. Neumann, and W. Zingg. *Biochimica et Biophysica Acta* 670 (1981): 64.
109. B. D. Ratner and A. S. Hoffman. In *Hydrogels for Medical and Related Applications.* Edited by J. D. Andrade, 1. A.C.S. Symposium Series, No. 31, American Chemical Society, Washington, DC, 1977.
110. D. Li and A. W. Neumann. *Colloids and Surfaces* 43 (1990): 195–206.
111. D. Y. Kwok and A. W. Neumann. *Advances in Colloid and Interface Science* 81 (1999): 167–249.

7 Thermodynamic Status of Contact Angles

Hossein Tavana

CONTENTS

7.1   INTRODUCTION

From daily experience we know that a drop of water deposited onto the surface of a plastic plate will form a sessile drop. Also, we often see that many liquids can climb up in a capillary tube to a certain height and form a meniscus at the top. In these phenomena, the angle formed between the liquid–vapor interface and the liquid–solid interface at the solid–liquid–vapor three-phase contact line is defined as the *contact angle*. It should be noted that, at the molecular level, the three phases do not meet in a line but within a zone of small but finite dimensions in which the three interfacial regions merge. Therefore, the microscopic contact angles may be different from their macroscopic counterparts measured by techniques such as axisymmetric drop shape analysis (ADSA). Discussion of microscopic contact angles lies beyond the scope of this chapter. For our purposes, the macroscopic contact angles will be the center of attention. The systems considered in this chapter consist of three bulk phases: solid, liquid, and vapor; and three interface phases: solid–liquid, solid–vapor, and liquid–vapor.

The interest in contact angles is twofold: they play a major role in a number of technological, environmental, and biological phenomena and processes; they are also a manifestation of the surface tension of the solid on which the contact angle is formed. Presently, the interpretation of contact angles is one of the better prospects to determine surface tensions of solids (see Chapters 8 and 9). Unfortunately, in spite of their seeming simplicity, contact angle phenomena are complex. This complexity is most easily appreciated by invoking the classical theory of capillarity [1]. Minimizing the overall free energy of a system consisting of a liquid in contact with a solid yields the Laplace equation of capillarity (see Chapters 1 and 2) [2]

$$\gamma_{lv}\left(\frac{1}{R_1}+\frac{1}{R_2}\right)=\Delta\rho gz+c=\Delta P, \tag{7.1}$$

and Young's equation [3]

$$\gamma_{lv}\cos\theta_e = \gamma_{sv} - \gamma_{sl}, \tag{7.2}$$

where R_1 and R_2 are the principal radii of curvature at a point of the liquid surface, $\Delta\rho$ is the density difference between the liquid and vapor phases, g is the local gravitational acceleration, z is the ordinate of a point of the liquid surface at which the principal radii of curvature are R_1 and R_2, c is a constant, ΔP is the capillary pressure or pressure of curvature, γ_{lv} is the liquid–vapor interfacial tension, γ_{sv} is the solid-vapor interfacial tension, γ_{sl} is the solid–liquid interfacial tension, and θ_e is the equilibrium contact angle.

The derivation of these relations assumes that the solid surface in contact with the liquid is smooth, homogeneous, isotropic, and nondeformable. This assumption is of little or no consequence for the range of validity of the Laplace equation, since Equation 7.1 essentially describes the shape of the liquid-vapor or liquid–liquid interface, away from the solid–vapor and solid–liquid interfaces. The Young equation, on the other hand, is an equilibrium condition involving properties that are a function of the solid surface; that is, γ_{sv} and γ_{sl}. The validity and applicability of Young's equation and the thermodynamic status of contact angles have been the subject of debates in the past, mainly due to the difficulty of preparing solid surfaces meeting the above conditions.

The above statement that the Laplace equation describes the shape of the liquid–vapor interface requires some further elaboration. To be more specific, the Laplace equation prescribes, for a given liquid of surface tension, γ_{lv}, a value of the mean curvature

$$J = \frac{1}{R_1} + \frac{1}{R_2},$$

as a function of the ordinate z at any point on the liquid–vapor interface. In addition, in many cases it will be necessary to introduce, as a boundary condition, what we shall term the phenomenological contact angle, θ. It is the angle between the solid–liquid interface and the tangent to the liquid–vapor interface at the three-phase line measured in a plane normal to the solid–liquid and the liquid–vapor interfaces. The phenomenological contact angle, θ, may or may not be the equilibrium contact angle, θ_e, and it may or may not be a dynamic contact angle. It will, in any case, represent the appropriate boundary condition that determines the shape of the liquid meniscus.

The equilibrium contact angle, θ_e, on the other hand, is a unique function of the interfacial tensions, γ_{lv}, γ_{sv}, and γ_{sl}, given by Young's equation. In the vast majority of cases, however, experimentally observed contact angles are not uniquely determined by the surface tensions of the solid and the liquid; there usually exists a range of contact angles in which any one contact angle gives rise to a mechanically stable liquid meniscus. The largest and the smallest of these angles are termed the advancing contact angle, θ_a, and the receding contact angle, θ_r, respectively. The difference between the advancing and the receding contact angles is called contact angle hysteresis. Contact angle hysteresis may be conveniently observed by first advancing and then receding a liquid drop over a solid surface. If the surface tension of the solid, γ_{sv}, and the interfacial tension γ_{sl} were modified due to contact with the liquid, we would, in fact, expect a change of the equilibrium contact angle, θ_e. If this were so, then the fact that the observed contact angles are not a unique property for any given solid–liquid system would not necessarily conflict with Young's equation. It will be demonstrated below (Section 7.3) that in addition to traditionally considered causes for contact angle hysteresis, for example, roughness and heterogeneity, such a true hysteresis is indeed operative in many solid–liquid systems.

It was only in 1936 that Wenzel recognized that Young's equation (Equation 7.2) may not be a universal equilibrium condition for the physical interaction between a solid and a liquid where the solid surface is rough [4]. He argued that, if the solid surface is rough, the interfacial tensions γ_{sv} and γ_{sl} should not be referred to the geometric area, but to the actual surface area. If we let

$$r = \frac{\text{actual surface area}}{\text{geometric surface area}}, \tag{7.3}$$

the so-called Wenzel equation

$$\gamma_{lv}\cos\theta_W = r(\gamma_{sv} - \gamma_{sl}) \tag{7.4}$$

results, where θ_W may be called the Wenzel contact angle, which we shall see later is, in fact, the equilibrium contact angle, θ_e, on a rough solid surface. Equation 7.4 was subsequently also derived more rigorously by Good [5].

The same type of reasoning was applied to heterogeneous and porous surfaces by Cassie and Baxter [6–8]. A heterogeneous solid surface is one that contains domains of different surface tension. Examples of heterogeneous solid surfaces are surfaces on which patches of a monomolecular film are adsorbed, or a surface of a polycrystalline material that exposes different crystallographic planes at its surface. For a solid surface consisting of two domains with the intrinsic contact angles θ_{e1} and θ_{e2} with respect to a given liquid, Cassie obtained

$$\cos\theta_C = a_1\cos\theta_{e1} + a_2\cos\theta_{e2}, \tag{7.5}$$

where a_1 and a_2 are the fractional surface areas of the two types of surfaces such that

$$a_1 + a_2 = 1, \tag{7.6}$$

and θ_C may be called the Cassie contact angle [6]. We shall see below that θ_C defined by Equation 7.5 is, like θ_W for a rough surface, the appropriate equilibrium contact angle for a heterogeneous solid surface.

One of the difficulties with relations such as the Cassie or the Wenzel equations is that they seemingly conflict with experimental observation. For example, the Wenzel equation predicts that the contact angle on a rough surface should be smaller than the contact angle on the smooth surface if the contact angle on the latter is smaller than 90°. If the contact angle on the smooth surface of the same material is larger than 90°, then a larger contact angle is predicted on the rough surface. Experimentally, however, one finds that with increasing roughness, the advancing contact angle increases and the receding contact angle decreases in many situations [9,10]. In addition to these difficulties, the explanations of the relation between roughness and heterogeneity on the one hand, and the phenomenon of contact angle hysteresis on the other, made further investigations necessary. The first more quantitative results linking contact angle hysteresis and heterogeneity, as well as roughness, are due to Johnson and Dettre [11,12]. Neglecting gravity, they considered a drop of liquid, centered on an alternating array of two types of smooth and narrow concentric rings of constant width, one type having an equilibrium contact angle θ_{e1} and the other θ_{e2} [11]. This arrangement represents their model heterogeneous surface. In order to obtain a homogeneous but rough model surface, they considered a surface of the same symmetry as their model heterogeneous surface by combining grooves and hills in such a way that a cut normal to the surface through the origin results in a sinusoidal waveform [12]. Neglecting gravity and assuming the local validity of Young's equation, Johnson and Dettre demonstrated, by minimization of the overall free energy of the system, the existence of a large number of metastable states.

In spite of the considerable insight into contact angle hysteresis provided by Johnson and Dettre, there remain three immediate reservations with respect to their model. First, a treatment of the capillarity phenomenon neglecting gravity is not general. Second, the concentric ring model is not very realistic physically; although it corresponds well to the symmetry of sessile drops, it is not easily modified to conform to actual patterns of heterogeneous solid surfaces. Third, it assumes the local validity of Young's equation. Although we shall see below (Section 7.2) that this latter assumption is essentially correct, it will be obtained as a result of an analysis based solely on the Laplace equation rather than on postulating the validity of the Young equation. Nevertheless, the important point remains that these models show that heterogeneity and roughness can produce contact angle hysteresis. The analysis in Sections 7.2.1–7.2.4 below presents a mathematically straightforward free energy analysis of contact angles on model heterogeneous and rough surfaces based solely on the Laplace equation. This type of analysis is then generalized to the flotation of heavy particles at the liquid/fluid interface in Section 7.2.5. Section 7.2 concludes with a discussion of models different from the Gibbsian one, Section 7.2.6. In that section, rather than considering the interface as a sharp mathematical plane and postulating the existence of surface excess quantities as in Chapter 1, the solid–vapor interface will be assumed to carry a thin film of the liquid that forms the drop.

Significant progress has been made in the past few years with respect to the preparation and characterization of solid surfaces. This has facilitated the fabrication of extremely smooth and chemically homogeneous surfaces that were not easily achievable a few decades ago. Despite the smoothness and homogeneity of such solid surfaces, contact angle hysteresis is still present, indicating that causes other than those related to the surface topography also give rise to hysteresis. More recently explored causes of hysteresis will be discussed in detail in Section 7.3, below.

The Gibbs phase rule is an often-used tool to gain insight into thermodynamic systems and their descriptions. That topic is pursued in Section 7.4 considering particularly possible implications for the formalism developed in Chapters 8 and 9 for the determination of solid surface tension from experimental contact angles.

Contact angle hysteresis is not the only difficulty in establishing equilibrium in contact angle systems. In many instances a phenomenon called "stick-slip" occurs: When, say, liquid is fed continuously into a sessile drop by means of a capillary, the three phase line does not necessarily move smoothly. Instead it may hinge at a certain point, while the drop volume and the contact angle increase continuously. The three phase line then jumps to a new position, where the contact angle is markedly smaller, and the whole process repeats. Such phenomena are discussed in Section 7.5.

Probably more often than not, the phenomenological contact angle is not the thermodynamic equilibrium contact angle. One particular group of contact angle phenomena on certain rough solids has found widespread interest in recent years: Certain rough surfaces cause very large contact angles, approaching 180°. This phenomenon, called superhydrophobicity, is introduced in Section 7.6. Finally, contact angles in the presence of an electric double layer are discussed in Section 7.7, and Section 7.8 presents a glossary of contact angle terminology.

7.2   THERMODYNAMIC MODELING AND FREE ENERGY ANALYSIS OF SOLID-LIQUID-FLUID SYSTEMS

7.2.1   THE VERTICAL PLATE MODEL

To avoid difficulties associated with the Johnson and Dettre model, Neumann and Good developed a vertical plate model [13]. They considered a vertical plate consisting of two types of parallel strips to represent the smooth but heterogeneous model surface. In their ideal rough surface model [14], they considered a vertical plate that consists of a number of smooth and homogeneous inclined surfaces, the angle of inclination changing discontinuously at constant increments in the vertical direction. Gravity was taken into account in both models.

In the following section, Neumann and Good's work will be discussed. The starting point of these models is a smooth and homogeneous vertical plate dipping into a pool of liquid, as illustrated in Figure 7.1. Depending on the equilibrium contact angle, the liquid will either rise or be depressed near the vertical wall.

The thermodynamic model of the wetting behavior of the vertical plate in contact with a liquid can be set up by calculating the change in free energy for any change in the configuration of the system. In order to perform the calculations, it is necessary to choose a reference state for which the free energy will be arbitrarily defined as zero. For convenience, the reference configuration is chosen to correspond to an instantaneous value of contact angle $\theta = 90°$. Moving from the reference state to an adjacent configuration, there will be a change in the total free energy of the system due to the following three terms.

1. ΔF_1; this is due to a change in the solid–vapor interfacial area and a corresponding change in the solid–liquid interfacial area.

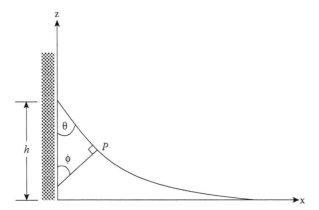

FIGURE 7.1 Capillary rise of a liquid at a vertical plate.

2. ΔF_2; this is due to a change in the liquid–vapor interfacial area.
3. ΔF_3; this is due to the work that has to be done against gravity for such a change in configuration.

Therefore, the overall free energy change, ΔF, for any hypothetical change in the configuration of the system is given by

$$\Delta F = \Delta F_1 + \Delta F_2 + \Delta F_3. \qquad (7.7)$$

To calculate the three free energy terms in Equation 7.7, the validity of the Laplace equation will be assumed, since it describes the equilibrium shapes of the liquid surface independent of the properties of the solid surface. On the other hand, no a priori assumption will be made regarding the validity of the Young equation. The surface area will be calculated based on a unit length, L (e.g., 1 cm), in the y-direction perpendicular to the plane of the paper (Figure 7.1).

7.2.1.1 The Driving Force Term, ΔF_1

The driving force term, ΔF_1, represents the work done by the system in replacing the area, Lh, having interfacial free energy γ_{sv}, by the same area having interfacial free energy γ_{sl}:

$$\Delta F_1 = -Lh(\gamma_{sv} - \gamma_{sl}). \qquad (7.8)$$

The capillary rise, h, may be obtained from an integration of the Laplace equation that may be written for the geometry of this model, that is, for a cylindrical liquid surface, as

$$\frac{1}{R_1} = -\frac{d\cos\phi}{dz} = \frac{\Delta\rho gz}{\gamma_{lv}} = Cz, \qquad (7.9)$$

where $C \equiv \Delta \rho g / \gamma_{lv}$ and ϕ is defined as the angle between the normal at a point P and the z axis, as shown in Figure 7.1. When P is on the z axis, we have $z = h$ and $\phi = 90°$ – θ, where θ is the instantaneous contact angle. Integrating Equation 7.9 yields

$$-\cos\phi = \frac{Cz^2}{2} + C'.$$

(7.10)

Since $z = 0$ at $\phi = 0$, it follows that $C' = -1$, so that

$$h = \sqrt{\frac{2\gamma_{lv}}{\Delta \rho g}} \sqrt{1 - \sin\theta}.$$

(7.11)

Thus, Equation 7.8 becomes

$$\Delta F_1 = -L(\gamma_{sv} - \gamma_{sl})\sqrt{\frac{2\gamma_{lv}}{\Delta \rho g}} \sqrt{1 - \sin\theta}.$$

(7.12)

7.2.1.2 The Free Energy Change of the Liquid–Vapor Interface, ΔF_2

Going from the reference state to any other configuration, work has to be done on the system to expand the liquid surface; therefore

$$\Delta F_2 = L\gamma_{lv}\Delta l,$$

(7.13)

where Δl is the increase in length of a line along the surface of the liquid, lying in the x-z plane (Figure 7.1)

$$\Delta l = \int_{x=0}^{x=\infty} (ds - dx).$$

Since $ds^2 = dx^2 + dz^2$, then

$$\Delta l = \int_{x=0}^{x=\infty} [(dx^2 + dz^2)^{1/2} - dx].$$

(7.14)

From Equation 7.9

$$dz = -\frac{d\cos\phi}{Cz},$$

(7.15)

and from elementary calculus

$$dx = -\frac{dz\cos\phi}{\sin\phi}.$$

(7.16)

Using Equations 7.15 and 7.16 and also Equation 7.10, Equation 7.14 becomes

$$\Delta l = \frac{1}{\sqrt{2C}} \int_{\phi=\frac{\pi}{2}-\theta}^{\phi=0} \frac{d\cos\phi}{(1+\cos\phi)^{1/2}}.$$ (7.17)

Using the result of integrating Equation 7.17, Equation 7.13 can be written as

$$\Delta F_2 = L\gamma_{lv}\sqrt{\frac{2\gamma_{lv}}{\Delta\rho g}}[\sqrt{2} - \sqrt{1+\sin\theta}]$$ (7.18)

7.2.1.3 Work Done Against Gravity, ΔF_3

Work must also be done on the system against gravity to raise the liquid near the vertical plate above the undisturbed level. Consider a small column of liquid of rectangular cross section $L\,dx$; the column is composed of successive increments of volume $L\,dx\,dz$. The work done in lifting each of these elements to its proper position, z, is $\Delta\rho g L z\,dx\,dz$. Integrating over all elements of the column, one obtains

$$\Delta F_{3\,column} = \frac{\Delta\rho g L z^2 dx}{2},$$ (7.19)

and integrating over all the columns

$$\Delta F_3 = \frac{1}{2}L\Delta\rho g \int_{x=0}^{x=\infty} z^2 dx.$$ (7.20)

Using Equations 7.15, 7.16, and 7.10 to integrate Equation 7.20 will yield

$$\Delta F_3 = \frac{1}{3}L\gamma_{lv}\sqrt{\frac{2\gamma_{lv}}{\Delta\rho g}}[(2-\sin\theta)\sqrt{1+\sin\theta} - \sqrt{2}].$$ (7.21)

The sum of ΔF_2 and ΔF_3 can be written in the following simpler form

$$\Delta F_2 + \Delta F_3 = \frac{1}{3}L\gamma_{lv}\sqrt{\frac{2\gamma_{lv}}{\Delta\rho g}}[2\sqrt{2} - (1+\sin\theta)^{3/2}].$$ (7.22)

From $d\Delta F/d\theta = 0$, one recovers $\gamma_{sv}-\gamma_{sl}-\gamma_{lv}\cos\theta_e = 0$; that is, Young's equation. Therefore ΔF_1 (Equation 7.12) can be rewritten as

$$\Delta F_1 = -L\gamma_{lv}\cos\theta_e\sqrt{\frac{2\gamma_{lv}}{\Delta\rho g}}\sqrt{1-\sin\theta}.$$ (7.23)

Combining Equations 7.22 and 7.23 gives the overall free energy change, ΔF

$$\Delta F = \frac{1}{3}L\gamma_{lv}\sqrt{\frac{2\gamma_{lv}}{\Delta\rho g}}[2\sqrt{2} - (1+\sin\theta)^{3/2} - 3\cos\theta_e\sqrt{1-\sin\theta}].$$ (7.24)

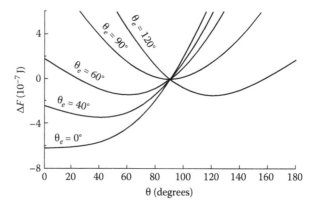

FIGURE 7.2 Free energy changes as a function of instantaneous contact angle, θ, on smooth, homogeneous solid surfaces of various equilibrium contact angles, θ_e; $\gamma_{lv} = 50.0$ mJ/m²; $\Delta\rho = 1000$ kg/m³. (Reprinted from Neumann, A. W. and Good, R. J., *Journal of Colloid and Interface Science*, 38, 341, 1972. With permission from Academic Press.)

Thus, one may characterize any specific system by giving the values of γ_{lv} and θ_e, instead of γ_{lv}, γ_{sv}, and γ_{sl}. Assuming a hypothetical liquid with $\gamma_{lv} = 50.0$ mJ/m² and $\Delta\rho = 1000$ kg/m³, free energy curves ΔF versus θ, for a number of θ_e values, are reproduced in Figure 7.2 using Equations 7.7, 7.12, 7.18, and 7.21. It can be seen that each curve passes through a minimum at $\theta = \theta_e$.

7.2.2 CONTACT ANGLES ON A SMOOTH BUT HETEROGENEOUS SURFACE CONSISTING OF HORIZONTAL STRIPS

In practice, various processing conditions can give rise to surface heterogeneity. For example, increasing the annealing temperature of films of polystyrene/poly(vinyl methyl) ether blend from 130 to 170°C induces phase separation between polystyrene (PS) and PVME moieties, resulting in a heterogeneous surface film [15]. Annealing of thin films of a polystyrene-block-poly(methyl methacrylate) diblock copolymer causes the PS and polymethylmethacrylate (PMMA) blocks to undergo microphase-separation [16]. Varying the annealing time changes the morphology of the film from a disordered state to a nanoscale depression morphology and finally to a striped morphology. Chemical incompatibility between the hydrocarbon and the fluorocarbon side chains of poly(octadecene-*alt*-N-(4-(perfluoroheptylcarbonyl) aminobutyl)maleimide) results in microscale phase-separation between the side chains and domains with different chemical compositions [17]. For similar reasons, the films of polymer blends of polybutadiene/polystyrene [18] as well as polybutadiene/poly(methyl methacrylate) [19] contain heterogeneous domains. Due to the difference in the chemistry of heterogeneous patches, there exist domains with different surface energies on a solid surface. The following energetic analysis demonstrates how surface heterogeneity causes contact angle hysteresis.

To avoid the limitations of the concentric rings model developed by Johnson and Dettre [11,12], Neumann and Good [13] considered a different but more realistic heterogeneous surface model: a vertical plate with two different horizontal strips of approximately equal width, one type having an equilibrium contact angle θ_{e1}, and the other θ_{e2}. For convenience, the strip widths, Δh_1 and Δh_2, are assumed to vary systematically so that constant increments of θ can be used as input (see Equation 7.11). Consider that this vertical plate is dipped into a hypothetical liquid with $\gamma_{lv} = 50.0$ mJ/m^2 and $\Delta\rho = 1000$ kg/m^3. Since the ΔF_2 and ΔF_3 terms depend only on the instantaneous contact angle θ, and not on the properties of the two types of solid, these two terms can be calculated, as in the previous section, from Equations 7.18 and 7.21. Moving from the reference state $\theta = 90°$ to other configurations, the solid-liquid-vapor three-phase line will traverse a number of type-1 and type-2 strips with equilibrium contact angles θ_{e1} and θ_{e2}, respectively. Therefore, to calculate ΔF_1 for a particular capillary rise, h, it is necessary to calculate the change in free energy for passage across every strip of type-1 and type-2, up to the specified value of h.

The general procedure to calculate ΔF_1 can be easily illustrated by a specific example. For convenience, the strips chosen are fairly wide so that the movement of the three-phase line from a θ_{e1}/θ_{e2} boundary to the next θ_{e2}/θ_{e1} boundary corresponds to a 2° increment in contact angle (according to Equation 7.11). The 90° configuration is located so that the three-phase line lies in the center of a θ_{e2} strip. Thus, going from the $\theta = 90°$ configuration to the 89° configuration, the three-phase line traverses only θ_{e2} surface:

$$\Delta F_1^{90-89°} = -L\gamma_{lv}\cos\theta_{e2}\sqrt{\frac{2\gamma_{lv}}{\Delta\rho g}}(\sqrt{1-\sin 89°} - \sqrt{1-\sin 90°}). \qquad (7.25)$$

Equation 7.25 is essentially equivalent to Equations 7.12 and 7.23, except that the term due to the lower limit of integration (cf. Equations 7.9 through 7.12), $\sqrt{1-\sin 90°}$, is retained. It should be noted that this term vanishes only if the lower limit of integration is $\theta = 90°$.

Now going from the configuration $\theta = 89°$ to $\theta = 88°$, the contact line traverses θ_{e1} surface area:

$$\Delta F_1^{89-88°} = -L\gamma_{lv}\cos\theta_{e1}\sqrt{\frac{2\gamma_{lv}}{\Delta\rho g}}(\sqrt{1-\sin 88°} - \sqrt{1-\sin 89°}). \qquad (7.26)$$

The free energy change, ΔF_1, for the process of going from $\theta = 90°$ to $\theta = 88°$ will be

$$\Delta F_1^{90-88°} = \Delta F_1^{90-89°} + \Delta F_1^{89-88°}. \qquad (7.27)$$

From $\theta = 88°$ to $\theta = 87°$, the three-phase line will still move across the θ_{e1} surface area; hence, the free energy change can be calculated as

$$\Delta F_1^{88-87°} = -L\gamma_{lv}\cos\theta_{e1}\sqrt{\frac{2\gamma_{lv}}{\Delta\rho g}}(\sqrt{1-\sin 87°} - \sqrt{1-\sin 88°}), \qquad (7.28)$$

and

$$\Delta F_1^{90-87°} = \Delta F_1^{90-88°} + \Delta F_1^{88-87°}.$$ (7.29)

If continuing to $\theta = 86°$, the three-phase line will traverse θ_{e2} surface area again. Thus

$$\Delta F_1^{87-86°} = -L\gamma_{lv}\cos\theta_{e2}\sqrt{\frac{2\gamma_{lv}}{\Delta\rho g}}(\sqrt{1-\sin86°} - \sqrt{1-\sin87°}),$$ (7.30)

and

$$\Delta F_1^{90-86°} = \Delta F_1^{90-87°} + \Delta F_1^{87-86°}.$$ (7.31)

As shown above, these calculations can be continued to any value of instantaneous contact angle θ.

For the range of contact angles from $\theta = 90°$ to $\theta = 0°$, the computations of $\Delta F_1 = \Delta F_1 + \Delta F_2 + \Delta F_3$ were performed for $\theta_{e1} = 40°$, $\theta_{e2} = 30°$, and a strip width equivalent to $2°$. The same type of computation was also performed for a strip width equivalent to $(2/3)°$. These two ΔF versus θ curves are given in Figure 7.3a and b, respectively.

The most significant feature of Figure 7.3 is that both curves have local minima. Between $30°$ and $40°$, the free energy curves have a sawtooth structure that corresponds to a number of metastable equilibrium configurations. Outside the contact angle range from $30°$ to $40°$, the sawtooth structure is still present but there are no more local minima: hence, there are no more metastable equilibrium configurations. Further away from the range of $30°-40°$, the free energy curves become almost smooth.

From the above thermodynamic model and Figure 7.3, several conclusions can be drawn as follows:

1. If strips with a much smaller width are chosen for the model and the corresponding envelope of the sawtooth structure is examined, it will be found that the absolute minimum is given by the Cassie Equation 7.5. Therefore, it can be concluded that the thermodynamic equilibrium contact angle θ_{ES} of the system under consideration is the Cassie angle θ_C; that is, $\theta_{ES} = \theta_C$. It should be noted that, however, experimental determination of θ_C might be difficult because of the existence of a large number of metastable contact angles.
2. Due to the presence of the metastable equilibrium states shown in Figure 7.3, if the system initially stays in any hypothetical configuration outside the range of contact angles from $30°$ to $40°$, when released, the system will spontaneously decrease its free energy and hence change its contact angle θ until it reaches $\theta_{e1} = 40°$ or $\theta_{e2} = 30°$, depending on whether the initial contact angle θ is greater than $40°$ or less than $30°$. To decrease the free energy of the system further, the free energy barriers shown in Figure 7.3 would

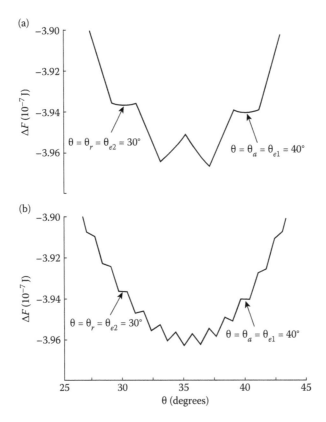

FIGURE 7.3 Free energy changes as a function of instantaneous contact angle, θ, on two surfaces consisting of two types of horizontal strips of equal width. The equilibrium contact angles of type-1 and type-2 strips are $\theta_{e1} = 40°$ and $\theta_{e2} = 30°$, respectively. (a) Strip width 2°; (b) strip width (2/3)°. (Reprinted from Neumann, A. W. and Good, R. J., *Journal of Colloid and Interface Science*, 38, 341, 1972. With permission from Academic Press.)

have to be overcome. It can be seen that the particular heterogeneity of this model predicts contact angle hysteresis.

3. From Figure 7.3, it can be seen that there are a number of metastable states within the contact angle range from 30° to 40°; the first local minimum on the lower angle side corresponds to the intrinsic contact angle θ_{e2} of the higher energy strips, and the first local minimum on the higher angle side corresponds to the intrinsic contact angle θ_{e1} of the lower energy strips. Therefore on such a heterogeneous surface, the three-phase line will advance only if the experimental contact angle becomes equal to θ_{e1}, and will recede if the experimental contact angle becomes equal to θ_{e2}; that is,

$$\text{advancing angle, } \theta_a = \theta_{e1}, \quad \text{receding angle, } \theta_r = \theta_{e2}. \quad (7.32)$$

Although neither θ_a nor θ_r is equal to the equilibrium contact angle, θ_{ES}, of this system, they are the intrinsic equilibrium contact angles of the two types of strips,

θ_{e1} and θ_{e2}, respectively; hence, they may be used in the Young equation. In order to keep in mind the fact that the advancing and receding contact angles are not the equilibrium angle θ_{ES}, we designate them as

$$\theta_a = \theta_{Y1}, \quad \theta_r = \theta_{Y2}. \tag{7.33}$$

In other words, the subscript Y indicates that the contact angles so designated may be used in conjunction with Young's equation. Thus, the equilibrium angle $\theta_{ES} = \theta_e$ on a smooth and homogeneous solid surface is a Young angle, whereas the Cassie angle, θ_C, in spite of being the thermodynamic equilibrium angle, is not a Young contact angle, θ_Y. This result implies the validity of Young's equation in the following forms

$$(\gamma_{sv})_1 - (\gamma_{sl})_1 = \gamma_{lv}\cos\theta_a, \tag{7.34}$$

$$(\gamma_{sv})_2 - (\gamma_{sl})_2 = \gamma_{lv}\cos\theta_r. \tag{7.35}$$

It is apparent from the above discussion that if an ideal heterogeneous surface is modeled as a smooth surface consisting of two types of horizontal strips, the heterogeneity of the model surface will cause metastable states or contact angle hysteresis, because moving from a type-1 strip to the next type-2 strip will cause a fluctuation of the free energy of the system [13].

7.2.3 CONTACT ANGLES ON A SMOOTH BUT HETEROGENEOUS SURFACE CONSISTING OF VERTICAL STRIPS

Above, we considered a specific type of heterogeneity (i.e., horizontal strips). In practice, heterogeneity is often patchwise; that is, the solid surface is composed of one type of surface patches distributed over a second type of surface. We will consider now a surface composed of vertical strips of fractional widths a_1 and a_2. This model will shed light on the behavior of patchwise heterogeneous surfaces.

The three-phase line is assumed first to be a horizontal straight line. That is to say, we assume the Laplace equation to hold, but do not insist that the Young equation must hold locally. Again, the ΔF_2 and ΔF_3 terms will be independent of any properties of the solid surface and, therefore, can be computed from Equations 7.18 and 7.21. The ΔF_1 term, however, will be broken into two terms:

$$\Delta F_{11} = -a_1 L\gamma_{lv} \cos\theta_{e1}\sqrt{\frac{2\gamma_{lv}}{\Delta\rho g}}\sqrt{1-\sin\theta}, \tag{7.36}$$

and

$$\Delta F_{12} = -a_2 L\gamma_{lv} \cos\theta_{e2}\sqrt{\frac{2\gamma_{lv}}{\Delta\rho g}}\sqrt{1-\sin\theta}, \tag{7.37}$$

so that

$$\Delta F_1 = \Delta F_{11} + \Delta F_{12}. \tag{7.38}$$

It is a simple matter to demonstrate that the thermodynamic equilibrium contact angle is again the angle θ_C given by the Cassie Equation 7.5. Differentiation of ΔF as defined by Equation 7.7, where ΔF_1 is now given by Equation 7.38, leads to

$$\frac{d(\Delta F)}{d\theta} = \frac{1}{2} L \sqrt{\frac{2\gamma_{lv}}{\Delta \rho g}} \frac{\cos\theta}{\sqrt{1 - \sin\theta}} (a_1 \gamma_{lv} \cos\theta_{e1} + a_2 \gamma_{lv} \cos\theta_{e2} - \gamma_{lv} \cos\theta). \tag{7.39}$$

The equilibrium condition

$$\frac{d(\Delta F)}{d\theta} = 0, \tag{7.40}$$

leads immediately to the Cassie Equation 7.5.

Retaining the parameters used in Figure 7.3 (i.e., $\theta_{e1} = 40°$, $\theta_{e2} = 30°$, $\gamma_{lv} = 50.0$ mJ/m^2, and $\Delta\rho = 1000$ kg/m^3), and choosing $a_1 = a_2 = 0.5$, ΔF was calculated using Equations 7.7, 7.18, 7.21, and 7.38. The result is shown in Figure 7.4. The minimum occurs at the Cassie angle θ_C, as proven above.

From Figure 7.4, we can reach a preliminary conclusion regarding patchwise heterogeneous surfaces. If the three-phase line is always straight and if the average

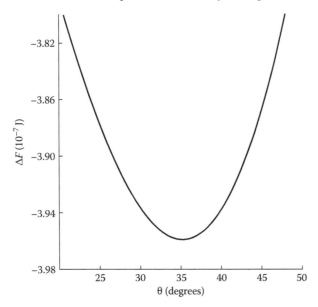

FIGURE 7.4 Free energy changes as a function of instantaneous contact angle on a surface consisting of vertical strips assuming a straight line of contact. $\theta_{e1} = 40°$, $\theta_{e2} = 30°$, $\gamma_{lv} = 50.0$ mJ/m^2, and $\Delta\rho = 1000$ kg/m^3. (Reprinted from Neumann, A. W. and Good, R. J., *Journal of Colloid and Interface Science*, 38, 341, 1972. With permission from Academic Press.)

composition of the surface is constant, there will be no metastable state and hence no contact angle hysteresis, and the capillary rise will be exactly that corresponding to the Cassie angle.

From experimental observations with real solid surfaces, we know that the three-phase line is not always straight. The cause of contortions of the three-phase line may most reasonably be considered to be the need for local compliance with Young's equation. The free energy arguments developed so far also make it plausible that contortions of the three-phase line may produce contact angle hysteresis. Let us envisage a patchwise heterogeneous solid surface where the two types of surface have equilibrium contact angles θ_{e1} and θ_{e2} with a certain liquid of surface tension γ_{lv}. Let us further consider a hypothetical configuration such that the line of contact is straight and the instantaneous contact angle has a value intermediate between θ_{e1} and θ_{e2}. If we now relax the constraint of the straightness of the line of contact, the system may be able to decrease its overall free energy by forming a contorted line of contact such that the capillary rise on θ_{e1} patches decreases and that on θ_{e2} increases. If we then apply a force to the system, which moves the contorted line of contact, say, upward, the portions of the line of contact on the θ_{e2} area will not move onto the θ_{e1} area since this process would increase the free energy, ΔF. Therefore, the externally imposed motion of the line of contact will tend to straighten out this line, a process that will also increase the free energy of the system. If the external force keeps building up, the line of contact will eventually be able to overcome the energy barrier connected with the crossing of the type-1 patches.

The movement of the three-phase line will thus be connected with fluctuations in the corrugation of the line of contact, and these fluctuations will also lead to fluctuations in the free energy curve ΔF versus θ. The latter fluctuations, if sufficiently pronounced, will represent free energy barriers of the type shown in Figure 7.3. It follows that contact angle hysteresis on a patchwise heterogeneous solid surface can occur if the three-phase line has the ability to deviate from a straight line. Similar conclusions were drawn by other researchers regarding hysteresis on heterogeneous surfaces [20,21].

The discussion given so far shows that heterogeneity of the solid surface may give rise to contact angle hysteresis. On an atomic or molecular scale, most surfaces are heterogeneous. Although the occurrence of contact angle hysteresis is, in fact, the rule rather than the exception, contact angle hysteresis is not observed on certain well-prepared solid surfaces [22–26]. We therefore conclude that there must be a lower limit for the lateral dimensions of patchwise heterogeneities, above molecular dimensions, below which heterogeneities do not contribute to contact angle hysteresis. This limit has been estimated to be of order 0.1 μm [13].

7.2.4 Contact Angles on Homogeneous but Rough Surfaces

The basic thermodynamic model described above has been applied to idealized rough surfaces [14]. Essentially, this application was made by considering, instead of a smooth and homogeneous vertical surface, a smooth and homogeneous inclined surface and changing the angle of inclination discontinuously at constant increments

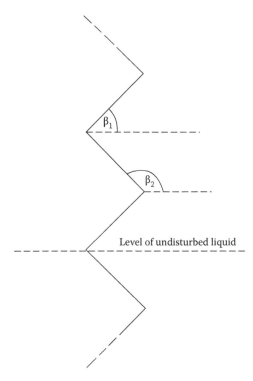

FIGURE 7.5   Vertical section through the model rough surface. The angles β_1 and β_2 are chosen symmetrically about 90° so that, when we switch from the microscopic picture given in this Figure to the macroscopic picture, we have a flat, vertical surface. (Reprinted from Eick, J. D., Good, R. J., and Neumann, A. W., *Journal of Colloid and Interface Science*, 53, 235, 1975. With permission from Academic Press.)

of the ordinate z. A cut through such a surface, parallel to the $x - z$ plane, is shown schematically in Figure 7.5. The first step toward the evaluation of the free energy ΔF for a liquid in contact with an idealized rough surface, as depicted in Figure 7.5, is to compute the ΔF_1, ΔF_2, and ΔF_3 terms for a unit length of an infinitely wide inclined plate (Figure 7.6).

7.2.4.1   The Driving Force Term, ΔF_1

The term ΔF_1 again represents the work done by the system in replacing the area Lh, having interfacial tension γ_{sv}, by the same area having interfacial tension γ_{sl}. The capillary rise, h, is measured along the inclined solid surface, in the $x - z$ plane. Thus, ΔF_1 is again given by Equation 7.8 or, when inserting Young's Equation 7.2, by

$$\Delta F_1 = -Lh\gamma_{lv}\cos\theta_e. \tag{7.41}$$

Integration of the Laplace Equation 7.1 leads again to

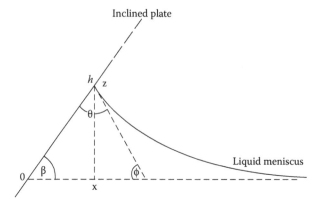

FIGURE 7.6 Capillary rise of a liquid at an inclined wall. (Reprinted from Eick, J. D., Good, R. J., and Neumann, A. W., *Journal of Colloid and Interface Science*, 53, 235, 1975. With permission from Academic Press.)

$$z = \sqrt{\frac{2\gamma_{lv}}{\Delta \rho g}}\sqrt{1-\cos\phi}. \tag{7.42}$$

It is to be noted that for the configuration given in Figure 7.6, the relation between ϕ and θ is given by

$$180° = \beta + \theta + \phi, \tag{7.43}$$

where β is the angle of inclination. From Figure 7.6, it can be seen that

$$h = \frac{z}{\sin\beta}, \tag{7.44}$$

so that

$$\Delta F_1 = -L\gamma_{lv}\frac{\cos\theta_e}{\sin\beta}\sqrt{\frac{2\gamma_{lv}}{\Delta \rho g}}\sqrt{1-\cos(180° - \beta - \theta)}. \tag{7.45}$$

Equation 7.45 reduces to Equation 7.23 for the special case of a vertical plate (i.e., for $\beta = 90°$).

7.2.4.2 The Free Energy Change of the Liquid–Vapor Interface, ΔF_2

Going from the reference state to any other configuration, work has to be done on the system to expand the liquid surface; the resulting ΔF_2 term consists of two parts:

$$\Delta F_2 = \Delta F_{21} + \Delta F_{22}. \tag{7.46}$$

The term ΔF_{21} is analogous to the ΔF_2 term in the analysis of the homogeneous, smooth solid surface (see Equation 7.18):

$$\Delta F_{21} = L\gamma_{lv}\sqrt{\frac{2\gamma_{lv}}{\Delta\rho g}}[\sqrt{2} - \sqrt{1+\cos(180° - \beta - \theta)}]. \tag{7.47}$$

The ΔF_{22} term arises from the fact that by going from the reference state, that is, the undisturbed liquid surface, to some other configuration, for example, the one depicted in Figure 7.6, the liquid surface area $L(OX)$ is annihilated. We therefore have

$$\Delta F_{22} = -L\gamma_{lv}x, \tag{7.48}$$

where

$$x = \overline{OX}. \tag{7.49}$$

Since

$$x = \frac{z}{\tan\beta}, \tag{7.50}$$

it follows from Equation 7.42 that

$$x = \sqrt{\frac{2\gamma_{lv}}{\Delta\rho g}}\sqrt{1-\cos\phi}\cot\beta. \tag{7.51}$$

Since $\cot\beta$ will change sign when β passes through 90°, and since ΔF_{22} has to change sign at this same point, it follows that Equation 7.51 is valid for $0 < \beta < 180°$. We therefore have

$$\Delta F_{22} = -L\gamma_{lv}\sqrt{\frac{2\gamma_{lv}}{\Delta\rho g}}\sqrt{1-\cos(180° - \beta - \theta)}\cot\beta, \tag{7.52}$$

and finally

$$\Delta F_2 = L\gamma_{lv}\sqrt{\frac{2\gamma_{lv}}{\Delta\rho g}}[\sqrt{2} - \sqrt{1+\cos(180° - \beta - \theta)} - \cot\beta\sqrt{1-\cos(180° - \beta - \theta)}]. \tag{7.53}$$

In the special case $\beta = 90°$, Equation 7.53 reduces to Equation 7.18.

7.2.4.3 Work Done Against Gravity, ΔF_3

Similar to the ΔF_2 term, the work done against gravity, ΔF_3, also will consist of two parts:

$$\Delta F_3 = \Delta F_{31} + \Delta F_{32}. \tag{7.54}$$

The first one of these terms represents the work done in lifting the liquid between X and ∞ and corresponds to the term given by Equation 7.21:

$$\Delta F_{31} = \frac{1}{3} L\gamma_{lv} \sqrt{\frac{2\gamma_{lv}}{\Delta\rho g}} [(2 - \cos(180° - \beta - \theta))\sqrt{1 + \cos(180° - \beta - \theta)} - \sqrt{2}]. \tag{7.55}$$

For the ΔF_{32} term we obtain (cf. Equation 7.20)

$$\Delta F_{32} = \frac{1}{2} L\Delta\rho g \int_0^x z^2 dx. \tag{7.56}$$

The relation between z and x needed to evaluate the integral in Equation 7.56 is given by Equation 7.50. It follows that

$$\Delta F_{32} = \frac{\sqrt{2}}{3} L\gamma_{lv} \sqrt{\frac{\gamma_{lv}}{\Delta\rho g}} \cot\beta[1 - \cos(180° - \beta - \theta)]\sqrt{1 - \cos(180° - \beta - \theta)}. \tag{7.57}$$

For the special case $\beta = 90°$, Equation 7.54 again reduces to Equation 7.21. To check for the internal consistency of the model, free energy calculations ΔF versus θ were performed for a number of angles of tilt β. The minima of these curves did, in fact, in all cases occur at $\theta = \theta_e$; that is, the model is in agreement with Young's equation. It is also apparent that Young's equation can be derived for an inclined plate from the preceding analysis in a way similar to the one used to derive it for the vertical wall.

7.2.4.4 Application to Idealized Rough Surfaces

We shall now apply the analysis given above to idealized rough surfaces. As indicated above, this may be accomplished by changing the angle of tilt β (Figure 7.6) at some specific value $h = H$ discontinuously to some other value, and after the same distance H back again to the original angle of tilt. Continuing in this way, if the two angles of tilt are symmetrical about 90° (i.e., if $\beta_1 = 180° - \beta_2$), we generate an idealized rough vertical surface, as shown in Figure 7.5. At this point, we also have to introduce a new contact angle concept. If H becomes very small, the optical systems normally employed to observe contact angle phenomena will no longer reveal that we are dealing with a surface having two types of area that differ from each other by their inclination. Operationally, we are dealing with a vertical surface on which we may measure a capillary rise $\bar{Z}$ that we may relate to a phenomenological or macroscopic contact angle, θ_M, by (cf. Equation 7.11)

$$\sin\theta_M = 1 - \frac{\Delta\rho g\bar{Z}^2}{2\gamma_{lv}}. \tag{7.58}$$

It is important to note that this phenomenological contact angle is of physical significance. Although it will, in general, not be identical with the microscopic contact angle θ, and therefore will not satisfy the Young Equation 7.2, it determines the shape of the liquid meniscus and therefore the pressure of curvature. Wetting kinetics will be determined entirely by this phenomenological contact angle, without regard to its thermodynamic status.

It will be apparent by now that the thermodynamics of contact angles on rough surfaces is considerably more involved than the thermodynamics of contact angles on heterogeneous surfaces. A substantial additional difficulty arises when we consider the mechanism of the capillary rise or capillary depression at a surface of the kind given in Figure 7.5. Assuming again that the Laplace equation is satisfied at all times, there will be a smooth change in the macroscopic contact angle θ_M, as we increase the capillary rise across a number of ridges. The instantaneous microscopic contact angle θ, however, will change discontinuously whenever the line of contact moves across the boundary between two strips of different inclinations. Adhering to the restrictions imposed by the Laplace equation, situations may arise in which the liquid meniscus would have to intersect the next lower ridge. Such cases are not considered here. This occurrence would obviously introduce conceptual and practical difficulties. In order to calculate, for the model surface shown in Figure 7.5, the changes in free energy ΔF as a function of the macroscopic contact angle θ_M, only the free energy terms ΔF_{22} and ΔF_{32} given in the analysis of the smooth inclined plate (Equations 7.52 and 7.57) have to be modified. The ΔF_{22} term will cycle between zero and some finite value that will be reached for the first time when $h = H$. The ΔF_{32} term comprises the work done in lifting the liquid elements to fill all the horizontal crevices of the model surface. The correctness of the procedure was checked by setting $H = 100$ μm, $\theta_e = 40°$, $\gamma_{lv} = 50$ mJ/m², $\Delta\rho = 1000$ kg/m³, and $L = 1$ cm, and letting $\beta_1 = \beta_2 = 90°$. The results were identical with those previously obtained for a smooth and homogeneous vertical plate.

Two typical examples of ΔF versus θ_M curves for two idealized rough surfaces are shown in Figures 7.7 and 7.8. The specific parameters used are the same as above except that $\beta_1 = 45°$ and $\beta_2 = 135°$ for both Figures, and $\theta_e = 70°$ for Figure 7.7, and $\theta_e = 40°$ for Figure 7.8. We note that this model rough surface produces metastable states, as do the horizontal strips. It is interesting to note that only the system in Figure 7.7 has an equilibrium contact angle θ_{ES}; that is, the envelope of the curve in Figure 7.7 does exhibit a minimum at the Wenzel angle θ_W (i.e., $\theta_{ES} = \theta_W$). However, in the case of $\theta_e = 40°$, there is no minimum (Figure 7.8). Therefore, the most favorable configuration for the system in Figure 7.8 is the lowest angle that is physically possible (i.e., $\theta_M = 0°$). The highest metastable state is at $\theta_M = 115°$ in Figure 7.7 and $\theta_M = 85°$ in Figure 7.8; that is, it occurs where Young's equation is satisfied locally. This may be seen by noting first that, for a given geometry, there exists the following

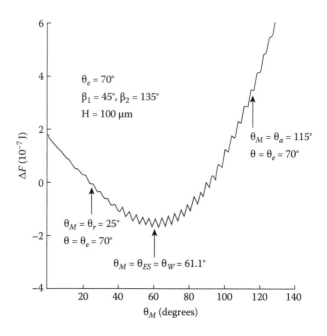

FIGURE 7.7 Free energy change as a function of the macroscopic contact angle; that is, the contact angle between the liquid and the envelope of the model solid surface. $\theta_e = 70°$; $H = 100$ μm, $\gamma_{lv} = 50$ mJ/m^2, $\Delta\rho = 1000$ kg/m^3; $L = 1$ cm; $\beta_1 = 45°$ and $\beta_2 = 135°$. (Reprinted from Eick, J. D., Good, R. J., and Neumann, A. W., *Journal of Colloid and Interface Science*, 53, 235, 1975. With permission from Academic Press.)

purely geometric relation between the macroscopic contact angle θ_M and the microscopic contact angle θ (cf. Figure 7.6)

$$\theta_M = \theta + \beta - 90°. \tag{7.59}$$

It should be noted that θ_M and θ represent any pair of instantaneous contact angles. The metastable state with the largest θ_M is reached when $\theta = \theta_e$, where θ_e designates a microscopic contact angle, which is identical with the equilibrium contact angle of the corresponding smooth solid surface. In the case of Figure 7.7, we have $\theta_e = 70°$, and θ_M reaches its first metastable state at $\theta = \theta_e$, or, since $\beta_2 = 135°$, at $\theta_M = 115°$. The last metastable state occurs at $\beta_1 = 45°$ (i.e., at $\theta_M = 25°$). Assuming that hysteresis on this model surface is only due to surface roughness, clearly $\theta_M = 115°$ and $\theta_M = 25°$ represent advancing and receding contact angles, respectively. In the case of Figure 7.8, where there is no equilibrium contact angle θ_{ES}, the receding contact angle is equal to $\theta_M = \theta_r = 0$. Obviously, this particular idealized rough surface predicts contact angle hysteresis.

The analysis of the heterogeneous surface consisting of horizontal strips has shown that the advancing contact angle is equal to the equilibrium contact angle on low energy portions of the solid surface. This result led us to the introduction of the concept of the Young contact angle, θ_Y, which, although not being the equilibrium

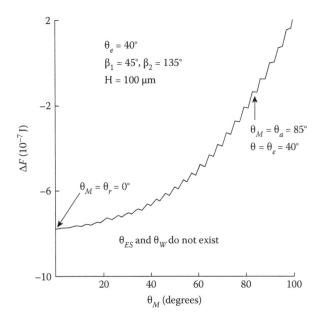

FIGURE 7.8 Free energy change as a function of the macroscopic contact angle; that is, the contact angle between the liquid and the envelope of the model solid surface. $\theta_e = 40°$, $H = 100$ μm, $\gamma_{lv} = 50$ mJ/m², $\Delta\rho = 1000$ kg/m³, $L = 1$ cm, $\beta_1 = 45°$ and $\beta_2 = 135°$. (Reprinted from Eick, J. D., Good, R. J., and Neumann, A. W., *Journal of Colloid and Interface Science*, 53, 235, 1975. With permission from Academic Press.)

contact angle of the system, could be inserted into Young's equation. It is apparent from the above results that this is not so for a model rough surface; that is,

$$(\theta_M)_a \neq \theta_Y, \tag{7.60a}$$

and

$$(\theta_M)_r \neq \theta_Y. \tag{7.60b}$$

In spite of representing the thermodynamic equilibrium condition, the Wenzel contact angle cannot easily be determined experimentally, because of the adjacent metastable states. Furthermore, there is no easy procedure to correlate it to measured (advancing or receding) contact angles. The Wenzel contact angle is also not a Young contact angle; that is, it may not be inserted into Young's Equation 7.2.

Further analysis of idealized rough surfaces using other angles of tilt, other microscopic equilibrium contact angles θ, and other values of H is available elsewhere [14].

7.2.5 FLOTATION OF CYLINDRICAL PARTICLES AT LIQUID–FLUID INTERFACES

The interfacial energetics and wettability of small particles are of technological interest in many areas of applied science. Examples are preparation of stable suspensions of particles, adhesion of particles to solid surfaces, and the dispersion of particles into a liquid or melt of a polymer. In this context, one of the important properties of particles directly related to wettability and surface free energy is floatability at liquid–fluid interfaces. This section will give a brief review of a free-energy analysis for particles with a particular geometry; that is, cylindrical particles, floating at liquid–vapor interfaces [27,28].

A solid particle, under the influence of gravity, may approach an interfacial region between two fluids. Depending on the relative densities of the two fluids and the particle, the size of the particle, and the nature of the interface, either the particle will pass through the interface and be engulfed, or it will come to some equilibrium position at the interface.

If the density of the solid particle (ρ_1) is intermediate in magnitude between that of the lower fluid (ρ_2), usually a liquid, and that of the upper fluid (ρ_3), usually a gas (but possibly another liquid), the solid particle will always take up an equilibrium position and float. In this instance, when $\rho_3 < \rho_1 < \rho_2$, buoyancy effects are sufficient to ensure equilibrium at the interface. When ρ_1 is greater than either of ρ_2 and ρ_3 (and still $\rho_2 > \rho_3$), buoyancy effects are not sufficient to ensure that the particle be supported at the interface. In this case, surface energies (or alternatively, surface forces) will begin to have an effect on the possible equilibrium position of the particle at the interfacial region. If the interfacial properties of the system are such that the surface energies contribute positively to the support of the particle, then there is a possibility that an appropriately sized particle will not pass completely through interface 23 (between fluids 3 and 2) into fluid 2.

This section discusses the equilibrium of cylindrical particles with their long axes parallel to the interface. Traditionally, studies of the equilibrium position of particles at fluid interfaces have been undertaken by means of a force analysis that predicts equilibrium when the net vertical force acting on the particle at an interface is zero. The drawback of the force analysis is that it does not clarify the status of the contact angle concepts employed.

In addition, all of the analytical work so far available refers to ideal (i.e., homogeneous and smooth) solid surfaces. Work on contact angles on idealized heterogeneous and idealized rough solid surfaces is in terms of a more general free-energy analysis [11,13,14,29]. This body of work cannot be readily transformed into a force analysis. It therefore appears desirable to formulate a flotation theory for particles in terms of a free-energy analysis as well, in order to generate a broader basis for the study of the behavior of nonideal particles at fluid interfaces. In a sense, the free-energy analysis proves to be the global description of the behavior of the system because it is sensitive to the possible states that the whole system may assume. Also, it has been shown that the free-energy analysis clarifies the status of the contact angle in terms of standard thermodynamic theory [13,29].

Figure 7.9 illustrates one possible configuration of a system consisting of a solid particle at a fluid interface. The diagram serves to define properties of the system that appear in the subsequent analysis. The radius of the particle is R, Z_0 is the capillary rise, and α is the angle from the vertical that locates the extended undeformed free surface on the solid surface. The angle ϕ is a position coordinate that locates the three-phase line on the solid surface, θ is the contact angle measured through fluid

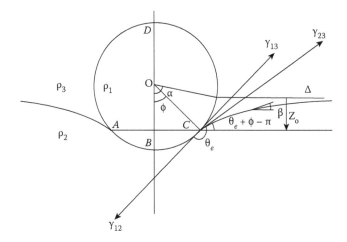

FIGURE 7.9 Stationary state of a small solid particle at a fluid interface. (Reprinted from Rapacchietta, A. V., Neumann, A. W., and Omenyi, S. N., *Journal of Colloid and Interface Science*, 59, 541, 1977. With permission from Academic Press.)

2 between tangents of surfaces 12 and 23 at the three-phase line. The free-energy analysis shows that the contact angle, θ, equals the equilibrium contact angle, θ_e. The angle β is the parameter of the profile of the meniscus; it is the angle between the horizontal and the tangent of the meniscus at any point on the profile. The following conditions apply: angle $\beta > 0$ if measured counterclockwise and $\beta < 0$ if measured clockwise. At the three-phase line, $\beta = \beta_0 = \theta_c + \phi -180°$.

Consider a cylindrical particle, with its long axis perpendicular to the plane of the paper, entirely within fluid 3 and that, under the influence of gravity, approaches interface 23 between fluids 2 and 3. Immediately after the particle touches the interface, a meniscus is formed to the surface of the particle and, as a consequence, the free energy of the system changes. Figure 7.9 illustrates a general stationary state assumed by the particle after such a process.

The reference state of the system will be chosen as that stationary state of the system when the bottom of the cylinder just touches the undeformed interface 23. Since it is desired to study the free energy as a function of the position coordinate, ϕ, which locates the three-phase line at the surface of the particle, the particle is allowed, in successive mechanically enforced stationary states, to penetrate the interface by specific increments of the angle α (locating the extended line of the free surface at the solid) such that ϕ changes in increments of 1°, and is allowed to achieve static conditions characterized by the cessation of capillary action. With reference to the case of the fixed cylindrical shell at a fluid interface, it has been shown that the capillary action will cease under these circumstances only when the instantaneous contact angle, θ, equals the thermodynamic equilibrium contact angle, θ_e [27]. This establishes the validity of the Young equation at the three-phase line for any mechanically enforced stationary state after transitory phenomena have stopped.

Of the full range of such conceptual mechanically enforced stationary states, there will be, at most, one for which the free energy of the system is a local minimum;

there will be none when it is impossible for the particle to stabilize at the interface. Such a state of minimum free energy corresponds to the thermodynamic and physical equilibrium state of the system. No conceptual or mechanical devices have to be employed to maintain the system in this state.

The total free energy change for the system undergoing a process that takes it from the reference state to some stationary state may be expressed as the sum of four distinct parts as follows.

7.2.5.1   The Driving Potential, ΔF_1

The driving potential, ΔF_1, represents the work done by the system in replacing the area ABC (Figure 7.9) having interfacial free energy γ_{12}. The dimensionless form of this driving potential is given by

$$\Delta F_1 = -2\phi\cos\theta_e. \tag{7.61}$$

7.2.5.2   Work to Alter Interface 23, ΔF_2

Work has to be done on the system to alter interfacial area 23 as the system goes from the reference state to some other configuration. The change in free energy associated with this amount of work is ΔF_2. For reasons that will become immediately apparent, it is convenient to consider this change in the free energy as made up of two parts; that is,

$$\Delta F_2 = \Delta F_{21} + \Delta F_{22}, \tag{7.62}$$

where, in dimensionless form,

$$\Delta F_{21} = 2\sqrt{\frac{2}{C}}\{\sqrt{2} - [1 - \cos(\theta_e + \phi)]^{1/2}\}\ \text{and}\ \Delta F_{22} = -2\sin\phi, \tag{7.63}$$

where

$$C = R^2(\rho_2 - \rho_3)g/\gamma_{23}. \tag{7.64}$$

When the liquid-vapor interface is altered to form a meniscus, its area changes; the change in free energy associated with this change in area is ΔF_{21}. During the deformation of interface 23, the geometry of the cylindrical particle will necessitate, simultaneously with and independently of establishing the meniscus, the liquid–vapor interface contract or expand horizontally to retain contact with the solid surface. The term ΔF_{22} represents the change in free energy associated with this phenomenon.

7.2.5.3 Work Done Against Gravity (on the Liquid), ΔF_3

Work must be performed either to depress or to raise the volume of fluid bounded by the meniscus, the extended line of the undeformed free surface, and the solid surface of the particle. This work is given by

$$\Delta F_3 = \Delta F_{31} + \Delta F_{32}, \qquad (7.65)$$

where ΔF_{31} is the work required to displace the fluid bounded by the meniscus, the undeformed free surface level, and the vertical plane through the three-phase line. In dimensionless form, it is

$$\Delta F_{31} = \frac{2}{3}\sqrt{\frac{2}{C}}\{[2 + \cos(\theta_e + \phi)] \times [1 - \cos(\theta_e + \phi)]^{1/2} - \sqrt{2}\}. \qquad (7.66)$$

The term ΔF_{32} is the amount of work required to displace the volume of fluid bounded by the undeformed free surface level, the solid surface, and the vertical plane through the three-phase line; ΔF_{32} is negative or positive depending on the particular configuration. In dimensionless form

$$\Delta F_{32} = C\{\cos^2\alpha(\sin\phi - \sin\alpha) - \cos\alpha[\phi - \alpha + 0.5(\sin 2\phi - \sin 2\alpha)]\}$$
$$= +C\left\{\sin\phi - \sin\alpha - \frac{1}{3}(\sin^3\phi - \sin^3\alpha)\right\}. \qquad (7.67)$$

This expression is dependent solely on the geometry of the particle.

When the top surface of the particle is below the level of the undeformed free surface, the latter expression does not apply because the extraneous volume of fluid cannot be bounded in the manner described. In this case, the appropriate, dimensionless expression is

$$\Delta F_{32} = C\sin\phi\left(\frac{Z_0}{R}\right)^2 + C(\pi - \phi + \sin\phi\cos\phi) \times \left(\frac{CG2}{R} + \frac{Z_0}{R}\right),$$

where CG2 is the centroid distance of volume ADC (Figure 7.9) from its flat base [28].

7.2.5.4 Work Done Against Gravity (on the Particle), ΔF_4

The work done against gravity, in dimensionless form, to displace the cylindrical particle vertically from the reference state to a stationary state is

$$\Delta F_4 = -\left(\frac{CG1}{R}\right)(DC - C)(\alpha - \sin\alpha\cos\alpha) + \left(\frac{CG2}{R}\right)DC(\pi - \alpha + \sin\alpha\cos\alpha) - \pi DC,$$

$$(7.68)$$

where CG1 is the centroid distance of volume ABC (Figure 7.9) from its flat side [27], and

$$D = (\rho_1 - \rho_3)/(\rho_2 - \rho_3). \tag{7.69}$$

The angle α must be used in these terms because the only changes in position of the particle giving rise to work against gravity are those with respect to the undeformed free surface.

When the entire particle is below the level of the undeformed free surface, Equation 7.68 does not apply. In a new and approximate form, ΔF_4 is

$$\Delta F_4 = \pi\{(DC - C)\times[\cos\phi - (Z_0/R)] - DC\}, \tag{7.70}$$

which is applicable when $Z_0 < 0$ and $|Z_0| > r(1 + \cos\phi)$.

The total free-energy expression is then given by

$$\Delta F = \Delta F_1 + \Delta F_2 + \Delta F_3 + \Delta F_4, \tag{7.71}$$

where each term has been properly defined. The expression has been nondimensionalized with division by $LR\gamma_{23}$, where L is the length of the cylinder.

Figure 7.10 is an illustration of the free-energy diagrams of several systems, plotted against ϕ with the equilibrium contact angle, θ_e, of each system. The curves in Figure 7.10 begin at successively higher values of ϕ as the parameter θ_e decreases

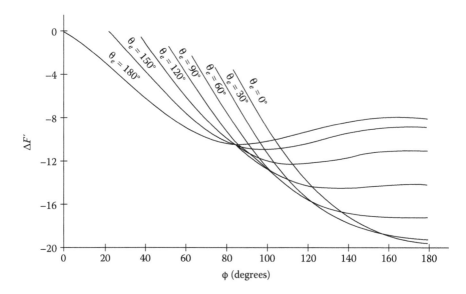

FIGURE 7.10 Free-energy diagrams for a cylindrical particle at fluid interfaces with $C = 2.8$ and $D = 1.25$. The ordinate is the ratio of the free energy of the system and the energy factor $LR\gamma_{23}$. (Reprinted from Rapacchietta, A. V., Neumann, A. W., and Omenyi, S. N., *Journal of Colloid and Interface Science*, 59, 541, 1977. With permission from Academic Press.)

for the following reasons. Recall that the reference state of the system is the one in which the bottom of the particle just touches interface 23 while still immersed in fluid 3. Immediately after contact, a meniscus takes shape until a stationary state is achieved (with the bottom of the particle still level with the undeformed free surface) having $\theta = \theta_e$. The value of ϕ compatible with this stationary state is determined by the surface properties of the system. It is, in general, greater than zero, so that any stationary state with a value of ϕ less than this first value is impossible.

Figure 7.10 indicates that the more complete the wetting, the larger the first permissible value of ϕ. For $\theta_e = 180°$, when there is no wetting, the first value of ϕ is $0°$; for $\theta_e = 0°$, when there is complete wetting, the first value of ϕ is about $80°$. In any case, intermediate values of ϕ between zero and the first realizable value would be possible only if necking were permitted, a state of the system in which a particle elevated off the undeformed free surface level is connected to fluid 2 by means of a raised column of the fluid. These states, however, violate the model of the system based, in part, on the assumption that the particle can have only a downward motion as it seeks equilibrium; such states would imply that the particle first makes contact with interface 23 and then travels upward against the influence of gravity. An occurrence of this sort would require that the free energy of the system increase, violating the condition that the system is free only to achieve states with lower free energies. For these reasons, the stationary states that involve necking are ignored.

For systems with $D > 1$ (that is, when $\rho_1 > \rho_2 > \rho_3$) and with constant material properties, the size, R, of a cylindrical particle at a fluid interface could become so large that equilibrium at the interface would be impossible for any value of ϕ. It is evident that there exists some limiting value of R, above which stability at an interface is impossible. This limiting size is called the critical radius and is directly dependent on the character of the rest of the system, as determined by $(\rho_2 - \rho_3)g/\gamma_{23}$, $(\rho_1 - \rho_2)/(\rho_2 - \rho_3)$, and θ_e. The critical radius is of practical importance in problems of flotation and engulfment, and in separation techniques. The correlation between the critical radius and other parameters of the system will be addressed below.

The free-energy analysis of cylindrical particles at fluid interfaces may be modified to produce the results appearing in Figures 7.11 and 7.12. Since equilibrium states are identified as the minima of ΔF versus ϕ curves, to develop a relation from the free-energy analysis that applies only to equilibrium states, it is necessary only to form the first derivative of the free energy with respect to ϕ and to set the result equal to zero.

Figure 7.11 represents five systems that have different values of D but the same contact angle ($\theta_e = 180°$). The maxima of these curves correspond to equilibrium states in which R is the largest possible value that may be tolerated without the cylindrical particle passing through interface 23. All points on the plots to the left of the maxima are stable equilibrium states (since R is always less than critical size).

Figure 7.11 reveals that the critical size of the cylindrical particle decreases as the value of D increases. This is expected because a denser particle (with larger D) will be under the influence of a proportionally larger sinking force than less dense particles. If the surface properties (which provide support for the particle at the interface) of the system remain constant, it is evident that particles of smaller critical sizes

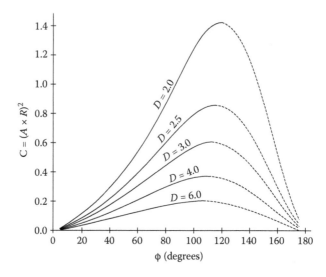

FIGURE 7.11 Equilibrium states of variable-sized cylindrical particles at fluid interfaces with $\theta_e = 180°$. (Reprinted from Rapacchietta, A. V., Neumann, A. W., and Omenyi, S. N., *Journal of Colloid and Interface Science*, 59, 541, 1977. With permission from Academic Press.)

will be supported at the interface as D increases. Figure 7.11 also indicates that the critical position of equilibrium (the value of ϕ when R is the critical size) occurs for smaller values of ϕ as D increases. Thus, as D increases there is a smaller number of stable equilibrium states available for systems with differently sized particles. The limit would be a point particle of infinite density at a fluid interface; such a particle could not achieve a stable equilibrium position at the interface.

The perspective of the analysis may be modified slightly so that systems with the same value of D may be investigated by plotting C versus ϕ with θ_e as the parameter. Figure 7.12 reveals that the critical radius, once again represented by C with constant $(\rho_2 - \rho_3)g/\gamma_{23}$, decreases with decreasing value of the equilibrium contact angle. For complete wetting (i.e., when $\theta_e = 0°$), no supportive surface tension force exists and so any particle in a system with $D > 1$ will sink. Figure 7.12 also shows that the critical equilibrium state occurs later; that is, for larger values of ϕ, as θ_e decreases. This is so because more of the particle needs to be submerged so that the other supportive force, buoyancy, will be proportionally larger than the decreasing surface tension.

A similar analysis was conducted for spherical particles. Assuming that Figure 7.9 represents a spherical particle, the total free energy change of the system was derived similarly to that of cylindrical particles. It is noted that unlike the analysis of cylindrical particles that is performed analytically, solving the resulting equations for spherical particles involves numerical methods [28]. The free energy change, ΔF, was plotted as a function of ϕ for fluid systems of specified properties with different θ_e values. The critical radius of spherical particles, that is, the largest radius that can be supported at the interface, was inferred from the maxima of these curves. Similar to the case of cylindrical particles, the critical radius of spherical particles at a fluid

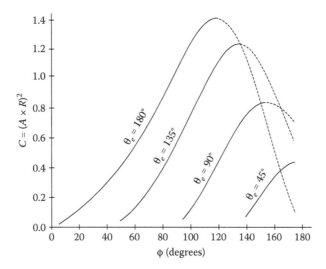

FIGURE 7.12 Equilibrium states of variable-sized cylindrical particles at fluid interfaces with $D = 2.0$. (Reprinted from Rapacchietta, A. V., Neumann, A. W., and Omenyi, S. N., *Journal of Colloid and Interface Science*, 59, 541, 1977. With permission from Academic Press.)

interface decreases as θ_e decreases, and the critical equilibrium state occurs at larger values of ϕ. A key finding was that for very large contact angles (>135°), a fluid system can support a sphere of larger radius than a cylinder. For contact angles less than 90°, the opposite is true. Details are available elsewhere [28].

7.2.6 CONTACT ANGLE PHENOMENA IN THE PRESENCE OF A THIN LIQUID FILM

It should be realized that the Young equation in the form of Equation 7.2 is not a universal formulation in all situations. As an illustration, it will be shown in this section that the equilibrium condition for the three-phase contact line will be quite different from Equation 7.2 in the presence of a thin liquid film.

It is well known that the adsorption of the liquid on the surface of a solid substrate may affect the magnitude of the contact angle. Generally, there are two ways to consider adsorption on a solid surface: one is the classical surface thermodynamics approach established by Gibbs [1], which takes adsorption into account by considering a solid–vapor interface with a surface tension γ_{sv}, rather than a pure solid–vacuum interface with a surface tension γ_s. The difference between γ_s and γ_{sv} is defined as the equilibrium spreading pressure

$$\pi = \gamma_s - \gamma_{sv}. \tag{7.72}$$

A second way to consider the adsorption of the liquid component on solid surfaces is given in the theory of thin liquid films [30–41]. When the adsorbate on the surface of a solid substrate forms a thin film with a thickness ranging from 10 to a few hundred Angstroms, the above interface model may become untenable, and a liquid film can probably no longer be modeled as a single, two-dimensional interface. The thickness of the thin film may be much greater than the thickness of a heterogeneous interfacial region between two uniform bulk phases, which is usually not more than a few molecular diameters. Although such a film is relatively thick, the long-range intermolecular interactions of the two heterogeneous boundary regions of the thin film overlap, causing the adsorbate inside this film to have different properties from those of the bulk phase and those of an interface.

Thin liquid films have been studied extensively; the modern theory of thin liquid films has been developed by Derjaguin, Churaev, and many other researchers [30–41]. An excellent summary can be found in Derjaguin, Churaev, and Muller's monograph *Surface Forces* [40]. To account for the intermolecular forces in a thin liquid film, Derjaguin introduced "disjoining pressure" as a distinctive property of the thin liquid film. In addition to the microscopic approach to thin liquid films, there have also been several thermodynamic descriptions [35,42–51].

For a system consisting of a liquid–vapor meniscus in equilibrium with a thin liquid film on a solid surface, an immediate concern is the effect of the presence of the thin liquid film on the basic thermodynamic relations, such as the Young equation. This section will present a model for a thin liquid film/contact angle system [48] and then focus only on the effects of the thin liquid film on some of the fundamental aspects of contact angle phenomena.

7.2.6.1 Thermodynamic Model

Consider a sessile drop of liquid resting on an ideal solid substrate, in equilibrium with the vapor of the liquid and a thin film, as illustrated in Figure 7.13. The solid surface is assumed to be isotropic, homogeneous, smooth, rigid, and insoluble in the liquid.

Physically, between the flat film and the free liquid–vapor meniscus, there is a transition zone. For the sake of simplicity, such a small transition zone will not be considered in the present model. In other words, only a flat film intersecting with a liquid–vapor meniscus governed by the Laplace equation will be considered, as shown in Figure 7.13. Since the thermodynamic fundamental equations for all the bulk phases and the liquid–vapor and the solid-liquid interface phases are well known from Chapters 1 and 2, attention will be focused only on the thin liquid film.

In this model, the film will be considered to be a simple thermodynamic phase, as are all the other phases in the system. That is, all the phases are assumed to be homogeneous and not subject to any chemical reactions or interactions with external fields. Since the main difference in the physical nature of a thin film and that of an interface separating two bulk phases stems from the overlap of long-range intermolecular interactions of the two interfacial regions of a thin liquid film, the thickness of the film is a characteristic parameter; hence, the thin liquid film will be considered as a three dimensional phase. Within Gibbsian thermodynamics, a simple thermodynamic system can be characterized by three types of fundamental variables: a thermal variable, geometric variables, and chemical variables. Generally, the thermal variable is the entropy of the system; the chemical variables are the mole numbers of the independent chemical components in the system; and the geometric variables

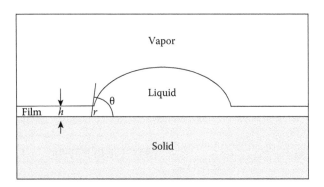

FIGURE 7.13 A sessile drop in equilibrium with a thin liquid film. The transition zone between the flat film and the liquid–vapor interface is neglected. (Reprinted from Li, D. and Neumann, A. W., *Advances in Colloid and Interface Science*, 36, 125, 1991. With permission from Elsevier.)

must be the appropriate mechanical work coordinates: for instance, volume, for a bulk phase, or surface area, for a moderately curved interface phase. As mentioned above, the distinct physical properties of a thin liquid film result from the overlap of the long-range intermolecular interactions of the two boundary regions of the film. The film thickness, thus, is critical to the physical properties and the energy of a thin liquid film phase. The film thickness, h, must be considered as a geometric variable, in addition to the film area, A_f. Therefore, the energy form of the fundamental equation for a thin liquid film can be written as

$$U_f = U_f(S_f, A_f, h, N_{1f}, ..., N_{rf}),$$ (7.73)

where U_f is the internal energy of a simple, uniform, thin liquid film phase, S_f is the entropy of the simple, uniform, thin liquid film phase, A_f is the film area, h is the film thickness, and N_{if} is the mole number of the ith independent chemical component in this thin film.

The differential form of the fundamental equation for such a film is

$$dU_f = \left(\frac{\partial U_f}{\partial S_f}\right)_{A_f,h,N_{if}} dS_f + \left(\frac{\partial U_f}{\partial A_f}\right)_{S_f,h,N_{if}} dA_f + \left(\frac{\partial U_f}{\partial h}\right)_{S_f,A_f,N_{if}}$$

$$dh + \sum_{i=1}^{r}\left(\frac{\partial U_f}{\partial N_{if}}\right)_{S_f,A_f,h,N_{if}\neq if} dN_{if} = TdS_f + \gamma_f dA_f - A_f \Pi dh + \sum_{i=1}^{r}\mu_{if}dN_{if},$$ (7.74)

where the intensive variables are defined by

$$T = \left(\frac{\partial U_f}{\partial S_f}\right)_{A_f,h,N_{if}}$$ (temperature) (7.75a)

$$\gamma_f = \left(\frac{\partial U_f}{\partial A_f}\right)_{S_f,h,N_{if}} \qquad \text{(film tension)} \qquad (7.75b)$$

$$\Pi = -\frac{1}{A_f}\left(\frac{\partial U_f}{\partial h}\right)_{S_f,A_f,N_{if}} \qquad \text{(disjoining pressure)} \qquad (7.75c)$$

$$\mu_{if} = \left(\frac{\partial U_f}{\partial N_{if}}\right)_{S_f,A_f,h,N_{if\neq if}} \qquad \text{(chemical potential).} \qquad (7.75d)$$

From the perspective of the classical Gibbsian interface model, the film tension, γ_f, and the disjoining pressure, Π, are novel quantities. The microscopic origin of the film tension and the disjoining pressure basically includes the contributions of all intermolecular forces—resulting from the thermal motion, hydrogen bonds, Born repulsion forces, van der Waals forces, electrostatic forces, and so forth [30–41]. However, only the macroscopic aspects of these two quantities will be discussed here. In addition, as implied by this simple thermodynamic model, the effects of electrostatic forces are not considered (see Section 7.7). Generally the film tension, γ_f, will depend on the molecular nature of both the liquid and the solid, and will be different from both the liquid–vapor interfacial tension, γ_{lv}, and the solid–liquid interfacial tension, γ_{sl}, since the intermolecular interactions from the two "interfacial" regions (film-vapor and solid-film) are overlapping. The disjoining pressure, Π, is the force required to separate a unit area of the two "interfaces" of the film. In other words, in order to compress the film thickness by dh, an amount of work, $dW = A_f\Pi dh$, must be done on the film. There may be a difference between the definition of the disjoining pressure given by Equation 7.75c and that given by Derjaguin et al. [40]. Further discussion of this matter, however, is beyond the scope of this chapter.

By the Euler theorem, the integral form of the fundamental equation is

$$U_f = TS_f + \gamma_f A_f - A_f\Pi h + \sum_{i=1}^{r}\mu_{if}N_{if}. \qquad (7.76)$$

The corresponding Gibbs-Duhem relation for the thin liquid film is

$$S_f dT + A_f d\gamma_f - A_f h d\Pi + \sum_{i=1}^{r}N_{if}d\mu_{if} = 0, \qquad (7.77)$$

or, dividing the above equation by the film area, A_f

$$s_f dT + d\gamma_f - h d\Pi + \sum_{i=1}^{r}\Gamma_{if}d\mu_{if} = 0. \qquad (7.78)$$

From Equation 7.78, one can readily see that

$$\left(\frac{\partial \gamma_f}{\partial \Pi}\right)_{T,\mu_{if}} = h > 0. \tag{7.79}$$

That is, the film tension always increases as the disjoining pressure increases. The grand canonical free energy of the thin film can be defined as follows:

$$\Omega_f = \Omega_f (T, A_f, h, \mu_{1f}, \dots, \mu_{rf}) = U_f - TS_f - \sum_{i=1}^{r} \mu_{if} N_{if} = (\gamma_f - \Pi h) A_f \tag{7.80}$$

$$d\Omega_f = -S_f dT + \gamma_f dA_f - A_f \Pi dh - \sum_{i=1}^{r} N_{if} d\mu_{if}$$

$$\Pi = -\frac{1}{A_f}\left(\frac{\partial \Omega_f}{\partial h}\right)_{T,A_f,\mu_{if}}.$$

It is noted that strictly speaking, it is always the grand canonical potential that should be discussed in the analysis of surface thermodynamic systems, but it is customary to just speak about "the free energy ΔF" such as in earlier sections of this chapter.

7.2.6.2 Mechanical Equilibrium Conditions

For the system illustrated in Figure 7.13, it can be proven that the thermal and chemical equilibrium conditions are the same as those in the absence of a liquid film; that is, the temperature, T, and the chemical potentials, μ_i ($i = 1, 2, \dots, r$), are constant through all phases in the system. However, the mechanical equilibrium conditions are different and are of more interest here. They can be derived by minimizing the total grand canonical free energy of the system while the intensive parameters, T and μ_i, in the grand canonical free energy, are held constant.

The total grand canonical free energy can be written as

$$\Omega = \Omega_l + \Omega_v + \Omega_s + \Omega_{lv} + \Omega_{sl} + \Omega_f$$

$$= -\int P_l dV_l - \int P_v dV_v - \int P_s dV_s + \int \gamma_{lv} dA_{lv} + \int \gamma_{sl} dA_{sl} + \int \gamma_f dA_f - \int \Pi h dA_f, \tag{7.81}$$

where the subscripts l, v, s, lv, sl, and f denote the liquid phase, the vapor phase, the solid phase, the liquid–vapor interface phase, the solid–liquid interface phase, and the thin film phase, respectively.

The equilibrium condition is that the variation of the total grand canonical free energy is zero; that is,

$$\delta\Omega = 0. \tag{7.82}$$

In the above, the symbol δ refers to a variation of the system in the sense of a virtual work variation; the symbol d refers to an element or differential of a quantity. In this model, the solid substrate is considered as an ideal, rigid solid phase and hence its volume and pressure are constant (i.e., independent of the interaction with a liquid drop). Therefore, combining Equation 7.81 with Equation 7.82 will give

$$\delta\Omega = -\delta\int P_l dV_l - \delta\int P_v dV_v - \delta\int P_s dV_s + \delta\int \gamma_{lv} dA_{lv}$$
$$+ \delta\int \gamma_{sl} dA_{sl} + \delta\int \gamma_f dA_f - \delta\int \Pi h dA_f = 0. \tag{7.83}$$

A virtual variation of the system is illustrated in Figure 7.14. The virtual variation terms in Equation 7.83 can then be obtained as follows.

The variation of the volume of the liquid drop consists of two parts: one due to the variation of the liquid–vapor interface position, the other due to the variation of the position of the three-phase (liquid drop/vapor/thin film) intersection; therefore,

$$\delta\int_{V_l} P_l dV_l = \int_{A_{lv}} P_l \delta N_{lv} dA_{lv} + \int_L P_l h \delta T dL. \tag{7.84}$$

The variation of the volume of the vapor phase also consists of two parts: one due to the variation of the liquid–vapor interface position, the other due to the variations of the film thickness. Therefore,

$$\delta\int_{V_v} P_v dV_v = -\int_{A_{lv}} P_v \delta N_{lv} dA_{lv} - P_v A_f \delta h. \tag{7.85}$$

The variation of the liquid–vapor interface area also has two parts: one due to the variation of the position of the three-phase intersection, the other due to the variation of curvature, or the shape, of the liquid–vapor interface [51]. Hence,

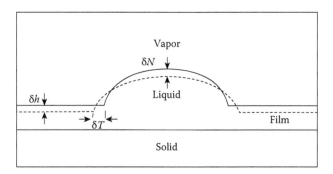

FIGURE 7.14 An illustration of a virtual variation from an equilibrium state. (Reprinted from Li, D. and Neumann, A. W., *Advances in Colloid and Interface Science*, 36, 125, 1991. With permission from Elsevier.)

$$\delta \int_{A_{lv}} \gamma_{lv} dA_{lv} = \int_L \gamma_{lv} \cos\theta \, \delta T dL + \int_{A_{lv}} \gamma_{lv} \left(\frac{1}{R_1} + \frac{1}{R_2} \right) \delta N_{lv} dA_{lv}. \tag{7.86}$$

The variations of the solid–liquid interface area and the film area are due to the variation of the three-phase intersection, but they are in opposite directions, so that

$$\delta \int_{A_{sl}} \gamma_{sl} dA_{sl} = \int_L \gamma_{sl} \delta T dL \tag{7.87a}$$

$$\delta \int_{A_{sl}} \gamma_f dA_f = - \int_L \gamma_f \delta T dL. \tag{7.87b}$$

Realizing that $hdA_f = dV_f$, one can express the variation of the film volume by two terms: one due to the variation of the film thickness, the other due to the variation of the position of the three-phase intersection. Therefore,

$$\delta \int_{A_{SL}} \Pi h dA_f = \Pi A_f \delta h - \int_L \Pi h \delta T dL. \tag{7.88}$$

In Equations 7.84 through 7.88, δN_{lv} is the normal component of motion of the element of the liquid–vapor interface into the vapor, dL is the element length of the three-phase (liquid drop/vapor/thin film) intersection, δT is the virtual motion of this line normal to dL along the solid and in a direction that increases with A_{sl}, θ is the contact angle, and R_1 and R_2 are the principal radii of curvature of dA_{lv}.

Substituting Equations 7.84 through 7.88 into Equation 7.83 and collecting similar terms yields

$$\delta\Omega = \int_{A_{lv}} \left[(P_v - P_l) + \gamma_{lv} \left(\frac{1}{R_1} + \frac{1}{R_2} \right) \right] \delta N_{lv} dA_{lv}$$
$$+ \int_L [\gamma_{lv} \cos\theta + \gamma_{sl} - \gamma_f - P_l h + \Pi h] \, \delta T dL + (P_v A_f - \Pi A_f) \delta h = 0. \tag{7.89}$$

Since the variations are arbitrary, it is necessary and sufficient that each integrand in the first two terms and the coefficient of the last term in Equation 7.89 be zero. This gives the following three equations:

$$P_l - P_v = \gamma_{lv} \left(\frac{1}{R_1} + \frac{1}{R_2} \right) \tag{7.90}$$

$$\Pi = P_v \tag{7.91}$$

$$\gamma_{lv} \cos\theta = (P_l - \Pi)h + \gamma_f - \gamma_{sl}. \tag{7.92}$$

Equation 7.90 is the well-known Laplace equation of capillarity, governing the liquid–vapor meniscus. Equation 7.91 is the mechanical equilibrium condition between the flat film and the bulk vapor phase. It clearly states that the disjoining pressure in the flat thin film must balance the external compressive force exerted by the vapor. It should also be noted that, as indicated by Equation 7.92, the disjoining pressure, Π, is a positive quantity just like the vapor pressure, P_v, for such a simple system (i.e., without considering, say, electrical effects). Equation 7.92 is the mechanical equilibrium condition at the intersection of the solid–liquid interface, liquid–vapor interface, and the thin liquid film. Although Equation 7.92 is similar to the classical Young equation (Equation 7.2), here γ_{sv} is replaced by $(P_l - \Pi)h + \gamma_f$. Equation 7.92 accounts for the effects of a thin liquid film on the equilibrium contact angle in terms of the disjoining pressure, Π, the film thickness, h, and the film tension, γ_f. It should be noted that γ_f, intrinsically, is also a function of h. When $h \to 0$, that is, the thin liquid film vanishes from the solid surface, the first term, $(P_l - \Pi)h$, will drop off Equation 7.92 and the second term, γ_f, can be expected to revert to the solid surface tension, γ_s. That is, Equation 7.92 becomes the Young equation for the case of no adsorption on the solid surface.

It should be noted that there is generally a transition zone between the free liquid–vapor meniscus and the flat thin liquid film that was not considered in the above model. A thermodynamic model has been developed to address this issue and details can be found elsewhere [52,53].

7.3 CONTACT ANGLE HYSTERESIS PHENOMENA: OVERVIEW AND CURRENT VIEW

The central relation used for the interpretation of contact angles is the Young equation. For a given solid–liquid–vapor system, Young's equation implies a unique contact angle. In practice, however, the contact angle of a liquid drop advancing on a solid surface (θ_a) is usually different from that obtained when it is receding (θ_r) and the difference represents the contact angle hysteresis (θ_{hyst}). Because of the existence of hysteresis, the interpretation of contact angles in the context of Young's equation is contentious. Numerous studies have been conducted in the past several decades to understand contact angle hysteresis; nonetheless, the underlying causes and origins of this phenomenon are not completely understood.

Earlier studies of contact angles attributed the discrepancy between theory and practice to imperfections of solid surfaces due to surface roughness and surface heterogeneity. Wenzel was the first to recognize that Young's equation might not be a universal equilibrium condition and the contact angle measured on a rough surface could not be used as the Young contact angle [4]. The same type of reasoning was applied to heterogeneous and porous surfaces by Cassie and Baxter [6–8]. They argued that a heterogeneous solid surface consists of domains of different solid surface tensions. Therefore, the contact angle on a heterogeneous surface should not be the same as the Young contact angle.

In 1946, Derjaguin introduced the concept of metastable states in conjunction with heterogeneous surfaces [54]. Further developments of this important concept by Johnson and Dettre (concentric ring model) [11,12] and Neumann and Good (vertical plate model) [13] have provided a framework for the understanding of contact angle hysteresis. This notion was pursued further by Marmur [55], Joanny and de Gennes [56], and Chen et al. [57]. The presence of metastable states yields more than one mechanically stable contact angle and, therefore, contact angle hysteresis.

In 1948, Shuttleworth and Bailey explained how roughness might produce contact angle hysteresis [58]. Bartell and Shepard examined the effect of roughness on contact angle hysteresis in 1953 [59]. Johnson and Dettre were the first to illustrate the relation of contact angle hysteresis, heterogeneity, and roughness with quantitative experimental results [11,12,60]. In 1985, Schwartz and Garoff examined the effects of a particular geometry of surface heterogeneity on contact angle behavior [61]. They found that contact angle hysteresis is strongly dependent on the patch structure, not just chemical heterogeneity. From their calculations, it was suggested that contact angle hysteresis on heterogeneous surfaces might vanish for patch sizes of the order of micrometers. Nevertheless, a study by McCarthy and coworkers in 1999 suggested that roughness on a molecular scale may also cause hysteresis [62]. Overall, these studies provide ample evidence that roughness and heterogeneity of solid surfaces can indeed cause contact angle hysteresis, as shown earlier in this chapter too.

Progress made in the past several years in the area of characterization of solid surfaces using spectroscopy and microscopy techniques has allowed analysis of surface topography and constituent moieties at the surfaces of solids and tuning them in order to prepare high quality smooth and homogeneous surfaces. It has been extensively reported that even on molecularly smooth surfaces, contact angle hysteresis does not vanish [22,26,63,64]. It is interesting to note that, for example, two solid surfaces that are prepared using two Teflon-type fluoropolymers, both homogeneous and with a mean surface roughness below one nanometer, exhibit considerably different hysteresis when exposed to a given liquid; that is, a few degrees on one surface but several times larger on the other [22]. Such evidence suggests that surface roughness and heterogeneity are not the only contributors to hysteresis and that there must be additional causes. Many studies have been conducted in recent years to resolve the complexities of the hysteresis phenomenon. It is the purpose of this section to report recent advances in understanding of contact angle hysteresis.

7.3.1 Interpretation of Time-Dependent Receding Angles

One of the earliest studies that discussed hysteresis more in the sense of a true hysteresis and not in terms of metastable states is that of Sedev et al. [65,66]. Two key findings of this work are as follows: (i) Contact angle measurements with three n-alkanes (n-octane, n-dodecane, n-hexadecane) on well-prepared FC-722 fluoropolymer surfaces showed that the receding angles decrease monotonically with the time of contact between solid and liquid. As a result, contact angle hysteresis increased with the contact time. (ii) Contact angle hysteresis was found to be related to the chain length of n-alkane molecules such that liquids with shorter molecular chains yielded larger contact angle hysteresis. These preliminary findings suggested that the polymer surface is modified over time due to contact with the liquid, giving rise to contact angle hysteresis. To understand the underlying mechanisms, Lam et al. performed a systematic study of the contact angles of a homologous series of n-alkanes (n-hexane to n-hexadecane) and 1-alcohols (1-ethanol to 1-undecanol) on films of FC-732 fluoropoymer [26,63]. It is noted that FC-732 contains the same film-forming chemical as the FC-722 fluorocarbon, but is provided in perfluorobutyl methylether rather than 1,1,2,-trichloro-1,2,2,-tri-fluoroethane (FC-722). The solid surfaces were prepared by a dip-coating technique (cf. Chapter 6). Atomic force microscopy (AFM) analysis showed that the rms mean roughness of the FC-732 surfaces is about 0.4 nm and that the films are quite homogeneous. The experiments were performed at low rates of advancing of the three-phase line using a motor-driven syringe mechanism, as described in Chapter 6. To ensure reproducibility, the measurements were repeated at least five times for each liquid. The contact angles of n-alkanes are plotted on the same time scale and are shown in Figure 7.15.

For the sake of discussion, the graphs were divided into three domains. The first domain (I) ranges from the beginning of the experiment to time t_0 when the motor was switched to the reverse mode and the liquid started to flow back into the syringe. Up to t_0, advancing contact angles are measured. Since the contact angles were constant upon advancing of the three-phase line, they were averaged to yield a mean advancing contact angle ($\bar{\theta}_a$) for each n-alkane, as given in Figure 7.15. The second domain (II) ranges from time t_0 to time $t_r = 0$, starting upon reversal of the motor. This second domain is characterized by a rapid decrease in the contact angles from advancing to receding with the three-phase line remaining stationary. This pattern continues until $t_r = 0$ at which the periphery starts to recede—the beginning of the third domain. The third domain (III) ranges from time $t_r = 0$ to the end of the experiment. Receding contact angles are measured during this third period. Unlike the advancing contact angles, the receding angles of all n-alkanes change with time. These results are consistent with the earlier findings of Sedev et al. [65,66]. From the contact angle graphs of Figure 7.15, it is conceivable that there are two contributing mechanisms to hysteresis: (i) A fast process that takes place in the second domain and brings about a fast decrease in the contact angles with time. The short time frame of this domain implies that the operative mechanism involves a surface effect. Retention of liquid molecules on the solid surface (surface retention) is the most likely explanation. (ii) A slow process that is operative in the third domain and causes the receding angles to decrease slowly but continuously with time. This process must involve a bulk effect. Penetration of liquid into the bulk of the polymer film is the most likely cause. The following detailed consideration of the receding angles supports this view.

Since the receding contact angles decrease with time, mean receding angles cannot be obtained. Unfortunately, it is conceptually impossible to measure a receding contact angle on a dry solid surface. Therefore, it was proposed to extrapolate the continuously decreasing receding angles of n-alkanes back to time t_0, when the motor was reversed [63]. The result is shown in Figure 7.15. One might expect that the extrapolation of the receding contact angles back to zero contact time t_0 should lead to the advancing contact angle. The fact that this is not so indicates that there is indeed more than one mechanism of solid–liquid interaction, and a very fast one would obviously not be detected by extrapolation of a slower one back to time zero (t_0). It is noted that contact angle experiments

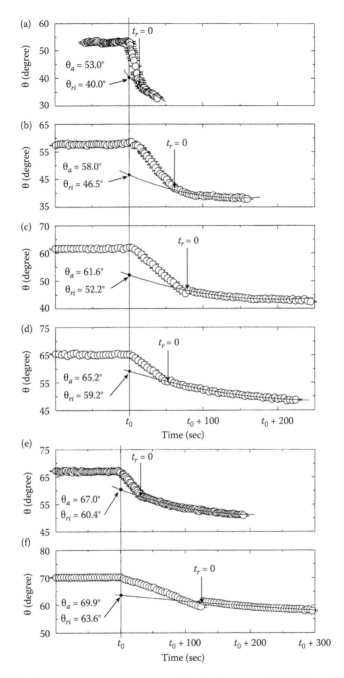

FIGURE 7.15 Dynamic contact angle plots of the 11 alkanes on FC-732 films: (a) *n*-hexane, (b) *n*-heptane, (c) *n*-octane, (d) *n*-nonane, (e) *n*-decane, (f) *n*-undecane, (g) *n*-dodecane, (h) *n*-tridecane, (i) *n*-tetradecane, (j) *n*-pentadecane, (k) *n*-hexadecane. (Reprinted from Lam, C. N. C., Kim, N., Hui, D., Kwok, D. Y., Hair. M. L., and Neumann, A. W., *Colloids and Surfaces A*, 189, 265, 2001. With permission from Elsevier.)

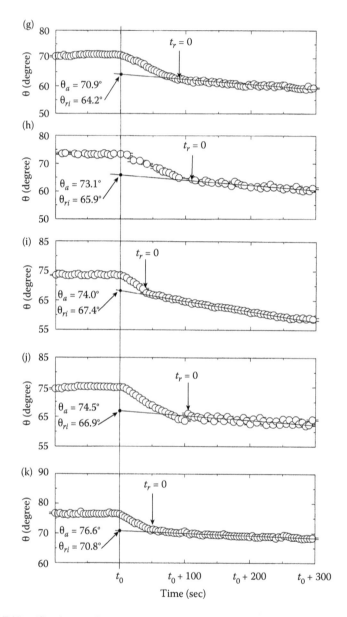

FIGURE 7.15 (Continued) Dynamic contact angle plots of the 11 alkanes on FC-732 films: (a) *n*-hexane, (b) *n*-heptane, (c) *n*-octane, (d) *n*-nonane, (e) *n*-decane, (f) *n*-unde-cane, (g) *n*-dodecane, (h) *n*-tridecane, (i) *n*-tetradecane, (j) *n*-pentadecane, (k) *n*-hexa-decane. (Reprinted from Lam, C. N. C., Kim, N., Hui, D., Kwok, D. Y., Hair. M. L., and Neumann, A. W., *Colloids and Surfaces A*, 189, 265, 2001. With permission from Elsevier.)

with 1-alcohols on FC-732 surfaces yielded results similar to those of *n*-alkanes and are not presented here [63].

To demonstrate the influence of these two processes on contact angle hysteresis, contact angles of *n*-alkanes were studied on the films of four other fluoropolymers with known properties as follows:

i. Poly(4,5-difluoro-2,2-bis(trifluoromethyl)-1,3-dioxole-*co*-tetrafluoro-ethylene), 65 mole% dioxole (Teflon AF 1600). This polymer has a glass transition temperature (T_g) of 160°C [67].

ii. Poly(2,2,3,3,4,4,4-heptafluorobutyl methacrylate) (EGC-1700) is a fluorinated acrylate polymer [68] with $T_g = 30$°C.

iii. Poly(ethene-*alt*-N-(4-(perfluoroheptylcarbonyl)aminobutyl)maleimide) (ETMF) is a maleimide copolymer possessing a long fluorinated side chain [69], with $T_g = 100$°C.

iv. Poly(octadecene-*alt*-N-(4-(perfluoroheptylcarbonyl)aminobutyl)maleimide) (ODMF) is a maleimide copolymer similar to ETMF but possesses an additional *n*-hexadecyl side chain [69]. The T_g of ODMF is 65°C.

The repeat unit of each polymer is shown in Figure 7.16. From the study of advancing contact angles with these polymers it was known that they behave very differently

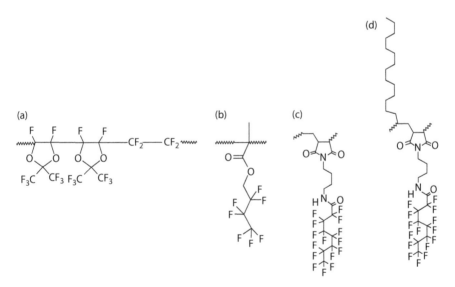

FIGURE 7.16 Repeat unit of: (a) Poly(4,5-difluoro-2,2-bis(trifluoromethyl)-1,3-dioxole-*co*-tetrafluoroethylene), 65 mole% dioxole (Teflon AF 1600), (b) Poly(2,2,3,3,4,4,4-heptafluorobutyl methacrylate) (EGC-1700), (c) Poly(ethene-*alt*-N-(4-(perfluoroheptylcarbonyl)aminobutyl) maleimide) (ETMF), (d) Poly(octadecene-*alt*-N-(4-(perfluoroheptylcarbonyl)aminobutyl) maleimide) (ODMF). (Reprinted from Tavana, H., Jehnichen, D., Grundke, K., Hair, M. L., and Neumann, A. W., *Advances in Colloid and Interface Science*, 134–135, 236, 2007. With permission from Elsevier.)

when exposed to one and the same probe liquid [17,70], due to the differences in the structural properties of these fluoropolymers (see Chapter 8).

Advancing contact angles of n-alkanes on the surfaces of Teflon AF 1600, EGC-1700, ETMF, and ODMF were constant over time and an average value was obtained for each liquid on a given solid surface (see Table 7.1). But similar to the findings of previous studies [26,63–66], the receding angles of n-alkanes depended on the contact time between solid and liquid; that is, they decreased continuously with the time of contact. As an example, the contact angles of n-undecane on the films of four polymers are illustrated in Figure 7.17. The time-dependence of the receding angles varies from polymer to polymer. To quantify this effect, the receding angles were extrapolated to time t_0 and the slope of the regression line of the receding angles on each polymer was calculated ($\Delta\theta_r/\Delta t$). As seen, $\Delta\theta_r/\Delta t$ is significantly different from one fluoropolymer to another, translating into various degrees of contact angle hysteresis (see Table 7.1; note: the contact angle hysteresis for each solid–liquid system was determined as the difference between the advancing and the extrapolated receding angles). Overall, the contact angle hysteresis of n-alkanes on Teflon AF 1600 and EGC-1700 films is significantly larger than that on ODMF and particularly ETMF surfaces. Considering that all solid surfaces are well-prepared and smooth, the difference in the receding angle patterns and the hysteresis results must be caused by different structural properties of the polymers, as discussed below.

Teflon AF 1600 is an amorphous polymer and comprises randomly distributed molecular chains that do not pack into ordered crystals [67,71,72]. To demonstrate this point, the x-ray diffraction pattern of a Teflon AF 1600 film was examined using wide angle x-ray scattering (WAXS) that is often used to determine polymer crystallinity and arrangement of polymer chains through Bragg's law:

$$n\lambda = 2d\sin\theta, \tag{7.93}$$

where n is a positive integer, λ is the x-ray wavelength, d represents net plane distances of the crystalline structure, and θ is the angle of the incident x-ray beam with respect to the vertical. The WAXS pattern of the Teflon AF 1600 showed only a broad maximum at $2\theta = 10°$ rather than a sharp intensity peak, suggesting lack of any crystalline structure. Teflon AF 1600 is optically transparent [i.e., contains no crystallites to scatter light and, hence, the refractive index is very low (1.31)] [73,74]; and it contains large fractional free volumes [67,75] and is highly permeable to gases and vapors [72,76–78], facts that also support the lack of crystallinity. EGC-1700 is also an amorphous fluoropolymer and is expected to present a WAXS pattern somewhat similar to that of Teflon AF 1600; that is, without any sharp peak.

Unlike Teflon and EGC-1700, the H-bonding interactions within the perfluorinated amide groups of the ETMF polymer support formation of a layered structure in the bulk and in the top layer of the surface [69]. The WAXS pattern of an annealed ETMF film displayed a sharp peak at the low angle $2\theta = 2.62°$ and less intense peaks at higher angles. The presence of this sharp peak indicates that the polymer chains are well-ordered and form a layered structure with distances of ~3.34 nm between

TABLE 7.1
Advancing Contact Angle (θ_a), Extrapolated Receding Contact Angles (θ^0_{ri}), and Contact Angle Hystersis (θ_{hyst}) of n-Alkanes on Teflon AF 1600, EGC-1700, ETMF, and ODMF Surfaces

Liquid	Teflon AF 1600			EGC-1700			ETMF			ODMF		
	θ_a	θ^0_{ri}	θ_{hyst}	θ_a	θ^0_{ri}	θ_{hyst}	θ_a	θ^0_{ri}	θ_{hyst}	θ_a	θ^0_{ri}	θ_{hyst}
n-Heptane	47.17	39.47	7.7	48.70	35.10	11.6	60.30	58.20	2.1	—	—	—
n-Octane	52.45	45.80	6.7	53.37	46.15	7.6	64.01	61.71	2.3	—	—	—
n-Nonane	56.22	49.88	6.3	57.03	50.00	7.0	66.63	64.43	2.2	—	—	—
n-Decane	59.29	53.58	5.7	59.84	52.60	7.2	68.82	66.72	2.1	67.14	63.14	4.0
n-Undecane	61.60	56.00	5.6	62.10	55.05	7.1	70.67	68.27	2.4	68.67	64.17	4.5
n-Dodecane	63.78	58.43	5.4	63.99	58.00	6.0	72.15	69.75	2.4	70.02	65.15	4.6
n-Tridecane	65.68	59.71	6.0	65.63	59.13	6.5	73.36	71.46	1.9	71.13	66.93	4.2
n-Tetradecane	67.15	61.56	5.6	67.04	60.75	6.3	74.44	72.21	2.2	72.72	68.31	4.4
n-Pentadecane	68.06	62.47	5.6	68.33	62.22	6.1	75.70	73.70	2.0	74.25	69.95	4.3
n-Hexadecane	69.48	64.19	5.3	69.33	63.50	5.8	76.87	73.67	3.2	74.72	70.02	4.7

Note: All data are in degrees. The three-phase contact line of the first three n-alkanes showed stick-slip on the films of ODMF polymer. Hence useful contact angles could not be obtained.

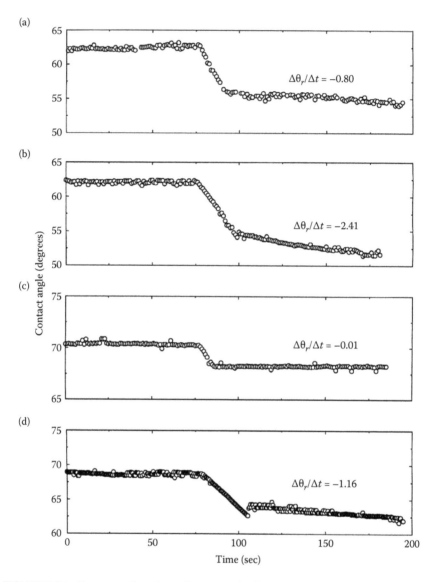

FIGURE 7.17 Contact angles of *n*-undecane on the films of four fluoropolymers: (a) Teflon AF 1600, (b) EGC-1700, (c) ETMF, (d) ODMF. Rates of change in deg/min.

the main chains. This organization of ETMF chains was also confirmed by molecular modeling [69].

ODMF has a main chain structure similar to that of ETMF with a 4-*N*-(perfluoroheptylcarbonyl)aminobutyl side chain. In addition, ODMF possesses an *n*-hexadecyl side chain. According to the angle-dependent XPS investigations of the annealed ODMF surfaces, the presence of *n*-hexadecyl side chains suppresses good self-organization of the perfluoroalkyl side chains on the outermost layer of the surface [69]. This specific structure only supports a weak self-organization of the perfluoroalkyl side chains on the top layer of ODMF films. Additional x-ray diffraction analysis at low angles showed only a very broad signal from $2\theta = 1.2°$ to $2\theta = 2.52°$ (d values of 7.4–3.5 nm, respectively). These results suggest a remarkable difference between the structural properties of ODMF and ETMF films.

Considering these differences in the chain configuration of the polymers, the receding contact angle patterns and the contact angle hysteresis of n-alkanes are plausible in terms of surface retention (fast process) and liquid penetration (slow process), as discussed below.

The primary cause of contact angle hysteresis is operative in region II. As the motor is reversed to decrease drop volume, the three-phase line remains hinged and the contact angles decrease rapidly over a relatively short time period. Examining the contact angle graphs of n-undecane in Figure 7.17 shows that the smallest decrease in the contact angles in region II takes place for ETMF with ~3° over 8 seconds whereas the largest decrease happens for EGC-1700 with ~9° over 25 seconds (both experiments were performed at a similar volume flow rate of the liquid). Taking into account that the rms roughness of the polymer films are 0.4 nm for ETMF and 1.4 nm for EGC-1700, it is conceivable that the larger decrease of contact angles on the latter polymer is due to more extensive surface retention of the liquid by the solid surface.

As discussed above, liquid penetration is the most likely event in region III and the slope of the regression line for the receding angles ($\Delta\theta_r/\Delta t$) represents the extent of this process. The larger slope of n-alkanes on Teflon AF 1600 and EGC-1700 surfaces compared to that on ETMF films is immediately plausible in terms of liquid penetration into the solid surface. The amorphous structure of these two polymers facilitates this process, whereas the layered molecular chains of ETMF significantly reduce liquid penetration. With the ODMF polymer, $\Delta\theta_r/\Delta t$ of n-alkanes is intermediate between those for the two amorphous polymers. This may seem counterintuitive considering that ODMF has a superior chain organization and is therefore expected to allow less extensive liquid penetration. However, AFM analysis of ODMF surfaces showed that the surface film comprises microscale hydrocarbon islands distributed over a fluorocarbon matrix. Existence of hydrocarbon patches facilitates adsorption of hydrocarbon molecules from the liquid phase. Most likely, it is the cooperative action of adsorption and liquid penetration that increases hysteresis on ODMF films beyond that expected for a homogeneous fluorocarbon surface of similar chain configuration.

Comparing the receding contact angle patterns of n-alkanes on the films of the two amorphous polymers, EGC-1700 and Teflon AF 1600, also sheds light on this discussion. The rate of decrease in the receding angles is larger on the surfaces of the former polymer, especially for those n-alkanes with shorter molecular chains [22]. For example, $\Delta\theta_r/\Delta t$ for n-heptane, n-undecane, and n-hexadecane on EGC-1700 films are −6.06, −2.41, and −1.63, respectively, whereas the corresponding values on Teflon AF 1600 surfaces are −1.14, −0.80, and −0.67 (all in degrees/minute). The data can be related to the difference in structural properties of the polymers. EGC-1700 contains mobile and flexible chains that can assume different configurations, whereas Teflon consists of stiff molecular chains. This is supported by the very different glass transition temperature of the two polymers ($T_{g,\text{Teflon AF 1600}} = 160°C$

versus $T_{g,\ \text{EGC-1700}} = 30°C$). The mobility and flexibility of EGC-1700 chains allows liquid molecules to be more easily accommodated. Consequently, liquid penetration will be more extensive and the polymer film is modified to a greater extent by the hydrocarbon molecules. This is manifested by the larger rate for decrease in the receding angles.

Overall, the results in Table 7.1 indicate that contact angle hysteresis correlates well with the configuration of the polymer chains. For a given liquid, the smallest hysteresis is measured with well-packed ETMF surfaces. The contact angle hysteresis on ODMF surfaces, whose chains have a poorer packing, is almost twice as large as that on ETMF surfaces. The largest hysteresis is obtained with the two amorphous polymers, Teflon AF 1600 and EGC-1700.

7.3.2 DYNAMIC CYCLING CONTACT ANGLE (DCCA)

The influence of liquid penetration into the solid surface as the cause of the slower contact angle hysteresis was further studied by dynamic cycling contact angle (DCCA) measurements with n-alkanes (n-nonane, n-dodecane, n-hexadecane) and 1-alcohols (1-hexanol, 1-nonanol, 1-undecanol) on the films of fluoropolymer FC-732 [64]. Each experiment consisted of at least 12 cycles, all on the same solid surface. Figure 7.18 illustrates a typical result obtained from a cycling experiment

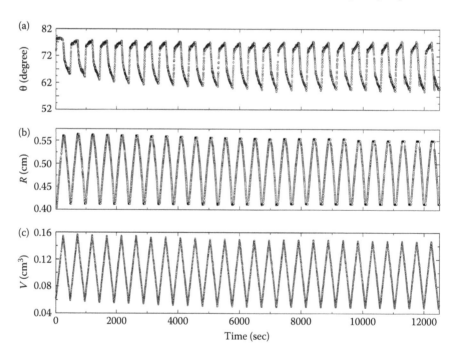

FIGURE 7.18 Dynamic cycling contact angles with n-hexadecane on an FC-732 surface: (a) contact angle, (b) contact radius, (c) drop volume. (Reprinted from Lam, C. N. C., Ko, R. H. Y., Yu, L. M. Y., Li, D., Hair, M. L., and Neumann, A. W., *Journal of Colloid and Interface Science*, 243, 208, 2001. With permission from Academic Press.)

with *n*-hexadecane. In this experiment, 26 cycles of advancing and receding contact angle measurements were performed consecutively in approximately three and a half hours. A complete cycle consists of expansion of the liquid drop from the initial volume/radius to the final (maximum) volume/radius and contraction back to the initial point, which can be identified from both the radius and the volume plots in Figure 7.18b and c. As expected, the receding contact angles are time-dependent in all cycles and the minimum values decrease with increasing number of cycles. Furthermore, beyond the first cycle, the advancing angles also depend on the solid–liquid contact time and increase as the drop radius increases. The observation is quite plausible considering the contact time between the liquid and the solid surface: At the end of the first cycle, wetted circular domains of the solid surface nearest the point where liquid retraction starts will have had the shortest contact time with the liquid and, therefore, are least modified by the liquid, yielding the maximum value in each cycle. On the other hand, domains closer to the center of the drop will have had longer contact times with the liquid and therefore are modified most, giving the minimum value. It also appears that both maximum and minimum values decrease over time. It is suggested that the contact angle at any given drop radius should follow a similar trend. This is illustrated in Figure 7.19 where both the advancing contact angle θ_a and receding contact angle θ_r at two different radii; that is, $R = 0.45$ and $R = 0.50$ cm, are plotted as a function of the number of cycles (the mean θ_a value

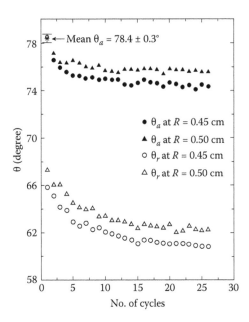

FIGURE 7.19 The advancing and receding contact angles of *n*-hexadecane on a FC-732 surface obtained for the two radii ($R = 0.45$ and 0.50 cm) versus the number of cycles; mean θ_a denotes the average value of advancing contact angles measured from the first cycle. (Reprinted from Lam, C. N. C., Ko, R. H. Y., Yu, L. M. Y., Li, D., Hair, M. L., and Neumann, A. W., *Journal of Colloid and Interface Science*, 243, 208, 2001. With permission from Academic Press.)

for the first cycle is also shown). Evidently, both θ_a and θ_r values at the two radii decrease with increasing number of cycles and, therefore, the solid-liquid contact time. Furthermore, it can be seen from Figure 7.19 that values of θ_a and θ_r obtained at the outer radius ($R = 0.50$ cm) are larger than those obtained at the inner radius ($R = 0.45$ cm). This is expected since the time of solid–liquid contact is longer at the inner radius than at the outer radius. These results indicate that the solid surface is modified due to contact with the liquid in a relatively slow process, presumably liquid penetration.

Direct evidence was found for penetration of liquid into polymer surfaces by Hennig et al. [79]. Dynamic cycling contact angle experiments with water on polyimide films yielded results similar to those in Figures 7.18 and 7.19 in terms of time-dependence of the contact angles, change in the minimum and maximum values of contact angles with time, and different contact angle values at different radii. To quantify the effect of liquid on the polymer film, a variable angle spectroscopic ellipsometry (VASE) technique was employed. The initial thickness of the dry polyimide film was first determined as 4952 nm. Then a sessile drop of water was deposited on the sample and left for 60 minutes. The volume and the contact area of the drop were 0.3 cm^3 and 0.1 cm^2, respectively. The drop was then removed from the surface using a motorized syringe. Immediately afterward, the ellipsometric data at three angles of incidence (65°, 70°, 75°) were collected from the dewetted surface near the central point of the drop contact area. Using an optical model; that is, the Bruggeman effective medium approximation (EMA), the ellipsometric data were translated into optical constants and layer thickness. Comparing the data from dry and dewetted polyimide surfaces, it was concluded that modification of the surface due to contact between the polymer surface and water can be very well described by a thickness increase of 10 nm of an EMA layer consisting of 98.6% polyimide and 1.4% water. This study verified the conclusions that the time-dependent receding angles are caused by a slow process described above as liquid penetration into the polymer film.

7.3.3 CONTACT ANGLE HYSTERESIS FOR LIQUIDS WITH BULKY MOLECULES

The above discussion highlighted the fact that the molecular properties of probe liquids play a role as important as those of solid surfaces in contact angle hysteresis phenomena. To further elucidate this issue, a number of liquids consisting of nonlinear molecules were considered. They are labeled liquids with "bulky" molecules and will be described in detail in the subsequent chapter, Chapter 8. The molecules of these liquids are round and nonflexible with a diameter of 0.6–1.0 nm and are therefore quite distinct from n-alkanes [70,80]. The liquids are: octamethylcyclotetrasiloxane (OMCTS), decamethylcyclopentasiloxane (DMCPS), p-xylene, o-xylene, cis-decalin, trans,trans,cis-1,5,9-cyclododecatriene, methyl salicylate, lepidine (4-methylquinoline), 1-fluoronaphthalene, 1-chloronaphthalene, 1-bromonaphthalene, and 1-iodonaphthalene.

Contact angle measurements with liquids consisting of bulky molecules on the films of the above four fluoropolymers of Figure 7.16 resulted in different receding angle patterns: (i) The receding angles of all these liquids on Teflon AF 1600 surfaces were independent of solid–liquid contact time [71,81]. (ii) On EGC-1700 and

ETMF surfaces, the receding angles were constant only for the siloxane-based liquids OMCTS and DMCPS but were time-dependent for other liquids [71]. (iii) With ODMF films, only OMCTS and DMCPS yielded useful contact angles. The other liquids either dissolved the polymer film or showed a stick-slip behavior [82]. As an example, the contact angles of DMCPS and *cis*-decalin are plotted in Figures 7.20 and 7.21, respectively. For all four polymers, the receding angles of DMCPS are almost constant. The receding angles of *cis*-decalin, however, are constant only on

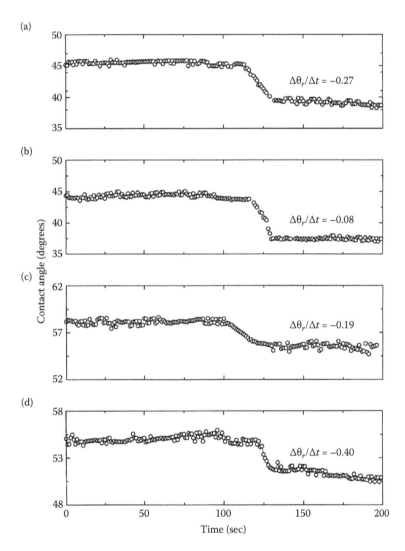

FIGURE 7.20 Advancing and receding contact angles of DMCPS on: (a) Teflon AF 1600, (b) EGC-1700, (c) ETMF, and (d) ODMF surfaces. The receding angles are fairly constant and the contact angle hysteresis is small. Rates of change in deg/min. (Reprinted from Tavana, H., Jehnichen, D., Grundke, K., Hair, M. L., and Neumann, A. W., *Advances in Colloid and Interface Science*, 134–135, 236, 2007. With permission from Elsevier.)

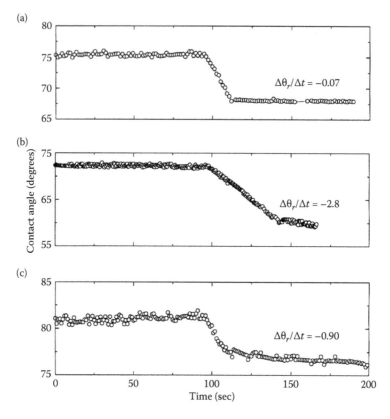

FIGURE 7.21 Advancing and receding contact angles of *cis*-decalin on: (a) Teflon AF 1600, (b) EGC-1700, and (c) ETMF surfaces. The receding angles are constant only on the Teflon AF 1600 surface. Rates of change in deg/min. (Reprinted from Tavana, H., Jehnichen, D., Grundke, K., Hair, M. L., and Neumann, A. W., *Advances in Colloid and Interface Science*, 134–135, 236, 2007. With permission from Elsevier.)

Teflon AF 1600 surfaces. They decrease with time on the EGC-1700 and ETMF films. The slope of the regression line ($\Delta\theta_r/\Delta t$) in these Figures shows the extent of the time-dependence of the receding angles.

Table 7.2 lists the advancing and receding angles of the liquids with bulky molecules on the films of all four polymers. For the systems showing time-dependence, the receding angles were obtained by the extrapolation technique, as described above. The contact angle hysteresis corresponding to each solid–liquid system is also given. Likely explanations for the receding angles and contact angle hysteresis on each polymer surface are given below.

(a) Teflon AF 1600: The receding angles of liquids with bulky molecules are independent of solid–liquid contact time. This suggests that the Teflon films are not modified by liquid penetration. As discussed above, the penetration process involves a continuous migration of liquid molecules into the solid matrix that would cause the receding angles to decrease over time. Most probably, the bulky molecules of

TABLE 7.2

Advancing (θ_a) and Receding (θ_r or θ^0_{ri}) Contact Angles and Contact Angle Hysteresis (θ_{hyst}) of the Liquids with Bulky Molecules Measured on the Films of Four Different Fluoropolymers

Liquid	Teflon AF 1600			EGC-1700			ETMF			ODMF		
	θ_a	θ_r	θ_{hyst}	θ_a	θ^0_{ri}	θ_{hyst}	θ_a	θ^0_{ri}	θ_{hyst}	θ_a	θ^0_{ri}	θ_{hyst}
OMCTS	43.68	38.40	5.2	42.33	36.83	5.5	57.70	54.70	3.0	53.60	50.20	3.4
DMCPS	45.65	39.45	6.2	44.47	37.47	7.0	58.16	54.66	3.5	55.19	51.29	3.9
p-Xylene	68.65	62.31	6.3	64.78	—	—	75.60	71.50	4.1	—	—	—
o-Xylene	71.17	64.74	6.4	66.35	—	—	77.50	74.00	3.5	—	—	—
cis-Decalin	75.53	67.90	7.6	72.75	63.50	9.3	80.87	77.20	3.7	—	—	—
trans,trans,cis-1,5,9 -Cyclododecatriene	78.29	70.79	7.5	73.70	65.00	8.7	82.73	78.52	4.2	—	—	—
Methyl salicylate	83.69	76.52	7.2	—	—	—	86.38	76.80	9.4	—	—	—
Lepidine	89.49	83.10	6.4	—	—	—	91.36	81.90	9.5	—	—	—
1-Bromonaphthalene	89.80	82.60	7.2	84.04	68.05	16.0	92.05	83.00	9.1	—	—	—

Note: Except for Teflon AF 1600, receding angles on other polymers were time dependent and hence extrapolated receding angles (θ^0_{ri}) are given. All data are in degrees.

the probe liquids eliminate this process. Thus, the main contributor to contact angle hysteresis in these systems is surface retention. The fact that contact angle hysteresis on Teflon AF 1600 films is almost the same for all liquids (i.e., ~6°–7°), supports this conclusion because the molecules of the liquids are similar in size and are expected to be retained on the Teflon films to a similar extent.

(b) EGC-1700: The receding angles of OMCTS and DMCPS on EGC-1700 surfaces are independent of time (cf. Figure 7.20b). Thus, surface retention is the likely cause for the contact angle hysteresis of these two liquids. On the other hand for other probe liquids, the receding angles are time-dependent and the contact angle hysteresis differs significantly from one liquid to the next (9°–16°). The molecules of this series of liquids are similar in terms of shape and size and surface retention is expected to be fairly similar with these liquids. Bearing this in mind, if liquid penetration was the only process causing time-dependent receding angles, one would expect similar hysteresis for all liquids due to their similar shape and size. The fact that this is not so indicates that other form(s) of solid–liquid interaction must be operative.

EGC-1700 contains mobile and flexible molecular chains. When it is exposed to a liquid whose molecules contain double bonds or exposable electronegative moieties (i.e., all liquids except OMCTS and DMCPS), EGC-1700 chains undergo major reorientation, exposing groups other than CF_2 and CF_3 at the surface (i.e., methyl or ester groups). The time-dependence of the receding angles suggests that the extent of this process increases with the contact time between the solid and the liquid, causing the solid–liquid interfacial tension (γ_{sl}) to decrease over time (see Chapter 8 for details of this process). Support for this explanation comes from the following simple experiment: Contact angles of *cis*-decalin were measured on a fresh EGC-1700 film and the average advancing angle was 72.75° ± 0.12. The sample was then blown dry with nitrogen and kept under vacuum for 24 hours. The contact angle measurement was repeated. The advancing contact angle of *cis*-decalin was now only 66.30° ± 0.09 (i.e., 6.45° lower). This simple test indicates that the conformation of the polymer changes due to contact with the test liquid and the contribution of the CF_2 and CF_3 groups at the outermost layer of the polymer surface decreases. It appears that the perturbations in the EGC-1700 chains are not reversible and the moieties that moved away from the top surface layer cannot easily relax back to the surface. A similar test with Teflon AF 1600 films showed that the contact angles were accurately reproducible, after thorough drying. This provides a clear indication of the fact that, depending on the structural properties of polymer films, different mechanisms contribute to the hysteresis phenomenon. Evidence also exists in the literature regarding change in the organization of chains of a polymer surface upon contact with a liquid [83–89]. These changes can take place through short-range motions of chains such as rotation around the chain axis or even long-range motions such as diffusion of specific moieties into the bulk. Therefore, reorientation of polymer chains due to contact with liquid molecules is the most likely cause of time-dependent receding angles on EGC-1700 surfaces.

(c) ETMF: The smallest hysteresis for a given liquid is obtained on ETMF surfaces. Even those liquids that show stick-slip on the EGC-1700 and ODMF films due to strong solid–liquid interactions yield a smooth motion on ETMF surfaces. These results imply that the long fluorinated side chains in ETMF provide a protective shield around the hydrophilic backbone. This is supported by the presence of the H-bonding interactions within the perfluorinated amide groups in ETMF that give rise to the formation of a layered structure both in the bulk and in the top layer of the surface film. The ordered molecular chains of ETMF as well as the bulky molecules of the probe liquids eliminate liquid penetration as the cause (*Note:* Even with *n*-alkanes that possess smaller and chain-like molecules, liquid penetration was insignificant; see Section 7.3.1). Therefore, time-dependence of the receding angles must have a different cause. The particular conformation of ETMF surface films makes major reorientation/perturbation of chains an unlikely event. Only minor changes in the arrangement of side chains are probable. Thus, surface retention and mobility of side chains are suggested as the main contributors to the contact angle hysteresis on ETMF surfaces.

(d) ODMF: On ODMF surfaces, only the two siloxane liquids (OMCTS and DMCPS) yielded useful contact angles. The more or less constant receding angles of these two liquids again suggest surface retention as the most likely cause of the hysteresis. Penetration of liquid molecules into the polymer film is less likely in these systems because it would be expected to cause the receding angles to decrease with time.

In conclusion, the contact angle results for liquids with bulky molecules on four different fluoropolymers strongly suggest the following causes for contact angle hysteresis: Surface retention, liquid penetration into the solid surface, major reorientation of polymer chains due to contact with certain probe liquids, and minor changes in the arrangement of the side chains of a polymer surface upon contact with a test liquid. The operative mechanism is determined by the morphological properties of the polymer film, configuration of the chains at the surface film, and chemistry of the solid surface and the test liquid. Furthermore, OMCTS and DMCPS show the smallest contact angle hysteresis on all four fluoropolymers, suggesting that they are the most inert liquids with respect to these, and most likely, other fluoropolymers.

7.3.4 SIZE OF *N*-ALKANE MOLECULES AND CONTACT ANGLE HYSTERESIS

An interesting trend can be observed in the contact angle hysteresis of the *n*-alkanes on the Teflon AF 1600 and EGC-1700 surfaces (Table 7.1). As the size of the hydrocarbon molecules increases from *n*-heptane to *n*-hexadecane, the contact angle hysteresis decreases by ~2.2° for Teflon AF 1600 and ~5.8° for EGC-1700 [22]. Again, the results are plausible in terms of surface retention and liquid penetration. Larger molecules do not fit into the morphological patterns of the polymer surface as easily as shorter molecular chains, reducing surface retention. Furthermore, longer chain *n*-alkanes penetrate the polymer films less readily than those with shorter molecular chains. Thus, both the fast and the slow mechanisms of contact angle hysteresis diminish significantly. It is noted that Zisman and coworkers reported a similar finding in the 1960s. They measured contact angles of various liquids on condensed monolayers of 17-(perfluoroheptyl)heptadecanoic acid adsorbed on polished chromium [90]. Hysteresis was shown to be related to the molecular volume of the liquid and to result from the penetration of liquid molecules into the porous monolayer. Contact angle hysteresis was negligible when the average diameter of liquid molecules was larger than the average cross-sectional diameter of the intermolecular pores. It was concluded that liquid penetration, even into pores of molecular dimensions, is a cause of significant hysteresis. Based on this investigation, they even proposed to estimate the intermolecular pore dimensions of such adsorbed monolayers from contact angle hysteresis data.

In order to test the expectation that contact angle hysteresis will become negligible for very long *n*-alkane chains, Lam et al. plotted the advancing contact angles and extrapolated initial receding angles versus the inverse number of carbon atoms for *n*-alkanes/FC-732 systems, as shown in Figure 7.22 [26]. The advancing contact angles and the extrapolated receding angles were approximated by two straight lines. When subjected to linear regression, it was found that the two straight lines merge near the zero inverse number of carbon atoms; that is, at infinite chain length. The contact angle hysteresis of an *n*-alkane, which would have an infinite number of carbon atoms, would be 0.7°; that is, essentially zero. In other words, the receding contact angle equals the advancing contact angle when the *n*-alkane molecules have

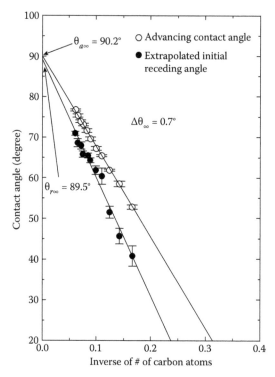

FIGURE 7.22 Advancing and extrapolated initial receding contact angles of 11 alkanes on FC-732 surfaces versus inverse number of carbon atoms. (Reprinted from Lam, C. N. C., Wu, R., Li, D., Hair, M. L., and Neumann, A. W., *Advances in Colloid and Interface Science*, 96, 169, 2002. With permission from Elsevier.)

infinite length. A similar result was obtained from the analysis of the contact angles of 1-alcohols on FC-732 surfaces [26]. The plausible explanation is that very large molecules are unlikely to be retained on the solid surface or penetrate it. Thus contact angle hysteresis vanishes.

In conclusion, the following picture emerges with respect to the contact angle hysteresis phenomena from the investigation of the contact angles of various liquids with different molecular properties, that is, n-alkanes and liquids with bulky molecules, on the surfaces of several fluoropolymers. There are two criteria that favor low or near zero contact angle hysteresis: (i) The solid surface is homogeneous, extremely smooth and consists of well-ordered and closely packed molecular chains. The smoothness of the film diminishes surface retention whereas packing of chains eliminates liquid penetration. (ii) The probe liquid possesses very large molecules; that is, they are significantly bulkier than the roughness scale on the solid surface and distances between molecular chains of semi-crystalline surfaces (or the size of pores in films of amorphous solids). From a practical standpoint, the latter condition may not be easily satisfied as materials with a high molecular weight are normally in solid form. The former criterion, on the other hand, might be met by selecting a crystalline film-forming material and using a reliable technique for surface preparation. This is indeed the case for the hysteresis-free surfaces of n-hexatriacontane ($C_{36}H_{74}$), which is a crystalline n-alkane, and siliconized glass that were prepared by Neumann et al. in the 1960s [29]. Thermal evaporation under vacuum was employed as the surface coating technique. A slow deposition rate apparently allowed the molecules to pack into ordered layers. The chains were so well packed that water molecules could not penetrate into the films and zero contact angle hysteresis was obtained.

Finally, from the perspective of the Young equation and the thermodynamic status of the contact angles in all the systems discussed here, it appears that in essence all advancing and receding contact angles are equilibrium contact angles on homogeneous and smooth surfaces and are compatible with Young's equation. The key point is that, upon receding, the three phase line leaves behind a solid–vapor surface that is different from the original dry surface. This newly generated surface presumably has been modified by any of the mechanisms introduced above; that is, surface retention, liquid penetration into the polymer film and molecular re-arrangement of the polymer chains at the surface. All these processes would be expected to change γ_{sv}, and hence the contact angle. The fact that the receding contact angle is always smaller than the advancing angle is also compatible with the above mechanisms: All the liquid surface tensions are larger than the solid surface tension (cf. Chapter 8) so that liquid retention and penetration processes will obviously increase γ_{sv}. The energetic effect of polymer chain rearrangement is less clear-cut. However, considering the production of these films also suggests an effect in the same direction: The polymer films were generated from solution by solvent evaporation. Hence the polymer chains have the opportunity to arrange themselves such that their free energy will be minimized. It is not conceivable that contact with the fairly inert contact angle liquids would cause a further decrease in the solid–vapor surface tension.

7.3.5 Rate of Motion of the Three-Phase Contact Line and Contact Angle Hysteresis

The influence of the rate of motion of the three-phase line on advancing and receding contact angles has been studied extensively; nevertheless, there is no general consensus among researchers in the surface science community on this issue. Following are some examples: Riddiford and coworkers showed that the advancing contact angles of water on siliconed glass plate and PTFE surfaces increase linearly with increasing rate of motion of the three-phase line from 1 to 7 mm/min [91,92]. Rate-dependence of advancing angles was argued in terms of disorientation of water molecules at the drop periphery from the equilibrium state. Outside this velocity range, two plateau regions were observed. The contact angles at the upper plateau were ascribed to the completion of disorientation, whereas those in the lower plateau region were considered the thermodynamic equilibrium angles. It was also suggested that after the "impressed drive" is removed from the liquid, contact angles relax rapidly back to equilibrium within 2–3 seconds of the cutoff time [93]. Johnson et al. investigated the advancing and receding angles of water and *n*-hexadecane on four different surfaces using a Wilhelmy plate technique [94]. Contrary to the above findings, this study showed no particular dependence of advancing or receding angles on the rate of motion of the three-phase line on homogeneous surfaces. No evidence was found in favor of the hypotheses of orientation/disorientation of water molecules at the three-phase line region or a preferred orientation of water molecules on a PTFE surface. It was argued that the velocity effect observed by Riddiford et al. is associated with their apparatus (two parallel plates confining the liquid drop) and is not caused by the motion of the three-phase line *per se*. Blake reported that advancing angles increase with increasing the rate of motion whereas the receding angles show an opposite trend [95]. He suggested that the rate-dependence of contact angles is more pronounced for viscous liquids. Cain et al. investigated this problem by contact angle measurements with water on siliconized glass slides by a capillary rise technique [96]. On smooth surfaces, a rate dependence of advancing contact angles below 0.1–0.2 mm/min was reported. This was followed by a plateau region for rates up to 0.5 mm/min. At higher rates, contact angles started increasing again. The initial increase in the contact angles was attributed to diffusion or spreading of liquid into the surface film. It was argued that above 0.2 mm/min, the rate of motion of the three-phase line overtakes the rate of diffusion such that the contact angles are rate-independent. The contact angles in the plateau region were considered the equilibrium advancing angles that can be used in the Young equation for the determination of solid surface tensions. Garoff and coworkers investigated the effect of inertia on the contact angles by measuring the shape of liquid–vapor interfaces within a few microns of the contact line for the immersion of a plate in PDMS solutions of different viscosities. The velocity range studied was 1–1300 μm/sec [97]. It was suggested that on a microscopic scale, inertia decreases the curvature of the free surface near the contact line, compared to the case of negligible Reynolds number. On a macroscopic scale, this translates into a decrease in the apparent contact angles.

Fluid mechanics theoreticians have also studied this problem. Generally, they suggest that the macroscopic advancing contact angles increase with the velocity of the three-phase line [95,98–102]. Some researchers proposed that the velocity dependence of the apparent contact angles follow cubic laws [103,104], and that the effect is more pronounced at fairly large capillary numbers [99] and for viscous liquids [95]. Mostly in these studies, the velocity is in the order of mm/sec to m/sec, which is far beyond the values suggested to obtain thermodynamically significant contact angles [96].

The short summary above highlights the inconsistencies encountered in the literature on this subject. Recently, a preliminary study of the contact angles of n-alkanes and 1-alcohols on the surfaces of FC-732 fluoropolymer by sessile drop experiments at drop front velocities of ~0.1–2.0 mm/min showed that neither advancing nor receding angles of the liquids depended on the rate of motion of the three-phase line [26]. To broaden this finding, advancing and receding contact angle measurements with several other probe liquids with a range of surface tension from 18.20 to 72.29 mJ/m^2 and a range of dynamic viscosity from 0.76 to 17.65 cP were performed on films of Teflon AF 1600 [105].

The advancing contact angles (θ_a) of seven different liquids on Teflon AF 1600 surfaces are presented in Table 7.3. For the first four liquids, that is, o-xylene to methyl salicylate, the advancing angles are essentially constant and independent of the rate of motion of the three-phase line in the range from ~0.2 to ~5–6 mm/min. For distilled water, even a wider range of velocity (i.e., 0.09–10.50 mm/min), was studied. Again, the advancing angles at different rates are similar and do not show a trend with respect to the velocity of the drop front.

As mentioned above, it has been suggested that the contact angles of viscous liquids are influenced more significantly by the velocity of the three-phase line [95]. This point was examined by choosing benzyl benzoate and ethylene glycol as probe liquids. Benzyl benzoate is about four times more viscous than the most viscous liquid among the first five liquids in Table 7.3 (i.e., methyl salicylate). The viscosity of ethylene glycol is about twice that of benzyl benzoate (see Table 7.3). The advancing contact angles at different rates of motion are given in Table 7.3. For benzyl benzoate, by increasing the rate of motion from 0.21 to 10.50 mm/min, the advancing angles show an increase of approximately 0.8°. For ethylene glycol, the advancing contact angles show an even more pronounced dependence on the rate of motion. Increasing the rate from 0.44 to 10.68 mm/min changes the advancing angles from ~103.8° to ~105.5°. The increase in the advancing angles of viscous liquids means that the drop becomes deformed (the curvature increases) due to hydrodynamic forces and the slope of the Laplacian profile at the contact point with the solid increases. Therefore, these results indicate that on smooth and homogeneous solid surfaces, the velocity dependence of advancing contact angles is an issue only for fairly viscous liquids. For liquids with a dynamic viscosity of well below 10 cP, the advancing angles are mainly determined by surface forces. As the viscosity of the probe liquid increases to $\mu \geq$ ~10 cP, viscous forces significantly affect the contact angles when increasing the rate of motion. Conversely, there is no evidence for a rate dependence at sufficiently low rates.

TABLE 7.3

Advancing (θ_a) and Receding (θ_r) Contact Angles of Test Liquids on Teflon AF 1600 Surfaces at Different Rates of Motion of the Three-Phase Line

Liquid	Rate (mm/min)	θ_a (°)	θ_r (°)	θ_{hyst} (°)
o-Xylene ($\mu = 0.76$ cP)				
	0.24	71.64 ± 0.04	65.68 ± 0.10	5.96
	0.81	71.71 ± 0.25	65.70 ± 0.03	6.01
	1.17	71.37 ± 0.29	65.61 ± 0.05	5.76
	1.26	71.57 ± 0.25	65.73 ± 0.10	5.84
	2.40	71.73 ± 0.19	65.92 ± 0.09	5.81
	5.12	71.68 ± 0.17	65.78 ± 0.14	5.90
trans,trans,cis-1,5,9-Cyclododecatriene (μ = N/A)				
	0.21	78.66 ± 0.11	71.38 ± 0.40	7.28
	0.35	78.50 ± 0.17	71.61 ± 0.05	6.89
	0.87	78.86 ± 0.16	71.09 ± 0.07	7.77
	1.40	78.67 ± 0.20	71.54 ± 0.06	7.13
	3.20	78.86 ± 0.54	71.56 ± 0.09	7.30
	3.50	78.96 ± 0.66	71.81 ± 0.14	7.15
	5.51	78.74 ± 0.51	71.68 ± 0.17	7.06
OMCTS ($\mu = 2.05$ cP)				
	0.18	44.42 ± 0.11	38.51 ± 0.03	5.91
	0.39	44.13 ± 0.08	38.75 ± 0.04	5.38
	0.72	44.21 ± 0.05	38.91 ± 0.06	5.30
	1.50	44.50 ± 0.27	39.02 ± 0.06	5.48
	2.46	44.38 ± 0.06	39.04 ± 0.08	5.34
	5.52	44.32 ± 0.31	39.04 ± 0.12	5.28
Methyl salicylate ($\mu = 2.34$ cP)				
	0.21	85.02 ± 0.10	77.76 ± 0.12	7.26
	0.32	84.78 ± 0.07	77.87 ± 0.09	6.91
	0.78	84.96 ± 0.12	77.77 ± 0.19	7.19
	1.10	84.58 ± 0.12	77.45 ± 0.08	7.13
	2.00	84.78 ± 0.28	77.29 ± 0.13	7.49
	3.80	84.58 ± 0.14	77.27 ± 0.08	7.31
	6.10	84.55 ± 0.11	77.37 ± 0.12	7.18
Distilled water ($\mu = 0.91$ cP)				
	0.09	126.54 ± 0.02	113.0 – 111.5	13.54–15.04
	0.20	126.56 ± 0.05	111.69 ± 0.14	14.87
	1.20	126.70 ± 0.07	114.5 – 112.5	12.20–14.20
	2.20	126.84 ± 0.25	115.60 ± 0.22	11.24
	5.90	126.82 ± 0.18	115.56 ± 0.51	11.26
	9.90	126.33 ± 0.18	116.07 ± 0.47	10.26
	10.50	126.44 ± 0.18	115.86 ± 0.71	10.58

TABLE 7.3 (Continued)

Advancing (θ_a) and Receding (θ_r) Contact Angles of Test Liquids on Teflon AF 1600 Surfaces at Different Rates of Motion of the Three-Phase Line

Liquid	Rate (mm/min)	θ_a (°)	θ_r (°)	θ_{hyst} (°)
Benzyl benzoate (μ = 9.62 cP)				
	0.21	89.13 ± 0.21	82.21 ± 0.04	6.92
	0.40	89.46 ± 0.33	82.12 ± 0.07	7.34
	0.92	89.45 ± 0.53	82.15 ± 0.08	7.30
	2.48	89.38 ± 0.28	82.00 ± 0.09	7.38
	4.15	89.50 ± 0.51	82.07 ± 0.14	7.43
	6.53	89.69 ± 0.45	81.94 ± 0.28	7.75
	10.50	89.92 ± 0.55	81.88 ± 0.34	8.04
Ethylene glycol (μ = 17.65 cP)				
	0.44	103.81 ± 0.41		
	0.79	103.85 ± 0.51		
	2.49	104.48 ± 0.54		
	3.67	104.48 ± 0.47		
	8.70	105.24 ± 0.43		
	10.68	105.50 ± 0.78		

Note: The 95% confidence limits for the contact angles and the values of contact angle hysteresis (θ_{hyst}) for each run are also given.

The receding angles of the test liquids at similar rates of motion of the three-phase line as for the advancing angles together with the corresponding contact angle hysteresis (θ_a–θ_r) data are given in the third and fourth columns of Table 7.3. For *o*-xylene, *trans,trans,cis*-1,5,9-cyclododecatriene, OMCTS, and methyl salicylate, the receding angles on Teflon AF 1600 films at different rates are similar and do not show a trend with respect to the velocity of the drop front. Consequently, identical contact angle hysteresis is obtained at different rates of motion. It is noted that for OMCTS at the slow rates of 0.18 and 0.39° mm/min, the receding angles are slightly smaller (by approximately 0.5°) than those obtained at the higher rates (see Table 7.3). It might be that the long solid–liquid contact time in these cases, that is, up to approximately 1500 seconds, causes the polymer film to be slightly modified due to liquid contact. Hence smaller receding angles are obtained.

For water, the receding angles show some variations. At the low rates of motion, the receding angles are overall smaller than those at the higher rates. This does not translate into a true hydrodynamic rate-dependence for the receding angles. If there were such an effect, the receding angles would have to decrease with increasing rate. However, the opposite effect is observed here. Unlike for the high rates of motion, the receding angles depend on the contact time between solid and liquid at the low rates. This is shown in Figure 7.23 for the contact angles of water at 1.20

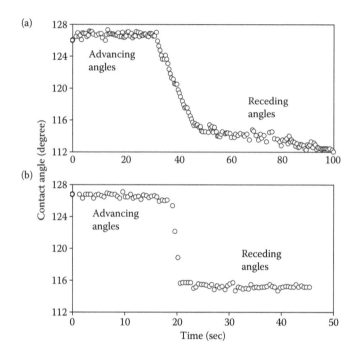

FIGURE 7.23 Contact angles of water at: (a) 1.20 mm/min, and (b) 5.90 mm/min. At the low rate of motion, the receding angles decrease with the solid-liquid contact time, but at the high velocity of the drop front, receding angles are fairly constant. (Reprinted from Tavana, H. and Neumann, A. W., *Colloids and Surfaces* A, 282–283, 256, 2006. With permission from Elsevier.)

and 5.90 mm/min. The causes of such effects are now known and were discussed in previous sections. At higher rates of motion, receding angles are not time-dependent indicating that water molecules do not find the opportunity to modify the polymer film significantly. Therefore, larger receding angles and hence smaller contact angle hysteresis are obtained. However, it appears that the primary mechanism of hysteresis (surface retention), which gives rise to the transition zone from advancing to receding angles, is not affected by increasing the rate of motion to values as high as ~10 mm/min. It is noted that the first four liquids in Table 7.3 consist of "bulky" molecules (molecular diameter from ~0.6 to ~1.0 nm, see Chapter 8), eliminating the likelihood of liquid penetration. Hence their receding angles are independent of solid–liquid contact time.

The receding angle of the viscous liquid, benzyl benzoate, decreases slightly by ~0.3°, as the rate of motion increases from 0.21 to 10.50 mm/min. Due to a more significant change of advancing angles within this velocity range, the contact angle hysteresis shows an overall increase of ~1.1°. For ethylene glycol, the receding angles were dependent on the solid–liquid contact time (similar to the receding angles of water at low rates of motion, i.e., Figure 7.23b), and hence are not reported here because they were not useful to study the rate-dependence of contact angles.

From the results presented above, it appears that rate-dependence of contact angles is an issue only for viscous liquids. In the case of liquids with low viscosity, surface forces are dominant and fairly constant contact angles are obtained at different rates of motion of the three-phase contact line. For liquids with high viscosity, increasing the rate of motion causes the advancing angles to increase and the receding angles to decrease, translating into larger contact angle hysteresis.

7.3.6 EFFECT OF A THIN LIQUID FILM ON CONTACT ANGLE HYSTERESIS

In the above discussions of contact angle hysteresis, it was assumed that no significant adsorption of liquid or its vapor occurs onto the solid surface. However, it should be noted that intermolecular forces might allow formation of a film of a submicrometer thickness extending beyond the macroscopic three-phase contact line [20]. A precursor film might be formed due to condensation of the vapor of the liquid ahead of the contact line [106], even in the absence of a significant vapor adsorption [107,108]. The thermodynamics of a solid–liquid–vapor system in the presence of a thin film was discussed above in Section 7.2.6. Below, we shall see how a thin liquid film can actually cause contact angle hysteresis. To simplify the analysis, other possible causes of hysteresis are ignored here.

Thin liquid film effects on contact angle hysteresis have been discussed by Churaev [109] and Zorin and Churaev [110]. Basically, contact angle hysteresis is considered to be caused by either the contact angle dependence on the capillary pressure or by coexistence of two uniform films with different thickness. For the latter case, Zorin and Churaev [110] reported an experiment with water films on quartz where metastable films were observed. First, a bubble was pressed against a quartz plate for 10–15 hours; the liquid–vapor meniscus had a finite contact angle in equilibrium with a uniform film of 10 nm thickness. When the meniscus was receded, a uniform wetting film with a thickness of about 40 nm was left behind. There was a sharp boundary between the two films and the situation did not change for several hours. On the thicker wetting film no finite contact angles were observed.

Such observations can be rationalized on the basis of the simple model discussed in this section. We can see how a thin film may cause the advancing contact angle, θ_a, to be different from the receding contact angle, θ_r. Replacing the term $(P_l - \Pi)$ in Equation 7.92 by $\gamma_{lv}J$ gives

$$\gamma_{lv}\cos\theta = \gamma_{lv}Jh + \gamma_f - \gamma_{sl}. \tag{7.94}$$

From this relation, we see that the contact angle depends on the Laplace pressure or the mean curvature, J, at the meniscus-film intersection, and on the film thickness, h. Consider a solid–liquid system of only two immiscible components; that is, solid and liquid. The solid–liquid interfacial tension, γ_{sl}, is assumed here to be a constant, independent of movement of the liquid–vapor meniscus, advancing or receding. At constant temperature and constant chemical potential, the Gibbs-Duhem equation for a uniform thin film phase yields

$$d\gamma_f = hd\Pi. \tag{7.95}$$

If we consider only the van der Waals interactions, the disjoining pressure has the following general form [40]

$$\Pi(h) = Ah^{-B}, \tag{7.96}$$

where A and B are constant for a given system. Integrating Equation 7.95 and using Equation 7.96, we will have

$$\gamma_f = \int hd\Pi = \frac{-B}{-B+1}h\Pi, \tag{7.97a}$$

or, by using Equation 7.91,

$$\gamma_f = \int hd\Pi = \frac{-B}{-B+1}hP_v. \tag{7.97b}$$

Now Equation 7.94 can be rewritten as

$$\gamma_{lv}\cos\theta = \gamma_{lv}Jh + \frac{-B}{-B+1}hP_v - \gamma_{sl}. \tag{7.98}$$

Equation 7.98 further shows an explicit dependence of contact angle on the thin film thickness: the contact angle, θ, will decrease as the film thickness, h, increases.

Let us denote the wetting film resulting from a receding liquid–vapor meniscus as the "receding film," and the film in front of an advancing liquid–vapor meniscus as the "advancing film." Since we know that $\theta_a > \theta_r$, we wish to compare the thickness of the receding film with that of the advancing film by using Equation 7.99. By $\cos\theta_a < \cos\theta_r$ and Equation 7.98, we have

$$\gamma_{lv}J_a h_a + \frac{-B}{-B+1}h_a P_{v,a} < \gamma_{lv}J_r h_r + \frac{-B}{-B+1}h_r P_{v,r}.$$

Rearranging the above equation yields

$$\frac{\gamma_{lv}J_a + \left(\dfrac{-B}{-B+1}\right)P_{v,a}}{\gamma_{lv}J_r + \left(\dfrac{-B}{-B+1}\right)P_{v,r}} < \frac{h_r}{h_a}.$$

Generally, the mean curvature of the receding meniscus is smaller than that of the advancing meniscus (i.e., $J_r < J_a$). For instance, when the bulk liquid is receded from an initial advancing position, due to the viscosity, the three phase contact line may show resistance against moving with the bulk liquid. This usually results in a lower curvature of the receding meniscus. Consequently, the lower curvature results in a lower equilibrium vapor pressure (i.e., $P_{v,r} < P_{v,a}$). Therefore, the above inequality can be further written as

$$1 < \frac{\gamma_{lv} J_a + \left(\dfrac{-B}{-B+1} \right) P_{v,a}}{\gamma_{lv} J_r + \left(\dfrac{-B}{-B+1} \right) P_{v,r}} < \frac{h_r}{h_a}. \tag{7.99}$$

Thus,

$$1 < \frac{h_r}{h_a} \quad \text{or} \quad h_a < h_r. \tag{7.100}$$

That is, the receding film is thicker than the advancing film, as observed in Zorin's experiment.

7.4 FURTHER THERMODYNAMIC CONSIDERATION OF THIN FILM PHENOMENA

7.4.1 THE NUMBER OF DEGREES OF FREEDOM

It has been shown that there are two degrees of freedom for a two-component solid–liquid–vapor system with a moderately curved interface [111] (see also Chapter 9). It is of interest to establish whether this is true also for a two-component solid–liquid–vapor system with a moderately curved interface and a flat thin film. Consider the sessile drop/thin film system illustrated in Figure 7.13. The solid substrate is considered an ideal, rigid solid phase with constant properties (independent of the interaction with a liquid drop) at a given temperature. Therefore, the solid phase will not be considered for the purpose of counting the number of degrees of freedom. Thus, for the system illustrated in Figure 7.13, there are a total of five phases: bulk liquid, bulk vapor, liquid–vapor interface, solid–liquid interface, and the thin liquid film. For each phase $(r + 1)$ intensive variables are required to describe the equilibrium state where r represents the number of components of the system; these variables may be chosen as follows:

$$
\begin{array}{lll}
\text{Bulk liquid phase} & T_l, P_l, x_{1l}, x_{2l}, \ldots, x_{(r-1)l} & \\
\text{Bulk vapor phase} & T_v, P_v, x_{1v}, x_{2v}, \ldots, x_{(r-1)v} & \\
\text{Liquid–vapor interface} & T_{lv}, \gamma_{lv}, x_{1lv}, x_{2lv}, \ldots, x_{(r-1)lv} & (7.101) \\
\text{Solid–liquid interface} & T_{sl}, \gamma_{sl}, x_{1sl}, x_{2sl}, \ldots, x_{(r-1)sl} & \\
\text{Thin liquid film} & T_f, \gamma_f, \Pi, x_{1f}, x_{2f}, \ldots, x_{(r-1)f} &
\end{array}
$$

where T is the temperature, P is the pressure, γ_{ij} is the surface tension of the ij interface, γ_f is the film tension, Π is the disjoining pressure of the film, and x_i represents the mole fraction of the ith component. When these five coexisting phases are in equilibrium, the thermal and chemical equilibrium conditions are given by

$$T_l = T_v = T_{lv} = T_{sl} = T_f \qquad \text{4 equations} \tag{7.102}$$

$$\mu_{il} = \mu_{iv} = \mu_{ilv} = \mu_{isl} = \mu_{if} \quad i = 1, 2,..., r \qquad 4r \text{ equations} \tag{7.103}$$

and the mechanical equilibrium conditions are given by

$$P_l - P_v = \gamma_{lv} J \tag{7.90}$$

$$\Pi = P_v \tag{7.91}$$

$$\gamma_{lv} \cos\theta = (P_l - \Pi)h + \gamma_f - \gamma_{sl}. \tag{7.92}$$

Note that Equation 7.90 and Equation 7.92 contain new variables, J and θ, which are not in the set listed in Equation 7.101; hence Equations 7.90 and 7.92 do not represent constraints among γ_{lv}, γ_{sl}, γ_f, Π, P_l, and P_v [111]. Only Equation 7.91 provides a constraint relation between Π and P_v. Therefore, for such an r component system in equilibrium, the total number of constraint equations is

$$4 + 4r + 1 = 4(r + 1) + 1. \tag{7.104}$$

As shown in Equation 7.101, there are a total of $5(r + 1) + 1$ variables. Therefore, the number of degrees of freedom for such a system is

$$f = 5(r + 1) + 1 - 4(r + 1) - 1 = r + 1. \tag{7.105}$$

If one considers a system of two components, that is, a solid component and a liquid component, then $r = 2$, and the number of degrees of freedom $f = 3$. This implies that there are three independent, intensive variables for a two-component interface/thin film system, such as the sessile drop/thin liquid film system illustrated in Figure 7.13. If the liquid–vapor surface tension, γ_{lv}, the film tension, γ_f, and the disjoining pressure, Π, or the vapor pressure, P_v ($\Pi = P_v$) are chosen to be the three independent variables, the solid–liquid surface tension, γ_{sl}, can then be expected to be a function of γ_{lv}, γ_f, and P_v; that is,

$$\gamma_{sl} = f(\gamma_{lv}, \gamma_f, \Pi) \quad \text{or} \quad \gamma_{sl} = f(\gamma_{lv}, \gamma_f, P_v). \tag{7.106}$$

Equation 7.106 may be referred to as an equation of state for the interfacial tensions and the film tension. A detailed description of an equation of state approach for a two-component solid–liquid–vapor system is given in Chapter 9.

7.4.2 An Equation of State Approach to Evaluate Film Tension

In principle, γ_f and γ_{sl} can be determined either by a microscopic approach involving calculations of intermolecular interactions, or by a macroscopic approach using a thermodynamic equation of state for the interfacial tensions and the film tension. Combining the equation of state in the form of Equation 7.106 with Equation 7.94, one will have

$$\gamma_{lv}\cos\theta = \gamma_{lv}Jh + \gamma_f - f(\gamma_{lv},\gamma_f,P_v). \qquad (7.107)$$

Knowing the values of the measurable quantities (γ_{lv}, J, h, θ), γ_f may be obtained from Equation 7.107 and then the solid–liquid interfacial tension, γ_{sl}, from either Equation 7.106 or Equation 7.94. An approximate, explicit functional form of such a thermodynamic equation of state, Equation 7.106, can be derived as follows.

The free energy of adhesion for a unit area of a solid–liquid pair is equal to the work required to separate a unit area of solid–liquid interface. In the presence of a thin liquid film, the grand canonical free energy of a thin film is given by Equation 7.80. The free energy of adhesion now can be written as

$$W_{sl(f)} = \Delta\Omega_{sl(f),adh} = \Omega_{final} - \Omega_{initial} = (\gamma_{lv} + \gamma_f - \Pi h) - \gamma_{sl}. \qquad (7.108)$$

Combining Equation 7.108 with Equation 7.94 gives

$$W_{sl(f)} = \Delta\Omega_{sl(f),adh} = \gamma_{lv}(1+\cos\theta) - \Pi h - \gamma_{lv}Jh. \qquad (7.109)$$

In analogy with the Berthelot combining rule for the attractive constant in the van der Waals equation of state [112–114], the free energy of adhesion, $W_{sl(f)}$, may be approximated as the geometric mean of the free energy of cohesion of the liquid pair, W_{ll}, and the free energy of cohesion of a pair of the thin liquid films, W_{ff}; that is,

$$W_{sl(f)} = \sqrt{W_{ll}W_{ff}} \qquad (7.110)$$

$$W_{ll} = 2\gamma_{lv} \qquad (7.111)$$

$$W_{ff} = 2(\gamma_f - \Pi h). \qquad (7.112)$$

Combining Equations 7.111 and 7.112 with Equation 7.110 yields

$$W_{sl(f)} = 2\sqrt{\gamma_{lv}(\gamma_f - \Pi h)}. \qquad (7.113)$$

Inserting Equation 7.113 into Equation 7.108, we have

$$\gamma_{sl} = \gamma_{lv} + \gamma_f - \Pi h - 2\sqrt{\gamma_{lv}(\gamma_f - \Pi h)},$$

or, in the light of Equation 7.81,

$$\gamma_{sl} = \gamma_{lv} + \gamma_f - P_v h - 2\sqrt{\gamma_{lv}(\gamma_f - P_v h)}. \tag{7.114}$$

Equation 7.114 is an equation of state for interfacial tensions and film tension. Combining Equation 7.114 with Equation 7.94 yields

$$\gamma_f = P_v h + \frac{0.25}{\gamma_{lv}}[\gamma_{lv}(\cos\theta + 1 - Jh) - P_v h]^2. \tag{7.115}$$

Clearly, as shown in Equation 7.115, the film tension, γ_f, can be estimated if P_v, J, h, γ_{lv}, and θ are known.

Without considering adsorption on solid surfaces, an equation of state for interfacial tensions has been developed [112,115,116] (see Chapters 8 and 9). By a phase rule approach, it has been shown that there are two degrees of freedom for a two-component solid–liquid–vapor surface system; hence, there exists an equation of state for interfacial tensions in the form of $\gamma_{sl} = f(\gamma_{lv}, \gamma_{sv})$. Combining such an equation of state with the Young equation will yield an equation relating the solid surface tension, γ_{sv}, to the measurable quantities—liquid surface tension, γ_{lv}, and contact angle, θ. An explicit form of such an equation of state for interfacial tensions has been derived as follows [116]:

$$\gamma_{sl} = f(\gamma_{lv}, \gamma_{sv}) = \gamma_{lv} + \gamma_{sv} - 2\sqrt{\gamma_{lv}\gamma_{sv}}\, e^{-\beta\ (\gamma_{lv} - \gamma_{sv})^2}, \tag{7.116}$$

where $\beta = 0.000125$ (mJ/m²)$^{-2}$. Combining Equation 7.116 with the classical Young Equation 7.2 produces

$$\cos\theta = g(\gamma_{lv}, \gamma_{sv}) = -1 + 2\sqrt{\frac{\gamma_{sv}}{\gamma_{lv}}}\, e^{-\beta\ (\gamma_{lv} - \gamma_{sv})^2}. \tag{7.117}$$

It is interesting to compare the interpretation of contact angles using the equation of state of interfacial tensions (see Chapter 8), that is, Equation 7.116, with that given by the equation of state of the interfacial tensions and film tension, Equation 7.114. For this purpose, the contact angles of six liquids on siliconized glass are presented in Table 7.4 [117]. It is significant that these data are free of contact angle hysteresis. The solid–vapor surface tension, γ_{sv}, of siliconized glass without appreciable adsorption is approximately 18.6 mJ/m² (see Table 7.4). This value is obtained from the water and glycerol contact angles on the siliconized glass by Equation 7.117. However, for the n-alkanes/siliconized glass systems, the adsorption on the

TABLE 7.4

Comparison of the Solid–Vapor Interfacial Tension, γ_{sv} (mJ/m²), of Siliconized Glass with the Film Tension, γ_f (mJ/m²), of n-Alkanes on Siliconized Glass

Liquid	γ_{lv}	$\theta_{measured}$ (°)	$\theta_{Equation\ 7.117}$ (°)	$\gamma_{sv,\ Equation\ 7.117}$	γ_f	$\gamma_{sv,\ Equation\ 7.119}$
Water	72.8	107.2		18.7		
Glycerol	63.4	99.2		18.5		
n-Hexadecane	27.6	27.2	51.3	24.7	24.6	24.6
n-Tetradecane	26.7	21.4	49.0	24.9	24.9	24.9
n-Dodecane	25.4	19.3	45.4	24.0	24.0	24.0
n-Decane	23.9	14.2	40.7	23.2	23.2	23.2

solid surface is quite prominent. The measured contact angles for n-hexadecane, n-tetradecane, n-dodecane, and n-decane are listed in Table 7.4. Comparing these with the contact angle values predicted by the equation of state approach, Equation 7.117, using $\gamma_{sv} = 18.6$ mJ/m² and the corresponding liquid surface tensions, it is seen that the measured contact angles are much lower than what they should be if there were no adsorption. Then, if we assume that adsorption is in the form of a thin liquid film on the siliconized glass, the corresponding film tensions can be calculated using Equation 7.115. To calculate γ_f from Equation 7.115, in addition to γ_{lv} and θ, the values of P_v, h, and J are needed. The film thickness, h, is generally of the order of $10^{-9} \sim 10^{-8}$ m. Therefore, in the case of moderate curvature, for example, the radius of curvature of the liquid–vapor interface being in the order of millimeters, the Jh term in Equation 7.115 will have a much smaller effect on γ_f than the other terms, unless the radius of curvature is less than a micrometer or so. Thus, for the low curvature case the exact value of J will not affect the γ_f value appreciably. Since these contact angles of n-alkanes on siliconized glass were measured by the technique of capillary rise at a vertical plate [117], the curvature at the three-phase intersection can be assumed to be $J = 1000$ m⁻¹ in the following calculations. The vapor pressures are as follows [118]: for n-hexadecane, $P_v = 0.2213$ N/m²; for n-tetradecane, $P_v = 1.866$ N/m²; for n-dodecane, $P_v = 7.599$ N/m²; and for n-decane, $P_v = 179.9$ N/m². It seems that, therefore, the $P_v h$ terms in Equation 7.115 also have only a minor effect on γ_f. Without affecting the essential results, we will assume that $h = 1 \times 10^{-9}$ m, approximately. The γ_f values for n-hexadecane, n-tetradecane, n-dodecane, and n-decane on siliconized glass are thus calculated and the results are listed in Table 7.4. As seen, the film tensions are larger than the solid–vapor surface tension without significant adsorption ($\gamma_{sv} = 18.6$ mJ/m²), and hence are the cause of the lower contact angles. However, there are essentially no differences between the γ_{sv} values calculated by Equation 7.117 and the γ_f values calculated by Equation 7.115 for the n-alkane/siliconized glass systems. However, it may be noted that Equation 7.115 (or Equation 7.114) is derived by using the geometric mean combining rule, as shown before, while Equation 7.116 or Equation 7.117 are derived by using a modified geometric mean combining rule [112,116] (see Chapters 8 and 9). Therefore, in order to compare γ_{sv} values with γ_f values at the same level, a simple form of the equation of state for interfacial tensions, derived also by using the geometric mean combining rule in a similar fashion to that for Equation 7.114 [112,114]; that is,

$$\gamma_{sl} = (\sqrt{\gamma_{lv}} - \sqrt{\gamma_{sv}})^2, \tag{7.118}$$

should be used. Combining Equation 7.118 with the classical Young Equation 7.2, yields

$$\gamma_{sv} = \gamma_{lv} \left(\frac{1 + \cos\theta}{2} \right)^2. \tag{7.119}$$

The γ_{sv} values calculated by Equation 7.119 are listed in the last column in Table 7.4. Again, there are no differences between γ_f and γ_{sv}.

It has been shown that the simple form of the equation of state derived by using the geometric mean combining rule is not applicable when the difference between γ_{lv} and γ_{sv} is large [112,116]. Therefore, we do not use Equations 7.114 and 7.115 to calculate γ_f for water/siliconized glass and glycerol/siliconized glass systems. For the same reason, we do not apply Equations 7.118 and 7.119 to water/siliconized glass and glycerol/siliconized glass systems to calculate γ_{sv} values.

It should be noted that the results for the film tension are based on a model that ignores the transition zone between the liquid–vapor meniscus and the flat thin liquid film, as illustrated in Figure 7.13. Therefore, the reliability of the above prediction depends on the validity of this model. Furthermore, within this model, these results are calculated for low curvature cases only. For high curvature situations, say, when the curvature of the liquid–vapor meniscus at the three-phase intersection is $J > 1 \times 10^6$ m^{-1}, one can expect that the curvature term in Equation 7.94 and hence in Equation 7.115 will have a significant effect; γ_f values in Table 7.4 will be different from the corresponding γ_{sv} values calculated by Equations 7.117 and 7.119. For example, if $J = 1 \times 10^7$ m^{-1} while the other parameters are the same, γ_f will be a few mJ/m^2 smaller than the γ_{sv} values listed in Table 7.4. Nevertheless, it can be concluded that the values of solid–vapor surface tension in Chapters 8 and 9 do not depend crucially on the validity of the Gibbsian interfacial model.

7.5 STICK-SLIP OF THE THREE-PHASE CONTACT LINE IN MEASUREMENTS OF DYNAMIC CONTACT ANGLES

Despite their seeming simplicity, contact angles have proven to entail many complexities both from measurement and interpretation points of view [119]. One such problem involves solid–liquid systems where the three-phase contact line does not move smoothly on the solid surface but shows stick-slip [17,119]: As liquid is pumped into the sessile drop and the volume increases, the contact angles increase at a constant contact radius; that is, the drop front remains pinned. Then the three-phase line slips abruptly on the solid surface as more liquid is supplied. This is accompanied by a sudden decrease in the contact angle and by a sudden increase in the contact radius. Supplying more liquid, the three-phase line sticks again to the solid surface at a new location and the radius again remains constant. This pattern is repeated as the measurement continues. Figure 7.24 shows an example for the contact angles of n-octane on a maleimide copolymer surface [17]. In Figure 7.25, two images of an n-octane drop (a) before and (b) after slipping are shown, corresponding to points (1) and (2) in Figure 7.24. Due to the abrupt slipping, the contact diameter of the drop increases by 0.33 mm in just 0.5 seconds and the contact angle shows a sudden decrease of ~6.7°. For normal motion of the three-phase line (no stick-slip), an increase of 0.33 mm in diameter would take more than 60 seconds with a low rate of motion of 0.3 mm/min for the drop front.

Stick-slip of the contact line, which is also referred to as pinning-depinning of the contact line, has been observed in different solid–liquid systems [17,20,82,119–122], including in microchannels with well-patterned superhydrophobic surfaces [123]. A few studies have investigated this phenomenon; however, the underlying mechanism is not well understood. Overall, stick-slip has been associated with noninertness of

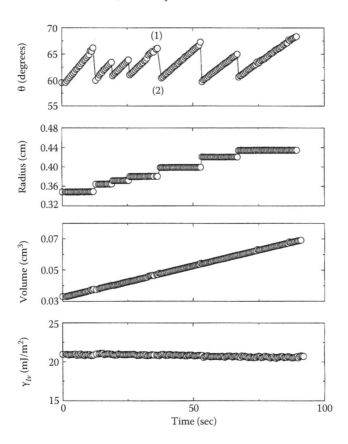

FIGURE 7.24   Contact angles, contact radius, drop volume, and surface tension of n-octane on a surface of ODMF. Stick-slip of the three-phase line and the abrupt changes in the drop radius by increase in the volume are seen. Points (1) and (2) in the contact angle graph correspond to images (a) and (b) in Figure 7.25, respectively. (Reprinted from Tavana, H.,Yang, G. C., Yip, C. M., Appelhans, D., Zschoche, S., Grundke, K., Hair, M. L., and Neumann, A. W., *Langmuir*, 22, 628, 2006. With permission from American Chemical Society.)

(a)   (b)

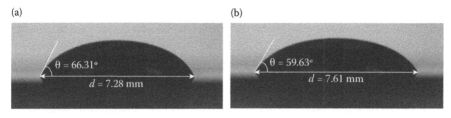

FIGURE 7.25   Images of an *n*-octane drop: (a) before, and (b) after slipping. As a result, the contact angle shows a sudden decrease of ~6.7° and the contact diameter increases abruptly by 0.33 mm. (Reprinted from Tavana, H.,Yang, G. C., Yip, C. M., Appelhans, D., Zschoche, S., Grundke, K., Hair, M. L., and Neumann, A. W., *Langmuir*, 22, 628, 2006. With permission from American Chemical Society.)

the solid surface and existence of defects on the solid surface. Defects might be in the form of physical roughness or domains with surface chemical differences from the chemistry of the solid matrix. The surface defects give rise to metastable states and create energy barriers against the smooth motion of the three-phase line. The contact line remains pinned onto the solid surface until the drop front possesses enough energy to overcome the energy barriers, resulting in slipping. For sufficiently large size of the defects, the three-phase line might also be contoured. It was suggested that the state of the three-phase contact line is determined by a competition between two opposing forces: On the one hand pinning forces that result from the existence of defects on the solid surface and tend to contour the contact line, and on the other restoring elastic forces; that is, surface tension forces, as well as the gravitational force that tend to straighten up the contact line [20,120]. For the strength of defects smaller than some limiting value, it was suggested that thermal fluctuations and vibrational noise cause the three-phase contact line to average the defects and avoid being pinned. In such situations, if defects are in the form of heterogeneous patterns, the macroscopic contact angle represents the thermodynamic equilibrium and will correspond to the Cassie contact angle. A threshold value of 1 μm was suggested for defect size [20]. It is the purpose of this section to present and discuss, as a case study, stick-slip of *n*-alkanes on the surface of a maleimide copolymer.

Contact angles of a homologous series of *n*-alkanes were measured on surfaces of ODMF, (see Figure 7.16 for the polymer repeat unit). Short-chain *n*-alkanes; that is, *n*-heptane to *n*-nonane, showed a stick-slip pattern as described above. From *n*-decane to *n*-tridecane, only one to two periods of stick-slip were observed at the beginning of the measurements, and then the drop front moved smoothly on the solid surface. For the last three *n*-alkanes, the pattern vanished and an even motion of the drop front was obtained [17]. The transition from stick-slip to a smooth motion of the three-phase line with increasing chain length of the *n*-alkanes is presented in Figure 7.26, where the contact angles of the liquids on the ODMF surfaces are given as a function of time. The average advancing contact angle for each liquid is given in this Figure.

It will be shown in Chapter 8 that on the films of a maleimide copolymer, ETMF, which has a similar molecular structure to that of ODMF except that it does not contain the *n*-hexadecyl side chain (see Figure 7.16), the three-phase line of all *n*-alkanes yields a smooth motion. This indicates that existence of an additional

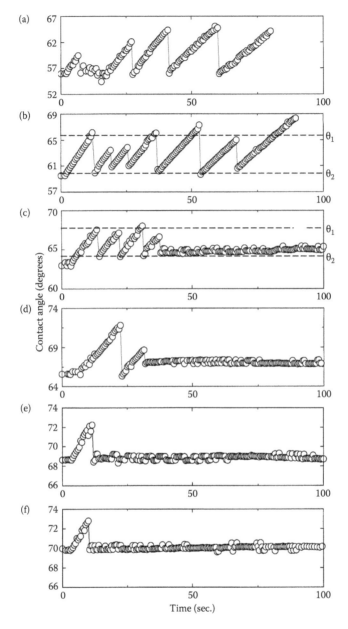

FIGURE 7.26 Advancing contact angles of *n*-alkanes on ODMF surfaces: (a) *n*-heptane, (b) *n*-octane, (c) *n*-nonane, (d) *n*-decane, (e) *n*-undecane, (f) *n*-dodecane, (g) *n*-tridecane, (h) *n*-tetradecane, (i) *n*-pentadecane, (j) *n*-hexadecane. The short-chain *n*-alkanes show a regular stick-slip, there is a transition from stick-slip to a smooth motion for the *n*-alkanes in the middle of the series, and the three-phase line of the long-chain *n*-alkanes moves smoothly. (Reprinted from Tavana, H.,Yang, G. C., Yip, C. M., Appelhans, D., Zschoche, S., Grundke, K., Hair, M. L., and Neumann, A. W., *Langmuir*, 22, 628, 2006. With permission from American Chemical Society.)

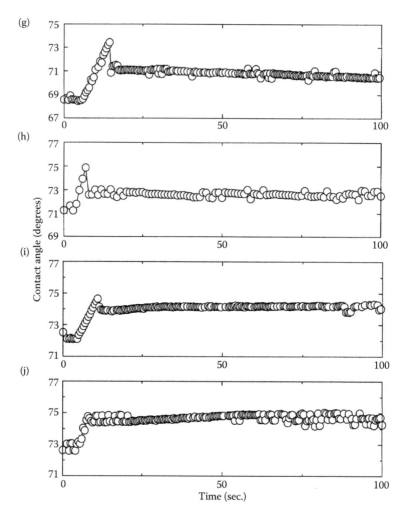

FIGURE 7.26 (Continued) Advancing contact angles of *n*-alkanes on ODMF surfaces: (a) *n*-heptane, (b) *n*-octane, (c) *n*-nonane, (d) *n*-decane, (e) *n*-undecane, (f) *n*-dodecane, (g) *n*-tridecane, (h) *n*-tetradecane, (i) *n*-pentadecane, (j) *n*-hexadecane. The short-chain *n*-alkanes show a regular stick-slip, there is a transition from stick-slip to a smooth motion for the *n*-alkanes in the middle of the series, and the three-phase line of the long-chain *n*-alkanes moves smoothly. (Reprinted from Tavana, H.,Yang, G. C., Yip, C. M., Appelhans, D., Zschoche, S., Grundke, K., Hair, M. L., and Neumann, A. W., *Langmuir*, 22, 628, 2006. With permission from American Chemical Society.)

n-hexadecyl side chain in ODMF compared to ETMF causes a significant difference in their surface properties. The morphology of the polymer films and variations in the compositions of ODMF and ETMF films were examined by atomic force microscopy (AFM). The measurements were performed through a simultaneous monitoring of both the amplitude and the phase of the oscillating cantilever in the tapping mode. Figures 7.27 and 7.28 illustrate the morphology and phase contrast images for

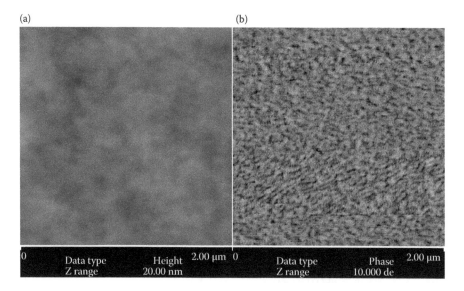

FIGURE 7.27   (a) Surface morphology, and (b) the corresponding phase contrast AFM images of an ETMF film. (Reprinted from Tavana, H.,Yang, G. C., Yip, C. M., Appelhans, D., Zschoche, S., Grundke, K., Hair, M. L., and Neumann, A. W., *Langmuir*, 22, 628, 2006. With permission from American Chemical Society.)

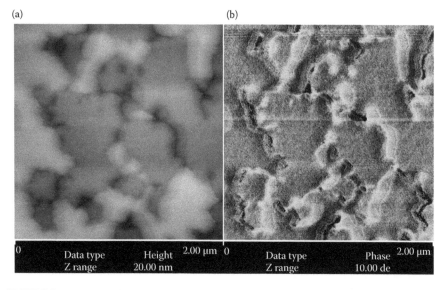

FIGURE 7.28   (a) Surface morphology, and (b) the corresponding phase contrast AFM images of an ODMF film. (Reprinted from Tavana, H.,Yang, G. C., Yip, C. M., Appelhans, D., Zschoche, S., Grundke, K., Hair, M. L., and Neumann, A. W., *Langmuir*, 22, 628, 2006. With permission from American Chemical Society.)

a scanned domain of $2 \times 2 \ \mu m^2$ on ETMF and ODMF films, respectively. From the phase contrast images, it is seen that unlike ETMF films, there are domains with noncontinuous structures on the ODMF surface film. It was demonstrated that these structures correspond to different compositions on the surface [17]. Due to the chemical incompatibility of n-hexadecyl ($C_{16}H_{33}$) and perfluoroalkyl (C_7F_{15}), the two side chains of ODMF segregate and give rise to microscale hydrocarbon patches of ~0.2–0.5 μm distributed over a fluorocarbon matrix. Thus, the ODMF films become heterogeneous in terms of distribution of surface energy. Additional angle-resolved XPS measurements showed that there is a preferential weak phase separation between the n-hexadecyl and the perfluoroheptyl side chains at the surface film of ODMF [69]. It will be shown in Chapter 8 that the surface tension of ODMF is 0.7 mJ/m² larger than that of ETMF. Assuming a surface tension of ~20 mJ/m² for a hydrocarbon film, an increase of 0.7 mJ/m² in the surface tension of ODMF (11.7 mJ/m²) over that of ETMF (11.0 mJ/m²) corresponds to a contribution of ~8% hydrocarbons on the ODMF film.

7.5.1 CONNECTION BETWEEN STICK-SLIP AND VAPOR ADSORPTION ONTO ODMF FILMS

As seen in the plots of Figure 7.26 for the contact angles of n-alkanes, the liquids with short-chain molecules show stick-slip on ODMF surfaces. There is a transition from stick-slip to a smooth motion for the liquids in the middle of the series. For the long-chain n-alkanes, that is, n-pentadecane and n-hexadecane, the patterns vanish and the three-phase line moves smoothly on the surface. Similarly, the vapor pressure of n-alkanes decreases significantly from short-chain to long-chain liquids; that is, from ~5.5 kPa for n-heptane to ~0.0001 kPa for n-hexadecane. This suggests that stick-slip might be connected to vapor pressure of the liquids.

The analysis of the contact angles of short-chain n-alkanes reveals that vapor molecules are adsorbed even onto pure fluoropolymer films (see Chapter 8). Bearing this in mind, vapor adsorption of short-chain n-alkanes onto the films of ODMF will be even more significant both due to the similar chemistry of the vapor molecules and the hydrocarbon patches on the surfaces and the fairly high surface tension of these patches compared to a pure fluorocarbon surface. It is also expected that vapor molecules have a preference for the hydrocarbon patches of the surface, yielding a nonuniform adsorption pattern.

The fact that the vapor pressure of n-alkanes decreases significantly with increasing chain length of liquid molecules (by two orders of magnitude from n-heptane to n-undecane) suggests that vapor adsorption onto the ODMF surfaces is less extensive for the liquids in the middle of the series compared to the first three liquids. This and the diminishing stick-slip suggest that the vapor of these liquids cannot reach far beyond the periphery of the initial sessile drop; that is, vapor adsorption occurs onto the solid surface only close to the drop deposited initially on the surface. The contact angles of any of the liquids in the middle of the series confirm this view as well. For instance for n-nonane, the contact angles in the "constant contact angle regime" are almost the same as the value predicted by Equation 7.117 for $\gamma_{sv} = 11.70$ mJ/m², which is the ODMF surface tension and is determined in Chapter 8. Knowing that Equation

7.117 does not take into account the effect of adsorption-type processes, it would have to predict a value different from the measured contact angles if n-nonane vapor were adsorbed far from the initial contact line. The possibility of unequal adsorption and modification of the solid surface near the contact line due to vapor adsorption was also reported in the literature [21].

All the evidence strongly suggests that stick-slip is caused cooperatively by vapor adsorption and existence of heterogeneous patches on ODMF films. Neither vapor adsorption nor heterogeneity alone can explain stick-slip: Vapor adsorption of short-chain n-alkanes onto homogeneous films of ETMF does not cause stick-slip nor do the long-chain n-alkanes (with negligible vapor pressure) show stick-slip on the heterogeneous ODMF surfaces. A tentative explanation is proposed below based on the analysis of the contact angles in Figure 7.26.

Suppose the sessile drop is in an equilibrium position; that is, the Cassie Equation 7.5 is satisfied ($\cos\theta_C = a_1\cos\theta_{e1} + a_2\cos\theta_{e2}$; a_1 and a_2 are the fractional surface areas of the two types of heterogeneous patches and θ_{e1} and θ_{e2} are the corresponding intrinsic contact angles). It is also assumed that the drop periphery maintains its circular shape. To understand why the three-phase line remains pinned as liquid is pumped into the sessile drop, it is useful to consider the size of the n-alkane-covered hydrocarbon patches and the optical resolution of the microscope-camera arrangement: This resolution is approximately two orders of magnitude below the size of the patches (20 μm vs. ~0.2–0.5 μm). This means that there might be a very small domain over which the three-phase line could move, but still retain contact with the patches, without being detected by the optical system. As the patches cover almost 10% of the solid surface, they provide, apparently, a sufficiently strong anchor to keep the three-phase line pinned. This state of affairs also has a favorable thermodynamic aspect: The three-phase line remaining stationary while the drop volume increase causes an increase in the contact angle on the more hydrophobic fluorocarbon matrix. This increase in contact angle is accompanied by a decrease in surface free energy, tending to a minimum value as the equilibrium contact angle on the fluorocarbon matrix is reached. It is significant that this increase in contact angles stops at a point where the advancing contact angle corresponds to $\gamma_{sv} = 11.0$ mJ/m²; that is, the surface tension of the fluorocarbon matrix, as obtained from ETMF. This limiting value (i.e., θ_1), is calculated from Equation 7.117 using $\gamma_{sv} = 11.0$ mJ/m² and is shown by the uppermost dashed line in Figure 7.26b and c for n-octane and n-nonane. Because of the presence of the much more hydrophilic hydrocarbon patches, there will be an ever increasing force opposing the continuous increase of the contact angles. The increasing opposing force caused by the hydrocarbon patches becomes predominant as the contact angle approaches the equilibrium value for the fluorocarbon matrix. Possibly due to the lubricity of the adsorbed film, slipping is initiated. As the three-phase line slips, the contact angles show a fast decrease. But, because of the large percentage of fluorocarbon area, slippage stops well short of the equilibrium contact angle on the hydrocarbon patches, which might be as low as 10–20°. An average contact angle line corresponding to the end of the slipping process is given by θ_2 in Figure 7.26b and c (lower dashed line). The question arises whether this limiting lower contact angle θ_2 has also a surface thermodynamic cause. The most likely explanation is again that the Cassie equation plays a role. The Cassie angle is given by the weighted average of the two types of solid surface; that is, the fluorocarbon matrix ($\gamma_{sv} \approx 11.0$ mJ/m²) and the hydrocarbon patches ($\gamma_{sv} \approx 20.0$ mJ/m²). To explore this venue, an equilibrium contact angle θ_3, belonging to the hydrocarbon patches was calculated from Equation 7.117, again on the assumption that $\gamma_{sv} \approx 20.0$ mJ/m² for the hydrocarbon patches.

Next, the ratio $(\cos\theta_2 - \cos\theta_1)/\cos\theta_3$ was defined as δ and calculated for each liquid. δ varies slightly from one liquid to another, between 5.5 and 9.7%, suggesting that this is the percentage area of the hydrocarbon patches (see Table 7.5). The average value of δ is 7.2%, consistent with the above approximation of 8% as the percentage area of hydrocarbons in the surface. It should be stressed that these two findings are independent of each other. The above estimate of 8% is based on systems that do not show stick-slip. Overall: It appears that the upper limit of the stick-slip pattern is given by the equilibrium contact angle of the fluorocarbon matrix and the lower limit by the Cassie contact angle.

These findings are illustrated by the *n*-nonane results in Figure 7.26: The upper limit of sticking is given by the equilibrium contact angle on the fluorocarbon patches; that is, the matrix with $\gamma_{sv} \approx 11.0$ mJ/m². The lower limit θ_2 is approximately equal to the global equilibrium contact angle; that is, the Cassie angle θ_2, as given by the constant contact angle after cessation of stick-slip. This cessation is likely caused by the absence or reduction of adsorption well away from the initial static drop [21].

Thus, the available evidence suggests that stick-slip is caused by processes occurring at the solid-vapor interface, and not, for example, at the solid–liquid interface. Processes in addition to adsorption such as liquid penetration may well occur, but

TABLE 7.5

Analysis of the Contact Angles of *n*-Alkanes on ODMF Surfaces

Liquid	$\theta_1(°)$	$\theta_2(°)$	$\theta_3(°)$	δ
n-Heptane	62.15	55.75	3.14	9.6
n-Octane	65.80	60.00	22.01	9.8
n-Nonane	68.24	64.26	28.58	7.3
n-Decane	70.09	65.80	32.79	8.3
n-Undecane	71.88	69.20	36.52	5.5
n-Dodecane	73.75	71.14	40.10	5.7
n-Tridecane	74.71	72.14	41.86	5.8
n-Tetradecane	75.63	73.09	43.25	5.9

θ_1: Contact angle on a fluoropolymer with $\gamma_{sv} = 11.0$ mJ/m²
θ_2: Contact angle reached at the end of slipping of the drop front
θ_3: Contact angle on a hydrocarbon film with $\gamma_{sv} = 20.0$ mJ/m²
$\delta = [(\cos\theta_2 - \cos\theta_1)/\cos\theta_3] \times 100$
δ represents the percentage area of the hydrocarbon patches.
Note: Reprinted from Tavana, H., Amirfazli, A., and Neumann, A. W., *Langmuir*, 22, 628, 2006. With permission from American Chemical Society.

they would be subsequent to adsorption. It appears that it is adsorption on the hydrocarbon patches that causes stick-slip, in the present systems.

In conclusion, stick-slip in the systems considered above is caused cooperatively by surface heterogeneity and vapor adsorption. The phenomenon has been observed in many other solid–liquid systems. For example, methyl salicylate and *p*-xylene on EGC-1700 surfaces [71] as well as *cis*-decalin and 1-bromonaphthalene on films of ODMF [82], diiodomethane on poly(methyl methacrylate/ethyl methacrylate) films [124], 3-pyridyl carbinol, ethyl cinnamate, methyl salicylate, and diiodomethane on fluoropolymer FC-725 films [125], and 2,2´-thiodiethanol on poly(styrene-*alt*-(hexyl/10-carboxydecyl(90/10)maleimide)) surfaces [126]. The explanation provided above may not be true for these systems because of the low vapor pressure of the probe liquids. It appears that elucidating the underlying causes in each of these cases would require a thorough investigation of the individual solid-liquid systems.

7.6 PHENOMENOLOGICAL CONTACT ANGLES: CONTACT ANGLES ON SUPERHYDROPHOBIC SURFACES

It has been demonstrated throughout this chapter that in many situations, experimental contact angles are not the same as the equilibrium contact angle for the solid–liquid system under consideration. One further example involves certain very rough surfaces on which a liquid drop tends to approach a spherical shape, exhibiting extremely large contact angles. Such surfaces are known as "superhydrophobic surfaces," and have found widespread interest in recent years. Examples of superhydrophobicity were discovered in nature, for example, the leaves of plants such as Lotus (*Nelumbo nucifera*), Taro (*Colocasia esculenta*), and Lady's mantle (*Alchemilla mollis*) as well as the wings of some insects (e.g., *Pflatoda claripennis*). On such biological surfaces, a water drop beads off completely and efficiently removes dirt and debris as it rolls off the surface. This "self-cleaning" property, known as the "Lotus effect" [127], is particularly vital for marsh and water plants. In aquatic habitats, high humidity and the presence of water supports many pathogenic organisms [128,129]. The water-repellent surface of plants eliminates the risk by hindering the adhesion of water needed for the germination of pathogens [129,130].

In 1996, Barthlott and Neinhuis studied this phenomenon in detail by examining the wettability and surface properties of various plant leaves [127]. It was shown that the water-repellent leaves exhibit a static contact angle of about 160°. Furthermore, water drops ran off the surface of the leaves very easily at inclination angles of < 5° without leaving any residue. Close examination of the surface of Lotus and Taro leaves by scanning electron microscopy (SEM) revealed that they consist of textures of dual-size roughness: bumps of about 10–20 μm covered with submicrometric wax crystals. The low surface energy of the wax and the unique hierarchical topography were suggested as the key characteristics of these leaves that make them water-repellent. Concurrent with this work, Onda et al. reported the fabrication of a "super water-repellent" fractal surface of alkylketene dimer that yielded a contact angle of 174° for water [131].

These studies attracted a great deal of interest mainly due to the potential of such surfaces for various applications, for example as waterproof coatings, stain-resistant finishes, anti-fog mirrors/lenses, and self-cleaning windshields/window panes. Extensive research has been conducted on the subject in the past decade and different techniques have been used to fabricate superhydrophobic surfaces: Plasma etching [132–135], electrodeposition [136], laser treatment [137], sol-gel processing [138], anionic oxidation [139], chemical etching [140–143], catalytic polymerization [144], plasma enhanced chemical vapor deposition [145], physical vapor deposition [146], micropatterning with templates [147–149], and so on. Superhydrophobic surfaces both with random and ordered surface textures have been fabricated as shown in Figures 7.29a and 7.29b [146,149].

It is known that roughness of a hydrophobic surface enhances its hydrophobicity [4]. For example, the contact angle of water on a smooth surface of a hydrophobic material such as those possessing CH_3 or CF_3 groups is typically of the order of $100° - 130°$ [29,70], but when the surface is made rough, the contact angle reaches values as large as ~170° [146]. Figure 7.30 shows a sessile drop of water on a rough surface of *n*-hexatriacontane (a crystalline *n*-alkane) [146]. The corresponding contact angle is 171°, whereas on the smooth surface of this material, the contact angle

(a)

(b)

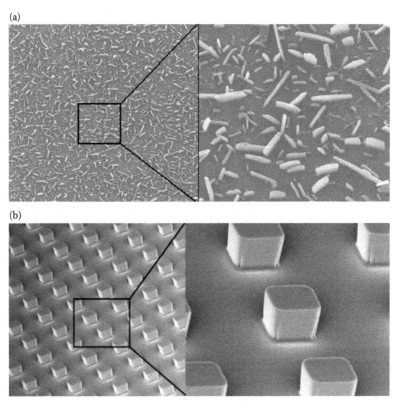

FIGURE 7.29 Two superhydrophobic surfaces with: (a) random, and (b) ordered surface textures. (Courtesy of He, B., Patankar, N. A., and Lee, J., *Langmuir,* 19, 4999, 2003.)

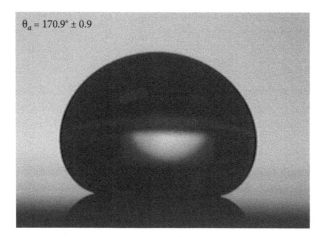

$\theta_a = 170.9° \pm 0.9$

FIGURE 7.30 An advancing sessile drop of water on a superhydrophobic surface of *n*-hexatriacontane (corresponding to Figure 7.29a). The contact diameter of the drop is 0.33 cm. (Reprinted from Tavana, H., Amirfazli, A., and Neumann, A. W., *Langmuir*, 22, 628, 2006. With permission from American Chemical Society.)

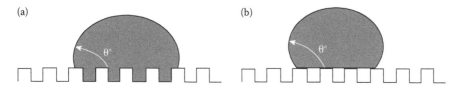

(a) $\theta°$

(b) $\theta°$

FIGURE 7.31 Sessile drops on two superhydrophobic surfaces: (a) in the Wenzel regime, and (b) in the Cassie regime.

of water is only ~105° [29]. Therefore, the superhydrophobic state results from a combination of surface roughness on the micrometer and/or nanometer scale and an intrinsically hydrophobic material.

The physics underlying superhydrophobicity is simple and well-understood [127,131,149–154]. On a superhydrophobic surface, two distinct types of wetting behavior might be observed depending on surface morphology and chemistry: (i) Wenzel model [4]; or (ii) Cassie model [8]. In the Wenzel regime, the liquid penetrates into the troughs of the surface texture and generally yields a large contact angle hysteresis (Figure 7.31a). On the other hand in the Cassie regime, the liquid drop sits on a composite surface that comprises solid and air pockets (Figure 7.31b). Normally a small hysteresis is observed in this regime. The Wenzel model is described by the following relation:

$$\cos\theta_W^r = r\cos\theta_Y, \qquad (7.120)$$

where θ_W^r is the apparent contact angle on the rough surface (Wenzel angle), θ_Y represents the intrinsic contact angle of the same material, and r is the roughness ratio;

that is, the ratio of actual to apparent surface area. The Cassie model is described by the following equation:

$$\cos\theta_C^r = -1 + \Phi_s(1 + \cos\theta_Y),\tag{7.121}$$

where θ_C^r is the apparent contact angle on the rough surface (Cassie angle) and Φ_s is the fraction of solid surface in contact with the liquid. Note that Equations 7.120 and 7.121 are identical with Equations 7.4 and 7.5, respectively, and are slightly rewritten here.

It has been demonstrated that both wetting regimes can be obtained on a super-hydrophobic surface depending on the geometry of the surface texture [149,154,155]. From a thermodynamic viewpoint, however, only one of these regimes corresponds to the minimum free energy of the system. Assume that a drop of water is initially in the Cassie state; that is, it sits on a composite of solid and air pockets. If external disturbances such as applying pressure on the drop or increasing the internal pressure of the drop do not cause water to penetrate into the spaces between protrusions on the solid, the Cassie regime is of a lower energy and the drop will remain on the composite solid–air surface. On the other hand, if such external disturbances cause water to wet the air-filled spaces, transition occurs from the Cassie regime to the Wenzel regime; that is, to a state of lower energy. It is important to note that the surface loses its water-repellency as a result of this switch in the wetting regime.

Figure 7.32 illustrates both models of superhydrophobicity, represented by solid lines. The dashed line represents the domain where the Cassie regime is metastable

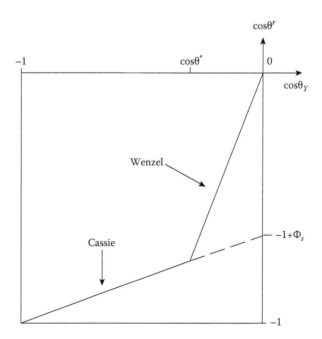

FIGURE 7.32 The Wenzel and the Cassie superhydrophobicity models and the threshold contact angle (θ^*) at which the switch from one regime to the other occurs.

and given that sufficient energy is provided to overcome the energy barriers, it would switch to the Wenzel regime. The switch takes place at a certain threshold contact angle (θ^*), which is obtained by equating Equations 7.120 and 7.121:

$$\theta^* = \cos^{-1}\left(\frac{\Phi_s - 1}{r - \Phi_s}\right). \tag{7.122}$$

It is inferred from Figure 7.32 that the likelihood of such an irreversible transition can be reduced if the texture is designed such that the corresponding threshold contact angle is as small as possible. Therefore, the metastability domain of the Cassie regime (the length of the dashed line in Figure 7.32) shrinks. Note that the Cassie regime stabilizes for $\theta > \theta^*$. This is particularly important from a practical standpoint since it guarantees that the surface will retain its water-repellency. It is, however, not trivial to design and fabricate such a surface.

In conclusion, superhydrophobic surfaces are generally fragile and the surface patterns are easily damaged due to contact with an object. The surfaces might also be vulnerable to oxidation, heat, and different solvents, depending on what material they are made of. Furthermore, the surfaces are prone to contamination that adversely affects both contact angle hysteresis and the roll-off angle, diminishing their water-repellent property. Such durability problems have hindered the commercialization of superhydrophobic surfaces, and to date only very few products have been launched. It still remains a challenge to fabricate long-lasting superhydrophobic surfaces with stable wetting characteristics.

7.7 CONTACT ANGLES IN THE PRESENCE OF ELECTRIC DOUBLE LAYERS

When an aqueous solution is brought in contact with a charged solid surface, electrostatic charges redistribute at the surface due to interactions between the solid and the liquid. The charging of a surface in a liquid occurs due to the ionization of surface groups or by the adsorption of ionic species from the liquid phase onto the surface, which might initially be uncharged or charged [156]. The surface charges attract ions of opposite charge (counterions) from the liquid phase but repel the co-ions. As a result, a small region in the vicinity of the solid carries higher concentrations of the counterions but lower concentrations of the co-ions compared to the bulk liquid. This net excess of counterions is considered to reside in two regions near the surface: the Stern (compact) layer, where most ions are transiently bound to the surface, and the Gouy-Chapman (diffuse) layer in which ions undergo rapid thermal motions close to the surface [156,157]. The two layers are collectively known as the electric double layer (EDL) whose thickness is represented by the Debye length, $1/\kappa$. The EDL thickness depends only on the properties of the liquid such as the ionic concentration and can range from less than a nanometer up to a micron [156,158]. These two layers are separated by a "shear plane." The electrical potential of the shear plane is referred to as the zeta potential (ζ) and can be determined experimentally [159]. Figure 7.33 shows the schematic of an EDL and the distribution of the electrical potential (ψ). The

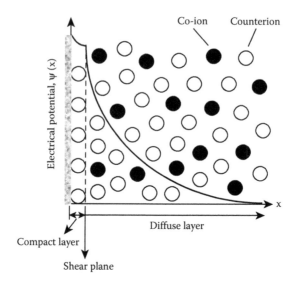

Co-ion   Counterion

Electrical potential, ψ (x)

x

Diffuse layer

Compact layer

Shear plane

FIGURE 7.33 Schematic of the electric double layer for a flat surface in contact with an aqueous solution and the corresponding electrical potential.

electrical potential decreases away from the Stern layer and the bulk liquid is assumed to carry no net charge.

The effect of EDL and the zeta potential on contact angles and solid surface tensions was studied by Chai et al. [160]. Films of poly(methyl methacrylate) (PMMA) were subjected to a DC pulsed oxygen plasma for various time periods up to 50 seconds. The plasma oxidation of a solid surface is known to introduce charged and polar functional groups at the surface and is a popular technique for the binding of layered microfluidic devices [161]. Immediately after the plasma treatment, contact angles of water were measured on the polymer films. The measurements were performed using ADSA-P. Figure 7.34 illustrates the advancing contact angle as a function of plasma exposure time. Contact angles decrease rapidly with the treatment time in the first 5 seconds. Longer exposure of polymer films to oxygen plasma does not change the contact angles significantly. Zeta potentials were also determined from the measurements of streaming potential and streaming current. Figure 7.35 shows the variation of ζ with the plasma treatment time. Zeta potential increases sharply within the first 5 seconds, and then the curve flattens out and reaches a value of −82.5 mV for 50 seconds of plasma treatment.

The zeta potential is related to the surface charge density (σ) through the following equation, which is obtained by solving the nonlinear Poisson–Boltzmann equation [159]

$$\sigma = \frac{2\varepsilon\varepsilon_0 K_B T \kappa}{e} \sinh \frac{e\zeta}{2K_B T}, \qquad (7.123)$$

where ε and ε_0 are permittivity of the liquid and vacuum, respectively, K_B represents the specific bulk conductivity of the liquid, κ is the inverse of the Debye length,

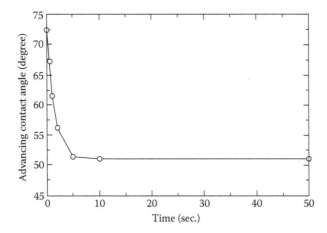

FIGURE 7.34   Contact angles of water on plasma-modified PMMA films. (Reprinted from Chai, J., Lu, F., Li, B., and Kwok, D. Y., *Langmuir*, 20, 10919, 2004. With permission from American Chemical Society.)

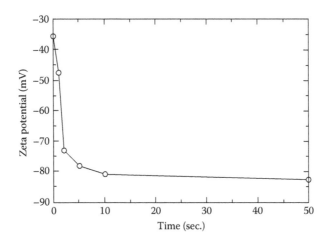

FIGURE 7.35   Variation of the zeta potential with plasma treatment time of PMMA surfaces. (Reprinted from Chai, J., Lu, F., Li, B., and Kwok, D. Y., *Langmuir*, 20, 10919, 2004. With permission from American Chemical Society.)

and T is the absolute temperature. The interfacial charge density increases from -0.0089 $\mu C/cm^2$ for untreated PMMA in contact with water to -0.028 $\mu C/cm^2$ for 50 seconds-treated PMMA in contact with water.

The contact angles in Figure 7.34 and the zeta potential in Figure 7.35 as well as the calculated surface charge densities follow a similar trend with respect to the duration of the plasma treatment and it might be argued that change in the contact angle is caused, at least in part, by the increase in the zeta potential and the interfacial charge density and therefore, by a stronger EDL. This point is examined below.

When the surface has a zeta potential ζ, a charge density σ builds up in the EDL of the liquid phase and causes the solid–liquid interfacial tension (γ_{sl}) to be different from that without the EDL. Using Lippmann's equation, a solid–liquid interfacial tension due solely to the EDL, γ_{sl}^{EDL}, is defined as [162]

$$\gamma_{sl}^{EDL} = \int \sigma d\phi, \tag{7.124}$$

where σ and ϕ are the surface charge density and electrical potential in the EDL, respectively. Using the expression for the charge density from Equation 7.123 and integrating Equation 7.124 from $\phi = 0$ to $\phi = \zeta$ gives the following relation for γ_{sl}^{EDL}

$$\gamma_{sl}^{EDL} = \frac{8 n_\infty K_B T}{\kappa} \left(\cosh \frac{e\zeta}{2 K_B T} - 1 \right). \tag{7.125}$$

Thus, in the presence of an EDL effect, γ_{sl} should be written as

$$\gamma_{sl} = \gamma_{sl}^0 - \gamma_{sl}^{EDL}, \tag{7.126}$$

where γ_{sl}^0 and γ_{sl}^{EDL} denote the solid–liquid interfacial tensions without and due to the EDL effect, respectively. Hence, a modified Young equation can be written as [163]

$$\gamma_{lv} \cos\theta = \gamma_{sv} - (\gamma_{sl}^0 - \gamma_{sl}^{EDL}). \tag{7.127}$$

The γ_{sl}^{EDL} term can be calculated from Equation 7.124 above. To obtain γ_{sl}^0, the following equation of state for the interfacial tensions can be used (see Chapters 8 and 9 for details)

$$\gamma_{sl}^0 = \gamma_{lv} + \gamma_{sv} - 2\sqrt{\gamma_{lv}\gamma_{sv}} \; e^{-\beta(\gamma_{lv}-\gamma_{sv})^2}, \tag{7.128}$$

where $\beta = 0.000125$ (mJ/m²)$^{-2}$. To determine γ_{sl}^0 from Equation 7.128, γ_{sv} should be calculated first. Eliminating the solid–liquid interfacial tension term (γ_{sl}^0) from the original Young equation ($\gamma_{lv} \cos\theta = \gamma_{sv} - \gamma_{sl}^0$) and the above equation of state yields

$$\cos\theta = -1 + 2\sqrt{\frac{\gamma_{sv}}{\gamma_{lv}}} e^{-\beta(\gamma_{lv}-\gamma_{sv})^2}. \tag{7.129}$$

For a given pair of γ_{lv} and θ (surface tension of water and its contact angle on PMMA surfaces in the present case), Equation 7.129 yields the γ_{sv} term. Finally, substituting γ_{sv} into the modified Young's equation (Equation 7.127) gives γ_{sl}^0.

Table 7.6 lists the calculated values of γ_{sv}, γ_{sl}^{EDL}, γ_{sl}^0, and γ_{sl} corresponding to PMMA surfaces treated with oxygen plasma for different time periods. For comparison, the data for nontreated films are also shown. Note that the surface tension

TABLE 7.6

Contact Angle of Water on PMMA Surfaces Modified by Oxygen Plasma for Different Time Periods, Corresponding Solid–Vapor Interfacial Tension (γ_{sv}), and Solid–Liquid Interfacial Tensions Without (γ_{sl}^0) and Due to the EDL Effect (γ_{sl}^{EDL})

Time (sec.)	θ (°)	γ_{sv} (mJ/m²)	γ_{sl}^{EDL} (mJ/m²)	γ_{sl}^0 (mJ/m²)	γ_{sl} (mJ/m²)
0	72.5 ± 0.6	40.09	0.0015	18.23	18.23
1	61.5 ± 0.6	46.85	0.0028	12.16	12.16
2	54.2 ± 0.7	51.22	0.0072	8.70	8.69
5	51.5 ± 1.4	52.81	0.0084	7.56	7.55
10	51.2 ± 1.4	52.98	0.0091	7.44	7.43
50	51.0 ± 0.6	53.09	0.0096	7.35	7.34

Note: Chai, J., Lu, F., Li, B., and Kwok, D. Y., *Langmuir*, 20, 10919, 2004. With permission from American Chemical Society.

of water is taken as $\gamma_{lv} = 72.70$ (mJ/m²). It is seen that the values of γ_{sl}^0 and γ_{sl} are essentially the same, differing only in the second or third significant digit after the decimal point. Thus, the values of γ_{sl}^{EDL} for surfaces with various plasma treatment times are extremely small. The key conclusion from the above set of data is that the contribution of EDL to the solid–liquid interfacial tension, and because of Equation 7.127 to the solid–vapor interfacial tension, is insignificant. This validates the assumption made in deriving the equation of state (Equation 7.128) to safely neglect the effect of EDL on the contact angles.

It is inferred that the enhanced hydrophilicity of plasma-treated PMMA surfaces is not due to the EDL; rather, it is the change of the overall surface tensions γ_{sv} and γ_{sl} due to the incorporation of polar groups at the surface that makes the surfaces more hydrophilic. In this study, PMMA was used as the substrate polymer material. It is expected that using polymers other than the substrate material will lead to a similar conclusion.

7.8 GLOSSARY OF CONTACT ANGLE CONCEPTS

θ: Contact angle in general: the phenomenological contact angle; that is, any contact angle that is observed, is often referred to simply as the contact angle, θ. For practical purposes, particularly for the study of surface tensions of solids, the validity or at least the applicability of Young's equation is required. This does not mean that contact angles, which do not satisfy the Young equation, are of no consequence. The phenomenological contact angle governs phenomena in which the Laplace pressure plays a role, such as the penetration of a liquid into a capillary.

θ_{ES}: Equilibrium contact angle of the system: the contact angle corresponding to the absolute minimum of the total free energy of the system, irrespective of whether the system is homogenous, heterogeneous, smooth, or rough. Thus, the equilibrium angle, θ_e, the Cassie angle, θ_C, and the Wenzel angle, θ_W, are all equilibrium angles, θ_{ES}, under the appropriate circumstances.

θ_e: Equilibrium contact angle: for a smooth and homogeneous solid surface, the equilibrium contact angle is given by Young's equation. The contact angle θ_e, is virtually a material property of an ideal solid surface. The conventional Young equation is not at all a universal equilibrium condition. For nonsimple systems, such as solids covered with a thin liquid film or an elastic liquid–vapor interface, the contact angle equilibrium condition will be different in form and content.

θ_a: Advancing contact angle: the contact angle of the liquid tending to advance across a solid surface.

θ_r: Receding contact angle: the contact angle of the liquid tending to recede over a solid surface.

$\theta_{hyst.}$: Contact angle hysteresis: the difference between the advancing and receding contact angles (i.e., $\theta_{hyst.} = \theta_a - \theta_r$).

θ_{dyn}: Dynamic contact angle: the contact angle associated with moving solid–liquid–vapor three-phase contact lines, as opposed to a static contact angle; at relatively low rates of motion of the three-phase line, static and dynamic contact angles tend to be identical on well-prepared solid surfaces (cf. Chapter 6).

θ_C: The Cassie angle: the equilibrium contact angle determined by the Cassie equation for a heterogeneous, but smooth surface. The Cassie angle is not readily accessible experimentally. Even if it could be found, it would not be useful for the determination of surface energetics, since the ratio of the two different surface components of the solid will not normally be known. However, in a simple model of a heterogeneous surface, the advancing contact angle, θ_a, is equal to the equilibrium contact angle that would be observed on the smooth homogeneous surface of the low energy surface component. Therefore, in such situations, the advancing contact angle represents an angle that is meaningful in the context of Young's equation. It may be called a Young contact angle, θ_Y (see below).

θ_W: Wenzel contact angle: the equilibrium contact angle for a rough surface; it is given by the Wenzel equation. For practical purposes, it is similarly inaccessible and unsuitable for energetic interpretations as the Cassie angle. Unlike the situation of some heterogeneous surfaces, the advancing angle on a rough surface cannot be readily interpreted in terms of the solid surface tension. Thus, contact angle hysteresis due to roughness precludes an interpretation in terms of solid surface tension, at least at the present state of knowledge.

θ_Y: Young contact angle: those contact angles that may be used in conjunction with the Young equation. For a smooth and homogeneous solid surface, the Young contact angle is just the equilibrium contact angle; for a smooth but heterogeneous surface the advancing and/or receding contact angles may represent Young contact angles, although they are clearly not equilibrium contact angles.

θ_M: The macroscopic contact angle: the contact angle observable with the usual optical means. In the case of, say, a sessile drop, the three-phase line may be contorted on a microscopic scale, causing the contact angle to vary along the three-phase line on that same scale. The observable, macroscopic contact angle may well be constant.

REFERENCES

1. J.W. Gibbs. *The Scientific Papers of J. Willard Gibbs.* Vol. 1. Dover, New York, 1961.
2. P. S. de Laplace. *Mechanique Celeste, Supplement to Book 10.* J. B. M. Duprat, Paris, 1808.
3. T. Young. In *Miscellaneous Works.* Edited by G. Peacock. J. Murray, London, 1855.
4. R. N. Wenzel. *Industrial & Engineering Chemistry* 28 (1936): 988.
5. R. J. Good. *Journal of the American Chemical Society* 74 (1952): 5041.
6. A. B. D. Cassie. *Discussions of the Faraday Society* 3 (1948): 11.
7. S. Baxter and A. B. D. Cassie. 36 (1945): T67.
8. A. B. D. Cassie and S. Baxter. *Transactions of the Faraday Society* 40 (1944): 546.
9. A. W. Neumann, D. Renzow, R. Reumuth, and I. E. Richter. *Fortschr. Kolloide Polym.* 55 (1971): 49.
10. J. J. Jasper and E. V. Kring. *Journal of Physical Chemistry* 59 (1955): 1019.
11. R. E. Johnson and R. H. Dettre. *Journal of Physical Chemistry* 68 (1964): 1744.
12. R. E. Johnson and R. H. Dettre. *Advances in Chemistry Series* 43 (1964): 112.
13. A. W. Neumann and R. J. Good. *Journal of Colloid and Interface Science* 38 (1972): 341.
14. J. D. Eick, R. J. Good, and A. W. Neumann. *Journal of Colloid and Interface Science* 53 (1975): 235.
15. A. Karim, T. M. Slawecki, S. K. Kumar, J. F. Douglas, S. K. Satija, C. C. Han, T. P. Russell, et al. *Macromolecules* 31 (1998): 857.
16. J. Peng, Y. Xuan, H. F. Wang, Y. M. Yang, B. Y. Li, and Y. C. Han. *Journal of Chemical Physics* 120 (2004): 11163.
17. H. Tavana, G. C. Yang, C. M. Yip, D. Appelhans, S. Zschoche, K. Grundke, M. L. Hair, and A. W. Neumann. *Langmuir* 22 (2006): 628.
18. D. Raghavan, X. Gu, T. Nguyen, M. VanLandingham, and A. Karim. *Macromolecules* 33 (2000): 2573.
19. D. Raghavan, M. VanLandingham, X. Gu, and T. Nguyen. *Langmuir* 16 (2000): 9448.
20. P. G. de Gennes. *Reviews of Modern Physics* 57 (1985): 827.
21. E. L. Decker, B. Frank, Y. Suo, and S. Garoff. *Colloids and Surfaces A: Physicochemical Engineering Aspects* 156 (1999): 177.
22. H. Tavana, D. Jehnichen, K. Grundke, M. L. Hair, and A. W. Neumann. *Advances in Colloid and Interface Science* 134–135 (2007): 236.
23. G. E. H. Hellwig and A. W. Neumann. *Kolloid-Z. Z. Polym.* 40 (1969): 229.
24. A. W. Neumann and D. Renzow. *Zeitschrift fur Physikalische Chemie* 11 (1969): 68.
25. W. J. Herzberg, J. E. Marian, and T. Vermeulen. *Journal of Colloid and Interface Science* 33 (1970): 1.
26. C. N. C. Lam, R. Wu, D. Li, M. L. Hair, and A. W. Neumann. *Advances in Colloid and Interface Science* 96 (2002): 169.
27. A. V. Rapacchietta, A. W. Neumann, and S. N. Omenyi. *Journal of Colloid and Interface Science* 59 (1977): 541.
28. A. V. Rapacchietta and A. W. Neumann. *Journal of Colloid and Interface Science* 59 (1977): 555.
29. A. W. Neumann. *Advances in Colloid and Interface Science* 4 (1974): 105.
30. A. N. Frumkin. *Zhurnal Fizicheskoi Khimii* 12 (1938): 337.
31. B. V. Derjaguin. *Journal of Physical Chemistry* 3 (1932): 29.
32. B. V. Derjaguin. *Journal of Physical Chemistry* 5 (1934): 379.
33. B. V. Derjaguin. *Zhurnal Fizicheskoi Khimii* 14 (1940): 137.
34. B. V. Derjaguin. *Kolloidnyi Zhurnal* 17 (1955): 207.
35. B. V. Derjaguin and N. V. Churaev. *Journal of Colloid and Interface Science* 54 (1976): 157.

36. B. V. Derjaguin, V. M. Starov, and N. V. Churaev. *Kolloidnyi Zhurnal of the USSR* 38 (1976): 875.
37. V. M. Starov and N. V. Churaev. *Kolloidnyi Zhurnal of the USSR* 42 (1980): 703.
38. N. V. Churaev, V. M. Starov, and B. V. Derjaguin. *Journal of Colloid and Interface Science* 89 (1982): 16.
39. B. V. Derjaguin and N. V. Churaev. *Wetting Films.* Nauka, Moscow, 1984.
40. B. V. Derjaguin, N. V. Churaev, and V. M. Muller. *Surface Forces.* Plenum Press, New York, 1987.
41. A. I. Rusanov. *Colloid Journal of the USSR* 29 (1967): 118.
42. B. V. Derjaguin and L. M. Shcherbakov. *Colloid Journal of the USSR* 23 (1961): 33.
43. J. A. Kitchener. *Endeavour* 22 (1962): 118.
44. A. Sheludko. *Advances in Colloid and Interface Science* 1 (1967): 391.
45. B. V. Derjaguin. *Journal of Colloid and Interface Science* 24 (1967): 357.
46. A. I. Rusanov. *Phase Equilibria and Interfacial Phenomena.* Khimia, Leningrad, 1967.
47. A. Sheludko, B. Radoev, and T. Kolarov. *Transactions of the Faraday Society* 64 (1968): 2213.
48. D. Li and A. W. Neumann. *Advances in Colloid and Interface Science* 36 (1991): 125.
49. H. B. Callen. *Thermodynamics and an Introduction to Thermostatistics.* 2nd ed. John Wiley, New York, 1985.
50. L. Boruvka and A. W. Neumann. *Journal of Chemical Physics* 66 (1977): 5464.
51. J. Rice. In *Commentary on the Scientific Writings of J. Willard Gibbs, Vol. I, Thermodynamics.* Edited by F. G. Donnan and A. Haas. Yale University Press, New Haven, 1936.
52. D. Li. In *Proceedings of A.S.M.E. Annual Winter Meeting.* New Orleans, LA, 1993.
53. D. Li and A. W. Neumann. *Journal of Colloid and Interface Science* 49 (1994): 147.
54. B. V. Derjaguin. *Comptes Rendus de L Academie des Sciences of the USSR* 51 (1946): 361.
55. A. Marmur. *Advances in Colloid and Interface Science* 50 (1994): 121.
56. J. F. Joanny and P. G. de Gennes. *Journal of Chemical Physics* 81 (1984): 552.
57. J. Long, M. N. Hyder, R. Y. M. Huang, and P. Chen. *Advances in Colloid and Interface Science* 118 (2005): 121.
58. R. Shuttleworth and G. L. Bailey. *Discussions of the Faraday Society* 3 (1948): 16.
59. F. E. Bartell and J. W. Shepard. *Journal of Physical Chemistry* 57 (1953): 211, 455, 458.
60. R. E. Johnson, Jr., and R. H. Dettre. In *Surface and Colloid Science, Wettability and Contact Angles.* Edited by E. Matijevic. Wiley, New York, 1969.
61. L. W. Schwartz and S. Garoff. *Langmuir* 1 (1985): 219.
62. A. Y. Fadeev and T. J. McCarthy. *Langmuir* 15 (1999).
63. C. N. C. Lam, N. Kim, D. Hui, D. Y. Kwok, M. L. Hair, and A. W. Neumann. *Colloids and Surfaces A: Physicochemical Engineering Aspects* 189 (2001): 265.
64. C. N. C. Lam, R. H. Y. Ko, L. M. Y. Yu, D. Li, M. L. Hair, and A. W. Neumann. *Journal of Colloid and Interface Science* 243 (2001): 208.
65. R. V. Sedev, J. G. Petrov, and A. W. Neumann. *Journal of Colloid and Interface Science* 180 (1996): 36.
66. R. V. Sedev, C. J. Budziak, J. G. Petrov, and A. W. Neumann. *Journal of Colloid and Interface Science* 159 (1993): 392.
67. W. H. Buck and P. R. Resnick. *Teflon AF Amorphous Fluoropolymers Technical Information: Properties of Amorphous Fluoropolymers Based on 2,2-Bistrifluoromethyl-4,5-Difluoro-1,3-Dioxole.* In 183rd meeting of the Electrochemical Society, Honolulu, HI, 1993.
68. 3M Inc., Product information. 3M Novec Electronic Coating EGC-1700, 2003.

69. D. Appelhans, Z.-G. Wang, S. Zschoche, R.-C. Zhuang, L. Häussler, P. Friedel, F. Simon, et al. *Macromolecules* 38 (2005): 1655.
70. H. Tavana and A. W. Neumann. *Advances in Colloid and Interface Science* 132 (2007): 1.
71. H. Tavana, F. Simon, K. Grundke, D. Y. Kwok, M. L. Hair, and A. W. Neumann. *Journal of Colloid and Interface Science* 291 (2005): 497.
72. T. C. Merkel, V. Bondar, K. Nagai, B. D. Freeman, and Y. P. Yampolskii. *Macromolecules* 32 (1999): 8427.
73. L. H. Tanner. *Journal of Physics D: Applied Physics* 12 (1979): 1473.
74. D. Y. C. Chan and R. G. Horn. *Journal of Chemical Physics* 83 (1985): 5311.
75. I. Pinnau and L. G. Toy. *Journal of Membrane Science* 109 (1996): 125.
76. P. R. Resnick. *Polymers of Fluorinated Dioxoles*. Patent No. 3,978,030, 1976.
77. M. Hung. *Macromolecules* 26 (1993): 5829.
78. S. M. Nemser and I. C. Roman. *Perfluorinated Membranes*. Patent No. 5,051,114, 1991.
79. A. Hennig, K.-J. Eichhorn, U. Staudinger, K. Sahre, M. Rogalli, M. Stamm, A. W. Neumann, and K. Grundke. *Langmuir* 20 (2004): 6685.
80. H. Tavana, R. Gitiafroz, M. L. Hair, and A. W. Neumann. *Journal of Adhesion* 80 (2004): 705.
81. H. Tavana, C. N. C. Lam, K. Grundke, P. Friedel, D. Y. Kwok, M. L. Hair, and A. W. Neumann. *Journal of Colloid and Interface Science* 279 (2004): 493.
82. H. Tavana, D. Appelhans, R.-C. Zhuang, S. Zschoche, K. Grundke, M. L. Hair, and A. W. Neumann. *Colloid and Polymer Science* 284 (2006): 497.
83. L. Lavielle and J. Schultz. *Journal of Colloid and Interface Science* 106 (1985): 438.
84. E. Ruckenstein and S. V. Gourisankar. *Journal of Colloid and Interface Science* 107 (1985): 488.
85. S. R. Holmes-Farley, R. H. Reamey, R. Nuzzo, T. J. McCarthy, and M. Whitesides. *Langmuir* 3 (1987): 799.
86. T. Yasuda, T. Okuno, K. Yoshida, and H. Yasuda. *Journal of Polymer Science* 26 (1988): 1781.
87. T. Yasuda, M. Miyama, and H. Yasuda. *Langmuir* 8 (1992): 1425.
88. T. Yasuda, M. Miyama, and H. Yasuda. *Langmuir* 10 (1994): 583.
89. J.-H. Wang, P. M. Claesson, J. L. Parker, and H. Yasuda. *Langmuir* 10 (1994): 3897.
90. C. O. Timmons and W. A. Zisman. *Journal of Colloid and Interface Science* 22 (1966): 165.
91. G. E. P. Elliott and A. C. Riddiford. *Journal of Colloid and Interface Science* 23 (1967): 389.
92. A. C. Lowe and A. C. Riddiford. *Chemical Communications* 106 (1970): 387.
93. M. C. Philips and A. C. Riddiford. *Journal of Colloid and Interface Science* 41 (1972): 77.
94. R. E. Johnson, Jr., R. H. Dettre, and D. A. Brandreth. *Journal of Colloid and Interface Science* 62 (1977): 205.
95. T. D. Blake. In *Wettability*. Edited by J. C. Berg, 251. Marcel Dekker, New York, 1993.
96. J. B. Cain, D. W. Francis, R. D. Venter, and A. W. Neumann. *Journal of Colloid and Interface Science* 94 (1983): 123.
97. K. Stoev, E. Ramé, and S. Garoff. *Physics of Fluids* 11 (1999): 3209.
98. E. B. Dussan V. *Annual Reviews of Fluid Mechanics* 1 (1979): 371.
99. R. G. Cox. *Journal of Fluid Mechanics* 357 (1998): 249.
100. R. L. Hoffman. *Journal of Colloid and Interface Science* 50 (1975): 228.
101. E. B. Gutoff and C. E. Kendrick. *AIChE Journal* 28 (1982): 459.
102. R. T. Foister. *Journal of Colloid and Interface Science* 136 (1990): 266.
103. L. H. Tanner. *Journal of Physics D: Applied Physics* 12 (1979): 1473.
104. V. V. Kalinin and V. M. Starov. *Kolloidnyi Zhurnal* 48 (1986): 907.
105. H. Tavana and A. W. Neumann. *Colloids and Surfaces A: Physicochemical Engineering Aspects* 282–283 (2006): 256.

106. W. Hardy. *Philosophical Magazine* 38 (1919): 49.
107. D. Bangham and S. Saweris. *Transactions of the Faraday Society* 34 (1938): 554.
108. W. V. Chang, Y. M. Chang, L. J. Wang, and Z. G. Wang. *Organic Coatings and Applied Polymer Science Proceedings.* In American Chemical Society, Washington, DC, 1982.
109. N. V. Churaev. *Revue de Physique Appliquée* 23 (1988): 975.
110. Z. Zorin and N. V. Churaev. *Colloid Journal of the USSR* 30 (1968): 279.
111. D. Li, J. Gaydos, and A. W. Neumann. *Langmuir* 5 (1989): 1133.
112. D. Li and A. W. Neumann. *Journal of Colloid and Interface Science* 137 (1990): 304.
113. D. Berthelot. *Comptes Rendus* 126 (1857): 1703.
114. L. A. Girifalco and R. J. Good. *Journal of Physical Chemistry* 61 (1957): 904.
115. A. W. Neumann, R. J. Good, C. J. Hope, and M. Sejpal. *Journal of Colloid and Interface Science* 49 (1974): 291.
116. D. Li and A. W. Neumann. *Advances in Colloid and Interface Science* 39 (1992): 299.
117. A. W. Neumann and D. Renzow. *Zeitschrift für Physikalische Chemie Neue Folge* 68 (1969): 11.
118. R. C. Wilhoit and B. J. Zwolinski. *Handbook of Vapor Pressure and Heats of Vaporization of Hydrocarbons and Related Compounds.* Thermodynamics Research Center, Department of Chemistry, Texas A&M University, TX, 1971.
119. D. Y. Kwok and A. W. Neumann. *Advances in Colloid and Interface Science* 81 (1999): 167.
120. B. Frank and S. Garoff. *Langmuir* 11 (1995): 87.
121. H. M. Princen, A. M. Cazabat, M. A. C. Stuart, F. Heslot, and S. Nicolet. *Journal of Colloid and Interface Science* 126 (1988): 84.
122. E. Schäffer and P.-Z. Wong. *Physical Review Letters* 80 (1998): 3069.
123. J. Zhang and D. Kwok. *Langmuir* 22 (2006): 4998.
124. D. Y. Kwok, D. Li, and A. W. Neumann. *Journal of Polymer Science Part B: Polymer Physics* 37 (1999): 239.
125. D. Y. Kwok, C. N. C. Lam, D. Li, A. Leung, R. Wu, E. Mok, and A. W. Neumann. *Colloids and Surfaces A: Physicochemical Engineering Aspects* 142 (1998): 219.
126. D. Y. Kwok, D. Li, C. N. C. Lam, R. Wu, S. Zschoche, K. Pöschel, T. Gietzelt, K. Grundke, H.-J. Jacobasch, and A. W. Neumann. *Macromolecular Chemistry and Physics* 200 (1999): 1121.
127. W. Barthlott and C. Neinhuis. *Planta* 202 (1997): 1.
128. C. L. Campbell, J. S. Huang, and G. A. Payne. In *Plant Disease: An Advanced Treatise.* Edited by J. G. Horsfall and E. B. Cowling. Academic Press, London, 1980.
129. E. A. Allen, H. C. Hoch, J. R. Steadman, and R. J. Stavely. In *Microbial Ecology of Leaves.* Edited by J. H. Andrews and S. S. Hirano. Springer, New York, 1991.
130. C. Neinhuis and W. Barthlott. *Annals of Botany* 79 (1997): 667.
131. T. Onda, S. Shibuichi, N. Satoh, and K. Tsujii. *Langmuir* 12 (1996): 2125.
132. M. B. O. Riekerink, J. G. A. Terlingen, G. H. M. Engbers, and J. Feijen. *Langmuir* 15 (1999): 4847.
133. S. Minko, M. Müller, M. M. M. Nitschke, K. Grundke, and M. Stamm. *Journal of the American Chemical Society* 125 (2003): 3896.
134. X. Zhang, M. Jin, Z. Liu, S. Nishimoto, H. Saito, T. Murakami, and A. Fujishima. *Langmuir* 22 (2006): 9477.
135. R. P. Garrod, L. G. Harris, W. C. E. Schofield, J. McGettrick, L. J. Ward, D. O. H. Teare, and J. P. S. Badyal. *Langmuir* 23 (2007): 689.
136. N. J. Shirtcliffe, G. McHale, M. I. Newton, and C. C. Perry. *Langmuir* 21 (2005): 937.
137. M. T. Khorasani, H. Mirzadeh, and Z. Kermani. *Applied Surface Science* 242 (2005): 339.
138. A. Nakajima, A. Fujishima, K. Hashimoto, and T. Watanabe. *Advanced Materials* 11 (1999): 1365.

139. K. Tsujii, T. Yamamoto, T. Onda, and S. Shibuichi. *Angewandte Chemie-International Edition in English* 36 (1997): 1011.
140. M. Thieme, R. Frenzel, S. Schmidt, F. Simon, A. Hennig, H. Worch, K. Lunkwitz, and D. Scharnweber. Advanced Engineering Materials 3 (2001): 691.
141. B. Qian and Z. Shen. *Langmuir* 21 (2005): 9007.
142. T. N. Krupenkin, J. A. Taylor, E. N. Wang, P. Kolodner, M. Hodes, and T. R. Salamon. *Langmuir* 23 (2007): 9128.
143. T. N. Krupenkin, J. A. T. T. M. Schneider, and S. Yang. *Langmuir* 20 (2004): 3824.
144. W. Han, D. Wu, W. Ming, H. Niemantsverdriet, and P. C. Thüne. *Langmuir* 22 (2006): 7959.
145. K. K. S. Lau, J. Bico, K. B. K. Teo, M. Chhowalla, G. A. J. Amaratunga, W. I. Milne, G. H. McKinley, and K. K. Gleason. *Nano Letters* 3 (2003): 1701.
146. H. Tavana, A. Amirfazli, and A. W. Neumann. *Langmuir* 22 (2006): 5556.
147. D. Öner and T. J. McCarthy. *Langmuir* 16 (2000): 7777.
148. L. Cao, H.-H. Hu, and D. Gao. *Langmuir* 23 (2007): 4310.
149. B. He, N. A. Patankar, and J. Lee. *Langmuir* 19 (2003): 4999.
150. J. Bico, C. Marzolin, and D. Quéré. *Europhysics Letters* 47 (1999): 220.
151. J. P. Youngblood and T. J. McCarthy. *Macromolecules* 32 (1999): 6800.
152. M. Callies and D. Quéré. *Soft Matter* 1 (2005): 55.
153. N. A. Patankar. *Langmuir* 20 (2004): 7097.
154. A. Lafuma and D. Quéré. *Nature Materials* 2 (2003): 457.
155. J. Bico, U. Thiele, and D. Quéré. *Colloids and Surfaces A: Physicochemical Engineering Aspects* 206 (2002): 41.
156. J. N. Israelachvili. *Intermolecular and Surface Forces*. 2nd ed. Academic Press, London, 1991.
157. A. W. Adamson. *Physical Chemistry of Surfaces*. 6th ed. John Wiley & Sons, Inc., New York, 1997.
158. D. Li. *Electrokinetics in Microfluidics*. Elsevier Academic Press, Boston, 2004.
159. R. J. Hunter. *Zeta Potential in Colloid Science*. Academic Press, London, 1981.
160. J. Chai, F. Lu, B. Li, and D. Y. Kwok. *Langmuir* 20 (2004): 10919.
161. D. C. Duffy, J. C. McDonald, O. J. A. Schueller, and G. M. Whitesides. *Analytical Chemistry* 70 (1998): 4974.
162. R. J. Hunter. *Foundations of Colloid Science*. Oxford University Press, New York, 2001.
163. G. Lippmann. *Annales de Chimie et de Physique* 5 (1875): 494.

8 Interpretation of Contact Angles

Hossein Tavana and A. Wilhelm Neumann

CONTENTS

8.1 INTRODUCTION

Contact angle phenomena have been the subject of numerous studies in surface science for several decades. Contact angles play a major role in phenomena such as wetting and adhesion as well as in a wide range of technological and biological systems. For systems involving contact between a liquid phase and a solid phase, contact angles are a manifestation of the energetics of the solid surface. Thus, a complete understanding of contact angles requires precise information about solid surface tensions. Interest has been enhanced by the recent advances in the field of micro/nanotechnology and by the ability to fabricate miniaturized devices. Due to the large ratio of surface area to volume in such micro/nanodevices, surface forces including surface tensions become extremely important and control various processes. Examples of such devices are: surface tension-based bubble valves and pumps that are being used in microfluidic and nanofluidic systems of inkjet printers, micro total analysis systems (µTAS) that facilitate the study of chemical species and living cells by fabricating microdrops from a liquid jet under the action of surface tension, microfluidic-based flow cytometers utilizing surface tension effects to control the flow configuration of a liquid sample containing living cells and chemical species in order to detect, sort, and analyze their morphological and biochemical properties, and nanoelectromechanical relaxation oscillators made of carbon nanotubes that take advantage of surface tension to oscillate droplets back and forth using electric current. To properly design such devices, one would certainly require a priori knowledge of the surface tensions of both liquid and solid phases and the corresponding solid–liquid interfacial tension.

Direct measurement of surface tension of liquids is straightforward and several techniques such as the Wilhelmy plate technique [1–4], the drop weight method [2–4], the oscillating jet method [3,4], the capillary wave method [2,3], the spinning drop method [2–4], and drop shape techniques [2–3] have been developed for this purpose. In contrast, due to the immobility of molecules in a solid phase, it is quite difficult to measure solid surface tensions directly. Nevertheless, a few experimental techniques have been proposed for this purpose. Cleavage of a crystal is one example [5–7]. During the cleavage process, if the material undergoes a reversible isothermal process, the solid–vapor interfacial tension (γ_{sv}) is approximately given by the free energy of cleavage. But, if irreversibilities occur during the process, the solid surface tension can no longer be defined as the free energy of cleavage. Thus, this conceptually simple treatment is suitable only for crystals with brittle structure, such as alkali halides and graphite. Historically, the interpretation of the measurements has been associated with ambiguity because cleavage usually involves effects such as plastic deformation of the sample as well as adsorption of atmospheric gases, which are believed to be responsible for variations of up to two orders of magnitude in the reported values for solid surface tensions in the literature [6]. To minimize plastic deformation, cleavage should be carried out at low temperatures. To avoid formation of adsorption layers from the atmosphere, such experiments will have to be carried out in ultra high vacuum.

Due to such uncertainties and difficulties associated with direct techniques for the determination of solid surface tensions, several indirect approaches, both experimental and theoretical, are often used. They include contact angle measurements with different liquids [8–13], direct force measurements [14–20], solidification front techniques [21–26], film flotation [27–30], sedimentation techniques [31–33], capillary penetration into columns of particle powder [34–37], gradient theory [38,39], and the Lifshitz theory of van der Waals forces [40–42]. Among these methods, the contact angle approach is the most broadly applicable one.

Estimating solid surface tensions from contact angles is based on a relation first recognized by Young [43]. In principle, the contact angle of a liquid drop on a solid surface is determined by the mechanical equilibrium under the action of three interfacial tensions, the liquid–vapor surface tension, γ_{lv}, the solid–vapor surface tension, γ_{sv}, and the solid–liquid interfacial tension, γ_{sl}. This equilibrium relation is known as the Young equation

$$\gamma_{lv}\cos\theta = \gamma_{sv} - \gamma_{sl}, \tag{8.1}$$

where θ is the Young contact angle. The only measurable quantities in the Young equation are γ_{lv} and θ. Therefore to obtain γ_{sv} and γ_{sl}, an additional relation is required.

Seeking an explicit relation between the three interfacial tensions is not novel and has a long history. Numerous studies have been conducted to meet this need. An early attempt is due to Antonow [44] who proposed the following simple relation to obtain the interfacial tension between two condensed phases from the interfacial tensions of these phases against air, in the present case:

$$\gamma_{sl} = \left|\gamma_{lv} - \gamma_{sv}\right|. \tag{8.2}$$

However, relations relying on this equation have not been able to provide a secure basis to estimate solid surface tensions (Chapter 9).

A significant contribution to the field of contact angles was made with the pioneering work of Zisman in the 1950s [45,46]. He performed an experimental study of contact angles on various low energy surfaces such as polytetrafluoroethylene, polystyrene, and polyethylene terephthalate (PET) using a large number of liquids of relatively high surface tension (i.e., $\gamma_{lv} > \gamma_{sv}$). Considering the measurement techniques available at the time, his contact angles were of good quality, having an accuracy of ~±2–3°. Zisman was the first to observe that if the cosine of contact angles of the liquids is plotted as a function of liquid surface tension ($\cos\theta$ vs. γ_{lv}), the data points fall within a linear band. From such patterns, he introduced the critical surface tension of wetting (γ_c) for a solid surface. Although he stated that γ_c behaves as one would expect γ_{sv} to behave, he took great care not to identify γ_{sv} with γ_c. Zisman advanced the understanding of contact angles and their interpretation in terms of energetics of solid surfaces remarkably. His work was a milestone in this field and set the stage for subsequent developments.

There were also studies attempting the determination of solid surface tensions using contact angles but involving certain theoretical aspects. Pioneered by Fowkes in the 1960s [9], this approach is known as the surface tension components approach. He argued that dispersive forces are always present in a material but other types of intermolecular forces may or may not be present depending on the specific chemical nature of the material. With this view, he postulated that the total surface tension of a material can be divided into two parts, the part due to dispersive forces (γ^d) and the part due to nondispersive forces (γ^r), and suggested that when two immiscible phases are brought into contact, only those intermolecular forces that are common to both phases act across the interface. This means that for a solid–liquid system, the contact angle not only depends on the total liquid and solid surface tensions (γ_{lv} and γ_{sv}), but also on the specific intermolecular forces. This approach was later extended by others including Owens, Wendt, Kaelble, and van Oss to take into account other types of intermolecular interactions [11–13,47]. However, it has been shown in detail that these approaches are not compatible with experimental data and do not reflect physical reality [48]. Furthermore, they do not satisfy the thermodynamic principle of the phase rule for capillary systems [49]. Of course, strictly speaking, phase rule arguments are not applicable for nonthermodynamic quantities like γ^d and γ^r. The failure of the surface tension component approaches to predict solid surface tensions will be discussed in Chapter 9.

8.2   CONTACT ANGLE MEASUREMENTS

The success of Zisman's work was an indication that more significant progress might be achieved on the experimental side of the study of contact angles. This was pursued further by Neumann and his coworkers. An important step in this direction was to scrutinize the measurement technique and to obtain contact angles with accuracy better than that achievable before.

Over the past two decades, advances of computational tools and numerical schemes have allowed for the development of new methodologies for the measurement of surface tension of liquids and contact angles, such as Axisymmetric Drop Shape Analysis (ADSA). ADSA determines liquid–fluid interfacial tensions and contact angles from the shapes of axisymmetric menisci; that is, sessile and pendant drops (Chapter 3). Assuming the experimental drop profile to be axisymmetric and Laplacian, ADSA finds the theoretical profile that best matches the profile extracted from the image of a real drop. From the best match, several parameters such as the liquid–vapor interfacial tension and the contact angle are determined. ADSA is a robust technique that yields liquid surface tensions and contact angles with a reproducibility of ±0.1 mJ/m^2 and ±0.2°, respectively. ADSA was developed in the 1980s [50] and was optimized and reformulated later in the 1990s [51,52]. A detailed account of ADSA is given in Chapters 3–6.

For the purpose of interpretation of contact angles in terms of solid surface tensions, contact angles that are thermodynamically significant and compatible with the Young equation have to be used. Experience shows that the measurement of such contact angles is not trivial. For example, simply depositing a droplet on a solid surface and measuring the corresponding static contact angle using a goniometer technique is not a suitable strategy. Processes such as evaporation of liquid and creeping of liquid into the solid can cause the static contact angle to be different from the equilibrium contact angle. Unfortunately, the literature of contact angle research contains a large amount of such inaccurate data. It has been argued that equilibrium contact angles compatible with the Young equation can be obtained if the three-phase line is advanced slowly (i.e., below 0.5 mm/min) during the contact angle experiment (see Chapter 6). The measurement can be facilitated by forming an initial small drop on the surface and then pumping more liquid into the drop from a small hole on the surface using a motorized syringe. Thus, drop fronts advancing on the solid surface and equilibrium advancing angles are obtained. Figure 8.1 illustrates the difference between static and low-rate dynamic contact angles for an experiment with distilled water on an FC-722 fluoropolymer surface. It is seen that carefully depositing an initial drop from above on the solid surface results in a contact angle of ~108°. Addition of a certain amount of liquid is required for the initial drop front to start advancing. Increasing the drop volume (V) linearly from 0.18 to 0.22 cm³, by a motorized-syringe mechanism, increases the contact angle (θ) from approximately 108° to 119° at constant three-phase contact radius (R). Further increase in the drop volume causes the three-phase line to advance, with essentially constant advancing contact angle (θ) as R increases. The rate of motion of the three-phase line in this particular example is 0.14 mm/min.

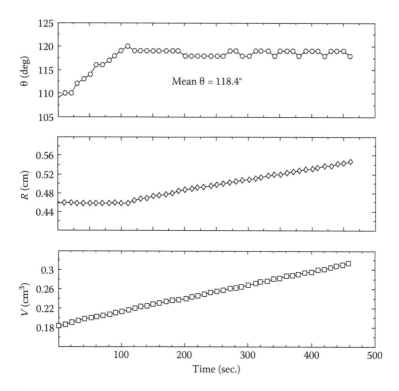

FIGURE 8.1   Low-rate dynamic contact angle θ, drop radius R, and drop volume V of water on the fluoropolymer FC-722 surface. (Reprinted from Kwok, D. Y., Lin, R., Mui, M., and Neumann, A. W., *Colloids and Surfaces* A, 116, 63, 1996. With permission from Elsevier.)

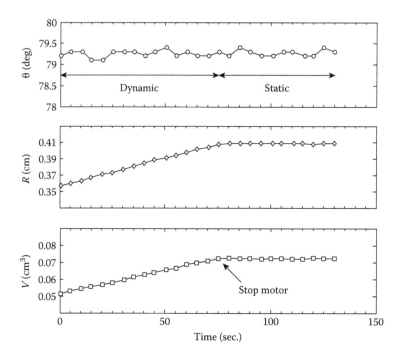

FIGURE 8.2 Dynamic and static contact angles of *cis*-decalin on the fluoropolymer FC-722 surface. The low-rate contact angles are identical to properly measured static angles. (Reprinted from Kwok, D. Y., Lin, R., Mui, M., and Neumann, A. W., *Colloids and Surfaces A*, 116, 63, 1996. With permission from Elsevier.)

It is noted that if supply of liquid into the sessile drop is stopped, the resulting static contact angles will be identical with the low-rate advancing angles. Figure 8.2 illustrates this point for a *cis*-decalin/FC-722 system. This and similar experiments indicate that the contact angles are independent of the slow rate of advancing, suggesting that the low-rate advancing contact angles and properly measured static contact angles are identical and may be used interchangeably. Details about the influence of the rate of motion on contact angles are given in Chapters 6 and 7.

8.3 GENERAL CONTACT ANGLE PATTERNS

Li et al. employed an experimental protocol similar to that explained above to measure contact angles of liquids with different properties (polar as well as nonpolar) on the films of three different polymers; that is, fluoropolymer FC-721, fluorinated ethylene propylene (FEP), and PET [53]. In order to follow Young's equation more closely, $\gamma_{lv}\cos\theta$ was plotted versus γ_{lv} for the contact angles of various liquids obtained on one and the same polymeric solid surface (i.e., FC-721). It was shown that this procedure yields data points that vary systematically with liquid surface tension, regardless of properties of the test liquids (Figure 8.3). Using the solid surfaces of FEP and PET (and hence changing γ_{sv}) shifted the contact angles in a regular manner. From contact

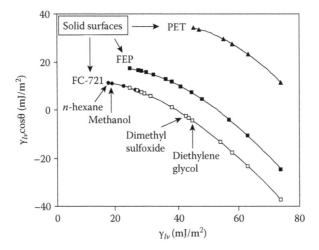

FIGURE 8.3 Plot of $\gamma_{lv}\cos\theta$ versus γ_{lv} for liquids with different molecular properties on three different solid surfaces (FC-721, FEP, and PET). The contact angles measured on each solid surface fall on a smooth curve. (Reprinted from Li, D. and Neumann, A. W., *Journal of Colloid and Interface Science*, 148, 190, 1992. With permission from Academic Press.)

angle results with these three solid surfaces, it was concluded that $\gamma_{lv}\cos\theta$ depends on liquid surface tension (γ_{lv}) and solid surface tension (γ_{sv}) only.

To ensure the universality of such contact angle patterns, contact angle measurements were extended to numerous other solid–liquid systems. The solid surfaces were: (i) fluoropolymer FC-721, (ii) fluoropolymer FC-722, (iii) fluoropolymer FC-725, (iv) Teflon FEP, (v) *n*-hexatriacontane, (vi) cholesteryl acetate, (vii) poly (propene-*alt*-*N*-(*n*-hexyl)maleimide), (viii) poly (*n*-butyl methacrylate), (ix) polystyrene (PS), (x) poly (styrene-*alt*-(*n*-hexyl/10-carboxydecyl(90/10)maleimide)), (xi) poly (methyl methacrylate/*n*-butyl methacrylate), (xii) poly (propene-*alt*-*N*-(*n*-propyl) maleimide), (xiii) poly (methyl methacrylate) (PMMA), and (xiv) poly (propene-*alt*-*N*-methylmaleimide). The contact angle experiments with these solids were conducted by an automated axisymmetric drop shape technique or a capillary rise technique at low rates of advancing of the three-phase line [48,54]. Both these techniques yield highly reproducible contact angles. Table 8.1 summarizes the results. The third column shows the liquid surface tension values measured by independent pendant drop experiments and the fourth column presents the advancing contact angles. The plot of $\gamma_{lv}\cos\theta$ versus γ_{lv} was constructed for the contact angle data obtained with different liquids on each of the above solid surfaces. Figure 8.4 encompasses the results for all the solid surfaces. The contact angles change smoothly and systematically with liquid surface tension on each solid. Changing the solid surface shifts the contact angles in a regular manner.

Overall, these contact angle patterns suggest that $\gamma_{lv}\cos\theta$ depends on liquid surface tension and solid surface tension only; that is,

$$\gamma_{lv}\cos\theta = f(\gamma_{lv}, \gamma_{sv}), \tag{8.3}$$

TABLE 8.1
Summary of Advancing Contact Angles of Various Liquids on Different Solid Surfaces

Solid Surface/Technique	Liquid	γ_{lv} (mJ/m²)	θ (degree)	γ_{sv}
FC-721-coated mica/capillary	Dodecane	25.03	70.4	11.7
rise [103]	2-Octanol	26.00	73.5	11.3
	Tetradecane	26.50	73.5	11.6
	1-Octanol	27.28	75.1	11.5
	Hexadecane	27.31	75.6	11.3
	1-Hexadecene	27.75	74.0	12.0
	1-Decanol	28.29	76.6	11.5
	1-Deodecanol	29.53	79.2	11.3
FC-722-coated mica/ADSA-P	Decane	23.88	67.36	11.9
[104]	1-Pentanol	26.01	72.95	11.5
	trans-Decalin	27.19	73.38	11.9
	Hexadecane	27.62	75.94	11.4
	1-Decanol	28.99	78.84	11.2
	cis-Decanol	32.32	79.56	12.4
	Ethyl cinnamate	37.17	86.54	12.2
	Dibenzylamine	40.80	90.70	12.2
	Dimethyl sulfoxide (DMSO)	42.68	90.95	12.9
	1-Bromonaphthalene	44.31	93.81	12.4
	Diethylene glycol	44.68	94.22	12.4
	Ethylene glycol	47.55	97.87	12.1
	Diiodomethane	49.98	101.18	11.7
	2,2'-Thiodiethanol	56.26	104.56	12.7
	Formamide	59.08	108.49	12.0
	Glycerol	65.02	111.73	12.8
	Water	72.70	118.69	12.2
FC-722-coated silicon wafer/	Hexane	18.50	50.83	12.4
ADSA-P [105]	2-Octanol	26.42	74.74	11.2
	Hexadecane	27.62	75.64	11.5
	Glycerol	65.02	111.89	12.7
FC-725-coated silicon wafer/	Dodecane	25.64	71.02	11.8
ADSA-P [106]	Hexadecane	27.62	73.41	12.1
	3,3-Thiodipropanol	39.83	90.48	11.9
	Diethylene glycol	45.16	94.47	12.5
	Ethylene glycol	48.66	100.05	11.6
	2,2'-Thiodiethanol	53.77	101.07	13.2
	Formamide	59.08	106.89	12.7
	Glycerol	63.13	110.21	12.7
	Water	72.70	119.31	11.9

TABLE 8.1   (Continued)
Summary of Advancing Contact Angles of Various Liquids on Different Solid Surfaces

Solid Surface/Technique	Liquid	γ_{lv} (mJ/m²)	θ (degree)	γ_{sv}
Teflon (FEP)/capillary rise [103]	Dodecane	25.03	47.8	17.7
	2-Octanol	26.00	52.3	17.2
	Tetradecane	26.50	52.6	17.5
	1-Octanol	27.28	54.4	17.5
	Hexadecane	27.31	53.9	17.7
	1-Hexadecene	27.75	54.2	17.9
	1-Dodecanol	29.53	55.7	18.6
	Dimethylformamide	35.21	68.6	17.7
	Methyl salicylate	38.85	72.2	18.4
Hexatriacontane/capillary rise [89,107]	Ethylene glycol	47.70	79.2	20.3
	2,2'-Thiodiethanol	54.00	86.3	20.3
	Glycerol	63.40	95.4	20.6
	Water	72.80	104.6	20.3
Cholesteryl acetate/capillary rise [89,108]	Ethylene glycol	47.70	77.0	21.3
	2,2'-Thiodiethanol	54.00	84.3	21.3
	Glycerol	63.40	94.0	21.3
	Water	72.80	103.3	21.1
Poly(propene-*alt*-N-(*n*-hexyl) maleimide)/ADSA-P [109,110]	*cis*-Decalin	32.32	28.81	28.5
	Triacetin	35.52	39.45	28.3
	Diethylene glycol	44.68	61.04	26.7
	Glycerol	65.02	82.83	28.6
	Water	72.70	92.26	27.8
Poly(*n*-butyl methacrylate)/ADSA-P [111]	Diethylene glycol	45.16	58.73	28.0
	3-Pyridylcarbinol	47.81	60.30	29.2
	2,2'-Thiodiethanol	53.77	68.00	29.4
	Formamide	59.08	76.41	28.5
	Glycerol	65.02	82.11	29.0
	Water	72.70	90.73	28.7
Polystyrene/ADSA-P [112]	Dimethyl sulfoxide (DMSO)	42.68	50.67	29.7
	Diethylene glycol	44.68	52.41	30.5
	Ethylene glycol	48.66	61.20	29.3
	Formamide	59.08	74.76	29.4
	Glycerol	63.11	78.38	30.0
	Water	72.70	88.42	30.2
Poly(styrene-*alt*-(hexyl/10-carboxydecyl(90/10) maleimide))/ADSA-P [113]	Diethylene glycol	45.16	51.32	31.3
	Ethylene glycol	48.66	59.72	30.2
	Formamide	58.45	70.28	31.4
	Glycerol	63.13	76.51	31.0
	Water	72.70	87.13	31.0

(*Continued*)

TABLE 8.1   (Continued)

Summary of Advancing Contact Angles of Various Liquids on Different Solid Surfaces

Solid Surface/Technique	Liquid	γ_{lv} (mJ/m^2)	θ (degree)	γ_{sv}
Poly(methyl methacrylate/n-	1-Iodonaphthalene	42.92	35.67	35.7
butyl methacrylate)/ADSA-P	3-Pyridylcarbinol	47.81	49.22	34.2
[114]	2,2'-Thiodiethanol	53.77	57.84	34.6
	Formamide	59.08	66.33	34.0
	Glycerol	65.02	74.72	33.3
	Water	72.70	81.33	34.6
Poly(propene-alt-N-(n-propyl)	1-Iodonaphthalene	42.92	35.19	35.9
maleimide)/ADSA-P	1-Bromonaphthalene	44.31	30.75	38.6
[109,110]	1,3'-Diiodopropane	46.51	39.98	37.1
	2,2'-Thiodiethanol	53.77	54.04	36.5
	Glycerol	65.02	70.67	35.7
	Water	72.70	77.51	37.0
Poly(methyl methacrylate)/	1,3-Diiodopropane	46.51	36.95	38.3
ADSA-P [115]	3-Pyridylcarbinol	47.81	39.47	38.4
	Diiodomethane	49.98	42.25	39.0
	2,2'-Thiodiethanol	53.77	50.35	38.3
	Formamide	59.08	57.73	38.6
	Glycerol	65.02	66.84	37.9
	Water	72.70	73.72	39.3
Poly(propene-alt-N-	Diiodomethane	49.98	30.71	43.7
methylmaleimide)/ADSA-P	Glycerol	65.02	60.25	41.7
[105]	Water	72.70	69.81	41.8

Note: The γ_{sv} values are calculated from Equation 8.6. Reprinted from Kwok, D. Y. and Neumann, A. W., *Advances in Colloid and Interface Science*, 81, 167, 1999. With permission from Elsevier.

where f is as yet an unknown function. Combining this relation with the Young equation yields

$$\gamma_{sl} = F(\gamma_{lv}, \gamma_{sv}), \tag{8.4}$$

where F is another as yet unknown function. It was concluded that an equation of state for the interfacial tensions exists. In view of this, unlike what is claimed by the proponents of the surface tension component approaches, intermolecular forces of liquids and solids do not appear to have any independent major effect on contact angles. Existence of an equation of state is also supported by a thermodynamic phase rule for capillary systems. For a solid–liquid system that is in contact with the vapor of the liquid (two-component system), there are only two degrees of freedom; that is, the contact angle can be changed by changing either the liquid or the solid (see Chapter 7). It is emphasized that the above relationship between $\gamma_{lv}\cos\theta$ and γ_{lv} and γ_{sv} may not be obtainable from inadequate procedures or methods, such as measurement on static drops. The above contact angle patterns can only be inferred from advancing contact angles carefully measured at low rates of motion of the three-phase line [48,53].

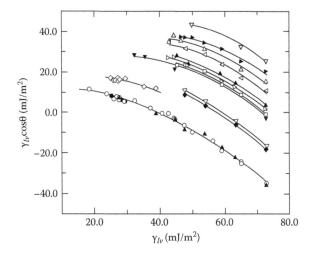

FIGURE 8.4 Plot of $\gamma_{lv}\cos\theta$ vs. γ_{lv} for various solid surfaces. The contact angles measured on each solid surface fall on a smooth curve. These patterns suggest that $\gamma_{lv}\cos\theta$ is only a function of γ_{lv} and γ_{sv}. ● FC-721-coated mica; ○ FC-722-coated mica and wafer; ▲ FC-725-coated wafer; ◇ Teflon FEP; ◆ hexatriacontane; ▽ cholesteryl acetate; ▼ poly(propene-*alt*-*N*-(*n*-hexyl)maleimide); □ poly(*n*-butyl methacrylate); ▷ polystyrene; ▲ poly(styrene-*alt*-(hexyl/10-carboxydecyl(90/10)maleimide)); ◁ poly(methyl methacrylate/*n*-butyl methacrylate); △ poly(propene-*alt*-*N*-(*n*-propyl)maleimide); ▶ poly(methyl methacrylate); ▿ poly(propene-*alt*-*N*-methylmaleimide). (Reprinted from Kwok, D. Y. and Neumann, A. W., *Advances in Colloid and Interface Science*, 81, 167, 1999. With permission from Elsevier.)

It was shown that the above functional relationship, called an "equation of state for interfacial tensions" may be sought in the form (see Chapter 9 for details)

$$\gamma_{sl} = \gamma_{lv} + \gamma_{sv} - 2\sqrt{\gamma_{lv}\gamma_{sv}}\, e^{-\beta(\gamma_{lv}-\gamma_{sv})^2},\qquad(8.5)$$

where β is an empirical constant. Combining this relation with the Young equation yields

$$\cos\theta = -1 + 2\sqrt{\frac{\gamma_{sv}}{\gamma_{lv}}}\, e^{-\beta(\gamma_{lv}-\gamma_{sv})^2}.\qquad(8.6)$$

Calculating the solid surface tension from this equation is straightforward. For a given set of γ_{lv} and θ for different liquids measured on one and the same type of solid surface, the constant β and γ_{sv} values can be determined by a multivariable optimization using a least-squares technique [55]. Starting out with arbitrary values for β and γ_{sv}, an iterative procedure can be used to identify a pair of β and γ_{sv} that provides the best fit of Equation 8.6 to the experimental γ_{lv} and θ values. The solid lines in Figures 8.3 and 8.4 have been obtained by applying this strategy to the experimental contact angles [48,53]. The β and γ_{sv} values corresponding to each solid surface of Figure 8.3 are given in Table 8.2. Since the values of β and γ_{sv} did not show any correlation, a weighted average of 0.000125 $(mJ/m^2)^{-2}$ was calculated for β.

TABLE 8.2

The Values of γ_{sv} and β for Three Solid Surfaces Obtained from Equation 8.6 Using a Multivariable Optimization Technique

Solid Surface	FC-721	FEP	PET
γ_{sv} (mJ/m²)	11.78	17.85	35.22
β (mJ/m²)$^{-2}$	0.000121	0.000134	0.000111

Note: β has a weighted average of 0.000125 (mJ/m²)$^{-2}$. Reprinted from Li, D. and Neumann, A. W., *Journal of Colloid and Interface Science*, 148, 190, 1992. With permission from Academic Press.

TABLE 8.3

γ_{sv} and β Values for Various Solid Surfaces Obtained from Equation 8.6 Using a Multivariable Optimization Technique

Solid Surface/Technique	Number of System	Least-Square Fit β	γ_{sv}
FC-721-coated mica/capillary rise	8	0.000124	11.7
FC-721-coated mica/ADSA-P	17	0.000111	11.8
FC-721-coated silicon wafer/ADSA-P	4	0.000111	11.8
FC-721-coated silicon wafer/ADSA-P	9	0.000114	11.9
Teflon FEP/capillary rise	9	0.000142	18.0
Hexatriacontane/capillary rise	4	0.000124	20.3
Cholesteryl acetate/capillary rise	4	0.000128	21.5
Poly(propene-*alt*-N-(*n*-hexyl)maleimide)/ADSA-P	5	0.000122	27.9
Poly(*n*-butyl methacrylate)/ADSA-P	6	0.000124	28.8
Polystyrene/ADSA-P	6	0.000120	29.7
Poly(styrene-*alt*-(hexyl/10-carboxydecyl(90/10) maleimide))/ADSA-P	5	0.000120	30.8
Poly(methyl methacrylate/*n*-butyl methacrylate)/ADSA-P	6	0.000136	34.7
Poly(propene-*alt*-N-(*n*-propyl)maleimide)/ADSA-P	6	0.000133	36.9
Poly(methyl methacrylate)/ADSA-P	7	0.000113	38.3
Poly(propane-*alt*-N-methylmaleimide)/ADSA-P	3	0.000167	43.4

Note: Reprinted from Kwok, D. Y. and Neumann, A. W., *Advances in Colloid and Interface Science*, 81, 167, 1999. With permission from Elsevier.

Using this β the γ_{sv} value corresponding to the contact angle of each liquid in Table 8.1 was calculated from Equation 8.6 and the results are shown in the last column of Table 8.3. For each solid surface, the γ_{sv} values are essentially independent of the choice of the probe liquid. This large body of data confirms the validity of the equation of state approach to determine solid surface tensions. A similar multivariable optimization can be carried out with the contact angles of different liquids on each of the solid surfaces of Figure 8.4. Table 8.3 shows the resulting β and γ_{sv} values. The value of β ranges from 0.000111 (mJ/m^2)$^{-2}$ to 0.000167 (mJ/m^2)$^{-2}$. A statistical analysis based on different β and γ_{sv} values again showed no correlation between them; that is, β did not change systematically with the solid surface. Thus an average value was determined for β as 0.000123 $\pm$ 0.000010 (mJ/m^2)$^{-2}$ [53]. This β is in excellent agreement with the original β value of 0.000125 (mJ/m^2)$^{-2}$ and such small differences in β do not influence the calculated γ_{sv} values significantly. This will be shown below.

These studies confirmed that the equation of state is compatible with experimental data and is capable of predicting consistent γ_{sv} values. Investigation of the contact angles of a large number of solid–liquid systems showed that the choice of probe liquids does not affect the results dramatically.

8.4 CONTACT ANGLE DEVIATIONS FROM SMOOTH CURVES OF $\gamma_{LV}\cos\theta$ VERSUS γ_{LV}

Close scrutiny of the smooth curves of $\gamma_{lv}\cos\theta$ versus γ_{lv} shows that contact angles usually do not fall perfectly on the curves and that there is a typical scatter of 1–3° [48,56]. An example is shown in Figure 8.5 for the contact angles of several

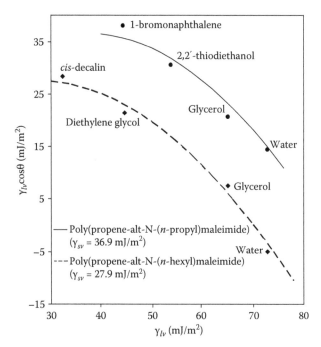

FIGURE 8.5 Plot of $\gamma_{lv}\cos\theta$ versus γ_{lv} for different liquids on two polymeric surfaces. There are deviations in the contact angles from the smooth curves. (Reprinted from Kwok, D. Y., Gietzelt, T., Grundke, K., Jacobasch, H.-J., and Neumann, A. W., *Langmuir*, 13, 2880, 1997. With permission from American Chemical Society.)

liquids on the films of two different polymers. The existence of the deviations introduces an element of uncertainty in the determination of solid surface tensions. For instance, a deviation of ±3° in the contact angles typically results in an error of ~±1.5 mJ/m² in the calculated value for the solid surface tension (γ_{sv}). A number of factors have been suggested as possible causes of the scatter [48,57]. Following are two examples:

(i) Adsorption: The equilibrium spreading pressure is given by

$$\pi = \gamma_s - \gamma_{sv},$$ (8.7)

where γ_s is the solid surface tension in vacuum. It was argued that for one and the same solid surface, γ_{sv} is expected to be constant if the vapor pressure of the liquid is negligible. But for liquids of fairly high vapor pressure, vapor adsorption onto the solid could cause γ_{sv} to be different from the surface tension of the bare solid, γ_s, giving rise to contact angle deviations of the type shown in the above figure. An equilibrium spreading pressure of ~1 mJ/m² has been suggested as a reasonable estimate for low-energy solid surfaces due to vapor adsorption [57].

(ii) Impurities: The liquids and solids selected for the contact angle experiments were usually of high purity, usually > 99%. Nevertheless, minor impurities in a solid–liquid system are unavoidable, for example, in the form of swelling of the polymer film. Such minor impurities were speculated to possibly account for the contact angle deviations.

To investigate the problem of deviation in contact angles from smooth curves and to tighten the determination of solid surface tensions from contact angles using the equation of state approach, the remainder of this chapter will focus on the question of contact angle deviations. In this context, it is important to explore the underlying causes for the contact angle deviations and to find out how solid surface tensions can be determined with accuracy better than that achievable so far. It is emphasized upfront that the existence of contact angle deviations does not weaken the status of the equation of state, as it was shown above that this approach yields fairly consistent values for solid surface tensions (Table 8.1) without the subsequent considerations.

In the present chapter, a comprehensive examination of contact angles of two diverse groups of liquids on the surfaces of four different fluoropolymers will be presented to explore the processes underlying contact angle deviations from smooth curves of $\gamma_{lv}\cos\theta$ versus γ_{lv}. This analysis will lead to a series of criteria that further scrutinize the equation of state for the determination of solid surface tensions.

A series of n-alkanes as well as liquids with "bulky" molecules were selected as test liquids for contact angle measurements. Table 8.4 lists the test liquids and shows their surface tension and purity. The surface tensions of the liquids were determined from

TABLE 8.4
Surface Tension of the Test Liquids from Pendant Drop Experiments at 24 ± 0.5°C and Purity of Each Liquid

Test Liquid	γ_{lv} (mJ/m²)	Purity (%)
n-Alkanes		
n-Hexane	18.32	99 + %
n-Heptane	20.03	99 + %
n-Octane	21.53	99 + %
n-Nonane	22.64	99 + %
n-Decane	23.54	99 + %
n-Undecane	24.47	99 + %
n-Dodecane	25.49	99 + %
n-Tridecane	26.04	99 + %
n-Tetradecane	26.58	99 + %
n-Pentadecane	27.07	99 + %
n-Hexadecane	27.30	99 + %
Liquids with Bulky Molecules		
OMTS	16.72	98%
OMCTS	18.20	98%
DMCPS	18.77	≥ 97%
p-Xylene	27.90	99 + %
o-Xylene	29.30	98%
cis-Decalin	32.16	99%
trans,trans,cis-1,5,9- Cyclododecatriene	33.88	98%
Tetralin	36.15	99%
Ethyl trans-cinnamate	36.60	99%
Diethyl phthalate	36.67	99.50%
Methyl salicylate	38.71	99 + %
Dibenzylamine	39.70	98%
Benzyl benzoate	41.75	99%
Lepidine	43.20	99%
4-Benzylisothiazole	44.03	98%
2-Pyridyl carbinol	47.55	98%
Naphthalene Compounds		
1-Fluoronaphthalene	36.12	≥ 99%
1-Methylnaphthalene	38.10	≥ 97%
1-Chloronaphthalene	40.65	97%
1-Bromonaphthalene	43.70	98%
1-Iodonaphthalene	46.59	98%
Distilled water	72.29	

Note: Reprinted from Tavana, H., et al., *Advances in Colloid and Interface Science*, 132, 1, 2007. With permission from Elsevier.

pendant drop experiments. Four fluoropolymers with distinct molecular structure and chemical composition were selected as the coating materials, as listed below:

a. poly(4,5-difluoro-2,2-bis(trifluoromethyl)-1,3-dioxole-*co*-tetrafluoroethylene), 65 mole% dioxole (Teflon AF 1600), a fluoropolymer with a glass transition temperature of $T_g = 160°C$ [58].
b. poly(2,2,3,3,4,4,4-heptafluorobutyl methacrylate) (EGC-1700), a fluorinated acrylate polymer with $T_g = 30°C$ [59].
c. poly(ethene-*alt*-N-(4-(perfluoroheptylcarbonyl)aminobutyl)maleimide) (ETMF), which is a maleimide copolymer possessing a long fluorinated side chain and has a T_g of 100°C [60].
d. poly(octadecene-*alt*-N-(4-(perfluoroheptylcarbonyl)aminobutyl)maleimide) (ODMF) is also a maleimide copolymer fairly similar to ETMF but possesses an additional *n*-hexadecyl side chain and has a $T_g = 65°C$ [60].

The repeat unit of each polymer is shown in Figure 8.6. The polymer films were prepared by spin-coating and dip-coating techniques, as described in Chapter 6. The morphology of the polymer surfaces was characterized by atomic force microscopy (AFM) in the tapping mode. Figure 8.7 illustrates an example of the morphology of Teflon AF 1600 over a scanned area of 4×4 μm². The films are very smooth with root-mean-square (RMS) roughness of ~0.4 nm and maximum peak-to-valley distances of ~2.0 nm. Similar measurements for EGC-1700, ETMF, and ODMF surfaces yielded RMS mean roughness of ~1.4 nm, ~0.4 nm, and ~1.3 nm, with corresponding

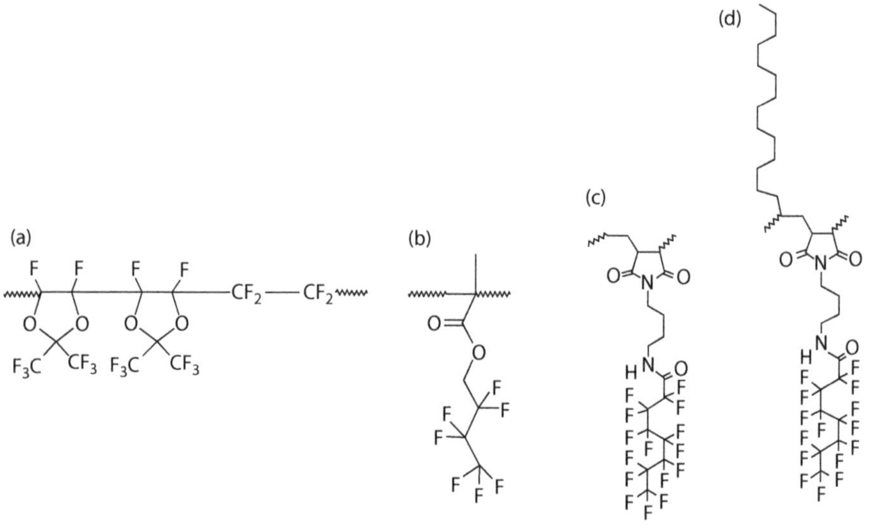

FIGURE 8.6 Repeat unit of: (a) Poly(4,5-difluoro-2,2-bis(trifluoromethyl)-1,3-dioxole-*co*-tetrafluoroethylene), 65 mole% dioxole (Teflon AF 1600); (b) poly(2,2,3,3,4,4,4-heptafluorobutyl methacrylate) (EGC-1700); (c) poly(ethene-*alt*-N-(4-(perfluoroheptylcarbonyl)aminobutyl)maleimide) (ETMF); (d) poly(octadecene-*alt*-N-(4-(perfluoroheptylcarbonyl)aminobutyl)maleimide) (ODMF).

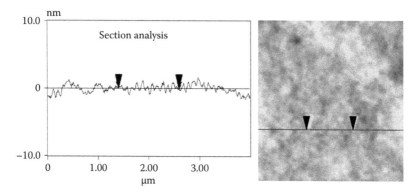

FIGURE 8.7   Morphology AFM image of a Teflon AF 1600 film for a scanned domain of $4 \times 4 \ \mu m^2$, a section analysis diagram, and the corresponding data are shown. The surface is quite smooth with RMS roughness of ~0.4 nm and maximum peak-to-valley distances ~2.0 nm.

maximum peak-to-valley distances of ~8.5 nm, ~3.5 nm, and ~4.0 nm, respectively. It is known that roughness on this scale does not influence the contact angles significantly [61].

8.5   REPRODUCIBILITY OF CONTACT ANGLE MEASUREMENTS

To address the problem of contact angle deviations from smooth curves of $\gamma_{lv}\cos\theta$ versus γ_{lv} in detail, it should be first established whether the contact angles are solely a material property or are influenced by other factors such as the thickness of the polymeric films and film preparation techniques. Langmuir [62] and then Zisman [63] were the first to state that a single monolayer should be sufficient to determine the wetting properties of a solid film. However data in the literature seemingly suggest that this is not necessarily the case. Following are two examples: (i) Cho et al. reported contact angles of water on different thicknesses of Teflon films produced by spin-coating and thermal evaporation techniques [64]. On spin-coated films of ~5 nm (produced by a 0.01% concentration solution), the contact angle of water was 105°. Similar values were measured on thermally evaporated films thicker than 3 nm. However, the contact angle decreased to as low as 51° when the thickness of a thermally evaporated film was reduced to 1.5 nm. (ii) Extrand investigated the contact angles of water and ethylene glycol on three different polymer surfaces: natural rubber (NR), polystyrene (PS), and poly(methyl methacrylate) (PMMA) [65]. The films were produced by a spin-casting technique on both heated silicon wafers and ozone-treated wafers. In the case of NR films, the "critical thickness"; that is, thickness above which contact angles do not depend on the film thickness, was found to be 9 nm for heated wafers and 30 nm for ozone-treated wafers. For both PS and PMMA films spin-cast on heated wafers, the critical thickness was as low as 2 nm. In these studies, inhomogeneity of thin films below a critical thickness was claimed as the cause for variation of contact angles with film thickness. Obviously, for the determination of the surface tension of such polymers, films in excess of such low thickness have to be studied.

8.5.1 CONTACT ANGLES AND FILM THICKNESS

To study the possible effect of coating film thickness on contact angles [66], a spin-coating technique can be used for film preparation, as described in Chapter 6. In this technique, film thickness is primarily influenced by spinning speed and concentration of polymeric solution. While increasing spinning rate tends to decrease the film thickness, a higher concentration of the coating solution would increase it. Different combinations of these two parameters were chosen to produce films of different thickness. Teflon AF 1600 was dissolved in FC-75 at volumetric ratios of 1:1, 1:4, and 1:8 to produce different concentrations of the solution, and spinning rates of 1000, 4000, and 8000 rpm were selected for the coating process. Contact angles of distilled water were measured on the surfaces of Teflon AF 1600 with different thickness at a slow rate of motion of 0.3–0.4 mm/min. For each thickness at least three experiments were performed, each on a freshly prepared solid surface. An example is shown in Table 8.5 for contact angles of water on films with a thickness of 420 nm resulting from the 1:1 (v/v) solution and a spinning rate of 1000 rpm in the spin-coating process. Since the contact angles were constant for all experiments, they were averaged and yielded a mean value of 127.05 ± 0.08°.

Similar contact angle measurements were performed with distilled water on Teflon AF 1600 films of different thickness prepared by spin-coating and the results are summarized in Table 8.6. This table lists the thickness of the coated films as determined by ellipsometry measurements and the corresponding combinations of the spinning rate and concentration of the polymeric solution. Applying different ratios of Teflon AF 1600 and FC-75 and using a range of spinning rates resulted in coated films with thicknesses ranging from 27 to 420 nm. It can also be seen that

TABLE 8.5

Advancing Contact Angles of Water on Teflon AF 1600 Films of 420 nm Thickness Prepared by a 1/1 (v/v) Polymer/Solvent Ratio and a Spinning Rate of 1000 rpm

Run	Three-Phase Line Velocity (mm/min)	Advancing Contact Angle (°)
a	0.32	126.90 ± 0.08
b	0.35	126.81 ± 0.06
c	0.31	126.87 ± 0.06
d	0.32	127.02 ± 0.08
e	0.37	127.42 ± 0.10
f	0.35	127.00 ± 0.08
g	0.32	127.31 ± 0.11
	mean:	127.05 ± 0.08

Note: The three-phase line velocity corresponding to each contact angle experiment is also given.

TABLE 8.6

Contact Angles of Water on Teflon AF 1600 Films of Different Thickness

Teflon AF 1600: FC-75 (v/v)	Spinning Speed (rpm)	Film Thickness (nm)	Replications	Mean Advancing Angle (°)
1:1	1000	420	7	127.05 ± 0.08
1:4	1000	150	5	127.11 ± 0.11
1:4	4000	72	3	127.04 ± 0.19
1:4	8000	72	3	126.86 ± 0.13
1:8	4000	27	4	127.10 ± 0.23
1:8	8000	27	4	127.07 ± 0.20
			mean:	127.04 ± 0.19

Note: The effect of concentration and spinning speed is also illustrated.

at fairly low concentrations of the coating solution (1:4 and 1:8), the film thickness does not depend on the spinning speed at rates above 4000 rpm. It is suggested that concentration of the polymeric solution has a more important effect in producing the final thickness of the Teflon AF 1600 layer than the rate of spinning. The number of contact angle measurements for each combination of concentration of the solution and spinning rate is also given.

The key result of Table 8.6 is that contact angles do not depend on the thickness of the Teflon AF 1600 films in the range from 27 to 420 nm. Within the 95% confidence limits, the contact angles are the same. These results imply smoothness and homogeneity of the polymer films. On such surfaces, except for the very first molecular layers, the subsequent layers are not in contact with the substrate, and presumably do not interact with it. Therefore the configuration of the polymer molecules at the solid–liquid interface is essentially the same for films of different thickness.

8.5.2 CONTACT ANGLES AND FILM PREPARATION TECHNIQUES

To find out whether contact angles are somehow influenced by film preparation techniques, contact angle measurements were performed with water, *n*-hexadecane, and 1-bromonaphthalene on dip-coated and spin-coated Teflon AF 1600 surfaces [66]. The thickness of the polymer film on dip-coated surfaces is 470 nm, determined from ellipsometry measurements. The experiments were all performed at low rates of advancing of the three-phase line. The mean value of contact angle from these experiments and the corresponding number of measurements are given in Table 8.7. The mean contact angle of water on the spin-coated surfaces is the grand average value from Table 8.6. The contact angles are essentially the same on both types of surfaces for each liquid, indicating that for solid surfaces of high quality, the coating technique does not have a dramatic effect on advancing contact angles. However, it is interesting that the mean values are consistently slightly higher for the dip-coated surfaces than for spin-coated surfaces. If this effect is

TABLE 8.7
Comparison of Advancing Contact Angles of Distilled Water, 1-Bromonaphthalene, and n-Hexadecane on Teflon AF 1600 Films Prepared by Dip-Coating and Spin-Coating Techniques

Liquid	Dip-Coating θ (°)	Replications	Spin-Coating θ (°)	Replications
Water	127.58 ± 0.19	6	127.04 ± 0.19	26
1-Bromonaphthalene	89.90 ± 0.19	4	89.51 ± 0.17	2
n-Hexadecane	69.68 ± 0.11	6	69.48 ± 0.09	2

TABLE 8.8
Reproducibility of the Advancing Contact Angles of n-Hexadecane from Five Different Measurements on Fresh Films of Teflon AF 1600

Run Number	θ (°)
1	69.45 ± 0.07
2	69.31 ± 0.12
3	69.62 ± 0.08
4	69.65 ± 0.15
5	69.37 ± 0.09
mean:	69.48 ± 0.16

Note: Reprinted from Tavana, H., et al., *Advances in Colloid and Interface Science*, 132, 1, 2007. With permission from Elsevier.

real, it might indicate that, since film formation in the dip-coating technique is slower, the polymer chains might be able to assume a surface configuration of slightly lower surface tension.

The results presented above assure that contact angles are solely a property of the polymer material and do not depend on the film thickness and film production techniques. The key remaining issue is the reproducibility of contact angle measurements and whether or not contact angle deviations from smooth curves can be a consequence of measurement errors.

8.5.3 REPRODUCIBILITY OF CONTACT ANGLES OF N-ALKANES

Contact angle measurements were performed with a homologous series of n-alkanes on the dip-coated films of Teflon AF 1600 [56]. To ensure reproducibility, at least five experiments were carried out for each liquid, each on a fresh solid surface. Table 8.8 is a typical example and shows the advancing contact angles of *n*-hexadecane obtained from five measurements. Since contact angles were constant during each run, they were averaged to yield a mean value for the experiment. Thus the contact angle of *n*-hexadecane on Teflon AF 1600 films is 69.48° ± 0.16. This procedure was repeated for each *n*-alkane. The results, presented in Table 8.9, are compared with those measured in 2000 by another operator on surfaces prepared similarly. The first column represents the older data and the second column shows the contact angles of each liquid from the 2004 work [53]. Comparing each pair of contact angles shows that the measurements have been accurately reproduced. The average reproducibility (the deviation between the two sets) is just 0.2°, which is inside the error limits of ADSA for contact angle measurements. This result implies consistency and reproducibility of ADSA. For one and the same liquid, carefully measured contact angles by different operators on well-prepared solid surfaces are identical.

In conclusion, the thickness of the coating films does not influence the contact angles as long as the polymer films are smooth and homogeneous, showing that film preparation technique, that is, dip-coating and spin-coating, does not have a dramatic effect on advancing contact angles. The comparison of contact angles of a series of *n*-alkanes on surfaces of Teflon AF 1600 by different operators at different times assure that contact angle deviations from smooth curves of $\gamma_{lv}\cos\theta$ versus γ_{lv} are not experimental errors and must be physically real. This calls for an elucidation of the underlying physical causes of the contact angle deviations.

TABLE 8.9
Older and Recent Advancing Contact Angles of
n-Alkanes on Teflon AF 1600 Surfaces

Liquid	Contact Angle Data (°) (2000)	Contact Angle Data (°) (2004)
n-Hexane	40.87 ± 0.37	40.31 ± 0.10
n-Heptane	47.49 ± 0.20	47.17 ± 0.10
n-Octane	52.37 ± 0.17	52.45 ± 0.15
n-Nonane	56.26 ± 0.09	56.22 ± 0.14
n-Decane	59.28 ± 0.11	59.29 ± 0.18
n-Undecane	61.78 ± 0.06	61.60 ± 0.22
n-Dodecane	63.59 ± 0.12	63.78 ± 0.32
n-Tridecane	65.30 ± 0.13	65.68 ± 0.05
n-Tetradecane	66.84 ± 0.10	67.15 ± 0.14
n-Pentadecane	68.25 ± 0.17	68.06 ± 0.14
n-Hexadecane	69.33 ± 0.06	69.48 ± 0.16

Note: Reprinted from Tavana, H., Lam, C. N. C., Friedel, P., Grundke, K., Kwok, D. Y., Hair, M. L., and Neumann, A. W., *Journal of Colloid and Interface Science*, 279, 493, 2004. With permission from Elsevier.

8.6 IDENTIFICATION OF THE CAUSES OF CONTACT ANGLE DEVIATIONS ON FLUOROPOLYMER SURFACES

8.6.1 GEOMETRICAL PROPERTIES OF LIQUID MOLECULES

The need to resolve the problem of deviation in the contact angles from smooth curves of $\gamma_{lv}\cos\theta$ versus γ_{lv} became apparent from the observation of different contact angle patterns for liquids with different molecular structures. Liquids consisting of chain-like molecules such as n-alkanes on polymer films usually show time-dependent receding contact angles; that is, the receding angles keep decreasing as the solid–liquid contact time increases (see Figure 8.8a for contact angles of n-tridecane on Teflon AF 1600 surface). It was suggested that in such systems phenomena such as liquid penetration modify the solid surface (see Chapter 7). Consequently the solid surface and the liquid become more alike, causing the contact angles to decrease. On the other hand, contact angle measurements with octamethylcyclotetrasiloxane (OMCTS), which consists of bulky molecules, yielded time-independent receding angles [56], suggesting that such liquids do not appear to modify the solid

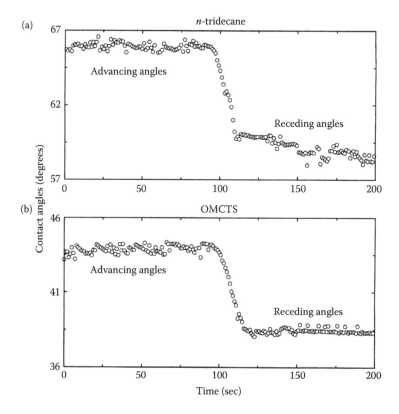

FIGURE 8.8 Contact angles of liquids with different molecular structures on Teflon AF 1600 films. (a) Liquids with chain-like molecules (e.g., n-tridecane) show time-dependent receding angles; (b) Liquids consisting of bulky molecules (e.g., OMCTS) yield constant receding angles.

surface significantly (Figure 8.8b). To test whether: (i) the shape and size of liquid molecules indeed affects the contact angles; and (ii) deviations in the contact angles from smooth curves are a consequence of sorption-like processes, we selected liquids with different geometrical properties for the contact angle measurements: A series of n-alkanes that is, n-hexane to n-hexadecane, and liquids with bulky molecules. The term bulky refers to molecules that are: (i) fairly round; (ii) rigid due to the presence of rings; and (iii) less flexible than n-alkane molecules.

To visualize the shapes of the molecules of the test liquids and to obtain quantitative information about geometrical properties of bulky molecules compared to the molecules of n-alkanes, all these molecules were modeled using computational chemistry software HyperChem 7.5 that provides optimized geometries [67]. The results are presented in Figure 8.9. Molecules of n-alkanes (e.g., n-hexane and n-hexadecane) have a worm-like structure with a mean diameter of only 0.2 nm. On the other hand, the mean diameter of the bulky molecules ranges from 0.61 nm for cis-decalin to 0.9 nm for OMCTS [68]. Our modeling results are in agreement with

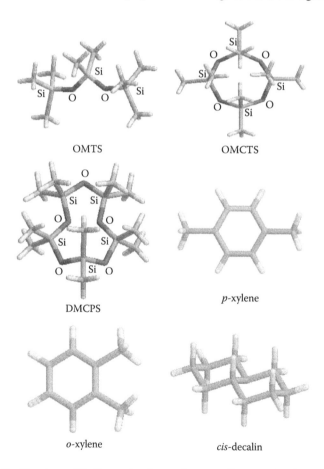

OMTS

OMCTS

DMCPS

p-xylene

o-xylene

cis-decalin

FIGURE 8.9 Structure of bulky molecules and two n-alkane molecules obtained from geometry optimization calculation using computational chemistry software HyperChem 7.5.

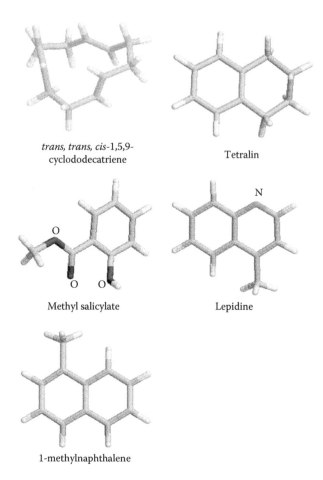

trans, trans, cis-1,5,9-
cyclododecatriene

Tetralin

Methyl salicylate

Lepidine

1-methylnaphthalene

FIGURE 8.9 (Continued) Structure of bulky molecules and two *n*-alkane molecules obtained from geometry optimization calculation using computational chemistry software HyperChem 7.5.

the data available in the literature, for example 0.75 to 1.08 nm for OMCTS [69–71]. It is therefore expected that the use of liquids with bulky molecules for contact angle measurements reduces the extent of interactions such as liquid sorption by the solid.

8.6.2   Interpretation of Contact Angles of Liquids Consisting of Bulky Molecules

8.6.2.1   Liquids with Bulky Molecules/Teflon AF 1600 Systems

The contact angle results for liquids with bulky molecules on Teflon AF 1600 surfaces are presented in two sections. This distinction is made based upon the different contact angle patterns the liquids yielded and also the difference in their molecular structures, as explained below. It is noted that the distinction is made for the contact angle results with Teflon AF 1600 only.

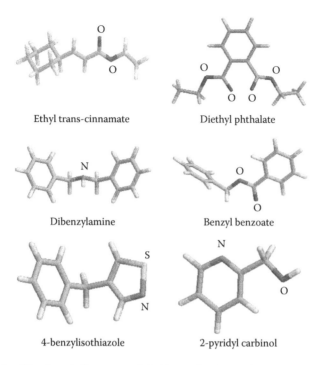

Ethyl trans-cinnamate   Diethyl phthalate

Dibenzylamine   Benzyl benzoate

4-benzylisothiazole   2-pyridyl carbinol

FIGURE 8.9 (Continued) Structure of bulky molecules and two *n*-alkane molecules obtained from geometry optimization calculation using computational chemistry software HyperChem 7.5.

8.6.2.1.1 First Group of Liquids with Bulky Molecules

Advancing contact angles of the liquids from the first group (11 liquids) were measured on Teflon AF 1600 films and the results are given in Table 8.10 [56,68]. Each contact angle in this table is an average value from at least five measurements on fresh polymer films. A multivariable optimization was used to determine γ_{sv} and β by finding the best fit of Equation 8.6 to the contact angle data. This yielded $\gamma_{sv} = 13.61$ mJ/m^2 and $\beta = 0.000116$ (mJ/m^2)$^{-2}$. The corresponding plot of $\gamma_{lv}\cos\theta$ versus γ_{lv} is shown in Figure 8.10. The contact angles of liquids with bulky molecules fall perfectly on this curve. The error analysis showed that the deviations are indeed small, averaging only $\pm0.24°$. The contact angle deviation ($\Delta\theta$) from this curve for each liquid is given in Table 8.10 (see the footnote of Table 8.10 for calculation of $\Delta\theta$). The minus sign for deviations means that the contact angle point is above the curve, while the plus sign shows that the contact angle falls below the curve. The largest deviation is $-0.69°$ for OMTS. It is noted that the β value is in good agreement with the literature value of $\beta = 0.000125$ (mJ/m^2)$^{-2}$, and such a small difference does not affect the γ_{sv} calculation significantly [61]. This is illustrated in Table 8.10 where the values of γ_{sv} calculated from each contact angle are given for both β values. Therefore $\gamma_{sv} = 13.61 \pm 0.07$ mJ/m^2 can be taken as the actual surface tension of the Teflon AF 1600 surface.

The consistency of the calculated γ_{sv} values from the contact angles of liquids with bulky molecules as well as the corresponding advancing and receding contact

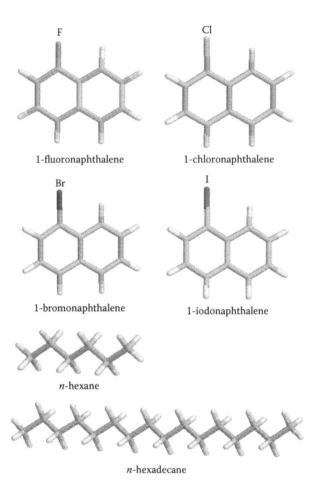

FIGURE 8.9 (Continued) Structure of bulky molecules and two *n*-alkane molecules obtained from geometry optimization calculation using computational chemistry software HyperChem 7.5.

angle patterns suggest that these liquids do not show significant interaction with the polymer film, for example, through liquid penetration. For example, the contact angles of *cis*-decalin and DMCPS are shown in Figure 8.11. Both the advancing and receding angles are constant over time, implying that the polymer film is not modified due to contact with the test liquids. It is concluded that these liquids with bulky molecules are "inert" with respect to Teflon AF 1600 film, and therefore are suited to characterize its surface tension. Most likely, the special geometry of bulky molecules is an important factor in making them inert with respect to Teflon AF 1600.

8.6.2.1.2 Second Group of Liquids with Bulky Molecules

If size and shape of the molecules of a liquid were the only parameters affecting the contact angle of a solid–liquid system, then contact angles of any liquid with bulky molecules would fall perfectly on the $\gamma_{sv} = 13.61$ mJ/m^2 curve in Figure 8.10.

TABLE 8.10

Advancing Contact Angles of the First Group of Liquids with Bulky Molecules on the Films of Teflon AF 1600, Contact Angle Deviations from the Smooth Curve of γ_{sv} = 13.61 mJ/m², the γ_{sv} Calculated Using Two Different β Values, and the Vapor Pressure of the Liquids

Liquid	θ (°)	Δθ (°)	γ_{sv} (mJ/m²)ᵃ	γ_{sv} (mJ/m²)ᵇ	p_v (kPa)
OMTS	35.75 ± 0.12	− 0.69	13.75	13.75	0.346
OMCTS	43.68 ± 0.13	+ 0.17	13.58	13.58	0.108
DMCPS	45.65 ± 0.15	+ 0.10	13.63	13.64	—
p-Xylene	68.65 ± 0.09	0.00	13.61	13.65	1.186
o-Xylene	71.17 ± 0.13	+ 0.12	13.57	13.63	0.884
cis-Decalin	75.53 ± 0.22	+ 0.01	13.61	13.68	0.091
trans,trans,cis-1,5,9-Cyclododecatriene	78.29 ± 0.19	+ 0.33	13.50	13.58	0.003
Tetralin	81.06 ± 0.20	+ 0.09	13.58	13.68	0.042
Methyl salicylate	83.69 ± 0.23	−0.44	13.77	13.90	0.003
Lepidine	89.49 ± 0.44	+ 0.30	13.49	13.66	0.001
1-Bromonaphthalene	89.80 ± 0.44	+ 0.07	13.58	13.76	0.001
Mean:		± 0.24	13.61 ± 0.07	13.68 ± 0.05	

Notes: Procedure to calculate deviation Δθ in the contact angle of a liquid from the smooth curve:

1. γ_{sl} is calculated from equation of state with γ_{lv} of the liquid and γ_{sv} = 13.61 mJ/m² as the surface tension of Teflon AF 1600:

$$\gamma_{sl} = \gamma_{lv} + \gamma_{sv} - 2\sqrt{\gamma_{lv}\gamma_{sv}}\,e^{-\beta(\gamma_{lv}-\gamma_{sv})^2}.$$

2. θ′ (ideal contact angle that would fall on the smooth curve) is obtained from Young's equation:

$$\theta' = \cos^{-1}\left(\frac{\gamma_{sv}-\gamma_{sl}}{\gamma_{lv}}\right).$$

3. Contact angle deviation is the difference between actual and ideal contact angles (i.e., Δθ = θ − θ′).

   (a) γ_{sv} values calculated by β = 0.000116 (mJ/m²)⁻².
   (b) γ_{sv} values calculated by β = 0.000125 (mJ/m²)⁻² (literature value).

However, this is not the case. Table 8.11 gives the contact angles of six other liquids with bulky molecules, labeled as the second group. In the $\gamma_{lv}\cos\theta$ versus γ_{lv} plot for Teflon AF 1600, these data deviate somewhat from this curve (see Figure 8.12; the circles show the contact angles of the liquids in the first group whereas the diamonds represent the liquids from the second group). Table 8.11 gives the deviations in contact angle from the "ideal" curve of Figure 8.12, calculated according to the procedure given in the footnote of Table 8.10. It is seen that the deviations are up to ~3°. Taking into account that the shape and size of molecules of these liquids and those belonging to the first group are similar, it is suggested that there must be additional factors affecting contact angles.

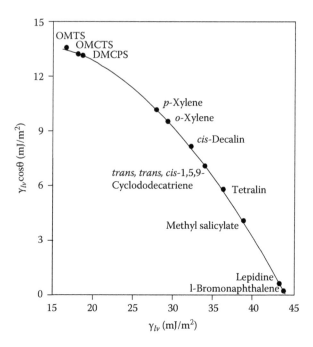

FIGURE 8.10 $\gamma_{lv}\cos\theta$ versus γ_{lv} for the contact angles of liquids with bulky molecules (first group) on Teflon AF 1600 surfaces. The smooth curve corresponds to $\gamma_{sv} = 13.61$ mJ/m^2 and $\beta = 0.000116$ (mJ/m^2)$^{-2}$.

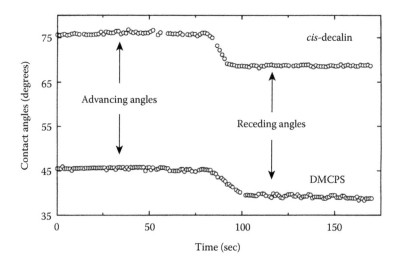

FIGURE 8.11 Advancing and receding contact angles of *cis*-decalin and DMCPS on Teflon AF 1600 films. The receding angles do not show time-dependence, confirming that sorption-type processes are not operative. (Reprinted from Tavana, H., Simon, F., Grundke, K., Kwok, D. Y., Hair, M. L., and Neumann, A. W., *Journal of Colloid and Interface Science*, 291, 497, 2005. With permission from Elsevier.)

TABLE 8.11

Advancing Contact Angles of the Second Group of Liquids with Bulky Molecules on the Films of Teflon AF 1600 and the Contact Angle Deviations from the Smooth Curve of $\gamma_{sv} = 13.61$ mJ/m²

Liquid	θ (°)	$\Delta\theta$ (°)	p_v (kPa)
Ethyl trans-cinnamate	84.68 ± 0.14	$+3.13$	0.037
Diethylphthalate	84.73 ± 0.27	$+3.09$	0.00006
Dibenzylamine	88.25 ± 0.29	$+2.96$	0.072
Benzyl benzoate	89.20 ± 0.19	$+1.58$	0.133
4-Benzylisothiazole	92.46 ± 0.36	$+2.54$	–
2-Pyridyl carbinol	96.70 ± 0.11	$+3.03$	–

Note: The plus sign in the deviation $\Delta\theta$ indicates that a data point lies below the ideal curve of Figure 8.12.

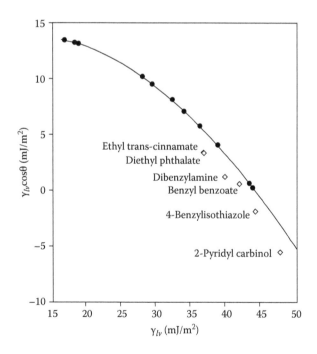

FIGURE 8.12 $\gamma_{lv}\cos\theta$ versus γ_{lv} for the contact angles of the second group of liquids with bulky molecules on Teflon AF 1600 surfaces. The smooth curve is identical with that of Figure 8.10 and represents $\gamma_{sv} = 13.61$ mJ/m² and $\beta = 0.000116$ (mJ/m²)⁻².

The consideration of Young's equation, which is assumed to be valid and applicable, is useful to explain the observed deviations. In this equation both γ_{lv} and θ are correctly measured quantities. Therefore, the contact angle deviations must be connected with processes that affect γ_{sv} and/or γ_{sl}. Because of the extremely low vapor pressure of these liquids (listed in Table 8.11), significant adsorption of their vapor onto the solid surface is not likely to take place. Hence, γ_{sv} is expected to remain constant during the contact angle experiments. Therefore, the reason must be sought in the solid–liquid interfacial tension, γ_{sl}.

The main difference between the two groups of liquids lies in their chemical structure. It is seen in Figure 8.9 that unlike most of the liquids in the first group, all liquid molecules in the second group contain exposable nitrogen (N) and/or oxygen atoms (O). Due to the fairly high electronegativity of these two atoms, their presence in a molecule causes a nonuniform electron density distribution over the molecule [68]. As a result, the part of the molecule including these atoms becomes more negatively charged. A similar hypothesis is plausible for the chains of Teflon AF 1600 due to the existence of CF_2 and CF_3 moieties. It is speculated that when liquid molecules approach a Teflon AF 1600 surface, since like charges repel each other, the negatively charged parts of the liquid molecules are repelled by the polymer chains, whose fluorine atoms are directed outward. This causes liquid molecules to be reoriented at the solid–liquid interface with their negatively charged parts directed away from the solid. As a result the solid–liquid interfacial tension becomes different from that given by the equation of state, Equation 8.5, since this equation does not take into account such restructuring effects. Therefore, the corresponding contact angles deviate from the $\gamma_{sv} = 13.61$ mJ/m^2 smooth curve.

It might be argued that reorientation could take place for the polymer chains rather than in the liquid molecules because polymer chains are known to be fairly mobile. For instance in response to the change in the contacting medium (e.g., from air to liquid), polymer chains might be restructured by a disappearance of hydrophobic CF_2 and CF_3 groups from the top layer of the surface [72–76]. However, migration of these moieties into the bulk solid phase would decrease the solid–liquid interfacial tension. This would yield a contact angle value lower than that measured for each of the liquids in the second group, and the experimental points would fall above the curve of $\gamma_{sv} = 13.61$ mJ/m^2. Therefore, change in the configuration of the Teflon films cannot be the reason for contact angle deviations of these liquids.

A second possibility exists for dibenzylamine and 2-pyridyl carbinol that contain N–H and O–H bonds, respectively. Because of the higher electronegativity of fluorine atoms compared to oxygen and nitrogen atoms, the hydrogen atom from the liquid molecules might be attracted to a fluorine atom from the polymer chains, forming a hydrogen bond with it. Formation of H-bonds between solid and liquid molecules upon their contact has also been reported [77]. If such interactions indeed take place at the solid–liquid interface, the solid–liquid interfacial tension (γ_{sl}) will be different from the value predicted by Equation 8.5, giving rise to contact angle deviations from the $\gamma_{sv} = 13.61$ mJ/m^2 curve. This process is less likely to occur than reorientation of liquid molecules at the surface; however, it cannot be excluded totally.

8.6.2.2 Liquid with Bulky Molecules/EGC-1700 Systems

Similar contact angle measurements were performed with liquids consisting of bulky molecules on the films of EGC-1700 [78]. It is a semifluorinated acrylic polymer whose molecular structure is shown in Figure 8.6.

Table 8.12 contains the advancing contact angles of liquids with bulky molecules and the solid surface tension values, γ_{sv}, calculated from Equation 8.6. Again, the contact angles in this table are the average values obtained from at least five experiments. It is pointed out that lepidine and methyl salicylate did not yield useful contact angle values. Lepidine dissolved the surface film or initiated hydrolyzation reactions while the contact line of methyl salicylate showed stick-slip on the solid surface.

The results presented in this table differ in certain respects from those presented above for Teflon AF 1600 (Table 8.10). The γ_{sv} values are not consistent and vary from 13.82 mJ/m^2 calculated for OMCTS to 16.10 mJ/m^2 for 1-bromonaphthalene. The immediate question that arises is: what mechanisms are responsible for the different behavior of these liquids on the two polymers (i.e., EGC-1700 and Teflon AF

TABLE 8.12

Advancing Contact Angles of Liquids with Bulky Molecules on EGC-1700 Surfaces and the Corresponding 95% Confidence Limits, the Solid Surface Tension Values Calculated from the Contact Angles, and Deviation in the Contact Angle of Each Liquid from the Smooth Curve of $\gamma_{sv} = 13.84$ mJ/m^2 and $\beta = 0.000125$ (mJ/m^2)$^{-2}$

Liquid	θ (°)	γ_{sv} (mJ/m^2)	$\Delta\theta$ (°)
OMCTS	42.33 ± 0.13	13.82	−0.08
DMCPS	44.47 ± 0.11	13.86	+ 0.09
p-Xylene	64.78 ± 0.15	14.77	−3.20
o-Xylene	66.35 ± 0.18	15.07	−4.09
cis-Decalin	72.75 ± 0.12	14.56	−2.25
trans,trans,cis-1,5,9-Dodecatriene	73.70 ± 0.11	15.08	−3.82
Tetralin	76.63 ± 0.10	15.17	−3.96
Diethyl phthalate	75.38 ± 0.14	15.89	−5.52
1-Methylnaphthalene	80.67 ± 0.15	14.62	−2.40
Methyl salicylate		(stick-slip)	
Dibenzylamine	81.32 ± 0.15	15.12	−3.36
Lepidine		(polymer dissolution)	
1-Bromonaphthalene	84.04 ± 0.17	16.10	−5.53

Note: Reprinted from Tavana, H., Simon, F., Grundke, K., Kwok, D. Y., Hair, M. L., and Neumann, A. W., Journal of Colloid and Interface Science, 291, 497, 2005. With permission from Elsevier.

1600), which were expected to be fairly similar in that they both expose mainly CF_3 groups at the outermost layer of the surface.

This question can be answered by considering the angle-resolved x-ray photo-electron spectroscopy (XPS) analysis of EGC-1700, its molecular configuration and those of the test liquids. The XPS analysis shows that the (F):(C) and (O):(C) elemental ratios do not depend on the take-off angle and similar values are obtained at 0°, 60°, and 75° [78]. It is suggested that the bulk of EGC-1700 polymer film has nearly the same composition as the top surface layers. This finding implies that the semi-fluorinated butyl side chains are not able to form an ordered molecular structure at the surface film. Furthermore, EGC-1700 has a low glass transition temperature ($T_g = 30°C$), which means that the chains are very flexible. It is therefore quite plausible that groups of a moderate polarity from EGC-1700 chains (e.g., ester groups) become accessible to certain test liquids.

If we consider the molecular structure of the liquids listed in Table 8.12, except for OMCTS and DMCPS, they all contain unsaturated bonds and/or electronegative substituents that are potential interaction sites of the molecule. It is speculated that upon advancing of the drop front in the contact angle measurement, some groups of the polymer chains less hydrophobic than CF_2 and CF_3, such as methyl and ester groups, come in contact with the liquid molecules so that the solid–liquid interfacial tension is not determined solely by the fluorine-containing moieties. On the other hand, OMCTS and DMCPS molecules do not contain unsaturated bonds or exposable electronegative atoms (molecular modeling showed that the oxygen atoms cannot be exposed). It is suggested that the interactions between these molecules and the molecular chains of the polymer must be nonspecific; that is, no significant reorientation of polymer chains or liquid molecules at the surface takes place. Hence, only OMCTS and DMCPS are inert with respect to EGC-1700 films. Following are three pieces of evidence in favor of this explanation.

(a) Receding contact angle patterns of the probe liquids on EGC-1700 compared to Teflon AF 1600 [78]: As shown above in Figures 8.8 and 8.11, the receding contact angles of liquids with bulky molecules on Teflon AF 1600 surfaces were independent of the contact time between solid and liquid. However, the case is different for EGC-1700. Here it is found that except for OMCTS and DMCPS, all the liquids yield time-dependent receding angles. Figure 8.13 shows contact angles of *cis*-decalin and DMCPS on EGC-1700 films. The receding angles of *cis*-decalin strongly depend on liquid-polymer contact time and decrease as the solid–liquid contact time increases, but DMCPS receding angles are constant. The contact time dependence of receding angles suggests that methyl and/or ester groups of the polymer might be exposed to the solid–liquid interface upon contact with the liquid and that the extent of the exposure increases with the solid–liquid contact time. Consequently, the solid–liquid interfacial tension decreases over time. This process is thermodynamically favorable. The driving force is the tendency of the system to decrease the overall free energy, which is facilitated by a decrease in the solid–liquid interfacial tension: Appearance of methyl and/or ester groups at

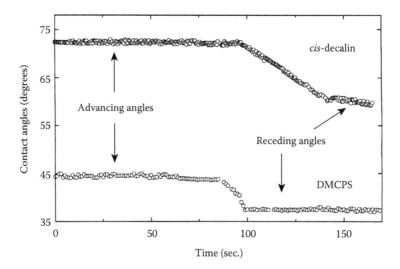

FIGURE 8.13 Advancing and receding contact angles of *cis*-decalin and DMCPS on EGC-1700 films. The receding angles of DMCPS are constant while *cis*-decalin shows time-dependent receding angles. (Reprinted from Tavana, H., Simon, F., Grundke, K., Kwok, D. Y., Hair, M. L., and Neumann, A. W., *Journal of Colloid and Interface Science*, 291, 497, 2005. With permission from Elsevier.)

the polymer surface would increase the solid surface tension from that of the original surface where only CF_2 and CF_3 groups were present at the surface film. Given that liquid surface tension is constant, an increase in the solid surface tension translates into a decrease in the solid–liquid interfacial tension.

Change in configuration of polymer surfaces upon contact with a liquid is well established [72–76]. These changes can take place through short-range motions of chains such as rotation around the chain axis or even long-range motions such as diffusion of specific moieties into the bulk. The extent of the perturbation in the configuration of the chains depends on the microstructure of the polymer and also the contacting liquid.

It should be mentioned that one cannot simply exclude the possibility of sorption-type processes as the cause of time-dependent receding angles. However, because of the shape and relatively large size of the molecules of the liquids used here, this seems less likely.

(b) Reproducibility of contact angles on EGC-1700 surfaces compared to Teflon AF 1600 films [78]: Advancing contact angles of *cis*-decalin on fresh surfaces of EGC-1700 and Teflon AF 1600 were measured as $72.75 \pm 0.12°$ and $75.53 \pm 0.22°$, respectively. Then both solid samples were blown dry with nitrogen and kept under vacuum for 24 hours. The advancing contact angle measurements were repeated with the same probe liquid on both films. While the contact angles were accurately reproduced on the Teflon AF 1600 surface, the second set of advancing angles on the EGC-1700 surface was $66.30 \pm 0.09°$; that is, $6.45°$ lower than the first set measured on the fresh sample. This implies that unlike Teflon AF 1600, the conformation of EGC-1700 polymer chains changes upon contact with the test liquid in a more or less irreversible manner. This conclusion is supported by considering the glass transition temperature of the two polymers; that is, $T_{g,\text{Teflon AF 1600}} = 160°C$ versus $T_{g,\text{EGC-1700}} = 30°C$.

(c) Contact angle hysteresis of OMCTS and DMCPS on EGC-1700 films in comparison with Teflon AF 1600 surfaces [78]: The receding angles of OMCTS and DMCPS were constant on both polymers. Therefore the corresponding contact angle hysteresis was simply obtained as the difference between advancing and receding angles. These data are given in Table 8.13. It is seen that hysteresis for OMCTS and DMCPS is essentially the same on the two polymers. Moreover, the γ_{sv} values obtained from the contact angles of these two liquids for both polymers are very close, implying that the same groups from the two polymer surfaces are "seen" by the liquid molecules. The values of γ_{sv} (see Table 8.12) suggest that these groups are mainly CF_3. It is inferred that the configurations of the polymer chains remain unchanged upon contact with these two liquids.

All of the above strongly suggest that OMCTS and DMCPS do not cause any significant change in the arrangement of polymer chains and the EGC-1700 film retains its original configuration upon contact with these two liquids. Hence the surface tension of EGC-1700 films can only be inferred from the contact angles of OMCTS and DMCPS. The average is $\gamma_{sv} = 13.84$ mJ/m^2.

Figure 8.14 shows the $\gamma_{lv}\cos\theta$ versus γ_{lv} plot for the liquids with bulky molecules on EGC-1700 surfaces. The smooth curve corresponds to $\gamma_{sv} = 13.84$ mJ/m^2 and $\beta = 0.000125$ (mJ/m^2)$^{-2}$. The two points on the curve shown by circles represent OMCTS and DMCPS. There are deviations in the contact angles of all other liquids

---

TABLE 8.13

Contact Angle Hysteresis for OMCTS and DMCPS on Teflon AF1600 and EGC-1700 Films

	Teflon AF 1600	EGC-1700
Liquid	θ_{hyst} (°)	θ_{hyst} (°)
OMCTS	5.3	5.5
DMCPS	6.2	7.0

Note: Reprinted from Tavana, H., Simon, F., Grundke, K., Kwok, D. Y., Hair, M. L., and Neumann, A. W., *Journal of Colloid and Interface Science*, 291, 497, 2005. With permission from Elsevier.

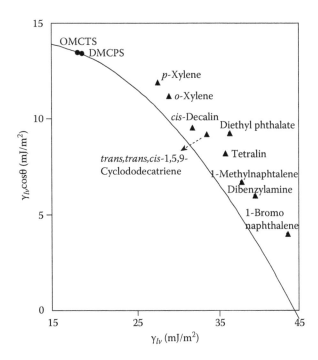

FIGURE 8.14 $\gamma_{lv}\cos\theta$ versus γ_{lv} for the contact angles of liquids with bulky molecules measured on EGC-1700 surfaces. (Reprinted from Tavana, H., Simon, F., Grundke, K., Kwok, D. Y., Hair, M. L., and Neumann, A. W., *Journal of Colloid and Interface Science*, 291, 497, 2005. With permission from Elsevier.)

from this curve, varying from about –2.3° for *cis*-decalin to about –5.5° for 1-bromonaphthalene, shown in the last column of Table 8.12. (The deviations were calculated using the procedure in the footnote of Table 8.10, using $\gamma_{sv} = 13.84$ mJ/m².) It is interesting to note that contact angles of all the liquids fall above the $\gamma_{sv} = 13.84$ mJ/m² curve, implying that the solid–liquid interfacial tension (γ_{sl}) is less than the value predicted by Equation 8.5. This agrees well with the explanation provided above that the molecular chains of EGC-1700 reorient upon contact with "noninert" liquids and expose groups less hydrophobic than CF_3 and CF_2 at the solid–liquid interface, resulting in a decrease in γ_{sl}.

For the purpose of determination of solid surface tension from contact angles, the following picture emerges from the results with EGC-1700: should the polymer chains contain hydrophilic moieties that can be exposed to the solid–liquid interface upon contact with the test liquid, the liquid should be completely inert; that is, without unsaturated bonds or exposable electronegative substituents. Then, the specific interactions of liquid with polymer molecules that is, reorientation/perturbation of polymer chains, are not stimulated. OMCTS and DMCPS appear to be two such liquids. This point seems to be crucial for amorphous polymers such as EGC-1700 whose chains can be perturbed easily because of the poor packing.

8.6.2.3   Liquids with Bulky Molecules/ETMF Systems

To further investigate the problem of contact angle deviations, ETMF (Figure 8.6c) was chosen as the next solid surface for contact angle measurements [79]. Although the overall structure of this polymer is similar to that of EGC-1700 (Figure 8.6b), two major differences should be noted: (a) In addition to the butyl and amide groups, ETMF possesses a much longer side chain, perfluoroheptyl compared to the perfluoropropyl side chain in EGC-1700; (b) Unlike EGC-1700, which is an amorphous fluorinated acrylate polymer, the H-bonding interactions within the perfluorinated amide groups in ETMF support formation of a layered structure in the bulk and in the top layer of the surface film. Molecular modeling and wide angle x-ray scattering (WAXS) measurements have confirmed the existence of a layered structure for ETMF [60]. Therefore, unlike EGC-1700, a major reorientation/perturbation of ETMF chains is less likely to occur. The long side chain in ETMF is also expected to shield the hydrophilic maleimide backbone.

Table 8.14 presents the contact angles of liquids with bulky molecules on ETMF surfaces. The solid surface tension value calculated from each contact angle is also given. The γ_{sv} values vary from 10.88 mJ/m^2 calculated for OMCTS to 12.94 mJ/m^2 for methyl salicylate. These results are similar to those obtained for EGC-1700 surfaces in terms of variations in the γ_{sv} values, suggesting that the mechanism responsible may be similar. Therefore an argument similar to that made in the case of EGC-1700 polymer can be used to infer the surface tension of ETMF: Upon advancing of the drop front of a noninert liquid on an ETMF film, the aliphatic

---

TABLE 8.14
Advancing Contact Angles of Liquids with Bulky Molecules on ETMF Surfaces and the Corresponding 95% Confidence Limits, the Solid Surface Tension Values Calculated from the Contact Angles, Deviation in the Contact Angles from the Smooth Curve of $\gamma_{sv} = 11.00$ mJ/m^2 and $\beta = 0.000125$ (mJ/m^2)$^{-2}$, and Error in the Calculation of Solid Surface Tension Due to Deviations

Liquid	θ (°)	γ_{sv} (mJ/m^2)	$\Delta\theta$ (°)	$\Delta\gamma_{sv}$ (mJ/m^2)
OMCTS	57.70 ± 0.08	10.88	+ 0.71	0.12
DMCPS	58.16 ± 0.09	11.11	−0.53	0.11
p-Xylene	75.60 ± 0.23	11.62	−2.17	0.62
o-Xylene	77.50 ± 0.17	11.71	−2.40	0.71
cis-Decalin	80.87 ± 0.29	11.95	−3.04	0.95
Tetralin	84.67 ± 0.12	12.42	−4.21	1.42
Diethyl phthalate	86.04 ± 0.18	12.05	−3.09	1.05
Methyl salicylate	86.38 ± 0.11	12.94	−5.41	1.94
Lepidine	91.36 ± 0.22	12.91	−5.14	1.91
1-Bromonaphthalene	92.05 ± 0.17	12.87	−4.95	1.87

Note: Reprinted from Tavana, H., et al., *Colloid and Polymer Science*, 284, 497, 2006. With permission from Springer-Verlag.

hydrocarbon segment (butyl) or possibly the amide group of the side chain can be exposed toward the liquid phase. This is quite likely considering that the long side chains of ETMF are flexible and can fluctuate, exposing groups other than fluorine-containing moieties. The contribution of butyl or amide groups at the three-phase line region causes the actual γ_{sl} to be less than the value predicted by Equation 8.5. Therefore the measured contact angles deviate from the ideal contact angle pattern of $\gamma_{lv}\cos\theta$ versus γ_{lv}. Such interactions appear to be insignificant when OMCTS and DMCPS are used as probe liquids due to the inertness of their molecules.

The receding contact angle patterns and contact angle hysteresis of the probe liquids support this proposition [79]. Except for OMCTS and DMCPS, the receding angles of all the liquids on ETMF films depend on the contact time between solid and liquid. This is shown in Figure 8.15 for (a) methyl salicylate, and (b) DMCPS. Change in the alignment of the side chain segments, penetration of liquid molecules into the polymer matrix, and retention of liquid molecules on the surface might all influence the time-dependence of the receding angles. The receding angles of these liquids were obtained by an extrapolation technique [80], and the results are given in Table 8.15. On the other hand, OMCTS and DMCPS show constant receding angles and the corresponding contact angle hysteresis was simply obtained as the difference between advancing and receding angles (see Table 8.16). The fact that OMCTS and DMCPS yield constant receding angles and also a very small contact angle hysteresis indicates that: (i) No

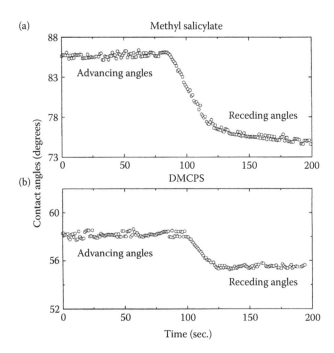

FIGURE 8.15 Contact angles of (a) methyl salicylate and (b) DMCPS on ETMF films. The receding angles of DMCPS are constant while methyl salicylate shows time-dependent receding angles. (Reprinted from Tavana, H., et al., *Colloid and Polymer Science*, 284, 497, 2006. With permission from Springer-Verlag.)

TABLE 8.15

Extrapolated Receding Contact Angles and Contact Angle Hysteresis for Those Liquids with Bulky Molecules that Showed Time-Dependent Receding Angles on ETMF Surfaces

Liquid	θ^0_r (°)	θ^0_{hyst} (°)
p-Xylene	71.5	4.1
o-Xylene	74.0	3.5
cis-Decalin	77.2	3.7
Tetralin	80.0	4.7
Methyl salicylate	76.8	9.4
Lepidine	81.9	9.5
1-Bromonaphthalene	83.0	9.1

Note: Reprinted from Tavana, H., et al., *Colloid and Polymer Science*, 284, 497, 2006. With permission from Springer-Verlag.

TABLE 8.16

Receding Contact Angles and Contact Angle Hysteresis for OMCTS and DMCPS on ETMF Films

Liquid	θ_r (°)	$\theta_{hyst.}$ (°)
OMCTS	54.70	3.0
DMCPS	54.66	3.5

Note: Reprinted from Tavana, H., et al., *Colloid and Polymer Science*, 284, 497, 2006. With permission from Springer-Verlag.

significant change in the arrangement of the polymer side chains occurs upon contact with these two liquids, (ii) The molecules of these two liquids do not penetrate into the polymer film significantly. Thus the surface tension of ETMF can be determined only from OMCTS and DMCPS measurements. The average is 11.00 mJ/m^2.

Figure 8.16 shows the $\gamma_{lv}\cos\theta$ versus γ_{lv} plot for liquids with bulky molecules/ ETMF systems. The smooth curve corresponds to $\gamma_{sv} = 11.00$ mJ/m^2 and $\beta = 0.000125$ (mJ/m^2)$^{-2}$. The two points on the curve shown by circles represent the OMCTS and DMCPS contact angles. The deviations in the contact angles of other liquids from this curve range from ~−2.2° for p-xylene to ~−5.4° for methyl salicylate, as given in Table 8.14. The minus sign means that the experimental points are located above the curve. The error in the calculated solid surface tension values due to the deviations

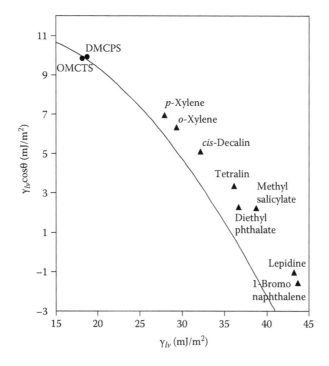

FIGURE 8.16 $\gamma_{lv}\cos\theta$ versus γ_{lv} for the contact angles of liquids with bulky molecules measured on ETMF surfaces. (Reprinted from Tavana, H., et al., *Colloid and Polymer Science*, 284, 497, 2006. With permission from Springer-Verlag.)

is also given in each case. Similar to the results for EGC-1700, contact angles of the test liquids fall above the $\gamma_{sv} = 11.00$ mJ/m^2 curve, implying that the solid–liquid interfacial tension (γ_{sl}) in these systems is less than the ideal value given by Equation 8.5. This is in agreement with the explanation presented above that the arrangement of chains in the polymer film may change due to contact with a noninert liquid. For the purpose of determination of the surface tension of polymer films from contact angles, this finding necessitates use of inert probe liquids, for example, OMCTS and DMCPS, to avoid specific solid–liquid interactions.

8.6.2.4 Liquids with Bulky Molecules/ODMF Systems

Contact angles of the probe liquids on ODMF surfaces are given in Table 8.17. Only OMCTS and DMCPS yielded useful contact angles. All other liquids either dissolved the polymer film or showed stick-slip of the three-phase line [79]. As shown in Figure 8.6, ODMF has a molecular structure fairly similar to that of ETMF. In addition, ODMF possesses a second *n*-hexadecyl side chain that contributes to the different contact angle response by the two polymer surfaces. Contrary to ETMF, the angle-dependent XPS investigations of the annealed ODMF surfaces showed that the presence of *n*-hexadecyl side chains suppresses good self-organization of the perfluoroalkyl side chains on the topmost layer of the film [60]. There is only a weak self-organization of the perfluoroalkyl side chains in the outermost layer of ODMF

TABLE 8.17

Advancing Contact Angles (θ^a), Receding Contact Angles (θ_r), and Contact Angle Hysteresis (θ_{hyst}) of Liquids with Bulky Molecules on ODMF Surfaces

Liquid	θ^a (°)	γ_{sv} (mJ/m²)	θ_r (°)	θ_{hyst} (°)
OMCTS	53.60 ± 0.16	11.68	50.20	3.4
DMCPS	55.19 ± 0.08	11.72	51.29	3.9
p-Xylene	(polymer dissolution)			
o-Xylene	(polymer dissolution)			
cis-Decalin	(stick-slip)			
Tetralin	(polymer dissolution)			
Methyl salicylate	(polymer dissolution)			
Lepidine	(polymer dissolution)			
1-Bromonaphthalene	(stick-slip)			

Note: Reprinted from Tavana, H., et al., *Colloid and Polymer Science*, 284, 497, 2006. With permission from Springer-Verlag.

films. AFM analysis revealed further information about the surfaces of ETMF and ODMF films. The AFM results are shown in Figures 8.17 and 8.18 for ETMF and ODMF, respectively. From the phase contrast image in Figure 8.17b, it is seen that the ETMF films are quite homogeneous. In contrast, domains with noncontinuous structures exist on the ODMF surface film, as shown in the phase contrast image of ODMF (Figure 8.18b). It has been shown that these structures correspond to different compositions on the surface [81]. Due to the chemical incompatibility of *n*-hexadecyl and perfluoroalkyl, the two side chains of ODMF segregate and give rise to micro-scale phase-separated fluorocarbon matrix and hydrocarbon islands, causing the ODMF films to be chemically heterogeneous.

The contact angle results indicate that the siloxane-based probe liquids, OMCTS and DMCPS, do not show specific interactions with the side chains of ODMF. In contrast, other liquids that are hydrocarbon derivatives interact with the nonfluorinated parts of the film, resulting in stick-slip or dissolution of the ODMF films. This again suggests that the surface tension of ODMF films should be calculated from contact angles of OMCTS and DMCPS. As a check, the receding contact angles and contact angle hysteresis of the two liquids were also investigated. Both liquids yield time-independent receding angles and small contact angle hysteresis on the ODMF films (see Table 8.17) indicating that they do not penetrate into the polymer film nor do they cause a change in the configuration of the polymer chains. Thus, the surface tension of ODMF was determined as $\gamma_{sv} = 11.70$ mJ/m² [79]. The surface tension of ODMF films is 0.7 mJ/m² larger than that of ETMF surfaces, presumably because of the presence of the *n*-hexadecyl side chain in the surface film, in addition to the perfluoroalkyl side chain.

It is important to note that for all the above solid–liquid systems, the contact angle deviation of liquids with bulky molecules from the smooth curves of $\gamma_{lv}\cos\theta$ versus γ_{lv} can be readily translated into a deviation of the solid–liquid interfacial tension

(a) (b)

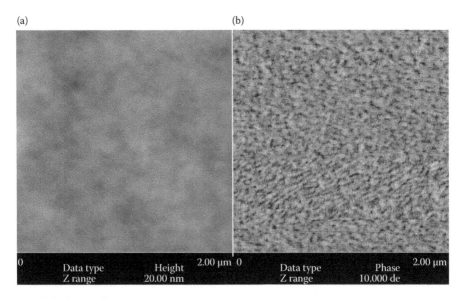

FIGURE 8.17 (a) Surface morphology and (b) the corresponding phase contrast AFM images of an ETMF film. (Reprinted from Tavana, H., Yang, C., Yip, C. Y., Appelhans, D., Zschoche, S., Grundke, K., Hair, M. L., and Neumann, A. W., *Langmuir*, 22, 628, 2006. With permission from American Chemical Society.)

(a) (b)

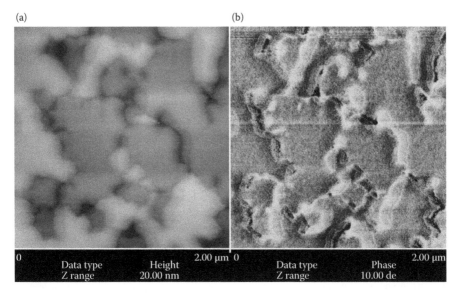

FIGURE 8.18 (a) Surface morphology and (b) the corresponding phase contrast AFM images of an ODMF film. (Reprinted from Tavana, H., Yang, C., Yip, C. Y., Appelhans, D., Zschoche, S., Grundke, K., Hair, M. L., and Neumann, A. W., *Langmuir*, 22, 628, 2006. With permission from American Chemical Society.)

(γ_{sl}) from the ideal value given by Equation 8.5; that is, from the value that would be operative if the contact angles would fall on the smooth curve. This will be shown in the following section that presents the contact angles of *n*-alkanes.

8.6.3 INTERPRETATION OF CONTACT ANGLES OF A HOMOLOGOUS SERIES OF *N*-ALKANES

8.6.3.1 *n*-Alkanes/Teflon AF 1600 Systems

Contact angle measurements were performed with a series of *n*-alkanes on the films of Teflon AF 1600 and the data (θ) are shown in Figure 8.19 [56]. The smooth curve is identical with that of Figure 8.10 and corresponds to $\gamma_{sv} = 13.61$ mJ/m^2 and $\beta = 0.000116$ (mJ/m^2)$^{-2}$. There are significant deviations in the contact angles from this curve. The deviation of each *n*-alkane from this curve ($\Delta\theta$) was calculated according to the procedure given in the footnote of Table 8.10. The results are presented in Table 8.18, with the same definition for the sign of the deviations as above. It is seen that *n*-hexane has the largest contact angle deviation with –3.52°. The deviations then decrease and the contact angles of *n*-decane, *n*-undecane, and *n*-dodecane fall very close to the smooth curve. From *n*-tridecane to *n*-hexadecane, the deviations increase again up to + 2.07°. The fact that the experimental contact angles for the *n*-alkanes

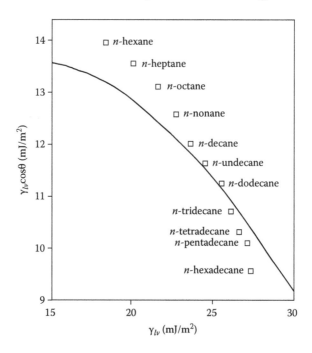

FIGURE 8.19 $\gamma_{lv}\cos\theta$ versus γ_{lv} for the contact angles of *n*-alkanes measured on Teflon AF 1600 surfaces. The smooth curve is identical to that of Figure 8.10, which represents $\gamma_{sv} = 13.61$ mJ/m^2 and $\beta = 0.000116$ (mJ/m^2)$^{-2}$. (Reprinted from Tavana, H., Lam, C. N. C., Friedel, P., Grundke, K., Kwok, D. Y., Hair, M. L., and Neumann, A. W., *Journal of Colloid and Interface Science*, 279, 493, 2004. With permission from Elsevier.)

TABLE 8.18
Experimental Contact Angles (θ) of *n*-Alkanes on Teflon AF 1600 Surfaces, Ideal Contact Angles (θ'), Contact Angle Deviations (Δθ) from the Smooth Curve, Actual and Ideal Solid–Liquid Interfacial Tensions (γ_{sl}^{θ} and $\gamma_{sl}^{\theta'}$), and Vapor Pressure of the Liquids at 23°C

Liquid	θ (°)	θ' (°)	Δθ (°)	γ_{sl}^{θ} (mJ/m²)	$\gamma_{sl}^{\theta'}$ (mJ/m²)	P_v (kPa)
n-Hexane	40.31 ± 0.10	43.83	−3.52	−0.33	0.42	18.69
n-Heptane	47.17 ± 0.10	49.90	−2.73	0.04	0.75	5.50
n-Octane	52.45 ± 0.15	54.50	−2.05	0.52	1.14	1.68
n-Nonane	56.22 ± 0.14	57.44	−1.22	1.05	1.46	0.52
n-Decane	59.29 ± 0.18	59.63	−0.34	1.62	1.74	0.17
n-Undecane	61.60 ± 0.22	61.74	−0.14	2.00	2.05	0.05
n-Dodecane	63.78 ± 0.32	63.90	−0.12	2.38	2.42	0.015
n-Tridecane	65.68 ± 0.05	65.01	+ 0.67	2.92	2.64	0.006
n-Tetradecane	67.15 ± 0.14	66.07	+ 1.08	3.32	2.85	0.001
n-Pentadecane	68.06 ± 0.14	66.99	+ 1.07	3.53	3.06	0.0003
n-Hexadecane	69.48 ± 0.16	67.41	+ 2.07	4.07	3.15	0.0001

Note: Reprinted from Tavana, H., Lam, C. N. C., Friedel, P., Grundke, K., Kwok, D. Y., Hair, M. L., and Neumann, A. W., *Journal of Colloid and Interface Science*, 279, 493, 2004. With permission from Elsevier.

fall in some cases above the curve and below in others suggests that more than one mechanism is operative. Tentative explanations of the complex pattern of the contact angle deviations of *n*-alkanes are given below.

Assuming the validity and applicability of Young's equation, since γ_{lv} and θ are correctly measured quantities, the deviation from the smooth curve can only be caused by deviations of γ_{sv} and/or γ_{sl} from their ideal values. The ideal γ_{sv} is the surface tension of bare Teflon AF 1600 film and is assumed to be indeed the value suggested by the contact angles of liquids with bulky molecules (i.e., 13.61 mJ/m²). Furthermore, two sets of solid–liquid interfacial tension (actual vs. ideal) are defined: The first set corresponds to experimental contact angles, θ, and is shown by γ_{sl}^{θ}. For each *n*-alkane/Teflon AF 1600 system, this value is obtained from Young's equation by substituting γ_{lv} and $\gamma_{sv} = 13.61$ mJ/m² as the surface tension of the bare solid, and experimental contact angles, θ. Thus, γ_{sl}^{θ} is the true, operative interfacial tension. The second set corresponds to the ideal contact angles, θ', shown by $\gamma_{sl}^{\theta'}$. It is obtained from Young's equation by substituting γ_{lv}, $\gamma_{sv} = 13.61$ mJ/m², and the ideal contact angles, θ', of each *n*-alkane on Teflon AF 1600. $\gamma_{sl}^{\theta'}$ represent the values that would exist if the experimental points would all fall on the smooth curve. Clearly, from the above definitions, if there was no deviation in the contact angles of *n*-alkanes from the smooth curve of $\gamma_{sv} = 13.61$ mJ/m², that is, θ = θ', the value of γ_{sl}^{θ} would be the same as $\gamma_{sl}^{\theta'}$ for each given *n*-alkane/Teflon AF 1600 system. Table 8.18 contains the two sets of the solid–liquid interfacial tension for each *n*-alkane/Teflon

AF 1600 system. Using the values of $\gamma_{sl}{}^{\theta}$, $\gamma_{sl}{}^{\theta'}$, and $\gamma_{sv} = 13.61$ mJ/m^2, the contact angle deviations of n-hexane and n-hexadecane, that is, the two liquids that fall furthest above and below the smooth curve, respectively, are explained below.

8.6.3.1.1   Contact Angle Deviation of n-Hexadecane

As implied by Young's equation, only a γ_{sv} and/or γ_{sl} different from their ideal values (described above) can cause deviation in contact angle from the smooth curve. The main cause for such a difference in the solid–vapor interfacial tension (γ_{sv}) is adsorption of vapor onto the solid. Considering the very low vapor pressure of n-hexadecane (see Table 8.18), it is not likely that a significant vapor adsorption onto the solid surface takes place. Therefore, the reason for deviation in the contact angle of n-hexadecane from the smooth curve must be sought in the solid–liquid interfacial tension (γ_{sl}). In this context, penetration of the liquid into the solid and/or surface retention might be thought of as a likely cause (see Chapter 7). It is obvious that this process would make the two bulk phases more alike, resulting in a decrease in solid–liquid interfacial tension. However, as shown in Table 8.18, $\gamma_{sl}{}^{\theta}$ of n-hexadecane/Teflon AF 1600 system is larger than $\gamma_{sl}{}^{\theta'}$. Therefore sorption of n-hexadecane cannot be the mechanism causing the contact angle deviation. The following explanation is proposed.

Molecules of n-alkanes consist of two CH$_3$ groups and a number of CH$_2$ groups. The higher the percentage of CH$_2$ groups, the higher the surface tension of the liquid. Furthermore, it is well-known that n-hexadecane molecules are flexible chains that can assume different shapes. The increase in the value of $\gamma_{sl}{}^{\theta}$ from $\gamma_{sl}{}^{\theta'}$ implies that n-hexadecane behaves like a liquid with higher surface tension than its actual value. This can be interpreted as an interaction between n-hexadecane and the solid surface such that n-hexadecane molecules are forced by the substrate to rearrange themselves somewhat flat and parallel to the polymer surface at the solid–liquid interface. This means a greater contribution of the CH$_2$ groups at the interface, translating into a larger solid–liquid interfacial tension.

Evidence can be found in the literature supporting this explanation. Klein et al. investigated the structural properties of liquid n-hexadecane films adsorbed on a flat metal (Au) surface using a Monte Carlo (MC) method [82]. The solid–liquid system was modeled by defining potential functions and parameters for all types of interactions such as bond stretching, bond bending, and torsional motion. It was found that the interactions between the hydrocarbons and the substrate are stronger than those between the hydrocarbons themselves. Consequently, the liquid molecules are forced to rearrange themselves parallel to the solid surface. This results in the formation of a "dense" monolayer at the solid–liquid interface. Interestingly, the layer density remains independent of the thickness of the liquid above it, confirming that the liquid molecules-substrate interactions are stronger than the intermolecular interactions of the liquid. Landman et al. performed a molecular dynamics study of interfacial n-alkane films on solid substrates [83]. It was shown that molecular chains of n-hexadecane close to the interface orient themselves parallel to the substrate and that a steady-state boundary layer of n-hexadecane is formed. Overney et al. compared the friction force between two silicon surfaces, where the gap between them was filled with n-hexadecane as well as with OMCTS [84]. A smaller friction force was measured with n-hexadecane as the lubricant. It was concluded that superior lubricity of this liquid is a result of interfacial liquid structuring; that is, a substrate-induced parallel orientation of liquid molecules in close vicinity to the substrate. This result is quite plausible considering that OMCTS consists of nonflexible bulky molecules that do not allow the substrate to restructure them significantly, as confirmed by the contact angle results of liquids with bulky molecules. Israelachvili et al. performed a series of experiments to measure the surface forces of n-alkanes between mica surfaces [85]. Oscillatory force profiles were obtained as a function of surface separation, implying a layering of the chains of liquid molecules under confinement parallel to the substrate.

All of the above agrees well with our postulate that a substrate-induced parallel reorientation of n-hexadecane molecules close to the surface occurs, making the solid–liquid interfacial tension larger than the ideal value predicted by Equation 8.5 and giving rise to a deviation in the contact angle of n-hexadecane from the $\gamma_{lv}\cos\theta$ versus γ_{lv} smooth curve.

8.6.3.1.2 Contact Angle Deviation of n-Hexane

As seen from Table 8.18, the value of $\gamma_{sl}{}^\theta$ for n-hexane/Teflon AF 1600 is negative. It has been previously argued that a negative solid–liquid interfacial tension is not possible and zero is the minimum possible value. Considering that γ_{lv} and θ are correctly measured quantities, Young's equation suggests that only an incorrect γ_{sv} value can lead to a negative $\gamma_{sl}{}^\theta$. Therefore the value of $\gamma_{sv} = 13.61$ mJ/m^2 cannot be correct for the n-hexane/Teflon AF 1600 system and must be higher to produce a positive (or zero) $\gamma_{sl}{}^\theta$. A higher γ_{sv} in turn suggests that once a drop of n-hexane is formed on the solid surface, its vapor is adsorbed onto the surface, resulting in an increase in the solid–vapor interfacial tension (γ_{sv}) from that of the bare polymer film. This means that the contact angle is measured on a surface that has been modified by the vapor of the test liquid, and therefore deviates from the curve that represents the surface tension of the original polymer film. Vapor adsorption onto Teflon AF 1600 films is quite plausible considering that: (i) n-hexane has a large vapor pressure (18.69 kPa); and (ii) the polymer consists of 65% bulky dioxole rings in its backbone that give rise to microvoids in the structure of the polymer [91], promoting the adsorption of vapor molecules of short-chain n-alkanes onto the polymer film. This finding can be confirmed as follows: Suppose $\gamma_{sl}{}^\theta$ has its minimum possible value (i.e., zero). Solving Young's equation for γ_{sv} with known γ_{lv} and θ of n-hexane gives $\gamma_{sv} = 13.97$ mJ/m^2. Since this is a higher value than the surface tension of Teflon AF 1600 film (13.61 mJ/m^2), it can be concluded that vapor adsorption indeed takes place and modifies the solid surface.

With the existence of adsorption of vapor of n-hexane onto the Teflon AF 1600 surface, the equilibrium spreading pressure can be determined. Using the values of $\gamma_s = 13.61$ mJ/m^2 (surface tension of the polymer film) and $\gamma_{lv} = 18.15$ mJ/m^2 (surface tension of n-hexane), Equation 8.5 yields a value of $\gamma_{sl} = 0.39$ mJ/m^2 as the solid–liquid interfacial tension of the n-hexane/Teflon system. Substituting this value and $\theta = 40.3°$ (experimental contact angle of n-hexane on Teflon AF 1600 surface) into Young's equation, a value of 14.23 mJ/m^2 is obtained for γ_{sv}. This leads to $\Pi = 0.62$ mJ/m^2 as the equilibrium spreading pressure. This calculation assumes that $\gamma_{sl} = 0.39$

mJ/m^2 is indeed the ideal value; that is, there are no additional interactions such as swelling of the solid by the liquid, penetration of liquid into the solid surface, or substrate-induced alignment of liquid molecules in close vicinity of the solid. In the case of swelling of solid and penetration of liquid into the solid, one would expect a γ_{sl} smaller than the corresponding ideal value ($\gamma_{sl}^{\theta'}$), whereas in the case of parallel alignment process a value larger than $\gamma_{sl}^{\theta'}$ would be expected.

It should be noted that the deviations in contact angles of the liquids in the middle of the n-alkane series, that is, n-decane, n-undecane, and n-dodecane, are inside the error limits of ADSA and these data fall close to the smooth curve of $\gamma_{sv} = 13.61$ mJ/m^2. This could result from a combination of effects responsible for deviations in contact angles of short-chain and long-chain n-alkanes.

8.6.3.2 *n*-Alkanes/EGC-1700 Systems

The contact angles of the series of n-alkanes on EGC-1700 surfaces are shown in Figure 8.20. The smooth curve is identical with the one in Figure 8.14 and corresponds to the surface tension of EGC-1700 that was determined above ($\gamma_{sv} = 13.84$ mJ/m^2) [78]. Short-chain n-alkanes deviate above the curve and the long-chain ones below it. This indicates that the mechanisms of vapor adsorption onto the surface prior to advancing of the drop and the parallel alignment of liquid molecules

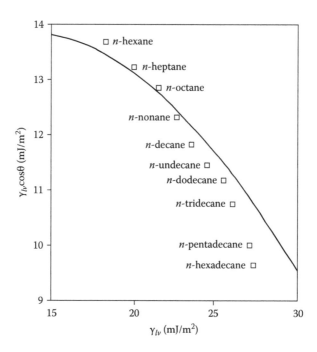

FIGURE 8.20 $\gamma_{lv}\cos\theta$ versus γ_{lv} for the contact angles of n-alkanes measured on EGC-1700 surfaces. The smooth curve is identical to that of Figure 8.14, which represents $\gamma_{sv} = 13.84$ mJ/ m^2 and $\beta = 0.000125$ (mJ/m^2)$^{-2}$. (Reprinted from Tavana, H., Simon, F., Grundke, K., Kwok, D. Y., Hair, M. L., and Neumann, A. W., *Journal of Colloid and Interface Science*, 291, 497, 2005. With permission from Elsevier.)

in the vicinity of the solid surface are operative here as well. In Table 8.19, the measured contact angles (θ) and the ideal contact angles (θ'), that is, the angle that would fall on the γ_{sv} = 13.84 mJ/m^2 curve, are given. The difference between θ and θ' gives the contact angle deviation for each liquid. The longest-chain n-alkane; that is, n-hexadecane shows the largest deviations with +2.56°. Moving from the longest to the shortest chains, the deviations decrease, change sign, and increase again.

Comparison of the contact angle deviations of the n-alkanes/Teflon AF 1600 with those of the n-alkanes/EGC-1700 systems; that is, Figures 8.19 and 8.20 reveals that in the former systems, the contact angles of more liquids fall above the curve (minus sign) and in the latter systems, more below it. This suggests that the relative importance of the two mechanisms differs for the two polymers: vapor adsorption is more pronounced on Teflon AF 1600 while the effect of parallel alignment of liquid molecules for the long-chain n-alkanes seems to be more dominant in the case of EGC-1700. For instance, the deviation for n-hexane is over 2° smaller on EGC-1700 than on Teflon AF 1600. The less extensive vapor adsorption on EGC-1700 is most probably due to the absence of any bulky groups in its backbone (in contrast to Teflon AF 1600). The larger contact angle deviations for the long-chain n-alkanes on EGC-1700 compared to Teflon AF 1600 is probably due to the same differences in their structural properties.

TABLE 8.19

Actual and Ideal Advancing Contact Angles of n-Alkanes/EGC-1700 Systems and the Deviations in the Contact Angle of Each Liquid from the Smooth Curve of γ_{sv} = 13.84 mJ/m^2 and β = 0.000125 (mJ/m^2)$^{-2}$

Liquid	θ (°)	θ' (°)	$\Delta\theta$ (°)
n-Hexane	41.68 ± 0.11	42.75	−1.07
n-Heptane	48.70 ± 0.07	49.08	−0.37
n-Octane	53.37 ± 0.12	53.67	−0.29
n-Nonane	57.03 ± 0.09	56.64	+0.38
n-Decane	59.84 ± 0.14	58.88	+0.96
n-Undecane	62.10 ± 0.11	61.02	+1.08
n-Dodecane	63.99 ± 0.08	63.21	+0.78
n-Tridecane	65.63 ± 0.12	64.34	+1.29
n-Pentadecane	68.33 ± 0.16	66.43	+1.90
n-Hexadecane	69.33 ± 0.14	66.77	+2.56

Note: Reprinted from Tavana, H., Lam, C. N. C., Friedel, P., Grundke, K., Kwok, D. Y., Hair, M. L., and Neumann, A. W., *Journal of Colloid and Interface Science*, 279, 493, 2004. With permission from Elsevier.

8.6.3.3 *n*-Alkanes/ETMF Systems

Contact angle measurements were also performed with the series of *n*-alkanes on the films of ETMF [80], just as for Teflon AF 1600 and EGC-1700. The results are shown in the plot of $\gamma_{lv}\cos\theta$ versus γ_{lv} in Figure 8.21. The smooth curve is identical with the one in Figure 8.16 and corresponds to the surface tension of ETMF that was determined from OMCTS and DMCPS contact angles ($\gamma_{sv} = 11.00$ mJ/m²). The contact angle points show an overall shift with respect to the smooth curve, compared to results obtained with the Teflon AF 1600 and EGC-1700 polymers. This indicates that the extent of the two mechanisms of vapor adsorption and parallel alignment of liquid molecules at the surface are different for ETMF. To quantify this point, the actual contact angles (θ), ideal contact angles (θ'), and the contact angle deviations from the $\gamma_{sv} = 11.00$ mJ/m² curve are given in Table 8.20 for each *n*-alkane. The deviation is up to about –2° for the short-chain *n*-alkanes and decreases toward the end of the series. The contact angle deviation of short-chain *n*-alkanes suggests that the process of vapor adsorption onto the solid is still operative. However the contact angles of the long-chain *n*-alkanes fall very close to this curve, suggesting that no significant parallel alignment of liquid molecules occurs near the surface.

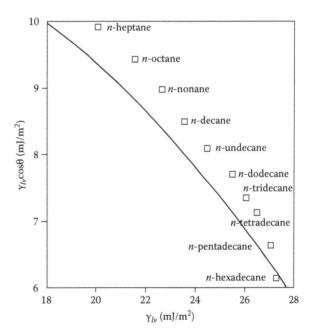

FIGURE 8.21 $\gamma_{lv}\cos\theta$ versus γ_{lv} for the contact angles of *n*-alkanes measured on ETMF surfaces. The smooth curve is identical to that of Figure 8.16, which represents $\gamma_{sv} = 11.00$ mJ/m² and $\beta = 0.000125$ (mJ/m²)⁻². (Reprinted from Tavana, H., Yang, C., Yip, C. Y., Appelhans, D., Zschoche, S., Grundke, K., Hair, M. L., and Neumann, A. W., *Langmuir*, 22, 628, 2006. With permission from American Chemical Society.)

TABLE 8.20

Actual and Ideal Contact Angles of *n*-Alkanes on ETMF Films and the Corresponding Deviations in the Actual Contact Angles from the Smooth Curve of $\gamma_{sv} = 11.00$ mJ/m² and $\beta = 0.000125$ (mJ/m²)$^{-2}$

Liquid	θ (°)	θ' (°)	$\Delta\theta$ (°)
n-Heptane	60.30 ± 0.29	62.25	−1.95
n-Octane	64.01 ± 0.24	65.90	−1.89
n-Nonane	66.63 ± 0.19	68.24	−1.61
n-Decane	68.82 ± 0.21	70.09	−1.27
n-Undecane	70.67 ± 0.22	71.88	−1.21
n-Dodecane	72.15 ± 0.07	73.75	−1.60
n-Tridecane	73.36 ± 0.12	74.71	−1.35
n-Tetradecane	74.44 ± 0.18	75.62	−1.18
n-Pentadecane	75.70 ± 0.25	76.44	−0.74
n-Hexadecane	76.87 ± 0.14	76.81	+0.06

Note: Reprinted from Tavana, H., Yang, C., Yip, C. Y., Appelhans, D., Zschoche, S., Grundke, K., Hair, M. L., and Neumann, A. W., *Langmuir*, 22, 628, 2006. With permission from American Chemical Society.

8.6.3.4   *n*-Alkanes/ODMF Systems

It was shown in Chapter 7 that not all *n*-alkanes yield useful contact angles on ODMF surfaces. The short-chain *n*-alkanes (e.g., *n*-heptane) show a stick-slip pattern. This pattern diminishes for those liquids in the middle of the series. For the last three *n*-alkanes, stick-slip almost vanishes and an even motion of the drop front is obtained. Due to stick-slip, the contact angles should be excluded from interpretation in terms of solid surface tensions. A more complete discussion of stick-slip was given in Chapter 7.

8.6.4   CONTACT ANGLE DEVIATIONS DUE TO STRONG MOLECULAR INTERACTIONS AT THE SOLID–LIQUID INTERFACE

It was shown that the deviations in the contact angles from smooth curves of $\gamma_{lv}\cos\theta$ versus γ_{lv} are due to specific interactions at solid–vapor and/or solid–liquid interfaces that are not taken into account by the equation of state. The use of short-chain *n*-alkanes as probe liquids causes the polymer film to be modified due to vapor adsorption so that γ_{sv} becomes different from the ideal value for the solid–liquid system under consideration. On the other hand, the extremely low vapor pressure of long-chain *n*-alkanes and liquids with bulky molecules eliminates the likelihood of vapor adsorption onto the solid surface. The contact angle deviations from smooth curves in systems involving such liquids are caused by interactions between liquid and solid molecules at the solid–liquid interface; that is, γ_{sl} becomes different from the ideal value [56,78,79]. The discussion presented so far reflects the fact that the processes affecting γ_{sl} are indeed very complicated. To understand such processes more fully and in more depth, solid–liquid systems with strong molecular interactions will be examined in detail below. It will be shown how such interactions actually cause contact angle deviations.

Contact angle measurements with two liquids consisting of bulky molecules, that is, 1-methylnaphthalene and 1-bromonaphthalene, on EGC-1700 surfaces yielded deviations from the $\gamma_{sv} = 13.84$ mJ/m^2 smooth curve (surface tension of the EGC-1700 polymer). The deviations are $-2.40°$ and $-5.53°$, respectively (see Table 8.12) [78]. The two liquids have similar molecular structures but the methyl group of the former is replaced by a bromine atom in 1-bromonaphthalene. This result suggests that the larger deviation for 1-bromonaphthalene might be due to electronegativity effects associated with the molecules, causing stronger interactions with the EGC-1700 chains. To test this proposition, a homologous series of naphthalene compounds that contain halogen moieties with different electronegativities were selected as test liquids for contact angle measurements on EGC-1700 surfaces (the liquids are listed in Table 8.4). If the above proposition is indeed correct, one should expect a larger contact angle deviation for the liquids with stronger electronegativity. The results of contact angle measurements on Teflon AF 1600 surfaces, which are inert with respect to many liquids, are reported for comparison with the results for EGC-1700 [87].

Table 8.21 presents the contact angles of the above liquids on Teflon AF 1600 films. The corresponding plot of $\gamma_{lv}\cos\theta$ versus γ_{lv} is shown in Figure 8.22. The smooth curve is identical to that of Figure 8.10; that is, $\gamma_{sv} = 13.61$ mJ/m^2 and $\beta = 0.000116$ (mJ/m^2)$^{-2}$. The contact angles of liquids with bulky molecules are shown by circles in Figure 8.22. The contact angles of the naphthalene compounds (triangles), that is, 1-fluoronaphthalene to 1-iodonaphthalene and 1-methylnaphthalene, all fall on the smooth curve and the corresponding deviations from this curve

---

TABLE 8.21

The Contact Angles (θ) of a Series of Naphthalene Compounds on Teflon AF 1600 Surfaces and the Calculated Values of Solid Surface Tension (γ_{sv}) from Each Contact Angle

Liquid	θ (°)	γ_{sv} (mJ/m^2)	$\Delta\theta$ (°)	$\Delta\gamma_{sv}$ (mJ/m^2)
1-Fluoronaphthalene	80.49 ± 0.28	13.86	-0.73	0.25
1-Chloronaphthalene	86.70 ± 0.17	13.64	-0.23	0.03
1-Bromonaphthalene	89.80 ± 0.19	13.75	-0.37	0.14
1-Iodonaphthalene	93.00 ± 0.22	13.71	-0.22	0.10
1-Methylnaphthalene	83.62 ± 0.24	13.65	-0.11	0.04

Source: Reprinted from Tavana, H., Hair, M. L., and Neumann, A. W., Journal of Physical Chemistry B, 110, 1294, 2006. With permission from American Chemical Society.

Note: The contact angle deviations ($\Delta\theta$) from the smooth curve of $\gamma_{sv} = 13.61$ mJ/m^2 and the corresponding error in the solid surface tension values ($\Delta\gamma_{sv}$) are also given.

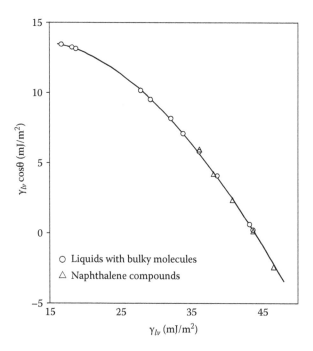

FIGURE 8.22 $\gamma_{lv}\cos\theta$ versus γ_{lv} for liquids with bulky molecules (circles) and naphthalene compounds (triangles), on Teflon AF 1600 films. The smooth curve is identical to that in Figure 8.10. (Reprinted from Tavana, H., Hair, M. L., and Neumann, A. W., *Journal of Physical Chemistry B*, 110, 1294, 2006. With permission from American Chemical Society.)

are small, averaging only $-0.33°$ (Table 8.21). Therefore, the contact angles can be used to determine the surface tension (γ_{sv}) of Teflon AF 1600 films, as given in Table 8.21. The γ_{sv} values are fairly constant and the variations ($\Delta\gamma_{sv}$) are negligible. It is suggested that Teflon AF 1600 is inert with respect to these liquids and does not interact specifically with the liquid molecules; that is, there is no significant change in the configuration of molecules of the test liquids or the polymer chains upon solid–liquid contact.

Contact angles of the naphthalene compounds on EGC-1700 films are given in Table 8.22. The corresponding $\gamma_{lv}\cos\theta$ versus γ_{lv} smooth curve ($\gamma_{sv} = 13.84$ mJ/m^2), which is identical to that of Figure 8.14, is shown in Figure 8.23. The circles represent OMCTS and DMCPS results. The contact angles of the naphthalene compounds show considerable deviations ($\Delta\theta$) from this curve, as given in Table 8.22. 1-Methylnaphthalene yields the smallest deviation ($-2.40°$). Deviations for the compounds containing halogen atoms increase from $-5.27°$ for 1-iodonaphthalene to $-7.29°$ for 1-fluoronaphthalene. This would correspond to an error of ~1–3 mJ/m^2 in the calculation of solid surface tension (γ_{sv}). These results and the fact that electronegativity of halogens also increases from iodine to fluorine confirms the proposition that specific solid–liquid interactions due to electronegativity effects correlate with the contact angle deviations. This point is illustrated in Table 8.22 where the Pauling electronegativities (EN) of the elements are given [88].

TABLE 8.22

Contact Angles (θ) of a Series of Naphthalene Compounds on EGC-1700 Surfaces, Contact Angle Deviations (Δθ) from the Smooth Curve of $\gamma_{sv} = 13.84$ mJ/m², Electronegativity (EN) of the Elements from the Pauling Scale, Actual (γ_{sl}^{θ}) and Ideal ($\gamma_{sl}^{\theta'}$) Solid–Liquid Interfacial Tensions, and the Percentage Difference Between the Actual and Ideal γ_{sl} Values

Liquid	θ (°)	Δθ (°)	EN	$\gamma_{sl}^{\theta a}$ (mJ/m²)	$\gamma_{sl}^{\theta'a}$ (mJ/m²)	$\%\delta\gamma_{sl}^{b}$
1-Fluoronaphthalene	73.29 ± 0.15	−7.29	3.98	3.5	7.9	56
1-Chloronaphthalene	80.09 ± 0.21	−6.23	3.16	6.8	11.1	39
1-Bromonaphthalene	84.04 ± 0.12	−5.53	2.96	9.3	13.5	31
1-Iodonaphthalene	87.38 ± 0.17	−5.27	2.66	11.7	16.0	27
1-Methylnaphthalene	80.67 ± 0.10	−2.40	—	7.7	9.3	17

[a] $\gamma_{sl}^{\theta'}$ is obtained from Equation 8.12 by substituting the liquid surface tension (γ_{lv}) and the actual surface tension of the polymer surface ($\gamma_{sv} = 13.84$ mJ/m²). γ_{sl}^{θ} is calculated from Young's equation by substituting the actual surface tension of the polymer film ($\gamma_{sv} = 13.84$ mJ/m²), the liquid surface tension (γ_{lv}), and the corresponding contact angle (θ).

[b] $\%\delta\gamma_{sl} = (\gamma_{sl}^{\theta'} - \gamma_{sl}^{\theta})/\gamma_{sl}^{\theta'}$

Note: Reprinted from Tavana, H., Hair, M. L., and Neumann, A. W., *Journal of Physical Chemistry B*, 110, 1294, 2006. With permission from American Chemical Society.

For the naphthalene compounds/EGC-1700 systems, the actual and ideal solid–liquid interfacial tensions were calculated. The ideal values were obtained by substituting γ_{lv} of each liquid and the actual solid surface tension of EGC-1700; that is, $\gamma_{sv} = 13.84$ mJ/m², into Equation 8.5. The actual values were calculated from the equilibrium condition, that is, Young's equation, using γ_{lv} and θ of each liquid and $\gamma_{sv} = 13.84$ mJ/m². The actual and ideal solid–liquid interfacial tensions (γ_{sl}^{θ} and $\gamma_{sl}^{\theta'}$) for each system are given in Table 8.22. In all cases the actual value is less than the corresponding ideal value. The 1-fluoronaphthalene/EGC-1700 system shows the most significant difference while the smallest difference occurs for the 1-methylnaphthalene/EGC-1700 system, in agreement with the corresponding contact angle deviations. The most likely explanation for the difference in the solid–liquid interfacial tensions from corresponding ideal values is that the polymer chains are reorganized at the uppermost layer of the film due to contact with the liquid molecules and expose groups less hydrophobic than CF_2 and CF_3 toward the liquid phase. This phenomenon was discussed in detail above.

The electronegativities shown in Table 8.22 are atomic properties. It can be shown that electronegativity correlates with the electronic properties of the corresponding molecule such as electrostatic potential, dipole moment, and electronic polarizability. For example, 1-fluoronaphthalene has the largest dipole moment and the most negative electrostatic potential. These properties were calculated using computational chemistry software, HyperChem 7.5 [67]. Thus, dipole–dipole and dipole-induced dipole interaction energies between liquid molecules and polymer

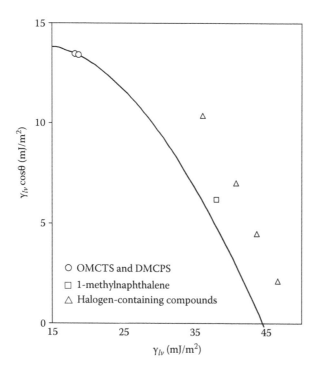

FIGURE 8.23 $\gamma_{lv}\cos\theta$ versus γ_{lv} for contact angles of naphthalene compounds containing halogen atoms (triangles) and 1-methylnaphthalene (square) on EGC-1700 films. The smooth curve is identical to that in Figure 8.14 obtained from OMCTS and DMCPS contact angles (circles). (Reprinted from Tavana, H., Hair, M. L., and Neumann, A. W., *Journal of Physical Chemistry B*, 110, 1294, 2006. With permission from American Chemical Society.)

chains were calculated. The results showed that, overall, contact angle deviations correlate with solid–liquid intermolecular interactions [87]. This is illustrated in Figure 8.24 where contact angle deviations are plotted versus the total interaction energy (i.e., the summation of dipole-dipole and dipole-induced dipole energies) for both Teflon AF 1600 and EGC-1700. In the case of Teflon AF 1600, the interaction energies are very small, regardless of dipole moment and polarizability of the liquid molecules. This is because the Teflon chains have a fairly symmetric and circular cross-section with a small dipole moment. On the other hand the large dipole moment of the perturbed molecular chains of EGC-1700 causes much stronger interactions with the molecules of the probe liquids. The interaction energies increase significantly from 1-methylnaphthalene to 1-fluoronaphthalene, in agreement with the corresponding contact angle deviations. Details of this study can be found elsewhere [87].

8.6.5 INERTNESS OF PROBE LIQUIDS WITH RESPECT TO A SOLID

An important picture emerges from contact angles of the naphthalene compounds on Teflon AF 1600 and EGC-1700 surfaces. Although both are fluoropolymers, Teflon

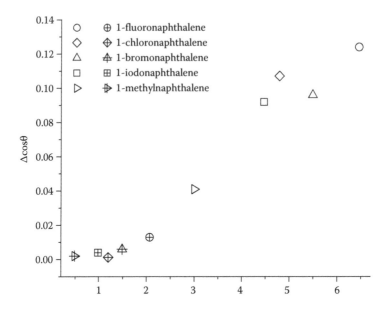

FIGURE 8.24 Deviations in the contact angles of naphthalene compounds measured on Teflon AF 1600 and EGC-1700 surfaces as a function of the corresponding energy of dipole-dipole and dipole-induced dipole interactions (J)(e-20). The crossed and open symbols represent data for Teflon AF 1600 and EGC-1700, respectively. (Reprinted from Tavana, H., Hair, M. L., and Neumann, A. W., *Journal of Physical Chemistry B*, 110, 1294, 2006. With permission from American Chemical Society.)

AF 1600 and EGC-1700 exhibit very different behavior when exposed to one and the same liquid. Teflon AF 1600 has stiff and nonflexible chains of molecules that do not reorganize upon contact with a liquid. On the other hand, EGC-1700 consists of very flexible chains of molecules that are easily perturbed due to contact with a noninert liquid. As the degree of inertness of the liquid decreases, for example, from 1-methylnaphthalene to 1-fluoronaphthalene, the interactions with EGC-1700 chains become stronger. Therefore, it is crucial to select appropriate probe liquids for contact angle measurements on a given polymer film. If the polymer is amorphous and possesses chains that are flexible and mobile with a nonsymmetrical electronic structure, then the probe liquid must be completely inert so that the strong interactions at the solid–liquid interface are eliminated.

If the solid surface were highly crystalline, that is, with chains packing in an orderly fashion, and possessing a uniform electron cloud over the constituent chains, then almost any liquid might be used for contact angle measurement to determine the solid surface tension. Such a situation has been reported with n-hexatriacontane surfaces [89]. The surfaces were of high quality with the chains so well packed that no contact angle hysteresis was observed. The contact angles of a number of liquids with different properties on n-hexatriacontane surfaces are given in Table 8.23. These data are shown in the $\gamma_{lv}\cos\theta$ versus γ_{lv} plot in Figure 8.25. The contact angles conform well to the smooth curve of $\gamma_{sv} = 20.60$ mJ/m^2 and $\beta = 0.000129$ (mJ/m^2)$^{-2}$, which result from the application of multivariable optimization to the experimental

TABLE 8.23

Contact Angles of Liquids with Different Properties on n-Hexatriacontane Surfaces

Liquid	θ (°)	θ' (°)	Δθ (°)	γ_{sv} (mJ/m²)
n-Nonane	25	26.4	+ 1.4	20.8
n-Decane	28	31.3	+ 3.3	21.2
n-Dodecane	38	37.3	−0.7	20.4
n-Tetradecane	41	41.5	+ 0.5	20.7
n-Hexadecane	46	44.2	−1.8	20.1
Ethylene glycol	79.2	78.7	−0.5	20.4
Thiodiglycol	86.3	86.0	−0.3	20.4
Glycerol	95.4	95.7	+ 0.2	20.7
Water	104.6	104.5	−0.1	20.6

Source: Reprinted from Tavana, H., et al., *Advances in Colloid and Interface Science*, 132, 1, 2007. With permission from Elsevier.

Note: n-Alkanes data are from Zisman, W. A., *Advances in Chemistry*, American Chemical Society, Washington, DC, 1964; the last four liquids are from Neumann, A. W., *Advances in Colloid and Interface Science*, 4, 105, 1974. The ideal contact angle of each liquid, the contact angle deviations from the smooth curve of $\gamma_{sv} = 20.60$ mJ/m² and $\beta = 0.000129$ (mJ/m²)⁻², and the solid surface tensions calculated from surface tension and contact angle of each liquid (both from literature) with $\beta = 0.000129$ (mJ/m²)⁻² using Equation 8.6 are also given.

contact angles obtained on n-hexatriacontane surfaces. To determine the extent of the contact angle deviations from this curve, the ideal contact angle (θ') for each liquid was also calculated, as given in Table 8.23. The deviations for n-alkanes are larger than the rest of the liquids. It is pointed out that the contact angles of the last four liquids in this table are from Neumann's work [89] and were measured with capillary rise, which is known as a reliable technique. The contact angles fall on the smooth curve, indicating that even these highly polar liquids do not show specific molecular interactions with n-hexatriacontane molecular chains. On the other hand, the n-alkanes data are taken from Zisman's work [8]. Due to the error associated with the measurement technique, these data have an error limit of at least ±2°. Most likely, the corresponding deviations reflect experimental errors rather than any physical phenomena. Since the scatter in the contact angles does not follow a trend, it is probable that the molecules of long-chain n-alkanes do not experience interfacial structuring at the n-hexatriacontane surface. In view of this, the contact angles of almost any of these liquids could be used to determine n-hexatriacontane surface tension (see Table 8.23). If the contact angles of n-alkanes on n-hexatriacontane films were of the accuracy obtainable with capillary rise or sessile drop techniques, they would not be expected to deviate significantly from the $\gamma_{sv} = 20.60$ mJ/m² smooth curve.

The importance of molecular chain configuration of a solid surface in the contact angle deviations context becomes more evident by comparing the contact angles of water/Teflon AF 1600 and water/n-hexatriacontane systems. Our preliminary

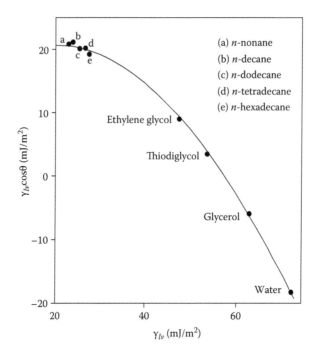

FIGURE 8.25 $\gamma_{lv}\cos\theta$ versus γ_{lv} for the contact angles of different liquids measured on the surfaces of n-hexatriacontane represents $\gamma_{sv} = 20.60$ mJ/m^2 and $\beta = 0.000129$ (mJ/m^2)$^{-2}$.

measurements with water yielded a deviation of ~7° from the $\gamma_{sv} = 13.61$ mJ/m^2 curve for Teflon AF 1600 [66]. It was shown above that Teflon AF 1600 is the most inert polymer used in this study in the sense that the interactions between many liquids and Teflon films are not specific. Nevertheless, the relatively large contact angle deviation of water indicates that even Teflon AF 1600 films are not inert with respect to water and strong interactions are operative between water molecules and Teflon films. The precise mechanisms responsible for the deviations are not known as yet. Possibly the anomalous interfacial properties of water are involved [90]. On the other hand, the negligible deviation of water contact angles from the smooth curve of n-hexatriacontane (see Table 8.23) implies that the close packing of the constituent chains plays an important role in eliminating specific molecular interactions even with water molecules.

8.7 CONTACT ANGLE DEVIATIONS ON SELF-ASSEMBLED MONOLAYERS (SAMs)

The preceding discussions of contact angles and their interpretation were essentially all based on contact angle measurements on films of a thickness well in excess of that of molecular dimensions. However, contact angles have also been studied extensively on monolayers, especially on self-assembled monolayers (SAMs), in particular of octadecanethiol [91–96]. Partly for historical reasons, these measurements were performed with methods and protocols not necessarily compatible with the strategy of contact angle measurement on the slowly advancing drop front, as discussed in previous chapters. Therefore Kwok et al. [97] performed a series of measurements with octadecanethiol adsorbed onto wafers coated with gold, using ADSA in conjunction with a slowly moving advancing drop front. The gold layers had been produced by a sputtering technique. Advancing and receding contact angles for five liquids including water are given in columns 3 and 4 of Table 8.24. Plotting these data in the usual fashion as $\gamma_{lv}\cos\theta$ versus γ_{lv} provides a plot with considerable scatter, of the same magnitude as the scatter with the thick polymeric films, as discussed above. The contact angle of water is $119°$ (i.e., extremely large). As the actual surface of the SAM is expected to consist essentially of CH_3 groups, one would expect a similar value of the contact angle as that on a hexatriacontane film formed by slow vapor deposition in vacuum. The contact angle on such a film against water was found to be $105°$, see Table 8.23. It should be noted that the contact angle hysteresis in the measurements for the SAMs in Table 8.24 was considerable, much larger than that for the polymeric films considered above.

To investigate the source of these discrepancies, Kwok et al. [97] investigated further the possible effect of different structures of the gold film on which the SAMs are formed. As it is known that thermally evaporated gold yields smoother and better polycrystalline structures than those obtained by sputtering [98,99], the gold surfaces were annealed prior to SAM formation. The advancing and receding contact angles are also given in Table 8.24.

Obviously, the contact angles are quite different due to annealing. The advancing angles now provide γ_{sv} values that are virtually independent of the liquid used. To illustrate this fact, the advancing contact angles on the annealed gold surfaces as substrate are plotted in Figure 8.26, together with the literature values for contact angles on hexatriacontane. There are two immediate conclusions to be drawn

TABLE 8.24
Experimental Advancing and Receding Contact Angles on SAMs of Octadecanethiol $CH_3(CH_2)_{17}SH$ Adsorbed onto Evaporated (Nonannealed) and Annealed Gold

Liquid	γ_{lv}	Nonannealed		Annealed	
		θ_a (deg.)	θ_r (deg.)	θ_a (deg.)	θ_r (deg.)
Water	72.7	119.1 ± 0.8	100.2 ± 0.7	106.9 ± 0.5	92.3 ± 0.9
Formamide	59.1	88.7 ± 0.8	63.0 ± 1.4	92.4 ± 1.5	69.2 ± 1.9
Ethylene glycol	47.6	81.5 ± 0.6	66.4 ± 1.1	81.6 ± 2.4	68.2 ± 1.6
Bromonaphthalene	44.3	67.2 ± 0.8	44.1 ± 0.8	76.1 ± 0.9	64.3 ± 1.3
Decanol	28.9	50.7 ± 0.5	38.2 ± 1.1	53.2 ± 0.9	45.1 ± 1.3
Hexadecane	27.6	45.4 ± 0.4	< 20.0	45.7 ± 0.8	35.4 ± 2.2

Note: Error bars are the 95% confidence limits. Reprinted from Yang, J., Han, J., Isaacson, K., and Kwok, D. Y., *Langmuir*, 19, 9231, 2003. With permission from American Chemical Society.

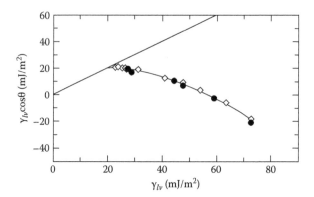

FIGURE 8.26 $\gamma_{lv}\cos\theta$ versus the liquid–vapor surface tension γ_{lv} for hexatriacontane ($\Diamond$) and SAMs of octadecanethiol $CH_3(CH_2)_{17}SH$ adsorbed onto thermally annealed Au ($\bullet$). (Reprinted from Yang, J., Han, J., Isaacson, K., and Kwok, D. Y., *Langmuir*, 19, 9231, 2003. With permission from American Chemical Society.)

from this figure [89]. One, the data points for the SAM surfaces merge perfectly with those for hexatriacontane, in agreement with the expectation that both types of surfaces are composed of closely packed CH_3 groups. And both sets of data fall on a continuous smooth line in a pattern identical with that discussed for polymeric surfaces above. The contact angle hysteresis on the SAMs with annealed gold substrate is also affected. Generally, the hysteresis is smaller in the case of the annealed substrates, although the very low values of the contact angle hysteresis in the case of the polymeric surfaces discussed above are not reached. This suggests that retention of the liquid in the case of the SAMs is more extensive than in the case of polymeric surfaces. A detailed Infrared Reflectance (IR) and ellipsometric study [97] indeed showed clearly that the two types of SAMs had different structure.

Independent AFM images shown in Figure 8.27 suggest that the annealed Au has larger terraces (as much as 200 nm), while the nonannealed Au has much smaller steps. From the interpretation of the IR and AFM results, a model was constructed in Figure 8.28 that illustrates a possible arrangement of octadecanethiol adsorbed onto nonannealed and annealed Au. From the schematic, it is expected that there are more methylenes per unit projected area on the nonannealed Au than on the annealed Au, in general agreement with the IR results. (The slightly larger errors for the contact angle data of octadecanethiol adsorbed onto annealed Au were presumably due to the variability of the annealing procedures.)

A definitive explanation of the different contact angle patterns on the nonannealed surfaces is not obvious. It is apparent from Table 8.24 that the contact angle hysteresis $H = \theta_a - \theta_r$ is indeed smaller for the annealed Au surfaces suggesting better surface quality. From the surface energetics and Young's equation standpoint, a key point is that for the annealed SAM, the water contact angle of 107° yields a γ_{sv} value in accordance with other contact angle measurements on solid surfaces consisting predominately of CH_3-groups, in the neighborhood of 20 mJ/m². A contact angle of 119° would imply a solid surface tension of approximately 12 mJ/m², which seems impossible for any hydrocarbon surface. A stipulation that in the case

(a)

(b)

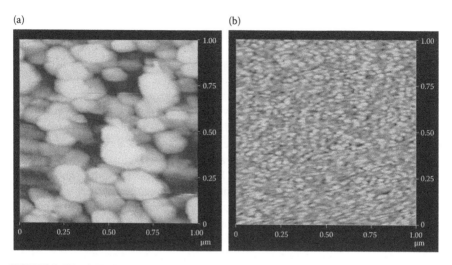

FIGURE 8.27 AFM images of (a) annealed Au (b) nonannealed Au for a scan size of 1 µm. (Reprinted from Yang, J., Han, J., Isaacson, K., and Kwok, D. Y., *Langmuir*, 19, 9231, 2003. With permission from American Chemical Society.)

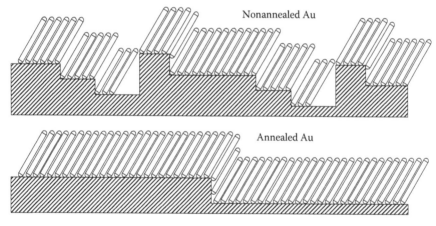

FIGURE 8.28 Schematic illustration of SAM assembly on two different Au substrates. The upper Figure illustrates SAM assembly of octadecanethiol adsorbed onto nonannealed Au with smaller gold steps. The lower Figure illustrates SAM assembly of octadecanethiol adsorbed onto annealed Au with larger terraces. (Reprinted from Yang, J., Han, J., Isaacson, K., and Kwok, D. Y., *Langmuir*, 19, 9231, 2003. With permission from American Chemical Society.)

of the nonannealed gold substrates the SAMs might exhibit a certain fraction of CH_2 groups on the surface would increase the solid surface tension above that of the pure CH_3 surface and hence would cause a smaller contact angle, not the larger observed value. All these facts and stipulations strongly suggest that the contact angle of water on SAMs with nonannealed substrates is not a Young contact angle and hence cannot be used for the determination of solid surface tension. Considerations common in matters of ultrahydrophobicity in Chapter 7 may provide further insight.

8.8   IMPACT OF RECENT WORK ON APPLICABILITY OF THE EQUATION OF STATE

The efficacy of the equation of state to estimate solid surface tensions was established in recent decades. Using this approach, surface tensions with an accuracy of ~1–2 mJ/m² could be obtained for polymer films. Nevertheless there were deviations in the contact angles from the smooth curves of $\gamma_{lv}\cos\theta$ versus γ_{lv} that were not well understood. The deviations were disregarded in the past. Because the contact angle points deviated both above and below the curve, the application of the multivariable optimization technique to these data averaged out the effect of processes such as vapor adsorption and parallel alignment of liquid molecules. Thus, reasonable values for the solid surface tension (γ_{sv}) and the constant β (Equation 8.6) were obtained.

In the major case study on fluorinated polymers of this chapter, the contact angle deviations were rigorously investigated and several causes were identified. Furthermore, the equation of state was evaluated to determine how large the errors in the calculation of solid surface tensions can be. The maximum error found from the extensive study of many S-L systems is ~2.6 mJ/m², which corresponds to the 1-fluoronaphthalene/EGC-1700 system, and is caused by strong intermolecular interactions.

It should be noted that the existence of the deviations does not weaken the status of the equation of state. To illustrate this point, the contact angles of all probe liquids; that is, first and second group of liquids with bulky molecules, n-alkanes, and naphthalene compounds, on Teflon AF 1600 surfaces are plotted in Figure 8.29. If the solid surface tension (γ_{sv}) is calculated from the contact angle of each liquid and the

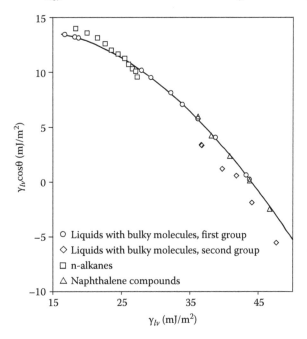

FIGURE 8.29 $\gamma_{lv}\cos\theta$ versus γ_{lv} for the contact angles of all probe liquids on Teflon AF 1600 films. The smooth curve represents $\gamma_{sv} = 13.61$ mJ/m² and $\beta = 0.000116$ (mJ/m²)⁻².

results are then averaged, a value of $\gamma_{sv} = 12.95$ mJ/m^2 is obtained. This value is close to $\gamma_{sv} = 13.61$ mJ/m^2, which was obtained from our scrutiny of the contact angles of these liquids. It is suggested that simply averaging the solid surface tension values obtained from contact angles of all the probe liquids using Equation 8.6 gives a fairly accurate value for the solid surface tension. This is because the effects of different processes that cause the contact angle deviations (e.g., vapor adsorption, alignment of liquid molecules at the solid surface, etc.) are averaged out. Using the averaging strategy, reasonable solid surface tension values were obtained for a large number of solid surfaces with different molecular properties [48].

The above highlights the fact that the equation of state approach is indeed a reliable formulation to predict the surface tension of solids. In particular for fluoropolymer surfaces, surface tension values with an accuracy of $\sim\pm0.2$ mJ/m^2 can be obtained if OMCTS and DMCPS are used as probe liquids for contact angle measurements. This is because these two liquids do not provoke specific interactions with fluoropolymers. The extremely low vapor pressure of these liquids and their special molecular structure eliminate the processes associated with n-alkanes as well as the specific intermolecular interactions at the S–L interface.

8.9 SUMMARY OF THE PHYSICAL CAUSES OF DEVIATIONS FROM THE SMOOTH CURVES

By accurately reproducing the contact angles of a series of n-alkanes on the surfaces of Teflon AF 1600 (average reproducibility was $\pm0.2°$), it was established that the contact angle deviations from smooth curves are not experimental errors but have physical causes. Therefore a systematic study of the contact angles of a large number of liquids on the surfaces of four different fluoropolymers was conducted to identify the causes of the deviations.

It was shown that specific interactions at solid–vapor (S–V) and/or solid–liquid (S–L) interfaces are responsible for the contact angle deviations. Vapor adsorption is the only obvious process that can affect the S–V interface. With respect to the S–L interface, specific interactions take place in different forms: parallel alignment of liquid molecules at the solid surface, reorganization of liquid molecules at the S–L interface, change in the configuration of polymer chains at the surface layer due to contact with the test liquids, and strong intermolecular interactions between solid and liquid. The mechanisms causing the contact angle deviations are summarized below.

1. *Vapor adsorption onto the solid surface*: The analysis of the contact angles of short-chain n-alkanes (e.g., n-hexane) showed that vapor of the test liquid is adsorbed onto the solid surface once an initial sessile drop is formed on the solid. Given that the liquid has a larger surface tension than the fluoropolymer films, vapor adsorption makes the solid surface tension (γ_{sv}) larger than that of the bare polymer film. Therefore the measured contact angles represent a modified surface, and hence deviate from the curve that represents the surface tension of the original polymer film.

2. *Parallel alignment of liquid molecules at the solid–liquid interface*: A detailed study of the contact angles of long-chain n-alkanes (e.g.,

n-hexadecane) revealed that liquid molecules are aligned parallel to the solid surface at the S–L interface. Such substrate-induced reorientation of liquid molecules is also reported in the literature, for example, in studies of friction between two surfaces with *n*-hexadecane as the lubricant between them and in the measurements of forces between surfaces with *n*-alkanes as the intervening medium. Due to the interfacial structuring of liquid molecules, the S-L interfacial tension (γ_{sl}) is not a precise function of γ_{lv} and γ_{sv}, as given by the equation of state. Parallel alignment of *n*-alkane molecules causes γ_{sl} to be somewhat larger than the value predicted by the equation of state. These effects are manifested by deviations of the experimental contact angles from the ideal contact angle pattern.

3. *Reorientation of liquid molecules at the solid–liquid interface*: Interpretation of the contact angles of liquids with exposable oxygen and nitrogen moieties (noninert liquids) on the films of a fluoropolymer that comprises inflexible molecular chains (e.g., Teflon AF 1600) led to the conclusion that liquid molecules can undergo a reorientation at the S–L interface. The presence of oxygen and nitrogen atoms causes a nonuniform electron density distribution over the liquid molecules. Such a nonuniform charge density is likely for the polymer chains that contain CF_2 and CF_3 groups. Based on the physical law that like charges repel each other, it was speculated that the negatively charged parts of the liquid molecules are repelled by the polymer chains. Such a reorientation of liquid molecules makes the S-L interfacial tension (γ_{sl}) larger than the value predicted by the equation of state. Therefore the measured contact angles fall below the smooth curve.

4. *Reorientation of polymer chains due to contact with probe liquids*: It was shown that if a fluoropolymer film consists of flexible and mobile chains of molecules (e.g., EGC-1700), the configuration of polymer chains changes upon contact with a noninert liquid (a liquid with exposable oxygen and nitrogen atoms). Due to the reorganization of polymer chains, groups less hydrophobic than CF_2 and CF_3 are exposed to the S–L interface. This causes the S–L interfacial tension (γ_{sl}) to become less than the value given by the equation of state. As a result, the corresponding contact angles fall above the $\gamma_{lv}\cos\theta$ versus γ_{lv} curve. This process is thermodynamically favorable. A decrease in the solid–liquid interfacial tension is expected to cause the overall free energy of the system to decrease.

5. *Intermolecular interactions at the solid–liquid interface*: The investigation of contact angles of a series of naphthalene compounds on EGC-1700 surfaces showed that contact angle deviations correlate with the electronic properties of liquid molecules. The results suggest that a larger dipole moment of liquid molecules and hence stronger dipolar interactions between solid and liquid causes a significant perturbation of the polymer chains. Due to the exposure of nonfluorinated moieties to the surface, the S–L interfacial tension (γ_{sl}) becomes less than the value predicted by the equation of state. As a result, the measured contact angles fall well above the smooth curve.

6. *Contact angles on self-assembled monolayers (SAMs)*: The findings above are all connected with physicochemical interactions between liquid and smooth polymer surfaces. In the case of SAMs on gold, on the other hand, the morphology of the gold surface is significant. The indications are that on sputtered, nonannealed surfaces the contact angles may be affected by roughness and hence not be compatible with Young's equation. On the other hand, SAMs on annealed gold surfaces yield contact angles that do not correspond specifically with any chemical or structural feature of the liquid, very much like highly crystalline hexatriacontane; that is, all contact angles fall on the smooth curve.

8.10 THE THERMODYNAMIC STATUS OF EXPERIMENTAL CONTACT ANGLES AND APPLICABILITY OF THE EQUATION OF STATE

The applicability of the equation of state depends on the applicability of Young's equation, as a necessary condition. Contact angles on rough surfaces, such as the SAMs on nonannealed sputtered gold surfaces are an obvious example where Young's equation is not satisfied. But validity of Young's equation does not guarantee the applicability of the equation of state, or at least not strictly. Examples are the deviations due to vapor adsorption, orientation of liquid molecules at the solid–liquid interface and reorientation of polymer chains due to contact with the liquid. The consequence of such processes at the solid–liquid interface is change in γ_{sl} from the ideal value that corresponds to the smooth curve in the standard contact angle curve.

Pure liquids should be used for contact angle measurements. The contact angles of mixtures of liquids and surfactant solutions are expected to involve more complexities. It has been found that contact angles of mixtures on one and the same solid surface do not follow smooth patterns [100]. Thermodynamically, such systems have more than two degrees of freedom [49,101,102] due to the additional liquid components. There is no reason to think that contact angles measured with a two-component liquid should satisfy the Young equation any less than one component liquids on the same smooth and homogenous surface. However, there is no independent way of establishing the applicability of the Young equation for any given solid surface-liquid system. Thus, the applicability of the Young equation had to be assumed a priori for all the smooth solid surfaces considered in this chapter. Fortunately, the fairly large array of results and conclusions presented strongly suggests that under these circumstances the Young equation is indeed applicable. The exceptions are the contact angles of SAMs on nonannealed gold surfaces. It is readily apparent that in this case, Young's equation is at variance with known physical facts.

REFERENCES

1. L. Wilhelmy. *Annals of Physics* 119 (1863): 177.
2. S. Hartland. In *Surface and Interfacial Tension: Measurement, Theory, and Applications.* Dekker, New York, 2004.

3. A. W. Adamson. In *Physical Chemistry of Surfaces*. 5th ed. John Wiley & Sons Inc., New York, 1990.
4. A. I. Rusanov and V. A. Prokhorov. In *Interfacial Tensiometry*. Edited by D. Möbius and R. Miller. Elsevier, Amsterdam, 1996.
5. J. C. Eriksson. *Surface Science* 14 (1969): 221.
6. J. J. Gilman. *Journal of Applied Physics* 31 (1960): 2208.
7. A. I. Rusanov. *Surface Science Reports* 23 (1996): 173.
8. W. A. Zisman. In *Advances in Chemistry*. American Chemical Society, Washington, DC, 1964.
9. F. M. Fowkes. *Industrial & Engineering Chemistry* 12 (1964): 40.
10. O. Driedger, A. W. Neumann, and P. J. Sell. *Kolloid-Z. Z. Polym.* 201 (1965): 52.
11. D. K. Owens and R. C. Wendt. *Journal of Applied Polymer Science* 13 (1969): 1741.
12. C. J. van Oss, M. K. Chaudhury, and R. J. Good. *Chemical Reviews* 88 (1988): 927.
13. R. J. Good and C. J. van Oss. In *Modern Approaches to Wettability: Theory and Applications*. Edited by M. Schrader and G. Leob. Plenum Press, New York, 1992.
14. B. V. Derjaguin, V. M. Muller, and Y. P. Toporov. *Journal of Colloid and Interface Science* 73 (1980): 293.
15. K. L. Johnson, K. Kendall, and A. D. Roberts. *Proceedings of the Royal Society of London* A324 (1971): 301.
16. A. Fogden and L. R. White. *Journal of Colloid and Interface Science* 138 (1990): 414.
17. H. K. Christenson. *Journal of Physical Chemistry* 90 (1986): 4.
18. P. M. Claesson, C. E. Blom, P. C. Horn, and B. W. Ninham. *Journal of Colloid and Interface Science* 114 (1986): 234.
19. P. M. Pashley, P. M. McGuiggan, and R. M. Pashley. *Colloids and Surfaces* 27 (1987): 277.
20. R. M. Pashley, P. M. McGuiggan, B. W. Ninham, and D. F. Evans. *Science* 229 (1985): 1088.
21. S. N. Omenyi and A. W. Neumann. *Journal of Applied Physics* 47 (1976): 3956.
22. A. E. Corte. *Journal of Geophysical Research* 67 (1962): 1085.
23. P. Hoekstra and R. D. Miller. *Journal of Colloid and Interface Science* 25 (1967): 166.
24. J. Cissé and G. F. Bolling. *Journal of Crystal Growth* 10 (1971): 67.
25. J. Cissé and G. F. Bolling. *Journal of Crystal Growth* 11 (1971): 25.
26. K. H. Chen and W. R. Wilcox. *Journal of Crystal Growth* 40 (1977): 214.
27. D. W. Fuerstenau and M. C. Williams. *Colloids and Surfaces* 22 (1987): 87.
28. D. W. Fuerstenau and M. C. Williams. *Particle Characterization* 4 (1987): 7.
29. D. W. Fuerstenau and M. C. Williams. *Journal of Mineral Processing* 20 (1987): 153.
30. D. W. Fuerstenau, J. Diao, and J. Hanson. *Energy Fuels* 4 (1990): 34.
31. E. I. Vargha-Butler, T. K. Zubovits, D. R. Absolom, and A. W. Neumann. *Journal of Dispersion Science and Technology* 6 (1985): 357.
32. E. I. Vargha-Butler, E. Moy, and A. W. Neumann. *Colloids and Surfaces* 24 (1987): 315.
33. E. I. Vargha-Butler, T. K. Zubovits, D. R. Absolom, and A. W. Neumann. *Chemical Engineering Communications* 33 (1985): 25.
34. H. G. Bruil. *Colloid and Polymer Science* 252 (1974): 32.
35. G. D. Cheever. *Journal of Coatings Technology* 55 (1983): 53.
36. H. W. Kilau. *Colloids and Surfaces* 26 (1983): 217.
37. K. Grundke, T. Bogumil, T. Gietzelt, H.-J. Jacobasch, D. Y. Kwok, and A. W. Neumann. *Progress in Colloid and Polymer Science* 101 (1996): 58.
38. S. J. Hemingway, J. R. Henderson, and J. R. Rowlinson. *Faraday Symposia of the Chemical Society* 16 (1981): 33.
39. R. Guermeur, F. Biquard, and C. Jacolin. *Journal of Chemical Physics* 82 (1985): 2040.

40. H. C. Hamaker. *Physica* 4 (1937): 1058.
41. J. N. Israelachvili. *Proceedings of the Royal Society of London* A331 (1972): 39.
42. A. E. v. Giessen, D. J. Bukman, and B. Widom. *Journal of Colloid and Interface Science* 192 (1997): 257.
43. T. Young. *Philosophical Transactions of the Royal Society of London* 95 (1805): 65.
44. G. Antonow. *Journal of Chemical Physics* 5 (1907): 372.
45. H. W. Fox and W. A. Zisman. *Journal of Colloid and Interface Science* 5 (1950): 514.
46. H. W. Fox and W. A. Zisman. *Journal of Colloid and Interface Science* 7 (1952): 428.
47. D. H. Kaelble. *Journal of Adhesion* 2 (1970): 66.
48. D. Y. Kwok and A. W. Neumann. *Advances in Colloid and Interface Science* 81 (1999): 167.
49. D. Li, J. Gaydos, and A. W. Neumann. *Langmuir* 5 (1989): 1133.
50. Y. Rotenberg, L. Boruvka, and A. W. Neumann. *Journal of Colloid and Interface Science* 93 (1983): 169.
51. P. Cheng, D. Li, L. Boruvka, Y. Rotenberg, and A. W. Neumann. *Colloids and Surfaces* 43 (1990): 151.
52. O. I. del Río. *On The Generalization of Axisymmetric Drop Shape Analysis.* M.A.Sc. Thesis, University of Toronto, 1993.
53. D. Li and A. W. Neumann. *Journal of Colloid and Interface Science* 148 (1992): 190.
54. D. Y. Kwok. *Contact Angles and Surface Energetics.* PhD Thesis, University of Toronto, 1998.
55. J. Adin Mann, Jr. In *Surface and Colloid Science.* Edited by E. Matijevic and R. J. Good, 213. Plenum Press, New York, 1984.
56. H. Tavana, C. N. C. Lam, P. Friedel, K. Grundke, D. Y. Kwok, M. L. Hair, and A. W. Neumann. *Journal of Colloid and Interface Science* 279 (2004): 493.
57. J. K. Spelt, D. R. Absolom, and A. W. Neumann. *Langmuir* 2 (1986): 620.
58. W. H. Buck and P. R. Resnick. In *183rd Meeting of the Electrochemical Society.* Honolulu, HI, 1993.
59. 3M Inc. In *Product Information, 3M Novec Electronic Coating EGC-1700.* Printed in U.S.A., 2003.
60. D. Appelhans, Z.-G. Wang, S. Zschoche, R.-C. Zhuang, L. Häussler, P. Friedel, F. Simon, et al. *Macromolecules* 38 (2005): 1655.
61. K. Grundke. In *Molecular Interfacial Phenomena of Polymers and Biopolymers.* Edited by P. Chen, 323. CRC Press LLC, Florida, 2005.
62. I. Langmuir. *Transactions of the Faraday Society* 15 (1920): 62.
63. W. C. Bigelow, D. L. Pickett, and W. A. Zisman. *Journal of Colloid and Interface Science* 1 (1946): 513.
64. C.-C. Cho, R. M. Wallace, and L. A. Files-Sesler. *Journal of Electronic Materials* 23 (1994): 827.
65. C. W. Extrand. *Langmuir* 9 (1993): 475.
66. H. Tavana, N. Petong, A. Hennig, K. Grundke, and A. W. Neumann. *Journal of Adhesion* 81 (2005): 29.
67. Hypercube Inc. In *HyperChem Release 7.0 for Windows Reference Manual.* Gainesville, FL, 2002.
68. H. Tavana, R. Gitiafroz, M. L. Hair, and A. W. Neumann. *Journal of Adhesion* 80 (2004): 705.
69. R. G. Horn and J. Israelachvili. *Journal of Chemical Physics* 75 (1981): 1400.
70. J. N. Israelachvili, P. M. McGuiggan, and A. M. Homola. *Science* 240 (1988): 189.
71. M. L. Gee, P. M. McGuiggan, and J. N. Israelachvili. *Journal of Chemical Physics* 93 (1990): 1895.
72. L. Lavielle and J. Schultz. *Journal of Colloid and Interface Science* 106 (1985): 438.

73. S. R. Holmes-Farley, R. H. Reamey, R. Nuzzo, T. J. McCarthy, and G. M. Whitesides. *Langmuir* 3 (1987): 799.

74. T. Yasuda, M. Miyama, and H. Yasuda. *Langmuir* 8 (1992): 1425.

75. T. Yasuda, M. Miyama, and H. Yasuda. *Langmuir* 10 (1994): 583.

76. J.-H. Wang, P. M. Claesson, J. L. Parker, and H. Yasuda. *Langmuir* 10 (1994): 3897.

77. S. H. Lee and P. J. Rossky. *Journal of Chemical Physics* 100 (1994): 3334.

78. H. Tavana, F. Simon, K. Grundke, D. Y. Kwok, M. L. Hair, and A. W. Neumann. *Journal of Colloid and Interface Science* 291 (2005): 497.

79. H. Tavana, D. Appelhans, R.-C. Zhuang, S. Zschoche, K. Grundke, M. L. Hair, and A. W. Neumann. *Colloid and Polymer Science* 291 (2005): 497.

80. C. N. C. Lam, N. Kim, D. Hui, D. Y. Kwok, M. L. Hair, and A. W. Neumann. *Colloids and Surfaces A: Physicochemical and Engineering Aspects* 189 (2001): 265.

81. H. Tavana, C. Yang, C. Y. Yip, D. Appelhans, S. Zschoche, K. Grundke, M. L. Hair, and A. W. Neumann. *Langmuir* 22 (2006): 628.

82. S. Balasubramanian, M. L. Klein, and J. I. Siepmann. *Journal of Chemical Physics* 103 (1995): 3184.

83. U. Landman, T. Xia, J. Ouyang, and M. W. Ribarsky. *Physical Review Letters* 69 (1992): 1967.

84. M. He, A. S. Blum, G. Overney, and R. Overney. *Physical Review Letters* 88 (2002): 154302.

85. H. K. Christenson, D. W. R. Gruen, R. G. Horn, and J. N. Israelachvili. *Journal of Chemical Physics* 87 (1989): 1834.

86. T. C. Merkel, V. Bondar, K. Nagai, B. D. Freeman, and Y. P. Yampolskii. *Macromolecules* 32 (1999): 8427.

87. H. Tavana, M. L. Hair, and A. W. Neumann. *Journal of Physical Chemistry* B 110 (2006): 1294.

88. N. N. Greenwood and A. Earnshaw. In *Chemistry of the Elements*. 2nd ed. Oxford, Boston, 2005.

89. A. W. Neumann. *Advances in Colloid and Interface Science* 4 (1974): 105.

90. J. Israelachvili. *Intermolecular and Surface Forces*. 2nd ed. Academic Press Ltd, San Diego, CA, 1992.

91. R. Bhatia and B. J. Garrison. *Langmuir* 13 (1997): 765.

92. A. T. Lusk and G. K. Jennings. *Langmuir* 17 (2001): 7830.

93. A. N. Parikh and D. L. Allara. *Journal of Chemical Physics* 96 (1992): 927.

94. M. H. Schoenfischand and J. E. Pemberton. *Journal of the American Chemical Society* 120 (1998): 4502.

95. A. Ulman. In *An Introduction to Ultrathin Organic Films: From Langmuir-Blodgett to Self-Assembly*. Academic Press, Boston, 1991.

96. F. P. Zamborini and R. M. Crooks. *Langmuir* 14 (1998): 3279.

97. J. Yang, J. Han, K. Isaacson, and D. Y. Kwok. *Langmuir* 19 (2003): 9231.

98. W. Guo and G. K. Jennings. *Langmuir* 18 (2002): 3123.

99. R. G. Nuzzo, F. A. Fusco, and D. L. Allara. *Journal of the American Chemical Society* 109 (1987): 2358.

100. D. Li, C. Ng, and A. W. Neumann. *Journal of Adhesion Science and Technology* 6 (1992): 601.

101. R. Defay. *Etude Thermodynamique de la Tension Superficielle*. Gauthier Villars, Paris, 1934.

102. D. Li and A. W. Neumann. *Advances in Colloid and Interface Science* 49 (1994): 147.

103. D. Y. Kwok, C. J. Budziak, and A. W. Neumann. *Journal of Colloid and Interface Science* 173 (1995): 143.

104. D. Y. Kwok, R. Lin, M. Mui, and A. W. Neumann. *Colloids and Surfaces A-Physicochemical and Engineering Aspects* 116 (1996): 63.

105. O. I. del Río, D. Y. Kwok, R. Wu, J. M. Alvarez, and A. W. Neumann. *Colloids and Surfaces A: Physicochemical and Engineering Aspects* 143 (1998): 197.
106. D. Y. Kwok, C. N. C. Lam, A. Li, A. Leung, R. Wu, E. Mok, and A. W. Neumann. *Colloids and Surfaces A-Physicochemical and Engineering Aspects* 142 (1998): 219.
107. G. H. E. Hellwig and A. W. Neumann. In *5th International Congress on Surface Activity, Section B*. 1968.
108. G. H. E. Hellwig and A. W. Neumann. *Kolloid-Z. Z. Polym.* 40 (1969): 229.
109. D. Y. Kwok, T. Gietzelt, K. Grundke, H.-J. Jacobasch, and A. W. Neumann. *Langmuir* 13 (1997): 2880.
110. D. Y. Kwok, C. N. C. Lam, A. Li, A. Leung, and A. W. Neumann. *Langmuir* 14 (1998).
111. D. Y. Kwok, A. Leung, A. Li, C. N. C. Lam, R. Wu, and A. W. Neumann. *Colloid and Polymer Science* 276 (1998): 459.
112. D. Y. Kwok, C. N. C. Lam, A. Li, K. Zhu, R. Wu, and A. W. Neumann. *Polymer Engineering and Science* 38 (1998): 1675.
113. D. Y. Kwok, A. Li, C. N. C. Lam, R. Wu, S. Zschoche, K. Poschel, T. Gietzelt, K. Grundke, H.-J. Jacobasch, and A. W. Neumann. *Macromolecular Chemistry and Physics* 200 (1999): 1121.
114. D. Y. Kwok, C. N. C. Lam, A. Li, and A. W. Neumann. *Journal of Adhesion* 69 (1998): 229.
115. D. Y. Kwok, A. Leung, C. N. C. Lam, A. Li, R. Wu, and A. W. Neumann. *Journal of Colloid and Interface Science* 206 (1998): 44.
116. H. Tavana and A. W. Neumann. *Advances in Colloid and Interface Science* 132 (2007): 1–32.
117. H. Tavana, D. Appelhans, R.-C. Zhuang, S. Zschoche, K. Grundke, M. L. Hair, and A. W. Neumann. *Colloid and Polymer Science* 284 (2006): 497–505.

9 Contact Angles and Solid Surface Tensions

Robert David, Jan Spelt,
Junfeng Zhang, and Daniel Kwok

CONTENTS

9.1   INTRODUCTION

The contact angle of a liquid drop on an ideal solid surface is determined by the mechanical equilibrium of the drop under the action of three interfacial tensions: solid–vapor (γ_{sv}), solid–liquid (γ_{sl}), and liquid–vapor (γ_{lv}). This equilibrium relation is known as Young's equation [1]:

$$\gamma_{lv} \cos\theta_Y = \gamma_{sv} - \gamma_{sl}, \qquad (9.1)$$

where θ_Y is the Young contact angle; that is, a contact angle that can be inserted into Young's equation (see Chapter 7).

Young's equation contains only two measurable quantities, the contact angle θ and the liquid–vapor surface tension, γ_{lv}. In order to determine γ_{sv} and γ_{sl}, an additional relation between these quantities must be sought. Nevertheless, Equation 9.1 suggests that the observation of the equilibrium contact angles of liquids on solids may be a starting point for investigating the solid surface tensions, γ_{sv} and γ_{sl}.

The determination of solid–vapor and solid–liquid interfacial tensions is of importance in a wide range of problems in pure and applied science. For example, the process of particle adhesion is dependent on the sign of the net free-energy change ΔF^{adh} of the system during the adhesion process, which depends explicitly on the solid (particle) surface tensions. Other applications include sedimentation of particles [2] and film flotation [3].

Fundamentally, liquid (γ_{lv}) and solid (γ_{sv}) surface tensions reflect the strength of molecular interactions within the bulk materials. It is therefore reasonable to expect that the solid–liquid interfacial tension γ_{sl}, reflecting cross-interactions between the two phases, may be derivable in terms of γ_{lv} and γ_{sv}. Such a relation—i.e., $\gamma_{sl} = f(\gamma_{lv}, \gamma_{sv})$—together with Young's equation (Equation 9.1), would indeed allow determination of γ_{sv} and γ_{sl} from measurements of γ_{lv} and θ.

This chapter presents a historical review of certain attempts to formulate such a relationship. These attempts fall into three broad categories: macroscopic approaches in which the only information needed to characterize the liquid and solid phases is γ_{lv} and γ_{sv} (i.e., equations of state); macroscopic approaches for which further information characterizing the phases is required (i.e., surface tension components (STCs)); and approaches that attempt to calculate surface and interfacial tensions directly from knowledge of molecular properties.

9.1.1   ZISMAN

Historically, the interpretation of contact angles in terms of solid surface energetics started with the pioneering work of Zisman [4]. While contact angles and contact angle measurements prior to the work of Zisman were somewhat suspect, particularly among physical chemists, they have since gained respectability to the extent that whole symposia are dedicated to contact angle phenomena. Zisman conducted numerous studies of contact angles on low-energy solid surfaces, such as Teflon, with liquids of relatively high surface tension (we know that this corresponds to

$\gamma_{lv} > \gamma_{sv}$). The key observation made by Zisman was that for a given solid, the measured contact angles did not vary randomly as the liquid was varied; rather, he found that for a homologous series of liquids, say alkanes, and a given solid, say Teflon, $\cos\theta$ changed smoothly with γ_{lv} in a fashion that suggested a straight-line relationship. The extrapolation of this straight line to the point where $\cos\theta = 1$ yielded "the critical surface tension γ_c"—the surface tension of a liquid that would just wet the solid surface completely. While Zisman stated that γ_c behaved as one would expect the surface tension of the solid, γ_{sv}, to behave, he took great care not to identify γ_c with γ_{sv}. When other types of liquids, say a homologous series of alcohols, were used instead, the contact angles changed with the liquid surface tension in a similar manner but did not superimpose completely on the alkane data. Such observations have been discussed in terms of a band in which all experimental points fall; or alternatively, in terms of different straight-line fits for different homologous series and hence different values of γ_c for one and the same solid depending on the types of liquids [4].

Subsequent to Zisman's work two schools of thought arose: surface tension components and equation of state.

9.2 SURFACE TENSION COMPONENT APPROACHES

9.2.1 FOWKES

The STC approach was pioneered by Fowkes [5], who proposed that surface tension can be expressed as a sum of surface tension components, each due to a particular type of intermolecular force:

$$\gamma = \gamma^d + \gamma^{di} + \gamma^h + \dots, \tag{9.2}$$

where γ is the total surface tension, and γ^d, γ^{di}, and γ^h are the STCs due to dispersion, dipole–dipole interactions, and hydrogen bonding, respectively. These components may vary according to the nature of the material. Such surface tension components lie outside the realm of thermodynamics (see the discussion of the phase rule in Section 9.3.5).

In practice, Equation 9.2 is often rearranged into the following form:

$$\gamma = \gamma^d + \gamma^n. \tag{9.3}$$

That is, the total surface tension γ is the sum of dispersive (γ^d) and nondispersive (γ^n) STCs. They are often said to be the apolar and polar STCs, respectively.

Within the framework of the Fowkes model, only dispersive cross-interface interactions are considered. It is therefore applicable only when at least one of the two phases is purely (or effectively) dispersive. For example, at a water-Teflon interface, since only dispersion forces are present in the Teflon, the large polar and hydrogen-bonding forces in the water are assumed not to act across the interface to affect the interfacial tension directly. Thus, a solid–liquid interfacial tension γ_{sl} can be expressed by means of a geometric mean relationship as

$$\gamma_{sl} = \gamma_s + \gamma_l - 2\sqrt{\gamma_s^d \gamma_l^d}. \tag{9.4}$$

Assuming that the solid phase is completely dispersive ($\gamma_s = \gamma_s^d$), Equation 9.4 reduces to

$$\gamma_{sl} = \gamma_s + \gamma_l - 2\sqrt{\gamma_s \gamma_l^d}. \tag{9.5}$$

It is this form of the Fowkes equation that is often used in conjunction with Young's equation for the determination of solid surface tensions of dispersive solids.

9.2.2 van Oss

Following the same line of thought as Fowkes, van Oss and coworkers [6] divided the surface tension into different components—the Lifshitz-van der Waals (LW), acid (+), and base (–) components—such that the total surface tension γ was proposed as

$$\gamma = \gamma^{LW} + 2\sqrt{\gamma^+ \gamma^-}, \tag{9.6}$$

for either a solid or a liquid phase. The interfacial tensions of the approach were postulated, based on an intuition of a new combining rule for the acid and base components, as

$$\gamma_{12} = \left(\sqrt{\gamma_1^{LW}} - \sqrt{\gamma_2^{LW}}\right)^2 + 2\left(\sqrt{\gamma_1^+} - \sqrt{\gamma_2^+}\right)\left(\sqrt{\gamma_1^-} - \sqrt{\gamma_2^-}\right). \tag{9.7}$$

Combining Equations 9.6 and 9.7 yields

$$\gamma_{12} = \gamma_1 + \gamma_2 - 2\sqrt{\gamma_1^{LW}\gamma_2^{LW}} - 2\sqrt{\gamma_1^+\gamma_2^-} - 2\sqrt{\gamma_1^-\gamma_2^+}. \tag{9.8}$$

Equation 9.8 is applicable to both liquid–solid and liquid–liquid systems. Since this approach endows the system with more degrees of freedom than provided for by the phase rule (see Section 9.3.5), like the Fowkes theory, it lies outside of thermodynamics. Table 9.1 summarizes the STCs of various commonly used liquids.

Experimental tests of the surface tension components approaches and the equation of state approach (Sections 9.3–9.4) are mainly discussed in Section 9.5; however, additional observations regarding the van Oss theory will be mentioned here. One concern with surface tension component theories has been discussed extensively in Chapter 8. It was shown that contact angles of different liquids on the same solid generally follow a single smooth curve. The minor deviations from this curve are caused by a variety of specific interactions between the solid and the liquid and cannot be attributed simply to the liquid *or* the solid. A test of Equation 9.8 for calculat-

TABLE 9.1

Surface Tensions and Components (mJ/m²)

Liquid	γ	γ^{LW}	γ^+	γ^-
Water	72.8	21.8	25.5	25.5
Glycerol	64	34	3.92	57.4
Formamide	58	39	2.28	39.6
Ethylene glycol	48.0	29	1.92	47.0
Dimethyl sulfoxide	44	36	0.5	32
Diiodomethane	50.8	50.8	0	0
1-Bromonaphthalene	44.4	44.4	0	0
Hexadecane	27.5	27.5	0	0
Tetradecane	26.6	26.6	0	0
Dodecane	25.4	25.4	0	0
Decane	23.8	23.8	0	0
Pentane	16.1	16.1	0	0

Note: From Good, R. J. and van Oss, C. J., *Modern Approaches to Wettability: Theory and Applications*, 1–27. Plenum Press, New York, 1992; van Oss, C. J., Good, R. J., and Chaudhury, M. K., *Journal of Colloid and Interface Science*, 111, 378, 1986; and Costanzo, P. M., Wu, W., Giese, Jr., R. F., and van Oss, C. J., *Langmuir*, 11, 1827, 1995.

ing solid STCs makes use of experimental contact angles through Young's equation (Equation 9.1). Combining Equation 9.8 with Young's equation yields:

$$\gamma_l\left(1+\cos\theta_Y\right)=2\sqrt{\gamma_l^{LW}\gamma_s^{LW}}+2\sqrt{\gamma_l^+\gamma_s^-}+2\sqrt{\gamma_l^-\gamma_s^+}. \tag{9.9}$$

Since the above equation contains three unknowns (γ_s^{LW}, γ_s^+, and γ_s^-) of a solid, it was suggested to use contact angle measurements of at least three different liquids (with known liquid STCs) on the same solid, and solve three simultaneous equations [7,8]. While these procedures imply the applicability of Young's equation and constancy of solid surface tension from one liquid to the next, this has sometimes been overlooked. For example, contact angle measurements on gels were used to determine solid STCs from Equation 9.9 [9]; such results may be misleading, since Young's equation may not be applicable on nonrigid surfaces. Water contact angles on noninert surfaces, such as films of human serum albumin [10], have been used in Equation 9.9. Water may dissolve such films upon contact, making the solid and/or liquid phases different from the original ones: if the operative solid surface tension is not constant from one liquid to the next due to such physical/chemical reactions, simultaneous solution of different equations (from contact angles of different liquids) will not be appropriate.

Kwok et al. [11] provided extensive tests of Equation 9.9, using experimental contact angles of triplets of polar and nonpolar liquids on inert solid surfaces, together with liquid STCs from the literature [7,8]. For an FC-721-coated mica surface, calculated γ_s values varied from −30.0 to 107.0 mJ/m², strongly dependent on the choice of the liquid triplets [11]. Results for solid surface tension components also varied by a very large margin. Performing the calculations with a proposed alternate method [12] based on Equation 9.9 with one nonpolar liquid in each triplet again resulted in inconsistent γ_s values, varying from −35.6 to 9.5 mJ/m² [11]. Similarly scattered results were also obtained for Teflon FEP and polyethylene terephthalate (PET) [11], and in another study for polystyrene (PS) and polymethyl methacrylate (PMMA) [13] surfaces.

In response, it has been argued that such results are a consequence of ill-conditioning of the simultaneous set of three equations (Equation 9.9), due to choices of liquid triplets with STCs that are too similar (from liquid to liquid) [14,15]. Appropriate choices must include one nonpolar liquid with high surface tension (i.e., high γ^{LW}), preferably diiodomethane; one liquid with high basicity (γ^+), of which water is the only known example; and one polar liquid (high γ), such as formamide, glycerol, or ethylene glycol [16]. Nevertheless, the extreme sensitivity of the approach to the liquids employed is of some concern; its practicality is also limited, with for example water-soluble surfaces excluded.

A further contentious issue with respect to the van Oss and other wetting theories has been the thermodynamic equivalence or nonequivalence of liquid–solid and liquid–liquid systems. Several investigators [17–19] have suggested that approaches for liquid–solid capillary systems must also be applicable to liquid–liquid systems. However, Young's equation itself runs counter to this assertion, being a special case of the Neumann triangle when a flat and rigid (i.e., solid) phase is present. In much the same way, while a comprehensive wetting theory that covers liquid–liquid systems (such as the STCs approach) should include liquid–solid systems as a special case, a more specialized theory (such as the equation of state approach) that applies only to liquid–solid systems need not be valid for liquid–liquid systems. In thermodynamic terms, the liquid–liquid interface has an extra degree of freedom relative to a liquid–solid interface, as detailed in Section 9.3.5.

In addition to solid–liquid systems, the van Oss STCs approach has also been evaluated for liquid–liquid systems. Interfacial tensions of a number of liquid–liquid pairs were measured [20] and compared with calculated values based on the STCs specified by van Oss et al. [7] (in Table 9.1). For liquid pairs that were immiscible, the predicted interfacial tensions ranged from 34% lower to 112% higher than the experimental values. Several liquid pairs were found to be miscible. The interfacial tensions of these systems should be zero or negative [7]. Equation 9.8, however, predicted these interfacial tensions to be all positive, varying from 2.0 to 7.0 mJ/m² [20]. This was ascribed by Della Volpe et al. [14] to inaccuracy in the suggested STCs.

The above tests make use of liquid STCs recommended in the literature. However, a number of different sets of these components have been proposed since 1986 [13]. The difficulty of measuring these quantities, postulated by Fowkes as unique material properties, is a drawback of the theory. Only after a set of liquid STCs is finalized can absolute testing be possible.

These and other issues regarding the validity of the acid–base theory and of the suggested STCs have been debated extensively in the literature by a number of other researchers [21–27].

9.3 EXISTENCE OF AN EQUATION OF STATE

9.3.1 INTRODUCTION

The calculation of solid surface tension γ_{sv} from the contact angle θ of a liquid of surface tension γ_{lv} starts with Young's equation (Equation 9.1). Of the four quantities in Young's equation, only γ_{lv} and θ are readily measurable. Thus, in order to determine γ_{sv}, further information is necessary. Conceptually, an obvious approach is to seek one more relation among the variables of Equation 9.1, such as an equation of state, possibly of the form

$$\gamma_{sl} = f\left(\gamma_{sv}, \gamma_{lv}\right). \tag{9.10}$$

The simultaneous solution of Equations 9.1 and 9.10 would solve the problem. Note that if the commonly used assumption of negligible liquid vapor adsorption is applied, then Equations 9.1 and 9.10 may be written in terms of γ_l and γ_s, rather than γ_{lv} and γ_{sv}.

9.3.2 GOOD'S INTERACTION PARAMETER

One simple equation of state that appears in the literature from time to time, despite never being derived, is Antonow's Rule

$$\gamma_{sl} = \left|\gamma_l - \gamma_s\right|. \tag{9.11}$$

Another old equation of state for solid–liquid interfacial tensions is that due to Rayleigh and later Good et al. (reviewed by Spelt [28])

$$\gamma_{sl} = \gamma_s + \gamma_l - 2\sqrt{\gamma_s \gamma_l}. \tag{9.12}$$

Combining Equation 9.12 with Young's equation (Equation 9.1) gives

$$\gamma_s = \frac{1}{4}\gamma_l\left(1 + \cos\theta\right)^2. \tag{9.13}$$

It is this equation that then allows the determination of solid surface tension γ_s from a pair of experimental liquid surface tension γ_l and contact angle θ values. The γ_{sl} value can then be determined either from Equation 9.1 or 9.12 once γ_s is known.

Early investigations by Good [29] showed that Equation 9.13 yields consistent values of γ_s when γ_s and γ_l are both relatively small (e.g., liquid alkanes on Teflon or paraffin wax). Contact angle data for liquids of higher surface tension, however, lead to values of solid surface tension that become progressively smaller as the liquid surface tension increases. The fourth column of Table 9.2 illustrates this for the contact angles of a wide range of liquids on solid hexatriacontane. On the assumption that the surface tension of the hexatriacontane should be approximately constant for all these liquids, it is evident that Equation 9.12 is inadequate as an equation of state.

TABLE 9.2

Solid Surface Tension (mJ/m²) of
***n*-Hexatriacontane at 20°C**

Liquid	γ_{lv}	θ	γ_s	γ_s^{EQS}
Water	72.8	104.6	10.2	19.8
Glycerol	63.4	95.4	13.0	20.0
Thiodiglycol	54.0	86.3	15.3	19.8
Ethylene glycol	47.7	79.2	16.8	19.8
Hexadecane	27.6	46	19.8	20.1
Tetradecane	26.7	41	20.6	20.7
Dodecane	25.4	38	20.3	20.4
Decane	23.9	28	21.2	21.2
Nonane	22.9	25	20.8	20.8

Note: γ_s calculated using Equation 9.13 and γ_s^{EQS} calculated using Equation 9.46. Contact angle data (degrees) from Neumann, A. W., *Advances in Colloid and Interface Science*, 4, 105, 1974.

Good [29] proposed that Equation 9.12 could be modified so that it yields constant values of γ_s for all γ_l values by incorporating an adjustable parameter, Φ, the "Good interaction parameter." Equation 9.12 is then written as

$$\gamma_{sl} = \gamma_s + \gamma_l - 2\Phi\sqrt{\gamma_s\gamma_l}, \tag{9.14}$$

and Equation 9.13 becomes

$$\gamma_s = \frac{1}{4}\frac{\gamma_l^2}{\Phi^2\gamma_l}(1+\cos\theta)^2, \tag{9.15}$$

with the understanding (from experimental observation) that Φ approaches 1 whenever γ_s and γ_l are both relatively small. However, it should be noted that Good [29] believed that it is not the magnitudes of the solid and liquid surface tensions that govern the condition $\Phi = 1$, but rather the similarity in the types of intermolecular forces in the solid and liquid. These two points of view are often coincidental, since as surface tension decreases below, say, 30 mJ/m², it is generally found that London dispersion forces are predominant in both solids and liquids.

At this point, three options present themselves for the further development of Equations 9.12 or 9.14.

1. Attempts can be made to calculate Φ using statistical mechanics, which is the very complex path followed by Good [29].
2. As described previously, Fowkes's approach starts with the conviction that the total surface tension can be decomposed into STCs:

$$\gamma = \gamma^d + \gamma^h + \ldots,\tag{9.16}$$

where γ^d and γ^h are considered unique physical properties, called dispersion and hydrogen-bonding components of surface tension.

In the context of Equation 9.12, this latter approach interprets the decrease in γ_s with increasing γ_l (cf. Table 9.2) as a reflection of the decrease in the relative importance of dispersion forces in the higher-surface-tension liquids found in Table 9.2. Equation 9.12 is thus thought to be deficient because it does not take into account the nature of the inter-molecular forces present in these various liquids. This reasoning led to the Fowkes equation (Equation 9.4). Equation 9.4 is supposed to be valid only for cases where at least one phase is completely dispersive. With respect to Table 9.2, this would mean, for example, that in the case of water where $\gamma_l^d < \gamma_l$, γ_{sl} would be larger than the value obtained from Equation 9.12; hence, in view of Young's equation, γ_{sv} would also be larger, potentially equal to the values obtained with low-surface tension liquids, i.e., the liquids having only dispersion forces. Combining Equation 9.4 with Young's equation (Equation 9.1) yields

$$\gamma_s^d = \frac{1}{4}\frac{\gamma_l^2}{\gamma_l^d}\left(1+\cos\theta\right)^2.\tag{9.17}$$

Thus, within the context of the STCs approach, γ_s^d can be determined from the liquid surface tension γ_l, its dispersive component γ_l^d and the contact angle θ.

3. The approach of Neumann et al. [30] begins with the observation that Equation 9.10 implies that in Equation 9.14, Φ must in fact be a function of the other variables. That an equation of the form of Equation 9.10 must exist is demonstrated thermodynamically in two ways in the following sections. The original formulation of the equation of state was based on the curve-fitting of contact angle data to

$$\Phi = \Phi\left(\gamma_{sl}\right).\tag{9.18}$$

The use of Φ as a correlating variable was essentially arbitrary and a more recent formulation of the equation of state presented below does not use it at all.

9.3.3 CONTACT ANGLE DATA

Historically, the equation of state approach for interfacial tensions may be seen as a development based on the pioneering work of Zisman [4]. The equation of state approach started out by asking if the $\cos\theta$ versus γ_{lv} relation as obtained by Zisman might not be more or less universal, and whether the deviations of the experimental points from a smooth curve could not have causes that are different from a lack

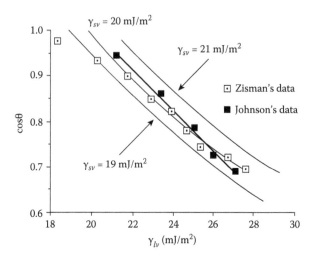

FIGURE 9.1 Contact angle data for Teflon. Least-squares straight line through Johnson's data; curved lines calculated from equation of state for γ_{sv} = 19, 20, 21 mJ/m². (From Li, D., Moy, E., and Neumann, A. W., *Langmuir,* 6, 885, 1990; Data from Fox, H. W., and Zisman, W. A., *Journal of Colloid Science,* 5, 514, 1950 and Johnson, R.E., and Dettre, R.H., *Langmuir,* 5, 293, 1989.)

of uniqueness of γ_c. A universal pattern would imply that $\theta = f(\gamma_{sv},\gamma_{lv})$, which is equivalent to an equation of state (i.e., a second relationship among the variables in Young's equation). To test this hypothesis, Figure 9.1 shows a plot of contact angle data for Teflon in $\cos\theta$ versus γ_{lv}. In this plot, a solid line calculated from the equation of state by Neumann (see below) for γ_{sv} = 20 mJ/m² was drawn. Two lines corresponding to γ_{sv} = 19 mJ/m² and 21 mJ/m² were drawn as well. One can readily see that nearly all experimental points fall within the band defined by these two lines. Thus, at least in this case, it appears that the deviations from a unique value for γ_{sv} are less than 1 mJ/m², and the question that should be considered is whether such deviations could have reasons different from lack of universality of Equation 9.10. There are a number of factors that could account for the scatter of the points in Figure 9.1. Some of the more important ones are discussed below (see also Chapter 8).

1. *Adsorption.* The equilibrium spreading pressure π is given by the equation

$$\pi = \gamma_s - \gamma_{sv}, \qquad (9.19)$$

where γ_s is the solid surface tension in a vacuum. This issue has been studied by Spelt et al. [31], and is discussed more fully in Section 9.4.1. Although there is no general consensus on the magnitude of the equilibrium spreading pressure, it would seem that a value of the order of 1 mJ/m² is not too high an estimate. Since the equilibrium spreading pressure will depend *inter alia* on vapor pressure, it will vary from liquid to liquid within a homologous series, as well as from one type of liquid to the next. Thus, a variety of patterns of experimental points is possible, without conflicting with the idea of a unique equation of state. Scatter due to this cause would not interfere with the possibility of calculating γ_{sv} from individual pairs of (γ_{lv}, θ) data: Changes of γ_{sv} with changes of liquid and liquid surface tension would simply reflect changes in the equilibrium spreading pressure and might be used to determine the latter.

2. *Contact angle measurement.* In Fox and Zisman's contact angle measurements [32,33], the sessile drop was formed by depositing the liquid from above onto the solid surface, and the contact angle was measured by a goniometer. In this procedure, certain vibrations or oscillations of the drop are inevitable. This may produce a value of contact angle θ between the true advancing contact angle θ_a and the receding contact angle θ_r. Furthermore, the error of the contact angles measured by a goniometer may be as large as $\pm 3°$; but a $3°$ difference in contact angle at $\gamma_{lv} = 27$ mJ/m^2 and $\theta = 50°$ leads to, approximately, a 1 mJ/m^2 difference in γ_{sv}.

3. Another possible error in contact angle measurements can arise from the drop-size dependence of contact angles, a possibility that has not yet been considered adequately. The dependence of contact angles on drop size may be caused by line tension [34]. There are a number of studies of contact angle drop-size dependence. As examples, several investigations [34] have reported that the contact angle changed by approximately $3°$–$5°$ while the radius of the three-phase contact circle increased from 1 to 5 mm (see Chapter 13).

Only after clarification of these points will we know whether, experimentally, Equation 9.10 is unique or universal. While it may not be possible to anticipate the answers to all these questions, our experience has shown that the more careful the experimentation (including the preparation of the solid surface), the closer do the experimental points fall to a smooth curve. An example is given in Figure 9.2, where two $\gamma_{lv} \cos\theta$ versus γ_{lv} curves are reproduced for hexatriacontane [35] and cholesteryl acetate [36] surfaces. Both surfaces were so smooth and homogeneous that contact angle hysteresis with water was zero. The measurements were made dynamically with the method of capillary rise at a vertical plate at very low rate of advance of the three-phase line. Figure 9.2 suggests that, in these cases, the equation of state is indeed universal, the contact angles are meaningful, and the equilibrium spreading pressure is negligible. Clearly, more high-quality contact angle data are sorely needed.

The ability to determine γ_{sv} and γ_{sl} from a single contact angle measurement depends on having an equation of state of the form of Equation 9.10. While Figures 9.1 and 9.2 display experimental evidence of the existence of such an equation, existence can also be demonstrated based on thermodynamic principles. This is done in two ways in the following sections. The explicit formulation of an empirical equation of state is then discussed in Section 9.4.

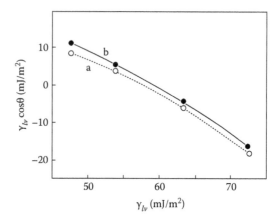

FIGURE 9.2 Contact angles of different liquids on the same solid lie on smooth curves. Curve (a), hexatriacontane; curve (b), cholesteryl acetate.

9.3.4 INTERFACIAL GIBBS–DUHEM EQUATIONS

Following the approach of Ward and Neumann [37], consider the system shown in Figure 9.3, where three phases are in equilibrium under the following conditions.

1. The surface of the solid is smooth and homogeneous.
2. There is no dissolution of the solid nor is there any absorption by the solid of the components of the liquid or gaseous phases.
3. The solid is assumed sufficiently rigid so that its state of strain is unaffected by movements of the three-phase line.

For simplicity, consider the case in which the liquid is pure and the gas is just the vapor phase of the liquid. The three interfacial Gibbs–Duhem equations are then

$$d\gamma_{sv} = -s_{(1)}^{sv}dT - \Gamma_{2(1)}^{sv}d\mu_2 \qquad (9.20a)$$

$$d\gamma_{sl} = -s_{(1)}^{sl}dT - \Gamma_{2(1)}^{sl}d\mu_2 \qquad (9.20b)$$

$$d\gamma_{lv} = -s^{lv}dT - \Gamma_2^{lv}d\mu_2, \qquad (9.20c)$$

where the subscript 2 indicates the liquid component and the subscript (1) refers to the definition of the Gibbs dividing surface chosen to eliminate adsorption of the solid component at the particular interface. The surface entropy of the solid–vapor interface is $s_{(1)}^{sv}$, T is the absolute temperature, $\Gamma_{2(1)}^{sv}$ is the surface excess concentration of component 2 (the liquid) at the solid–vapor interface, and μ_2 is the chemical potential of the liquid component. Similarly, $\Gamma_{2(1)}^{sl}$ is the surface excess concentration of component 2 at the solid–liquid interface, and Γ_2^{lv} is the surface excess concentration of component 2 at the liquid–vapor interface. The surface entropies at the solid–liquid and liquid–vapor interfaces are, respectively, $s_{(1)}^{sl}$ and s^{lv}.

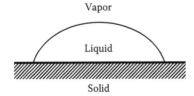

FIGURE 9.3   Ideal solid–liquid–vapor system.

Equations 9.20 indicate that each of the surface tensions is a function of T and μ_2; that is,

$$\gamma_{sv} = \gamma_{sv}(T,\mu_2) \tag{9.21a}$$

$$\gamma_{sl} = \gamma_{sl}(T,\mu_2) \tag{9.21b}$$

$$\gamma_{lv} = \gamma_{lv}(T,\mu_2). \tag{9.21c}$$

Thus, there are three equations in terms of the two variables T and μ_2, implying that any one of the Equations 9.21 may be expressed as a linear combination of the other two. In other words, there must exist an equation of the form of Equation 9.10.

9.3.5   PHASE RULE FOR INTERFACIAL SYSTEMS

The existence of an equation of state for interfacial tensions has also been proven [39] by determining the number of degrees of freedom in the equilibrium state of the system shown in Figure 9.3.

The Gibbs phase rule

$$f = r + 2 - M, \tag{9.22}$$

gives the number of degrees of freedom f in a system of r independent chemical components and M phases under the following restrictions.

1. The system must have negligible boundary effects and all boundaries between phases must be thermally conducting, deformable, and permeable to all components.
2. No chemical reactions occur.
3. Volume is the only work coordinate; that is, PdV work is the only mode of work.

These conditions are not satisfied by the system of Figure 9.3, and a different form of the phase rule must be used to determine the number of independent intensive variables or degrees of freedom.

The required phase rule for surface systems may be derived by subtracting the number of equilibrium constraint equations from the number of variables required to describe the system. For a surface system of M phases (bulk and surface), each with r independent components, each bulk phase, α, may be described by the variables T^α, P^α, x_1^α, x_2^α, ..., x_{r-1}^α, where T^α and P^α are, respectively, the temperature and pressure of phase α, and $x_i^\alpha (i = 1, 2, ..., r-1)$ is the mole fraction of the ith component in phase α. The surface phases in the system may be described by a similar set of independent variables, only replacing P^α by $\gamma_{\alpha\beta}$, the interfacial tension between adjacent bulk phases α and β. Thus, the total number of intensive variables describing the surface system is $M(r + 1)$.

Considering the number of constraint equations, the equilibrium of the surface system is defined by the following conditions.

Thermal equilibrium conditions:

$$T^\alpha = T^\beta = ... = T^M \qquad (M - 1)\text{equations.} \tag{9.23}$$

Chemical equilibrium conditions:

$$\mu_i^\alpha = \mu_i^\beta = ...\mu_i^M \qquad r(M - 1)\text{equations,} \tag{9.24}$$

where $i = 1, 2, ... , r$.

Mechanical equilibrium conditions of three possible types:

1. Laplace equations,

$$P^\alpha - P^\beta = \gamma_{\alpha\beta}J^{\alpha\beta}, \tag{9.25}$$

where α and β represent adjacent bulk phases separated by a curved liquid-fluid interface and $J^{\alpha\beta}$ is the mean curvature of the $\alpha\beta$ interface. If this interface is planar, then $J^{\alpha\beta}$ equals zero and Equation 9.25 reduces to $P^\alpha = P^\beta$, which is the mechanical equilibrium condition used in the derivation of the Gibbs phase rule, Equation 9.22.

2. Young equations, Equation 9.1.
3. Neumann triangle relations,

$$2\gamma_{12}\gamma_{23}\cos\theta = \left(\gamma_{12}\right)^2 - \left(\gamma_{23}\right)^2 - \left(\gamma_{13}\right)^2, \tag{9.26}$$

where the phases are defined as in Figure 9.4a, and θ is the angle within phase 3 between the 1–3 and 2–3 interfaces.

It should be emphasized that Equations 9.1, 9.25, and 9.26 do not, in general, serve as constraint equations since neither $J^{\alpha\beta}$ nor θ are members of the set of intensive variables describing the state of the surface system. For example, Equation 9.25 can

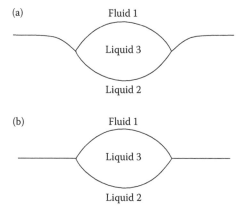

(a)   Fluid 1

Liquid 3

Liquid 2

(b)   Fluid 1

Liquid 3

Liquid 2

FIGURE 9.4 (a) Liquid–liquid–fluid system, (b) liquid–liquid–fluid system with a planar interface.

always be satisfied by adjusting $J^{\alpha\beta}$ while the values of the pressures and the surface tension in Equation 9.25 can be assigned freely. Also, in the Young equation and the Neumann triangle relation, the presence of the contact angle θ introduces a new unknown variable. Therefore, it is not possible to use these equations to calculate one of the interfacial tensions from knowledge of the other two; that is, these are not constraint equations for the set of independent intensive variables. It is only when $J^{\alpha\beta} = 0$ (a planar interface) and, hence, $P^{\alpha} = P^{\beta}$, that a mechanical constraint is imposed by any of the Equations 9.1, 9.25, and 9.26. With this in mind, let N be the number of distinct $P^{\alpha} = P^{\beta}$ type relations among the mechanical equilibrium conditions; then, for a surface system with M phases and r independent chemical components, the total number of constraint equations is given by

$$
\underset{\substack{\text{Thermal}\\\text{Equilibrium}}}{(M-1)} \quad + \quad \underset{\substack{\text{Chemical}\\\text{equilibrium}}}{r(M-1)} \quad + \quad \underset{\substack{\text{Mechanical}\\\text{equilibrium}}}{N}
$$

Remembering that the number of intensive variables of the system is $M(r+1)$, the degrees of freedom f are given by

$$
f = M(r+1) - \left[(M-1) + r(M-1) + N \right] = r + 1 - N, \tag{9.27}
$$

where r is the number of independent chemical components in each phase of the system, and N is the number of $P^{\alpha} = P^{\beta}$ relations among the mechanical equilibrium conditions; that is, the number of distinct planar interfaces between adjacent bulk phases. Equation 9.27 is the phase rule for surface systems.

Consider now the application of Equation 9.27 to the two component (solid and liquid–vapor) surface system shown in Figure 9.3, which has three bulk phases and three surface phases. Note that such a system is typically called a two-component, three-phase system. Assuming that the solid phase is isotropic and may be characterized by a single hydrostatic pressure, P^S (see Münster [38] for a discussion of other cases), one of the mechanical equilibrium conditions is $P^\alpha = P^\beta$ and Equation 9.27 gives $f = 2$ indicating that any two of the intensive variables describing the system may be independently varied. Any of the other variables is then a function of these two arbitrarily chosen independent ones. Thus if we choose γ_{sv} and γ_{lv} as the two independent variables out of the complete set of $M(r + 1)$ variables, then γ_{sl} may be expressed as a function of these variables; that is, Equation 9.10.

The inclusion of linear phases in the development of the phase rule for surface systems does not affect this result [39].

It is also important to note that for a system composed of three bulk fluid phases, since, in general, all of the interfaces are curved, there are no mechanical equilibrium constraints of the type $P^\alpha = P^\beta$ and therefore $N = 0$. For a two-component liquid-lens system as shown in Figure 9.4a, Equation 9.27 predicts three degrees of freedom and no equation of state relation can exist among the three interfacial tensions. Such a relation can be formed only if one of the interfaces is planar (Figure 9.4b) so that $N = 1$ and Equation 9.27 gives $f = 2$.

We have thus demonstrated thermodynamically that an equation of state must exist relating γ_{sl}, γ_{sv}, and γ_{lv} in the system of Figure 9.3 comprising a pure liquid, its vapor, and a rigid, insoluble solid on which there is no liquid or vapor absorption or adsorption, and which is smooth and homogeneous. While such an equation would strictly be applicable to just one system as it experienced different values of, for example, temperature and pressure, we expect a single equation to in fact describe a large class of systems, as is found with, for example, the Ideal Gas Law or equations of state for bulk liquids. Assuming that the presence of air may be ignored, the explicit form of such a relation has been determined empirically by curve-fitting large sets of contact angle data [30,40].

9.4 FORMULATION OF AN EQUATION OF STATE

In the previous sections, the existence of an equation of state of the form of Equation 9.10 was proven in two different ways. The explicit formulation of Equation 9.10 can be done empirically through the interpretation and curve fitting of contact angle data [30] or, in principle, through statistical mechanics, although at present this remains beyond our capabilities. This section will examine two equivalent empirical methods that have been used to explicitly formulate the equation of state for interfacial tensions [30].

9.4.1 ROLE OF ADSORPTION

In order to obtain an explicit formulation of $\gamma_{sl} = f(\gamma_{lv}, \gamma_{sv})$, it is desirable to keep one of the three variables γ_{sl}, γ_{lv}, and γ_{sv} constant, to subject a second one to a known change, and to register the effect of this change on the third variable. Of these three quantities, only γ_{lv} can be readily measured and serve as the independent variable; γ_{sl} will change with γ_{lv}, and γ_{sv} remains unchanged if the adsorption of the liquid's vapor on the solid–vapor interface can be neglected. The validity of this assumption is demonstrated as follows.

Consider a solid of surface tension γ_s against vacuum and liquid of surface tension γ_{lv}. Let us assume that the vapor of the liquid is initially prevented from contacting the solid surface so that the vapor pressure near the solid surface is equal to zero; then, the vapor of vapor pressure P_v is allowed to contact the solid surface. In order to obtain an expression for the resulting solid–vapor interfacial tension γ_{sv}, we perform a Taylor series expansion and retain only the first-order term

$$\gamma_{sv} = \gamma_s + \left(\frac{\partial \gamma_{sv}}{\partial P_v}\right)_T \Delta P_v, \tag{9.28}$$

which, since the original vapor pressure was equal to zero, can be written as

$$\gamma_{sv} = \gamma_s + \left(\frac{\partial \gamma_{sv}}{\partial P_v}\right)_T P_v. \tag{9.29}$$

From the Gibbs–Duhem equation for the solid–vapor interface, Equation 9.20a, we will have

$$-\Gamma_{2(1)}^{sv} = \left(\frac{\partial \gamma_{sv}}{\partial \mu_2}\right)_T, \tag{9.30}$$

and

$$\left(\frac{\partial \gamma_{sv}}{\partial \mu_2}\right)_T = \left(\frac{\partial \gamma_{sv}}{\partial P_v}\right)_T \left(\frac{\partial P_v}{\partial \mu_2}\right)_T. \tag{9.31}$$

Assuming for simplicity that the vapor is an ideal gas, we have

$$\mu_2 = \mu_2^0(T) + RT \ln P_v, \tag{9.32}$$

so that

$$\left(\frac{\partial \mu_2}{\partial P_v}\right)_T = \frac{RT}{P_v}. \tag{9.33}$$

It follows that

$$\left(\frac{\partial \gamma_{sv}}{\partial P_v}\right)_T = -\Gamma_{2(1)}^{sv} \frac{RT}{P_v}, \tag{9.34}$$

and

$$\gamma_{sv} = \gamma_s - RT \Gamma_{2(1)}^{sv}. \tag{9.35}$$

Since $\Gamma_{2(1)}^{sv} > 0$, we infer from Equation 9.35 that adsorption will decrease γ_s. On the other hand, if we only consider systems for which $\gamma_s < \gamma_{lv}$, the adsorption of the vapor of the liquid would increase the solid surface tension. Since this would contradict Equation 9.35, it may be expected that adsorption will not play a major role for systems having $\gamma_s < \gamma_{lv}$.

Adsorption from the vapor has also been investigated experimentally. The few experimental data available at present [41,42] seem to indicate that the equilibrium spreading pressure is normally less than approximately 1 mJ/m^2 if the contact angle is not too low, for instance, greater than 20 or 30°.

Based on the above arguments, the solid–vapor surface tension γ_{sv} will be considered as a constant, independent of the wetting liquid. For a discussion of the implications for an equation of state when adsorption is significant enough to produce a thin liquid film coating the solid surface, see Chapter 7.

9.4.2 EQUATION OF STATE: ORIGINAL FORMULATION

An equation of state relation $\gamma_{sl} = f(\gamma_{lv}, \gamma_{sv})$ can be formulated from contact angle data on low-energy solids. Such a formulation was first attempted in the 1960s [43,44] and continued in the 1970s, using extensive contact angle data on eight polymeric solids [30]. In Figure 9.5, the data for the eight solids are plotted in terms of $\gamma_{lv} \cos\theta_Y$ versus γ_{lv}. As seen in these diagrams, all experimental points fall reasonably close to smooth curves, which have, moreover, the same general shape in all cases. In view of the Young equation (Equation 9.1), these continuous curves are consistent with the hypothesis that for any constant γ_{sv}, γ_{sl} is a unique function of γ_{lv}. It should be noted that the contact angles of Figure 9.5 are advancing contact angles that were measured on carefully prepared solid surfaces with pure liquids. Such contact angles are denoted "Young contact angles" θ_Y because they are thermodynamically significant and satisfy the Young equation for a given solid and liquid. The measurement of a Young contact angle is complicated by the influence of surface roughness, vapor adsorption, and liquid impurities [45]. Two general conclusions can be drawn from the plots in Figure 9.5.

1. As γ_{lv} decreases $\gamma_{lv}\cos\theta_Y$ increases and, by the Young equation, since γ_{sv} is assumed constant, γ_{sl} decreases.
2. The slope

$$\frac{d(\gamma_{lv} \cos\theta)}{d\gamma_{lv}},$$

is equal to zero at $\theta_Y = 0$.

This second point was demonstrated quantitatively by computer curve-fitting the experimental data to a second-order polynomial in each case:

$$\gamma_{lv} \cos\theta_Y = a\gamma_{lv}^2 + b\gamma_{lv} + c. \tag{9.36}$$

The 45° line

$$\gamma_{lv} \cos\theta_Y = \gamma_{lv}, \tag{9.37}$$

that is, the limiting condition $\theta_Y = 0$, is also shown in each case. The intercept of the computed curve (Equation 9.36) with the 45° line is given by

$$a\gamma_{lv}^2 + (b-1)\gamma_{lv} + c = 0. \tag{9.38}$$

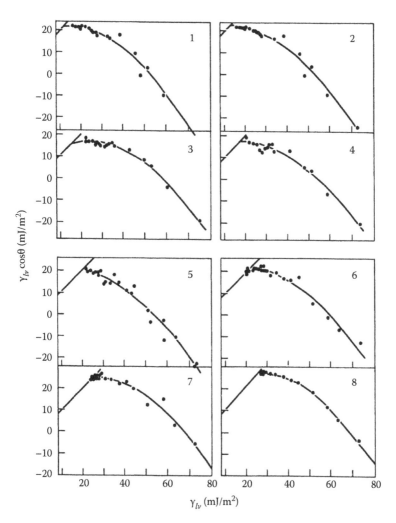

FIGURE 9.5 Plot of $\gamma_{lv}\cos\theta$ as a function of the surface tension, γ_{lv}, of various liquids. 1, Methacrylic polymer A with fluorinated side chain (3M, Inc.); 2, methacrylic polymer S with fluorinated side chain (3M, Inc.); 3, 17-(perfluoropropyl)-heptadecanoic acid; 4, 17-(perfluoroethyl)-heptadecanoic acid; 5, polytetrafluoroethylene; 6, 80–20 copolymer of tetrafluoroethylene and chlorotrifluoroethylene; 7, 60–40 copolymer of tetrafluoroethylene and chlorotrifluoroethylene; and 8, 50–50 copolymer of tetrafluoroethylene and polyethylene.

The intercepts and the limiting slopes at the intercepts are given in Table 9.3. The average limiting angle of inclination, calculated from the average limiting slope given, is $0.1 \pm 4.2°$. Thus, it is reasonable to conclude that

$$\lim_{\theta_Y \to 0} \frac{d\left(\gamma_{lv} \cos\theta_Y\right)}{d\gamma_{lv}} = 0. \tag{9.39}$$

TABLE 9.3

Limiting Slopes and Intercepts for the Eight Systems in Figure 9.5

No.	Solid	Slope	Intercept
1	3M, Inc. methacrylic polymer A with fluorinated side chain	0.0611	11.66
2	3M, Inc. methacrylic polymer S with fluorinated side chain	0.0089	12.44
3	17-(Perfluoropropyl)-heptadecanoic acid	0.1714	15.87
4	17-(Perfluoroethyl)-heptadecanoic acid	−0.0520	17.38
5	Polytetrafluoroethylene	−0.1759	20.23
6	80–20 copolymer of tetrafluoroethylene and chlorotrifluoroethylene	−0.0207	20.93
7	60–40 copolymer of tetrafluoroethylene and chlorotrifluoroethylene	0.0379	24.63
8	50–50 copolymer of tetrafluoroethylene and ethylene	−0.0443	26.90

From the above fact that γ_{sl} decreases as $\gamma_{lv}\cos\theta$ increases, and from Equation 9.39, we conclude that γ_{sl} has its minimum value when $\theta = 0$.

From our knowledge of liquid–liquid interfaces, zero is the lower limit for the interfacial tension between two liquid phases at equilibrium. It would then be very difficult to understand this if arbitrarily small solid–liquid interfacial free energies were not possible. Therefore, the minimum value of the solid–liquid interfacial tension is zero as the contact angle approaches zero; that is,

$$\lim_{\theta_Y \to 0} \gamma_{sl} = \gamma_{sl}^* = 0. \tag{9.40}$$

This will be discussed further in Section 9.4.4, where the possibility of negative interfacial tensions will be considered.

The formulation of the equation of state is essentially an empirical curve-fit to contact angle data. As such, there are a variety of ways of proceeding and in this instance [30] it was decided to correlate the data in terms of Good's interaction parameter

$$\Phi = \frac{\gamma_{sv} + \gamma_{lv} - \gamma_{sl}}{2\sqrt{\gamma_{lv}\gamma_{sv}}}. \tag{9.41}$$

This can be done as follows:

1. Assuming γ_{sv} is a constant and $\gamma_{sl}^* = 0$, determine γ_{sv} graphically from Figure 9.5 using

$$\gamma_{sv} = \lim_{\theta_Y \to 0} \gamma_{lv} = \gamma_{lv}^*. \tag{9.42}$$

2. Using this constant γ_{sv} and experimental values of γ_{lv} and $\cos\theta_Y$, obtain γ_{sl} as a function of θ_Y from the Young equation (Equation 9.1).
3. Using the values of γ_{sv} and γ_{sl} as obtained in steps 1 and 2, compute Φ using Equation 9.41.

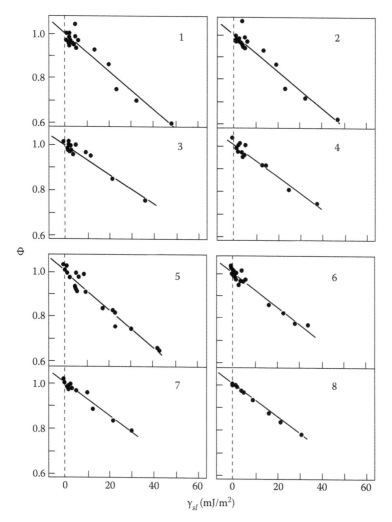

FIGURE 9.6 Interaction parameter, Φ, as a function of γ_{sl} for the eight systems given in Figure 9.5.

Figure 9.6 shows graphs of Φ versus γ_{sl}, for the eight systems given in Figure 9.5. Clearly, the fit of the data to straight lines is satisfactory and we conclude that:

1. As a good approximation, Φ is a linear function of γ_{sl} for a particular solid with a series of liquids. The straight lines shown in Figure 9.6 are least-square fits.
2. All data points of all eight systems can be fitted to a single straight line

$$\Phi = \alpha\gamma_{sl} + \beta. \tag{9.43}$$

From this, one may conclude that the same relationship $\Phi = f(\gamma_{sl})$ holds for all low-energy surfaces.

The values of the two constants α and β in Equation 9.43 were found by best fit to the experimental data [30]:

$$\alpha = \frac{d\Phi}{d\gamma_{sl}} = -0.0075 \text{ m}^2/\text{mJ} \qquad \beta = 1.000. \tag{9.44}$$

Combining Equation 9.41 with Equations 9.43 and 9.44, an explicit form of the equation of state can be obtained as

$$\gamma_{sl} = \frac{\left(\sqrt{\gamma_{lv}} - \sqrt{\gamma_{sv}}\right)^2}{1 - 0.015\sqrt{\gamma_{lv}\gamma_{sv}}}. \tag{9.45}$$

Combining this equation of state with the Young equation, we have

$$\cos\theta_Y = \frac{(0.015\gamma_{sv} - 2.00)\sqrt{\gamma_{lv}\gamma_{sv}} + \gamma_{lv}}{\gamma_{lv}\left(0.015\sqrt{\gamma_{lv}\gamma_{sv}} - 1\right)}. \tag{9.46}$$

Notice that difficulties may arise for liquids with relatively large surface tensions γ_{lv} since the denominator of Equation 9.45 can become zero. This limitation of the equation of state formulation is due to the use of Good's interaction parameter Φ and is purely mathematical. Physical reasoning was used to "mend" Equation 9.45 and, in practice, Equation 9.45 is implemented as a computer program [30] or a set of tables [46].

However, as was mentioned above, the use of Good's interaction parameter Φ is not the only way to find an explicit expression for the equation of state. In the next section, we present a different formulation of the equation of state [40], giving the same results but being free of the shortcomings of the above development. This work of Li and Neumann [40] used the contact angle data of Neumann et al. [30]. Contact angle data of greater accuracy were used shortly thereafter [47] to refine the formulation.

9.4.3 EQUATION OF STATE: ALTERNATE FORMULATION

The solid–liquid free energy of adhesion is equal to the work required to separate a unit area of the solid–liquid interface [48]; that is,

$$W_{sl} = \gamma_{lv} + \gamma_{sv} - \gamma_{sl}. \tag{9.47}$$

Usually, in analogy with the Berthelot combining rule for the attractive constants in the van der Waals equation of state, the free energy of adhesion, W_{sl}, is taken as the geometric mean of the free energy of cohesion of the solid W_{ss} and the free energy of cohesion of the liquid W_{ll} [49]; that is,

$$W_{sl} = \sqrt{W_{ll}W_{ss}}. \tag{9.48}$$

By the definitions $W_{ll} = 2\gamma_{lv}$ and $W_{ss} = 2\gamma_{sv}$, Equation 9.48 becomes

$$W_{sl} = 2\sqrt{\gamma_{lv}\gamma_{sv}}. \tag{9.49}$$

Therefore, combining Equation 9.47 with Equation 9.49, the solid–liquid interfacial tension γ_{sl} can be written as

$$\gamma_{sl} = \gamma_{lv} + \gamma_{sv} - 2\sqrt{\gamma_{lv}\gamma_{sv}} = \left(\sqrt{\gamma_{lv}} - \sqrt{\gamma_{sv}}\right)^2. \tag{9.50}$$

Note that this is the same as Equation 9.12. As previously discussed, it has been found that this simple equation of state works only for situations where γ_{lv} values are close to the values of γ_{sv}. This is because the geometric mean combining rule Equation 9.48 is valid only for $W_{ll} \approx W_{ss}$. To modify the geometric mean combining rule, Girifalco and Good [49] introduced the above-mentioned interaction parameter Φ as the ratio of the free energy of adhesion between two phases to the geometric mean of the free energies of cohesion of these two phases; that is,

$$W_{sl} = \Phi\sqrt{W_{ll}W_{ss}}. \tag{9.51}$$

An alternative form of the interaction parameter, Φ, is given in Equation 9.41, which was derived by combining Equation 9.51 with Equation 9.47. It was found that [30]

$$\Phi \leq 1. \tag{9.52}$$

In other words, the geometric mean combining rule, Equation 9.48, generally overestimates the value of W_{sl}.

Actually, the above pattern holds generally for bulk systems, too. In the theory of intermolecular interactions and the theory of mixtures, the combining rule is used to evaluate the parameters of unlike-pair interactions in terms of those of like-pair interactions. It should be pointed out that, as for many other combining rules, the Berthelot rule, that is, the geometric mean combining rule,

$$\varepsilon_{ij} = \sqrt{\varepsilon_{ii}\varepsilon_{jj}}, \tag{9.53}$$

where ε_{ij} is the energy parameter for unlike-pair interactions and ε_{ii}, ε_{jj} are the energy parameters for like-pair interactions, is only a useful approximation and does not provide a secure basis for the understanding of the unlike-pair interactions. Finding better combining rules to characterize unlike-pair interactions in terms of like ones has been the subject of much research related to equations of state of liquid mixtures. Reviews on this subject can be found elsewhere [50,51]. An important question is how far the interactions of unlike molecules can be expressed in terms of the two like interactions, as in the Berthelot rule, Equation 9.53. By London's theory of dispersion forces, it has been shown [50] that the geometric mean combining rule Equation 9.53 is applicable only for similar molecules, because implicit in this rule is the condition that the two energy parameters of like-pair interactions must be very close to each other; that is, $\varepsilon_{ii} \approx \varepsilon_{jj}$. However, for the interactions between two very dissimilar molecules or materials, where there is an apparent difference between ε_{ii} and ε_{jj}, it has been demonstrated [52,53] that the geometric mean combining rule generally overestimates the strength of the unlike-pair interactions. This is the case even for the interactions between similar molecules [50], although the extent of the overestimation is smaller than that between dissimilar molecules. Similar conclusions may be expected to hold in the context of interactions across surfaces.

In the study of mixtures, it has become common practice to introduce a factor $(1 - K_{ij})$ to the geometric mean combining rule:

$$\varepsilon_{ij} = \left(1 - K_{ij}\right)\sqrt{\varepsilon_{ii}\varepsilon_{jj}}, \tag{9.54}$$

where K_{ij} is an empirical parameter quantifying deviations from the geometric mean combining rule. Since the geometric mean combining rule overestimates the strength of the unlike-pair interactions, the modifying factor $(1 - K_{ij})$ should decrease with the difference $(\varepsilon_{ii} - \varepsilon_{jj})$ and be equal to unity when the difference $(\varepsilon_{ii} - \varepsilon_{jj})$ is zero. Based on this thought, we may consider a modified combining rule of the form

$$\varepsilon_{ij} = \sqrt{\varepsilon_{ii}\varepsilon_{jj}}\, e^{-\alpha\left(\varepsilon_{ii}-\varepsilon_{jj}\right)^2}, \tag{9.55}$$

where α is an empirical constant. The square of the difference $(\varepsilon_{ii} - \varepsilon_{jj})$ rather than the difference itself reflects the symmetry of this combining rule, and hence the anticipated symmetry of the equation of state [54].

Correspondingly, for the cases of large differences $|W_{ll} - W_{ss}|$ or $|\gamma_{lv} - \gamma_{sv}|$, the combining rule for the free energy of adhesion of a solid–liquid pair can be written as

$$W_{sl} = \sqrt{W_{ll}W_{ss}}\, e^{-\alpha\left(W_{ll}-W_{ss}\right)^2}, \tag{9.56}$$

or, more explicitly, by using $W_{ll} = 2\gamma_{lv}$ and $W_{ss} = 2\gamma_{sv}$,

$$W_{sl} = 2\sqrt{\gamma_{lv}\gamma_{sv}}\, e^{-\beta\left(\gamma_{lv}-\gamma_{sv}\right)^2}. \tag{9.57}$$

In the above equations, α and β are as yet unknown constants. Clearly, when the values of γ_{lv} and γ_{sv} are close to each other, Equation 9.57 will revert to Equation 9.49, the geometric mean combining rule. Coupling Equation 9.57 with Equation 9.47, an equation of state for interfacial tensions can be written as

$$\gamma_{sl} = \gamma_{lv} + \gamma_{sv} - 2\sqrt{\gamma_{lv}\gamma_{sv}}\, e^{-\beta\left(\gamma_{lv}-\gamma_{sv}\right)^2}. \tag{9.58}$$

Obviously, Equation 9.58 will not have the difficulty of a singularity as Equation 9.45 does. Combining Equation 9.58 with the Young equation (Equation 9.1) will yield

$$\cos\theta = -1 + 2\sqrt{\frac{\gamma_{sv}}{\gamma_{lv}}}\, e^{-\beta\left(\gamma_{lv}-\gamma_{sv}\right)^2}, \tag{9.59}$$

where θ is understood to be the Young contact angle θ_Y.

By fitting Equation 9.59 to the experimental data [30] used for deriving the original equation of state (Equation 9.45), the constant β in Equations 9.58 and 9.59 was obtained as $\beta = 0.000115$ $(m^2/mJ)^2$ [40]. A more accurate value of β has been obtained using newer contact angle data [47], as discussed below. It is apparent that Equation 9.59 has three variables, γ_{lv}, θ, and γ_{sv}, and thus will enable us to determine the solid surface tension γ_{sv} when we have experimental data for the liquid surface tension γ_{lv} and the contact angle θ. Solutions of Equation 9.59 with a handheld calculator [40] and with a FORTRAN program [55] are available in the literature.

Equation 9.46 was compared with Equation 9.59 by calculating the solid surface tension γ_{sv} with hypothetical values of liquid surface tension γ_{lv} and contact angle θ. Examples of these results are given in Table 9.4. It is evident that Equation 9.59 yields essentially the same results as those of Equation 9.46. The larger discrepancies

TABLE 9.4

Comparison Between the Equations of State
(Equations 9.46 and 9.59)

		γ_{sv} (mJ/m²)	
γ_{lv} (mJ/m²)	θ (degrees)	Equation 9.46	Equation 9.59
70.0	20.0	66.0	66.1
70.0	30.0	61.7	61.9
70.0	40.0	56.5	56.8
70.0	50.0	50.8	51.2
70.0	60.0	45.5	45.3
70.0	70.0	39.1	39.2
70.0	80.0	33.2	33.0
70.0	90.0	27.1	26.9
70.0	100.0	21.1	20.8
70.0	110.0	15.2	15.1
50.0	20.0	47.1	47.1
50.0	30.0	44.0	43.9
50.0	40.0	40.0	39.9
50.0	50.0	35.5	35.4
50.0	60.0	30.7	30.7
50.0	70.0	25.7	25.8
50.0	80.0	20.9	20.9
50.0	90.0	16.2	16.2
50.0	100.0	11.9	11.9
30.0	20.0	28.2	28.2
30.0	30.0	26.2	26.2
30.0	40.0	23.6	23.6
30.0	50.0	20.6	20.7
30.0	60.0	17.5	17.5
30.0	70.0	14.3	14.3
30.0	80.0	11.3	11.2

at high values of γ_{lv} occur due to the linear interpolation used to overcome the mathematical difficulty associated with Equation 9.46. We suspect that, if extreme accuracy matters, the results of Equation 9.59 are more reliable than those of Equation 9.46, particularly if β is obtained from more recently published contact angle data [47] as described below.

The accuracy of either Equation 9.59 or Equation 9.46 is, of course, limited by the accuracy of the contact angle data used to determine the equation parameters. Three sets of contact angle data were produced with a wide range of liquids on very smooth, homogeneous solid surfaces of PET, fluorinated ethylene propylene (FEP), and mica coated with the fluoropolymer FC-721 (3M, Inc.) [47,56]. The contact angles were measured using Axisymmetric Drop Shape Analysis-Profile (ADSA-P, Chapter 6) and are thus more accurate than the goniometer contact angle measurements used to formulate Equation 9.46 and to give $\beta = 0.000115$ $(m^2/mJ)^2$ in Equation 9.59. These data are listed in Tables 9.5 through 9.7 and are plotted in Figure 9.7. Since all three curves are smooth, it can be concluded that the adsorption on these surfaces is indeed negligible [37], and hence the assumption that γ_{sv} is approximately constant is valid. This represents a confirmation of the analysis of Section 9.4.1.

TABLE 9.5
Liquid Surface Tensions and Contact Angles Measured on FC-721/Mica Surface

Liquid	γ_{lv} (mJ/m²)	θ (°)	γ_{sv} (mJ/m²)
Decane	23.43	65.97	11.98
Dodecane	25.44	69.82	12.02
Tetradecane	26.55	73.31	11.61
Hexadecane	27.76	75.32	11.62
trans-Decalin	29.50	76.71	12.02
cis-Decalin	31.65	79.87	12.02
Tetralin	35.96	83.14	12.84
Ethylcinnamate	38.37	88.20	12.08
Dibenzylamine	40.63	92.06	11.60
DMSO	43.58	94.47	11.84
1-Bromonaphthalene	44.01	95.29	11.70
Diethylene glycol	45.04	96.84	11.50
Ethylene glycol	47.99	99.03	11.75
Thiodiglycol	54.13	103.73	12.11
Formamide	57.49	107.32	11.80
Glycerol	63.11	111.38	12.04
Water	72.75	119.05	11.88

Source: Li, D. and Neumann, A. W., *Journal of Colloid and Interface Science,* 148, 190, 1992.

Note: Solid surface tension calculated with Equation 9.59 using $\beta = 0.0001247$ $(m^2/mJ)^2$.

TABLE 9.6
Liquid Surface Tensions and Contact Angles
Measured on FEP Surface

Liquid	γ_{lv} (mJ/m²)	θ (°)	γ_{sv} (mJ/m²)
Decane	23.43	43.70	17.54
Dodecane	25.44	47.96	17.97
Tetradecane	26.55	52.51	17.52
Hexadecane	27.76	53.75	17.99
trans-Decalin	29.50	58.14	17.80
cis-Decalin	31.65	62.60	17.69
Dimethylformamide	35.57	66.84	18.76
Tetralin	35.96	68.52	17.91
Ethylcinnamate	38.37	72.61	17.92
Dibenzylamine	40.63	75.99	17.80
DMSO	43.58	80.35	17.53
1-Bromonaphthalene	44.01	79.70	18.03
Diethylene glycol	45.04	81.48	17.79
Ethylene glycol	47.99	85.56	17.48
Formamide	57.49	95.38	17.45
Glycerol	63.11	100.63	17.45
Water	72.75	111.59	15.96

Source: Li, D. and Neumann, A. W., *Journal of Colloid and Interface Science,* 148, 190, 1992.

Note: Solid surface tension calculated with Equation 9.59 using $\beta = 0.0001247$ (m²/mJ)².

TABLE 9.7
Liquid Surface Tensions and Contact Angles
Measured on PET Surface

Liquid	γ_{lv} (mJ/m²)	θ (°)	γ_{sv} (mJ/m²)
Diethylene glycol	45.04	41.19	35.58
Ethylene glycol	47.99	47.52	35.07
Thiodiglycol	54.13	55.57	35.95
Formamide	57.49	61.50	35.35
Glycerol	63.11	68.10	35.72
Water	72.75	79.09	35.86

Source: Li, D. and Neumann, A. W., *Journal of Colloid and Interface Science*, 148, 190, 1992.

Note: Solid surface tension calculated with Equation 9.59 using $\beta = 0.0001247$ (m²/mJ)².

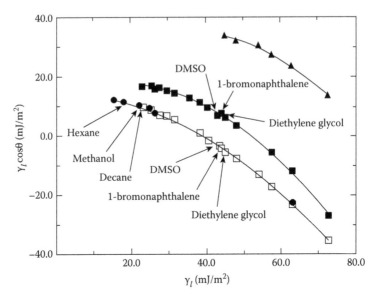

FIGURE 9.7 A plot of $\gamma_l \cos\theta$ versus γ_l for three well-prepared solid surfaces, FC-721 dip-coated on mica, Teflon (FEP) heat-pressed against mica, and polyethylene terephthalate (PET). Note that polar and nonpolar liquids lie on the same smooth curve. □ FC-721 (Li, D. and Neumann, A. W., *Journal of Colloid and Interface Science*, 148, 190, 1992); ● FC-721 (Li, D., Xie, M., and Neumann, A. W., *Colloid and Polymer Science* 271, 573, 1993); ■ FEP (Li, D. and Neumann, A. W., *Journal of Colloid and Interface Science*, 148, 190, 1992); ▲ PET (Li, D. and Neumann, A. W., *Journal of Colloid and Interface Science*, 148, 190, 1992.)

The contact angles and liquid surface tensions listed in each of Tables 9.5 through 9.7 were fitted separately using a nonlinear least-square technique to give three values for β, the weighted average of which is β = 0.0001247 (m²/mJ)² [47]. Figure 9.8 shows a comparison between Equation 9.59 with β = 0.0001247 (m²/mJ)² and the original equation of state Equation 9.46. In most cases, the differences in γ_{sv} are small. The larger discrepancies at higher values of γ_{lv} are due to the linear interpolation used to overcome the singularity associated with Equation 9.45. Tables 9.5 through 9.7 also list the consistent values of γ_{sv} obtained from each individual contact angle measurement using β = 0.0001247 (m²/mJ)².

9.4.4 The Possibility of Negative Solid–Liquid Interfacial Tensions

As shown in Section 9.4.2, the formulation of the equation of state (Equations 9.45 or 9.58) is based on the assumption that the minimum value of solid–liquid interfacial tension γ_{sl} is zero (equivalently, $\phi \leq 1$). In this section, the possibility of negative solid–liquid interfacial tensions will be discussed. For this purpose, the experimental data obtained from a variety of essentially independent methodologies are examined in conjunction with the equation of state approach for calculating interfacial tensions. Contact angle measurements, liquid–liquid interfacial tensions, and advancing solidification front/particle interactions are employed to demonstrate that zero is the lower limit of solid–liquid interfacial tensions [58].

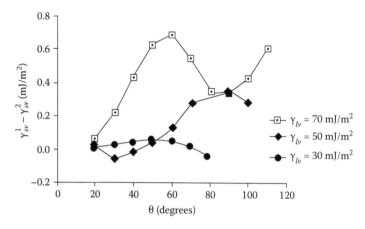

FIGURE 9.8 Difference between solid surface tension calculated by Equation 9.59 with $\beta = 0.0001247$ $(m^2/mJ)^2$ (γ_{sv}^1) and the original formulation of the equation of state Equation 9.46 (γ_{sv}^2).

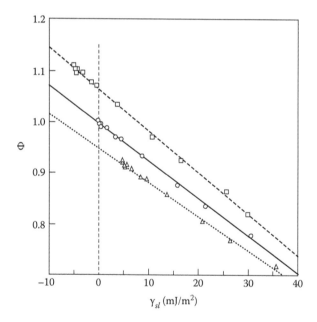

FIGURE 9.9 Copolymer (50-50) of tetrafluoroethylene and ethylene; plot of Φ as a function of γ_{sl} for different estimates of γ_{sv}. $\square$, $\gamma_{sv} = 21.9$ mJ/m^2; O, $\gamma_{sv} = 26.9$ mJ/m^2; $\triangle$, $\gamma_{sv} = 31.9$ mJ/m^2.

In order to study this question in more detail, let us consider the plots in Figure 9.9. For a given set of contact angle data on a given solid, three hypothetical γ_{sv} values were selected. The middle one ($\gamma_{sv} = 26.9$ mJ/m2) corresponds to the assumption that the lowest possible value of γ_{sl} is zero. The others correspond to nonzero lower limits for γ_{sl}. For each γ_{sv}, the corresponding hypothetical γ_{sl} and the Good interaction parameter Φ were calculated from the Young equation and the definition of Φ,

Equation 9.41, respectively. It was found that, in each case, there was a linear relation between the hypothetical Φ and γ_{sl} values, as shown in Figure 9.9. The problem of determining the correct γ_{sv} value was thus reduced to determining the correct straight line from such a family of curves. It has been argued [30] that the only possible choice was the straight line that intersects the Φ-axis at $\Phi = 1$, when $\gamma_{sl} = 0$. This argument was, however, partially based on the assumption that γ_{sl} could not be negative and the approach taken considered only situations of nonzero contact angles; that is, it excluded spreading situations throughout.

Regarding this aspect of the problem, let us consider pairs of polymer melts that are all mutually insoluble to a degree comparable with that of polymeric solids and low-molecular-weight liquids. In Figure 9.10, we reproduce a plot of Φ versus γ_{12}, the interfacial tension between pairs of polymer melts. In some cases, the free energy of spreading is negative while in others it is positive, so that conditions conducive to negative interfacial tensions should exist in some of these cases. However, it is clear from Figure 9.10 that, independent of spreading or nonspreading, all points fall close to a straight line, giving Φ as a function of γ_{12} with a limiting value of $\Phi = 1.0$ at $\gamma_{12} = 0$, in agreement with our choice for the solid–liquid case in Figure 9.9.

At this point, it is worth repeating that the use of Φ in the explicit (empirical) formulation of the equation of state is essentially arbitrary. As was noted in Section 9.4.2, since Φ was used as a correlating parameter in the original equation of state formulation, it is convenient to refer to it in the present context.

As a second test of the validity of a particular Φ versus γ_{sl} relation in Figure 9.9, consider the interaction of small particles embedded in a liquid with an advancing solidification front. Whether a particle, when encountered by the solidification front, is engulfed or swept along by the solidification front, is expected to depend on the sign of the free energy of adhesion

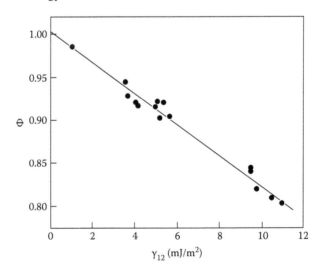

FIGURE 9.10 Polyethylene melt in contact with various polymer melts; plot of Φ as a function of the polymer–polymer interfacial tension, γ_{12}. All interfacial tensions in this case are readily measurable.

$$\Delta F^{adh} = \gamma_{ps} - \gamma_{pl} - \gamma_{sl}, \tag{9.60}$$

or, alternatively, the free energy of engulfment

$$\Delta F^{eng} = \gamma_{ps} - \gamma_{pl}, \tag{9.61}$$

where γ_{ps} is the particle-solid interfacial tension and γ_{pl} is the particle-liquid interfacial tension. At low rates of solidification, for $\Delta F^{adh} < 0$ or $\Delta F^{eng} < 0$ engulfment is predicted, whereas for $\Delta F^{adh} > 0$ or $\Delta F^{eng} > 0$ particle rejection should occur. Thus, the equation of state approach and its choice of Φ that allows the prediction of interfacial tensions can be verified experimentally through observations of particle behavior at solidification fronts [59]. To do so, the surface tension of the melt γ_{lv} and the contact angles on the solid matrix materials as well as the contact angles on the particle materials were measured. From the contact angle data, the surface tension of the solid matrix γ_{sv} was calculated by the equation of state. In a second step, the required interfacial tensions in Equations 9.60 and 9.61 were calculated.

The various straight lines in Figure 9.9, which correspond to different lower limits for γ_{sl}, may be represented by

$$\Phi = \beta' - \alpha\gamma_{sl}. \tag{9.62}$$

In the equation of state approach [30], the constants were chosen to be $\beta' = 1.00$ and $\alpha = 0.0075$ m²/mJ. Solidification front observations can then be used to test the validity of choices other than the straight line with the limiting value $\gamma_{sl} = 0$ at $\Phi = 1.00$. The results of comparisons of such calculations with experimental observations for several matrix-polymer systems are listed in Tables 9.8 and 9.9 [58]. Overall, it can be seen that for $\beta' > 1.00$, particle rejection is predicted in a number of cases where the microscopic observation is engulfment; for $\beta' < 1.00$, particle engulfment is predicted where the microscopic observation is rejection; only when $\alpha = 0.0075$ m²/mJ and $\beta' = 1.00$ are the predictions confirmed by the experimental

TABLE 9.8

Testing for Possible Values of $\alpha \geq 0.0075$ m²/mJ and $\beta' \geq 1.00$ Against Freezing-Front Observations

System	$\alpha = 0.0075$ $\beta' = 1.00$		$\alpha = 0.0075$ $\beta' = 1.05$		$\alpha = 0.0085$ $\beta' = 1.05$		Observation[a]
	ΔF^{adh}	ΔF^{eng}	ΔF^{adh}	ΔF^{eng}	ΔF^{adh}	ΔF^{eng}	
Naphthalene/ polystyrene	−1.46	−0.02	1.36	0.30	2.83	0.77	E
Biphenyl/polystyrene	−0.68	−0.10	2.45	0.15	3.94	0.57	E
Benzophenone/PMMA	−0.39	−0.01	5.08	0.55	7.01	1.07	E
Prediction[a]	E	E	R	R	R	R	

[a] E, particle engulfment; R, particle rejection.

TABLE 9.9

Testing for Possible Values of $\alpha \leq 0.0075$ m²/mJ and
$\beta' \leq 1.00$ Against Freezing-Front Observations

System	$\alpha = 0.0075$ $\beta' = 1.00$		$\alpha = 0.0065$ $\beta' = 0.95$		Observation[a]
	ΔF^{adh}	ΔF^{eng}	ΔF^{adh}	ΔF^{eng}	
Thymol/nylon-6,10	0.088	0.11	−0.98	−7.08	R
Prediction[a]	R	R	E	E	

[a] E, particle engulfment; R, particle rejection.

observations. For over 60 cases, $\beta' = 1.00$ has yielded agreement between experimental observations and thermodynamic predictions, with very few exceptions where ΔF^{adh} was very close to zero so that experimental error could not be ruled out [59]. Thus, $\beta' = 1.00$ corresponding to a minimum $\gamma_{sl} = 0$ produces the best agreement with these experiments. Further details on solidification front experiments can be found in Chapter 12.

Overall, it can be concluded that, considering a variety of systems and situations (including situations of spreading and nonspreading), there is no evidence for negative solid–liquid interfacial tensions. Instead, there is considerable evidence that zero is the lower limit of all solid–liquid interfacial tensions.

9.5 EXPERIMENTAL DATA

For many years, a persistent problem in surface science has been the direct measurement of interfacial tensions involving a solid phase. The many uncertainties associated with such measurements have dissuaded most authors from seeking independent, direct experimental support for their predictions of γ_{sv} and γ_{sl} obtained indirectly, for example, from contact angles. In this section, we will consider a number of experimental observations that serve to verify the γ_{sv} and γ_{sl} predictions of the equation of state (Equation 9.58) from contact angles. The data will also be considered in the light of two earlier equations of state—Antonow's Rule (Equation 9.11) and the unmodified Good equation (Equation 9.12)—as well as the surface tension components approach.

9.5.1 DIRECT FORCE MEASUREMENTS

The surface force apparatus of Israelachvili is capable of directly measuring the intermolecular forces between solid substrates separated by gases or liquids [60]. If such forces are recorded as a function of the separation distance, it is possible to integrate to obtain the energy of surface interaction (W_{ss} or W_{sl}), and thereby calculate the surface tensions (γ_{sv} or γ_{sl}). Results for direct force measurements for surfactant-coated mica sheets in different liquids have been published [61–64]. The surfactants used were hexadecyltrimethylammonium bromide (HTAB), a single-chain alkylammonium surfactant; and dimethyldioctadecylammonium bromide (DDOAB) and dihexadecyldimethylammonium acetate (DHDDA), two double-chained surfactants. The reported results for both HTAB and DDOAB are for force measurements performed in water [63,64] and in octamethylcyclotetrasiloxane (OMCTS), a nonpolar liquid [62]. For DHDDA surfaces, only results for interactions in water are available [61].

Contact angle data, from measurements with water on the surfactant monolayers, are also reported in the same publications. These values of contact angles can be used to calculate the solid surface tension γ_{sv} and the solid–liquid interfacial tensions γ_{sl} from either the equation of state or the Fowkes equation. The Fowkes equation in conjunction with the Young equation, as given by Equation 9.17, provides only the means to evaluate the dispersion component of γ_{sv} and the total solid surface tension remains unknown. However, all surfaces used in these studies can be considered to be purely dispersive so that $\gamma_{sv} = \gamma_{sv}^{d}$. The dispersion component of the surface tension of water is $\gamma_{lv}^{d} = 21.8$ mJ/m^2 [66], and the total surface tension is $\gamma_{lv} = 72.8$ mJ/m^2. The calculated values of γ_{sv} and γ_{sl} can be compared with the values obtained from direct force measurements, thereby providing a means of independently evaluating the accuracy of the equation of state and the Fowkes equation.

The contact angle of water on DDOAB monolayers was found to be 94° [63] and 93 ± 2° [62], in very good agreement. The reported contact angles of water on HTAB monolayers are, however, very different. Christenson [62] reports a value of 60 ± 2° while, in a later publication, Pashley et al. [64] give a value of 95°. The value of contact angles of water on DHDDA surfaces was 95° [61]. Moy [67] reported water contact angles on HTAB and DDOAB surfaces of 93 ± 1° and 94 ± 1°, respectively. These two values, together with the 95° water contact angle for DHDDA, were used in calculations of γ_{sv} and γ_{sl}. The equation of state estimate of the solid–liquid interfacial tension for OMCTS was obtained using the solid surface tension calculated with the water contact angle.

The calculated values for γ_{sv} and γ_{sl} obtained from the equation of state and from Equation 9.17, along with values obtained from direct force measurements, are given in Table 9.10. The direct force measurements are in much better agreement with the values calculated from the equation of state.

The equation of state predicts γ_{sl} values equally well for systems involving polar and nonpolar liquids, with the exception of HTAB-water. The discrepancy between the results for HTAB-water is puzzling since the measured contact angle is known to be accurate. A possible explanation for the differences, according to Pashley et al. [66], may be contamination of the sample of HTAB used for the force measurements. The differences between the values for interactions across OMCTS are most likely caused by the uncertainties in the measured forces. Another possible reason may be the limitations of the original semiempirical formulation of the equation of state, but not of the approach itself. Section 9.4.3 describes a new formulation that should be more accurate in cases of low interfacial tension.

The lack of agreement between the calculated values of γ_{sv} from the Fowkes equation and from direct force measurements results from the inherent assumption

TABLE 9.10

Solid (γ_{sv}) and Solid–Liquid (γ_{sl}) Surface Tensions Calculated with the Equation of State and the Fowkes Equation: Comparison with Direct Force Measurements

Surfactant Monolayer	Liquid	γ_{sv} (mJ/m²)			γ_{sl} (mJ/m²)		
		Equation of State	Fowkes	Direct Measurement	Equation of State	Fowkes	Direct Measurement
HTAB	Water	26.9	54.6	25 [64]	30.7	58.4	11 [64]
	OMCTS	—	—	—	1.2	9.6	0.5 [62]
DDOAB	Water	26.2	52.6	27 [63]	31.0	57.7	34 [63]
	OMCTS	—	—	—	0.9	8.8	0.4 [62]
DHDDA	Water	26.0	50.6	32 [65]	32.3	57.0	28 [61]

in the Fowkes theory that intermolecular forces of different origin do not cross-interact. By ignoring these cross-interactions, the Fowkes equation overestimates both the solid and solid–liquid interfacial tensions for systems involving a nondispersive material. The very existence of these cross-interaction terms indicates that it is not possible to obtain the so-called dispersion component of surface tension from contact angle data unless both the solid and the liquid are purely dispersive. Thus, in order to be complete, the Fowkes approach should include the explicit formulations for the various types of cross-interactions that, at the present time, are not known.

9.5.2 Solidification Fronts

The behavior of microscopic solid particles at advancing liquid solidification fronts can be explained in terms of interfacial free energy changes [68]. This fact was used by Omenyi and colleagues [59,69] to provide an independent test of predictions of solid and solid–liquid interfacial tensions determined by the equation of state from contact angle measurements. Chapter 12 provides a detailed discussion of solidification front experiments.

When a microscopic particle initially embedded in the liquid phase of a matrix material is approached by the solid–liquid interface of the solidifying matrix, its subsequent behavior is governed largely by the free energy of adhesion of the particle and the contacting interface. Particle engulfment must be preceded by adhesion, and particle rejection by repulsion. Thus, the initial rejection or engulfment of the particle may be predicted by the free energy of adhesion, given by Equation 9.60. If ΔF^{adh} is positive, the adhesion of the particle is thermodynamically unfavorable because of the predicted increase of the system's free energy. Particle adhesion is favored if ΔF^{adh} is negative, thereby resulting in a decrease in the overall system free energy. Therefore, knowledge of the three relevant interfacial tensions can be used to calculate ΔF^{adh} and predict the outcome of a particle-engulfing experiment. The accuracy of such predictions can, in turn, be used to judge the accuracy of the method used to obtain γ_{ps}, γ_{pl}, and γ_{sl}. This can be the basis of a test of the equation of state.

Solid surface tensions γ_{sv} and γ_{pv} were determined from contact angles with liquids of known surface tension, using the equation of state. The data were obtained at different temperatures thereby allowing the calculation of $d\gamma_{sv}/dT$ and $d\gamma_{pv}/dT$, which then could be used to find γ_{sv} and γ_{pv} at the melting points of the various matrix materials. The surface tension of each matrix liquid phase was measured with the Wilhelmy plate technique. Finally, in order to calculate the solid–solid interfacial tension γ_{ps}, it was necessary to treat the equation of state as a "generic" equation of the form

$$\gamma_{12} = f(\gamma_{1v}, \gamma_{2v}). \tag{9.63}$$

This represents an extrapolation to two solid phases because the explicit form of the equation of state was derived empirically from contact angles of liquids on solids. Although this lacks a rigorous justification, the fact that this approach does, in fact, accurately predict particle engulfing behavior, lends confidence to this generic usage of the equation of state.

As seen in Chapter 12, predictions of particle engulfing or rejection by the equation of state were observed to be accurate in almost all cases. Considering only those experiments in which $|\Delta F^{adh}| \geq 0.2$ mJ/m^2 to be representative of the most clear-cut predictions (unambiguous pushing or rejection), the equation of state correctly predicted 29 out of 31 cases (94%).

Other methods can also be used to predict particle engulfment or rejection by calculating ΔF^{adh}. Predictions using the Fowkes approach for interfacial tensions were found to contradict experimental data in most cases [70]. Predictions were also generated using Lifshitz theory (see Chapter 10), for cases in which the matrix material was dispersive. Lifshitz theory accounts for only dispersive forces. Its predictions also proved inaccurate, indicating that nondispersive interactions must be significant between a dispersive and a nondispersive material. This runs contrary to an underlying assumption of the STCs approaches [71].

The free energy of adhesion can also be calculated using other equations of state, such as Equations 9.11 and 9.12. For a hypothetical matrix material with $\gamma_{sv} = 35$ mJ/m^2 and $\gamma_{lv} = 40$ mJ/m^2, curves of ΔF^{adh} as a function of particle surface tension γ_{pv} are plotted in Figure 9.11. With Equations 9.12 and 9.58, ΔF^{adh} is negative at low γ_{pv} and becomes positive with increasing γ_{pv} when $\gamma_{pv} = \gamma_{lv}$. These characteristics of the curves imply that there is a transition from particle engulfment to particle rejection (pushing) when $\gamma_{pv} = \gamma_{lv}$. On the other hand, with Equation 9.11, ΔF^{adh} reaches zero at the same point as the curves from Equations 9.50 and 9.58, but remains zero. Thus, ΔF^{adh} never becomes positive, so that pushing of the particles would not be observed under any circumstances, in contradiction with the overwhelming experimental evidence to the contrary.

In Table 9.11, the results of relevant microscopic observations are listed for several types of polymer particles and two matrix materials: benzophenone and bibenzyl. Also, the surface tensions γ_{pv} are calculated from the contact angle measurements on

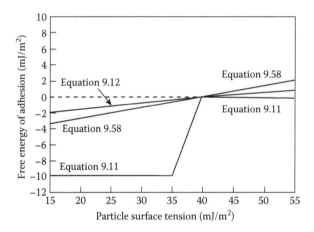

FIGURE 9.11   Free energy of adhesion calculated using three equations of state type relations.

TABLE 9.11

Observation of Particle Pushing or Engulfment in the Melt Materials Benzophenone and Bibenzyl

Particle Material	Benzophenone $\gamma_{lv} = 39.9$ mJ/m²		Bibenzyl $\gamma_{lv} = 24.9$ mJ/m²	
	γ_{pv} (mJ/m²)	Observation	γ_{pv} (mJ/m²)	Observation
Acetal	44.5	pushed	44.3	pushed
Nylon-6	43.7	pushed	43.4	pushed
Nylon-6,6	43.1	pushed	42.8	pushed
Nylon-12	40.8	pushed	40.6	pushed
Nylon-6,10	38.2	engulfed	37.8	pushed
PMMA	37.2	engulfed	36.8	pushed
Nylon-6,12	34.3	engulfed	34.0	pushed
Teflon	18.1	engulfed	17.8	engulfed

Note: The surface tensions are at the respective melting points of the matrix materials, 48°C for benzophenone and 52°C for bibenzyl.

smooth surfaces of the same polymeric materials using Equation 9.58. Equation 9.12 could not be used because it does not yield a consistent γ_{pv} value across the liquids with which the contact angles were measured. Table 9.11, in conjunction with the curves for Equations 9.58 and 9.12 in Figure 9.11, indicates that the surface tensions of Teflon, nylon-6,12, nylon-6,10, and PMMA are lower than 39.9 mJ/m² (γ_{lv} of benzophenone at its melting point) and the γ_{pv} of all the other polymer particles is greater than 39.9 mJ/m^2. From the observations with bibenzyl, we conclude similarly that the surface tension of Teflon is lower than 24.9 mJ/m^2 and that of all other polymer particles is greater than 24.9 mJ/m^2.

It is apparent that, given a sufficiently large number of matrix materials, one could put narrow limits on the surface tension of each particle, thus effectively determining these surface tensions. It is also clear that these results would not be a unique consequence of Equation 9.58. While Equation 9.11 is at variance with experimental observations of particle rejection, the above inferences can be made on the basis of Equation 9.12 as well as Equation 9.58. The surface tensions γ_{pv} as calculated from contact angles using Equation 9.58 are in complete agreement with the implications of both Equation 9.58 and Equation 9.12 with respect to the observations of engulfment and rejection.

9.5.3 SEDIMENTATION VOLUMES

Sedimentation experiments are a well-established technique to study the stability of powder dispersions in liquids. The behavior of such systems is governed largely by van der Waals and electrostatic interactions, although a complete model has yet to be developed. In this section, we shall consider sedimentation volume data that indicate the role of van der Waals forces as reflected by the relevant interfacial tensions of the solid particles suspended in a liquid. Chapter 11 provides further information on the sedimentation volume technique.

The relationship between van der Waals interactions and surface thermodynamics is evident when one realizes that the free energy of adhesion between two solids p and s in a liquid l is just the integral of the van der Waals forces from infinity to the equilibrium separation distance at adhesion, and is also equal to $\gamma_{ps} - \gamma_{pl} - \gamma_{sl}$. The free energy of cohesion for like particles p in a liquid is therefore

$$\Delta F_{plp}^{coh} = -2\gamma_{pl}.\tag{9.64}$$

which is maximum when $\gamma_{pl} = 0$ corresponding to $\gamma_{pv} = \gamma_{lv}$. In other words, for a given level of electrostatic repulsion, the degree of particle attraction and hence sedimentation volume will be a function of the surface tension of the suspending liquid γ_{lv}, reaching a minimum when $\gamma_{lv} = \gamma_{pv}$. In fact, depending on whether the sedimentation mechanism involves particle agglomeration, the condition $\gamma_{lv} = \gamma_{pv}$ may result in either a maximum or a minimum in the sedimentation volume (see Chapter 11 for details). In either case, this provides another method of independently measuring a solid surface tension γ_{pv}, and comparing it with a value obtained from contact angles and the equation of state.

Sedimentation volumes have been recorded for a number of polymer powders in both pure liquids and binary liquid mixtures of various surface tensions [2,72,73]. As detailed in Chapter 11, the equation-of-state contact angle predictions of γ_{pv} are in good agreement (average difference <3%) with those inferred independently from the thermodynamic model of sedimentation. Because the latter did not, in any way, involve the explicit form of the equation of state and depended only on liquid surface tension measurements, these data constitute another piece of evidence that the equation of state approach is correct.

The results can also be interpreted using alternative equations of state, Equation 9.11 and Equation 9.12. Figure 9.12 shows the free energy of cohesion ΔF^{coh} versus the liquid surface tension γ_{lv} for hypothetical particles having a surface tension $\gamma_{pv} = 20$ mJ/m². All three equations produce a minimum at $\gamma_{lv} = \gamma_{pv}$, suggesting that in a sedimentation experiment one should observe an extremum (presumably a minimum) in the sedimentation volumes at that point.

Thus, Equations 9.58, 9.12, and 9.11 yield identical results for the prediction of sedimentation volumes, suggesting that all three equations have common features. By inspection, we find that for all three equations

$$\gamma_{12} = f(\gamma_{1v}, \gamma_{2v}) = f(\gamma_{2v}, \gamma_{1v}), \tag{9.65}$$

and

$$\gamma_{12} = f(\gamma_{1v}, \gamma_{2v}) = 0 \quad \text{when} \quad \gamma_{1v} = \gamma_{2v}. \tag{9.66}$$

In other words, all three equations are symmetric in γ_{lv} and γ_{sv}, and all three predict zero interfacial tension when $\gamma_{lv} = \gamma_{sv}$.

The sedimentation experiments may be interpreted by referring only to the basic characteristics that are common to the three equations of state, Equations 9.58, 9.12, and 9.11. By Equation 9.66, when $\gamma_{lv} = \gamma_{pv}$ then $\gamma_{pl} = 0$ and ΔF^{coh} is a maximum. In fact, condition Equation 9.66 may be relaxed slightly. It can be shown that as long as γ_{12} is a minimum when $\gamma_{1v} = \gamma_{2v}$, and the minimum may not necessarily be zero, the minima for the free energy of cohesion ΔF^{coh} in Figure 9.12 and the extrema of the sedimentation volumes will occur at the same liquid surface tension. Hence, the predictions of the above three equations of state will remain the same.

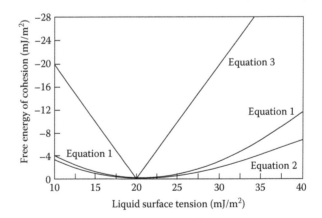

FIGURE 9.12   Free energy of cohesion calculated using three equation of state type relations: Equation 9.58 as Equation 1, Equation 9.50 as Equation 2, and Equation 9.11 as Equation 3.

9.5.4 Particle Suspension Layer Stability

When a dilute suspension of small particles in a liquid is carefully layered on a dense liquid as, for example, in zone electrophoresis, centrifugation, or isoelectric focusing, the suspension often forms a zone or layer of finite thickness with a well defined interface between the suspension layer and the supporting liquid. For a stable suspension layer, the suspension-layer/liquid-cushion interface is planar and "sharp." When the suspension layer becomes unstable, droplets of suspension form at the interface and fall through into the liquid cushion. This instability is generally referred to as "streaming" or "droplet sedimentation," and is affected by the initial particle concentration, the diffusion of solutes, particle charge, and van der Waals attractive forces [74]. The latter two factors were the focus of a study of suspension layer stability of fixed erythrocytes on a D_2O cushion, which is described below [74].

Under the hypothesis that colloidal stability theory was applicable to suspensions of biological cells, the total energy of interaction between like cells was modeled as the sum of a repulsive component due to electrostatic charge, and an attractive component due to van der Waals forces. It has been shown that the Hamaker coefficients, which give the van der Waals potential energy of attraction, may be expressed in terms of the surface tensions of the solid particle (the cell) and the suspending liquid [75,76]. The Hamaker coefficients for the cells in the liquid reach their minimum value of zero when the cell surface tension equals that of the suspending liquid. At this point, the attractive van der Waals forces are reduced to zero and the repulsive action of the electrostatic charge is maximized. This serves to prevent cell agglomeration and subsequent droplet sedimentation. In other words, the model predicts that the cell suspension is most stable when the cell surface tension γ_{cv} equals the liquid surface tension γ_{lv}. Thus, we can again measure a solid surface tension, by determining the liquid surface tension that gives maximum particle suspension stability.

The model was tested using five species (human, horse, chicken, canine, and turkey) of glutaraldehyde-fixed erythrocytes (i.e., red blood cells, treated to render the cells rigid). Because of their uniformity of size and shape, fixed erythrocytes make an excellent model particle in many physical studies. The cells were suspended in a saline solution and then layered onto a cushion of D_2O. Stability was recorded as the time required for the onset of droplet sedimentation as indicated by the distortion of the D_2O-suspension interface. The surface tension of the saline was varied by adding dimethylsulfoxide (DMSO), which had a relatively small effect on the cell surface charge potential as measured electrophoretically. It was therefore possible to regulate the van der Waals attraction forces while leaving the electrostatic repulsive forces relatively constant. Table 9.12 lists the surface tensions of the suspending liquids that produced greatest stability, as measured by the elapsed time before the interface became distorted. According to the model, this liquid surface tension should be equal to that of the cells.

So far, we have not made reference to the equation of state for interfacial tension. Unfortunately, it is not possible to measure contact angles on layers of fixed erythrocytes as it is with, say, platelets or bacteria [77]. However, the cell surface

TABLE 9.12

Comparison of the Surface Tension of Erythrocytes Obtained Via Two Independent Methods

	Technique	
Erythrocyte Species	Droplet Sedimentation (mJ/m²)	Freezing Front (mJ/m²)
Turkey	65.7	65.4
Chicken	65.2	65.1
Canine	64.4	64.2
Horse	65.4	64.5
Human	64.3	64.1

tensions could be compared with earlier values obtained with the solidification front technique (Section 9.5.2), which is itself based on the equation of state [78,79]. These values of cell surface tension, which are also listed in Table 9.12, are in remarkably good agreement with those obtained from suspension stabilities. Thus, we have further reassurance that the equation of state, which was used to develop and calibrate the solidification front technique, is indeed reliable.

9.5.5 TEMPERATURE DEPENDENCE OF CONTACT ANGLES

The equation of state can be used to calculate solid surface entropies from measurements of the temperature dependence of equilibrium contact angles. As a short aside, we first describe how heats of immersion can be used to verify that measured contact angles are in fact equilibrium values.

Harkins and Jura [80] were the first to point out that the heat of wetting, ΔH_w, may be obtained directly from the temperature dependence of contact angles. The heat of wetting is defined as

$$\Delta H_w = \left(\gamma_{sl} - T \frac{d\gamma_{sl}}{dT} \right) - \left(\gamma_{sv} - T \frac{d\gamma_{sv}}{dT} \right), \tag{9.67}$$

or, using Young's equation (Equation 9.1),

$$\Delta H_w = -\left(\gamma_{lv} \cos\theta - T \frac{d(\gamma_{lv} \cos\theta)}{dT} \right). \tag{9.68}$$

Thus, the only quantities to be determined experimentally are the surface tension, γ_{lv}, and the contact angle, θ, both as a function of temperature. Since calorimetric heats of immersion are available for polytetrafluoroethylene in contact with several n-alkanes [41], it is of interest to compare these calorimetric heats with those calculated from the temperature dependence of contact angles.

Temperature-dependent contact angles were recorded [81] with seven *n*-alkanes against Teflon (PTFE). Advancing contact angles between PTFE and the *n*-alkanes from decane to hexadecane were measured for temperatures from ambient to 70°C. From these data and the literature values [82] of the temperature dependence of γ_{lv}, the heats of wetting were calculated using Equation 9.68. These are shown in Figure 9.13 (filled symbols) together with the values calculated from *n*-heptane and *n*-nonane contact angles [83]. The calorimetric results of Whalen and Wade [41] are also given (open symbols). The agreement between the two types of results is remarkable. It appears that the rise of the heats of wetting (solid symbols) from *n*-decane to *n*-dodecane is real. Since the reproducibility of the calorimetric heats of immersion is somewhat poorer [41], it cannot be decided whether or not the calorimetric results show a similar behavior.

The agreement between the calorimetric heats of immersion and the contact angle heats of wetting indicates that the observed contact angles must be Young angles. We then cannot only expect to obtain the correct solid surface free energies γ_{sv}, but also the solid surface entropies $-d\gamma_{sv}/dT$ [mJ/m²·K] from the temperature dependence of the contact angles and the equation of state.

Indeed, from measurements of the temperature dependence of contact angles on Teflon, the average surface entropy of Teflon obtained from a series of *n*-alkane test liquids was 0.064 mJ/m²·K [81]. This value is in excellent agreement with literature data for Teflon melt [84]. There was a slight systematic change of the $-d\gamma_{sv}/dT$ data with the chain length of the liquids, presumably due to sorption of the vapor of the shorter chain length *n*-alkanes.

Surface entropies were also obtained for surfaces of two low molecular weight substances, cholesteryl acetate and hexatriacontane. Cholesteryl acetate undergoes an allotropic phase change at about 40°C. The measured surface entropies of the two solid phases of the cholesteryl acetate were 0.085 mJ/m²·K for the low temperature modification and 0.055 mJ/m²·K for the high temperature modification. It is interesting to note that this latter value is close to the value observed for liquid cholesteryl acetate, 0.048 mJ/m²·K [85]. The surface entropy of even the low temperature modification of hexatriacontane was 0.14 mJ/m²·K; that is, considerably

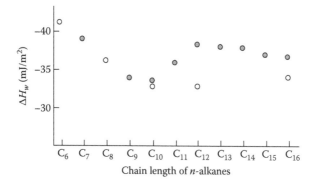

FIGURE 9.13   Heats of wetting (solid symbols) and calorimetric heats of immersion (open symbols) for *n*-alkanes on PTFE.

higher. This result is in line with the rather low value of the surface tension. As was pointed out by Fowkes [86], the low value of the surface tension γ_{sv} is due to the fact that the long chains are aligned in parallel array and the methyl end groups are exposed at the surface. We therefore expect surface tension and surface entropy of the hexatriacontane in the solid state to be closer to the values for short-chain alkanes in the liquid state (>0.1 mJ/m$^2\cdot$K) than to those for the longer-chain alkanes in the liquid state.

The surface tensions of siliconized glass as a function of temperature were measured and are plotted in Figures 9.14 and 9.15. The plot for the γ_{sv} values obtained from the water contact angles is perfectly straight with very little scatter below 40°C; above 40°C, the data points tend to fall above this straight line due to condensation of water vapor. The γ_{sv} value for 20°C is 18.3 mJ/m^2, in good agreement with a measurement performed with glycerol at 20°C, which yielded 18.1 mJ/m^2 [81]. Such a relatively small value is to be expected in view of the discussion given for the hexatriacontane surface. Chemisorption of the silicone oil on the siliconized glass

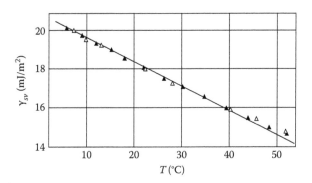

FIGURE 9.14 The temperature dependence of the surface tension of two samples of siliconized glass, calculated from water contact angles.

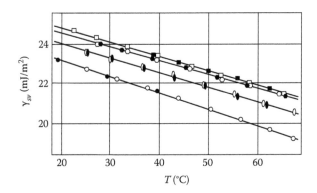

FIGURE 9.15 The temperature dependence of the surface tension of two samples of siliconized glass, calculated from n-alkane contact angles (from bottom to top: decane, dodecane, tetradecane, and hexadecane).

surface leads to a surface consisting of methyl groups. The large surface entropy of 0.126 mJ/m^2·K obtained from the water contact angles below 40°C corroborates this expectation.

The γ_{sv} values at 20°C obtained from the alkane contact angles are neither constant, nor consistent with the values obtained from the water and the glycerol data. Since condensation and hence adsorption become quite prominent in the measurements with water above 40°C, it is to be expected that adsorption plays an even more important role in the measurements with the alkanes. It is therefore reasonable to expect that the relatively high values of γ_{sv} in these cases are due to adsorption of the vapors of the measuring liquids. This conclusion would imply that the contact angle measurements were not performed on a surface consisting exclusively of methyl groups, but rather one that consisted to a considerable degree of CH$_2$ groups. This supposition is again corroborated by the surface entropies, which were rather small (0.068–0.084 mJ/m^2·K). They are in fact somewhat smaller than the surface entropies of the liquid alkanes. But even this is not surprising, since the chains of, say, decane when adsorbed on a solid surface will produce a surface of a larger ratio of CH$_2$ to CH$_3$ groups than when they are in a more or less randomly oriented three-dimensional liquid phase.

In summary, surface entropies measured for four solid surfaces (Teflon, cholesteryl acetate, hexatriacontane, and siliconized glass) with the aid of the equation of state turn out to be of the expected magnitudes based on surface chemistry and independent literature data, lending further support to the validity of the equation of state.

9.5.6 CONSISTENCY OF SOLID SURFACE TENSIONS

There is one immediate criterion that the results obtained with any approach for measuring solid surface tensions must satisfy. When measuring contact angles with a number of liquids on a low-energy solid, the surface tension γ_{sv} is expected to be constant, independent of the liquid surface tension γ_{lv}. This expectation is based on the fact that the equilibrium spreading pressure in such situations is generally low, of the order of 1 mJ/m^2 or less [42]. In this case, θ and γ_{lv} of the measuring liquid are the only two inputs into any of the approaches, and γ_{sv} and γ_{sl} are outputs.

Before we consider a series of contact angle experiments, it is appropriate to note the considerable care that must be taken to obtain good-quality contact angles. The measurement of contact angles is subject to many sources of error that are often overlooked. This lack of appreciation of the subtleties involved has caused many investigators to be misled by spurious data. Particularly important are the effects of surface roughness and the necessity of establishing a truly advancing contact angle. Scratches less than 0.5 μm deep can cause an advancing liquid drop to "hinge" at the scratch, rendering the contact angle essentially meaningless [28,87]. Surface roughness and homogeneity should always be assessed by measurement of both the advancing and receding contact angles to give the contact angle hysteresis. It should be noted, as well, that the common practice of adding liquid to an existing drop by touching it momentarily with a pendant drop suspended from a platinum wire or the tip of a pipet may be incorrect. Although the additional liquid indeed causes the drop to grow, the contact angle is often not the maximum advancing angle because of the vibration caused by the sudden addition of the pendant drop. It has been observed many times that this effect can produce intermediate, metastable contact angles as much as 7° below the advancing angle [28]. If a pipet or syringe is used to add liquid to a sessile drop, it must penetrate the interface and not be withdrawn prior to observation.

Combining Fowkes's equation (Equation 9.4) with the Young equation and assuming that vapor adsorption is negligible (i.e., $\gamma_{sv} = \gamma_s$), the dispersive component of a solid surface tension γ_s^d can be determined by contact angle measurements from

$$\gamma_s^d = \frac{1}{4}\frac{\gamma_l^2}{\gamma_l^d}(1+\cos\theta)^2. \tag{9.69}$$

Tables 9.5 through 9.6 list experimental contact angle data for both dispersive and nondispersive liquids on two dispersive solids, and Table 9.7 lists contact angle data for nondispersive liquids on a nondispersive solid. Employing these data (and data for a few other liquids from Li et al. [57]) in Equation 9.69, γ_s^d values were determined and are shown in Table 9.13. In the case of dispersive solids, $\gamma_s = \gamma_s^d$; for dispersive liquids, $\gamma_l = \gamma_l^d$; while for nondispersive liquids, γ_l^d values were taken from Fowkes and colleagues [88,89], if available. As can be seen in Table 9.13, the γ_s values from contact angle measurements of dispersive liquids on FC-721 and FEP surfaces tend to decrease with increasing γ_l. In the case of FC-721, γ_s values vary by approximately 25%. From the contact angle data of nondispersive liquids on each surface, non-constant, γ_s and γ_s^d values also arise, especially for water. In addition, consistent values of γ_s are not obtained between the groups of dispersive and nondispersive liquids.

When vapor adsorption is negligible, it is expected that the values of γ_s and γ_s^d should be approximately constant and independent of the liquids used, but the results obtained from the Fowkes approach do not bear this out. In order to bring the γ_s value from water on FC-721 in line with the other values, say the γ_s value from decane, the γ_l^d value for water would have to be 30.2 mJ/m². This, however, would disagree with Fowkes's γ_l^d value for water of about 21 mJ/m². The same calculation can be performed for FEP, using for example, the γ_s value from cis-decalin. The required γ_l^d value for water would be 31.3 mJ/m².

In contrast, the values for γ_s calculated from the equation of state (Equation 9.58) are consistent across the different liquids tested (see Tables 9.5 through 9.7). Many other solid surfaces have been used to test the equation of state, one example being hexatriacontane deposited by vacuum sublimation on clean glass [35]. These surfaces were so smooth and homogeneous that no contact angle hysteresis (difference between advancing and receding angles) was observed, even though the measurement technique was capillary rise at a vertical plate, which has a precision of about 0.1° (better than this if used with digital image analysis; see Chapter 6). It is significant that, regardless of the relative magnitudes of the various intermolecular forces within the different liquids, the equation of state, using only the measured contact

TABLE 9.13

Solid Surface Tensions and Surface Tension Components of FC-721, FEP, and PET Obtained from Contact Angle Measurements of Dispersive (D) and Nondispersive (ND) Liquids Using the Fowkes Approach

Type	Liquid	γ_l (mJ/m²)	γ_l^d (mJ/m²)	FC-721, $\gamma_s = \gamma_s^d$ (mJ/m²)	FEP, $\gamma_s = \gamma_s^d$ (mJ/m²)	PET, γ_s^d (mJ/m²)
D	Pentane	15.65	15.65	12.37	—	—
	Hexane	18.13	18.13	12.14	—	—
	Decane	23.43	23.43	11.60	17.39	—
	Dodecane	25.44	25.44	11.50	17.73	—
	Tetradecane	26.55	26.55	11.00	17.18	—
	Hexadecane	27.76	27.76	10.90	17.57	—
	cis-Decalin	31.65	31.65	10.94	16.87	—
	1-Bromonaphthalene	44.01	44.01[a]	9.07	15.29	—
ND	Methanol	22.30	17.4[b]	15.3	—	—
	Dimethylformamide	35.57	30.2[b]	—	19.55	—
	DMSO	43.58	29.0[b]	13.91	22.32	—
	Diethylene glycol	45.04	32.3[b]	12.18	20.70	48.22
	Ethylene glycol	47.99	29.3[c]	13.97	22.81	55.15
	Formamide	57.49	28.0[b]	14.55	24.24	64.39
	Glycerol	63.11	36[c]	11.17	18.40	52.14
	Water	72.75	21.1[b]	16.59	25.05	88.69

[a] Value obtained from van Oss, C. J., Good, R. J., and Chaudhury, M. K., *Journal of Colloid and Interface Science*, 111, 378, 1986.

[b] Value obtained from Fowkes, F. M., Riddle, Jr., F. L., Pastore, W. E., and Weber, A. A., *Colloids and Surfaces*, 43, 367, 1990.

[c] Value obtained from Israelachvili, J. N., *Proceedings of the Royal Society of London A*, 331, 39, 1972.

angle and the total liquid surface tension, correctly predicts a constant solid surface tension γ_{sv} (Table 9.14).

Either of the other equations of state considered—i.e., Equations 9.11 and 9.12—in conjunction with the Young equation (Equation 9.1), can be used to calculate values for the solid surface tensions γ_{sv} from the same contact angles. The results obtained with these equations are also given in Table 9.14. Only Equation 9.58 gives results that are essentially independent of γ_{lv}, suggesting that Equations 9.11 and 9.12 are deficient. The fact that all three equations predict γ_{sv} values near 20 mJ/m² as the contact angle decreases reflects the fact that all three equations imply $\gamma_{sl} \to 0$ as $\theta \to 0$.

Given this last observation, the analysis can be reversed by using the common value of γ_{sv} calculated from any of the equations of state for an almost-wetting liquid to predict contact angles of other liquids on the same solid surface. Results are shown in Figure 9.16 for a surface of Teflon AF 1600 [90] (see also Chapter 8), using

TABLE 9.14

Solid Surface Tensions of Hexatriacontane at 20°C

Liquid	γ_{sv} (mJ/m²)		
	Equation 9.11	Equation 9.12	Equation 9.58
Water	27.2	10.2	19.8
Glycerol	28.7	13.0	20.0
Thiodiglycol	28.7	15.3	19.8
Ethylene glycol	28.3	16.8	19.7
Hexadecane	23.3	19.8	20.0
Tetradecane	23.4	20.6	20.7
Dodecane	22.7	20.3	20.4
Decane	21.3	19.1	21.2
Nonane	21.8	20.8	20.8

Note: Liquid surface tensions and contact angles are given in Table 9.2.

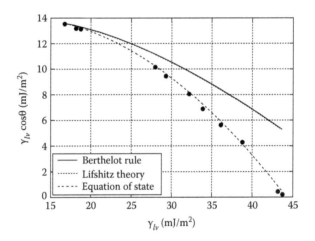

FIGURE 9.16 Contact angles for a series of liquids with bulky molecules on Teflon AF 1600. (From Tavana, H. and Neumann, A. W., *Advances in Colloid and Interface Science*, 132, 1, 2007.) Lines are the contact angle predictions using three approaches and the common γ_{sv} of 13.6 mJ/m² found for the low surface tension liquids. The line for Lifshitz theory is slightly below the line for Berthelot's rule (i.e., Equation 9.12).

the equations of state (Equations 9.50 and 9.12). It is seen that Equation 9.58 is far more accurate.

The consistency of solid surface tensions calculated by the equation of state (Equation 9.58) has been further examined using extensive contact angle data on low energy surfaces (mainly polymers). Liquid surface tensions have spanned the range from 19 to 73 mJ/m². Data for 15 different solid surfaces are collected in the review by Kwok and Neumann [55]. Measured solid surface tensions γ_{sv} are in all cases consistent across different probe liquids. Similarly consistent fits of the equation of state were found for contact angles measured decades earlier by Zisman and coworkers on 34 different solid surfaces [55].

The ability of the equation of state to correlate a wide range of measured $\theta - \gamma_{lv}$ data with a single γ_{sv} is a fundamental test of both the equation of state concept [that $\gamma_{sl} = f(\gamma_{lv}, \gamma_{sv})$] and the explicit formulation of the equation.

9.5.7 Contact Angles of Polar and Nonpolar Liquids

Perhaps the most explicit test of the equation of state concept is to compare the measured contact angles of polar and nonpolar liquids on the same solid surface [31]. Consider two different pure liquids that are chosen to have equal overall surface tensions. These same liquids are, however, also selected to have widely disparate compositions of intermolecular forces. In other words, one liquid may be an alkane (a liquid that has only dispersion forces) while another may be characterized by a large dipole moment. According to the theory of STCs, the contact angles of these two liquids on a single solid surface should differ in proportion to the differences in the makeup of the intermolecular forces. In contrast, the equation of state approach predicts that the contact angles will be equal since both the total liquid and solid surface tensions are constant. This simple experiment provides a direct test of the basic premise of each of the two theories, and, moreover, it is independent of the specific form of any Fowkes-type equation or of any particular equation of state.

Liquids used in such an experiment are listed in Table 9.15. In order to minimize the potential for vapor adsorption, the seven liquids were chosen to have relatively high boiling points, the lowest being that of heptaldehyde at 153°C. The first three liquids in Table 9.15 were selected to be significantly more dispersive than the remaining four liquids, which are characterized by much larger dipole moments and by relatively smaller dispersion components of the solubility parameters. In addition, the prediction of the Burrell hydrogen-bonding classification [91], which is another empirical aid for the prediction of solubility, is "moderate" for the last four liquids and "poor" for the rest. This is not meant to imply the actual existence of hydrogen bonding in our systems, but in the present context serves to indicate that independent experimental observation has established significant differences in the character of the intermolecular forces.

Furthermore, Beerbower [92] has developed a correlation between liquid surface tension and the dispersion, polar, and hydrogen-bonding components of the solubility parameter. The last column of Table 9.15 lists the percentage of the total predicted surface tension that is due to the dispersion component of the solubility parameter. It should be noted that, as Beerbower himself did [92], this dispersive fraction bears no relation to the dispersion component of surface tension from the Fowkes theory. Taken together, the information in Table 9.15 indicates that, relative to the last four liquids, dispersion forces in pentadecane, dibenzylamine, and 1-methylnaphthalene are responsible for a significantly larger fraction of the total intermolecular binding energy. For the purpose of the present investigation, the exact magnitudes of such differences are unimportant.

TABLE 9.15
Liquid Properties

Liquid	Dipole Moment[a] (Debye)	Surface Tension[b] (mJ/m²)	$\delta_d/\delta_T \times 100$[c]	$\gamma_{disp}/\gamma \times 100$[d]
Pentadecane	0.0	$28.93 - 0.08531T$	100	100
Dibenzylamine	0.97	$42.14 - 0.1054T$	71	94
1-Methylnaphthalene	0.23	$41.82 - 0.1188T$	79	97
Benzaldehyde	2.77	$43.24 - 0.1195T$	60	76
Ethyl caprylate	1.68	$29.12 - 0.1018T$	50	58
Heptaldehyde	2.58	$28.50 - 0.0766T$	47	56
Methyl salicylate	2.23	$41.84 - 0.1201T$	44	56

[a] From McClellan, A. L., *Tables of Experimental Dipole Moments*, W. H. Freeman, San Francisco, 1963.

[b] Measured by the Wilhelmy plate technique (From Spelt, J. K., *Solid Surface Tension: The Equation of State Approach and the Theory of Surface Tension Components*, PhD Thesis, University of Toronto, 1985.) with an uncertainty of ±0.15 mJ/m². T = temperature in °C.

[c] Percentage of total solubility parameter (δ_T) attributed to dispersion forces at 25°C. Calculated using the solubility parameters (From Barton, A. F. M., *CRC Handbook of Solubility Parameters and Other Cohesion Parameters*, CRC Press, Boca Raton, FL, 1983) and correlations (From Barton, A. F. M., *Chemical Reviews*, 75, 731, 1975). For details see Spelt, J. K., *Solid Surface Tension: The Equation of State Approach and the Theory of Surface Tension Components*, PhD Thesis, University of Toronto, 1985.

[d] Percentage of total predicted surface tension due to the dispersion component of the solubility parameter. (From Beerbower, A., *Journal of Colloid and Interface Science*, 35, 126, 1971.)

Finally, the predominance of dispersion forces may also be inferred from the Lifshitz theory using data for the refractive index n and the dielectric constant (static permittivity) ε_0 [93]. It is known that for electrically symmetric molecules, ε_0 is approximately equal to n^2. Permanent dipoles and hydrogen bonding tend to increase the static permittivity so that in the case of water, for example, $\varepsilon_0 = 78.4$ while $n^2 = 1.77$. Literature values for permittivity were found for only three of the present liquids [98]: dibenzylamine, $\varepsilon_0 = 3.6$, $n^2 = 2.47$; benzaldehyde, $\varepsilon_0 = 17.8$, $n^2 = 2.39$; methyl salicylate, $\varepsilon_0 = 9.72$, $n^2 = 2.36$. As expected, the relative importance of nondispersive interactions is significantly greater for the latter two liquids.

Contact angle measurements were performed on two surfaces. The first was heat-pressed Teflon (FEP; Dupont), a surface that is exceptionally smooth and homogeneous, with measured contact angle hysteresis of only 3° [28,95]. Two FEP samples (designated A and B) were employed, each having been prepared in a different way and each having a unique thermal history. The latter fact caused the surface tensions of the samples to be slightly different. A second solid surface, which was used for only one pair of liquids, was siliconized glass (details of dimethyldichlorosilane treatment given by Neumann [81]). The advancing contact angle on this surface with water was 105° while the receding angle was between

95 and 100°. Sessile drop contact angles were measured using Axisymmetric Drop Shape Analysis (ADSA) [96]. Chapter 3 gives a complete description of ADSA in its latest form.

Table 9.15 lists the measured surface-tension/temperature relation for each of the seven liquids. By controlling the temperature of the contact angle experiment, it was possible to match more exactly the total surface tensions of the various liquids. All of the experiments were therefore performed in a temperature-controlled chamber. Further experimental details are available in Spelt et al. [31].

The results of the contact angle experiments are reported in Tables 9.16 through 9.18 for the solid substrates Teflon (FEP) sample A, Teflon (FEP) sample B, and siliconized glass, respectively. The data are grouped in pairs according to the matched surface tensions of the liquids used. The first liquid in each pair is the one that is completely or overwhelmingly exhibits London dispersion forces. The fourth column in these tables shows the contact angle of the first liquid minus that of the second liquid. The average of the eight contact angle differences is +0.4°.

The Fowkes equation for solid–liquid interfacial tensions,

$$\gamma_{sl} = \gamma_s + \gamma_l - 2\sqrt{\gamma_s^d \gamma_l^d}, \tag{9.70}$$

is strictly applicable only to situations in which at least one phase is a saturated hydrocarbon (n-alkane, paraffin wax, etc.) since this ensures that only dispersion

TABLE 9.16
Advancing Contact Angles on Substrate Teflon (FEP) A

Liquids	Liquid Surface Tension (mJ/ m²)	Contact Angle (°)	Contact Angle Difference	Expt. Temp. (°C)	γ_l^d/γ_l for Liquid 2[a]
1-Methylnaphthalene	39.0	72.6 ± 1.4	–0.2	24	0.97
Methyl salicylate	39.0	72.8 ± 0.7			
Dibenzylamine	41.8	75.4 ± 0.6	+2.0	3	1.02
Benzaldehyde	42.9	73.4 ± 0.4			
Dibenzylamine	41.8	75.4 ± 0.6	+2.5	3	1.00
Methyl salicylate	41.5	72.9 ± 0.6			
Pentadecane	25.6	52.4 ± 0.7	–0.8	39	0.98
Heptaldehyde	25.5	53.2 ± 0.3			
Pentadecane	27.7	53.6 ± 0.3	+0.6	14	1.01
Ethyl caprylate	27.7	53.0 ± 0.4			

[a] Dispersive fraction of liquid 2 from Equation 9.72.

TABLE 9.17
Advancing Contact Angles on Substrate Teflon (FEP) B

Liquids	Liquid Surface Tension (mJ/ m²)	Contact Angle (°)	Contact Angle Difference	Expt. Temp. (°C)	γ_l^d/γ_l for Liquid 2[a]
Dibenzylamine	41.8	72.4 ± 3.2	+ 3.1	3	1.01
Methyl salicylate	41.5	69.3 ± 0.6			
Pentadecane	25.6	49.1 ± 0.5	−1.2	39	0.98
Heptaldehyde	25.5	50.3 ± 0.4			

[a] Dispersive fraction of liquid 2 from Equation 9.72.

TABLE 9.18
Advancing Contact Angles on Siliconized Glass

Liquids	Liquid Surface Tension (mJ/m²)	Contact Angle (°)	Contact Angle Difference	Expt. Temp. (°C)	γ_l^d/γ_l for Liquid 2[a]
1-Methylnaphthalene	39.0	58.3 ± 1.1	−2.7	24	0.92
Methyl salicylate	39.0	61.0 ± 0.3			

[a] Dispersive fraction of liquid 2 from Equation 9.72.

forces are operative within that phase. Equation 9.70 predicts that if γ_s and γ_l are fixed, then γ_{sl} will vary inversely with γ_l^d. With respect to the contact angle experiments, the dispersive liquid in each pair should therefore have the smaller contact angle on a surface that is interacting only through dispersion forces. As was mentioned above, however, the average contact angle difference for the eight liquid pairs was +0.4°, indicating that the opposite trend was more prevalent. Considering the four cases where the contact angle difference exceeds the error limits, in two of these the difference is positive (contrary to Equation 9.70), while in the other two cases it is indeed negative.

Equation 9.70 may be combined with the Young equation (Equation 9.1) to yield

$$2\sqrt{\gamma_s^d} = \sqrt{\frac{\gamma_l}{\gamma_l^d}}\sqrt{\gamma_l}\left(1+\cos\theta\right). \tag{9.71}$$

For a given pair of liquids (denoted "1" and "2") on a single substrate, the left-hand side of Equation 9.71 is constant so that

$$\frac{\gamma_{l_2}^d}{\gamma_{l_2}} = \frac{\gamma_{l_1}^d}{\gamma_{l_1}}\frac{\gamma_{l_2}}{\gamma_{l_1}}\left(\frac{1+\cos\theta_2}{1+\cos\theta_1}\right)^2. \tag{9.72}$$

Considering liquid "1" to be the dispersive liquid (the first liquid listed in each pair in Tables 9.16 through 9.18) and assuming for these liquids that the dispersion fraction is that listed in the last column of Table 9.15 (from the Beerbower correlation at 25°C), Equation 9.72 may be used to calculate the implied dispersion fraction of liquid "2" (the left-hand side of Equation 9.72). Note that Equation 9.72 can also be used to give the ratio of the dispersive fractions of liquids "1" and "2" without regard to the Beerbower correlation.

The last column of Tables 9.16 through 9.18 shows the results of these calculations; that is, the prediction of Fowkes theory for the dispersive fraction of the "2" liquid (the nondispersive one) within each liquid pair. In all cases, this dispersive fraction is very close to 1.00, indicating that Equation 9.70 predicts that the "2" liquids are just as dispersive as the "1" liquids. This contradicts the predictions of the Beerbower correlation and, in general, to the expectations based on solubility parameters, molecular structure, and molecular properties.

Since the total liquid surface tensions are constant within a given pair of liquids, the contact angles are predicted (by an equation of state) to be equal on a single solid substrate. This does appear to be largely the case, although explanations must be found for the small, but nonzero, contact angle differences that persist. One possibility is vapor adsorption.

The equation of state can be used to estimate the equilibrium spreading pressure required to make the contact angles equal in each liquid pair. As shown in Spelt et al. [31], in all eight cases this pressure was less than or equal to 1.4 mJ/m^2. As demonstrated by Good [44], it is not unreasonable to assume that such spreading pressures can occur on surfaces of Teflon (FEP) and siliconized glass. The contact angle data are thus seen to be consistent with the predictions of the equation of state approach and provide an experimental verification of this theory.

A related argument [97] based on liquid properties has been used to examine the Fowkes approach for interfacial tensions. By combining Equation 9.70 with the Young equation (Equation 9.1), and assuming that adsorption is negligible, it is evident that the Fowkes approach implies that $\gamma_l \cos\theta$ depends, in general, on four variables: γ_l, γ_s, $\gamma_l{}^d$, and $\gamma_s{}^d$. The equation of state approach, however, predicts that $\gamma_l \cos\theta$ is only a function of γ_l and γ_s. This discrepancy between the two approaches can be tested by plotting $\gamma_l \cos\theta$ versus γ_l for a wide range of liquids on different solid surfaces; if smooth curves of similar shape emerge for a number of solid surfaces, one can conclude that $\gamma_l \cos\theta$ depends only on γ_l and γ_s and not on the STCs, which are varying randomly from liquid to liquid.

In fact, as already shown in Figure 9.7, curves describing the contact angles of different liquids on the same solid surfaces are so smooth and similar that one has to conclude that $\gamma_l \cos\theta$ indeed depends only on γ_l and γ_s. If the Fowkes hypothesis was correct (i.e., $\gamma_l \cos\theta$ was also dependent on $\gamma_l{}^d$ and $\gamma_s{}^d$) then the variations in the dispersive character of each liquid would lead to contact angles that were arbitrarily scattered in Figure 9.7.

It is instructive to focus on the experimental contact angle data of hexane, methanol, and decane on FC-721 in Figure 9.7. The surface tension of methanol is intermediate between the surface tensions of hexane and decane, and methanol is polar while the alkanes are nonpolar. Because of the difference in polarity or nondispersive property, the Fowkes approach implies a contact angle for methanol that should be significantly different from those of hexane and decane. Within the framework of the Fowkes model, it stands to reason that for purely dispersive liquids such as alkanes, the contact angle changes smoothly as the liquid surface tension changes. By interpolation between hexane and decane, it can be established what the contact angle for a purely dispersive liquid of the same surface tension as methanol (i.e., $\gamma_l = 22.3$ mJ/m^2) would be. The interpolated contact angle is found to be $\theta = 62.67°$, in good agreement with the experimental results for methanol ($\theta = 62.39 \pm 0.25°$). The fact that methanol has the same contact angle as a nonpolar liquid of the same surface tension indicates that the Fowkes approach is not tenable. Similar calculations can be performed for other sets of liquids, with similar results [97].

It is apparent, and in principle obvious from Figure 9.7 that the surface tensions γ_s and γ_l determine the contact angle completely. Clearly, this does not mean that intermolecular forces are irrelevant; they determine the primary surface tensions, γ_{lv} and γ_{sv}. But intermolecular forces do not have an additional, independent effect on contact angles. There is therefore no obvious way to use interfacial tensions and contact angles to determine the intermolecular forces.

The overall conclusion that emerges from the experimental data presented in the last seven sections is that of all the approaches examined, only the equation of state given by Equation 9.58 consistently matches observations. Nevertheless, it is interesting to note that most of the indirect techniques are not sensitive to the exact form of the equation of state used.

The theories discussed in this chapter up to this point are semiempirical, top-down approaches to understanding contact angles. In the final section, we consider a bottom-up approach based on intermolecular forces.

9.6 INTERMOLECULAR THEORY

9.6.1 CALCULATION OF INTERFACIAL TENSIONS AND CONTACT ANGLES

From the theory of intermolecular forces, the surface tension of a given liquid or solid can be numerically estimated given the strength of molecular interactions. Similar calculations can also be performed to estimate interfacial tensions but a specific solid–liquid interaction is required and can be obtained by means of established combining rules. When the three surface and interfacial tensions are calculated, the anticipated Young contact angles can then be obtained from Young's equation, allowing direct comparison with those obtained from experiments.

A mean-field approximation has been employed [98,99] to numerically calculate surface and interfacial tensions from molecular interactions. In a simple van der Waals model, the fluid molecules are idealized as hard spheres interacting with each other through a potential $\phi_{ff}(r)$, where r is the distance between two interacting molecules. A Carnahan-Starling model [98,100,101] was adopted as the hard sphere reference system. For a planar interface formed by a liquid and its vapor, each of which occupies a semi-infinite space, $z > 0$ and $z < 0$ respectively, the surface tension is given by [98,102]

$$\gamma_{lv} = \min_{\rho} \int_{-\infty}^{+\infty} dz \left\{ F[\rho(z)] + \frac{1}{2}\rho(z) \int_{-\infty}^{+\infty} dz' \overline{\phi}_{ff}(z'-z)[\rho(z')-\rho(z)] \right\}, \qquad (9.73)$$

where the minimum is taken over all possible density profiles $\rho(z)$. The excess free energy density $F(\rho)$ can be found from the equation of state of the fluid, and $\overline{\phi}_{ff}$ represents the interaction potential that has been integrated over the whole $x'y'$ plane.

For the solid–fluid (i.e., solid–liquid or solid–vapor) interface, the solid can be modeled as a semi-infinite impenetrable wall occupying the domain of $z < 0$ and exerting an attraction potential $V(z)$ on the fluid molecule at a distance z from the solid surface. The interfacial tension of such an interface can be obtained from

$$\gamma_{sf} = \gamma_s + \min_{\rho} \int_0^{+\infty} dz \left\{ F[\rho(z)] + \rho(z)V(z) \right.$$
$$+ \frac{1}{2}\rho(z)\int_0^{+\infty} dz' \overline{\phi}_{ff}(z'-z)[\rho(z')-\rho(z)] - \frac{1}{2}\rho^2(z)\int_{-\infty}^0 dz' \overline{\phi}_{ff}(z'-z) \right\}, \qquad (9.74)$$

where γ_s is the solid–vacuum surface tension, a constant that exists in the calculations of both γ_{sv} and γ_{sl}. This constant (γ_s) will be canceled out in the calculations of the contact angles via Young's equation (Equation 9.1); for the purpose of contact angle determination, it has no impact on the final results since we are concerned only with the difference between γ_{sv} and γ_{sl}.

To carry out the calculations of interfacial tensions and hence the contact angles, an interaction potential is required. Here we assume a (12:6) Lennard-Jones potential model and consider only the attraction part. The Lennard-Jones potential function requires knowledge of two parameters: the potential strength ε and the collision diameter σ. The potential strength ε_{sf} for $\phi_{sf}(r)$ is obtained from the fluid ε_{ff} and solid ε_{ss} potential strengths via a combining rule, as discussed in the next section.

For the calculation of liquid surface tension γ_{lv}, the two parameters ε_{ff} and σ_f can be related to the critical temperature T_c and pressure P_c of the liquid via the following expressions for the Carnahan-Starling model [98,100]

$$kT_c = 0.18016\alpha / \sigma_f^3$$
$$P_c = 0.01611\alpha / \sigma_f^6, \qquad (9.75)$$

where k is the Boltzmann constant and α is the van der Waals parameter, given by

$$\alpha = -\frac{1}{2}\int \phi_{ff}(r)\, dr. \qquad (9.76)$$

The densities of the liquid ρ_l and vapor ρ_v were obtained by requiring the liquid and vapor to be in coexistence at a given temperature T [98,100]. In the calculations, 30 liquids of different molecular structures were selected and the following parameters were assumed: $T = 21°C$, $\sigma_s = 10$ Å, and $\rho_s = 10^{27}$ molecules/m³ for the solid surface.

Each liquid is modeled by the parameters ε_{ll} and σ_l. The solid is modeled by its potential parameter ε_{ss}, density ρ_s, and collision diameter σ_s. For the solid–vapor and solid–liquid interfacial tensions (γ_{sv} and γ_{sl}), in addition to the parameters of fluid and solid, a combining rule $g(\sigma_l/\sigma_s)$ was also required. Thus the three interfacial tensions γ_{lv}, γ_{sv}, and γ_{sl} at a given temperature T could be expressed as

$$\gamma_{lv} = \gamma_{lv}\left(\varepsilon_{ll},\sigma_l,T\right)$$
$$\gamma_{sv} = \gamma_{sv}\left(\varepsilon_{ll},\sigma_l,\varepsilon_{ss},\rho_s,\sigma_s,g,T\right) \quad (9.77)$$
$$\gamma_{sl} = \gamma_{sl}\left(\varepsilon_{ll},\sigma_l,\varepsilon_{ss},\rho_s,\sigma_s,g,T\right).$$

For a given solid surface at a given temperature with a selected combining rule g and different fluids, ε_{ss}, ρ_s, σ_s, T and the function $g(\sigma_l/\sigma_s)$ are fixed, so that

$$\gamma_{lv} = \gamma_{lv}\left(\varepsilon_{ll},\sigma_l\right)$$
$$\gamma_{sv} = \gamma_{sv}\left(\varepsilon_{ll},\sigma_l\right) \quad (9.78)$$
$$\gamma_{sl} = \gamma_{sl}\left(\varepsilon_{ll},\sigma_l\right).$$

The above theoretical framework can be utilized to determine interfacial tensions and hence the contact angle/adhesion patterns [99]. Table 9.19 lists the liquid–vapor interfacial tensions for 30 different liquids as calculated from the generalized van der Waals molecular theory described above together with the experimental values. In most cases, differences between the calculated and experimental liquid–vapor surface tensions are less than 20%. The largest discrepancy comes from water with a calculated γ_{lv} value of 93 mJ/m² instead of an experimental value of 72.8 mJ/m². Considering the simplified fluid model, larger deviations for hydrogen-bonding liquids are indeed expected as the Lennard-Jones potential should not completely reflect the complicated interactions of, for example, water. Once γ_{lv} and γ_{sv} are calculated from the above van der Waals theory, the determination of γ_{sl}, however, requires the value of a solid–liquid energy parameter that can be related to the individual solid and liquid energy parameters through a combining rule.

9.6.2 COMBINING RULES FOR SOLID–LIQUID INTERFACIAL TENSIONS

In the theory of molecular interactions and mixtures, combining rules are often used to evaluate the parameters of unlike-pair interactions in terms of those of the like

TABLE 9.19
Comparison Between the Calculated (γ_{lv}^{cal}) and Experimental (γ_{lv}^{exp}) Liquid–Vapor Surface Tensions

Liquid	γ_{lv}^{exp} (mJ/m²)	γ_{lv}^{cal} (mJ/m²)	Liquid	γ_{lv}^{exp} (mJ/m²)	γ_{lv}^{cal} (mJ/m²)
CH₃Cl	16.20[a]	13.72	Hexadecane	27.76	17.68
Pentane	16.65	13.07	CH₂Cl₂	27.84[a]	24.69
Hexane	18.13	15.00	Benzene	28.88[a]	26.04
Methylamine	19.89[a]	16.98	trans-Decalin	29.50	26.15
Methanol	22.30	29.28	cis-Decalin	31.65	26.77
Decane	23.43	17.83	CS₂	32.32[a]	33.44
Ethyl acetate	23.97[a]	18.85	Chlorobenzene	33.59[a]	30.95
Acetone	24.02[a]	20.14	Bromobenzene	35.82[a]	34.15
Ethyl methyl ketone	24.52[a]	21.29	Iodobenzene	39.27[a]	38.38
Methyl acetate	25.10	19.98	Aniline	42.67[a]	38.55
Dodecane	25.44	18.08	Diethylene glycol	45.04	36.01
Tetradecane	26.55	16.82	Ethylene glycol	47.99	45.72
CCl₄	27.04[a]	24.29	Glycerol	63.11	50.32
Fluorobenzene	27.26[a]	24.59	Hydrazine	67.60[a]	71.81
CHCl₃	27.32[a]	25.01	Water	72.75	92.92

Note: Experimental values were obtained from Kwok, D. Y. and Neumann, A. W., *Advances in Colloid and Interface Science,* 81, 167, 1999, at 21°C. (Reprinted from Zhang, J. and Kwok, D. Y., *J. Phys. Chem. B* 106, 12594, 2002. With permission from American Chemical Society.)

[a] From Jasper, J. J., *Journal of Physical Chemistry Reference Data*, 1, 841, 1972, measured at 20°C.

interactions [51,100,102,109]. As with many other combining rules, the Berthelot rule [115]

$$\varepsilon_{ij} = \sqrt{\varepsilon_{ii}\varepsilon_{jj}},\tag{9.79}$$

is a useful approximation, but does not provide a secure basis for the understanding of unlike-pair interactions; ε_{ij} is the potential energy parameter (well depth) of unlike-pair interactions, and ε_{ii} and ε_{jj} are for like-pair interactions.

Historically, from the London theory of dispersion forces, the attraction potential ϕ_{ij} between a pair of unlike molecules i and j is given by

$$\phi_{ij} = -\frac{3}{2}\frac{I_i I_j}{I_i + I_j}\frac{\alpha_i \alpha_j}{r^6},\tag{9.80}$$

where I is the ionization potential, α is the polarizability and r is the distance between the pair of unlike molecules. For like molecules Equation 9.80 becomes

$$\phi_{ij} = -\frac{3}{4}\frac{I_i\alpha_i^2}{r^6}. \tag{9.81}$$

The total intermolecular potential $V(r)$ expressed by the (12:6) Lennard-Jones potential is in the form

$$V(r) = 4\varepsilon_{ii}\left[\left(\sigma_i/r\right)^{12} - \left(\sigma_i/r\right)^6\right], \tag{9.82}$$

where σ is the collision diameter. The attractive potentials in Equations 9.81 and 9.82 can be equated to give

$$\frac{3}{4}I_i\alpha_i^2 = 4\varepsilon_{ii}\sigma_i^6. \tag{9.83}$$

Equation 9.83 can be used to derive α_i and α_j; substituting these quantities into Equation 9.80 yields

$$\phi_{ij} = -\frac{2\sqrt{I_iI_j}}{I_i+I_j}\frac{4\sigma_i^3\sigma_j^3}{r^6}\sqrt{\varepsilon_{ii}\varepsilon_{jj}}. \tag{9.84}$$

If we write ϕ_{ij} in the form $-4\varepsilon_{ij}\sigma_{ij}^6/r^6$ such that $\sigma_{ij} = (\sigma_i + \sigma_j)/2$, the energy parameter for two unlike molecules can be expressed as

$$\varepsilon_{ij} = \frac{2\sqrt{I_iI_j}}{I_i+I_j}\left[\frac{4\sigma_i/\sigma_j}{\left(1+\sigma_i/\sigma_j\right)^2}\right]^3\sqrt{\varepsilon_{ii}\varepsilon_{jj}}. \tag{9.85}$$

The above expression for ε_{ij} can be simplified: when $I_i = I_j$, the first term of Equation 9.85 becomes unity; when $\sigma_i = \sigma_j$ the second factor becomes unity. When both conditions are met, we obtain the well-known Berthelot rule; that is, Equation 9.79.

For the interactions between two very dissimilar types of molecules or materials where there is an apparent difference between ε_{ii} and ε_{jj}, it is clear that the Berthelot rule cannot describe the behavior adequately. It has been demonstrated [50,52,53] that the Berthelot geometric mean combining rule generally overestimates the strength of the unlike-pair interactions. In general, the differences in the ionization potential are not large, that is, $I_i \approx I_j$; thus the most serious error comes from the difference in the collision diameters σ for unlike molecular interactions.

The minimum of the solid–liquid interaction potential ε_{sl} is often expressed in the following manner [102,103,106]

$$\varepsilon_{sl} = g(\sigma_l/\sigma_s)\sqrt{\varepsilon_{ss}\varepsilon_{ll}}. \tag{9.86}$$

Several functional forms of $g(\sigma_l/\sigma_s)$ have been suggested. For example, by comparing ε_{sl} with the minimum in the (9:3) Lennard-Jones potential, one obtains, the (9:3) combining rule:

$$\varepsilon_{sl} = \frac{1}{8}\left(1+\frac{\sigma_l}{\sigma_s}\right)^3 \sqrt{\varepsilon_{ss}\varepsilon_{ll}}. \qquad (9.87)$$

An alternative function has been investigated by Steele [111] and others [112]:

$$\varepsilon_{sl} = \frac{1}{4}\left(1+\frac{\sigma_l}{\sigma_s}\right)^2 \sqrt{\varepsilon_{ss}\varepsilon_{ll}}. \qquad (9.88)$$

Further, from the (12:6) Lennard-Jones potential, Equation 9.85 implies a (12:6) combining rule of

$$\varepsilon_{sl} = \left[\frac{4\sigma_l/\sigma_s}{\left(1+\sigma_l/\sigma_s\right)^2}\right]^3 \sqrt{\varepsilon_{ss}\varepsilon_{ll}}. \qquad (9.89)$$

These equations are attempts for a better representation of solid–liquid interactions from solid–solid and liquid–liquid interactions. In general, these functions are normalized such that $g(\sigma_l/\sigma_s) = 1$ when $\sigma_l = \sigma_s$; in other words, they revert to the Berthelot geometric mean combining rule Equation 9.79 when $\sigma_l = \sigma_s$. Each of the combining rules given above was employed as to determine interfacial tensions and adhesion/contact angle patterns from the generalized van der Waals model.

9.6.3 Calculated Adhesion and Contact Angle Patterns

For comparison with computed values, experimental adhesion and contact angle patterns are available for a large number of polar and nonpolar liquids on a variety of carefully prepared low-energy solid surfaces [55,113,114]. Figure 9.17a illustrates that, for a given solid surface, the experimental solid–liquid work of adhesion W_{sl} increases up to a maximum value as γ_{lv} increases. Further increase in γ_{lv} causes W_{sl} to decrease from its maximum. The trend described here appears to shift systematically to the upper right for a more hydrophilic surface (such as PMMA) and to the lower left for a more hydrophobic surface.

Figure 9.17b shows the experimental contact angle patterns in $\cos\theta$ versus γ_{lv}. For a given solid surface, as γ_{lv} decreases, the cosine of the contact angle ($\cos\theta$) increases, intercepting at $\cos\theta = 1$ with a limiting γ_{lv} value. As γ_{lv} decreases beyond this limiting value, contact angles become more or less zero ($\cos\theta \approx 1$), representing the case of complete wetting. The trend described here appears to change systematically to the right for a more hydrophilic surface (such as PMMA) and to the left for a relatively more hydrophobic surface (such as fluorocarbon).

The 30 polar and nonpolar liquids from Table 9.19 were used to calculate the solid–vapor and solid–liquid surface tensions for the adhesion patterns using the

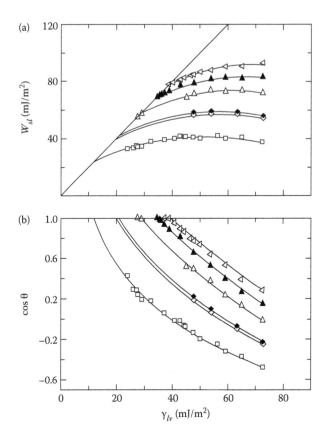

FIGURE 9.17 (a) The solid–liquid work of adhesion W_{sl} versus the liquid–vapor surface tension γ_{lv}; and (b) cosine of the contact angle $\cos\theta$ versus the liquid–vapor surface tension γ_{lv} for a fluorocarbon FC-722 ($\square$), hexatriacontane ($\diamondsuit$), cholesteryl acetate ($\blacklozenge$), poly(n-butyl methacrylate) ($\triangle$), poly(methyl methacrylate/n-butyl methacrylate) ($\blacktriangle$), and poly(methyl methacrylate) ($\triangleleft$) surfaces. (Reprinted from Zhang, J. and Kwok, D. Y., *Langmuir*, 19, 4666, 2003. With permission from American Chemical Society.)

above van der Waals theory together with the Berthelot, (9:3), Steele, and (12:6) combining rules [99] (Figure 9.18). The calculation results suggested that Berthelot's rule (Figure 9.18a) does not follow the general behavior of the other combining rules considered here. In fact, a larger discrepancy is seen as Berthelot's rule predicts the cosine of the contact angle to increase with larger γ_{lv}, contrary to the experimental patterns.

The calculated adhesion and contact angle patterns from the other combining rules generally predicted the trend of the patterns well. The Steele combining rule produced the least scatter; the (12:6) combining rule correctly predicted the local maximum in the solid–liquid work of adhesion that was observed in Figure 9.17a.

The question arises as to whether or not a more accurate combining rule could have been deduced by means of the experimental patterns. For this reason, a new

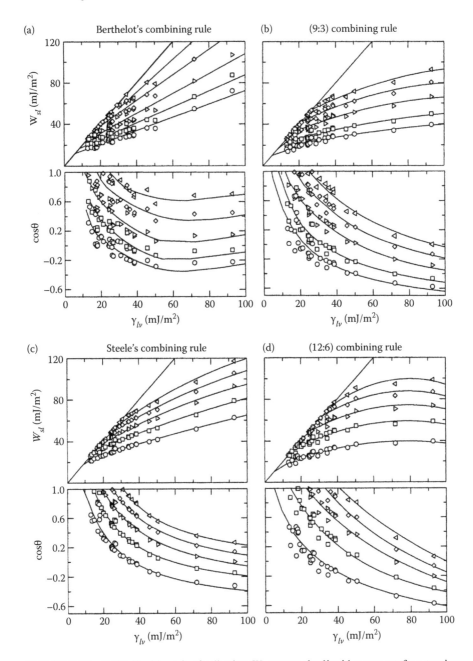

FIGURE 9.18 Solid–liquid work of adhesion W_{sl} versus the liquid–vapor surface tension γ_{lv} and cosine of the contact angle $\cos\theta$ versus γ_{lv} calculated from (a) Berthelot's rule, (b) the (9:3) combining rule, (c) Steele's rule, and (d) the (12:6) combining rule. The symbols are calculated data, and the curves are the general trends of the data points. (Reprinted from Zhang, J. and Kwok, D. Y., *Langmuir*, 19, 4666, 2003. With permission from American Chemical Society.)

combining rule has been formulated and was shown to generate a much smoother trend of $\cos\theta$ versus γ_{lv} curve, similar to the experimental trend observed [113].

Intermolecular theory based on the above generalized van der Waals description of intermolecular forces allows direct calculations of reasonable interfacial tensions and hence adhesion/contact angle patterns when accurate combining rules are used. As seen above, the same adhesion patterns can be described thermodynamically by an equation of state. While intermolecular forces obviously determine the interfacial tensions and hence the contact angle, a link between the two approaches has yet to be established.

9.6.4 LIFSHITZ THEORY

The equation of state, in either of its forms Equation 9.45 or Equation 9.58, contains one constant that is fit to experimental data. Since both forms are essentially equivalent, we restrict our attention in this closing note to Equation 9.58. In order to understand the physical origin of this equation of state, an explanation must be derived for the constant $\beta = 0.000125$ (m²/mJ)².

Molecular origins of contact angle patterns and interfacial tensions were explored in the previous sections in terms of density functional theory (Equations 9.73 and 9.74). Contact angles and interfacial tensions have also been studied using Lifshitz theory [115] (see Chapter 10). This theory accounts for only van der Waals forces.

Under certain approximations, the surface tension of a liquid or solid can be expressed in Lifshitz theory as [115]

$$\gamma = \frac{1}{24\pi d^2} \frac{3\hbar\omega}{16\sqrt{2}} \frac{(\varepsilon_0 - 1)^2}{(\varepsilon_0 + 1)^{3/2}}, \tag{9.90}$$

where d is the molecular separation, $\hbar$ is Planck's constant divided by 2π, ω is the dominant UV absorption frequency of the material, and ε_0 is its static dielectric constant.

Assuming $d = 0.165$ nm and $\omega = 1.9 \times 10^{16}$ rad/s to be relatively constant from one material to another [115], the work of adhesion between a solid and a liquid (subscripts 1 and 2) is

$$W_{12} = \frac{1}{12\pi d^2} \frac{3\hbar\omega}{8\sqrt{2}} \frac{(\varepsilon_{10} - 1)(\varepsilon_{20} - 1)}{(\varepsilon_{10} + 1)^{1/2}(\varepsilon_{20} + 1)^{1/2}} \frac{1}{(\varepsilon_{10} + 1)^{1/2} + (\varepsilon_{20} + 1)^{1/2}}. \tag{9.91}$$

As discussed above, the Berthelot or geometric mean combining rule is an approximation for intermolecular forces between unlike molecules. It corresponds to the equation of state 9.12. The exponential term in the equation of state 9.58 can be regarded as a correction to this combining rule. Equations 9.90 and 9.91 also deviate from the geometric mean combining rule. Making use of Equation 9.57 to equate these two corrections,

$$\frac{1}{2}\frac{(\varepsilon_{10}+1)^{1/2}+(\varepsilon_{20}+1)^{1/2}}{(\varepsilon_{10}+1)^{1/4}(\varepsilon_{20}+1)^{1/4}} = \exp\left[\beta(\gamma_1-\gamma_2)^2\right]. \tag{9.92}$$

Starting with any liquid and solid surface tensions, Equation 9.90 can be used to estimate values for the respective dielectric constants, which can then be substituted into Equation 9.92 to find a value for β. For various pairs of liquid and solid surface tensions in the range 10–70 mJ/m², the resulting β is typically about 2×10^{-6} (m²/mJ)², or around 60 times smaller than the experimental value (see also Figure 9.16, which shows predicted contact angles from Equation 9.91). Hence, the observed value of β cannot be understood solely in terms of the Lifshitz theory of van der Waals forces (subject to common approximations).

REFERENCES

1. T. Young. *Philosophical Transactions of the Royal Society of London* 95 (1805): 65.
2. E. I. Vargha-Butler, T. K. Zubovits, D. R. Absolom, and A. W. Neumann. *Journal of Dispersion Science and Technology* 6 (1985): 357.
3. D. W. Fuerstenau and M. C. Williams. *Colloids and Surfaces* 22 (1987): 87.
4. W. A. Zisman. In *Contact Angle, Wettability and Adhesion*. Edited by R. F. Gould, 2. Advances in Chemistry Series, No. 43, American Chemical Society, Washington, DC, 1964.
5. F. M. Fowkes. *Industrial & Engineering Chemistry* 56 (1964): 40.
6. C. J. van Oss, R. J. Good, and M. K. Chaudhury. *Langmuir* 4 (1988): 884.
7. R. J. Good and C. J. van Oss. In *Modern Approaches to Wettability: Theory and Applications*. Edited by M. Schrader and G. Loeb, 1–27. Plenum Press, New York, 1992.
8. C. J. van Oss, R. J. Good, and M. K. Chaudhury. *Journal of Colloid and Interface Science* 111 (1986): 378.
9. C. J. van Oss, L. Ju, M. K. Chaudhury, and R. J. Good. *Journal of Colloid and Interface Science* 128 (1989): 313.
10. C. J. van Oss and R. J. Good. *Journal of Macromolecular Science-Chemistry* A26 (1989): 1183.
11. D. Y. Kwok, D. Li, and A. W. Neumann. *Langmuir* 10 (1994): 1323.
12. W. Wu, R. F. Giese, Jr., and C. J. van Oss. *Langmuir* 11 (1995): 379.
13. D. Y. Kwok. *Colloids and Surfaces A* 156 (1999): 191.
14. C. Della Volpe, D. Maniglio, M. Brugnara, S. Siboni, and M. Morra. *Journal of Colloid and Interface Science* 271 (2004): 434.
15. A. Holländer. *Journal of Colloid and Interface Science* 169 (1995): 493.
16. C. J. van Oss. *Journal of Adhesion Science and Technology* 16 (2002): 669.
17. R. E. Johnson and R. H. Dettre. *Langmuir* 5 (1989): 293.
18. I. D. Morrison. *Langmuir* 7 (1991): 1833.
19. L. H. Lee. *Langmuir* 9 (1993): 1898.
20. D. Y. Kwok, Y. Lee, and A. W. Neumann. *Langmuir* 14 (1998): 2548.
21. J. Kloubek. *Langmuir* 5 (1989): 1127.
22. E. Chibowski, M. L. Kerkeb, and F. González-Caballero. *Journal of Colloid and Interface Science* 155 (1993): 444.
23. B. Jańczuk, E. Chibowski, J. M. Bruque, M. L. Kerkeb, and F. González-Caballero. *Journal of Colloid and Interface Science* 159 (1993): 421.

24. L. H. Lee. *Langmuir* 12 (1996): 1681.
25. J. C. Berg. In *Wettability*. Edited by J. C. Berg, 75. Marcel Dekker, New York, 1993.
26. M. Morra. *Journal of Colloid and Interface Science* 182 (1996): 312.
27. M. Greiveldinger and M. E. R. Shanahan. *Journal of Colloid and Interface Science* 215 (1999): 170.
28. J. K. Spelt. *Solid Surface Tension: The Equation of State Approach and the Theory of Surface Tension Components*. PhD Thesis, University of Toronto, 1985.
29. R. J. Good. *Journal of Colloid and Interface Science* 59 (1977): 398.
30. A. W. Neumann, R. J. Good, C. J. Hope, and M. Sejpal. *Journal of Colloid and Interface Science* 49 (1974): 291.
31. J. K. Spelt, D. R. Absolom, and A. W. Neumann. *Langmuir* 2 (1986): 620.
32. H. W. Fox and W. A. Zisman. *Journal of Colloid Science* 5 (1950): 514.
33. H. W. Fox and W. A. Zisman. *Journal of Colloid Science* 7 (1952): 428.
34. A. Amirfazli and A. W. Neumann. *Advances in Colloid and Interface Science* 110 (2004): 121.
35. G. E. H. Hellwig and A. W. Neumann. *5th International Congress on Surface Activity*. 687. Section B, Ediciones Unidas, Barcelona, 1969.
36. G. E. H. Hellwig and A. W. Neumann. *Kolloid-Z. Z. Polymere* 40 (1969): 229.
37. C. A. Ward and A. W. Neumann. *Journal of Colloid and Interface Science* 49 (1974): 286.
38. D. Li, J. Gaydos, and A. W. Neumann. *Langmuir* 5 (1989): 1133.
39. A. Münster. *Classical Thermodynamics*. Translated by E. S. Halberstadt. John Wiley, Toronto, 1970.
40. D. Li and A. W. Neumann. *Journal of Colloid and Interface Science* 137 (1990): 304.
41. J. W. Whalen and J. W. Wade. *Journal of Colloid and Interface Science* 24 (1967): 372.
42. R. J. Good. In *Adsorption at Interfaces*. Edited by K. L. Mittal. A.C.S. Symposium Series, No. 8, American Chemical Society, Washington, DC, 1975.
43. O. Driedger, A. W. Neumann, and P. J. Sell. *Kolloid-Z.Z. Polymere* 201 (1965): 52.
44. O. Driedger, A. W. Neumann, and P. J. Sell. *Kolloid-Z.Z. Polymere* 204 (1965): 101.
45. A. W. Neumann. In *Wetting, Spreading and Adhesion*. Edited by J. F. Padday, 3. Academic Press, New York, 1978.
46. A. W. Neumann, D. R. Absolom, D. W. Francis, and C. J. van Oss. *Separation and Purification Methods* 9 (1980): 69.
47. D. Li and A. W. Neumann. *Journal of Colloid and Interface Science* 148 (1992): 190.
48. A. Dupré. *Théorie Méchanique de la Chaleur*. 369. Gauthier-Villars, Paris, 1969.
49. L. A. Girifalco and R. J. Good. *Journal of Physical Chemistry* 61 (1957): 904.
50. G. C. Maitland, M. Rigby, E. B. Smith, and W. A. Wakeham. *Intermolecular Forces: Their Origin and Determination*. Clarendon Press, Oxford, 1981.
51. K. C. Chao and R. L. Robinson, Jr. *Equations of State: Theories and Applications*. American Chemical Society, Washington, 1986.
52. J. N. Israelachvili. *Proceedings of the Royal Society of London A* 331 (1972): 39.
53. J. Kestin and E. A. Mason. *AIP Conference Proceedings* 11 (1973): 137.
54. R. P. Smith, D. R. Absolom, J. K. Spelt, and A. W. Neumann. *Journal of Colloid and Interface Science* 110 (1986): 521.
55. D. Y. Kwok and A. W. Neumann. *Advances in Colloid and Interface Science* 81 (1999): 167.
56. D. Li and A. W. Neumann. *Advances in Colloid and Interface Science* 39 (1992): 299.
57. H. Tavana and A. W. Neumann. *Advances in Colloid and Interface Science* 132 (2007): 1.
58. A. W. Neumann, J. K. Spelt, R. P. Smith, D. W. Francis, Y. Rotenberg, and D. R. Absolom. *Journal of Colloid and Interface Science* 102 (1984): 278.
59. S. N. Omenyi, A. W. Neumann, and C. J. van Oss. *Journal of Applied Physics* 52 (1981): 789.

60. J. N. Israelachvili and G. E. Adams. *Journal of the Chemical Society-Faraday Transactions* 1, no. 74 (1978): 975.
61. R. M. Pashley, P. M. McGuiggan, B. W. Ninham, and D. F. Evans. *Science* 229 (1985): 1088.
62. H. K. Christenson. *Journal of Physical Chemistry* 90 (1986): 4.
63. P. M. Claesson, C. E. Blom, P. C. Herder, and B. W. Ninham. *Journal of Colloid and Interface Science* 114 (1986): 234.
64. R. M. Pashley, P. M. McGuiggan, R. G. Horn, and B. W. Ninham. *Journal of Colloid and Interface Science* 126 (1988): 569.
65. P. M. McGuiggan and R. M. Pashley. *Colloids and Surfaces* 27 (1987): 277.
66. F. M. Fowkes. *Journal of Adhesion Science and Technology* 1 (1987): 7.
67. E. Moy. *Approaches for Estimating Solid/Liquid Interfacial Tensions*. PhD Thesis, University of Toronto, 1989.
68. D. R. Uhlmann, B. Chalmers, and K. A. Jackson. *Journal of Applied Physics* 35 (1964): 2986.
69. S. N. Omenyi and A. W. Neumann. *Journal of Applied Physics* 47 (1976): 3956.
70. J. K. Spelt, R. P. Smith, and A. W. Neumann. *Colloids and Surfaces* 28 (1987): 85.
71. E. Moy and A. W. Neumann. *Colloids and Surfaces* 43 (1990): 349.
72. E. I. Vargha-Butler, E. Moy, and A. W. Neumann. *Colloids and Surfaces* 24 (1987): 315.
73. E. I. Vargha-Butler, T. K. Zubovits, M. K. Weibel, D. R. Absolom, and A. W. Neumann. *Colloids and Surfaces* 15 (1985): 233.
74. S. N. Omenyi, R. S. Snyder, D. R. Absolom, C. J. van Oss, and A. W. Neumann. *Journal of Dispersion Science and Technology* 3 (1982): 307.
75. A. W. Neumann, S. N. Omenyi, and C. J. van Oss. *Journal of Colloid and Polymer Science* 257 (1979): 413.
76. A. W. Neumann, S. N. Omenyi, and C. J. van Oss. *Journal of Physical Chemistry* 86 (1982): 1267.
77. A. W. Neumann, D. R. Absolom, D. W. Francis, S. N. Omenyi, J. K. Spelt, Z. Policova, C. Thomson, W. Zingg, and C. J. van Oss. *Surface Phenomena in Hemorheology*. Edited by A. L. Copley and G. V. F. Seaman. New York Academy of Sciences, New York, 1983.
78. J. K. Spelt, D. R. Absolom, W. Zingg, C. J. van Oss, and A. W. Neumann. *Cell Biophysics* 4 (1982): 117.
79. D. R. Absolom, Z. Policova, E. Moy, W. Zingg, and A. W. Neumann. *Cell Biophysics* 7 (1985): 267.
80. W. D. Harkins and G. Jura. *Journal of the American Chemical Society* 66 (1944): 1362.
81. A. W. Neumann. *Advances in Colloid and Interface Science* 4 (1974): 105.
82. J. J. Jasper and E. V. Kring. *Journal of Physical Chemistry* 59 (1955): 1019.
83. A. W. Neumann. *Deut. Akad. Wiss. Berlin, Kl. Chem., Geol. Biol.* 6b (1966): 811.
84. B. B. Sauer and G. T. Dee. *Macromolecules* 27 (1994): 6112.
85. A. W. Neumann and P. J. Sell. *Zeitschrift fur Physikalische Chemie (Frankfurt)* 65 (1969): 19.
86. F. M. Fowkes. *Society of the Chemical Industry of London, Monograph* 3. Society of Chemical Industry, London, 1967.
87. J. F. Oliver, C. Huh, and S. G. Mason. *Colloids and Surfaces* 1 (1980): 79.
88. F. M. Fowkes, F. L. Riddle, Jr., W. E. Pastore, and A. A. Weber. *Colloids and Surfaces* 43 (1990): 367.
89. F. M. Fowkes. *Journal of Adhesion* 4 (1972): 155.
90. D. Li, M. Xie, and A. W. Neumann. *Colloid and Polymer Science* 271 (1993): 573.
91. A. F. M. Barton. *CRC Handbook of Solubility Parameters and Other Cohesion Parameters*. CRC Press, Boca Raton, FL, 1983.

92. A. Beerbower. *Journal of Colloid and Interface Science* 35 (1971): 126.
93. D. B. Hough and L. R. White. *Advances in Colloid and Interface Science* 14 (1980): 3.
94. R. C. Weast. *Handbook of Chemistry and Physics.* 63rd ed. CRC Press, Boca Raton, FL, 1982.
95. J. K. Spelt, Y. Rotenberg, D. R. Absolom, and A. W. Neumann. *Colloids and Surfaces* 24 (1987): 127.
96. Y. Rotenberg, L. Boruvka, and A. W. Neumann. *Journal of Colloid and Interface Science* 93 (1983): 169.
97. D. Y. Kwok, D. Li, and A. W. Neumann. *Colloids and Surfaces A* 89 (1994): 181.
98. A. E. van Giessen, D. J. Bukman, and B. Widom. *Journal of Colloid and Interface Science* 192 (1997): 257.
99. J. F. Zhang and D. Y. Kwok. *Journal of Physical Chemistry B* 106 (2002): 12594.
100. D. E. Sullivan. *Physical Review B* 20 (1979): 3991.
101. N. F. Carnahan and K. E. Starling. *Physical Review A* 1 (1970): 1672.
102. D. E. Sullivan. *Journal of Chemical Physics* 74 (1981): 2604.
103. T. M. Reed. *Journal of Physical Chemistry* 59 (1955): 425.
104. T. M. Reed. *Journal of Physical Chemistry* 59 (1955): 428.
105. G. H. Hudson and J. C. McCoubrey. *Transactions of the Faraday Society* 56 (1960): 761.
106. B. E. F. Fender and G. D. Halsey, Jr. *Journal of Chemical Physics* 36 (1962): 1881.
107. D. V. Matyushov and R. Schmid. *Journal of Chemical Physics* 104 (1996): 8627.
108. J. S. Rowlinson and F. L. Swinton. *Liquids and Liquid Mixtures.* Butterworth Scientific, London, 1981.
109. W. A. Steele. *The Interaction of Gases with Solid Surfaces.* Pergamon Press, New York, 1974.
110. D. Berthelot. *Comptes Rendus* 126 (1898): 1857.
111. W. A. Steele. *Surface Science* 36 (1973): 317.
112. J. E. Lane and T. H. Spurling. *Australian Journal of Chemistry* 29 (1976): 8627.
113. D. Y. Kwok and A. W. Neumann. *Journal of Physical Chemistry B* 104 (2000): 741.
114. D. Y. Kwok, H. Ng, and A. W. Neumann. *Journal of Colloid and Interface Science* 225 (2000): 323.
115. J. N. Israelachvili. *Intermolecular and Surface Forces.* 2nd ed. Academic Press, London, 1991.

10 Theoretical Approaches for Estimating Solid–Liquid Interfacial Tensions

Elio Moy, Robert David, and A. Wilhelm Neumann

CONTENTS

10.1 INTRODUCTION

Solid–liquid interfacial tensions, γ_{sl}, play an important role in a wide range of problems in the fields of pure and applied science. Interfacial phenomena are responsible for the behavior and properties of commonly used materials, for example paints, adhesives, detergents, and lubricants. Also, the general patterns of the adhesion exhibited by biological cells to materials such as glass, metals, and polymers cannot be attributed to specific chemical interactions; rather, they are related to the change in the overall free energy for the process of adhesion, which, in turn, is dependent on the surface and interfacial tensions relevant to the process. Thus, it is not surprising to find that thermodynamic models based on interfacial tensions have been used to explain many phenomena such as cell adhesion [1,2], sedimentation [3], and the behavior of particles at solidification fronts [4].

In the field of materials science, the interfacial tension between a solid and its melt dictates to a large extent the temperatures at which solids nucleate from their liquids [5,6]. The solid–melt interfacial tension is often important in the determination of the morphology of growth, and may also lead to solidification taking place in preferred crystallographic orientations [7]. When liquid metals migrate to surface cracks in solids, γ_{sl} replaces the solid surface tension, γ_{sv}, as the effective energy of cohesion for the solid. The value of γ_{sl} will usually be considerably lower than γ_{sv} so that the strength of the solid is drastically reduced, thereby causing surface embrittlement [8].

Unlike liquid–fluid interfacial tensions, interfacial tensions involving a solid phase are difficult to measure directly. The most common approach for estimating these tensions involves the interpretation of contact angle data [9], which has been discussed in detail in Chapters 8 and 9. The approaches favored by materials science groups involve the direct or indirect application of the so-called Gibbs–Thomson equation [10]. These two approaches are unfortunately in considerable disagreement, as will be shown in Section 10.3. The values of γ_{sl} obtained by the two approaches, for the same systems, differ by almost two orders of magnitude.

The central purpose of this chapter is to determine what approach—contact angle interpretation or implementation of the Gibbs–Thomson equation—provides the best estimates of solid–liquid interfacial tensions. In order to do so, two independent theoretical approaches for estimating γ_{sl} are investigated: gradient theory and Lifshitz theory.

10.2   CONTACT ANGLE INTERPRETATION

The attractiveness of using contact angle data to estimate γ_{sl} and γ_{sv} arises from the fact that contact angles can be measured with relative ease on suitably prepared surfaces. In order to use contact angle data to estimate γ_{sl} and γ_{sv}, in addition to the Young equation

$$\gamma_{sv} - \gamma_{sl} = \gamma_{lv} \cos\theta, \tag{10.1}$$

an equation involving the three interfacial tensions (γ_{sl}, γ_{sv}, and γ_{lv}) is required.

The numerous interfacial tension equations that are found in the literature can be divided into two groups: (i) those based on the surface tension components approach; and (ii) those based on the equation of state approach. The existence of an equation of state for interfacial tensions and the validity of the surface tension components approach were discussed in detail in Chapters 8 and 9. While the differences in the calculated values of γ_{sl} and γ_{sv} from the two approaches may be as large as 100%, they are small compared with the differences between results from contact angle interpretation and from Gibbs–Thomson equation considerations, which are orders of magnitude.

In Chapters 8 and 9, it has been shown that the equation of state approach is superior for interpreting contact angle data on solid sufaces. Therefore, in this chapter, the equation of state, Equation 10.2, will be used to calculate γ_{sl} and γ_{sv} from contact angle considerations. Whenever possible, the values that are obtained from the surface tension components approach will also be included for comparative purposes. The equation of state for interfacial tensions can be written as (cf. Chapter 9)

$$\gamma_{sl} = \gamma_{lv} + \gamma_{sv} - 2\sqrt{\gamma_{lv}\gamma_{sv}}\,e^{-\beta(\gamma_{lv}-\gamma_{sv})^2}, \tag{10.2}$$

where $\beta = 0.000125 \ (\mathrm{m^2/mJ})^2$.

10.3   GIBBS–THOMSON EQUATION

The Gibbs–Thomson equation is widely used in the field of materials science for estimating solid–melt interfacial tensions. It is a relative of the Kelvin equation for liquid–vapor interfaces that links droplet curvature to vapor pressure.

The Gibbs–Thomson equation gives the relation between the change in equilibrium temperature and the curvature of the interface, for pure substances, at a constant external pressure. It shows that, for a given external pressure, the equilibrium temperature of a liquid droplet decreases with decreasing drop size. The Gibbs–Thomson equation can be used to explain the phenomenon of supercooling of a saturated vapor and of a saturated liquid. The extent of supercooling of a saturated liquid is the basis for obtaining γ_{sl} from nucleation experiments, which will be discussed later in this section.

The Gibbs–Thomson equation is deduced by combining the Laplace equation of capillarity and the condition of chemical equilibrium through the Gibbs–Duhem relations. The Gibbs–Thomson equation was originally developed for liquid–fluid systems and later applied to solid–liquid systems. The complete derivation of the original Gibbs–Thomson equation can be found in the book by Defay and Prigogine [11]. The simplified version of this equation, which is the one widely used in the field of materials science, is

$$\Delta T = \frac{2\gamma_{sl}}{r}\frac{T_0 v_s}{\Delta H_f}, \tag{10.3}$$

where $\Delta T = T_0 - T$ represents the degree of undercooling, v_s is the molar volume of the solid, r is the radius of curvature, and ΔH_f is the latent heat of fusion.

In the remainder of this section, the use of Equation 10.3 to determine γ_{sl}, directly and indirectly, will be described.

10.3.1 INDIRECT APPROACHES: HOMOGENEOUS
NUCLEATION AND MELTING IN PORES

Most of the older published values of solid–melt interfacial tensions of pure metals were obtained by means of homogeneous nucleation experiments first performed by Turnbull and coworkers [12,13]. In these experiments, the homogeneous nucleation rate, J, is measured as a function of the degree of undercooling $\Delta T = T_0 - T$.

The parameters J and ΔT are related via Boltzmann statistics and the Gibbs–Thomson equation (Equation 10.3) by

$$ J = K_v \exp\left(-\frac{16\pi}{3\kappa T} \frac{\gamma_{sl}^3 T_0^2 v_s^2}{\Delta H_f \Delta T^2} \right), \tag{10.4} $$

where K_v is an experimentally obtained constant, and κ is Boltzmann's constant. A plot of $\ln J$ versus $1/T\Delta T^2$ allows the determination of γ_{sl} from the slope of the straight line. Experimentally, the principal difficulty of the method is ensuring that true homogeneous nucleation occurs. Theoretically, the expected degree of undercooling can be several hundred degrees [14]; however, experimentally, undercooling of more than a few degrees is rarely attained, indicating that heterogeneous nucleation is taking place. The major difference between homogeneous and heterogeneous nucleation is that in the former the nucleus is formed via the aggregation of a number of molecules of the pure substance from the melt, while in the latter the formation of the nucleus is induced by a foreign material that is in contact with the melt.

The major difficulty in interpreting the data from nucleation experiments results from applying macroscopic thermodynamics to systems that are typically 1 nm in size and contain only a few hundred molecules [10,15]. Also, in situations where true homogeneous nucleation is apparently observed, the estimates of γ_{sl} are obtained at temperatures that are as much as 200 K below T_0. In order to relate these values to those expected at T_0, the temperature coefficient of γ_{sl} is necessary. These coefficients are, however, as yet unknown. Values of γ_{sl} for various materials, obtained by the homogeneous nucleation method, are given in Table 10.1. Typical results from nucleation experiments are those for n-alkanes by Uhlmann et al. [16].

A second indirect approach for determining γ_{sl} from Equation 10.3 is by calorimetric measurement of ΔH_f and T_0 for liquids confined in porous solids. From data for solids of different pore radii, γ_{sl} can be found via the slope of a plot of ΔT versus $1/r$. Using this method, Jackson and McKenna [17] measured values of γ_{sl} for seven organic liquids confined in controlled pore glass. Their values were smaller than those from other methods based on the Gibbs–Thomson equation (their result for naphthalene is included in Table 10.1). Uncertainties in their results were mainly due to lack of knowledge of the liquid/solid structure within the pores, and whether bulk modeling concepts were valid at a scale of tens of nanometers. For example, the authors found significant changes in the enthalpy of fusion ΔH_f in smaller pores.

TABLE 10.1
Comparison of Solid-Melt Interfacial Tensions (mJ/m²) Obtained by Gibbs–Thomson Equation Approaches and by Contact Angle Considerations

| System | Gibbs–Thomson Approaches | | | | Contact Angle Approaches | |
	Nucleation	Melting in Pores	Skapski	Grain Boundary	Equation of State	Fowkes' Equation
Ice water	26.1	—	120 ± 10 [19]	29.1 ± 0.8	0.38 [27]	—
Naphthalene	> 27.2	8.2 [17]	—	61 ± 11	1.44 [26]	0.10
Biphenyl	> 24.0	—	—	50 ± 10	0.67 [26]	0.01
n-Dodecane	12.1 [14]	—	—	—	0.15 [28]	0.09
n-Hexadecane	15.2 [14]	—	—	—	0.89 [28]	0.57

Note: Values for Gibbs–Thomson approaches were obtained from Jones, D. R. H., *Journal of Materials Science* 9, 1, 1974 unless otherwise stated. Skapski's result was incorrectly listed in Jones' reference.

10.3.2 DIRECT APPROACHES: SKAPSKI'S METHOD AND ANALYSIS OF GRAIN BOUNDARY GROOVES

The direct evaluation of γ_{sl} via the Gibbs–Thomson equation (Equation 10.3) entails equilibrium experiments. These experiments involve the analysis of the equilibrium shape of the solid–liquid interface for a system whose temperature is maintained at a value below the normal melting temperature, T_0.

The first successful attempts at measuring γ_{sl} directly from the Gibbs–Thomson equation were due to Skapski and coworkers [18–20]. A schematic of the apparatus used by Skapski et al. is shown in Figure 10.1. The experiments were performed at temperatures of 0.02–0.8 ± 0.002 K below the melting temperature of the materials. A wedge geometry was used to achieve equilibrium conditions and to observe the interfacial curvatures in transparent systems.

The equilibrium position of the solid–liquid interface for a given degree of under-cooling is given by the Gibbs–Thomson equation, now modified to account for the nonspherical geometry of the interface:

$$\Delta T = \frac{\gamma_{sl} T_0 v_s}{\Delta H_f}\left(\frac{1}{r_1}+\frac{1}{r_2}\right),\tag{10.5}$$

where r_1 and r_2 are the principal radii of curvature. For the geometry depicted in Figure 10.1, $r_1 = h/2$ and $r_2 = \infty$. Thus, by knowing the position of the solid–liquid interface, it is possible to obtain γ_{sl} directly from Equation 10.5. This method was

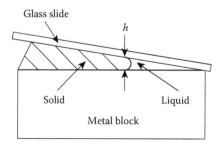

FIGURE 10.1 Schematic of Skapski's experimental setup for the determination of solid–melt interfacial tensions.

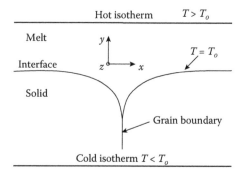

FIGURE 10.2 Schematic diagram of the profile of a solid–liquid interface that is intersected by a grain boundary groove.

used to study various transparent materials. The result for ice-water is given in Table 10.1.

The most attractive method of measuring γ_{sl} directly, for macroscopic systems, involves the observation of equilibrium shapes of grooves formed by the intersection of planar grain boundaries with an otherwise planar solid–liquid interface in a system that is subjected to a temperature gradient. The advantage of this method over the Skapski approach is that the whole grain boundary groove is used in the determination of γ_{sl} whereas in the latter, only the wedge thickness is used in the calculation of γ_{sl} via Equation 10.5.

The required geometry of the grain boundary groove is shown in Figure 10.2. Far from the groove, the solid–liquid interface is planar and is therefore in equilibrium at the normal melting temperature, T_0. Closer to the root of the groove, the curvature of the interface will increase steadily so as to compensate for the decreasing interfacial temperature imposed by the temperature gradient, according to Equation 10.5. At each point on the interface, the local curvature is $1/r_1$, where r_1 is the principal radius of curvature in the x-y plane (Figure 10.2) since it is assumed that the interface is invariant in the z-direction ($r_2 = \infty$).

The Gibbs–Thomson equation, for the geometry depicted in Fig. 10.2, has been solved analytically by Bolling and Tiller [7] by rewriting Equation 10.5 as a differential equation and assuming a linear temperature gradient. For a linear temperature gradient, G, along the y-axis, $\Delta T = Gy$ and Equation 10.5 can be rewritten as

$$y = \left(\frac{\gamma_{sl} T_0 v_s}{G \Delta H_f} \right)^2 y'' \left(1 + y'^2 \right)^{-3/2}, \tag{10.6}$$

where the primes denote differentiation with respect to x.

The assumption of a linear temperature gradient applies strictly to systems where the thermal conductivities of the solid and the melt are the same. A numerical solution to Equation 10.5 for situations where the thermal conductivities are different has been developed by Nash and Glicksman [21]. The grain boundary groove shape analysis technique has been applied to a variety of transparent materials [22,23]. A series of more recent measurements has also been published (see, e.g., Akbulut et al. [24]). Some results obtained by this technique are given in Table 10.1.

10.3.3 Comparison of Solid–Liquid Interfacial Tension Values

In Table 10.1, the estimated values of γ_{sl}, obtained from contact angle considerations and from the application of the Gibbs–Thomson equation, are compared. Because the contact angle approach is limited to low-energy systems ($\gamma < 100$ mJ/m^2) and the approaches based on the Gibbs–Thomson equation are limited to solid–melt systems, there are only a few cases for which the values of γ_{sl} can be compared.

The use of the term "contact angle" is slightly misleading in the context of solid–melt systems. While contact angles of the melts on the solids of the same materials are, in principle, measurable if γ_{lv} is larger than γ_{sv}, they are not essential in the determination of γ_{sl} for solid–melt systems. Experimentally, the simplest way of calculating γ_{sl} for such systems is to determine γ_{lv} and γ_{sv} independently and to use Equation 10.2 to determine γ_{sl} (cf. Chapter 9).

The value of γ_{lv} can be measured directly by the Wilhelmy plate method [25] or by means of Axisymmetric Drop Shape Analysis-Profile (ADSA-P, Chapter 3), at different temperatures so that γ_{lv} at the melting temperature can be obtained by extrapolation. The value of γ_{sv} can be obtained by measuring contact angles (of liquids other than the melt) as a function of temperature, allowing the estimation of γ_{sv} at the melting point, also by extrapolation. The values shown in Table 10.1 were obtained in this fashion by Omenyi et al. [26].

From the values of γ_{sl} shown in Table 10.1, it is readily apparent that the Gibbs–Thomson and contact angle approaches are in considerable disagreement. The values from contact angle considerations are about two orders of magnitude smaller than those from Gibbs–Thomson considerations. The differences in γ_{sl} values obtained from the application of the Fowkes equation and from the equation of state are relatively small when compared with those from the Gibbs–Thomson approaches.

The question that arises from Table 10.1 is what approach correctly estimates solid–liquid interfacial tensions. In the next sections, independent approaches for estimating γ_{sl} will be discussed: gradient theory and Lifshitz theory. It will be shown that the two independent approaches estimate γ_{sl} values that are consistent with those from contact angle considerations.

At this point, some physical implications of the values presented in Table 10.1 can be discussed. The γ_{sl} values for naphthalene and biphenyl, as obtained by the grain boundary groove method, are numerically larger than the corresponding γ_{lv} values. This is physically unrealistic. Interfacial tensions are manifestations of intermolecular interactions between bulk phases separated by an interface. The more different the two bulk phases are, the higher the interfacial tension should be; thus, a value of γ_{sl} that is larger than γ_{lv} implies that, improbably, there is a greater difference between the interactions in two condensed phases of the same material (solid and liquid) than there is between a condensed (liquid) and a vapor phase. While phase changes bring about surface tension changes, these changes should not be as drastic as the results from the grain boundary groove method seem to suggest.

10.4 THEORETICAL ESTIMATIONS OF SOLID–LIQUID INTERFACIAL TENSIONS

10.4.1 MICROSCOPIC APPROACH TO INTERFACES: GRADIENT THEORY

A system consisting of two fluids in contact, for example, a liquid and its vapor, must be described in terms of three regions: two bulk phases corresponding to the liquid and vapor, and a region between the bulk phases whose properties are different from either phase. These properties must, however, become equal to those of the bulk phases at the extremities of the interface layer. There are two ways in which to define the interfacial layer: (i) the macroscopic approach originally proposed by Gibbs [29], which replaces the interfacial region by a two-dimensional layer of zero thickness and of uniform properties; or (ii) a microscopic approach where the interfacial region is treated as having a finite volume and where the properties are continuously varying throughout the thickness. The common feature of both approaches is that surface quantities such as surface tension, internal energy, and entropy are definable. The macroscopic approach for interfaces, the generalized theory of capillarity, was the topic of Chapter 1. In this section, the microscopic approach will be discussed using the concepts of the gradient theory of van der Waals, and Cahn and Hilliard [30]. Gradient theory has been used by other workers to determine the density profile and surface tension of liquid drops [31–33], to study wetting transitions [34], and to study the structure of the solid–liquid–gas contact region [35]. In this section, gradient theory analysis will be used to calculate solid–liquid interfacial tensions.

From a molecular point of view, the interfacial region between two contiguous bulk phases in equilibrium is a zone of finite thickness (approximately 1 nm). Within this region, the density, composition, and pressure tensor vary rapidly from the characteristics of one bulk phase to the characteristics of the other phase. It will be assumed that gravity will have negligible effects on the bulk and interfacial regions, other than ordering the phases according to their density.

A microscopic formulation of the surface tension, from a mechanical point of view, can be obtained by means of the varying pressure tensor in the interfacial region as proposed by Kirkwood and Buff [36]. In order to clarify this concept, the equation of hydrostatics is used. This equation can be obtained from the momentum Equation 10.7, written using the Einstein summation convention:

$$\rho \frac{\partial u_j}{\partial t} + \rho u_k \frac{\partial u_j}{\partial u_k} = -\frac{\partial P_{ij}}{\partial x_i} + \rho f_j, \tag{10.7}$$

where ρ is the density, u_j are the components of the velocity vector, P_{ij} are the elements of the pressure tensor, and f_j are the components of body forces per unit mass. For a static system, the velocity vector $\mathbf{u}$ is equal to zero, resulting in the equation of hydrostatics:

$$-\frac{\partial P_{ij}}{\partial x_i} + \rho f_j = 0. \tag{10.8}$$

For a horizontal interface between two isotropic fluid bulk phases, the lighter phase will be on top because of gravity. Thus, Equation 10.8, in the absence of external forces other than gravity, reduces to

$$\frac{\partial P_{ij}}{\partial x_i} = 0. \tag{10.9}$$

For a planar geometry, the fluid density and pressure tensor in the interfacial region depend only on the distance x perpendicular to the flat interface. The equation of hydrostatics can therefore be expressed as

$$\frac{dP_{xv}}{dx} = 0, \qquad v = x, y, z. \tag{10.10}$$

Equation 10.10 implies that P_{xx}, P_{xy}, and P_{xz} are constant in a planar system. Far below and above the interface, the fluids are homogeneous and isotropic such that the pressure tensor is simply $\mathbf{P} = P_B \mathbf{I}$ where P_B is the bulk equilibrium thermodynamic pressure and $\mathbf{I}$ is the unit tensor. Thus, Equation 10.10 leads to the conclusion that $P_{xx} = P_B$ and $P_{xy} = P_{xz} = 0$; that is, the pressure tensor component normal to the interface is equal to the thermodynamic bulk pressure, P_B, and remains constant throughout the whole interfacial region.

The condition of symmetry of the pressure tensor at equilibrium implies that $P_{yx} = P_{zx} = 0$. Also, assuming that the interfacial region is isotropic and homogeneous in the transverse y-z direction, the fluid is isotropic in every layer parallel to the interface, and the final restrictions on the pressure tensor are obtained: $P_{yz} = P_{zy} = 0$, and $P_{yy} = P_{zz}$.

Thus, the planar interface has two distinct principal pressures: a normal one, $P_N = P_{xx}$; and a transverse one, $P_T = P_{yy} = P_{zz}$. The normal component remains constant throughout the interfacial region and is equal to the bulk thermodynamic pressure, P_B, and the transverse pressure can vary throughout the interfacial region but can only be a function of x. The pressure tensor can therefore be written as

$$\mathbf{P} = \mathbf{P}(x) = \begin{bmatrix} P_N & 0 & 0 \\ 0 & P_T(x) & 0 \\ 0 & 0 & P_T(x) \end{bmatrix}. \tag{10.11}$$

Because $\mathbf{P} = P_B \mathbf{I}$ above and below the interfacial region, it follows that $P_T = P_N$ outside of the interfacial region.

The pressure deficit in the interfacial layer $(P_N - P_T)$ manifests itself as the surface tension, γ. The relation between γ and the pressure deficit is

$$\gamma = \int_{-\infty}^{+\infty} (P_N - P_T) dx. \tag{10.12}$$

The limits of integration of Equation 10.12 are $\pm\infty$, instead of the thickness of the interfacial region, for the simple reason that the integrand is zero outside the interfacial region since $P_N = P_T$ in the bulk phases.

Thus, from a microscopic point of view, if the variation of the pressure tensor, $\mathbf{P}$, across the interfacial region is known, then the interfacial tension can be determined from Equation 10.12. The gradient theory model of the interface will be used to determine the pressure tensor, $\mathbf{P}(x)$, within the interfacial region. Gradient theory will first be developed for fluid–fluid systems with flat interfaces without any external potentials. In gradient theory, the symbol used for density is n, which represents the number of molecules per unit volume, as opposed to the conventional symbol ρ, which represents the number of moles per unit volume.

The pressure tensor $\mathbf{P}$, as will be shown below, is a function of the equilibrium density profile of the fluid across the interface. It is to be expected that if a solid wall is introduced to a fluid–fluid system, the equilibrium density profile for the new system will change. The wall will attract liquid molecules thereby increasing the liquid density above that of the bulk, near the solid–fluid interface. However, far away from the wall, the density of the fluid should not be affected. The difference between the equilibrium densities of the fluid–fluid system with and without the wall gives rise to solid–liquid interfacial tension. Therefore, in gradient theory, extension to solid–liquid (from fluid–fluid) systems is done by simply adding to the fluid–fluid system an external potential to model the solid–fluid interaction.

The pressure tensor, $\mathbf{P}$, in the interfacial region depends not only on the local density but also on the density distribution in the vicinity of $\mathbf{r}$. In gradient theory, it is assumed that the local pressure tensor is a function of the local density plus a finite number of local derivatives of the density. That is, the pressure tensor, $\mathbf{P}$, is of the form

$$\mathbf{P} = \mathbf{P}(n, \nabla n, \nabla\nabla n, \ldots). \tag{10.13}$$

It has been shown [37] that interfacial properties can be estimated accurately by keeping only the first two derivatives of n. A Taylor expansion of $\mathbf{P}$ about $\nabla n = \nabla\nabla n = 0$ yields, through second order in ∇,

$$\mathbf{P}(\mathbf{r}) = P_0 \mathbf{I} + l_{11} \nabla^2 n \mathbf{I} + l_{12} \nabla\nabla n + l_{21} (\nabla n)^2 \mathbf{I} + l_{22} \nabla n \nabla n, \tag{10.14}$$

where $P_0(n)$ is the homogeneous pressure of the fluid at density n, and l_{ij} are functions of the local density and are related to the correlation functions of the homogeneous fluids [38]. For a planar interface, the number density n is a function of x, the direction normal to the interface. The normal and transverse pressures for such interfaces are, from Equation 10.14,

$$P_N = P_0(n) + (l_{11} + l_{12}) \frac{d^2 n}{dx^2} + (l_{21} + l_{22}) \left(\frac{dn}{dx} \right)^2, \tag{10.15}$$

and

$$P_T = P_0(n) + l_{11} \frac{d^2 n}{dx^2} + l_{21} \left(\frac{dn}{dx} \right)^2. \tag{10.16}$$

Computing the interfacial tension from Equation 10.12 using Equations 10.15 and 10.16 results in

$$\gamma = \int_{-\infty}^{+\infty} \left(l_{12} \frac{d^2 n}{dx^2} + l_{22} \left(\frac{dn}{dx} \right)^2 \right) dx. \tag{10.17}$$

Integrating Equation 10.17 by parts, and knowing that the density gradients vanish as $x \to \pm\infty$ (i.e., outside the interface),

$$\gamma = \int_{-\infty}^{+\infty} c \left(\frac{dn}{dx} \right)^2 dx, \tag{10.18}$$

where the influence parameter c [38] is defined as

$$c = l_{22} - \frac{\partial l_{12}}{\partial n}. \tag{10.19}$$

Equation 10.18, first obtained by van der Waals, shows that the smaller the density gradient between the phases, the lower the interfacial tension. Thus, fluids near the critical point, where the density difference between the liquid and vapor phases is small, will have lower surface tensions. The parameter c is obtainable from properties of the homogeneous fluid [38], as will be discussed later.

In order to calculate the interfacial tension via Equation 10.18, it is necessary to determine the equilibrium density profile, $n(x)$, and the influence parameter, c, for the system. So far, only mechanical equations for the interfacial region have been discussed. Surface tension, from Equation 10.12, has been defined in terms of mechanical quantities apart from any thermodynamic considerations. Thermodynamic laws provide the criteria for determining the equilibrium density profile. In gradient theory, it is assumed that the entropy function exists for interfacial regions and that the Second Law of Thermodynamics is the same regardless of whether the system is homogeneous or nonhomogeneous. That is, for any system in equilibrium, the entropy must be a maximum.

For a system in thermal equilibrium subject to the constraint of a fixed number of molecules in the interfacial region, the determination of the equilibrium density profile is best formulated in terms of the grand canonical potential. For a homogeneous bulk phase, the grand canonical potential, or free energy, is defined as

$$\Omega = U - TS - \mu_E N = F - \mu_E N, \tag{10.20}$$

where the definition of the Helmholtz potential, $F = U - TS$, has been used, and μ_E is the constant equilibrium chemical potential (Lagrange multiplier) resulting from the constraint of a fixed number of molecules in the interfacial region. For the nonhomogeneous region, the free energy for a three-dimensional system will be given by the definite integral of the local specific free energy, ω:

$$\Omega = \int_V \omega(\mathbf{r}) d^3 r, \tag{10.21}$$

or, in terms of the specific Helmholtz potential f,

$$\Omega = \int_V [f(\mathbf{r}) - \mu_E n(\mathbf{r})] d^3 r. \tag{10.22}$$

Equation 10.22 is the form of the free energy equation that will be used since the gradient approximation to the specific Helmholtz potential, $f(\mathbf{r})$, has been derived independently by Cahn and Hilliard [30], Davis and Scriven [38], and Yang et al. [39].

Analogously to what was done for the pressure tensor, the function $f(\mathbf{r})$ in the interfacial region is obtained by assuming that $f(\mathbf{r})$ is a function of the local density, $n(\mathbf{r})$, and the local derivatives of $n(\mathbf{r})$. Expanding about the homogeneous state, the Cahn-Hilliard form of the local Helmholtz potential, which is correct to third order in gradients [30], is given by

$$f(n(\mathbf{r})) = f_0(n) + \frac{1}{2}c(\nabla n)^2, \qquad (10.23)$$

where $f_0(n)$ is the specific Helmholtz function at density n, and c is the influence parameter that also appears in Equations 10.18 and 10.19 and is also a function of the density n. The first- and third-order terms in gradients vanish because of the isotropy of the homogeneous fluid. Thus, the free energy, Ω, can be expressed as

$$\Omega = \int_V \left[f_0(n) + \frac{1}{2}c(\nabla n)^2 - \mu_E n \right] d^3r. \qquad (10.24)$$

For the special case of a planar interface, the number density, n, is only a function of x; that is, $n = n(x)$. For a system of cross-sectional area A, Equation 10.24 becomes

$$\Omega = A \int_a^b \left[f_0(n) + \frac{1}{2}c\left(\frac{dn}{dx}\right)^2 - \mu_E n \right] dx, \qquad (10.25)$$

where a and b are the boundaries of the interfacial region. The equilibrium density profile for a flat interface will be such that the integral in Equation 10.25 is minimized. The minimization of Equation 10.25 is a problem of the calculus of variations. Therefore, the equilibrium density profile must be such that it is a solution of the Euler–Lagrange equation [40]

$$\frac{\partial I}{\partial n} - \frac{d}{dx}\frac{\partial I}{\partial(dn/dx)} = 0, \qquad (10.26)$$

where I represents the integrand in Equation 10.25. Evaluation of Equation 10.26 yields

$$\mu_0(n) - \frac{1}{2}\frac{\partial c}{\partial n}\left(\frac{dn}{dx}\right)^2 - c\frac{d^2n}{dx^2} - \mu_E = 0, \qquad (10.27)$$

where μ_0 is given by

$$\mu_0 = \frac{\partial f_0}{\partial n}, \qquad (10.28)$$

and represents the chemical potential of a homogeneous fluid at density n. Equation 10.27 is the statement that the chemical potential throughout the interfacial region, given by the first three terms of Equation 10.27, remains constant and equal to the equilibrium chemical potential, μ_E. The derivative terms in Equation 10.27 give the corrections to the local chemical potential, μ_0, to account for the nonhomogeneous nature of the interface and to ensure that the chemical potential at any point within the interface remains constant. Note that Equation 10.27 is equally applicable in the bulk phases since the homogeneous nature of the fluids ensures that the density gradients are equal to zero in the bulk phases.

Solution of Equation 10.27 gives the equilibrium density profile, $n(x)$, which, in turn, can be used to calculate the surface tension from Equation 10.18. The boundary conditions for Equation 10.27 are $n(-\infty) = n_L$ and $n(+\infty) = n_B$; that is, outside the interfacial region the density of the fluid corresponds to that of the bulk phases.

The interfacial tension equation can also be obtained from Equation 10.25. First, Equation 10.27 must be rearranged. Multiplying Equation 10.27 by dn/dx the following is obtained:

$$[\mu_0(n) - \mu_E]\frac{dn}{dx} = \frac{dn}{dx}\left[\frac{1}{2}\frac{\partial c}{\partial n}\left(\frac{dn}{dx}\right)^2 + c\frac{d^2n}{dx^2}\right]. \tag{10.29}$$

Both sides of Equation 10.29 can be integrated by parts resulting in

$$f_0(n) - \mu_E n + K = \frac{1}{2}c\left(\frac{dn}{dx}\right)^2, \tag{10.30}$$

where K is a constant of integration. From the definition of the homogeneous specific free energy, $\omega = f_0(n) - \mu_E n$ and, by definition, $\omega = -P$, the thermodynamic pressure. The constant of integration can be evaluated at $x = a$, the lower boundary of the interface. At a, the pressure $P = P_N$ and $dn/dx = 0$, implying that $K = P_N$. Substituting Equation 10.30 in Equation 10.25 results in

$$\Omega = -P_N V + A\int_a^b c\left(\frac{dn}{dx}\right)^2 dx. \tag{10.31}$$

The interfacial tension is the difference, or excess, per unit area between Equation 10.31 and the free energy if the system were composed of a homogeneous phase. Since $\Omega = -PV$ for a homogeneous phase, the interfacial tension is just the second term of Equation 10.31 divided by A, the cross-sectional area of the interface. This integral part of Equation 10.31 is identical to what was obtained from mechanical considerations (Equation 10.18).

In summary, for systems with no external forces other than gravity, solution of Equation 10.27, subject to the appropriate boundary conditions, provides the equilibrium density profile $n(x)$ that, in turn, allows the determination of interfacial tensions from Equation 10.18. Before solving Equation 10.27, however, an explicit formula-tion of $\mu_0(n)$, the chemical potential for a homogeneous fluid at density n, is neces-sary and can be obtained from any of the numerous equations of state for fluids that are available in the literature.

Equation 10.31 provides the definition of the interfacial tension from thermody-namic considerations. The interfacial tension is seen as the energy in excess of that for a system composed only of a homogeneous phase. Thus, the concept of the excess energy associated with the Gibbs dividing surface of zero thickness is maintained in the gradient theory treatment of interfaces, even though the interfacial region is considered to have a finite thickness.

So far, the effects of external potentials have been neglected. The effects of grav-ity were considered negligible since the distances involved in the interfacial regions were so small that the hydrostatic pressure of the bulk fluids adjacent to the interface remained constant. If external potentials other than gravity exist, then the equilib-rium density profile must reflect such effects. The treatment that follows is for the one-dimensional density profile.

Consider an external potential energy function $\varphi(x)$. The one-dimensional equa-tion of hydrostatics for systems in the presence of external forces can be obtained from Equation 10.8

$$\frac{dP_N}{dx} = -n\frac{d\varphi}{dx},$$
(10.32)

where the definition of the external body force, $f = -d\varphi/dx$, has been employed.

The expression for the free energy of a planar interface of cross-sectional area A, in the presence of the external potential, can be obtained by extension of Equation 10.25 to include body forces:

$$\Omega = A\int_a^b \left[f_0(n) + \frac{1}{2}c\left(\frac{dn}{dx}\right)^2 - \mu_E n + n\varphi(x) \right] dx,$$
(10.33)

where $n\varphi(x)$ represents the energy potential density function. The solution of the Euler–Lagrange equation (Equation 10.26) results in

$$\mu_0(n) + \varphi(x) - \frac{1}{2}\frac{\partial c}{\partial n}\left(\frac{dn}{dx}\right)^2 - c\frac{d^2 n}{dx^2} - \mu_E = 0.$$
(10.34)

The implication of Equation 10.34 is that for systems under the influence of external potential fields, the chemical potential μ is not constant but is instead dependent on the location. This is true for both the homogeneous bulk phases and for the interfacial region. Solution of Equation 10.34 allows the determination of the equilibrium density profile for systems under the influence of an external potential.

Because the normal pressure is not constant but is a function of x, the equation for the interfacial tension is more complex and involves more than just the density gradients, as is the case in Equation 10.18 or in the integral from Equation 10.31. For systems in an external field, it is simpler to use the excess free energy concept to express the interfacial tension. Recalling that the interfacial tension is the difference, per unit area, between the free energy of a system that has an interfacial region and a system that does not, the following is obtained

$$\gamma = \int_a^b \left[f_0(n) + \frac{1}{2}c\left(\frac{dn}{dx}\right)^2 - \mu_E n + \varphi n + P \right] dx, \tag{10.35}$$

where $-P$ represents, again, the specific free energy, Ω/V, for a homogeneous system. Outside the interfacial region, the integrand in Equation 10.35 is identically zero and the integration limits can be extended to $\pm\infty$ without loss of generality. The correct boundary conditions for the solid–fluid systems are $n(0) = 0$ and $n(\infty) = n_L$, corresponding, respectively, to the assumption of no liquid adsorption on the surface of the wall and to the fact that the external potential has no effect on the bulk liquid far away from the wall.

In summary, the effects of an external potential on the equilibrium density profile have been examined and are given by Equation 10.34. Also, the interfacial tension Equation 10.35 for systems in an external potential field has been derived from excess free energy considerations.

An external potential will be used to model the interaction between a solid wall and a liquid molecule. The interfacial tension calculated from Equation 10.35 will, therefore, correspond to the solid–liquid interfacial tension, γ_{sl}. The interaction potential between a solid and a liquid molecule is modeled by a Lennard-Jones 6–12 potential, which, after integration for a semi-infinite wall, results in

$$\varphi(x) = 4\pi\varepsilon n_s \sigma^3 \left[\frac{1}{45}\left(\frac{\sigma}{x}\right)^9 - \frac{1}{6}\left(\frac{\sigma}{x}\right)^3 \right], \tag{10.36}$$

where ε represents the depth of the interaction potential well and σ represents the collision diameter. The parameter ε is related to the Hamaker constant, a bulk property of the system, via

$$A_{sl} = 4\pi^2\varepsilon\sigma^6 n_s n_l, \tag{10.37}$$

where A_{sl} is the solid–liquid Hamaker constant, and n_s and n_l are the number densities of the solid wall and of the bulk fluid, respectively. The collision diameter, on the other hand, cannot be obtained from the bulk properties of either the solid or the liquid. Its determination must be done empirically, as will be shown later in this section.

Before calculations of γ_{sl} from gradient theory considerations can proceed, it is necessary to specify the influence parameter c, and the equation of state for homogeneous fluids. The equation of state used to model the liquids is the Peng–Robinson [41] equation 10.38. Similar to the van der Waals equation of state, the Peng–Robinson (PR) equation is a two-parameter equation. The major difference between the two equations is that the attractive pressure term in the PR equation has been modified to generate more accurate density values of the liquid phase. The PR equation is of the form

$$P = \frac{n\kappa T}{1-nb} - \frac{n^2 a(T)}{1+nb(2-nb)},\tag{10.38}$$

where b and a are material properties, and κ is the Boltzmann constant. The constant b accounts for the excluded volume of the molecules, and $a(T)$ is the energy-related temperature function that accounts for the attraction between molecules. The two constants are defined as

$$a(T) = a(T_c)\alpha(T/T_c,\omega),\tag{10.39}$$

$$b(T) = b(T_c) = 0.07780\frac{\kappa T_c}{P_c},\tag{10.40}$$

where

$$a(T_c) = 0.452724\frac{\kappa^2 T_c^2}{P_c},\tag{10.41}$$

and

$$\sqrt{\alpha} = 1 + (0.37464 + 1.54226\omega - 0.26992\omega^2)(1 - \sqrt{T/T_c}).\tag{10.42}$$

In Equations 10.39 through 10.42, P_c and T_c represent the critical pressure and temperature, respectively, and ω represents the acentric factor. All the required parameters for the PR equation are available from the literature, such as in Reid et al. [42]. For room temperature, the PR equation is excellent at predicting densities, especially for nonpolar materials [41]. It should be stressed that any other equation of state for fluids, such as the Benedict–Webb–Rubin equation [43] or the Redlich–Kwong equation [44], can be used in the gradient theory analysis.

The specific Helmholtz potential, f_0, and the chemical potential, μ_0, can be obtained, from thermodynamic considerations, by Equation 10.38. They are

$$f_0 = f_0(T) - n\kappa T \ln\left(\frac{1-nb}{n}\right) + \frac{an}{2\sqrt{2}b}\ln\left(\frac{1+nb(1-\sqrt{2})}{1+nb(1+\sqrt{2})}\right),\tag{10.43}$$

and

$$\mu_0 = -\kappa T \ln\left(\frac{1}{n} - b\right) + \frac{\kappa T}{1-nb} - \frac{a}{2\sqrt{2}b}\ln\left(\frac{1+nb(1+\sqrt{2})}{1+nb(1-\sqrt{2})}\right)$$

$$-\frac{na}{1+2nb-(nb)^2} + \mu_0(T), \tag{10.44}$$

where the functions $f_0(T)$ and $\mu_0(T)$ represent the temperature-dependent parts of the Helmholtz potential and the chemical potential, respectively.

Equations 10.38, 10.43, and 10.44 are the necessary equations for gradient theory analysis. These equations represent, respectively, the pressure, the Helmholtz potential, and the chemical potential of a homogeneous fluid of density n.

The influence parameter, c, will affect the calculated γ_{sl} to a great extent, as can be seen from Equation 10.35. Carey et al. [33] calculated c from the homogeneous properties of fluids. They found that c is not strongly dependent on the density, n, of the fluid and can therefore be taken as a constant, obtained from the following empirical equation

$$c = 0.27ab^{2/3} + 2\times10^{-67} \text{ Jm}^5. \tag{10.45}$$

The value of c is very small because it represents the influence parameter per molecule (as do the parameters a and b).

The assumption that c is not strongly dependent on the density, n, greatly simplifies the differential Equation 10.34 that needs to be solved to obtain the equilibrium density profile. Also, the pressure component equations for the interfacial region are simplified to a great extent.

It is convenient to introduce nondimensional variables to simplify the pertinent equations. These dimensionless variables are

$$n^* = nb, \quad T^* = b\kappa T/a, \quad x^* = x(a/c)^{1/2}, \quad \mu^* = \mu b/a,$$

$$\gamma^* = b^2\gamma/(ac)^{1/2}, \quad \varphi^* = \varphi b/a, \quad P^* = Pb^2/a, \quad f^* = fb^2/2, \tag{10.46}$$

$$\omega^* = \omega b^2/2.$$

The parameter $(c/a)^{1/2}$ represents the diameter of the fluid molecule. Therefore, in dimensionless variables, distances are measured in terms of molecular diameter. In terms of dimensionless variables, Equation 10.34, with the assumption of constant c, becomes

$$\frac{d^2n^*}{dx^{*2}} = \mu_0^*(n^*) + \varphi^*(x^*) - \mu_E^*, \tag{10.47}$$

where

$$\mu_0^* = T^* \ln\left(\frac{n^*}{1-n^*}\right) + \frac{T^*}{1-n^*} - \frac{1}{2\sqrt{2}} \ln\left[\frac{1+n^*\left(1+\sqrt{2}\right)}{1-n^*\left(1-\sqrt{2}\right)}\right]$$
$$- \frac{n^*}{1+2n^*-n^{*2}} + \mu_0^*(T^*),$$

(10.48)

and

$$\varphi^* = W\left[\frac{1}{45}\left(\frac{d}{x^*}\right)^9 - \frac{1}{6}\left(\frac{d}{x^*}\right)^3\right].$$

(10.49)

In Equation 10.49, W and d are defined as

$$W = 4\pi b\varepsilon n_s \sigma^3 / a,$$

(10.50)

and

$$d = \sigma / (c/a)^{1/2}.$$

(10.51)

Using Equation 10.37, the W parameter, Equation 10.50, can be written in terms of Hamaker constants as

$$W = \frac{b^2 A_{sl}}{\pi a n_l^* \sigma^3}.$$

(10.52)

The appropriate boundary conditions for Equation 10.47 are $n^* = 0$ at $x^* = 0$, and $n^* = n_l^*$ at $x^* = \infty$. That is, there is no adsorption of liquid molecules on the solid wall ($x^* = 0$), and the density of the fluid approaches that of the bulk liquid far from the wall ($x^* = \infty$). Equation 10.47 was solved numerically using a finite difference approach [45] in conjunction with a nonlinear equation solver. The finite difference technique was chosen over more complex finite element techniques, such as the Galerkin method [46], because of its simplicity and ease of implementation.

All required nondimensional equations have now been defined. The solid–liquid interaction potential, φ^*, is given by Equation 10.49, and the equation of state for the fluids is given by Equation 10.48. However, as was mentioned above, the collision diameter, σ, cannot be obtained from the bulk properties of either the liquid or the solid. Its determination must therefore be made empirically. For the determination of σ, the solid–liquid interfacial tensions of various n-alkanes on Teflon (polytetrafluoroethylene) (PTFE) were used.

TABLE 10.2

Determination of Collision Diameters, σ, from Contact Angle Data for
n-Alkanes on Teflon (PTFE)

Liquid	Surface Tension, γ_{lv} (mJ/m²)	Contact Angle, θ (deg)	Hamaker Constant, A_{sl} ($\times 10^{-20}$ J)	Solid–Liquid Interfacial Tension, γ_{sl} (mJ/m²)	Collision Diameter, σ (nm)
n-Octane	21.62	26	4.11	0.57	0.121
n-Decane	23.83	35	4.40	0.80	0.118
n-Dodecane	25.35	42	4.55	1.16	0.115
n-Hexadecane	27.47	46	5.02	1.24	0.113
				Mean	0.117

The solid surface tension, γ_{sv}, of PTFE has been obtained by various techniques, the sedimentation volume method among them [3]. The estimated value of γ_{sv} is approximately 20 mJ/m². The choice of PTFE as the "calibrating" solid was motivated by three factors: (i) PTFE is considered to be purely dispersive, so that the Lennard-Jones potential (Equation 10.36) is a good model for the solid–liquid interactions; (ii) there are accurate dispersion data that are required to calculate the solid–liquid Hamaker constants [47]; and (iii) contact angle data for n-alkanes on PTFE are also available from the literature [48].

Using a value of 20 mJ/m² for γ_{sv} and using literature values for γ_{lv} and θ, it is possible to obtain the required solid–liquid interfacial tensions, γ_{sl}, by using Young's equation (Equation 10.1). These values are, in turn, used to determine the average collision diameter, σ. In Table 10.2, the contact angles of the n-alkanes on PTFE, the solid–liquid Hamaker constants, A_{sl}, the solid–liquid interfacial tensions, γ_{sl}, and the calculated σ values are given.

The collision diameters, σ, vary between 0.121 and 0.113 nm, the larger value corresponding to n-octane and the smaller value to n-hexadecane, with a mean value of 0.117 nm. Because the parameter σ is material dependent, exact values of the solid–liquid interfacial tensions for the other solid–melt pairs could not be obtained. Instead, assuming that the values of σ obtained for n-alkane/PTFE systems were representative of the range of values of σ for most materials, a range of solid–liquid interfacial tensions was determined for each system.

Solid–liquid interfacial tensions, γ_{sl}, were calculated from gradient theory considerations for four solid–melt systems: naphthalene, biphenyl, n-dodecane, and n-hexadecane. The calculated γ_{sl} values for the range of σ, as given in Table 10.2, for each system are shown in Table 10.4 of Section 10.4.3.

10.4.2 LIFSHITZ THEORY OF VAN DER WAALS FORCES

There is an interaction force between any two macroscopic bodies in a medium or in vacuum. This interaction between the bodies is of a much longer range than the interaction between two molecules. The interaction between two particles can be divided into long-range and short-range forces. The short-range forces will not be discussed here. The long-range forces are:

1. The electric double-layer force that results from the electrostatic interaction among the charges on the surfaces of the solids, screened by the intervening medium.
2. The van der Waals force that results from the electrodynamic interaction among the molecules of the solid and of the intervening medium.

It is important to stress that van der Waals forces are always present between all particles, regardless of the properties of the constituent molecules. The electric double-layer force, on the other hand, only occurs when charges are present.

This section deals with the van der Waals forces between macroscopic bodies. The concept of Hamaker constants, which have been used in gradient theory analysis, and how these parameters are obtained, represents the bulk of this section. An approach for estimating solid–liquid interfacial tensions based on Lifshitz theory considerations will be developed. The attractiveness of using Lifshitz theory for calculating such values is that bulk properties such as refractive index data and dielectric constants, which are easily measured in principle, are the only required information.

The relationship between dispersion interactions and interfacial tensions will be discussed in terms of Hamaker constants. Hamaker constants will then be used to determine the solid–melt interfacial tensions of the four systems that were also studied in the previous section. The use of Hamaker constants in this manner represents a novel way of estimating solid–liquid interfacial tensions. It is surprising to find that Lifshitz theory and Hamaker constants, although they are intimately related to surface and interfacial tensions, have not been used to estimate solid–liquid interfacial tensions. The reverse process, that is, using surface and interfacial tension data to obtain Hamaker constants, has been used by Omenyi [49] to show that repulsive van der Waals forces between macroscopic bodies can, and do, exist.

The interaction between macroscopic bodies can be written in terms of two parameters—the first to represent the geometry of the bodies and the second to represent the material properties of the interacting bodies. This second parameter is the Hamaker constant [50]. The major use of the Hamaker constant is to explain the stability of colloidal particles dispersed in liquid media in conjunction with the DLVO theory—Derjaguin-Landau-Verwey-Overbeek [51].

Since van der Waals interactions exist between neutral molecules, a force must also exist between macroscopic bodies. The calculation of this force can be accomplished by taking into account the interactions between all the molecules of the macroscopic bodies. This microscopic approach was first introduced by Hamaker [50]. Hamaker considered only the dispersion component of van der Waals interactions since, except for highly polar molecules, the dispersion term dominates. Hamaker also assumed pairwise additivity of the intermolecular dispersion interactions. That is, when a number of molecules are present, the total dispersion energy consists of the sum of the interaction energies between each pair of molecules.

The interaction energy, per unit area, for two half-spaces with molecular densities n_1 and n_2, respectively, and a separation distance, l, is given by

$$E = -\frac{A}{12\pi l^2}.$$ (10.53)

The Hamaker constant, A, is defined as

$$A = \pi^2 n_1 n_2 C_{12},$$ (10.54)

where C_{12} is a measure of the strength of the molecular interaction between a molecule of type 1 and one of type 2. The corresponding van der Waals force can be obtained by differentiation of the interaction energy equation. For the two half-spaces, the force per unit area is given by

$$F = \frac{A}{6\pi l^3}.$$ (10.55)

It is inherently assumed in Hamaker's approach that the energy of interaction of body 1 with the intervening medium is unaffected by the presence or absence of body 2. While this assumption may be valid for two bodies interacting across a vacuum, it will not be correct for two bodies separated by a medium. Thus, the method of pairwise summations cannot be used to calculate the interactions between two bodies separated by a condensed medium unless corrections for three-body and many-body effects are included.

The problem of additivity is completely avoided in Lifshitz theory, where the molecular structure of the bodies is ignored and the interactions between large bodies, now treated as continuous media, are derived in terms of bulk properties such as dielectric constants and indices of refraction. The interaction between the bodies is then considered to take place through a fluctuating electromagnetic field. However, it should be pointed out that the expression for the interaction energy between two half-spaces, Equation 10.53, as obtained by Hamaker, is still valid within the framework of Lifshitz theory. The only difference is the way in which the Hamaker constant, A, is calculated.

Within all media, the electrons are in continuous motion. This motion gives rise to a fluctuating electromagnetic field that still exists at absolute zero. At absolute zero, these fluctuations are of a purely quantum mechanical nature due to the random excitation of the electrons. At nonzero temperatures, there are additional contributions due to thermal excitation of the molecules [52]. On a macroscopic scale, the fluctuating electromagnetic field can be pictured as comprising oscillatory waves that act within the body but also extend beyond its boundaries. For two macroscopic bodies, the fields at a point in body 1 and a point in body 2 are at least partially in phase over distances of the order of magnitude of the absorption wavelengths. It is this phase correlation that gives rise to van der Waals interactions. Van der Waals interactions can be seen as being the change in the energy of the electromagnetic field due to the fluctuations of the field caused by the presence of condensed phases. The total interaction results from contributions from all the electromagnetic fluctuations whose wavelengths are large compared with intermolecular separations; however, the major contributions come from wavelengths that are of the same order of magnitude as the separation distance between the macroscopic bodies [53].

The first general macroscopic theory of van der Waals interactions is due to Lifshitz [54] who derived the van der Waals force between two nonmagnetic dielectric half-spaces separated by a vacuum. Later, Dzyaloshinskii et al. [53] extended Lifshitz theory to two half-spaces separated by a third medium. The mathematical difficulties involved in the solution of the Maxwell equations are formidable when the fluctuating magnetic fields are introduced. In this section, only the final result from the analysis of Dzyaloshinskii et al. and further simplifications are included. The interaction energy, at finite temperatures, for two parallel plates is given by

$$E(l) = \frac{\kappa T}{8\pi l^2} \sum_{n=0}^{\infty}{'} I(\xi_n, l), \tag{10.56}$$

where

$$
\begin{aligned}
I(\xi_n, l) = \left(\frac{2\xi_n l \varepsilon_3^{1/2}}{c} \right)^2 \int_l^{\infty} p\, dp \Bigg\{ &\ln\left[1 - \Delta_{13}^{-R} \Delta_{23}^{-R} \exp\left(-\frac{2p\xi_n l \varepsilon_3^{1/2}}{c} \right) \right] \\
&+ \ln\left[1 - \Delta_{13}^{R} \Delta_{23}^{R} \exp\left(-\frac{2p\xi_n l \varepsilon_3^{1/2}}{c} \right) \right] \Bigg\},
\end{aligned}
\tag{10.57}
$$

where

$$\Delta_{k3}^{-R} = \frac{s_k \varepsilon_3 - p \varepsilon_k}{s_k \varepsilon_3 + p \varepsilon_k}, \quad \Delta_{k3}^{R} = \frac{s_k - p}{s_k + p} \tag{10.58}$$

$$s_k = \sqrt{p^2 - 1 + \varepsilon_k / \varepsilon_3}, \quad \varepsilon = \varepsilon(i\xi_n), \quad \xi_n = 4\pi^2 n \kappa T / h,$$

where the prime on the summation symbol in Equation 10.56 indicates that the term $n = 0$ is given half weight, ε_j represents the dielectric response of the jth medium as a function of imaginary frequencies $i\xi_n$, κ is Boltzmann's constant, T is the temperature, and h is Planck's constant.

Fortunately, some simplifications of Equation 10.57 are possible, making it more tractable, and, as will be shown below, an unretarded Hamaker constant can be calculated when the separation distance, l, becomes small ($l < 5$ nm). The simplification results in

$$I(\xi_n,0) = (\xi_n / \xi_s)^2 \int_0^\infty p\,dp\left[1-\left(\frac{\varepsilon_1-\varepsilon_3}{\varepsilon_1+\varepsilon_3}\right)\left(\frac{\varepsilon_2-\varepsilon_3}{\varepsilon_2+\varepsilon_3}\right)e^{-(\xi_n/\xi_s)p}\right]$$
$$= \int_0^\infty x\,dx\ln\left[1-\left(\frac{\varepsilon_1-\varepsilon_3}{\varepsilon_1+\varepsilon_3}\right)\left(\frac{\varepsilon_2-\varepsilon_3}{\varepsilon_2+\varepsilon_3}\right)e^{-x}\right],$$

(10.59)

where a change of variable to $x = p\xi_n/\xi_s$ was performed. Thus, the interaction energy for two slabs, 1 and 2, separated by a medium, 3, for the unretarded case, can be obtained from Equation 10.56 where Equation 10.59 replaces Equation 10.57 as the summand.

For the unretarded case, the interaction energy for two half-spaces, 1 and 2, separated by a medium, 3, of thickness l can be written in the form obtained by Hamaker, Equation 10.53; that is,

$$E_{132}(l) = -\frac{A_{132}}{12\pi l^2},$$

(10.60)

which, when compared to Equation 10.56, defines the Hamaker constant, A_{132}, as

$$A_{132} = -\frac{3\kappa T}{2}\sum_{n=0}^\infty{}' \int_0^\infty x\,dx\ln\left(1-\Delta_{13}\Delta_{23}e^{-x}\right),$$

(10.61)

where

$$\Delta_{kj} = \frac{\varepsilon_k(i\xi_n)-\varepsilon_j(i\xi_n)}{\varepsilon_k(i\xi_n)+\varepsilon_j(i\xi_n)},$$

(10.62)

and ξ_n is as defined above.

The major difference between the Hamaker constant as defined by Equation 10.54 and by Equation 10.61 is that the former is calculated from a microscopic quantity (the intermolecular pair potential), whereas the latter is calculated from a bulk property of the condensed phase (the dielectric response).

In Equation 10.61, the term $\Delta_{13}\Delta_{23}e^{-x}$ is always less than 1, so that the integration may be performed by expanding the logarithmic term in a power series and integrating term-by-term. Bearing in mind that Δ_{kj} is not a function of x, one obtains

$$A_{132} = \frac{3\kappa T}{2}\sum_{n=0}^\infty{}'\sum_{s=1}^\infty \frac{\left(\Delta_{13}\Delta_{23}\right)^s}{s^3}.$$

(10.63)

Equation 10.63 allows the calculation of the unretarded Hamaker constant, from Lifshitz theory considerations, which can be used in the interaction energy equations that were derived from a microscopic approach. It must be stressed that Equation 10.63 can only apply for situations where the separation distance, l, is smaller than 5 nm. For larger separations, the Hamaker constant becomes a function of the separation distance; that is, $A = A(l)$.

The need to know the dielectric permittivity function, $\varepsilon(\omega)$, for all frequencies seems to be an obstacle to the application of the equations given in the previous section, because the complete data on the absorption spectra are not available for most materials of interest. Fortunately, however, only partial knowledge of the functions $\varepsilon(\omega)$ is sufficient to determine the force and energy of interaction between bodies [55].

The dielectric permittivity function, $\varepsilon(\omega)$, of a material at frequency ω represents the change in the strength of the electromagnetic field of the material due to an externally applied field of frequency ω. This external field, for the case of van der Waals interactions, arises from the presence of another condensed phase in the vicinity of a macroscopic body. The function $\varepsilon(\omega)$ is a complex function of the frequency, ω, which is itself complex. The function $\varepsilon(\omega)$ is usually written as

$$\varepsilon(\omega) = \varepsilon'(\omega) + i\varepsilon''(\omega), \tag{10.64}$$

where $\omega = \omega_R + i\xi$ represents the complex frequency. For the calculation of dispersion interactions, only the values of ε on the imaginary frequency axis, $\varepsilon(i\xi)$, are required. The real part of the dielectric response, ε', because of its relationship to the refractive index, gives a measure of the transmission properties of the material [47]. The imaginary part of the dielectric response, ε'', as a function of the frequency, ω, represents the absorption spectrum of the material. At certain frequencies, the function $\varepsilon''(\omega)$ has a zero value, meaning that the material does not absorb any energy at those frequencies; the material is said to be transparent to electromagnetic energy of those frequencies. In the frequency range where $\varepsilon''(\omega)$ is zero, the total dielectric response is real and is related to the refractive index, $n(\omega)$, of the medium via [56]

$$\varepsilon(\omega) = \varepsilon'(\omega) = n^2(\omega) \quad [\varepsilon''(\omega) = 0]. \tag{10.65}$$

The real part, $\varepsilon'(\omega)$, and imaginary part, $\varepsilon''(\omega)$, of the dielectric response are not independent but are related via the Kramers–Kronig relation

$$\varepsilon'(\omega) = 1 + \frac{2}{\pi} P \int_0^\infty \frac{x\varepsilon''(x)}{x^2 - \omega^2} dx, \tag{10.66}$$

where P is the polarization density (see Hough and White [47] for details). The dielectric response along the imaginary frequency axis, $\varepsilon(i\xi)$, is also related to the absorption spectrum via a second Kramers–Kronig relation [47]

$$\varepsilon(i\xi) = 1 + \frac{2}{\pi} \int_0^\infty \frac{x \varepsilon''(x)}{x^2 + \xi^2} dx. \tag{10.67}$$

Equations 10.66 and 10.67 mean that if the absorption spectrum, $\varepsilon''(\omega)$, for all frequencies is known, then the function $\varepsilon'(\omega)$ can be reconstructed. The function $\varepsilon''(\omega)$ can be represented by a discrete set of delta functions representing the absorption peaks at various frequencies, ω_j:

$$\varepsilon''(\omega) = \sum_{j=1}^N f_j \delta(\omega - \omega_j), \tag{10.68}$$

where f_j represents the height of the absorption peak at the absorption frequency ω_j.

As was mentioned previously, for the calculation of the van der Waals interactions, only the dielectric permittivity function evaluated along the imaginary frequency axis is required; that is, $\varepsilon = \varepsilon(i\xi)$. Also, for most materials of interest, the complete absorption spectrum, $\varepsilon''(\omega)$, is not available. Therefore, an approximation to the function $\varepsilon(i\xi)$ is required. The most common approximation used is that due to Ninham and Parsegian [55,57], which can be obtained by substituting Equation 10.68 into Equation 10.67:

$$\varepsilon(i\xi) = 1 + \frac{C_{mv}}{1 + \xi/\omega_{mv}} + \sum_j \frac{C_j}{1 + (\xi/\omega_j)^2}, \tag{10.69}$$

where ω_{mv} and ω_j represent the resonance frequencies, and the constants C_{mv} and C_j represent the oscillator strengths. The parameters C_j and f_j are related by

$$C_j = \frac{2}{\pi} \frac{f_j}{\omega_j}. \tag{10.70}$$

The first term in Equation 10.69 models the function in the microwave frequencies and the second term models the function in the infrared through ultraviolet frequencies.

From Equation 10.69, it is clear that the dispersion interaction is determined solely by the oscillator strengths, C, and absorption frequencies, ω, of the component materials. Equation 10.69 is the so-called Ninham–Parsegian approximation of the dielectric response function in terms of experimentally accessible quantities. The importance of the Ninham–Parsegian approximation cannot be overstated: it allows the determination of dispersion interactions, via Lifshitz theory, in systems where only partial absorption data are available.

It should be pointed out that the major contribution to the calculation of unretarded Hamaker constants, from Equation 10.63, comes from the ultraviolet region. The reason for this is that the frequencies ξ_n occur at equally spaced intervals of about 3×10^{14} rad/s, at $T \approx 300$ K. In the microwave frequencies ($\approx 10^{11}$ rad/s) and in the infrared frequencies ($\approx 10^{14}$ rad/s), there are very few sampling points. For example, in the frequency region $\xi < 10^{16}$ rad/s there are only 30 sampling points whereas in the ultraviolet region $10^{16} < \xi < 10^{17}$ rad/s there are about 300 terms. There are, however, materials for which contributions from the lower frequencies cannot be ignored. A good example is water; the dielectric response of water decreases from a value of 80 at zero frequency to approximately 4 at the infrared frequencies. For dispersive systems, the contributions from the microwave region are nonexistent since microwave absorption exists only for polar molecules [58].

For condensed phases, the errors introduced by using the Ninham–Parsegian approximation in the calculation of Hamaker constants are minimized because only the difference in the dielectric responses (Equation 10.63), at a given frequency, contributes to the calculated unretarded Hamaker constants. The errors in the Hamaker constants are slightly higher when the intervening medium, 3, is a vacuum, since for this case $\varepsilon_3 = 1$ for all frequencies.

For systems that are predominantly dispersive, only the ultraviolet oscillator strength, C_{uv}, and absorption frequency, ω_{uv}, need to be determined in order to obtain good estimates of Hamaker constants from Lifshitz theory. Thus, the Ninham–Parsegian construction, Equation 10.69, simplifies to

$$\varepsilon(i\xi) = 1 + \frac{C_{uv}}{1 + \left(\xi / \omega_{uv}\right)^2}.$$ (10.71)

The value of the ultraviolet absorption frequency, ω_{uv}, and of the oscillator strength, C_{uv}, can be obtained from refractive index data by means of so-called Cauchy plots [47]. The relation (simplified) between the refractive index, n, and the parameters ω_{uv} and C_{uv}, in the visible region, can be obtained from Equation 10.65 and by substituting Equation 10.68 into Equation 10.66:

$$\varepsilon'(\omega) = \varepsilon(\omega) = n^2(\omega) = 1 + \frac{C_{uv}}{1 - \left(\omega / \omega_{uv}\right)^2}.$$ (10.72)

Rearranging Equation 10.72, the following is obtained

$$n^2(\omega) - 1 = \left[n^2(\omega) - 1\right] \frac{\omega^2}{\omega_{uv}^2} + C_{uv}.$$ (10.73)

Therefore, a plot of $[n^2(\omega) - 1]$ versus $[n^2(\omega) - 1]\omega^2$ should yield a straight line of slope $1/\omega_{uv}^2$ and intercept C_{uv}. This is known as the Cauchy plot. Experimentally, it is relatively easy to measure the refractive index as a function of the wavelength λ ($= 2\pi c/\omega$) in the visible region, and this information is usually available in the literature for most common substances.

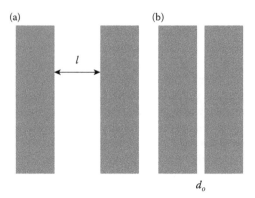

FIGURE 10.3 Schematic of two half-spaces, (a) initially at infinity that are then brought into contact (b).

In summary, from the refractive index data in the visible region, it is possible to approximate the dielectric response function along the imaginary axis, $\varepsilon(i\xi)$. For most materials, except in the cases of highly polar materials, the simple Ninham–Parsegian representation for $\varepsilon(i\xi)$, as given by Equation 10.71, with a single oscillator function corresponding to the ultraviolet region, is sufficient to compute Hamaker constants of the materials accurately. Once the Hamaker constants are known, surface and interfacial tensions can be estimated, as described below.

Consider two liquid half-spaces originally at infinite separation where the intervening medium is air, as shown in Figure 10.3a. The two half-spaces are now brought together to some separation distance such that the interfacial region is indistinguishable from the bulk phase, as in Figure 10.3b. The change in free energy of the system is given by the grand canonical potential for surfaces. For the situation depicted in Figure 10.3, the change in free energy of adhesion, per unit area, is

$$\Delta\omega^{adh} = -2\gamma_{lv} \tag{10.74}$$

since two liquid–air interfaces are destroyed and the "new" surface created is indistinguishable from the bulk phases. This change in free energy must be related somehow to the interaction energy as given by Equation 10.60.

Equation 10.60 diverges as the separation distance, l, approaches zero. This divergence does not reflect physical reality; rather, it is the result of treating the molecules as point-polarizable entities; thus, at zero separation distance, molecules would be overlapping. In reality, constituent molecules have a finite size and, therefore, macroscopic bodies cannot approach to $l = 0$. Thus, a cutoff distance, d_0, which is related to the collision diameter, σ, is often used to relate the interaction energy, as given by Equation 10.60, to the thermodynamic free energy, Equation 10.74. When two macroscopic bodies are at a distance d_0, they are considered to be in molecular contact, thus eliminating the divergence in Equation 10.60. Combining Equations 10.74 and 10.60 yields

$$\Delta\omega^{adh} = -2\gamma_{lv} = -\frac{A_{ll}}{12\pi d_0^2},\tag{10.75}$$

where A_{ll} represents the liquid–air–liquid Hamaker constant (it is an accepted convention to not assign a symbol to represent the intervening medium if it happens to be air). Equation 10.75 establishes the relationship between surface tensions and Hamaker constants when the concept of cutoff distances is employed. While such an approach is simplistic, the computed surface tensions for saturated hydrocarbon liquids, using Equation 10.75, agree very well with experimentally obtained values [59], so that such an agreement cannot be regarded as fortuitous.

The relationship between surface tensions and interaction energy is not restricted to either macroscopic bodies of the same material or systems where the intervening medium is air. If, in Figure 10.3, the two half-spaces were solid and the intervening medium were liquid, the following relationship would be obtained

$$\gamma_{sl} = \frac{A_{sls}}{24\pi d_0^2},\tag{10.76}$$

which relates the solid–liquid–solid Hamaker constant to the solid–liquid interfacial tension, γ_{sl}.

Solid–melt interfacial tensions, γ_{sl}, for the identical systems studied using gradient theory will now be calculated. The necessary dielectric response functions, $\varepsilon(i\xi)$, are approximated using the Ninham-Parsegian approach. The value used for the cutoff distance d_0 is 0.134 nm, which is related to the average collision diameter, σ, obtained from gradient theory considerations, as shown in Table 10.2. Before proceeding with the calculations, some questions regarding the limiting value for solid–liquid interfacial tensions, γ_{sl}, can be investigated by using Equation 10.76 as the starting point. In the development of the equation of state for interfacial tensions, described in Chapter 9, two assumptions were made regarding the possible range of values for γ_{sl}: the first assumption was that γ_{sl} would have a limiting value of zero, which would only occur when $\gamma_{lv} = \gamma_{sv}$; the second that γ_{sl} would always have a positive value. The first assumption has been criticized by Johnson and Dettre [60]. Equation 10.76 provides an answer to the controversy, at least in the case of purely dispersive solid–liquid systems, if it is assumed that the value of the cutoff distance, d_0, is universal. Equation 10.76 indicates that γ_{sl} is directly related to the Hamaker constant, A_{sls}. Thus, for the two assumptions to be valid, A_{sls} must always be greater than or equal to zero. However, A_{sls} is calculated from Equation 10.63, which, for two surfaces of the same material (1 = 2), is given by

$$A_{sls} = \frac{3\kappa T}{2}\sum_{n=0}^{\infty}{}'\sum_{s=1}^{\infty}\frac{\left(\Delta_{sl}\right)^{2s}}{s^3},\tag{10.77}$$

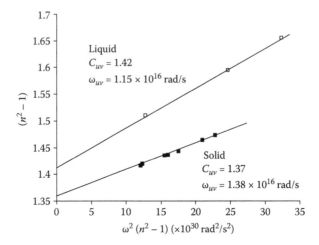

FIGURE 10.4 Cauchy plot for solid and liquid biphenyl at the melting temperature.

where the function

$$\Delta_{sl} = \frac{\varepsilon_s\left(i\xi\right) - \varepsilon_l\left(i\xi\right)}{\varepsilon_s\left(i\xi\right) + \varepsilon_l\left(i\xi\right)}.$$ (10.78)

In Equation 10.77, the function Δ_{sl} to the power of $2s$ will always be positive; therefore, it is obvious from Equation 10.77 that the Hamaker constant for a solid–liquid–solid system is always positive, implying a positive solid–liquid interfacial tension, γ_{sl}. Also, for the special case where the dielectric response of the liquid medium, $\varepsilon_l(i\xi)$, is similar to that of the solid, $\varepsilon_s(i\xi)$, the Hamaker constant is equal to zero and the corresponding value for γ_{sl}, from Equation 10.76, is zero. The special case of equal dielectric responses for the liquid and solid results in $A_{ll} = A_{ss}$, which, according to Equation 10.75, corresponds to the special case of $\gamma_{lv} = \gamma_{sv}$. Thus, the assumptions made in the development of the equation of state are in perfect agreement with Lifshitz theory considerations, at least in systems where Lifshitz theory is applicable, and assuming a constant value of d_0.

Cauchy plots for the solid and liquid phases of biphenyl and naphthalene were produced from refractive index data available from the literature [61–63], and are shown in Figures 10.4 and 10.5, respectively. For the n-alkanes, the Cauchy plots were obtained from Hough and White [47] for the liquid phase.

The Hamaker constant, A_{sls}, like the surface tension, has a temperature dependence. This temperature dependence comes about in two ways: the temperature, T, appears explicitly in Equation 10.63, and implicitly since the refractive index is also temperature dependent. Where possible, the temperature dependence of the refractive index, dn/dT, which is linear, was obtained for the different wavelengths from the available temperature data. Otherwise, a value of $dn/dT = -0.0005°C^{-1}$ was used; the value chosen is typical for most nonpolar organic

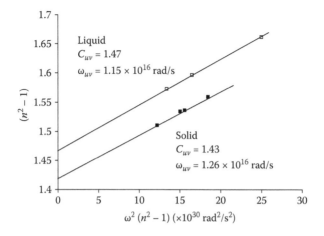

FIGURE 10.5   Cauchy plot for solid and liquid naphthalene at the melting temperature.

TABLE 10.3
Oscillator Strengths, C_{uv}, and Absorption Frequencies, ω_{uv}, for the Solid and Liquid Phases of the Various Materials at Their Melting Temperatures

		Phase				Hamaker
		Solid		Liquid		Constant,
	Melting					
	Temperature		ω_{uv}		ω_{uv}	A_{sls}
Material	(K)	C_{uv}	($\times 10^{16}$ rad/s)	C_{uv}	($\times 10^{16}$ rad/s)	($\times 10^{-22}$ J)
Naphthalene	353.5	1.40	1.26	1.41	1.15	2.22
Biphenyl	342.4	1.37	1.38	1.42	1.15	7.40
n-Dodecane	263.4	1.10	1.89	1.03	1.89	1.55
n-Hexadecane	291.2	1.13	1.89	1.02	1.82	5.11

materials at all wavelengths [64]. The values of ω_{uv} and C_{uv} obtained from the Cauchy plots are given in Table 10.3. In the case of the n-alkanes, no refractive index data are available for the solid phase so that Cauchy plots cannot be used to calculate the required parameters. However, the static dielectric constants, ε_0, for both solid n-dodecane and n-hexadecane at the freezing temperatures are known [65,66]. From these data, it is possible to estimate the oscillator strength, C_{uv}. From Equation 10.71, evaluated at zero frequency, the following is obtained

$$\varepsilon_0 = 1 + C_{uv}, \qquad (10.79)$$

TABLE 10.4

Solid–Melt Interfacial Tensions Calculated from Gradient and Lifshitz Theories

Material	Solid–Melt Interfacial Tensions, γ_{sl} (mJ/m²), From:				
	Lifshitz Theory	Gradient Theory	Equation of State	Fowkes' Equation	Gibbs–Thomson Equation
Naphthalene	0.17	0.8–4.6	1.44	0.10	61
Biphenyl	0.55	0.7–3.7	0.67	0.01	50
n-Dodecane	0.11	<2.0	0.15	0.09	12.1
n-Hexadecane	0.38	<0.4	0.89	0.57	15.2

where, for the reasons stated above, the oscillator strengths for infrared and microwave regions have been ignored. The values obtained from Equation 10.79 are given in Table 10.3. For the absorption frequency a value of $\omega_{uv} = 1.89 \times 10^{16}$ rad/s, which is typical for long-chained hydrocarbon crystals [67], was used.

The data given in Table 10.3 are sufficient for the determination of the solid–melt–solid Hamaker constants, A_{sls}, which, in turn, allow the calculation of the interfacial tensions, γ_{sl}, from Equation 10.76; A_{sls} values were calculated from Equation 10.63 using a computer program [28]. The calculated values of γ_{sl} are given in Table 10.4.

10.4.3 RESULTS AND DISCUSSION

The results shown in Table 10.4 indicate that both gradient theory and Lifshitz theory calculate solid–melt interfacial tension, γ_{sl}, values that are consistent with those from contact angle interpretation.

The accuracy of the values of γ_{sl} from gradient theory for n-dodecane may not be very high since the melting temperature of n-dodecane is well below room temperature and the PR equation of state may not describe the fluid behavior at such low temperatures. Nevertheless, gradient theory still predicts a γ_{sl} value that is in agreement with the value from contact angle considerations.

The values of γ_{sl} from the application of the Gibbs–Thomson equation are still about one order of magnitude larger than those calculated from gradient theory. While it is seen that the calculated γ_{sl} depends on the value chosen for σ, the range of γ_{sl} is still of the same order of magnitude as the values obtained from contact angle considerations. For the gradient theory results to agree with those from Gibbs–Thomson equation methods, the collision diameters would have to be much larger than those used in the present calculations. For example, for the gradient theory results for naphthalene to agree with that from the grain boundary groove measurement, a σ-value of 0.7 nm would be required. It is, however, not expected that the collision diameter could be larger than the diameter of the liquid molecule.

The values obtained from Lifshitz theory are within the range of values that were obtained from gradient theory considerations. The Lifshitz theory values, however, tend to be somewhat smaller than those from contact angle/equation of state considerations. Nevertheless, it is obvious from Table 10.4 that, among the four approaches, only the values obtained from the Gibbs–Thomson equation method are in any considerable disagreement.

The differences between Lifshitz theory results and those from equation of state considerations may be due to two possible reasons. First, the errors might be due to the need to use an approximate dielectric response function $\varepsilon(i\xi)$, instead of the full absorption spectrum. However, as was mentioned before, the errors introduced by the approximations tend to cancel out if the intervening medium is also a condensed phase, as is the case in a solid–liquid–solid system. Second, the need to introduce ad hoc cutoff distances d_0, to eliminate the divergence of the interaction energy as the separation distance approaches zero, is likely a cause of the discrepancy between Lifshitz theory and contact angle/equation of state approaches. At the present time, it is not possible to obtain the cutoff distance or collision diameter from the bulk properties of either the solid or the liquid.

Without more information regarding minimum separation distances, Equation 10.75 should be viewed as an empirical relationship between the surface or interfacial tension, γ, and the Hamaker constant, A, with d_0 being the adjustable parameter whose value is in the range of 0.1–0.2 nm. If the values of γ_{sl} obtained from techniques based on the Gibbs–Thomson equation were correct, then the value of d_0 would be in the range of 0.005–0.01 nm; that is, one to two orders of magnitude smaller than the commonly accepted values. It should be noted that most of the d_0 values available in the literature result from the correlation between the liquid surface tension, γ_{lv}, and the Hamaker constant, A_{ll}. Also, γ_{lv} and A_{ll} can be measured and calculated respectively, with a high degree of accuracy; thus, values of d_0 that are one to two orders of magnitude smaller than those obtained from liquid surface tension data can be considered to be incorrect. For the γ_{sl} values from Lifshitz theory to be comparable with those from contact angle interpretation, the d_0 values would range from 0.04 to 0.13 nm. The smaller value of d_0 corresponds to naphthalene and is smaller than the accepted values; however, the larger value of d_0 is certainly within the acceptable range.

In conclusion, the results from gradient theory and from Lifshitz theory are in good agreement with those from contact angle considerations. These results give further indication that the methods based on the Gibbs–Thomson equation cannot give correct estimates, not even in the order of magnitude, of solid–liquid interfacial tensions. The deficiencies of the Gibbs–Thomson equation will be discussed in the next section.

10.5   DEFICIENCIES OF THE GIBBS–THOMSON EQUATION

The main conclusion from the results of Section 10.4 is that the contact angle interpretation approach provides the best estimates of solid–melt interfacial tensions for low-surface-energy systems. The estimated solid–melt interfacial tensions for various systems, from both gradient theory and Lifshitz theory considerations, are in good agreement with those from the contact angle/equation of state approach. As was mentioned previously, approaches based on the Gibbs–Thomson equation give estimates of solid–melt interfacial tensions, γ_{sl}, which are almost two orders of magnitude larger than those obtained from contact angle interpretation. Results from both gradient and Lifshitz theories support the notion that the magnitudes of the γ_{sl} values for the various systems, as calculated by the contact angle/equation of state approach, are correct.

Because solid–liquid interfacial tensions cannot be measured directly, discrepancies between γ_{sl} values, as obtained by various independent approaches, are expected due to the assumptions and approximations inherent in any of the approaches discussed. Thus, it is not surprising that neither gradient theory nor Lifshitz theory provide γ_{sl} values that are in perfect agreement with those obtained from equation of state considerations. Nevertheless, it is still expected that the various independent approaches would estimate γ_{sl} values that are at least of the same order of magnitude.

There are many possible explanations for the lack of agreement between the Gibbs–Thomson equation approaches and the other approaches discussed in the previous section. Experimental difficulties, such as ensuring true homogeneous nucleation in the nucleation rate experiments, might give rise to such discrepancies. A more likely reason for the observed discrepancies is the assumption that the Gibbs–Thomson equation, developed originally for liquid–vapor systems, can be applied to solid–liquid systems in all situations.

This section examines the possible reasons for the failure of the Gibbs–Thomson equation to predict solid–melt interfacial tensions, γ_{sl}, correctly. The difficulties in using the Gibbs–Thomson equation to interpret data from nucleation experiments and melting in pores have already been discussed in Section 10.3. These difficulties are due to the small size of the solid that contains no more than a few molecules. For such small solids, it is not possible to ignore the dependence of surface tension on the curvature of the interface. Also, the approximation that the interface has zero thickness, when compared with the size of the bulk phase, is no longer valid for such small systems. The major portion of this section deals with the problems in using the Gibbs–Thomson equation to analyze the shapes of grain boundary grooves.

10.5.1 SURFACE STRESS AND SURFACE TENSION

The discrepancies in the calculated solid–melt interfacial tensions result in large part from the confusion between the quantities "surface tension" and "surface stress." Surface tension, γ, a scalar quantity, represents the work required by any reversible process to form a unit area of new surface. Surface stress, on the other hand, represents the work required to increase a unit area of surface by stretching. Surface stress, f_{ij}, is generally a tensor quantity. Much of the confusion between the two quantities arises from the fact that most of the earlier work on surfaces dealt only with fluid–fluid interfaces where the distinction between surface stresses and tension may often be irrelevant. If the surface of a fluid is distorted, then because it has no long-range order, there is no barrier that prevents molecules from entering or leaving the surface region. Thus, a new state of equilibrium can always be reached in which every one of the surface molecules occupies the same area as in the original undistorted surface. The process of stretching the surface is identical to creating more of the same surface, since, while the number of molecules in the surface changes, the area per surface molecule does not. In a solid crystal, on the other hand, there is a long-range correlation in atomic or molecular positions. Thus, a distortion of the surface represents a change in the surface area that cannot be accommodated by the migration of molecules to and from the surface. In the case of stretching a solid surface, the number of molecules in the surface remains constant but the area occupied by each of the molecules differs from the undistorted state. If the stress distorting the surface is removed, the surface area per molecule returns to the undistorted state.

The distinction between surface tension and surface stresses was originally pointed out by Gibbs [29], who argued that when solids are involved, "there is no such equivalence between stretching of the surface and the forming of new surface." Herring [68] also pointed out that, in a solid, the number of surface molecules and the state of strain of the surface are unrelated. Some authors [69–73] classify surface tension as the work required to strain the surface plastically, because the surface area occupied by the surface molecules remains constant; and surface stress as the work required to strain the surface elastically, because the number of surface molecules remains constant. The distinction between solid surface tension and surface stress, and measurement methods for both, were reviewed by Butt and Raiteri [74].

For normal pure liquids, only plastic deformation is possible and the total work in deforming the surface is just γ. For solids, on the other hand, both plastic and elastic deformations are possible. Herring [68] and Shuttleworth [75] have derived the relationship between surface stress, f, and surface tension, γ. This relation is given by

$$f_{ij} = \gamma \delta_{ij} + \frac{\partial \gamma}{\partial \varepsilon_{ij}}, \tag{10.80}$$

where ε_{ij} is the surface strain tensor and δ_{ij} is the Kronecker delta.

Makkonen has objected to Equation 10.80 by redefining γ to include elastic surface distortions [76]. He did, however, concur on the main point, which is the distinction between surface tension and surface stresses ("external stresses").

From Equation 10.80, it can be seen that surface stress and surface tension will be numerically equal only if γ is unaffected by deformation. As was explained above, this is evidently true in the case of pure liquids where molecules can move freely to and from the surface. The surface stress for liquids is isotropic with zero shear components so that it can be characterized by a single quantity, f, where

$$f = \gamma. \tag{10.81}$$

On the other hand, for solids, because of the low mobility of molecules, it is not possible to keep constant the local configuration around any particular molecule in the surface region where the deformation of the surface area is performed. Thus, for solids, γ will be altered; that is, $\partial \gamma / \partial \varepsilon_{ij} \neq 0$. This contribution to f_{ij} can also be seen as the result of the surface molecular configuration and/or molecular density being different from that which minimizes the free energy of the surface. It should also be pointed out that, even at elevated temperatures where the mobility of the molecules in solids is greater, it is expected that $\partial\gamma/\partial\varepsilon_{ij} \neq 0$ so that the surface tension and surface stresses will, generally, not be numerically equal in solids [77].

There are, nevertheless, situations in which the shape of the crystal will be determined solely by the surface tension of the solid rather than the surface stresses. An example is the shape of crystals growing from their melt. The process of crystal growth from its melt or solution can be viewed as a build-up of monolayers of molecules from the liquid phase onto the crystal surfaces. These monolayers are added to the existing surfaces such that each subsequent surface remains parallel during growth. The liquid molecules will deposit preferentially onto the solid surfaces such that the overall free energy of the crystal is minimized. Because the molecules of the liquid phase can freely adhere to the solid surface, the process of crystal growth represents the creation of new solid surface, and not the stretching or deformation of the existing surface; therefore, the representative work term in the process of crystal growth is γ and not f.

The equilibrium shape of the crystal will be such that the free energy of the system is minimized. In the case of fluid–fluid systems, where the surface tension is isotropic, the equilibrium shape for a liquid drop is a sphere. For crystals, on the other hand, surface tension will usually depend on the crystallographic orientation of the surface. A Laplace-type equation (Equation 10.82) of capillarity for solids can be derived from the minimization of the free energy of the system, provided that the state of stress within the solid reduces to the isotropic pressure, P_s. Only under a state of isotropic pressure is it possible to define a chemical potential, μ, for the solid phase. The definability of a chemical potential for the solid phase is essential in the derivation of the Gibbs–Thomson equation. The equilibrium equation

$$P_s - P_l = 2\frac{\gamma_{sl}^i}{\lambda^i}, \tag{10.82}$$

is very similar to the Laplace equation of capillarity for a liquid drop in equilibrium with its vapor. In Equation 10.82, $P_s - P_l$ will be constant for the system in equilibrium so that the right-hand side of the equation can be rewritten as

$$\gamma_{sl}^1 / \lambda^1 = \gamma_{sl}^2 / \lambda^2 = \ldots = \gamma_{sl}^i / \lambda^i = \text{constant}, \tag{10.83}$$

where λ^i denotes the length normal to the ith face extended into the crystal center (refer to Figure 10.6). Equation 10.83 represents the Gibbs–Wulff theorem [78], which states that the surfaces with the higher interfacial tensions will be the ones that are furthest away from the center of the crystal.

In summary, the existence of a Laplace-type equation of capillarity for solids requires that the stresses within the bulk be isotropic so that a thermodynamic pressure, P, can be defined for the solid. The shape of the crystal will then satisfy the

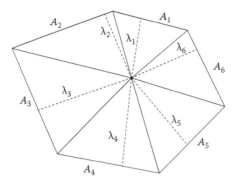

FIGURE 10.6 Geometrical variables defining the size and shape of a crystal. Terms A_i and λ_i represent the surface area and distance from the center of face i, respectively.

Gibbs–Wulff theorem, Equation 10.83, such that the solid–melt interfaces are planar. Also, and most importantly, it is necessary that the process of crystal growth be dictated solely by the interfacial tensions, $\gamma_{sl}{}^i$ (no elastic deformations of the surface are allowed). The implication of Equation 10.83 is that while surface tension, γ, plays an important role in the determination of the equilibrium shape of crystals growing from a melt, it cannot change the shape of the solid–melt interface from that of a plane. That is, growth of a crystal surface is accomplished by the addition of layers of molecules that are parallel to the original surface. If the planar surface of a crystal were to become curved, such a process would involve the deformation of the surface, which would inevitably introduce surface stresses so that Equation 10.83 would no longer be applicable.

The derivation of the Gibbs–Thomson equation inherently assumes that the Laplace equation of capillarity is applicable to solid–liquid systems. As was shown above, a Laplace-type equation does exist for solids and, therefore, the Gibbs–Thomson equation is correct for solids; however, its applicability is restricted to crystals whose bulk pressure is isotropic and whose shape satisfies the Gibbs–Wulff relation. In other situations, such as the case of a grain boundary intersecting with a surface in a temperature gradient, shown schematically in Figure 10.7, the profile of the solid surface near the grain boundary is not "Laplacian" (in the liquid–solid sense) and the Gibbs–Thomson equation cannot be used to describe the shape of the interface. Thus, the technique of grain boundary groove profile analysis to determine solid–melt interfacial tensions, γ_{sl}, is incorrect. Because equilibrium shapes of crystals involve only planar interfaces, the observed shape of the grain boundary grooves is most likely the result of stresses in the solid–melt interfaces due to the presence of the grain boundaries.

10.5.2 GRAIN BOUNDARY ENERGY AND SOLID–MELT INTERFACIAL TENSIONS

It is assumed that, when a grain boundary intersects a free solid–melt interface, the interface will be distorted so as to balance the grain boundary interfacial energy [14]. This balancing of interfacial energies is similar to what is observed when a

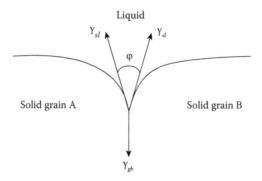

FIGURE 10.7 Schematic of a solid–liquid interface that is intersected by a grain boundary.

liquid drop is placed on a solid substrate. The contact angle formed by the liquid drop on the surface is related to the pertinent interfacial tensions via the Young equation (Equation 10.1). The equation for the energy balance for the system shown schematically in Figure 10.7 is expressed by

$$\gamma_{gb} = 2\gamma_{sl}\cos(\varphi/2),\qquad(10.84)$$

where γ_{gb} represents the grain boundary interfacial tension and φ is the dihedral angle defining the groove. In Equation 10.84, it is assumed that the surface tension of the solid is independent of orientation (isotropic) so that the shape of the groove is symmetric about the grain boundary. The presence of a temperature gradient, as is the case in grain boundary groove shape analysis experiments, can distort the shape of the groove, but at the intersection point between the grain boundary and the solid–melt interface, Equation 10.84 must still be satisfied [14].

The term γ_{gb} requires closer examination. The grain boundary region has often been treated as a thin layer of subcooled liquid separating two crystals and having a thickness of more than 10 atomic spacings [79]. Using this model, the associated grain boundary energy would be $\gamma_{gb} = 2\gamma_{sl}$. However, a grain boundary should be viewed as an interface where two crystals of different orientation contact, and where the thickness of the interface is much less than 10 atomic spacings [8]. Therefore, γ_{gb} should simply reflect the interfacial tension between two solids that have different solid surface tensions, γ_{sv}. It is often assumed that, near the melting point, the surface tension anisotropy of the crystal is minimal [8]; thus, as was discussed in Section 10.4.2, the interfacial tension between two solids that have the same surface tension, γ_{sv}, should be equal to zero; that is, $\gamma_{gb} = 0$ for this special case. Under these circumstances, for Equation 10.84 to be satisfied, the dihedral angle, φ, should have a value of 180° and the solid–liquid interface would be planar, as shown in Figure 10.8. Again, the effects of the solid–melt interfacial tension on the shape of the interface would not be noticeable under such circumstances.

More correct models for describing grain boundaries are the so-called dislocation models. In these models, the grain boundary is viewed as a transition lattice formed

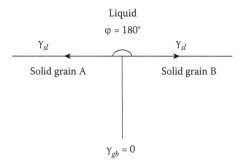

FIGURE 10.8   Energy balance at a grain boundary for the special case where the solid surface tensions, γ_{sv}, of both grains are identical.

by an array of dislocations [80]. Dislocations are defects in crystals that result from the misfit between atomic layers. The crystal lattice is distorted by the presence of dislocations. Dislocations are therefore a source of internal stresses and it is no longer possible to define a thermodynamic pressure, P, in crystals that have dislocations present. The presence of stresses also makes it difficult to define a chemical potential for the solid phase. Under these conditions, it is not possible to apply Equation 10.83 to describe the shape of the solid–melt interface. The energy associated with a grain boundary can be obtained from the work of Read and Shockley [81], who found that the energy of a low-angle dislocation grain boundary is given by

$$\gamma_{gb} = E_0\Theta\left(A_0 - \ln\Theta\right), \tag{10.85}$$

where E_0 is dependent on the elastic constants of the material and A_0 is related to the core energy of a dislocation; Θ represents the misorientation angle between the contacting crystals. It is obvious from Equation 10.85 that the quantity defined as grain boundary energy is not related in any way to surface tension. As was explained in Section 10.4.2, in terms of van der Waals interactions, surface tensions for dispersive systems arise from the fluctuating electromagnetic fields that extend beyond the boundaries of macroscopic bodies. Surface tension can be calculated from macroscopic quantities such as the refractive index and the dielectric constant of the material. The quantity that is calculated from Equation 10.85 depends on the elastic properties of the material and is therefore more related to the state of stress of the solid, due to the presence of defects, than to surface tensions. While the interfacial tension of a grain boundary in isotropic systems would be equal to zero, the grain boundary energy, as calculated from Equation 10.85, would certainly not be equal to zero.

   It has been argued to the contrary [82] that surface stresses cannot be active in grain boundary groove experiments since new molecules are freely added to the existing solid surface. However, due to the long range order that characterizes solids, it should be expected that the configuration of new molecules solidifying near the grain boundary will be affected by any stress already developed in the pre-existing material.

In summary, the equilibrium profile of a solid–melt interface that is intersected by a grain boundary is not determined by the solid–melt interfacial tension. The effect of surface tension on the equilibrium profile would be to maintain the interface flat. On the other hand, if surface stresses are involved then the equilibrium profile of the groove may take any shape so as to maintain the system in mechanical equilibrium. Therefore, the profile of the grain boundary groove cannot be described by the Gibbs–Thomson equation, which is a function of surface tension and not of surface stresses.

Surface stresses in solids can be significantly different from surface tension values. Flueli and Solliard [83] calculated the surface stress in gold particles by determining the change in lattice parameter with particle size. Values of $f = 3.9$ J/m^2 and 4.4 J/m^2 (for different crystal orientations) were obtained for surface stress, which are about two times larger than the surface tension of gold: $\gamma = 1.7$–2 J/m^2.

10.6　CONCLUSIONS

In this chapter, independent approaches were used to calculate solid–liquid interfacial tensions for the special case of a solid-melt system. The results from the independent approaches—the Lifshitz and gradient theories—are in relatively good agreement with those from the contact angle/equation of state approach. The relative agreement in the calculated γ_{sl} values among the three approaches suggests that the methods based on the Gibbs–Thomson equation are incorrect.

The results from the methods based on the Gibbs–Thomson equation are one to two orders of magnitude larger than those from the other methods discussed in this chapter. Such discrepancies are most likely the result of the interdependence of surface tension and surface stresses in the case of solids, which is not considered in the usual application of this equation. Also, the applicability of the Gibbs–Thomson equation to a solid–melt interface that is intersected by a grain boundary is questionable in view of the fact that the equation can only be used in crystals that obey the Gibbs–Wulff theorem.

REFERENCES

1. A. W. Neumann, D. R. Absolom, W. Zingg, and C. J. van Oss. *Cell Biophysics* 1 (1979): 79.
2. D. Chattopadhyay, J. F. Rathman, and J. J. Chalmers. *Biotechnology and Bioengineering* 48 (1995): 649.
3. E. I. Vargha-Butler, T. K. Zubovits, H. A. Hamza, and A. W. Neumann. *Journal of Dispersion Science and Technology* 6 (1985): 357.
4. S. N. Omenyi, A. W. Neumann, and C. J. van Oss. *Journal of Applied Physics* 52 (1981): 789.
5. M. J. Stowell. *Philosophical Magazine* 22 (1970): 1.
6. D. Turnbull. *Journal of Chemical Physics* 18 (1950): 198.
7. G. F. Bolling and W. A. Tiller. *Journal of Applied Physics* 31 (1960): 1345.
8. L. E. Murr. *Interfacial Phenomena in Metals and Alloys*. Addison-Wesley, Reading, MA, 1975.

9. H. Tavana and A. W. Neumann. *Advances in Colloid and Interface Science* 132 (2007): 1.
10. D. R. H. Jones. *Journal of Materials Science* 9 (1974): 1.
11. R. Defay and I. Prigogine. *Surface Tension and Adsorption*. Longmans, London, 1966.
12. D. Turnbull and R. E. Cech. *Journal of Applied Physics* 21 (1950): 804.
13. D. Turnbull. *Journal of Applied Physics* 20 (1950): 411.
14. D. P. Woodruff. *The Solid–Liquid Interface*. Cambridge University Press, London, 1973.
15. S. Balibar and F. Caupin. *C.R. Physique* 7 (2006): 988.
16. D. R. Uhlmann, G. Kritchevsky, R. Straff, and G. Scherer. *Journal of Chemical Physics* 62 (1975): 4896.
17. C. L. Jackson and G. B. McKenna. *Journal of Chemical Physics* 93 (1990): 9002.
18. R. C. Sill and A. S. Skapski. *Journal of Chemical Physics* 24 (1956): 644.
19. A. S. Skapski, R. Billups, and A. Rooney. *Journal of Chemical Physics* 26 (1957): 1350.
20. A. S. Skapski, R. Billups, and D. Casavant. *Journal of Chemical Physics* 31 (1959): 1431.
21. G. E. Nash and M. E. Glicksman. *Philosophical Magazine* 24 (1971): 577.
22. D. R. H. Jones. *Philosophical Magazine* 27 (1973): 569.
23. S. C. Hardy. *Philosophical Magazine* 35 (1977): 471.
24. S. Akbulut, Y. Ocak, U. Böyük, M. Erol, K. Keşioğlu, and N. Maraşli. *Journal of Applied Physics* 100 (2006): 123505.
25. A. W. Neumann, M. A. Moscarello, and R. M. Epand. *Biopolymers* 12 (1973): 1945.
26. S. N. Omenyi, A. W. Neumann, and C. J. van Oss. *Journal of Applied Physics* 52 (1981): 789.
27. J. K. Spelt, D. R. Absolom, W. Zingg, C. J. van Oss, and A. W. Neumann. *Cell Biophysics* 4 (1982): 117.
28. E. Moy. *Approaches for Estimating Solid/Liquid Interfacial Tensions*. PhD Thesis, University of Toronto, 1989.
29. J. W. Gibbs. *The Scientific Papers of J. Willard Gibbs*. Dover, New York, 1961.
30. J. W. Cahn and J. E. Hilliard. *Journal of Chemical Physics* 28 (1958): 258.
31. S. J. Hemingway, J. R. Henderson, and J. S. Rowlinson. *Faraday Symposia of the Chemical Society* 16 (1981): 33.
32. R. Guermeur, F. Biquard, and C. Jacolin. *Journal of Chemical Physics* 82 (1985): 2040.
33. B. S. Carey, L. E. Scriven, and H. T. Davis. *AICHE Journal* 26 (1980): 705.
34. G. F. Teletzke, L. E. Scriven, and H. T. Davis. *Journal of Colloid and Interface Science* 87 (1982): 550.
35. R. E. Benner, Jr., L. E. Scriven, and H. T. Davis. *Faraday Symposia of the Chemical Society* 16 (1981): 169.
36. J. G. Kirkwood and F. P. Buff. *Journal of Chemical Physics* 17 (1949): 338.
37. V. Bongiorno, L. E. Scriven, and H. T. Davis. *Journal of Colloid and Interface Science* 57 (1976): 462.
38. H. T. Davis and L. E. Scriven. *Advances in Chemical Physics* 49 (1982): 357.
39. A. J. M. Yang, P. D. Fleming, III, and J. H. Gibbs. *Journal of Chemical Physics* 64 (1976): 3732.
40. H. Margenau and G. M. Murphy. *The Mathematics of Physics and Chemistry*. D. van Nostrand, Princeton, 1943.
41. D.-Y. Peng and D. B. Robinson. *Industrial & Engineering Chemistry Fundamentals* 15 (1976): 59.
42. R. C. Reid, J. M. Prauswitz, and T. K. Sherwood. *The Properties of Gases and Liquids*. McGraw-Hill, New York, 1977.

43. M. Benedict, G. B. Webb, and L. C. Rubin. *Journal of Chemical Physics* 8 (1940): 334.
44. O. Redlich and J. N. S. Kwong. *Chemical Reviews* 44 (1949): 233.
45. C. F. Gerald. *Applied Numerical Analysis*. Addison-Wesley, Toronto, 1980.
46. O. C. Zienkiewicz and K. Morgan. *Finite Elements and Approximation*. John Wiley, New York, 1983.
47. D. B. Hough and L. R. White. *Advances in Colloid and Interface Science* 14 (1980): 3.
48. H. W. Fox and W. A. Zisman. *Journal of Colloid Science* 5 (1950): 514.
49. S. N. Omenyi. *Attraction and Repulsion of Particles by Solidifying Melts*. PhD Thesis, University of Toronto, 1978.
50. H. C. Hamaker. *Physica* 4 (1937): 1058.
51. G. Frens and J. Th. G. Overbeek. *Journal of Colloid and Interface Science* 38 (1972): 376.
52. J. N. Israelachvili and D. Tabor. *Progress in Surface and Membrane Science* 7 (1973): 1.
53. I. E. Dzyaloshinskii, E. M. Lifshitz, and L. P. Pitaevskii. *Advanced Physics* 10 (1961): 165.
54. E. M. Lifshitz. *Soviet Physics-JETP* 2 (1956): 73.
55. B. W. Ninham and V. A. Parsegian. *Biophysical Journal* 10 (1970): 646.
56. L. D. Landau and E. M. Lifshitz. *Electrodynamics of Continuous Media*. Pergamon Press, Oxford, 1960.
57. V. A. Parsegian and B. W. Ninham. *Nature* 224 (1969): 1197.
58. J. Mahanty and B. W. Ninham. *Dispersion Forces*. Academic Press, New York, 1976.
59. J. N. Israelachvili. *Journal of the Chemical Society, Faraday II* 69 (1973): 1729.
60. R. E. Johnson and R. H. Dettre. *Langmuir* 5 (1989): 294.
61. Beilstein Institute. *Beilsteins Handbuch der Organischen Chemie*. Springer-Verlag, New York, 1918–1983.
62. R. Nasini and O. Bernheimer. *Gazetta Chimica Italiana* 15 (1855): 59.
63. S. B. Hendricks and M. E. Jefferson. *Journal of the Optical Society of America* 23 (1933): 299.
64. J. Timmermans. *Physicochemical Constants of Pure Organic Compounds*. Elsevier, New York, 1965.
65. R. W. Dornte and C. P. Smyth. *Journal of the American Chemical Society* 52 (1930): 3546.
66. J. D. Hoffmann and C. P. Smyth. *Journal of the American Chemical Society* 72 (1950): 171.
67. J. N. Israelachvili. *Intermolecular and Surface Forces*. Academic Press, London, 1985.
68. C. Herring. In *The Physics of Powder Metallurgy*. Edited by W. E. Kingston, 143. McGraw-Hill, New York, 1951.
69. P. R. Couchman, W. A. Jesser, D. Kuhlmann-Wilsdorf, and J. P. Hirth. *Surface Science* 33 (1972): 429.
70. P. R. Couchman and W. A. Jesser. *Surface Science* 34 (1973): 212.
71. P. R. Couchman and D. H. Everett. *Journal of Electroanalytical Chemistry* 67 (1976): 382.
72. P. R. Couchman and D. H. Everett. *Journal of Colloid and Interface Science* 52 (1975): 410.
73. R. G. Linford. *Chemical Reviews* 78 (1978): 81.
74. H.-J. Butt and R. Raiteri. In *Surface Characterization Methods: Principles, Techniques, and Applications*. Edited by A. J. Milling, 1. Marcel-Dekker, New York, 1999.
75. R. Shuttleworth. *Proceedings of the Physical Society (London)* A63 (1950): 444.
76. L. Makkonen. *Langmuir* 18 (2002): 1445.

77. J. M. Blakely. *Introduction to the Properties of Crystal Surfaces.* Pergamon Press, Oxford, 1973.
78. G. Wulff. *Zeitschrift fur Kristallographie* 34 (1901): 449.
79. N. F. Mott. *Proceedings of the Royal Society* 60 (1948): 391.
80. J. M. Burgers. *Proceedings of the Royal Society* 52 (1940): 52.
81. W. T. Read and W. Shockley. *Physical Review* 78 (1950): 275.
82. L. Makkonen. *Langmuir* 16 (2000): 7669.
83. M. Flueli and C. Solliard. *Surface Science* 156 (1985): 487.

11 Wettability and Surface Tension of Particles

Yi Zuo, Dongqing Li, and A. Wilhelm Neumann

CONTENTS

11.1   INTRODUCTION

The interfacial energetics and wettability of small particles are of technological interest in many areas of applied science. Areas where such phenomena are important include the preparation of stable suspensions of particles (e.g., color pigments in paints), the adhesion of particles to solid surfaces in various scenarios (e.g., lubrication), the dispersion of particles into a liquid or melt of a polymer, and the modification of particle surface properties through the adsorption of polymeric macromolecules or surfactants. The successful manipulation of the process being considered is largely determined by the physicochemical surface properties of the interacting surface components, and particularly the wettability and the surface tension of the particles. The general complexities of contact angle phenomena and solid surface tension measurements have been discussed in Chapters 6 and 9.

Compared to bulk materials, the measurement of surface properties, such as wettability and surface tension, of small particles is an even more difficult problem. It becomes even more complex for particles that are highly heterogeneous with respect to size and shape. The main surface properties are surface tension and surface charge density of the interacting components. It is not the intent of this chapter to discuss the methods available to measure surface charge. Those methods, which generally rely on the determination of the electrophoretic mobility of particles when suspended in liquid, were extensively reviewed elsewhere [1,2]. The purpose of this chapter is to describe the behavior of particles at liquid–vapor interfaces and the various strategies used for measuring particle wettability. These methods can be broadly divided into two categories: qualitative and quantitative approaches. This distinction is based on whether or not the technique is capable of characterizing wettability in terms of the contact angle and/or the solid surface tension. The major emphasis in this chapter will be on the quantitative approaches; however, for the sake of completeness, a brief review of several qualitative approaches will also be given.

A general review of contact angles and contact angle measurement techniques has been given in Chapters 6–8, and by Neumann [3]. In some cases, contact angle measurements on small particles can be performed by the modification of methods used for the study of extended surfaces. A typical example is small fibers below 10 μm in diameter, the wettability of which can be studied by means of the Wilhelmy plate method [4] (see Chapter 6). While this method will not be discussed further in this chapter, it should be pointed out that surface tensions and contact angles on fibers can also be measured by some of the indirect methods described in this chapter.

11.2 QUALITATIVE APPROACHES

All of the qualitative techniques suffer from the disadvantage that they are not able to express the wettability in terms of the contact angle and/or the surface tension of the particles. Generally, they are relative in the sense that they provide information as to whether one population of particles is more or less wettable than another. In view of their widespread use, it is still worthwhile to discuss them here briefly. It should also be pointed out that many of these techniques have been developed by researchers in the biological sciences, which illustrates the widespread and diverse interest in characterizing the wettability of small particles. There is no a priori reason why these techniques could not be adapted to study nonbiological particles.

11.2.1 Liquid–Liquid Contact Angle Measurement

This technique requires the use of two immiscible liquids having different densities. Typically, eight solutions of polyethylene glycol (PEG) 6,000 and dextran t-500 are formed by dissolving the appropriate mass of the polymers in tissue culture media (e.g., RPMI 1640 or Hanks Balanced Salt Solution). Equal volumes of these two solutions are mixed together and then allowed to phase separate. The less dense upper phase is PEG-rich and the more dense lower phase is dextran rich. Layers of cells or bacteria are then deposited on some substrate material (e.g., anisotropic cellulose acetate membrane) by ultrafiltration, or through adhesion to siliconized glass slides (or tissue-culture plastic). The cell layers are then immersed in a bath of PEG-rich liquid and a drop of fixed volume ($\approx$ 2 μl) of the dextran-rich phase is deposited on the surface of interest. The contact angle that this droplet makes with the surface may be recorded by using a stereomicroscope attached to a camera. Initially the contact angle is close to 180°, as expected; however, as contact is made and the dextran droplet begins to interact with the cellular material, the contact angle is reduced considerably. After approximately 5 minutes, the observed contact angles reach a stable plateau value that is taken as the effective contact angle of the specimen. In many respects, this technique is similar to the well-known two-phase partition technique developed many years ago [5]. At this time, it is still not possible to interpret and quantify the solid–liquid–liquid contact angle in terms of the interfacial tensions involved. The best that can be done is to rank materials in terms of increasing or decreasing contact angle. Liquid–vapor contact angle measurements (Chapter 6), on the other hand, can be used to derive quantitative information about the surface tension of various substrates.

Despite this disadvantage, the two-phase technique has been employed to document major differences in the surface properties of various biological cell lines, including platelets, granulocytes, lymphocytes, macrophages [6,7], vascular endothelium [6], and various strains of bacteria [8,9]. The major advantage of this technique is its extraordinary sensitivity to even small differences in cellular surface properties that are manifested as large changes in the observed contact angles.

11.2.2 TWO-PHASE PARTITION METHODS

The partition of particles between a hydrophobic liquid and water was utilized first for the (qualitative) classification of bacterial surfaces according to their hydrophobicity, in the pioneering efforts of Mudd and Mudd [10]. All of these partitioning techniques suffer from the drawback that they cannot be used to provide quantitative information about the surface properties of the particles being investigated.

Albertsson [5] introduced the use of immiscible aqueous dextran and PEG solutions in partition separation. The extent to which the particles of interest partition into the hydrophobic PEG phase is taken as a relative measure of their hydrophobicity. Stendahl et al. [11] used this method to demonstrate that the more hydrophobic bacteria are the ones that become phagocytized to the greatest extent. Using the same strains of bacteria, these results were later correlated with the quantitative contact angle method by Cunningham et al. [12]. This method has the advantage that it is very sensitive to even small changes in surface properties. The sensitivity of the technique can be considerably enhanced through the use of either dextran or PEG polymers in which various chemical groups have been substituted.

Rosenberg et al. [13–16] have introduced a novel approach for measuring particle hydrophobicity. The technique has been employed primarily for the characterization of bacteria but can, in principle, be applied to other particles as well. The technique relies on quantifying the extent of Particle adhesion to hydrocarbon (PATH). This technique may be briefly summarized as follows. To a fixed volume of aqueous suspension, at a standard particle concentration, various volumes of hydrocarbon (e.g., n-hexadecane, n-octane, or p-xylene) are added. After incubation at 30°C for 10 minutes, the binary system is vortexed for 2 minutes to ensure complete mixing. This system is then allowed to phase separate, the aqueous phase is carefully removed, and its turbidity is measured through light absorbance at 400 nm. When hydrophobic particles are tested, they attach to the hydrocarbon droplets and rise with these less dense drops following mixing. The adherent cells are therefore removed from the aqueous phase. The proportion of adherent cells is then determined by comparing the decrease in light absorbance (following PATH) with the absorbance of the aqueous suspension (prior to PATH) of known particle concentration. This technique permits the ranking of bacterial hydrophobicity but it does not allow a quantitative assessment of surface tension.

11.2.3 HYDROPHOBIC INTERACTION CHROMATOGRAPHY

The chromatographic interaction of particles with various matrix materials (e.g., phenyl- and octyl-sepharose) originally developed for protein separation has found widespread use in many studies of bacterial hydrophobicity [17–19]. In this technique, aqueous suspensions of sepharose beads with covalently bound hydrophobic moieties (e.g., phenyl or octyl groups) are usually packed into small columns, and the bacterial suspension is subsequently applied. Retention may be determined by various means, such as turbidimetric readings, colony-forming units, or radiotracer techniques. In many cases, salting-out agents are added to promote adhesion to the gel. In some cases, adherent cells can be desorbed by lowering the ionic strength of the eluent, or by adding detergent. The retention time can be correlated with data obtained by other qualitative as well as quantitative methods.

11.2.4 SALTING-OUT AGGREGATION TEST

This technique is based on the premise that the same laws governing the precipitation of protein molecules from aqueous solution hold true for the aggregation of particles; that is, the more hydrophobic the particle, the greater its tendency to precipitate out of the solution (i.e., through particle–particle interactions) at a lower concentration of salting-out agents [20]. In the original report on the use of this method with particles, bacterial cells were suspended in dilute phosphate buffer, and ammonium sulfate was added until aggregation occurred. This technique appears to correlate well with other methods in most, but not all, instances [21]. Configurational changes of cell surface structures due to the high salt concentrations may introduce errors in measurement.

11.3 QUANTITATIVE APPROACHES

Many methods have been developed to quantitatively determine the wettability of different particles, such as polymer powders, coal powders, fibers, microporous membranes, self-assembled nanoparticles, liposomes, and biological cells.

One category of these methods relies on direct contact angle measurement on the compressed pellets of the particles. Once the contact angle is known, the solid surface tension of the particles, γ_{sv}, can be determined from Young's equation, for example in combination with the equation of state approach (Chapter 9). As will be discussed later, however, direct contact angle measurement may not be applicable for some particles and may introduce serious errors due to inappropriate surface preparation. To overcome these problems, many indirect methods have been developed for measuring contact angle and/or surface tension of particles. Four of these methods will be introduced following the direct contact angle method. They are: heat of immersion, film flotation, sedimentation volume, and capillary penetration methods. Focus will be given to the last two methods that determine the surface tension of particles, γ_{sv}, without requiring knowledge of the contact angle. Another novel approach, known as the solidification front technique, will be covered in Chapter 12.

11.3.1 DIRECT CONTACT ANGLE MEASUREMENTS

The theoretical aspects of contact angles have been discussed in Chapters 1 and 9. It was shown that the values of the solid–vapor surface tension, γ_{sv}, and the solid–liquid interfacial tension, γ_{sl}, can be determined by interpreting contact angle data in terms of Young's equation

$$\gamma_{lv} \cos \theta = \gamma_{sv} - \gamma_{sl}, \tag{11.1}$$

and an equation of state for interfacial tensions

$$\gamma_{sl} = f(\gamma_{lv}, \gamma_{sv}). \tag{11.2}$$

Therefore, direct contact angle measurement represents an important method in the quantitative approaches.

A general review of the techniques for contact angle measurement has been given in Chapter 6. The choice of method for measuring contact angles depends directly on the geometry of the system. For example, the sessile drop is the most convenient method for measuring contact angle on a smooth plane surface. However, this method is not applicable on the inner surface of a capillary tube, or for fine textile fibers and powders. For these systems, the direct observational problems, the optics, the mechanics of manipulation, and the manner in which the Laplace equation of capillarity is involved in the measurement, all vary widely.

Several approaches are available for measuring contact angles on layers of particles. The success of these methods appears to depend on the nature of the particles themselves. If the particles are reasonably pliable, then it is fairly easy to measure accurate and reproducible contact angles on layers of these particles. For example, contact angles have been measured successfully on a wide variety of biological and other highly hydrated materials. If, however, the particles are rigid, success generally has not been achieved due to the inevitable difficulty with surface roughness.

In all cases, the equipment required for direct contact angle measurements is a horizontally leveled surface on which the material of interest (i.e., layers of the particles) is placed. A liquid drop is deposited on the particle layer with a micropipette. It may be desirable that the pipet should remain in contact with the drop during the measurement process. This allows fluid to be slowly added to the drop, thereby creating a slowly advancing three-phase line. The advancing contact angle determined in this way is the correct value for use in Young's equation (see Chapter 7 for detailed explanation), provided that the measured angle is not affected by surface roughness. In all cases, both sides of the droplet should be measured in order to check the symmetry of the drop. If the drop is not symmetrical, the observed readings might have to be disregarded. The volume of the drop is then slowly increased by adding more liquid and the contact angles on both sides of the drop are again determined. Several techniques are available for measuring the contact angle of the sessile drop, such as the use of a telescope with a goniometer eyepiece, and Axisymmetric Drop Shape Analysis (ADSA) either from the side view or top view of the drop [22–24] (see Chapters 3, 4 and 6 for details of ADSA).

Preparing an appropriate surface of the test particles is the key to the direct contact angle measurements. Only a measurement on a carefully prepared, flat, smooth, homogeneous, rigid, and insoluble solid surface reveals the Young contact angle, thus permitting the determination of solid surface tension. For biological and other highly hydrated particles, layers of these particles may be prepared by deposition from suspension onto a hydrophobic surface, such as siliconized glass, or by ultrafiltration on anisotropic cellulose acetate membranes. More rigid particles can be processed to form flat smooth surfaces suitable for contact angle measurements. Methods for this purpose include heat pressing, solvent casting, vapor deposition, and compressing powder cakes. Clearly, such techniques should ensure that the prepared flat surface has the same surface properties as the particles themselves. Details of these and other surface preparation techniques are given in Chapter 6.

It should be noted that preparing an ideal surface of solid particles for Young contact angle measurement is not a trivial task and sometimes not even possible. Even closely packed polymer beds or pellets are usually rough and porous. It is well-known that serious problems arise when contact angles of sessile drops are measured on such surfaces of porous materials. Surface roughness, heterogeneity, and the penetration of the liquid drop into the porous material may affect the measured contact angles, causing meaningless results in the surface energetic interpretation of these contact angle data [3]. Under these circumstances, indirect methods as follows should be used.

11.3.2 HEAT OF IMMERSION

Immersion of a solid in a liquid is usually accompanied by the release of heat, called the heat of immersion (ΔH_i). The heat of immersion is defined as the heat liberated per square centimeter of the particles immersed in the liquid, and is related to the contact angle of the particles [25,26]. Therefore it is possible to characterize the wettability of particles by measuring the heat of immersion. The principles of this method are briefly given as follows.

The free energy of immersion (ΔF_i) of the particles is

$$\Delta F_i = \gamma_{sl} - \gamma_{sv},$$ (11.3)

where γ_{sl} and γ_{sv} are the solid–liquid and solid–vapor surface tensions of the particles, respectively.

Combining Young's equation (Equation 11.1) with Equation 11.3 gives

$$\Delta F_i = -\gamma_{lv} \cos\theta.$$ (11.4)

The enthalpy of immersion (ΔH_i), that is, the heat of immersion, at a constant temperature and volume, is related to ΔF_i by

$$\Delta H_i = \Delta F_i + T\Delta S = \Delta F_i - T\frac{d\Delta F_i}{dT},$$ (11.5)

where S is the entropy, and T is the absolute temperature. Substituting Equation 11.4 into Equation 11.5 gives

$$\Delta H_i = T\frac{d(\gamma_{lv}\cos\theta)}{dT} - \gamma_{lv}\cos\theta.$$ (11.6)

Equation 11.6 can be rewritten to

$$\cos\theta = \frac{\dfrac{d(\gamma_{lv}\cos\theta)}{dT}T - \Delta H_i}{\gamma_{lv}}.$$ (11.7)

Equation 11.7 can be solved numerically to determine the contact angle from the heat of immersion, ΔH_i, measured using a calorimeter. However, the heat of immersion method involves several serious complications: first, the specific surface area of the particles/powder must be known [e.g., from Brunauer-Emmett-Teller (BET) measurements]. Second, one needs to ensure that the experimental heat of immersion is indeed the heat of wetting, and that there are no contributions from the partial dissolution of the particles. Third, the heat of wetting (an enthalpy) is not only related to the contact angle (a free-energy type of quantity), but also to the temperature dependence of the contact angle (see Equation 11.7). Strictly speaking, one will not normally have information on the temperature dependence of contact angles without already knowing the contact angle as well. According to Young's equation, $d(\gamma_{lv}\cos\theta)/dT$ in Equation 11.7 can be approximated by the temperature dependence of solid surface tension; that is, $d\gamma_{sv}/dT$. It is known that $d\gamma_{sv}/dT$ can span a large range from -0.04 to -0.13 across different solid surfaces [27]. Thus, the heat of immersion method will normally only provide relative and semiquantitative information.

11.3.3 Film Flotation

It is a well-known fact that small particles can float on a liquid surface even if their density is greater than that of the liquid. The prerequisite for this flotation is a relatively large contact angle. For cylindrical and spherical particles, mathematical solutions are available to relate the depth of immersion of such particles to the contact angle [28–30](see also Chapter 7). For irregularly shaped particles, such an analysis is not possible. Nevertheless, some qualitative information can be obtained from the floatability of small particles. The simplest strategy is to dust the powder onto the liquid surface and record the time required until the powder sinks. This may be possible even in situations where the contact angle is not large. If the contact angle is large, the powder may not sink at all. In this case, one can compare the mobility of different powders sprinkled on the liquid surface when a stream of air is directed obliquely at the surface. The larger the contact angle, the larger the mobility will be, since with large contact angles the particles will be immersed less deeply [31]. Based on the floatability of particles, a technique called film flotation [32–42] has been developed by Fuerstenau and coworkers to characterize the wettability of solid particles.

Generally speaking, film flotation is an experimental technique designed to find the surface tension of a liquid that will just wet a solid particle; that is, the so-called critical surface tension of wetting. In a film flotation experiment, closely sized particles are sprinkled onto the surface of the wetting liquid (such as an aqueous methanol solution) and the fraction of particles that sink into the liquid is determined. Depending on the wetting characteristics of the material and the surface tension of the test liquid, the particles either remain at the liquid–vapor interface or are immediately engulfed into the liquid. At a particular surface tension, those particles that do not sink into the wetting liquid are considered to be hydrophobic, while those that are imbibed into the liquid are considered to be hydrophilic. After performing a film flotation test, the hydrophobic and hydrophilic fractions are recovered, dried, and weighed. The percentage by weight of the hydrophobic fraction of the particles for each solution is plotted as a function of the surface tension of the solution. From this curve, four parameters for defining the wetting characteristics of the particulate samples may be determined: the mean critical wetting surface tension, the minimum and the maximum wetting surface tensions, and the standard deviation of the wetting surface tension [36,43].

Recently, this method has been employed to measure the critical wetting surface tensions of particles of sulfur, silver iodide, methylated glass beads, quartz, paraffin-wax-coated coal, and surfactant-coated pyrite. Generally, Fuerstenau and coworkers [32–42] found that the film flotation technique is sensitive to the surface hydrophobicity and the heterogeneity of the particles. It was also found that particle size, particle shape, particle density, film flotation time, and the nature of the wetting liquids have negligible effects on the results of film flotation. Moreover, the liquid and the solid particles used in the experiments must not have any chemical interactions.

11.3.4   SEDIMENTATION VOLUME

11.3.4.1   Theory

Sedimentation experiments are a well-established technique to study the stability of dispersions of powders in liquids [44–46]. While, in many cases, the behavior of such systems is governed by van der Waals and electrostatic interactions, it is to be expected that for polymer particles, particularly in nonaqueous media, the effect of the electrostatic interactions may be considered negligible. It is also of importance to note that van der Waals interactions can be related to surface tensions [47–49]. The van der Waals interaction between two parallel, infinitely extended flat surfaces in a liquid medium was first calculated by Hamaker [48]. For the work done by the van der Waals force in bringing these surfaces from infinity to a distance d_0, Hamaker obtained

$$W = -\frac{A_{132}}{12\pi d_0^2}, \tag{11.8}$$

where the coefficient A_{132} has subsequently been called the "Hamaker coefficient." The indices 1, 2, and 3 refer, respectively, to the solid (1), solid (2), and liquid (3). If we assume that d_0 is so small that we, in fact, have contact between the two solid phases, this work is the thermodynamic free energy of adhesion

$$\Delta F^{adh} = \gamma_{12} - \gamma_{13} - \gamma_{23}, \tag{11.9}$$

where γ denotes interfacial tensions.

It has been shown that the free energy of adhesion can be positive, negative, or zero, implying that van der Waals interactions can be attractive as well as repulsive [47,50,51] (see Chapter 12 for an example of repulsive van der Waals interactions). While in the above, Equation 11.8 can, strictly speaking, be expected to hold only for systems that interact by means of dispersion forces only, there are no restrictions on Equation 11.9. Since Equation 11.9 describes fundamental patterns of the behavior of particles, including macromolecules, independent of the type of molecular interactions present, it was found to be convenient to define an "effective Hamaker coefficient" that reflects the free energy of adhesion [47].

While van der Waals interactions between unlike solids in a third medium may be attractive as well as repulsive, it is clear from the underlying thermodynamics [50,52] that like particles can only attract each other, with zero interaction in the limiting case. For interaction between particles of the same kind embedded in a liquid, this interaction is governed by the free energy of cohesion

$$\Delta F^{coh} = -2\gamma_{sl}. \tag{11.10}$$

Because solid–liquid interfacial tensions are always positive, or zero as a limiting case, it follows that $\Delta F^{coh} \leq 0$; implying that, in the absence of electrostatic forces, there will always be an attraction between like particles suspended in a liquid (with no interaction, as a limiting case, at $\gamma_{sl} = 0$).

It can be expected that the sedimentation volume (V_{sed}) of particles will show extrema in the special case of zero driving force; that is, $\Delta F^{coh} = 0$. There are at least two possible patterns of behavior, depending on whether or not agglomeration of the particles at the early stages of sedimentation is possible.

1. If there is no agglomeration at nonzero values of the free energy of cohesion, then, for zero free energy of cohesion, least close packing of the sediment and, hence, a maximum in the sedimentation volume V_{sed} is expected.
2. If there is agglomeration at nonzero values of the van der Waals attraction in the early stages of sedimentation, then this agglomeration would cease when the van der Waals attraction approaches zero. Since the irregularly shaped aggregates resulting from agglomeration do not pack well, one would expect minimum sedimentation volume at zero van der Waals attraction.

For systems of solid particles in a single component liquid, the solid–liquid interfacial tension, γ_{sl}, is a function of the liquid–vapor surface tension, γ_{lv}, and the solid–vapor surface tension, γ_{sv}; that is, $\gamma_{sl} = f(\gamma_{lv}, \gamma_{sv})$, as predicted by the equation of state (see Chapters 8 and 9). A basic feature of the equation of state is that $\gamma_{sl} = 0$ when $\gamma_{lv} = \gamma_{sv}$. In turn, this will result in a maximum value for ΔF^{coh}; that is, $\Delta F^{coh} = 0$ when $\gamma_{lv} = \gamma_{sv}$. To illustrate this point, Figure 11.1 shows a plot of the free energy of cohesion, ΔF^{coh}, against the liquid surface tension, γ_{lv}, for hypothetical particles having a surface tension $\gamma_{sv} = 20$ mJ/m². It is apparent that $\Delta F^{coh} = 0$ occurs at $\gamma_{lv} = \gamma_{sv} = 20$ mJ/m². In view of the possibility that the sedimentation volume of the particles may show extrema when $\Delta F^{coh} = 0$, such an extremum in the sedimentation volume may provide a means to determine the solid–vapor surface tension of the particles. The solid–vapor surface tension, γ_{sv}, of particles would be equal to γ_{lv}, the surface tension of the suspending liquid at which the sedimentation volume extremum occurs.

Therefore, to determine the particle surface tension by using the sedimentation volume technique, the required basic procedures in the experiments may be summarized as follows:

1. Prepare a series of liquids with a surface tension range covering the surface tension of the particles of interest in suitable graduated cylinders.
2. Put an equal amount of the particles into each liquid.
3. Determine the liquid surface tension, γ_{lv}^{*}, at which an extremum in the sedimentation volume occurs.

The surface tension of the particles can then be determined as $\gamma_{sv} = \gamma_{lv}^{*}$.

In practice, sedimentation volume experiments are performed with binary liquid mixtures as the suspending liquids, in order to have a sufficiently large range of surface tension and to be able to adjust the liquid surface tension to any specific value.

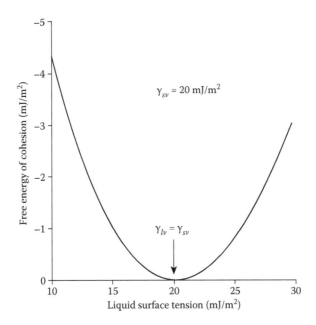

FIGURE 11.1 The free energy of cohesion, ΔF^{coh}, of hypothetical particles ($\gamma_{sv}= 20$ mJ/m^2) as a function of the surface tension of the suspending liquid, γ_{lv} .

However, it should be noted that the equation of state, $\gamma_{sl} = f(\gamma_{lv}, \gamma_{sv})$, is applicable only to single-component liquid systems, for the reasons discussed in Chapter 9. Nevertheless, the sedimentation volume technique has been found experimentally to be insensitive to the use of either single component liquids or binary liquid mixtures (detailed later). Therefore, it will become apparent that sedimentation volume is a simple and reliable method to determine particle surface tensions, and by implication, the wettability of particles.

11.3.4.2 Sedimentation of Polymer Particles in Binary Liquid Mixtures

Sedimentation experiments were performed [44] with the following polymer powders: (1) polytetrafluoroethylene (Teflon), PTFE [two samples with different particle size were used: Grade 1 (Polyscience Inc.) and No. 6 (DuPont)]; (2) polyvinylidenefluoride, PVDF (Polyscience Inc.); (3) polyvinylfluoride, PVF (Polyscience Inc.); (4) high-density polyethylene, HDPE (DuPont); (5) polyhexamethylene adipamide (nylon 6,6), PA 66 (Commercial Plastics); and (6) polysulfone, PSF (Union Carbide Corp.).

The sedimentation volume, V_{sed}, of the polymer powders was determined in mixtures of pairs of nonpolar as well as polar liquids. The liquid combinations were chosen such that the surface tension, γ_{l_1v}, of one liquid was lower, and that of the second one, γ_{l_2v}, was higher than that of the polymer particles, γ_{sv}.

There are several requirements that the suspending liquids must satisfy in order to be used for the sedimentation experiments: (1) they should be chemically inert with the solid particles; (2) the boiling temperature should be reasonably high to minimize evaporation; (3) the density of the liquid should be less than that of the particles; (4) liquids with zero or nearly zero dipole moment as well as those with higher dipole moment should be used in order to cover a wide polarity range; (5) the liquid components should be miscible in all ratios with each other; and (6) the surface tensions, γ_{lv}, of the mixtures should cover the surface tension of the particles, γ_{sv}.

The liquids selected according to the above criteria, together with their relevant boiling points [53,54], density [53], dipole moments [53,55,56], and surface tensions [56,57], are summarized in Table 11.1.

For the sedimentation experiments, the following seven pairs of liquids were chosen from the 10 liquids listed in Table 11.1: (1) n-hexane/n-hexadecane; (2) n-octane/ tetralin (1,2,3,4-tetrahydronaphthalene); (3) diethyl ether/tetralin; (4) diethyl ether/ n-hexanol; (5) diethyl ether/ethylene glycol (1,2-ethanediol); (6) n-propanol/cyclo-hexanone; and (7) n-propanol/2,2'-thiodiethanol. These liquids were Aldrich, Baker, and/or Fluka reagents (certified laboratory grade).

The liquid–vapor surface tension, γ_{lv}, of the different liquid mixtures, as well as that of the pure solvents, was determined by means of the modified Wilhelmy method [58] at 20°C prior to the sedimentation measurements.

Prior to the actual sedimentation experiments, the polymer powders were thoroughly agitated in the appropriate liquids in order to break up any aggregates and to displace air bubbles. For this purpose, a fixed amount of a given polymer powder was placed into small polyethylene microcentrifuge tubes with well-fitting lids. The amount of polymer powder used varied from 0.05 to 0.20 g, depending on the density and/or the particle size of the powder. To be specific, 0.20 g was used for both PTFE samples, 0.05 g for PVF, and 0.10 g for all other polymers. The powder was weighed with analytical accuracy. The powders in the centrifuge tubes were then suspended with approximately 0.1–0.2 ml of liquid by agitating the samples for at least 5 minutes using a test-tube mixer. In order to achieve complete displacement of air, the samples were left for at least 2 hours and then agitated again for another 5 minutes.

For the actual sedimentation experiments, graduated micro(test)tubes (100 mm high with an inner diameter of 3 mm) were used. The total volume, 1.00 ml, of these tubes is divided into 100 graduations, and the sedimentation volume was read (or estimated) to 0.1 of a graduation; that is, to 0.001 ml.

Each of the dispersions prepared in the centrifuge tubes was then transferred into a microtube using Pasteur micropipets. The selected liquid mixture was used as rinsing liquid in each case, so that the microtubes were filled up to 1.00 ml with the liquid mixture. Then, the graduated microtubes were sealed and shaken for approximately 5 minutes so that the polymer powder was totally and homogeneously suspended in the liquid medium. The reading of the sedimentation volume, V_{sed}, of the polymer powder was taken every day for three days to one week, depending on the sample, until no further change in V_{sed} occurred.

Plots of the sedimentation volume of PTFE (Grade 1) powder suspended in liquid pairs of n-hexane/n-hexadecane, diethyl ether/tetralin and diethyl ether/n-hexanol are shown in Figures 11.2 and 11.3. Since these experiments are time consuming,

TABLE 11.1
Physical Properties of Liquids Used for Sedimentation

Liquids	Chemical Formula	Boiling Point, b.p. (°C)	Density at 20°C, ρ_l (kg/m³)	Dipole Moment at 20°C, $\mu \times 10^{-30}$ (cm)	Surface tension at 20°C, γ_{lv} (mJ/m²) Literature	Surface tension at 20°C, γ_{lv} (mJ/m²) Measured
n-Hexane	C_6H_{14}	68.95	660.3	0.00	18.40	18.3
n-Octane	C_8H_{18}	125.66	702.5	0.00	21.62	21.8
n-Hexadecane	$C_{16}H_{34}$	286.80	773.3	0.00	27.47	27.0
Tetralin	$C_{10}H_{12}$	207.57	970.2	1.63	35.60	36.5
Diethyl ether	$C_4H_{10}O$	34.51	713.8	3.84	17.10	17.4
n-Propanol	C_3H_8O	97.40	803.5	5.60	23.71	23.8
n-Hexanol	$C_6H_{14}O$	157.85	813.6	5.17	26.21	25.2
Ethylene glycol	$C_2H_6O_2$	198.93	1108.8	7.16	48.43	48.9
Cyclohexanone	$C_6H_{10}O$	155.65	947.8	9.01	35.19	34.3
2,2'-Thiodiethanol	$C_4H_{10}O_2S$	282	994.0	Not available	54.00	52.5

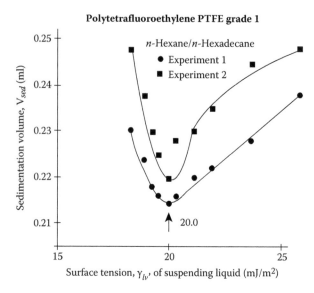

FIGURE 11.2 Reproducibility of the sedimentation volume measurements; results from two separate experiments with Grade 1 Teflon (PTFE) in *n*-hexane/*n*-hexadecane. Minima at $\gamma_{lv} = 20.0$ mJ/m².

reproducibility testing was limited to performing two independent experiments for one and the same system—Teflon/*n*-hexane/*n*-hexadecane—as shown in Figure 11.2. While the two curves do not coincide, they show the common relevant feature; that is, a minimum at $\gamma_{lv} = 20$ mJ/m² in each case. The sedimentation volumes as a function of the surface tension of the suspending liquids are given in Figures 11.4 through 11.8 for the other six polymers investigated in various liquid combinations.

It is apparent that the final sedimentation volume, V_{sed}, changes with composition and hence surface tension, γ_{lv}, of the liquid mixtures, such that it shows an extremum at a certain liquid surface tension. It turns out that these extrema are minima in the case of nonpolar and slightly polar liquid combinations such as *n*-hexane/*n*-hexadecane, or diethyl ether/tetralin. In the case of more polar liquids, such as mixtures of diethyl ether/*n*-hexanol and *n*-propanol/ethylene glycol, the extrema are maxima. The types of extrema of the sedimentation volume, V_{sed}, obtained for the seven polymers in the various liquid combinations are given in Table 11.2, together with the surface tension and dipole moment range of the liquid pairs.

The surface tensions, γ_{lv}, of the liquid mixtures at which the extrema occur are summarized in Table 11.3 for the various polymers. The contact angles with water, obtained for the various polymers (measured on smooth, homogenous, flat surfaces of the polymers), and the polymer surface tensions, γ_{sv}, calculated from these contact angles using the equation of state approach (Chapter 9), are listed in Table 11.4. The agreement between these data and those from the sedimentation volume extrema is so striking that it seems justifiable to suggest that the sedimentation technique might be considered as a means to determine the surface tension of solids in powder form, or for that matter, of any particles, including biological cells.

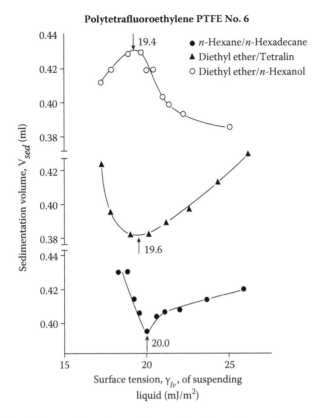

FIGURE 11.3 Sedimentation volume, V_{sed}, as a function of the surface tension, γ_{lv}, of the suspending liquid for PTFE (No. 6).

The fact that we have maxima in some cases and minima in others might suggest that both of the possible mechanisms predicted in the theory of the sedimentation volume method (Section 11.3.4.1) are operative. This suggestion is further corroborated by the fact that it is far easier to resuspend the systems where we had observed maxima, than those where we had minima. This observation is in keeping with the hypothesis of a correspondence between an absence of aggregation and a maximum in the sedimentation volume at $\gamma_{sv} = \gamma_{lv}$. This question will require further study by such means as sedimentation kinetics or direct observation of aggregation.

One further unanswered question is related to adsorption of the liquid components at the liquid–air interface as well as the solid–liquid interface. In the case of a single-component liquid in contact with a polymer particle, $\gamma_{sv} = \gamma_{lv}$ implies zero polymer-liquid interfacial tension, γ_{sl}. It is not guaranteed a priori that this is also true for binary liquid mixtures. However, it is interesting to note that the experimental results, as shown by the two bottom curves in Figure 11.3, one for a combination of nonpolar liquids and the other for polar liquids, also support the validity of this approach. In these experiments, significant preferential adsorption of one component in mixtures of n-hexane and n-hexadecane is not to be expected, because of the similarity of these two liquids. However, it is conceivable that this could be very different

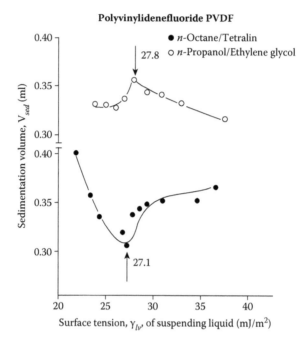

FIGURE 11.4 Sedimentation volume, V_{sed}, as a function of the surface tension, γ_{lv}, of the suspending liquid for PVDF.

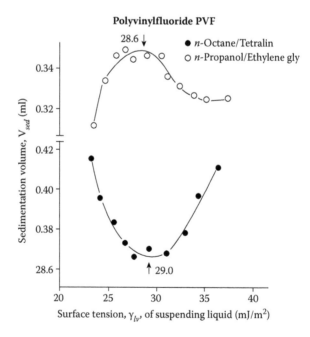

FIGURE 11.5 Sedimentation volume, V_{sed}, as a function of the surface tension, γ_{lv}, of the suspending liquid for PVF.

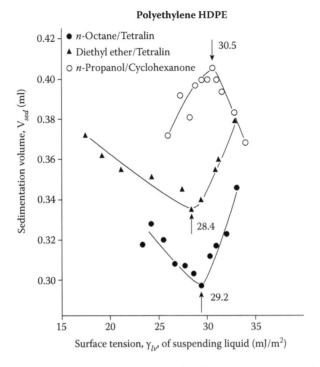

FIGURE 11.6 Sedimentation volume, V_{sed}, as a function of the surface tension, γ_{lv}, of the suspending liquid for HDPE.

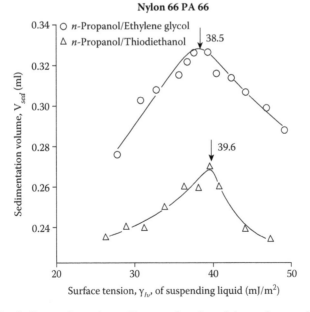

FIGURE 11.7 Sedimentation volume, V_{sed}, as a function of the surface tension, γ_{lv}, of the suspending liquid for PA66.

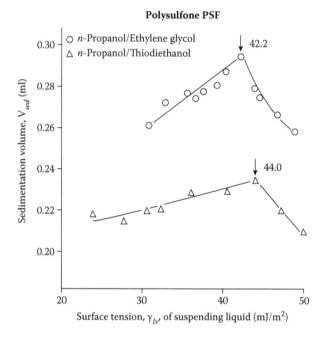

FIGURE 11.8 Sedimentation volume, V_{sed}, as a function of the surface tension, γ_{lv}, of the suspending liquid for PSF.

in the diethyl ether/tetralin system. Nevertheless, the two curves for the dependence of sedimentation volume on liquid–vapor surface tension are very similar, and the position of the minima is, within the experimental error, indistinguishable. This suggests that preferential adsorption of one component at the liquid–air interface, if it occurs, is mimicked by similar adsorption at the solid–liquid interface. More discussion of this matter will be given in the following section.

11.3.4.3 Sedimentation Behavior in Single-Component Liquids and in Binary Liquid Mixtures

In all the above sedimentation experiments, binary liquid mixtures were used as the suspending media. This is because the use of binary liquid mixtures allows a close control of the surface tension, γ_{lv}, of the suspending liquid, thus making the sedimentation technique a simple, inexpensive yet accurate method of determining the solid surface tension, γ_{sv}, of particulate matter. In the case of a single-component liquid in contact with a particle, $\gamma_{lv} = \gamma_{sv}$ implies zero γ_{sl}. However, as mentioned above, it is not guaranteed a priori that the same is also true for binary liquid mixtures. For this purpose, it is necessary to compare the sedimentation behavior of particles in both single-component liquids and binary liquid mixtures.

An experimental investigation was performed on nonpolar particles (different Teflon powders and a coal powder) using a series of nonpolar single-component liquids (n-alkanes) and also a series of nonpolar binary liquid mixtures [45]. The three polymer powders used in the experiment were Grade 1 Teflon; Aldrich Teflon

TABLE 11.2

Type of Sedimentation Volume, *Extremum* for the Polymers Investigated in Liquids of Different Polarity

Liquid Pairs Used for Sedimentation	Surface Tensions, γ_{lv} (ml/m²)	Dipole Moments, μ (10^{-30} cm)	Type of V_{sed} Extremum Obtained for:					
			PTFE	PVDF	PVF	HDPE	PA66	PSF
n-Hexane	18.3	0	Min					
n-Hexadecane	27.0	0						
n-Octane	21.8	0						
Tetralin	36.5	1.63		Min	Min	Min		
Diethyl ether	17.4	3.84	Min					
Tetralin	36.5	1.63				Min		
Diethyl ether	17.4	3.84						
n-Hexanol	25.2	5.17	Max					
n-Propanol	23.8	5.60						
Ethylene glycol	48.9	7.61		Max	Max		Max	Max
n-Propanol	23.8	5.60						
Cyclohexanone	34.3	9.01				Max		
n-Propanol	23.8	5.60						
Thiodiethanol	52.5	Not available					Max	Max

TABLE 11.3

Extrema of Sedimentation Volumes for Polymers in Different Liquid Combinations

		γ_{lv} of Liquid Mixtures at 20°C at V_{sed} Extrema						
		PTFE						
		Grade 1	Number 6	PVDF	PVF	HDPE	PA66	PSF
Liquid Combinations								
n-Hexane	n-Hexadecane	20.0	20.0					
n-Octane	Tetralin			27.1	29.0	29.2		
Diethyl ether	Tetralin	20.1	19.6			28.4		
Diethyl ether	n-Hexanol	20.2	19.4					
n-Propanol	Ethylene glycol			27.8	28.6		38.5	42.2
n-Propanol	Cyclohexanone					30.5		
n-Propanol	Thiodiethanol						39.6	44.0

TABLE 11.4

Comparison of the Surface Tension of Polymers Obtained from Contact Angles, θ, and from Sedimentation Volumes, V_{sed}

| Polymer | | θ_{H_2O} (deg) | Reference | Polymer Surface Tension, γ_{sv} (mJ/ m²) from: | |
				Contact Angle	Sedimentation
PTFE	Grade 1				20.2
	No.6	104.0± 2.0	49	20.0 ± 1.3	19.7
PVDF		94.8 ± 2.0	44	25.5 ± 1.3	27.5
PVF		88.6 ± 2.5	44	29.4 ± 1.7	28.8
HDPE		87.1 ± 2.0	44	30.3 ± 1.3	29.4
PA66		70.0 ± 3.0	44	41.1 ± 1.8	39.1
PSF		66.0 ± 2.0	44	43.1 ± 1.1	43.1

(PTFE); and DuPont fluorocarbon micropowder, DLX-6000, of unknown composition. The experiments were also performed with bituminous coal powder (Pittsburgh No. 8 coal). In dry form, the agglomerated size of the polymer powders varied from 150 μm (DLX-6000) to 600 μm (for the Teflon powders). The individual particle diameter for the Teflon powders was less than 60 μm. The particle diameter for

Pittsburgh No. 8 coal was less than 37 μm. The liquid combinations were n-hexane/n-hexadecane, n-hexane/decalin, and n-octane/tetralin. The single-component liquid series (n-alkanes) was chosen such that the surface tension of the liquids, γ_{lv}, spanned a range that included the surface tension of the particles, γ_{sv}.

The sedimentation volumes, V_{sed}, of the three polymer powders (Grade 1 Teflon, Aldrich Teflon, and DLX-6000) as a function of the surface tension, γ_{lv}, of the suspending liquids are plotted in Figures 11.9 through 11.11. The plot for Pittsburgh No. 8 coal is given in Figure 11.12. From these figures, it can be seen that the final sedimentation volume, V_{sed}, changes with varying liquid surface tension, γ_{lv}, such that V_{sed} exhibits a minimum both for single-component liquids and for binary liquid mixtures. The surface tensions, γ_{lv}, of the liquids at which the extrema occur are summarized in Table 11.5.

It is interesting to note that the sedimentation curves of each powder, obtained from different liquid series, are not superimposable. In the case of polymer powders, the curves from binary liquid mixtures have a higher final volume than the curves from the single-liquid series. In the case of the coal powder, the opposite is true. The differences are not understood at the present time.

The results in Table 11.5 show that, for one and the same powder, the liquid surface tension, γ_{lv}, at which the minimum in the sedimentation volume occurs is very similar for both the single-component liquids and the binary liquid mixtures. Better agreement in the position of the minima cannot be expected since, in the case of the homologous alkanes (single-component liquids), we are limited to a small number of points that are up to 2 mJ/m² apart. It should also be noted that the reproducibility of the position of the minima, identifying the particle surface tension, compares very favorably

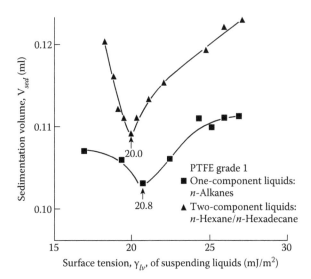

FIGURE 11.9 Sedimentation volume, V_{sed}, as a function of the surface tension, γ_{lv}, of the one-component (n-alkanes) and of the two-component (n-hexane/n-hexadecane) suspending liquids for Grade 1 Teflon (PTFE) powder.

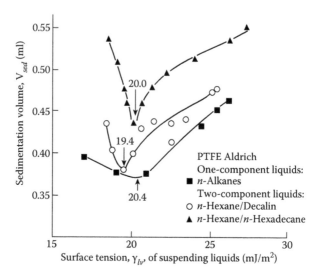

FIGURE 11.10   Sedimentation volume, V_{sed}, as a function of the surface tension, γ_{lv}, of the one-component (n-alkanes) and of the two-component (n-hexane/n-hexadecane) suspending liquids for Aldrich Teflon (PTFE) powder.

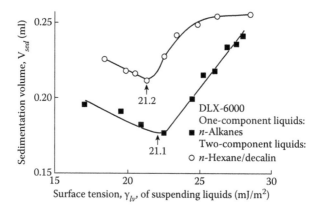

FIGURE 11.11   Sedimentation volume, V_{sed}, as a function of the surface tension, γ_{lv}, of the one-component (n-alkanes) and of the two-component (n-hexane/n-hexadecane) suspending liquids for a fluorinated Teflon (PTFE) micropowder (DLX-6000).

to the reproducibility of γ_{sv} values obtained from contact angle measurements with different liquids on a smooth polymer surface.

Independent γ_{sv} values for the powders used in these experiments are not available, so that a direct comparison to the solid surface tensions determined by the sedimentation volume method is not possible. Contact angle measurements performed on smooth Teflon films give a value of γ_{sv} of approximately 20 mJ/m^2, which compares very well to the results given in Table 11.5 for the Teflon powders. Freezing front experiments (see Chapter 12) performed with various bituminous coal

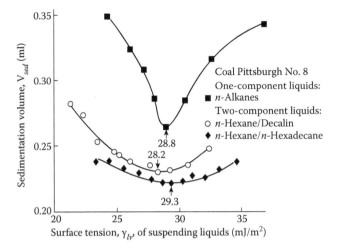

FIGURE 11.12 Sedimentation volume, V_{sed}, as a function of the surface tension, γ_{lv}, of the one-component (n-alkanes) and of the two-component (n-hexane/n-hexadecane) suspending liquids for the—400 mesh fraction of Pittsburgh No. 8 coal powder.

TABLE 11.5

Extrema of Sedimentation Volumes for Polymer and Coal Powders in One- and Two-Component Liquid Media

	Surface Tension, γ_{lv}, of Liquid Medium at which Minimum Occurs (mJ/m²)			
	Teflon			Coal
Dispersion Medium	Grade 1	Aldrich	DLX-6000	Pittsburgh No. 8
One-component liquids				
n-Alkanes	20.8	20.4	22.1	28.8
Two-component liquids				
n-Hexane/n-hexadecane	20.0	20.0	—	—
n-Hexane/decalin	—	19.4	21.2	28.2
n-Octane/tetralin	—	—	—	29.3

powders [59–61] give γ_{sv} values that are comparable to those reported in Table 11.5 for Pittsburgh No. 8 coal.

It appears that the position of the minimum in the sedimentation volume is determined by the surface tension of the liquid, be it binary or single component. These findings give further credence to the identification of the position of the sedimentation volume extremum as the particle surface tension.

As an application, the sedimentation behavior of coal particles has received much attention in recent years. A peculiar pattern of the interaction of coal with the suspending liquids has been found [62,63]. Under the same experimental conditions, inert materials, such as certain polymer particles, exhibit a unique surface tension regardless of the nature of the suspending liquids; but, the effective surface tension of coal seems to vary depending on the liquid medium with which it is in contact. In essence, coal is relatively hydrophobic in organic liquids, with a surface tension typically in the range of 30–45 mJ/m^2 whereas it is hydrophilic in water or in aqueous media, having a surface tension near 70 mJ/m^2. The duality of response to organic and aqueous liquids has an immediate bearing on practical aspects of coal processing and utilization. Therefore, the sedimentation behavior of coal fines was studied [46] to examine the effect of the polarity of the suspending liquids on the type of extremum in sedimentation volume, as well as on the resultant surface tension of coal particles. Particle-size dependence and the duality of the coal surface were also examined. Details of these studies can be found elsewhere [46,62–68].

11.3.5 Capillary Penetration

11.3.5.1 Theory

The method of capillary penetration is based on wicking of liquid into a porous material. It has attracted increasing interest in recent years for measuring wettability of solid particles. Quite often, the purpose of studying contact angles on particles/powders is to predict and control the penetration of a liquid into a powder bed or, conversely, the dispersion of the powder in the liquid. Rather than attempting contact angle measurements on the compressed powder, it may be better to consider the process of capillary penetration of the liquid into a suitably prepared powder bed. There is one important point to keep in mind. While a contact angle measured on a rough surface will normally not be meaningful in conjunction with Young's equation (i.e., it will not be possible to use it to determine the solid surface tension, γ_{sv}), it is indeed this phenomenological contact angle that will, together with the liquid surface tension, determine the Laplace pressure (ΔP) of the meniscus in a capillary, and hence capillary penetration [69]. One method, therefore, is to measure the pressure necessary to just balance the Laplace pressure, which drives liquid into a capillary; that is, the limiting pressure necessary to prevent further capillary penetration:

$$\Delta P = \frac{2\gamma_{lv}\cos\theta}{r},$$ (11.11)

where r is the radius of the capillary.

Washburn [70] first established the correlation between the Laplace pressure and the drop of the hydrostatic pressure as the liquid column travels in the capillary. The velocity (v) of the liquid–air meniscus along the capillary is predicted by the Hagen-Poiseuille equation for laminar flow,

$$v = \frac{dh}{dt} = \frac{r^2\Delta P}{8\eta h},$$ (11.12)

where h is the height of the liquid front, that is, the distance travelled along the tube by the meniscus in time t, and η is the dynamic viscosity of the liquid. Assuming that the driving pressure is solely the capillary pressure, one can combine Equations 11.11 and 11.12 to yield

$$\frac{dh}{dt} = \frac{r\gamma_{lv}\cos\theta}{4\eta h}. \tag{11.13}$$

After integration one obtains Washburn's equation,

$$h^2 = \frac{rt}{2\eta}\gamma_{lv}\cos\theta. \tag{11.14}$$

Washburn's equation is valid with the following assumptions: (1) laminar flow predominates in the pore spaces; (2) gravity can be neglected; and (3) the geometry of the capillary (i.e., r) is constant.

Given the fact that it is much easier and more accurate to experimentally determine the weight of the imbibed liquid than the penetration velocity of the liquid, Washburn's equation can be further modified by replacing h with the weight M of the liquid that penetrates into the capillary:

$$M = \rho V = \rho h A, \tag{11.15}$$

where A is the cross-sectional area of the capillary and ρ is the density of the liquid. It follows that

$$\frac{M^2}{\rho^2 A^2} = \frac{rt}{2\eta}\gamma_{lv}\cos\theta, \tag{11.16}$$

and by rearranging

$$\gamma_{lv}\cos\theta = \left[\frac{2}{A^2 r}\right]\left[\frac{\eta}{\rho^2}\right]\left[\frac{M^2}{t}\right], \tag{11.17}$$

where $[2/A^2 r]$ is a factor representing the geometry of the capillary, $[\eta/\rho^2]$ reflects the physical properties of the test liquid and $[M^2/t]$ is determined in the experiment.

In the case of powder packings or other porous solids, such as membranes, the geometry of the capillary system is not known. The value of $[2/A^2 r]$ in Equation 11.17 is therefore replaced by an unknown factor $1/K$; that is,

$$\gamma_{lv}\cos\theta = \frac{1}{K}\left[\frac{\eta}{\rho^2}\right]\left[\frac{M^2}{t}\right], \tag{11.18}$$

or

$$K\gamma_{lv}\cos\theta = \left[\frac{\eta}{\rho^2}\right]\left[\frac{M^2}{t}\right], \tag{11.19}$$

where K is an unknown parameter that depends on the geometry of the porous material.

In practice, the quantity $[M^2/t]$ can be determined by measuring the weight M of a penetrating liquid into a porous solid as a function of time using an electrobalance. If M is plotted versus $\sqrt{t}$ the experimental quantity $[M^2/t]$ can be obtained by determining the slope of the linear part of these plots.

Washburn's equation has been used to characterize the wettability of porous materials, such as polymer packings, in which polymer powders uniformly packed into a tube are modeled as a bundle of capillary tubes [71–73]. The geometric factor, K, or the average equivalent radius, $\bar{r}$, of the capillaries are determined using a liquid that completely wets the powder packing (i.e., by enforcing $\cos\theta = 1$). Then, for the same packing, values of $\cos\theta$ for other liquids can be determined. Subsequently, the solid surface tension of the porous materials, γ_{sv}, can be determined from the contact angle value. However, this procedure is questionable due to the concern that the contact angles determined directly from Washburn's equation are apparent contact angles that are affected by the geometry (roughness and porosity) of the porous materials, as for example the local inclination angles of the capillary walls. Only in a cylindrical capillary with smooth and homogeneous walls does the calculated θ coincide with the intrinsic Young contact angle. This is, however, obviously not the case of porous materials such as polymer packings, in which complicated pore geometry, varying cross-sectional areas, and rough surfaces are expected. In general, it has been found that capillary penetration experiments tend to overestimate the contact angles compared to directly measured contact angles on smooth surfaces of the same material [74–76]. For example, the contact angle of hexadecane was calculated to be $\theta = 88°$ for a polytetrafluoroethylene (PTFE) powder using Washburn's equation [76]. On a flat and smooth PTFE surface, however, this contact angle is well known to be 46°. Obviously, the former value only reflects the contact angle of hexadecane on "rough" PTFE powder, which is meaningless for energetics calculations in conjunction with Young's equation. Thus, it can be concluded that the contact angles of porous materials determined directly from Washburn's equation do not reflect material properties of the surfaces; rather, they reflect morphological ones. In addition, if the rate of motion of the three-phase line is relatively high it cannot be excluded that the "dynamic" contact angles calculated from capillary penetration experiments differ from static advancing contact angles. It is well-known that they can be quite different at higher velocities of the moving meniscus.

To circumvent the difficulties associated with the contact angle measurement, Grundke et al. [76] have developed a novel approach that allows characterization of the solid surface tension, γ_{sv}, directly from capillary penetration experiments without knowledge of contact angle and geometric factors of the porous materials. The principles of this approach are as follows. Experimental results obtained for powders, liposomes, and membranes, consisting of hydrophobic and hydrophilic materials show that there exists a maximum when the values $K\gamma_{lv}\cos\theta$ are plotted against the surface tension of the penetrating liquids, γ_{lv} [76–81]. This maximum occurs when the surface tension of the liquid, γ_{lv}, is equal to the surface tension of the solid particles; that is, at $\gamma_{lv}^* = \gamma_{sv}$. This phenomenon can be explained by the schematic shown in Figure 11.13, which plots $\gamma_{lv}\cos\theta$ as a function of γ_{lv} for a hypothetical solid surface under ideal conditions. It is found that a maximum exists in the plot.

1. To the left of the maximum, that is, at $\gamma_{lv} < \gamma_{lv}^* = \gamma_{sv}$, the liquids wet the solid completely, which leads to a zero contact angle and thus $\gamma_{lv}\cos\theta = \gamma_{lv}$.
2. To the right of the maximum, that is, at $\gamma_{lv} > \gamma_{lv}^* = \gamma_{sv}$, the liquids only partially wet the surface, which results in a nonzero contact angle (i.e., $\theta > 0$) and thus $\gamma_{lv}\cos\theta < \gamma_{lv}$. The curve of $\gamma_{lv}\cos\theta$ versus γ_{lv} at $\gamma_{lv} > \gamma_{sv}$ follows a smooth path for a given solid as predicted by the equation of state for interfacial tensions (see Chapters 8 and 9).

When $K\gamma_{lv}\cos\theta$, instead of $\gamma_{lv}\cos\theta$, is plotted against γ_{lv}, the curve can be distorted from the ideal shape as shown in Figure 11.13 [76]. However, as will be shown later, this distortion only affects the shape of the curve but not the location of the maximum. Therefore, based on the plot of $K\gamma_{lv}\cos\theta$ versus γ_{lv}, it is possible to directly determine the solid surface tension, γ_{sv}, using the capillary penetration method, without the necessity of knowing the contact angle (θ) and the geometric factor (K) of the system. There are two basic requirements for the test liquids used in these experiments: (1) they should be chemically inert with respect to the porous material of interest; and (2) the range of the surface tensions, γ_{lv}, of the test liquids should cover the anticipated surface tension of the solid particles, γ_{sv}.

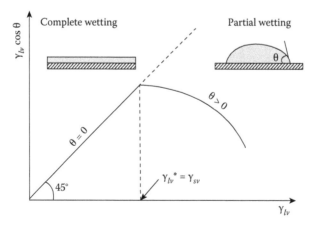

FIGURE 11.13 A schematic contact angle plot for an ideal solid surface. Along the 45° straight line, at $\gamma_{lv} < \gamma_{sv}$, $\theta = 0$ and hence $\gamma_{lv}\cos\theta = \gamma_{lv}$. As γ_{lv} increases beyond γ_{sv}, θ becomes nonzero and hence $\gamma_{lv}\cos\theta < \gamma_{lv}$.

11.3.5.2 Experiments

Figure 11.14 shows the experimental setup for the capillary penetration measurements. The powder is packed in a glass tube of which the lower end is closed with a glass filter. Considerable care is necessary to obtain a constant and homogeneous powder packing. A precisely weighed quantity of the powder has to fill up to the same height in the glass tube by manually tapping the powder. The filled columns are attached to an electrobalance and brought into contact with several test liquids. Their penetration velocities are determined by measuring the weight gain with the electrobalance as a function of time. The physical properties of the liquids used for the capillary penetration measurements are given in Table 11.6.

It is found that the main source of error of the measurement is the geometry of the porous system (K). It is difficult to reproduce K for each measurement due to (1) variations in the packing preparation; and (2) variations in the penetration of different liquids. Consequently, the shape of the curve $K\gamma_{lv}\cos\theta$ versus γ_{lv} may vary in an unpredictable way for each experiment. However, from experiments with microporous expanded PTFE membranes of different pore geometries it could be concluded that information about the geometric constant K is not needed. The position of the maximum, which is expected to reflect the solid surface tension γ_{sv} of the porous material, was not affected by the different geometries of the membranes [77].

Figure 11.15 shows a typical result of capillary penetration for a PTFE (Teflon 807-N) powder. The maximum occurs at $\gamma_{lv}* = 20.4$ mJ/m^2. Thus, the γ_{sv} value of the PTFE particles would be 20.4 mJ/m^2. When this γ_{sv} value and $\gamma_{lv} = 72.5$ mJ/m^2

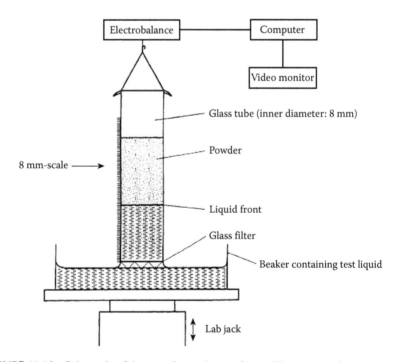

FIGURE 11.14   Schematic of the experimental setup for capillary penetration.

TABLE 11.6

Physical Properties of the Liquids Used for the Capillary Penetration Measurements

Liquids	γ_{lv} (mJ/m²)[a]	η (mPa sec)[b]	ρ (g/cm³)
Perfluoropolyether	14.31	1.427	1.720
Isopentan	17.13	0.223	0.624
Hexane	18.30	0.308	0.659
Heptane	20.50	0.413	0.683
Octane	21.42	0.546	0.702
Decane	23.22	0.907	0.730
Dodecane	24.69	1.383	0.751
Tetradecane	26.40	2.128	0.761
Hexadecane	27.90	3.032	0.775
Tetralin	34.54	2.020	0.976
Benzylalcohol	39.00	5.474	1.044
1-Bromonaphthalene	44.53	4.520	1.483
Ethylene glycol	48.00	19.900	1.113
Formamide	58.20	3.300	1.133
Water	72.50	1.001	0.998

[a] Measured by the ring method at room temperature.
[b] Tabulated values from the literature of Lide, D. R., CRC Handbook of Chemistry and Physics, CRC Press, Boca Raton.

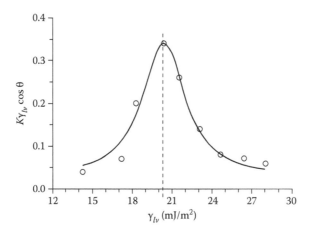

FIGURE 11.15 A plot of $K\gamma_{lv}\cos\theta$ versus γ_{lv} using nine liquids obtained from capillary penetration experiments for PTFE (Teflon 807-N) powder.

for water are used in conjunction with the equation of state approach for interfacial tensions, a water contact angle of 104° can be predicted. Remarkably, this is exactly what one would observe on a smooth Teflon surface.

Although the packed powder bed certainly does not represent a flat and smooth solid surface, the derived value for γ_{sv} is that obtained by direct contact angle

measurements on a flat and smooth surface. It is well known that even a highly compacted hydrophobic powder, presenting a seemingly flat and smooth solid surface,
does not yield the same contact angle as truly smooth and coherent solid surfaces. It
appears that an indirect method, such as the capillary penetration, can provide much
more relevant information concerning the solid surface energetics than direct contact
angle measurements on imperfect solid surfaces.

Table 11.7 shows the comparison of solid surface tensions, γ_{sv}, of three hydrophobic
polymers (polypropylene, polyethylene, and polystyrene), determined using the capillary penetration approach and direct contact angle measurements on smooth surfaces
of these polymers [78]. Perfect agreement is found between these two methods. Also
shown in Table 11.7 are the contact angles of benzylalcohol and 1-bromonaphthalene
directly determined from Washburn's equation (Equation 11.19), in which the geometric factor K was estimated using liquids that wet the solid completely. This procedure
has been used by many workers [71–73]. As can be seen, the contact angles determined
from Washburn's equation are distinctly higher than those directly measured on the
smooth polymer surfaces. This can result in an erroneously low solid surface tension
by about 15 mJ/m². Obviously, these calculated contact angles do not reflect the surface energetics of the powdered material but rather its morphological properties.

Figure 11.16 shows that the above described approach is not only applicable to
hydrophobic surfaces but also to very hydrophilic surfaces, such as cellulose membranes [76]. Similar to the case of the PTFE powder (Figure 11.15) we obtain curves
with a maximum when $K\gamma_{lv}\cos\theta$ versus γ_{lv} of the test liquids is plotted. It can be seen
that the unmodified cellulose fiber shows the highest γ_{sv} value (i.e., 48 mJ/m²) whereas
the modified types have a lower γ_{sv}; that is, they are less hydrophilic surfaces. Using
Young's equation, with γ_{sv} = 48 mJ/m² and γ_{lv} = 72.5 mJ/m² for water, a water contact
angle of 59° can be predicted on an ideally smooth unmodified cellulose hollow
fiber cuprophan. It has to be considered that processing agents are applied during the
manufacture of the cuprophan fibers, which can be expected to influence even the

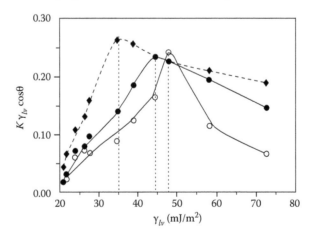

FIGURE 11.16   A plot of $K\gamma_{lv}\cos\theta$ versus γ_{lv} using 11 liquids for three different cellulose hollow fibers: ○ unmodified cellulose cuprophan; ● chemically modified cellulose M1; ◆ chemi-
cally modified cellulose M2.

TABLE 11.7

Comparison of the γ_{sv} Values Determined by the Capillary Penetration of Liquids into Polymer Powder Packings and by Contact Angle Measurements on Smooth Surfaces Prepared from the Same Polymer Powders

Polymers	γ_{sv} (mJ/m²)	Liquids[a]	Capillary Penetration of Liquids Into Polymer Powder Packings		Contact Angle Measurements on Smooth Polymer Surfaces	
			Contact Angle Predicted by EQS[b] (deg.)	Contact Angle Calculated by Washburn's Equation (deg.)	Directly Measured Contact Angle θ (deg.)	γ_{sv} Calculated by EQS from Measured θ (mJ/m²)
Polypropylene	30.2	Benzylalcohol	43	83	42.4 ± 3.0[c]	30.4
		1-Bromonaphthalene	47	81	46.7 ± 0.2[d]	30.4
Polyethylene	34.4	Benzylalcohol	30	71	35.1 ± 2.0[c]	33.0
		1-Bromonaphthalene	36	77	35.3 ± 2.0[c]	34.6
Polystyrene	27.5	Water	92	—	90.9 ± 0.3[d]	28.2

[a] Measured $\gamma_{lv} = 39.45$ mJ/m² for benzylalcohol and $\gamma_{lv} = 41.45$ mJ/m² for 1-bromonaphthalene.
[b] Equation of state approach for solid–liquid interfacial tensions (EQS).
[c] Static advancing contact angle measured by the goniometer technique.
[d] Low rate dynamic contact angle measured by the ADSA contact angle technique.

surface properties of the unmodified cellulose material. By chemical modification of the cellulose, the γ_{sv} could be decreased to 35 mJ/m^2, as can be seen from Figure 11.16; a water contact angle of 80° can be predicted for this modified cellulose material from Young's equation.

REFERENCES

1. C. J. van Oss. *Separation and Purification Methods* 8 (1979): 119.
2. J. Th. G. Overbeek and B. H. Bijsterbosch. In *Electrokinetic Separation Methods*. Edited by P. G. Righetti, C. J. van Oss, and J. W. Vanderhoff. Elsevier/North-Holland (Amsterdam), New York, 1980.
3. A. W. Neumann. *Advances in Colloid and Interface Science* 4 (1974): 105.
4. S. K. Li, R. P. Smith, and A. W. Neumann. *Journal of Adhesion* 17 (1984): 105.
5. P. A. Albertsson. *Partition of Cells, Particles and Macromolecules*. Wiley-Interscience, New York, 1971.
6. J. F. Boyce. PhD Thesis, University of Western Ontario, 1985.
7. D. F. Gerson, C. Cupo, A. M. Benoliel, and P. Bongrand. *Biochimica et Biophysica Acta* 692 (1982): 147.
8. D. F. Gerson and J. Akit. *Biochimica et Biophysica Acta* 602 (1980): 281.
9. D. F. Gerson and D. Scheer. *Biochimica et Biophysica Acta* 602 (1980): 506.
10. S. Mudd and E. B. H. Mudd. *Journal of Experimental Medicine* 40 (1924): 647.
11. O. Stendahl, C. Tagesson, and L. Edebo. *Infection and Immunity* 8 (1973): 36.
12. R. K. Cunningham, T. O. Soderstrom, C. F. Gillman, and C. J. van Oss. *Immunological Communications* 4 (1975): 429.
13. M. Rosenberg, D. Gutnick, and E. Rosenberg. *F.E.M.S. Microbiology Letters* 9 (1980): 29.
14. M. Rosenberg, A. Perry, E. A. Bayer, D. L. Gutnick, E. Rosenberg, and I. Ofek. *Infection and Immunity* 33 (1981): 29.
15. M. Rosenberg and E. Rosenberg. *Journal of Bacteriology* 148 (1981): 51.
16. E. Weiss, M. Rosenberg, H. Judes, and E. Rosenberg. *Current Microbiology* 7 (1982): 125.
17. C. J. Smyth, P. Johnsson, E. Olsson, O. Söderlind, J. Rosengren, S. Hjertén, and T. Wadström. *Infection and Immunity* 22 (1978): 462.
18. I. Stjernström, K. E. Magnusson, O. Stendahl, and C. Tagesson. *Infection and Immunity* 18 (1977): 261.
19. K. E. Magnusson, O. Stendahl, I. Stjernström, and L. Edebo. *Immunology* 36 (1979): 439.
20. H. E. Schutze and J. F. Heremans. In *Molecular Biology of Human Proteins*. Vol. I. Elsevier, Amsterdam, 1966.
21. M. Lindahl, A. Faris, T. Wadström, and S. Hjertén. *Biochimica et Biophysica Acta* 677 (1981): 471.
22. J. F. Boyce, S. Schürch, Y. Rotenberg, and A. W. Neumann. *Colloids and Surfaces* 9 (1984): 307.
23. P. Cheng, D. Li, L. Boruvka, Y. Rotenberg, and A. W. Neumann. *Colloids and Surfaces* 43 (1990): 151.
24. J. M. Alvarez, A. Amirfazli, and A. W. Neumann. *Colloids and Surfaces A-Physicochemical and Engineering Aspects* 156 (1999): 163.
25. W. D. Harkins and G. Jura. *Journal of the American Chemical Society* 66 (1944):1362.
26. R. J. Good and L. A. Girifalco. *Journal of Physical Chemistry* 62 (1958): 1418.
27. A. W. Adamson. *Physical Chemistry of Surfaces*. Wiley, New York, 1990.
28. A. V. Rapacchietta, A. W. Neumann, and S. N. Omenyi. *Journal of Colloid and Interface Science* 59 (1977): 541.
29. A. V. Rapacchietta and A. W. Neumann. *Journal of Colloid and Interface Science* 59 (1977): 555.

30. A. W. Neumann, O. Economopoulos, L. Boruvka, and A. V. Rapacchietta. *Journal of Colloid and Interface Science* 71 (1979): 293.
31. E. Weber. *Habilitationsschrift.* University of Stuttgart, 1968.
32. D. W. Fuerstenau and M. C. Williams. *Colloids and Surfaces* 22 (1987): 87.
33. D. W. Fuerstenau and M. C. Williams. *Particle Characterization* 4 (1987): 7.
34. J. Diao. MASc Thesis, University of California, Berkeley, 1987.
35. D. W. Fuerstenau, M. C. Williams, and J. Diao. Paper presented at AIME Annual Meeting, New Orleans, March, 1986.
36. D. W. Fuerstenau and M. C. Williams. *Colloids and Surfaces* 22 (1987): 87.
37. M. C. Williams and D. W. Fuerstenau. *International Journal of Mineral Processing* 20 (1987): 153.
38. D. W. Fuerstenau, K. S. Narayanan, R. H. Urbina, and J. L. Diao. In *1987 International Conference on Coal Science.* Elsevier, Amsterdam, 1987.
39. D. W. Fuerstenau, M. C. Williams, K. S. Narayanan, J. L. Diao, and R. H. Urbina. *Energy and Fuels* 2 (1988): 237.
40. D. W. Fuerstenau, J. Diao, and J. Hanson. Paper presented at the American Chemical Society National Meeting, Los Angeles, September, 1988.
41. D. W. Fuerstenau, J. Diao, and J. Hanson. *Energy and Fuels* 4 (1990): 34.
42. D. W. Fuerstenau. Private communication. 1988.
43. J. Laskowski and J. A. Kitchener. *Journal of Colloid and Interface Science* 29 (1969): 670.
44. E. I. Vargha-Butler, T. K. Zubovits, H. A. Hamza, and A. W. Neumann. *J. Dispersion Science and Technology* 6, (1985): 357.
45. E. I. Vargha-Butler, E. Moy, and A. W. Neumann. *Colloids and Surfaces* 24 (1987): 315.
46. E. I. Vargha-Butler, T. K. Zubovits, D. R. Absolom, and A. W. Neumann. *Chemical Engineering Communications* 33 (1985): 255.
47. A. W. Neumann, S. N. Omenyi, and C. J. van Oss. *Journal of Physical Chemistry* 86 (1982): 1267.
48. H. C. Hamaker. *Physica* 4 (1937): 1058.
49. S. N. Omenyi, A. W. Neumann, and C. J. van Oss. *Journal of Applied Physics* 52 (1981): 789.
50. A. W. Neumann, S. N. Omenyi, and C. J. van Oss. *Colloid and Polymer Science* 257 (1979): 413.
51. C. J. van Oss, D. R. Absolom, A. W. Neumann, and W. Zingg. *Biochimica et Biophysica Acta* 670 (1981): 64.
52. A. W. Neumann, R. J. Good, C. J. Hope, and M. Sejpal. *Journal of Colloid and Interface Science* 49 (1974): 291.
53. R. C. Weast, ed. *Handbook of Chemistry and Physics.* 63rd ed. CRC Press, Boca Raton, FL, 1982–1983.
54. J. Timmermans, ed. *Physico-Chemical Constants of Pure Organic Compounds.* Elsevier, Amsterdam, 1965.
55. D. W. van Krevelen, ed. *Properties of Polymers.* Elsevier, Amsterdam, 1976.
56. J. A. Dean, ed. *Lange's Handbook of Chemistry.* McGraw-Hill, New York, 1973.
57. D. R. Lide, ed. *Journal of Physical and Chemical Reference Data* I, no. 3 (1972). ACSAIP-NBS Publication, Washington, DC.
58. A. W. Neumann, M. A. Moscarello, and R. M. Epand. *Biopolymers* 12 (1973): 1945.
59. M. R. Soulard. MASc Thesis, University of Toronto, 1980.
60. M. R. Soulard, E. I. Vargha-Butler, H. A. Hamza, and A. W. Neumann. *Chemical Engineering Communications* 21 (1983): 329.
61. E. I. Vargha-Butler, T. K. Zubovits, R. P. Smith, I. H. L. Tsim, H. A. Hamza, and A. W. Neumann. *Colloids and Surfaces* 8 (1984): 231.
62. A. W. Neumann, E. I. Vargha-Butler, H. A. Hamza, and D. R. Absolom. *Colloids and Surfaces* 17 (1986): 131.

63. D. R. Absolom, K. Eom, E. I. Vargha-Butler, H. A. Hamza, and A. W. Neumann. *Colloids and Surfaces* 17 (1986): 143.
64. D. H. Lee. MASc Thesis, University of Toronto, 1984.
65. M. R. Soulard. MASc Thesis, University of Toronto, 1980.
66. M. R. Soulard, E. I. Vargha-Butler, H. A. Hamza, and A. W. Neumann. *Chemical Engineering Communications* 21 (1983): 329.
67. E. I. Vargha-Butler, M. R. Soulard, A. W. Neumann, and H. A. Hamza. *The Canadian Mining & Metallurgical Bulletin* 74 (1981): 54.
68. E. I. Vargha-Butler, M. R. Soulard, H. A. Hamza, and A. W. Neumann. *Fuel* 61 (1982): 437.
69. P. S. De Laplace. *Méchanique Céleste, Supplement to Book 10.* J. B. M. Duprat, Paris, 1908.
70. E. W. Washburn. *Physical Review* 17 (1921): 374.
71. P. M. Costanzo, R. F. Giese, and C. J. van Oss. *Journal of Adhesion Science and Technology* 4 (1990): 267.
72. C. J. van Oss, R. F. Giese, Z. Li, K. Murphy, J. Norris, M. K. Chaudhury, and R. J. Good. *Journal of Adhesion Science and Technology* 6 (1992): 413.
73. E. Chibowski and L. Holysz. *Langmuir* 8 (1992): 710.
74. A. Marmur and R. D. Cohen. *Journal of Colloid and Interface Science* 189 (1997): 299
75. G. E. Parsons, G. Buckton, and S. M. Chatham. *Journal of Adhesion Science and Technology* 7 (1993): 95.
76. K. Grundke, T. Bogumil, T. Gietzelt, H.-J. Jacobasch, D. Y. Kwok, and A. W. Neumann. *Progress in Colloid and Polymer Science* 101 (1996): 58.
77. J. Tröger, K. Lunkwitz, K. Grundke, and W. Bürger. *Colloids and Surfaces A* 134 (1998): 299.
78. K. Grundke, and A. Augsburg. *Journal of Adhesion Science and Technology* 14 (2000): 765.
79. T. R. Desai, D. Li, W. H. Finlay, and J. P. Wong. *Colloids and Surfaces B-Biointerfaces* 22 (2001): 107.
80. C. Bellmann, N. Petong, A. Caspari, W. Jenschke, F. Simon, and K. Grundke. *Surface Coatings International Part B-Coatings Transactions* 89 (2006): 69.
81. A. Synytska, L. Ionov, V. Dutschk, S. Minko, K. J. Eichhorn, M. Stamm, and K. Grundke. *Progress in Colloid and Polymer Science* 132 (2006): 72.

12 Behavior of Particles at Solidification Fronts

Dongqing Li, Yi Zuo, and A. Wilhelm Neumann

CONTENTS

12.1 INTRODUCTION

The behavior of insoluble particles at the solid–liquid interface of an advancing solidification front is a multifaceted phenomenon. Consider a channel containing an insoluble particle in a liquid. The channel is cooled at one end to a temperature below the melting point of the liquid. As schematically illustrated in Figure 12.1, the liquid at the cooling end will freeze and a solidification front will advance along the temperature gradient; that is, moving from the cold right end to the warm left end in the case of Figure 12.1. The rate of the advancing solidification front depends on the heat extraction from the cooling end for a given liquid–solid system. When the solidification front approaches the foreign particle (Figure 12.1a), one of three phenomena

633

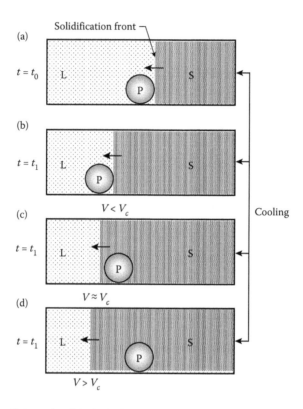

FIGURE 12.1 Schematic of the behavior of a particle at a solidification front moving from right to left. (a) At $t = t_0$; the solidification front approaches the particle. (b) At $t = t_1$ $(t_1 > t_0)$; the particle is rejected by the solidification front; that is, pushed ahead of the front, if the front moves with a rate lower than a critical velocity (V_c). (c) At $t = t_1$; the particle is pushed ahead for a short distance and then engulfed by the solidification front, if the front moves with a rate close to V_c. (d) At $t = t_1$; the particle is instantly engulfed by the solidification front, if the front moves with a rate greater than V_c.

may occur depending on the velocity of the solidification front [1,2]. (1) The particle can be rejected and pushed ahead of the solidification front (Figure 12.1b) if the rate of the front is below a critical velocity V_c. (2) At intermediate rates, the particle can be pushed for a distance before being engulfed (Figure 12.1c). (3) The particle can be engulfed instantly by the solidification front (Figure 12.1d) if the front moves sufficiently rapidly; that is, at a rate beyond V_c. V_c is dependent on the particle geometry, and thermophysical and surface properties of both the particle and the matrix material, as detailed in the next section.

The study of particle behavior at solidification fronts is of practical importance in metallurgy, material, soil, food and biological sciences; for example, in the fabrication of composite materials [3,4], casting of alloys [5], crystal growth [6], freezing of soils [7], phagocytosis [8,9], cryogenic preservation of cells, tissues and perishable foods [10,11], behavior of vegetation in permafrost and cold regions [12], and separation of solid particles by particle chromatography [13,14]. Some of these applications require the particles to be engulfed by the solidification fronts, thus resulting in a uniform distribution of the particles in the matrix material, while the other applications need the particles to be selectively rejected by the solidification fronts.

As a further example, in the processing of molten steel, deoxidizers are usually added to the melt prior to solidification. The deoxidation products tend to form particles in the 5–50 μm range. When steel is solidified in the form of an ingot or as a continuous cast product, the physical properties of the materials are dependent on whether these deoxidation products are engulfed, and are thus incorporated in the cast metal, or whether they are rejected. In other cases of materials processing, it is either necessary to purify the solid phase and to keep it free of second-phase inclusions, or it may be desirable to achieve a controlled incorporation of particles (e.g., ceramic), thus yielding better mechanical or electrical properties of the solid matrix.

Another important example may be the study of phagocytosis (ingestion) of bacteria by single white blood cells (neutrophils or macrophages). While no solidification front is present, the situation may be treated in a similar thermodynamic fashion. In many cases, the bacteria will be destroyed if engulfed by the white blood cells due to chemical reactions. Phagocytosis thus plays an important role in the ability of the human body to combat pathogenic microorganisms such as bacteria. Another biological application is the behavior of the advancing freezing front of aqueous salt solutions during the cryopreservation of biological cell suspensions [15–17].

The remainder of this chapter will be organized as follows. Section 12.2 reviews the existing theoretical models for studying the interactions between particles and solidification fronts. Section 12.3 presents the experimental data for the behavior of particles at solidification fronts with a variety of particle-matrix material combinations. Section 12.4 describes qualitatively the thermodynamic interpretation of these experimental observations. In Section 12.5, the critical velocities (V_c) of different particle-matrix material combinations are experimentally determined and the quantitative correlation between V_c and the surface free energy of adhesion (ΔF^{adh}) is established through dimensional analysis. Based on the $V_c - \Delta F^{adh}$ correlation established in Section 12.5, Section 12.6 introduces a novel method of determining solid surface tensions using the solidification front technique. In Section 12.7, this solidification front technique is applied to measure the surface tensions of fibers, biological cells, and coal particles as examples. Finally, Section 12.8 is dedicated to theoretical interpretation of the particle-front interactions in terms of the repulsive force that arises from van der Waals interactions within the particle-liquid–solid system. The critical repulsive forces in different systems are experimentally determined using solidification experiments on inclines. From these experimental data, new insight into the critical separation distance for particle engulfment is gained.

12.2   REVIEW OF THEORETICAL MODELS

12.2.1   Critical Velocity for Particle Engulfment

In the study of behavior of particles at solidification fronts, the particles can refer to solid particles [1,2,7,18], liquid droplets [19], or gas bubbles [20,21]. The solidification fronts can grow in either the vertical [22–27] or horizontal [1,2,8,12,28–31] direction.

It is generally accepted that the behavior of particles at solidification fronts is largely determined by the balance of forces acting on the particles, mainly between the attractive viscous drag force and the repulsive particle-front intermolecular force. Practically, the microscopic force balance is reflected by a critical velocity (V_c) at which engulfment may result, as indicated in Figure 12.1. Consequently, the study of the particle-front interactions can be converted to the characterization of V_c. Once V_c is defined for a particle-matrix material system, the behavior of the particles at the solidification fronts of the matrix material can be qualitatively predicted. In general, V_c is a function of the particle geometry (such as size, shape, and roughness), the thermophysical properties of the particles and the matrix material (such as the thermal conductivities of the particles and the matrix material), and the surface free energies of the liquid-particle, solid-particle, and solid–liquid interfaces [32]. In addition, when the solidification fronts grow in vertical directions, V_c can be affected by buoyancy forces. In the case of droplets or bubbles, surface tension driven drop/bubble migration, caused by the thermocapillary effect or uneven distribution of surfactants, can also play a role in altering V_c. Qualitative effects of these influencing factors on V_c are listed below. Detailed review of the existing theoretical models and the mathematical descriptions of these influencing factors can be found in Section 12.2.2.

1. V_c decreases with increasing particle size [1,2,12,22–30,33].
2. V_c depends upon the particle shape and roughness [22]. For spherical particles, V_c is inversely proportional to the particle diameter; while for particles in a disc shape, V_c is inversely proportional to the cube of the particle diameter [32]. The effect of increasing particle roughness appears to be equivalent to reducing the effective diameter compared to a smooth sphere, thus resulting in an increased V_c [24,25,28].
3. V_c decreases with increasing liquid matrix viscosity [1,2,12,28].
4. V_c depends upon the temperature gradient [24] and the relative thermal conductivities of the particles and matrix materials. Generally, V_c decreases with increasing thermal conductivity of the particles [26,34].
5. V_c depends upon the surface tensions (i.e., surface free energies) of the solid and liquid matrix material and the particles [1,2,28–30,33]. When the surface tension of the liquid is between that of the solid matrix material and that of the particle, the free energy of adhesion is positive and the particle is pushed. When the free energy of adhesion is negative, the particle is engulfed. V_c has been observed to increase with increasing free energy of adhesion. For organic matrix materials, where the surface tension of the solid is less than that of the liquid matrix, V_c has been observed to increase with increasing particle surface tension.
6. V_c may depend upon the body forces acting on the particle [27,28]. For example, in the case of countergravity solidification (i.e., solidification fronts moving upward), buoyancy forces will either favor engulfment (i.e., decreasing V_c) or impede engulfment (i.e., increasing V_c), depending on whether the density of the particles is larger or smaller than the density of the liquid phase. The effect of buoyancy forces tends to be negligible for fine particles and when the density differences are small [32].

7. Differently from solid particles, V_c of droplets or bubbles can be affected by surface tension driven drop/bubble migration due to the thermocapillary effect or uneven distribution of surfactants at the interfaces [19–21,35]. The thermocapillary effect refers to drop/bubble motion caused by the temperature gradient across the drop/bubble. The thermocapillary effect is especially pronounced for bubbles in which the surface tension of the air–water interface can be significantly affected by temperature. Being unevenly heated or cooled, a bubble tends to move toward the direction of decreasing its surface free energy; that is, the warmer end of the bubble. The thermocapillary migration will therefore impede engulfment of the bubble by the solidification fronts [19,20]. Surface tension driven migration can also be induced by surfactants unevenly distributed at the drop/bubble surface [35]. Its effects on V_c are dependent on the gradient of surfactant concentration along the drop/bubble.

12.2.2 AVAILABLE THEORETICAL MODELS

A number of theoretical models have been developed to study the interaction between particles and advancing solidification fronts. Uhlmann et al. [12] investigated, theoretically, the pushing of a particle ahead of an advancing solidification front. In order for the particle to be pushed, there must be a force acting on it and preventing its engulfment by the solid, and there must be mass transfer of the liquid matrix material to the region between the particle and the solid so that the solidification front can advance. Engulfment occurs when the mass cannot be transferred rapidly enough for the solidification front to advance at a uniform rate, and its shape therefore becomes unstable.

The shape of the solid–liquid interface in the region near the particle must allow sufficient mass transfer by diffusion for the solidification front to advance uniformly. This was determined [12] from the variation of the surface free energy with separation distance. The shape was found [12] to be dependent upon the velocity of the solidification front. The stability of the shape of the solid–liquid interface was then investigated using a modified form of the Gibbs–Thomson relation to determine an equilibrium condition for the shape of the solid–liquid interface. It was found that there was a maximum or critical velocity above which the shape of the solid–liquid interface was unstable. For velocities above this critical value, the particle would be engulfed by the advancing solid.

For smooth spherical particles of radius R_0, the critical velocity was found [12] to vary as $1/R_0^2$. Generally, particles are not ideally smooth, but are rough. For a rough particle of radius R_0 with irregularities of radius R, the critical velocity was determined to be:

$$V_c = \frac{d_s h La_0 d_1}{6\mu n R_0 R^2}\left[-1+\left(1+\frac{6\mu R_0 n(n+1)v_0 D}{d_s h d_1 kT}\right)^{1/2}\right],\qquad(12.1)$$

where L is the latent heat of fusion of the liquid, a_0 is a constant of the order of one molecular diameter, μ is the liquid viscosity, v_0 is the atomic volume of the liquid,

D is the diffusion coefficient of the liquid, k is the Boltzmann constant, T is the temperature, n is the exponent of the variation of the surface free energy with separation distance, d_1 is the minimum separation distance between the particle and the solid (assumed to be two molecular diameters), d_s is the minimum separation distance for fluid flow (assumed to be 10 molecular diameters), and h is the penetration depth of the particle into the solid (assumed to be 10 molecular diameters). Assuming the radius of the irregularities to be independent of the particle radius, R_0, the critical velocity varies as $1/R_0$ for small particles and as $1/\sqrt{R_0}$ for large particles.

The expression for the critical velocity, V_c, however, does not contain the free energy of adhesion, which is defined [12] as $\Delta F^{adh} = \gamma_{ps} - \gamma_{pl} - \gamma_{sl}$, where γ_{ps}, γ_{pl}, and γ_{sl} are the surface free energies (or the surface tensions) of the particle-solid, particle-liquid, and solid–liquid interfaces, respectively. The critical velocity is therefore predicted to depend only upon the properties of the matrix material and to be independent of the properties of the particle. This result, obtained using the Gibbs–Thomson relation, is not consistent with the experimental observations that the critical velocity is dependent upon the properties (such as surface tension) of the particle [1,2,29,30,33]. It is also not consistent with the initial assumption that there must be a repulsive force between the particle and the solid in order for pushing to occur.

The reason that the predicted critical velocity was independent of ΔF^{adh}, and therefore of the properties of the particle, is that the critical velocity was determined only from the stability of the solid–liquid interface. The model of Uhlmann et al. [12] is not adequate for the determination of the critical velocity since it does not consider the balance of forces acting on the particle and therefore predicts that the critical velocity is independent of the properties of the particle. However, several important contributions were made to the understanding of the pushing of particles by an advancing solidification front:

1. There must be a force acting on the particle and preventing its engulfment, and this force must be related to the free energy of adhesion.
2. Engulfment occurs when the liquid cannot be transferred rapidly enough to the region between the particle and the solid for the solidification front to advance at a uniform rate.
3. The drag force acting on a spherical particle being pushed by a planar solidification front was correctly determined.
4. Generally, particles are not ideally smooth, but are rough; and the critical velocity for a rough particle differs from that for a smooth particle.

The pushing of a particle by an advancing solidification front was also investigated by Bolling and Cissé [36]. The critical velocity for a smooth particle of radius R was determined from the stability of the solid–liquid interface in a manner similar to that of Uhlmann et al. [12]. Again, their expression for the critical velocity is a function of the properties of the matrix material only, and is independent of the properties of the particle. This is also a consequence of considering only the stability of the solid–liquid interface in the determination of the critical velocity. The balance of forces acting on the particle, or even the presence of a repulsive force preventing engulfment, was not considered. However, Bolling and Cissé also make several contributions to the solution of this problem:

1. The drag force acting on a spherical particle being pushed by a concave solidification front was correctly determined.
2. The effects of thermal conductivity and temperature gradient on the shape of the solid–liquid interface were qualitatively investigated. When the thermal conductivity of the particle is equal to that of the matrix material, the heat flow is unidirectional. However, when the thermal conductivity of the particle is greater than that of the matrix material, the heat flow is not unidirectional since the heat is more easily conducted through the particle than through the matrix. This causes the solid–liquid interface to become concave near the particle. Similarly, when the thermal conductivity of the particle is less than that of the matrix material, the solid–liquid interface is convex. The greater the temperature gradient in the matrix material, the greater the heat flow, and the more concave or convex the interface becomes.

In the investigation of Chernov and coworkers [37–40], the equilibrium conditions for both the particle and the shape of the solid–liquid interface were used to determine an expression for the critical velocity of the particle. There must be a balance of forces acting on the particle in order for it to be continuously pushed by the advancing solidification front. Two forces were assumed to act on the particle: a viscous drag force that tends to push the particle toward the solid, and a repulsive force between the particle and the solid. This repulsive force was considered to be an unretarded van der Waals interaction, which was related to the disjoining pressure between the particle and the solid.

Assuming that the solid–liquid interface was a paraboloid in the region near the particle, the repulsive interaction force for a spherical particle of radius R was

$$F_R = \frac{\pi B_3 R}{(1-v)h_0^2},$$ (12.2)

where h_0 is the minimum separation distance between the particle and the solid, v is the ratio of the particle-radius to the radius of curvature of the concave solid–liquid interface, and B_3 is related to the Hamaker coefficient by

$$B_3 = -\frac{A_{pls}}{6\pi}$$ (12.3)

In order for this force to be repulsive, B_3 must be greater than zero (or A_{pls} must be less than zero). This corresponds to a positive free energy of adhesion that has to be overcome during the porcess of particle engulfment. The drag force was the same as determined by Bolling and Cissé [36] and is given by

$$F_D = \frac{6\pi \mu V R^2}{(1-v)^2 h_0},$$ (12.4)

where μ is the liquid viscosity and V is the velocity of the solidification front.

For continuous pushing of the particle these forces must be equal, and the velocity was therefore given by

$$V = \frac{B_3(1-v)}{6\mu R h_0}. \tag{12.5}$$

This expression does not give the critical growth rate since the minimum separation distance, h_0, and the relative curvature of the solid–liquid interface, v, have not been determined. They were determined from the second equilibrium condition, the stability of the shape of the solid–liquid interface. The shape of the solid–liquid interface was assumed to be determined by three factors:

1. the disjoining pressure that tends to increase the curvature of the solid–liquid interface;
2. the Gibbs-Thomson effect that tends to decrease the curvature of the interface; and
3. the temperature gradient in the matrix material that also tends to decrease the curvature of the interface.

The solid–liquid interface was assumed to be a paraboloid in the region near the particle and a plane away from the particle. The shape of the solid–liquid interface was then determined from the relation

$$\Omega\gamma\left(\frac{1}{R_1} + \frac{1}{R_2}\right) - \frac{B_3\Omega}{h^3} + \Delta sG(z - H) = 0, \tag{12.6}$$

where Ω is the specific molecular volume of the liquid, γ is the solid–liquid interfacial tension, R_1 and R_2 are the principal radii of curvature of the interface, Δs is the entropy of melting, G is the temperature gradient, and H is the distance from the center of the spherical particle to the planar solid–liquid interface.

The critical velocity was determined from the stability of the shape of the solid–liquid interface. Two characteristic lengths, l and λ, were introduced:

$$l = \left(\frac{B_3\Omega}{\Delta sG}\right)^{1/4}, \lambda = \left(\frac{\gamma\Omega}{\Delta sG}\right)^{1/2}. \tag{12.7}$$

For small particles, $R < \lambda^2/l$, only the first two terms in Equation 12.6 are significant, and the critical velocity was determined to be

$$V_c = \frac{0.11B_3}{\mu R}\left(\frac{2\gamma}{B_3 R}\right)^{1/3}. \tag{12.8}$$

For large particles, $R > \lambda^2/l$, only the last two terms in Equation 12.6 are significant, and the critical velocity was determined to be

$$V_c = \frac{0.15B_3}{\mu R} \left(\frac{\Delta s G}{B_3 \Omega} \right)^{1/4}.$$ (12.9)

The critical velocity was found to vary as $R^{-4/3}$ for small particles and as R^{-1} for large particles. It was also found to vary inversely with the liquid viscosity, in agreement with the predictions of Uhlmann et al. [12] and Bolling and Cissé [36]. However, unlike the previous models, the critical velocity was also found to be dependent upon the properties of the particle material. For both large and small particles, the critical velocity was determined to be dependent upon B_3, which is related to the Hamaker coefficient, A_{pls}, which, in turn, is a function of the surface tensions of the solid and liquid matrix material and of the particle. These expressions for the critical velocity are therefore consistent with the experimental observations that the critical velocity is dependent upon the properties of the particle, such as its surface tension [1,2,30,33].

For large particles, the critical velocity was found to vary with the temperature gradient in the form

$$V_c \propto G^{1/4}.$$ (12.10)

The predicted variation of the minimum separation distance, h, with the particle radius is as follows. For small particles, this minimum separation distance was given by

$$h = 1.3 \left(\frac{B_3 R}{2\gamma} \right)^{1/3}.$$ (12.11)

For large particles, the minimum separation distance was found to be independent of the particle size, but dependent upon the temperature gradient:

$$h = \left(\frac{B_3 \Omega}{Gs} \right)^{1/4}.$$ (12.12)

The minimum separation distance was therefore predicted to increase with increasing particle radius for small particles until it reached a maximum value determined by the temperature gradient.

The shape of the solid–liquid interface was predicted by Bolling and Cissé [36] to be dependent upon the relative values of the thermal conductivities of the particle and the matrix material. As shown in Equation 12.5, the critical velocity was predicted by Chernov and coworkers [37–40] to be dependent upon the curvature of the solid–liquid interface and therefore must also be dependent upon the relative thermal conductivities. Chernov and coworkers [37–40] found that for large particles, those for which the temperature gradient was important, the critical velocity varied with the thermal conductivity in the form

$$V_c \propto \left(\frac{k_L}{k_p}\right)^{3/4} \left(\frac{3k_L}{2k_L + k_p}\right)^{1/4},$$ (12.13)

where k_L is the liquid thermal conductivity and k_p is the particle thermal conductivity.

The critical velocity was therefore predicted to decrease with increasing particle thermal conductivity. This occurs since the solid–liquid interface becomes more concave when the thermal conductivity of the particle is greater than that of the liquid, resulting in a lower critical velocity. When the thermal conductivity of the particle is less than that of the liquid, the interface is convex and the critical velocity is therefore larger.

The critical velocity predicted by Equation 12.8 can be compared with the experimental observations of Omenyi et al. [1,2,29,30,33] since all the parameters in this equation have been determined. Omenyi et al. measured the critical velocity of irregularly shaped polymer particles in organic matrix materials. For acetal particles in naphthalene [1,2], $B_3 = 1.73 \times 10^{-19}$ mJ, $\mu = 9.67 \times 10^{-4}$ kg/s·m, and $\gamma = 1.44$ mJ/m². For a particle with a radius of 50 μm, Equation 12.8 predicts a critical velocity of 0.03 μm/s. This is three orders of magnitude smaller than the observed velocity of 40 μm/s.

Chernov and Temkin [37] made the most complete investigation of the critical velocity for smooth spherical particles. The effects of particle thermal conductivity, impurities, and viscosity fluctuations on the critical velocity were investigated. The expressions that were derived for the critical velocity were in good qualitative agreement with experimental observations. However, since the effect of particle shape and roughness were not investigated, the predicted critical velocity did not agree quantitatively with experimental results for real particles, which are generally neither smooth nor spherical.

Gilpin [41] used an approach similar to that of Chernov and Temkin [37]. He considered both the equilibrium conditions for the particle and the shape of the solid–liquid interface in the determination of the critical velocity of the particle. The repulsive force between the particle and the solid was related to the chemical potential of the liquid layer between them. The chemical potential, μ_l, was given by

$$\mu_l = \mu_{lb} - g(h),$$ (12.14)

where $g(h)$ is some function and μ_{lb} is the bulk chemical potential. Two relations determined empirically from measurements of pressure- and temperature-gradient-induced regelation for ice-water were used for $g(h)$. They were

$$g_1(h) = a \frac{L}{T_a} \left(\frac{h}{h_1}\right)^{-2}$$ (12.15)

$$g_2(h) = a \frac{L}{T_a} \exp[-\beta(h - h_1)], \tag{12.16}$$

where L is the latent heat of fusion, a is a constant equal to 1, T_a is the absolute temperature, h is the liquid layer thickness, and h_1 and β are empirically determined quantities.

The values of β and h_1 were assumed to be constants for the matrix material, water, and independent of the particle material. As a result, the critical velocity was found to vary with particle density for vertical motion only and would be independent of the particle material for horizontal motion. Furthermore, this model predicted that particles of all materials would be pushed by the advancing solidification front. This is not consistent with experimental observations [1,2,29,30,33].

The independence of the critical velocity from the properties of the particle is a result of the relationship used for $g(h)$. If the repulsive force between the particle and the solid is assumed to be due to an unretarded van der Waals interaction, then $g(h)$ would be given by

$$g(h) = \frac{A_{pls} v_l}{6\pi h^3}, \tag{12.17}$$

where A_{pls} is the Hamaker coefficient and v_l is the specific volume of the liquid. Using this expression, Equation 12.17, instead of Equations 12.15 or 12.16, for $g(h)$, and following Gilpin's method, an approximate analytical expression can be obtained for the critical velocity. Assuming that the solid and liquid specific volumes are equal, the critical velocity for a spherical particle of radius R is given by

$$V_c = -\frac{A_{pls}}{36\pi\mu R} \left(\frac{12\pi\gamma}{-A_{pls}R} \right)^{1/3}. \tag{12.18}$$

This result is approximately 50% greater than that determined by Chernov and Temkin [37], Equation 12.8, for the same circumstances. The difference results from the determination of the shape of the solid–liquid interface.

Gilpin's contribution to the solution of this problem was the determination of the shape of the solid–liquid interface [41]; the equilibrium condition for the liquid layer thickness, h, was given by

$$-\Delta v p_{lh} - L\hat{T} / T_a + v_s \gamma \bar{k} - g(h) = 0, \tag{12.19}$$

where Δv is the specific volume difference $v_s - v_l$, p_{lh} is the pressure at the solid–liquid interface, $\hat{T}$ is the temperature difference from the bulk fusion temperature, and $\bar{k}$ is the mean curvature of the solid–liquid interface. By dividing the solid–liquid interface into three regions—inner, transition, and outer—and solving the equation for the shape of the interface for each region separately, Gilpin was able to determine numerically the shape of the solid–liquid interface, as opposed to the arbitrary surface chosen by Chernov and Temkin [37] consisting of a paraboloid for the inner region and a planar surface for the outer region. Similar numerical calculations of the shape of the solid–liquid interface have also been made by Aubourg [28].

In summary, the behavior of particles at solidification fronts; that is, engulfment or rejection, generally depends on the fluid mechanics and thermophysics of the solidification process. However, if the rate of solidification is low, as in the case of studying critical velocity (V_c), thermodynamic effects may dominate, as will be discussed in the remainder of this chapter. Readers who are interested in the fluid mechanics and heat transfer aspects of this study may refer to recent research papers [7,20,34,42–45].

12.3 EXPERIMENTAL

12.3.1 EXPERIMENTAL SETUP

An experimental apparatus to study the behavior of small inert particles at solidification fronts is illustrated in Figure 12.2 [2,46]. The particles and matrix material are contained in a small horizontal copper channel and are covered with a glass microscope slide. A temperature gradient is imposed on the matrix material by maintaining a temperature gradient in the copper. The solidification front is caused to move by lowering the temperature of the copper, and therefore the temperature of the matrix material. The particles and the solidification front are observed by reflection microscopy.

The copper cell is 78 mm long, 30 mm wide, and 5 mm high; the groove in the cell is 5 mm wide and 0.6 mm deep. The temperature of the copper, and therefore the matrix material, is varied by heating or cooling by means of thermoelectric devices

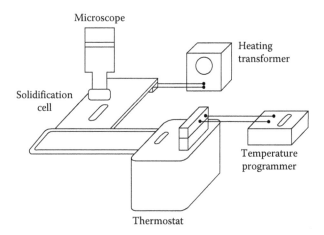

FIGURE 12.2 An experimental apparatus to study the behavior of small particles at solidification fronts.

at each end of the copper channel. The temperature at each end is measured by a thermistor and is independently controlled by a feedback controller. The measured value of each temperature is compared with its programed value, the difference determining the electric current provided by a bipolar controller (Cambion 809-3020-09). The rate of heat transfer by the thermoelectric device (Cambion 801-3958-01) to or from the copper channel is determined by this current. A water-cooled heat sink is used to cool the thermoelectric devices.

The apparatus is designed to operate in the temperature range from 0 to 100°C. The maximum temperature is limited by the maximum operating temperature of the thermoelectric devices. This apparatus is suitable for critical velocity experiments using matrix materials with a melting point between 20°C and 80°C, such as naphthalene, biphenyl, and thymol. The horizontal temperature gradient of the matrix material is regulated by controlling the temperature difference between each end of the copper channel using a temperature programmer.

The solidification channel is mounted on the moving stage of a Leitz Orthomat-Orthoplan photomicroscope so that particle interaction with the solidification front can be viewed. Two overall magnifications (through the eyepiece) of 64 × and 128 × with reflected light are used. The microscope can be used with a video camera, allowing the experiment to be recorded for analysis at a later time.

12.3.2 Matrix Materials

The requirements for the matrix materials to be used in the solidification experiments generally include the following:

1. The matrix materials should not react chemically with the particle materials.
2. The design of the experimental cell requires that melting temperatures of the matrix materials should be below 100°C. In the case where water is used as a thermostating liquid to regulate and control the temperature of the solid–liquid interface, the melting point of the matrix should not be below 30°C. However, if lower-melting-point matrix materials are used, the thermostating liquid can be replaced with a mixture of water and ethylene glycol and a temperature of about −60°C can be attained.
3. The matrix materials should be transparent in order to allow direct observation of particle interactions with the solidification fronts.
4. The matrix materials should not be toxic since the nature of the cell makes it possible for the vapor to escape into the environment.
5. The availability of the physical and thermal properties of the matrix materials is an important criterion for material selection.
6. Most importantly, the matrix materials should be able to form plane solid–liquid interfaces. Without a smooth interface, particle behavior at solidification fronts cannot be studied. The presence of dendritic growth leads to immediate entrapment of foreign particles, so this will not provide the desired results. One of the important factors for forming a plane solid–liquid interface is the purity of the matrix materials.

Finally, as the thermodynamic prediction of particle behavior at the solidification fronts will require the use of the equation of state for interfacial tensions, which is valid for low-energy solids, the materials of the particles and matrix should be low energy.

Tables 12.1 and 12.2 list the physical properties of several matrix materials that were found to meet essentially all the requirements.

12.3.3 EXPERIMENTS

In the experiment, the cell groove is polished with alumina polishing powders. The polishing powders are then washed away with distilled water, and the cell is rinsed with toluene. A few of the desired particles are sprinkled into the cell groove and then covered completely with the matrix material powder. A clean glass slide is placed on top of the matrix material. The cell is then connected to the heater and

TABLE 12.1
Physical and Chemical Properties of the Matrix Materials

Matrix Material	Chemical Formula	Melting Temperature T_{mp} (°C)	Mol. Weight	Source
Naphthalene	$C_{10}H_8$	80	128	Fisher Scientific (F.S.)
Biphenyl (phenyl benzene)	$C_{12}H_{10}$	70	154	F.S.
Thymol (2-hydroxy-1-isopropyl-4-methyl benzene)	$C_{10}H_{14}O$	51.5	150	Eastman Kodak (E.K.)
Salol (phenyl salicylate)	$C_{13}H_{10}O_3$	43	214	E.K.
o-Terphenyl (1,2-diphenyl benzene)	$C_{18}H_{14}$	58	230	E.K.
Pinacol (tetramethyl ethylene glycol)	$C_6H_{14}O_2$	43.5	118	E.K.
2-phenyl phenol (2-hydroxy biphenyl)	$C_{12}H_{10}O$	59	170	E.K.

TABLE 12.2
Thermal and Physical Properties of Organic Matrix Melts

Substance	Density ρ_l (kg/m³)	Viscosity $\mu \times 10^3$ (Pa s)	Specific Heat c_l (J/g °C)	Thermal Conductivity k_l (W/m K)	Latent Heat L (J/g)	Diffusion Coefficient $D_l \times 10^6$ (cm²/s)
Naphthalene	978	0.967	1.683	0.226	146.74	7.20
Biphenyl	992	1.490	1.842	0.142	109.20	4.30
Thymol	925		2.369	0.130	114.98	6.98
Salol	1200	0.746	1.637	0.084	89.07	7.50

the copper cooling coil as shown in Figure 12.2. With the thermostat turned off, the heater is turned on until all the matrix melts and the glass slide falls into place on the glass cell, thereby trapping the melt in the cell groove. Some melt may spread on top of the glass slide and may be drained off. When all the matrix material has melted, the temperature of the thermostating water is reduced at a very low rate until slow solidification is initiated. Using the temperature programmer, different solidification rates can be attained, and the interface can be advanced or retracted as desired, enabling a given particle to be studied in detail. The heating current to be supplied depends on the matrix being studied and the correct setting is obtained on a trial-and-error basis.

As the interface advances, the particle–interface interactions can be observed through the micrometer eyepiece of the microscope at a chosen magnification. When contact is made between particle and interface, if the particle is not pushed, a smaller interface speed is reinitiated and used to study the interaction. If, at a very low speed, say 2 μm/s or less, the particle is still not pushed, one may conclude that particle engulfment has occurred. In case of doubt, the melting front should be made to recede, and the process of solidification may be reinitiated, possibly at a lower rate. This procedure is of critical importance, as all particles, irrespective of thermodynamic properties of the system, will be engulfed at relatively large rates of solidification.

In the experiments, parts of the interface could be seen to be very sharp, some straight, and others ragged. It is important to ensure that the particle considered is positioned on a plane part of the interface.

The particle may float or lie on the cell floor, depending on whether the particle is lighter or heavier than the melt. By properly adjusting the focus of the lenses, the particles can be studied. Graticules are incorporated into the micrometer eyepiece for determining particle size and the advancing distance. The sizes of the squares of the graticules are 19×19 μm for the $128 \times$ magnification and 38×38 μm for the $64 \times$ overall magnification through the eyepiece. The graticules are precalibrated by matching the squares against a standard scale.

As mentioned earlier, not all the interfaces are smooth. Naphthalene, biphenyl, and thymol exhibit straight interfaces at low advancing rates whereas the rest of the matrix materials in Table 12.1 show different types of ragged interfaces. For a sawtooth interface, the particle is considered only if it is located on a plane part of the interface and observed in such a way that it is moving in a direction perpendicular to the interface.

Some large particles may be observed to rotate or jerk as they move forward. This could possibly be due to some contact with the cell floor. Most small particles, however, will not rotate or jerk, suggesting that no contact with the cell floor occurs. It has been shown that friction with the cell floor has no significant effect on the outcome of the engulfing experiments [1,46,47].

Some particles, when being pushed, are seemingly partially embedded in the solid phase. The explanation for such observations is that the solidification front may not be a vertical plane, due to a slight temperature difference between the bottom and the top of the cell. With the present setup it is not possible to see the gap between the particle and the interface as the particle is being pushed. Generally, a liquid film must exist between the particle and the interface for the particle to be pushed, and it is the properties of this liquid film that determine whether a particle should be engulfed or not. This separation distance will be discussed later in this chapter (Section 12.8.2) and more details can be found elsewhere [46,47].

Sometimes, the particles in the melt are hardly visible due to a similarity in the refractive index of the particle material and the melts; for example, nylon particles in the melts of thymol, pinacol, o-terphenyl, and 2-phenyl phenol are in this category. This difficulty may be overcome by using polarized light. Some typical results of particle pushing/engulfing experiments [1] are summarized in Table 12.3. The particle sizes and the speeds at which a variety of materials were pushed or engulfed are listed for the seven matrix materials of Tables 12.1 and 12.2.

12.4 THERMODYNAMIC INTERPRETATION

Particle engulfment and rejection may, in general, depend on the fluid mechanics of the solidification process; on thermodynamic properties such as interfacial tensions, temperature, and concentration; and also on the supersaturation or supercooling of the crystallizing liquid. If the rate of solidification is low and the solid-melt interface is smooth, thermodynamic effects may dominate.

Thermodynamically, the process of particle engulfment by the advancing solidification front (see Figure 12.3) can be modeled by the net free-energy change of the system during the engulfing process. As illustrated in Figure 12.3, the net free-energy change per unit surface area for the engulfment process is given by

$$\Delta F^{eng} = \gamma_{ps} - \gamma_{pl}, \tag{12.20}$$

where γ_{ps} and γ_{pl} are the particle-solid and particle-liquid interfacial tensions, respectively. Equation 12.20 reflects the fact that, for a particle of unit surface area, 1 cm^2 of particle-solid interface is generated and 1 cm^2 of particle-liquid interface is annihilated, as a result of particle engulfment.

The condition for particle engulfment is that the net change in the free energy of the system, ΔF^{eng}, is negative; that is,

$$\Delta F^{eng} < 0. \tag{12.21}$$

If ΔF^{eng} is positive, that is,

$$\Delta F^{eng} > 0, \tag{12.22}$$

there will be particle rejection. An intermediate step in the process of particle engulfment is particle adhesion. For a square particle, the associated free-energy change, the free energy of adhesion, is given by:

$$\Delta F^{adh} = \gamma_{ps} - \gamma_{pl} - \gamma_{sl}. \tag{12.23}$$

TABLE 12.3

Microscopic Observations of Particle Pushing and Engulfing

System	Particle Diameter (μm)	Remarks	Velocity (μm/s)
Naphthalene/acetal	47	Pushed	40.0
Nylon-6	63	Pushed	28.7
Nylon-6,6	78	Pushed	19.3
Nylon-12	60	Pushed	15.6
Nylon-6,10	59	Pushed	7.7
Nylon-6,12	32	Pushed	17.0
Polystyrene	51	Engulfed	3.2
Teflon	79	Engulfed	2.7
Siliconized glass	67	Engulfed	2.1
Biphenyl/acetal	31	Pushed	29.0
Nylon-6	59	Pushed	14.3
Nylon-6,6	70	Pushed	11.4
Nylon-12	51	Pushed	8.6
Nylon-6,10	88	Pushed	4.1
Nylon-6,12	69	Pushed	5.3
Polystyrene	43	Engulfed	1.2
Teflon	31	Engulfed	0.9
Siliconized glass	56	Engulfed	1.1
2-phenyl phenol/acetal	12	Pushed	0.20
Nylon-6	Could not be observed	Could not be observed	Could not be observed
Nylon-6,6	Could not be observed	Could not be observed	Could not be observed
Nylon-12	76	Reoriented and engulfed	0.56
Nylon-6,10	38	Reoriented and engulfed	0.90
Nylon-6,12	76	Reoriented and engulfed	0.30
Polystyrene	20	Engulfed	0.33
Teflon	57	Engulfed	0.93
Siliconized glass	51	Engulfed	0.70
Pinacol/acetal	50	Pushed	0.80
Nylon-6	10	Reoriented and engulfed	0.34
Nylon-6,6	8	Reoriented and engulfed	0.30
Nylon-12	8	Pushed	0.20
Nylon-6,10	10	Pushed	0.20
Nylon-6,12	47	Engulfed	0.40
Polystyrene	4	Pushed	0.41
Teflon	10	Pushed	0.39
Siliconized glass	56	Engulfed	0.40
Thymol/acetal	60	Pushed	9.60
Nylon-6	38	Pushed	0.75

(Continued)

TABLE 12.3   (Continued)
Microscopic Observations of Particle Pushing and Engulfing

System	Particle Diameter (μm)	Remarks	Velocity (μm/s)
Nylon-6,6	57	Pushed	1.20
Nylon-12	95	Pushed	1.35
Nylon-6,10	95	Pushed	1.00
Nylon-6,12	57	Pushed	1.00
Polystyrene	25	Pushed	2.00
Teflon	13	Reoriented and engulfed	0.30
Siliconized glass	20	Engulfed	0.90
o-Terphenyl/acetal	25	Pushed	1.70
Nylon-6	57	Pushed	0.70
Nylon-6,6	57	Pushed	0.48
Nylon-12	28	Pushed	0.42
Nylon-6,10	57	Pushed	0.50
Nylon-6,12	76	Pushed	0.42
Polystyrene	54	Engulfed	0.61
Teflon	72	Engulfed	0.50
Siliconized glass	13	Engulfed	0.80
Salol/acetal	50	Pushed	1.30
Nylon-6	38	Pushed	0.30
Nylon-6,6	25	Reoriented and engulfed	0.30
Nylon-12	19	Pushed	0.90
Nylon-6,10	19	Pushed	0.82
Nylon-6,12	13	Reoriented and engulfed	0.20
Polystyrene	28	Engulfed	0.90
Teflon	48	Engulfed	0.30
Siliconized glass	51	Engulfed	0.80

The process of particle adhesion (see Figure 12.3) involves generation of particle-solid interface and annihilation of particle-liquid interface as well as solid–liquid interface. It is not clear a priori which of the two free energies, ΔF^{eng} or ΔF^{adh}, should be used to study the interaction between a particle and a solidification front. A discussion of this point will be given later. Fortunately, γ_{sl} is usually quite small, so that there will not normally be a large difference between ΔF^{eng} and ΔF^{adh}.

The interfacial tensions involved in Equation 12.23 must be known in order to calculate the net free energy change. For a solid–liquid–vapor system, the solid-vapor and solid–liquid interfacial tensions can be determined from the experimental data for liquid surface tensions and contact angles by an equation of state approach [48–52]. In this approach, an equation of state for interfacial tensions is given by

$$\gamma_{sl} = \gamma_{lv} + \gamma_{sv} - 2\sqrt{\gamma_{lv}\gamma_{sv}}\ e^{-\beta(\gamma_{lv}-\gamma_{sv})^2}.\tag{12.24}$$

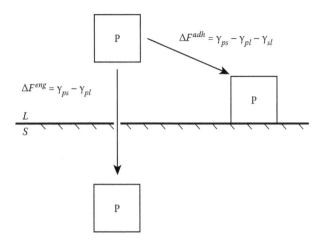

FIGURE 12.3 Free-energy changes during particle-solid adhesion and particle engulfment processes.

The constant β in Equation 12.24 is obtained as $\beta = 0.0001247$ $(m^2/mJ)^2$. Combining this equation with Young's equation

$$\gamma_{lv} \cos\theta = \gamma_{sv} - \gamma_{sl},\qquad(12.25)$$

will yield

$$\cos\theta = -1 + 2\sqrt{\frac{\gamma_{sv}}{\gamma_{lv}}}\,e^{-\beta(\gamma_{lv}-\gamma_{sv})^2}.\qquad(12.26)$$

It is apparent that Equation 12.26 has three variables, the liquid–vapor surface tension, γ_{lv}, the contact angle, θ, and the solid-vapor surface tension, γ_{sv}, and thus will enable us to determine solid surface tension, γ_{sv}, when we have experimental data γ_{lv} and θ. Finally, the solid–liquid interfacial tension, γ_{sl}, can be determined from either Young's equation, Equation 12.25, or the equation of state, Equation 12.24. Detailed description and recent progress in determining solid surface tensions using the equation of state approach can be found in Chapters 8 and 9.

The particle–liquid interfacial tension, γ_{pl}, in Equation 12.23 can be obtained by using the above equation of state approach, if contact angle information for the particle material is available. For instance, it may be possible to produce a smooth surface out of polymer particles by heat pressing (see Chapter 11 for details). However, for the particle-solid interfacial tension, γ_{ps}, one has to take Equation 12.24 as a generic correlation among the three interfacial tensions, γ_{12}, γ_{13}, and γ_{23}; that is,

$$\gamma_{12} = f(\gamma_{13}, \gamma_{23}).\qquad(12.27)$$

Then, γ_{ps} may be obtained from γ_{pv} and γ_{sv}, both of which can be determined by the contact angle/equation of state method. While not justified explicitly within the equation of state approach, this strategy has been found to be successful in predicting the engulfment or rejection of particles by solidification fronts [2,29,33,53].

Since particle engulfment and rejection occur at the melting temperature of the matrix material, the various interfacial tensions must refer to the melting temperature. However, measurements of γ_{lv} and θ at the melting point are not practical, so that measurements of the temperature dependence of these quantities and extrapolation to the melting point are called for.

In such studies [2,29,33], the temperature-dependent surface tensions of melts of the matrix materials were determined using the Wilhelmy plate technique [54], both on heating and on cooling. The surface tension at the melting point was obtained by extrapolation.

All matrix materials were zone refined and the purity was checked qualitatively by differential scanning calorimetry prior to final measurements and engulfing experiments. The polymeric particles were commercial products that were used as received. Flat and smooth surfaces of the matrix and particle materials were prepared for contact angle measurements by film casting on glass slides from solutions of the materials in suitable solvents. For Teflon and acetal, the surfaces were prepared by heat pressing of the polymer between clean glass slides. Siliconized glass slides as well as siliconized glass powder were prepared by heating clean glass spheres in silicone oil to 150°C for 2 hours. They were then allowed to cool to room temperature in the silicone oil. Eventually, the excess oil not bonded to the glass surface was rinsed off with toluene. Preliminary contact angles of water and glycerol were measured using the conventional sessile-drop technique to check the quality of the solid surfaces.

In order to determine γ_{sv} and γ_{pv} values; that is, the surface tension of the matrix and the particle materials, contact angles as a function of temperature were measured on all matrix and all particle materials. Temperature-dependent contact angles were obtained using the method of capillary rise at a vertical plate [55] (see Chapter 6).

The solid surface tensions, γ_{sv}, of the matrix materials at their respective melting points were calculated from the temperature dependence of the contact angles, using the equation of state relations. The solid surface tension, γ_{pv}, for each particle material was similarly calculated for the various temperatures of the matrix materials. The relevant interfacial tensions, γ_{sl}, γ_{pl}, and γ_{ps}, for the melting point of the matrix material were also calculated from the equation of state relations. As examples, some of these data are listed in Tables 12.4 and 12.5. The net free-energy changes for particle engulfment, ΔF^{eng}, and the free energy of adhesion, ΔF^{adh}, can then be calculated from these data.

Table 12.6 shows the comparison between the thermodynamic predictions of ΔF^{eng} and the results of actual engulfing experiments. In Table 12.6, the letter R designates rejection; the letter E, engulfing; and (E), conditional engulfing, for example, a slight reorientation of the particle prior to engulfment, but definitely not a transport of the particle by the solidification front. As clearly seen, there is very good agreement between thermodynamic prediction and microscopic observation.

The largest discrepancy between clear-cut observation (i.e., E or R) and thermodynamic predictions is for the two systems nylon-6,12/pinacol and Teflon/pinacol, with ΔF^{eng} values of $+ 0.03$ mJ/m^2 and -0.05 mJ/m^2, respectively. A detailed analysis [1] showed that errors of the magnitude of these data can easily arise from errors

TABLE 12.4

Surface Tensions (in mJ/m²) of Particle Materials, γ_{pv}, and Solid Matrix Materials, γ_{sv}, at the Melting Points of the Corresponding Matrix Materials

	Salol	Pinacol	Thymol	o-Terphenyl	2-Phenyl Phenol	Biphenyl	Naphthalene
Matrix materials (γ_{sv})	34.5	28.0	29.4	30.4	38.6	25.6	22.9
Particle materials (γ_{pv})							
Acetal	44.9	44.9	44.3	43.8	43.8	43	42.2
Nylon-6	44.2	44.1	43.4	42.9	42.8	41.8	40
Nylon-6,6	43.5	43.4	42.8	42.3	42.2	41.3	40.5
Nylon-12	41.2	41.2	40.6	40.1	40.0	39.2	38.5
Nylon-6,10	38.6	38.5	37.8	37.3	37.2	36.3	35.4
Nylon-6,12	34.7	34.6	34.0	33.5	33.4	32.6	31.8
Polystyrene	30.4	30.4	29.8	39.3	29.2	28.4	27.7
Teflon	18.4	18.4	17.8	17.3	17.2	16.4	15.7
Siliconized glass	16.0	15.9	14.9	14.1	14.0	12.6	11.3

in γ_{lv} alone. Errors due to errors in γ_{sv} might easily be larger, for example, 0.1–0.2 mJ/m², owing to errors in the contact angles and the fact that the equation of state relations used in interpreting the data, in turn, rely on experimental contact angles. Apart from these two cases, all other doubtful results are connected with the observations of particle reorientation and engulfment (E), which in itself might well be understood as being indicative of a small absolute value of ΔF^{eng}. Interpreting the observation (E) as engulfing, the largest discrepancy occurs for nylon-6,6/salol with $\Delta F^{eng} = + 0.18$ mJ/m²; and interpreting it as rejection, the largest discrepancy would occur for Teflon/thymol with $\Delta F^{eng} = -0.19$ mJ/m². The sign of either of these could be due to slight errors in γ_{lv}, γ_{sv}, and γ_{sl}. It was shown [1] that friction of the particles on the floor of the cell is not a major influence on the behavior of the particles. Overall, it can be concluded that particle engulfment by solidification fronts at low rates of solidification can be described by surface thermodynamic properties.

12.5 CRITICAL VELOCITY AND DIMENSIONAL ANALYSIS

As an experimental fact, particle engulfment depends on the rate of the advancing solidification front. Therefore, it is necessary to study rate-dependent phenomena. It was shown [2] that, at even the lowest rates, Teflon particles and siliconized glass spheres were engulfed by the advancing solid–liquid interfaces of biphenyl and naphthalene melts, while nylon and acetal particles were pushed. Polystyrene latex spheres were just reoriented without being pushed. As a result of these observations, nylon and acetal particles were chosen for rate studies. Rates of solidification were measured by timing the progress of the solidification front over distances measured

TABLE 12.5
Solid–Liquid (γ_{sl}), Particle–Liquid (γ_{pl}), and Particle–Solid (γ_{ps}) Interfacial Tensions at the Melting Points of Matrix Materials

System	γ_{sl} (mJ/m²)	γ_{pl} (mJ/m²)	γ_{ps} (mJ/m²)
Naphthalene/naphthalene	1.44		
Naphthalene/acetal		1.17	5.23
Naphthalene/nylon-6		0.87	4.56
Naphthalene/nylon-6,6		0.79	4.37
Naphthalene/nylon-12		0.43	3.45
Naphthalene/nylon-6,10		0.09	2.27
Naphthalene/nylon-6,12		0.01	1.17
Naphthalene/polystyrene		0.38	0.36
Naphthalene/Teflon		4.64	0.94
Naphthalene/siliconized glass		7.81	2.84
Biphenyl/biphenyl	0.67		
Biphenyl/acetal		1.46	4.12
Biphenyl/nylon-6		1.15	3.59
Biphenyl/nylon-6,6		1.04	3.59
Biphenyl/nylon-12		0.60	2.56
Biphenyl/nylon-6,10		0.20	1.59
Biphenyl/nylon/6,12		0.00	0.69
Biphenyl/polystyrene		0.23	0.13
Biphenyl/Teflon		4.03	1.46
Biphenyl/siliconized glass		6.56	3.12
Thymol/thymol	0.00		
Thymol/acetal		2.76	2.90
Thymol/nylon-6		2.44	2.63
Thymol/nylon-6,6		2.22	2.40
Thymol/nylon-12		1.52	1.67
Thymol/nylon-6,10		0.85	0.96
Thymol/nylon-6,12		0.23	0.29
Thymol/polystyrene		0.00	0.00
Thymol/Teflon		2.39	2.20
Thymol/siliconized glass		3.78	3.55
o-Terphenyl/o-terphenyl	0.33		
o-Terphenyl/acetal		0.95	2.40
o-Terphenyl/nylon-6		0.74	2.07
o-Terphenyl/nylon-6,6		0.63	1.88
o-Terphenyl/nylon-12		0.30	1.26
o-Terphenyl/nylon-6,10		0.05	0.64
o-Terphenyl/nylon-6,12		0.04	0.13
o-Terphenyl/polystyrene		0.49	0.02
o-Terphenyl/Teflon		4.89	2.70
o-Terphenyl/siliconized glass		7.09	4.40

TABLE 12.5   (Continued)
Solid–Liquid (γ_{sl}), Particle–Liquid (γ_{pl}), and Particle–Solid (γ_{ps})
Interfacial Tensions at the Melting Points of Matrix Materials

System	γ_{sl} (mJ/m²)	γ_{pl} (mJ/m²)	γ_{ps} (mJ/m²)
Salol/salol	0.01		
Salol/acetal		1.20	1.42
Salol/nylon-6		1.03	1.23
Salol/nylon-6,6		0.87	1.05
Salol/nylon-12		0.45	0.59
Salol/nylon-6,10		0.14	0.22
Salol/nylon-6,12		0.01	0.00
Salol/polystyrene		0.33	0.22
Salol/Teflon		4.25	3.86
Salol/siliconized glass		5.74	5.29
2-Phenyl phenol/2-phenyl phenol	0.00		
2-Phenyl phenol/acetal		0.33	0.34
2-Phenyl phenol/nylon-6		0.21	0.22
2-Phenyl phenol/nylon-6,6		0.16	0.17
2-Phenyl phenol/nylon-12		0.02	0.03
2-Phenyl phenol/nylon-6,10		0.03	0.02
2-Phenyl phenol/nylon-6,12		0.37	0.35
2-Phenyl phenol/polystyrene		1.21	1.18
2-Phenyl phenol/Teflon		6.81	6.75
2-Phenyl phenol/siliconized glass		9.39	9.31
Pinacol/pinacol	0.00		
Pinacol/acetal		3.77	3.85
Pinacol/nylon-6		3.44	3.52
Pinacol/nylon-6,6		3.15	3.22
Pinacol/nylon-12		2.29	2.35

with a micrometer eyepiece graticule. With the aid of a temperature programmer, it was possible to adjust the velocity of the interface at will.

Conventionally, such experiments were performed with the aim of determining what is called the critical velocity, V_c; that is, the velocity that separates pushing and engulfing of particles. However, from experimental observations, the transition from pushing to engulfing is not sharp so that three modes of particle behavior may be defined. As illustrated in Figure 12.1, at relatively high rates of solidification, the particles are engulfed instantly on contact with the solidification front (mode 1). At intermediate rates, the particles are pushed through various distances before being engulfed (mode 2). At relatively low rates of solidification, steady-state pushing of any individual particle can be observed without engulfment (mode 3). Typical examples of these three modes of engulfing and pushing for various particle diameters are represented in Figures 12.4 and 12.5.

TABLE 12.6

Theoretical Predictions (ΔF^{eng} in mJ/m²) and Microscopic Observations of Particle Behavior at Solidification Fronts

	Naphthalene	Biphenyl	Thymol	*o-*Terphenyl	Salol	2-Phenyl Phenol	Pinacol
Acetal	+4.06	+2.66	+0.20	+1.45	+0.22	+0.01	+0.08
	R[a]	R	R	R	R	R	R
Nylon-6	+3.69	+2.44	+0.19	+1.33	+0.20	+0.01	+0.08
	R	R	R	R	R	—	(E)
Nylon-6,6	+3.58	+2.35	+0.18	+1.25	+0.18	+0.01	+0.07
	R	R	R	R	(E)	—	(E)
Nylon-12	+3.02	+1.96	+0.15	+0.96	+0.14	+0.01	+0.06
	R	R	R	R	R	(E)	R
Nylon-6,10	+2.18	+1.39	+0.11	+0.59	+0.08	−0.01	+0.05
	R	R	R	R	R	(E)	R
Nylon-6,12	+1.16	+0.69	+0.06	+0.09	−0.01	−0.02	+0.03
	R	R	R	R	(E)	(E)	E
Polystyrene	−0.02	−0.10	+0.00	−0.47	−0.11	−0.03	+0.01
	E	E	R	E	E	E	R
Teflon	−3.70	−2.57	−0.19	−2.19	−0.39	−0.06	−0.05
	E	E	(E)	E	E	E	R
Siliconized glass	−4.97	−3.44	−0.23	−2.69	−0.45	−0.08	−0.06
	E	E	E	E	E	E	E

[a] R, rejection; E, engulfing; (E), reorientation and engulfing.

The three modes of particle pushing are clearly shown in both figures. The area between the broken lines in each figure represents mode 2. These transition velocities are clearly particle size dependent. This was to be expected, since dynamic effects (e.g., viscous drag, wall friction, etc.) retard particle motion more strongly the larger the particles. For both matrix materials, the transition velocities are higher for acetal particles than for nylon particles of the same size. Figures 12.4 and 12.5 show that the transition velocities are higher for naphthalene than for biphenyl, whereas the thermodynamic driving forces would predict the opposite behavior. This finding is therefore probably due to fluid mechanic effects. According to the experimental data (Figures 12.4 and 12.5), the critical velocity, V_c, is defined as the central line through the area representing mode 2.

The critical velocity of engulfing, V_c, can be understood as follows. If the free energy of adhesion between the solidification front and the solid particle is positive, then repulsion occurs, and the particle will be pushed along at low rates of solidification. This motion, however, sets up a viscous drag force acting on the particle and opposing the thermodynamic or van der Waals type of repulsion. As the rate of solidification increases, the viscous drag will increase and finally overpower the van der Waals type of repulsion; engulfment will take place, and the corresponding rate

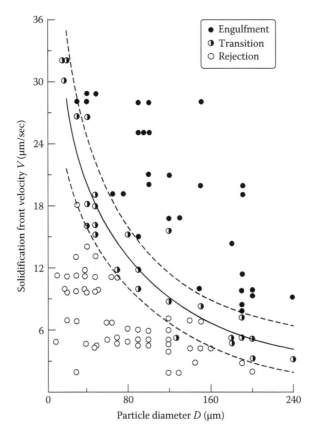

FIGURE 12.4 Pushing and trapping velocities as a function of the particle diameters for the biphenyl (matrix)-nylon (particle) systems.

of solidification is the critical velocity, V_c. The latter is thus a measure of the balance between van der Waals repulsion and viscous drag. Thus, in essence, measuring the critical velocity of engulfing V_c and knowing the viscous drag on the particle would allow one to determine the van der Waals interaction of the particle with the solidification front, and hence the free energy of adhesion, ΔF^{adh}, and other interfacial free-energy quantities, such as the solid-melt interfacial tension.

Unfortunately, even the description of viscous drag acting on a spherical particle near a flat and smooth solid is a complicated matter. Generally, particles used in the engulfment experiments are not spherical, but are irregularly shaped, and are therefore only characterized by a mean diameter. Furthermore, the solidification front is not necessarily smooth and flat. In addition, due to different thermal conductivities of solid, melt, and particle material, the particle itself may modify the shape of the solid-melt interface. It is in view of these complications that a scheme of dimensional analysis was developed [30] for the description of pushing and engulfing of particles by a solidification front. This analysis provides a correlation between the critical velocity, the related material properties, and the free energy of adhesion.

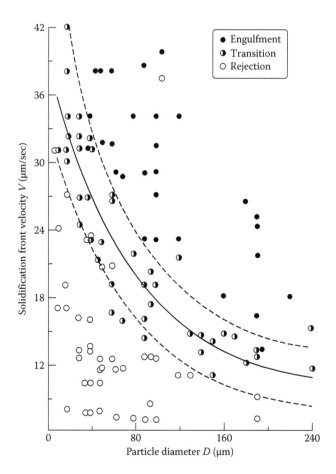

FIGURE 12.5 Pushing and trapping velocities as a function of the particle diameters for the naphthalene (matrix)-nylon (particle) systems.

As discussed above, the driving forces for particle pushing are provided by the interfacial free energies whereas the main retarding force is hydrodynamic drag. The various particle and matrix properties that appear in the expressions for the retarding and driving forces can thus be combined by the method of dimensional analysis. In engulfing experiments, the particles are wholly in the melt as they are being pushed, so that apart from the free-energy change at the interface, only melt and particle properties will be taken into account. Through a more complete physical analysis of the process, the following can be concluded [30]:

1. The shape of the interface behind the particles is a determining factor for particle engulfment; the shape is affected by the relative thermal conductivities of the particle and the melt (and, to a lesser extent, those of the solid phase). This means that thermal transfer governs the shape formation and that a relevant dimensionless parameter should be the ratio, $\bar{F}$, of the

interfacial free energy per unit area to the heat content of the particle per unit surface area of the particle, since this ratio compares the two potential energies of the problem:

$$\bar{F} = \frac{\Delta F^{adh}}{\rho_p c_p D T},$$ (12.28)

where T is the interface absolute temperature, ΔF^{adh} is the free energy of adhesion per unit area, and D, ρ_p, and c_p are the mean diameter, density, and specific heat of the particle, respectively.

2. The solidification front proceeds by mass diffusion: the diffusion sequence may not be adequately described from the point of view of a continuous fluid flow since the gap between solidification front and particle is very small; that is, of molecular dimension; but if one refers to kinetic theory, a solution is given by the Stokes–Einstein relation for spherical molecules diffusing very slowly in a stationary fluid, and the self-diffusion coefficient can be calculated. The appropriate dimensionless parameter is the Lewis number of the melt, which is the ratio between mass and heat diffusion; that is,

$$Le = \frac{D_l}{\alpha_l},$$ (12.29)

where D_l is the self-diffusion coefficient of the liquid phase, and α_l is the thermal diffusivity of the melt, $\alpha_l = k_l/(\rho_l c_l)$, with k_l, ρ_l, and c_l representing the thermal conductivity, the density, and the specific heat of the melt, respectively.

3. The various work terms that act on the particle are proportional to the following quantities: (a) the work done against the interface $\Delta F^{adh} D^2$; (b) the thermal energy exchanged (mainly through conduction since convection effects are improbable), $k_p T D^2/V_c$; (c) the work done by viscosity, $\mu V_c D^2$; and (d) the work of the inertia force, $\rho_p V_c^2 D^3$.

Conceptually, this analysis leads to three independent dimensionless parameters containing the above work terms, and it is convenient to introduce these as the Reynolds number,

$$Re = \rho_l V_c D/\mu,$$ (12.30)

the "capillary number,"

$$\Delta F^{adh} / \mu V_c,$$ (12.31)

and the ratio of the free energy of adhesion to the conducted heat,

$$V_c \Delta F^{adh} / k_p T.$$ (12.32)

On the basis of the above dimensional reasoning, the following general relationship holds:

$$\phi(\bar{F}, Le, Re, \Delta F^{adh}/\mu V_c, V_c \Delta F^{adh}/k_p T) = 0. \tag{12.33}$$

Correlation of the experimental data can therefore be sought in the form [30]

$$Re = h\bar{F}^l Le^n \left(\Delta F^{adh}/\mu V_c\right)^p \left(V_c \Delta F^{adh}/k_p T\right)^q. \tag{12.34}$$

Experimental determination of the four exponents, l, n, p, and q would involve four independent sets of measurements, but only three are easily accessible; that is, changes in the particle materials, in particle diameter, and in the matrix material. Therefore, a further hypothesis is necessary and it was decided arbitrarily to group two parameters (assuming implicitly that their exponents are equal) by replacing the two last ones by their geometric mean:

$$Q = \left[\frac{\Delta F^{adh}}{\mu V_c} \frac{V_c \Delta F^{adh}}{k_p T}\right]^{1/2} = \frac{\Delta F^{adh}}{\left(\mu k_p T\right)^{1/2}}, \tag{12.35}$$

which reduces Equation 12.34 to

$$Re = h\bar{F}^l Q^m Le^n. \tag{12.36}$$

For computing these exponents, one must notice that the Lewis number depends on the properties of the melt only, and that D appears only in Re and $\bar{F}$. Then, the determination of the four constants, h, l, m, and n can be based on the following strategy:

1. For given matrix and particle materials, all the variables will remain constant except V_c and D, so that

$$Re \propto \bar{F}^l, \tag{12.37}$$

   and l can be determined from the slope of a plot of log Re versus log $\bar{F}$.
2. For a given matrix material and fixed particle diameter (but different particle materials), only Q varies, and

$$Re \cdot \bar{F}^{-l} \propto Q^m, \tag{12.38}$$

   which gives m.
3. For a fixed particle material and fixed particle size, V_c and the melt properties will change, hence

$$Re \cdot \bar{F}^{-l} Q^{-m} = hLe^n, \tag{12.39}$$

and the constants h and n can be deduced from a logarithmic plot of

$$Re\bar{F}^{-l}Q^{-m} \text{ versus } Le.$$

Based on the above strategy and a large number of experimental data, Equation 12.36 finally becomes [30]:

$$Re = 1.29 \times 10^{-3} \bar{F}^{-0.52} Q^{1.70}, \tag{12.40}$$

for $D < 100$ μm; and

$$Re = 3.98 \times 10^{-29} \bar{F}^{-3.15 - 1.10 \log Re} Q^{-11.12}, \tag{12.41}$$

for $D > 100$ μm.

The formulation of Equation 12.36 assumes implicitly that the exponents l, m, and n are constants. The determination of l from the experimental data showed that l is not unique, but depends on particle size. While l is constant for small particles up to approximately 100 μm diameter, it depends on Re for larger particles. If one wishes to maintain a formulation with constant exponents, l may be rewritten in the following form: $l = -a - b \log Re$ with $a = 0.52$, $b = 0$ for $D < 100$ μm and $a = 3.15$, $b = 1.10$ for $D > 100$ μm. Equation 12.36 may then be written as

$$Re^{1 + b \log \bar{F}} = h\bar{F}^{-a} Q^m Le^n. \tag{12.42}$$

Although the formulation of Equation 12.42 makes all of the exponents constant, it must be realized that the existence of the two different particle-size regimes has not been included in the physical analysis used as a guide for the dimensional analysis. A possible explanation is that for small particles, surface effects are predominant; whereas for larger particles, volume effects start playing a significant role.

12.6 DETERMINATION OF PARTICLE OR SOLID SURFACE TENSIONS FROM THE CRITICAL VELOCITY

As discussed in the previous sections, at the critical velocity, V_c, the repulsive force, due to a positive ΔF^{adh}, and the retarding forces acting on the particle are equal, and hence engulfment results from any small disturbance. Through the dimensional analysis, the critical velocity, V_c, and the free energy of adhesion, ΔF^{adh}, are related through Equation 12.40 or Equation 12.41. Because the physical properties (involved in Equation 12.40 or Equation 12.41) of the various matrix materials are known or can be determined experimentally, it is thus possible, using V_c data, to calculate ΔF^{adh} by Equation 12.40 or Equation 12.41. Having thus obtained a value for ΔF^{adh}, one can, through the equation of state approach, obtain the various relevant interfacial tensions and in this manner determine the surface tension of small particles, γ_{pv}. Thus, the sequence of steps required to obtain, for example, γ_{pv} may be summarized as follows:

$$V_c \rightarrow \text{Dimensional analysis} \rightarrow \Delta F^{adh} \rightarrow \begin{array}{c} \text{Equation of state for} \\ \text{interfacial tensions} \end{array} \rightarrow \gamma_{pv}$$
$$\text{(by Equation 12.40 or}$$
$$\text{Equation 12.41)}$$

In such a procedure, the γ_{lv} values can be obtained experimentally by conventional methods such as the Wilhelmy plate technique or a drop shape method (see Chapter 3). The solid surface tension γ_{sv} of the respective matrix materials may be determined through the contact angle/equation of state approach as described in Section 12.4 (see also Chapters 8 and 9 for details). Because contact angle measurements at the melting point of the matrix material are not feasible, these measurements must be performed as a function of temperature at temperatures below the melting point, with the data at the melting point obtained by extrapolation. A suitable method for such temperature-dependent contact angle measurements is the technique of capillary rise at a vertical plate (see Chapter 6).

To illustrate this approach, let us consider particle-thymol systems with polymethyl methacrylate (PMMA) and polyvinyl chloride (PVC) particles as an example. As discussed before, the solid-vapor and liquid-vapor interfacial tensions of thymol have to be determined first. As literature data on the viscosity of thymol are not available, this quantity was measured with an Ostwald viscometer placed in a liquid bath at the melting point of the matrix material. The liquid–vapor surface tension, γ_{lv}, for thymol at its melting point was determined from temperature-dependent surface tension measurements at and above the melting point, using the Wilhelmy plate method [54]. These data are given in Figure 12.6. The surface tension at the melting point of thymol was taken from the curve fit of the data in Figure 12.6 using linear regression. The results of these viscosity and surface tension measurements are

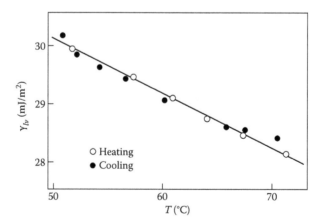

FIGURE 12.6 Temperature dependence of the liquid-vapor surface tension for thymol at and above its melting point, 51.5°C.

given in Table 12.7 with the liquid density and the melting temperature as obtained from the literature; this information is also given for three additional matrix materials: benzophenone, bibenzyl, and naphthalene.

The solid-vapor surface tension, γ_{sv}, of thymol at the melting point was obtained from the extrapolation of temperature-dependent contact angles measured with glycerol on thymol in the solid state. Smooth surfaces of thymol were prepared by solvent casting [1]. Contact angles were measured using the capillary rise at a vertical plate method [55,56] and then converted to solid surface tensions via the equation of state approach. A line fit to the data using linear regression was extrapolated to obtain the γ_{sv} value at the melting point. The measurements of capillary rise, h, for thymol-glycerol are given in Figure 12.7, together with the resulting contact angles and γ_{sv} values. The γ_{sv} values at the melting point are given for all four matrix materials in Table 12.7.

Having measured the required physical properties for thymol, it is now possible to determine particle surface tensions, γ_{pv}, from critical velocity measurements using thymol as the matrix. To illustrate the procedure for determining γ_{pv} from freezing front experiments, critical velocity measurements with PVC and PMMA particles in thymol will be used to calculate the surface tensions of these two particles. These surface tensions will then be compared with values obtained from contact angle measurements.

Solidification front experiments with PVC and PMMA particles must first be performed in order to obtain V_c data. Figure 12.8 shows the experimental observations of pushing, engulfment, and transition; that is, momentary pushing followed by engulfment for PVC particles in thymol [2]. The critical velocity curve for this system is then obtained from the center line of the band between the broken lines in Figure 12.8, as before. Critical velocity curves produced in this manner for PVC and PMMA in thymol are given in Figure 12.9. Points on each curve for a range of diameters were selected and ΔF^{adh} was calculated from the physical properties of PVC and PMMA given in Table 12.8. The particle surface tension, γ_{pv}, was calculated for each diameter ΔF^{adh} using the previously determined values of γ_{lv} and γ_{sv} for thymol (see Table 12.7). The results are summarized in Table 12.9. The slight variation in γ_{pv} for the different diameters is thought to be due to experimental error. An average value of γ_{pv} for the various particle sizes was calculated.

As a check on the γ_{pv} values obtained from V_c measurements, the surface tensions of PVC and PMMA were determined from contact angle measurements [1]. Since these measurements were performed at 20°C, a value of $(d\gamma_{pv})/(dT) = -0.075$ mJ/m^2 °C

TABLE 12.7

Physical Properties of Matrix Materials: Density, ρ_l, Viscosity, μ, Surface Tensions, γ_{lv}, γ_{sv}, and Melting Temperature, T_{mp}

Matrix Material	ρ_l (kg/m^3)	T_{mp} (°C)	$\mu \times 10^3$ (Pa s)	γ_{sv} (mJ/m^2)	γ_{lv} (mJ/m^2)
Thymol	925	51.5	3.97	29.4	29.9
Benzophenone	1140	48.0	5.15	34.5	39.9
Bibenzyl	958	52.0	2.46	20.8	24.9
Naphthalene	978	80.0	0.967	22.9	32.8

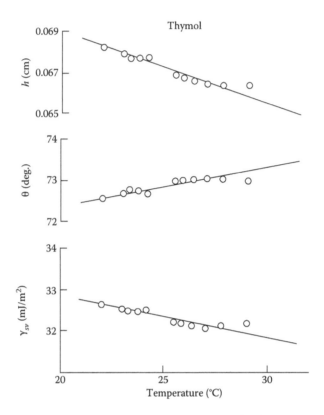

FIGURE 12.7 Temperature dependence of the contact angle, θ, of glycerol on solid thymol, determined from the capillary rise, h, on a flat vertical surface, and the temperature dependence of the resulting surface tension, γ_{sv}, of thymol in the solid state.

was assumed [1] for both PVC and PMMA in order to obtain surface tension values at 51.5°C. These results are also given in Table 12.9. It is apparent that there is good agreement between the two experimental strategies, supporting our contention that the freezing front technique is a valid tool to characterize surface properties of small particles.

In the foregoing discussions, particle-vapor interfacial tensions were obtained from V_c in situations where the matrix properties γ_{lv} and γ_{sv} were known. It stands to reason that it should be possible to determine the surface tension of the matrix material in the solid state, γ_{sv}, if its surface tension in the liquid state, γ_{lv}, as well as the surface tension of the particles, γ_{pv}, are known. This possibility may be of practical interest as it may be easier to perform freezing front experiments than precise temperature-dependent contact angle measurements. The procedure is illustrated below.

Surface tensions of the melt of benzophenone and bibenzyl as well as viscosities were determined as described above. Freezing front experiments were performed with particles of acetal and nylon, the surface tensions of which are known [29]. The results for acetal particles in benzophenone are shown in Figure 12.10. Again, the center line of the "transition" band was established as the critical velocity, V_c, and plotted in Figure 12.11, together with the corresponding curves obtained from

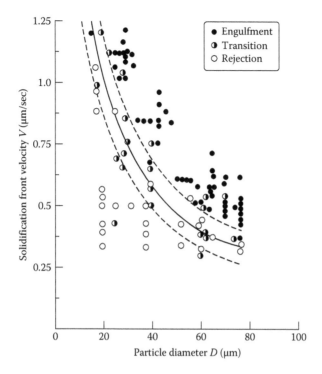

FIGURE 12.8 Solidification front velocity of thymol as a function of particle diameter for PVC particles.

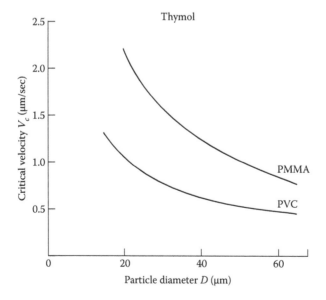

FIGURE 12.9 Critical velocity of thymol solidification front as a function of particle diameter obtained from the mean of the band of plots of the type given in Figure 12.8.

TABLE 12.8

Physical Properties of Particle Materials: Densities, ρ_p, Specific Heats, C_p, and Thermal Conductivities, k_p

Particle Material	ρ_p (kg/m³)	C_p (J/kg K)	k_p (W/m K)
PVC	1400	1250	0.163
PMMA	1250	1380	0.188
Acetal	1430	1520	0.260
Nylon-6	1090	1600	0.260
Nylon-6,6	1080	1590	0.260
Nylon-12	1060	1590	0.260
Nylon-6,12	1040	1590	0.260

TABLE 12.9

Particle Surface Tension, γ_{pv}, Determined from Critical Velocity Measurements in Thymol and Compared with γ_{pv} Values Obtained from Direct Contact Angle Measurements. γ_{pv} Given for 51.5°C

Particle Material	Diameter D (µm)	V_c (µm/s)	γ_{pv} (mJ/m²)	γ_{pv} (Average) (mJ/m²)	γ_{pv} (from θ Measurements) (mJ/m²)
	20	1.10	33.2		
PVC	40	0.60	32.5	32.7	31.8
	60	0.45	32.3		
	20	2.20	36.4		
PMMA	40	1.10	34.8	35.3	36.8
	60	0.80	34.8		

measurements with nylon particles. Corresponding curves for bibenzyl were similarly obtained and are given in Figure 12.12. The values of ΔF^{adh} for each particle-matrix system were then calculated as before, using the data in Tables 12.7 and 12.8. Table 12.10 shows the results of these calculations for particles in benzophenone. In the case of bibenzyl, only average ΔF^{adh} values are reported (see Table 12.11).

Finally, to obtain γ_{sv} for these materials, it is necessary to use the equation of state approach in conjunction with the known values of γ_{pv}, γ_{lv}, and ΔF^{adh}. The surface tension values, γ_{sv}, obtained from the measurements with each type of polymer particle are given in Tables 12.11 and 12.12, respectively. Final average γ_{sv} values are also given. The consistency of the results obtained for the different particle materials is remarkable.

In order to test the accuracy and reliability of these results, contact angles that a liquid of known surface tension (i.e., glycerol with $\gamma_{lv} = 63.4$ mJ/m²) would form on benzophenone and bibenzyl were predicted and compared with actual contact angle measurements. In order to avoid the complexities of the capillary rise measurements at elevated temperatures, contact angle measurements on solid benzophenone and

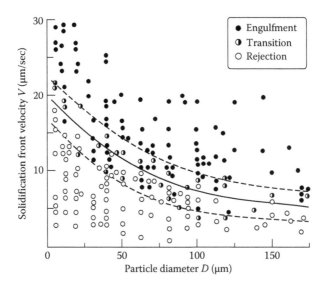

FIGURE 12.10 Solidification front velocity of benzophenone as a function of particle diameter for acetal particles.

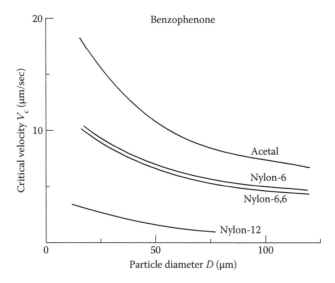

FIGURE 12.11 Critical velocity of benzophenone solidification front as a function of particle diameter obtained from the mean of the band of plots of the type given in Figure 12.10.

bibenzyl were simply performed at room temperature. However, the γ_{sv} values in Tables 12.11 and 12.12 refer to the respective melting points of the two matrix materials. Therefore, the γ_{sv} values for 20°C were calculated by assuming, as before, a temperature coefficient $(d\gamma_{sv})/(dT) = -0.075$ mJ/m² °C. Contact angles that glycerol would form on surfaces of benzophenone and bibenzyl were then predicted from the equation of state approach. Next, contact angles were measured with glycerol

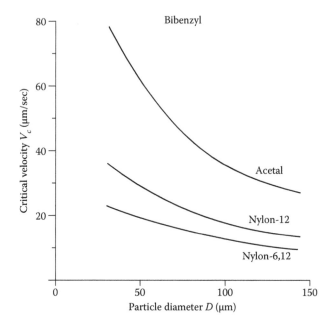

FIGURE 12.12 Critical velocity of bibenzyl solidification front as a function of particle diameter obtained from the mean of the band of plots of the type given in Figure 12.10.

TABLE 12.10
Free Energy of Adhesion, ΔF^{adh}, for Particles in Benzophenone Determined from Critical Velocity, V_c, Measurements

Particle Material	Diameter D (μm)	V_c (μm/s)	ΔF^{adh} (mJ/m²)	ΔF^{adh} (Average) (mJ/m²)
Acetal	30	14.0	0.684	
	60	9.5	0.652	0.655
	90	7.3	0.628	
Nylon-6	30	8.5	0.488	
	60	6.3	0.507	0.496
	90	5.0	0.492	
Nylon-6,6	30	8.1	0.477	
	60	5.9	0.484	0.475
	90	4.6	0.463	
Nylon-12	10	2.9	0.130	
	20	2.2	0.135	0.131
	30	1.7	0.129	

TABLE 12.11

Solid-Vapor Surface Tension, γ_{sv}, for Bibenzyl at 52°C, Calculated from ΔF^{adh} Obtained from V_c Measurements and Known Values of the Particle-Vapor Surface Tension, γ_{pv}, and the Liquid-Vapor Surface Tension, γ_{lv} ($\gamma_{lv} = 24.9$ mJ/m² at the Melting Point)

Particle Material	γ_{pv} (mJ/m²)	ΔF^{adh} (mJ/m²)	γ_{sv} (mJ/m²)	γ_{sl} (mJ/m²)
Acetal	44.3	2.59	20.3	0.345
Nylon-12	40.6	1.54	21.5	0.191
Nylon-6,12	34.0	1.14	20.7	0.294
Average values			20.8	0.280

TABLE 12.12

Solid-Vapor Surface Tension, γ_{sv}, for Benzophenone at 48°C, Calculated from ΔF^{adh} Obtained from V_c Measurements and Known Values of the Particle-Vapor Surface Tension, γ_{pv}, and the Liquid-Vapor Surface Tension, γ_{lv} ($\gamma_{lv} = 39.9$ mJ/m² at the Melting Point)

Particle Material	γ_{pv} (mJ/m²)	ΔF^{adh} (mJ/m²)	γ_{sv} (mJ/m²)	γ_{sl} (mJ/m²)
Acetal	44.5	0.655	34.5	0.381
Nylon-6	43.7	0.496	34.9	0.326
Nylon-6,6	43.1	0.475	34.3	0.410
Nylon-12	40.8	0.131	34.3	0.410
Average values			34.5	0.381

TABLE 12.13

Contact Angle, θ, of Glycerol ($\gamma_{lv} = 63.4$ mJ/m²) on Solid Matrix Materials at 20°C

Matrix Material	θ from V_c Measurements (deg)	θ from Direct Measurements (deg)
Benzophenone	64.0	64.9
Bibenzyl	88.3	86.5

on layers of benzophenone and bibenzyl obtained by solvent casting. Clean glass slides were dipped in a bath containing a 1% solution of benzophenone or bibenzyl dissolved in a suitable solvent. These glass slides were then stored in an evacuated desiccator for at least 4 hours to remove excess solvent. Contact angle measurements were performed using the vertical plate method [55,56]. The results of these measurements, together with the predictions from the critical velocity measurements, are given in Table 12.13. The agreement is excellent, indicating that the solid surface tensions, γ_{sv}, obtained for benzophenone and bibenzyl from freezing front experiments are reliable.

12.7 APPLICATIONS OF SOLIDIFICATION FRONT TECHNIQUE TO DETERMINE PARTICLE SURFACE TENSIONS

12.7.1 SURFACE TENSIONS OF FIBERS

The wettability of reinforcing fibers is of interest since this property has an effect on the fiber-resin bond strength in composite materials. The overall strength of a reinforced material depends heavily on how strong this bond is. Good adhesion is essential for the transfer of stresses, thus strengthening the composite material, while poor adhesion means a decrease in the composite tensile or compression strength or the lack of ability of the material to maintain its normal strength on exposure to water under adverse conditions. An important condition for achieving a good adhesive bond between the fiber and matrix resin is that the liquid has to wet the surface of the fiber to obtain complete and intimate contact. For a liquid resin to wet or spread on a fiber completely, the surface tension of the liquid resin must be less than that of the fiber itself. Therefore, in order to optimize the reinforcement of the composite material, it is necessary to know the surface tension of the liquid resin and the solid surface tension of the fiber. The fibers used for the study were a high-strength, high-modulus carbon fiber (Thornel 300 carbon fiber from Union Carbide Corp.) and an aromatic polyamide fiber (Kevlar, from DuPont de Nemours & Co.).

In the solidification front experiments, the fibers were first chopped into small segments from 10 to 300 μm in length before being put into the solidification cell. At low solidification front velocities, the fibers always lined up with the front and were then pushed further along. Such observations were considered to be rejection. Sometimes, after being pushed for a small distance (less than, approximately, 30 μm) they were engulfed by the advancing front. Since it was not certain whether these cases were engulfments or rejections, they were recorded as "transitions." If the fiber was engulfed immediately after it had lined up with the front, it was taken as an "engulfment." From the plots of the solidification front velocity versus the fiber particle length, the upper limit of the critical velocity was taken as the line above which all the fiber particles were engulfed and the lower limit was taken as that below which all the fiber particles were rejected. The average between these two limits was taken as the mean critical velocity, as before.

For interpreting the results of solidification front measurements, the effective "particle diameter" of the fiber was taken as the diameter of a sphere with equivalent projected area. The physical properties of the matrix materials are given in Table 12.7. From the measured mean critical velocity values, the average fiber surface tensions, γ_{pv}, were calculated [57] using the procedure described in Section 12.6. It was found [57] that the surface tensions of the fibers do not depend on the fiber length, and the average surface tensions over all matrix materials are 41.8 mJ/m² for carbon fiber and 46.4 mJ/m² for Kevlar, respectively, at 24°C. An error analysis was performed [57] by taking into account the errors in critical velocity. It was found that the error limits for the results where thymol was used as a matrix material were approximately ±1.5 mJ/m² while for those results with matrix materials other than thymol, the errors were substantially lower, between 0.1 and 0.7 mJ/m². Once the surface tensions of the fibers and the liquid melt are known, the corresponding contact angles can be calculated using the equation of state approach.

The solid surface tensions of the fibers measured from the solidification front technique were compared with those measured from the contact angle/equation of state approach [57]. The contact angles were determined from the Wilhelmy plate technique [58,59] (see also Chapter 6 for details). Briefly, due to capillary effects, when a vertical solid plate is partially immersed into a liquid, the liquid will either rise or be depressed along the vertical wall, thus exerting a force on the solid. The magnitude of the force, F, exerted on the plate by the liquid is given by

$$F = p\gamma_{lv} \cos \theta, \qquad (12.43)$$

where p is the perimeter of the solid, γ_{lv} is the liquid surface tension, and θ is the contact angle between the solid and the liquid. Therefore, if the perimeter of the immersed fiber and the surface tension of the liquid are known, the contact angle between the fiber and the liquid matrix can be calculated by measuring the pulling force, F, using an electrobalance.

To calculate the contact angle accurately, the precise value for the perimeter of the fiber must be known. The perimeter of the fiber was obtained by first dipping the fiber into a liquid that was known to wet the fiber completely. Under the complete wetting condition, $\cos\theta$ in Equation 12.43 is equal to 1; thus, the perimeter, p, is determined by

$$p = \frac{F}{\gamma_{lv}}. \qquad (12.44)$$

Since the two fibers, Thornel 300 and Kevlar, were expected to have surface properties close to common polymers, a liquid with low surface tension should wet the fibers completely. Toluene, which has a surface tension of 28.4 mJ/m² at 24°C, was chosen. It was found that the diameters of the carbon fibers were rather uniform from one fiber to another, ranging from 6.8 to 8.0 μm. The diameters of the Kevlar, which ranged from 10.8 to 13.4 μm, seemed to have a wider variation between fibers. The fiber diameter was further checked by using a digital image analysis system (Bausch & Lomb Omnicon 3000).

The gravimetric results obtained from the Wilhelmy plate method were used to calculate the contact angles of three different liquids on the fibers. The results are listed in Table 12.14. Using the equation of state for interfacial tensions, at 24°C, the average surface tension of the carbon fiber was found to be 42.4 mJ/m² and that for Kevlar was 43.7 mJ/m², as summarized in Table 12.14. The error limits in Table 12.14 include the errors in the electrobalance output, the liquid surface tensions, and the fiber perimeter measurements.

TABLE 12.14

The Static Advancing Contact Angle, θ_a, of Various Liquids on Carbon and Kevlar Fibers and the Surface Tension, γ_{pv}, of the Fibers, as Obtained from Wilhelmy Plate Measurements

Fiber	Liquid	θ_a (deg)	γ_{pv} (mJ/m²)
Carbon	Ethylene glycol	31.2 ± 8.8	41.6 ± 3.2
Carbon	Glycerol	55.7 ± 3.6	42.2 ± 1.6
Carbon	Water	66.0 ± 2.6	43.3 ± 1.5
Kevlar	Ethylene glycol	27.8 ± 5.7	42.8 ± 2.0
Kevlar	Glycerol	52.8 ± 2.7	43.8 ± 1.6
Kevlar	Water	63.8 ± 1.5	44.6 ± 0.9

TABLE 12.15

Summary of the Contact Angles, θ, and Surface Tension, γ_{pv}, of Carbon and Kevlar Fibers Obtained by the Wilhelmy Plate Technique and the Solidification Front Technique

	Wilhelmy Plate Technique			Solidification Front Technique		
	θ (deg)			θ (deg)		
Fiber	With Water	With Glycerol	γ_{pv} (mJ/m²)	With Water	With Glycerol	γ_{pv} (mJ/m²)
Carbon	68.4	50.8	42.4	68.9	51.2	41.8
Kevlar	66.2	49.0	43.7	61.2	45.1	46.4

A comparison of the results obtained from the Wilhelmy plate technique and the solidification front technique is shown in Table 12.15. The contact angles and surface tensions determined by these two techniques for the carbon fibers are essentially the same. For the Kevlar fibers there is a somewhat larger discrepancy in the results obtained by the two techniques. Overall, the comparison of the results illustrates that both techniques will produce reliable surface tension and contact angle data for small-diameter fibers.

It should be noted that, in the Wilhelmy plate experiment, it is not possible to measure the fiber diameter and the contact angle simultaneously; also, finding the correct perimeter of a less uniform fiber might be difficult. The fiber geometry is also very important; for example, a convoluted cross section would generate a wicking effect and thus give incorrect balance readings. Furthermore, a rough fiber surface would result in a large contact angle hysteresis, entailing the possibility of obtaining an incorrect contact angle measurement. Since the Kevlar fibers used in this investigation have a rougher surface than the carbon fibers, it may be expected that the solidification front experiments will give more reliable results than the Wilhelmy plate method, as far as γ_{pv} is concerned. As for the solidification front method, there are fewer limitations. Also, the consistency and reproducibility shown in different matrix systems provide credence to the technique.

12.7.2 SURFACE TENSIONS OF BIOLOGICAL CELLS

The solidification front technique has been applied to biological cells. Generally, the free energy of adhesion for the process of a cell coming into contact with a solid surface is given by

$$\Delta F^{adh} = \gamma_{cs} - \gamma_{cl} - \gamma_{sl}. \tag{12.45}$$

Here γ_{cs} is the interfacial tension of a cell-solid interface, while γ_{cl} and γ_{sl} are, respectively, the surface tensions between the cell and the suspending liquid and between the solid and the suspending liquid. If ΔF^{adh} is negative, the adhesion process is favored because it results in a net reduction of the free energy of the system. If ΔF^{adh} is positive, the adhesion of the cell to the substrate increases the system's free energy and is thus not favored. The validity of this type of analysis has been well established through adhesion studies with granulocytes [60] and platelets [61,62], as well as through experiments pertaining to the engulfment of bacteria of differing surface tensions by human granulocytes [63].

In order to compute the free energy of adhesion as illustrated in Equation 12.45, it is necessary to obtain the surface tension of the cell along with the surface tensions of the other interacting materials: for example, the suspending liquid and solid substrate. The surface tensions in Equation 12.45 can be calculated by means of the equation of state for interfacial tensions (see Section 12.4). In this way, Equation 12.45 may be rewritten as

$$\Delta F^{adh} = \gamma_{cs} - \gamma_{cl} - \gamma_{sl} = f(\gamma_{cv}, \gamma_{sv}) - f(\gamma_{cv}, \gamma_{lv}) - f(\gamma_{sv}, \gamma_{lv}), \tag{12.46}$$

where $f(\gamma_{1v}, \gamma_{2v})$ is an equation of state. The liquid surface tension, γ_{lv}, is usually easy to measure, for example, by the Wilhelmy plate method or the pendant drop method. The cell and solid substrate surface tensions, γ_{cv} and γ_{sv}, respectively, may be obtained by a variety of means including the freezing front technique. In the past, the surface tensions of biological cells have been determined using the equation of state approach, with contact angle measurements of drops of saline on monolayers of cells such as granulocytes, lymphocytes, and platelets [63–66]. This technique, however, cannot easily be employed with the more rigid types of cells, such as glutaraldehyde-fixed erythrocytes, which do not readily adhere to a substrate to form a relatively regular monolayer of cells on which to perform the contact angle measurement [64,67].

In view of these limitations, the solidification front technique was used to measure the surface tension of biological cells such as fresh human lymphocytes and granulocytes, as well as glutaraldehyde-fixed human erythrocytes [68,69]. The surface tensions of these cells have been measured previously using other methods [60,64,70,71] and thus it is also possible to compare the accuracy and suitability of the freezing front technique. The experimental procedures in these studies are essentially the same as explained for the determination of the critical velocity of engulfment in the previous sections. Once V_c is known, the mathematical procedures required to obtain the surface tension of the engulfed particles are also the same as given previously. In the cases where γ_{cv} is known, this scheme may be used to determine γ_{sv}.

The freezing front technique has been used previously only with organic matrix materials. With biological cells it is necessary, for physiological reasons, to use an aqueous matrix [64]. For this reason, as a first step, it was necessary to determine the surface tension of ice (i.e., γ_{sv}). Some preliminary results on the characterization of water as the matrix material have been previously reported by Shiffman [72]. For this purpose, it was necessary to obtain the critical velocity of a particle of known surface tension, γ_{cv}, in the water-ice matrix system. Then, the surface tension of biological cells may be determined from the critical velocities of these cells in a water-ice matrix system. Procedures of this study are as follows:

1. The surface tension of calibration cells (γ_{cv}) was determined with a matrix material of known γ_{sv}. The cells used for this purpose were glutaraldehyde-fixed human erythrocytes and the matrix material was thymol.
2. The critical velocity of engulfment (V_c) of these cells in a water-ice matrix system could then be established, after which the surface tension of ice (γ_{sv}) could be calculated.
3. V_c of the biological cells of interest in the water-ice system was determined and used together with the γ_{sv} of ice, determined in Step 2, to obtain the γ_{cv} of these cells. Fresh human granulocytes and lymphocytes were investigated. Details of the material preparation can be found elsewhere [69].

From a study [70,71] of the stability of cells suspended in various liquids (droplet sedimentation technique, Chapter 11), a value for the surface tension of glutaraldehyde-fixed human erythrocytes was obtained: $\gamma_{cv} = 65.0$ mJ/m^2 at 25°C. In order to check this value, the critical velocity of engulfment of glutaraldehyde-fixed human erythrocytes was determined in thymol [69], an organic matrix material that had been extensively zone refined. Then, human erythrocytes from the same preparation batch were used to determine V_c in the water-ice system. By analyzing the experimental results [69], the average critical velocity of erythrocytes in thymol at 51.5°C (the solidification temperature of this matrix) was found to be 18 μm/s. The diameter, D, of the erythrocytes was chosen as $D = 4.94$ μm; that is, the diameter of an idealized "spherical" erythrocyte of the same projected surface area as the real, disk-shaped one with its 8 μm diameter and 2.4 μm thickness. Using these values together with various matrix material and cell properties, the free energy of adhesion was calculated to be $\Delta F^{adh} = 0.458$ mJ/m^2.

Since the solid-vapor ($\gamma_{sv} = 29.4$ mJ/m^2) and the liquid-vapor ($\gamma_{lv} = 29.9$ mJ/m^2) interfacial tensions of thymol are known [73,74], it is possible to calculate, using the equation of state for interfacial tensions, the surface tension, γ_{cv}, of the fixed erythrocytes at 51.5°C to be 61.9 mJ/m^2.

It is reasonable to assume a surface tension temperature dependence of -0.1 mJ/ m^2 °C [75], yielding a value of 67.1 mJ/m^2 at 0°C, and 64.6 mJ/m^2 at 25°C, in excellent agreement with the reported value [70,71]; that is, 65.0 mJ/m^2 at 25°C. Having thus established the surface tension of the fixed erythrocytes, these cells were then used in a similar manner to determine the surface tension, γ_{sv}, of ice. It was found that fixed erythrocytes in water had a much higher critical velocity of 88 μm/s [69]. The corresponding free energy of adhesion is $\Delta F^{adh} = 1.77$ mJ/m^2.

The surface tension of water (γ_{lv}) at 0°C is known to be 75.6 mJ/m^2 [76]. Together with the γ_{cv} of erythrocytes measured above; that is, 67.1 mJ/m^2 at 0°C, it is possible to obtain the surface tension (γ_{sv}) of ice: $\gamma_{sv}^{ice} = 80.2$ mJ/m^2. This value is in reasonable agreement with the results obtained from similar freezing front experiments by Shiffman [72]. More discussions about this value for the surface tension of ice can be found elsewhere [69].

Having obtained a value for the surface tension of ice, it was possible to use this value to obtain the surface tension of human granulocytes and lymphocytes from the critical velocities. The average critical velocities in water for fresh human lymphocytes and granulocytes were 13.5 and 31.0 μm/s, respectively. Using the same physical parameters as employed previously for the fixed erythrocytes, and using a measured average diameter of 11 μm for lymphocytes and 13 μm for granulocytes, the free energies were found to be: $\Delta F^{adh} = 0.531$ mJ/m^2 for lymphocytes and $\Delta F^{adh} = 1.000$ mJ/m^2 for granulocytes.

Using the surface tensions of ice and water at 0°C (described above) with these free energies, the equation of state yields: $\gamma_{cv} = 72.8$ mJ/m^2 for lymphocytes at 0°C and $\gamma_{cv} = 70.5$ mJ/m^2 for granulocytes at 0°C. Correcting these values to 22°C, assuming, as before, a surface tension temperature dependence of -0.1 mJ/m^2 °C, results in: $\gamma_{cv} = 70.6$ mJ/m^2 for lymphocytes and $\gamma_{cv} = 68.3$ mJ/m^2 for granulocytes. To illustrate the relative behavior of the three types of cells in the water-ice matrix, the percentage rejection histograms for each cell have been plotted on a common velocity axis in Figure 12.13. The large differences in critical velocity reflect the tremendous sensitivity of the freezing front technique to small differences in the surface tension of the cells.

The average cell surface tension values determined by the freezing front technique agree well with the results obtained via alternative techniques, as indicated in Table 12.16. This suggests that the technique does indeed provide an alternate and valid method for the determination of cellular surface tensions. The outstanding feature of the freezing front technique is its sensitivity. A difference of only 2.3 mJ/m^2 in the surface tensions of lymphocytes and granulocytes produces a difference of 18 μm/s in the critical velocity. This pronounced sensitivity lends great promise to the method for detecting the small surface tension differences that have been reported to exist between healthy, normal cells, and pathological cells [77,78], and between normal and activated cells [79] or virus-transformed cells [80]. It is also possible that the freezing front technique may be able to discriminate between cell subpopulations, for example, between T and B lymphocytes. Phase partition chromatography [81] has been employed to determine qualitative differences in the hydrophobicity of various species of cells [82–85]. An advantage of the freezing front technique is that it is able to supply quantitative information about the surface tension of the cells studied.

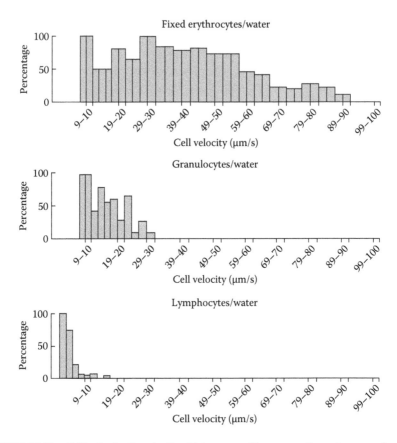

FIGURE 12.13   Critical velocity rejection histograms of human erythrocytes, granulocytes, and lymphocytes in water.

TABLE 12.16
Comparison of Cell Surface Tensions at 22°C, Obtained by Various Techniques

Cell Type	Technique	Surface Tension at 22°C (mJ/m²)
Granulocytes	Contact angle	69.1
	Freezing front	68.3
Lymphocytes	Contact angle	70.1, 70.4
	Freezing front	70.6
Fixed human erythrocytes	Droplet sedimentation	65.0
	Freezing front	64.9

12.7.3 Surface Tensions of Coal Particles

The solidification front technique has been applied to study coal particles [86,87] in order to provide information relevant to coal processing, such as flotation. The matrix materials include thymol, biphenyl, and naphthalene. Figure 12.14 is a set of photomicrographs illustrating the process of engulfment of coal particles by an advancing solidification front of thymol.

The results indicate some small dependence of coal particle surface tensions on the matrix material and on particle size [86,87]. The dependence on the matrix material is believed to be due to the complexities of coal composition [86–89]. The dependence on particle size can be explained in terms of a decreasing percentage of inorganic matter in the coal of smaller particle size [86,88,89]. Surface tension values for a number of different coals are given in Table 12.17 for 60 μm diameter particles at 20°C.

When the solidification front experiments were performed with water as the matrix material, the surface tensions of coal particles were found to be much greater than those obtained using organic matrix materials. That is, in the presence of water, the coal is significantly more hydrophilic, with a surface tension

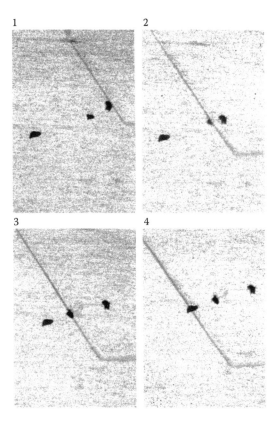

FIGURE 12.14 Photomicrographs illustrating an engulfing process of coal particles by an advancing solidification front of thymol.

TABLE 12.17

Particle Surface Tensions, γ_{pv}, from Solidification Front Experiments in Organic Matrix Materials for Various Bituminous Coal Samples; γ_{pv} are Given for 60 μm Particle Size

Coal Sample	Matrix Material	γ_{pv} at 20°C (mJ/m²)
No. 2	Thymol	40.8 ± 1.1
	Biphenyl	39.0 ± 0.3
	Naphthalene	39.4 ± 0.4
Devco	Thymol	39.7 ± 1.0
Minto	Thymol	39.8 ± 1.1
B.C.	Thymol	41.5 ± 1.4
Cardinal River	Thymol	42.0 ± 1.9

near that of water. This striking difference suggests that the same coal sample may have two effective surface tensions, or even a range of surface tensions, depending on the suspending liquid. It is likely that there is significant sorption of water by the coal particles so that they are perceived by other phases present as essentially "black water particles". The fact that the particle size dependence exhibited in organic matrix materials vanishes in the case of water matrix supports this contention.

12.8 MICROSCOPIC INTERPRETATION OF PARTICLE-FRONT INTERACTIONS

12.8.1 PARTICLE-FRONT BEHAVIOR IN VIEW OF VAN DER WAALS INTERACTIONS

As discussed in the previous sections, the thermodynamic predictions were found to be in excellent agreement with the microscopic observations of particle engulfing and rejection by a solidification front. The observations of particle engulfment or rejection by a solidification front may also be understood on the basis of van der Waals interactions between the particles and the solidification front [90,91]. Integration of the van der Waals force between the interacting bodies from infinity to the equilibrium separation should yield the thermodynamic free energy of adhesion [90,91]. Material parameters in the theories of van der Waals interactions, such as the macroscopic Lifshitz theory, should therefore be deducible from interfacial tensions and contact angles. It is desirable to know how well the predictions of the van der Waals interactions agree, or at least are consistent, with those of the free-energy approach and the observed particle behavior at the solidification fronts. For this purpose, the systems chosen in these studies are largely nonpolar so that dipole orientation effects, which may play a considerable role at small separations [92], can be neglected.

It has been shown [90,91] that the free energy of adhesion between a particle and a solidification front can be represented by an interaction potential energy between two semi-infinite slabs (i.e., the particle and the solidification front) in the presence of a third medium (i.e., the liquid):

$$\Delta F_{pls}^{adh}(d_0) = -\frac{A_{pls}}{12\pi d_{0pls}^2}, \tag{12.47}$$

where A_{pls} is the combined Hamaker coefficient for the particle (p), liquid (l), and solid (s). A_{pls} has energy units. d_{0pls} is the minimum separation distance in a third medium (i.e., the liquid phase) at the point of particle-front adhesion. It is generally accepted that d_{0pls} is close to the molecular distance and is usually given a general value of 2 Å. This extremely small separation distance lays the foundation for the geometrical assumption of two infinite slabs used in deriving Equation 12.47. At larger separation distances, however, the interactions between a particle and a solidification front should be more analogous to interactions between a sphere with a known radius and a semi-infinite slab.

It is known that the sign of the free energy change, ΔF^{eng} in Equation 12.20, determines whether engulfment or rejection of the solid particles should occur. As adhesion of the particle to the solidification front is a necessary intermediate step in the process of engulfing, the sign considerations should in principle apply similarly to the free energy of adhesion, ΔF^{adh} in Equation 12.23. In view of Equation 12.47, this means that for

$$A_{pls} > 0 \tag{12.48}$$

particle engulfment is expected, and for

$$A_{pls} < 0, \tag{12.49}$$

particle rejection should occur. Therefore, particle engulfment and rejection can be predicted in terms of Hamaker coefficients.

In order to predict particle engulfment and rejection according to Equations 12.48 and 12.49, the value of the Hamaker coefficient, A_{pls}, must be known. Generally, A_{pls} can be calculated from the values of Hamaker coefficients, A_{ij} $(i, j = p, l, s)$ as discussed below.

In the derivation of Equation 12.47, it is assumed that only dispersion forces are operative. If the minimum separation distance is d_0, and if Equation 12.47 is still applicable at such a small separation, then the Hamaker coefficient becomes, by combining Equation 12.47 with Equation 12.23,

$$A_{pls} = -12\pi d_{0pls}^2 (\gamma_{ps} - \gamma_{pl} - \gamma_{sl}). \tag{12.50}$$

Equation 12.23 yields, when written generally for two substances, i and j, and the vapor phase, v:

$$\Delta F_{ij}^{adh} = \gamma_{ij} - \gamma_{iv} - \gamma_{jv},$$ (12.51)

so that

$$\gamma_{ij} = \Delta F_{ij}^{adh} + \gamma_{iv} + \gamma_{jv}.$$ (12.52)

Substituting Equation 12.52 in Equation 12.50 yields

$$A_{pls} = -12\pi d_{0\,pls}^2 (\Delta F_{ps}^{adh} + \Delta F_{ll}^{coh} - \Delta F_{pl}^{adh} - \Delta F_{sl}^{adh}),$$ (12.53)

where the cohesion energy $\Delta F_{ll}^{coh} = -2\gamma_{lv}$.

From Equation 12.47, the adhesion energy of two surfaces in a vacuum or a gas phase, ΔF_{ij}^{adh}, can be written as

$$\Delta F_{ij}^{adh} = -\frac{A_{ij}}{12\pi d_{0ij}^2}.$$ (12.54)

Substituting Equation 12.54 into Equation 12.53 yields

$$A_{pls} = \frac{d_{0\,pls}^2}{d_0^2}\left(A_{ps} + A_{ll} - A_{pl} - A_{sl}\right),$$ (12.55)

where the assumption has been made that $d_{0ps} \approx d_{0pl} \approx d_{0sl} \approx d_{0ll} = d_0$. Equation 12.55 is similar to the one first derived by Haymaker [93]

$$A_{pls} = A_{ps} + A_{ll} - A_{pl} - A_{sl}.$$ (12.56)

Obviously, Equation 12.55 will reduce to Equation 12.56 if the equilibrium separation distance between solidification front and particle in the liquid phase (d_{0pls}) is the same as the equilibrium separation distance in a vacuum or a gas phase (d_0). While there is evidence [1] that for practical purposes the equilibrium separation distances for two-phase systems are the same, no information on d_{0pls} is available. The value of A_{pls} calculated from Equation 12.56 was found to be lower than the results obtained by the more accurate macroscopic approach of Lifshitz and coworkers [94,95]. The latter approach yields

$$A_{pls} = \frac{3}{4\pi}\hbar\,\bar{w}_{pls},$$ (12.57)

where $\hbar$ is Planck's constant divided by 2π, and $\bar{w}_{pls}$ is the average frequency obtained from the optical properties of the phases p, l, and s. Visser [96] therefore proposed that Equation 12.56 should be modified to

$$A_{pls} = C\left(A_{ps} + A_{ll} - A_{pl} - A_{sl}\right). \tag{12.58}$$

By using the A_{ii} values obtained from the Lifshitz theory, Visser calculated the A_{ij} values from the geometric mean combining rule,

$$A_{ij} = \left(A_{ii}A_{jj}\right)^{1/2}, \tag{12.59}$$

and then calculated A_{pls} from Equation 12.58. Visser compared this value of A_{pls} with that obtained directly from Equation 12.57 and deduced that for water, polystyrene, and gold, each as a third medium, $C = 1.6$, 1.7, and 3.1, respectively. Visser defined C as a correction for the transmission of a force through a third medium.

Returning to Equation 12.56, we note that two different approaches can be used to determine the A_{ij} values required to calculate A_{pls}. First, A_{ij} can be calculated from the free energy of adhesion by combining Equation 12.54 with Equation 12.51. Second, A_{ij} can be calculated from the free energy of cohesion using the following equations:

$$\Delta F_{ii}^{coh} = -2\gamma_{iv} \quad \text{and} \quad \Delta F_{ii}^{coh} = \frac{-A_{ii}}{12\pi d_{0ii}^2}, \tag{12.60}$$

in conjunction with the geometric mean combining rule, Equation 12.59. The Hamaker coefficients obtained from the free energy of adhesion are designated A_{ij}^a and those from the free energy of cohesion, A_{ij}^c. In these calculations, the interfacial tensions, γ_{pv}, γ_{sv}, γ_{pl}, γ_{sl}, and γ_{ps}, are determined by means of the equation of state for interfacial tensions (see Section 12.4) and are listed in Tables 12.4 and 12.5. Furthermore, in order to calculate Hamaker coefficients, A_{ij}, from relations of the type of Equation 12.54, one must have an explicit value of the equilibrium separation, d_0, to which the free energy data presumably apply. In the absence of adequate direct information, one has two options. Either one uses lattice and atomic size considerations, or else one considers calibration systems for which Hamaker coefficients, surface tension, and free energy data are available so that one can calculate d_0 from Equation 12.54. As it was desired to establish a connection between surface thermodynamics and theories of van der Waals interactions, the second alternative was chosen. To do so, the following materials of known interfacial tensions and Hamaker coefficients were used to determine d_0: polystyrene, Teflon, nylon-6,6, polymethyl methacrylate PMMA, n-decane, polyethylene, polyvinyl acetate (PVA), and polyhexafluoropropylene (PHFP). The surface tension information for these materials is summarized in Table 12.18. The work of cohesion was determined from Equation 12.60 and Hamaker coefficients, A_{ii}, were

TABLE 12.18

Surface Tensions of Calibration Systems

	Polystyrene	Teflon	Nylon-6,6	PMMA	n-Decane	Polyethylene	PVA	PHFP
γ_{sv} (or γ_{lv}) (mJ/m^2)	33.0	20.0	46.0	39.0	23.9	31.0	41.0	17.0

then calculated for a range of hypothetical values of d_0. These hypothetical Hamaker coefficients were compared, in each case, with the average of the literature values as compiled by Visser [96]. Agreement between the literature value and a specific hypothetical value identified the correct equilibrium separation, d_0 (see Table 12.19). These d_0 values are very similar in all cases; the mean value is $d_0 = 1.82$ Å. While it has been argued [97] on the basis of lattice and molecular size considerations that this separation distance should be approximately 4 Å, from the point of view of the macroscopic theory [90], implying a continuum model, an equilibrium separation of approximately 2 Å is reasonable. Moreover, Israelachvili [98] has argued that d_0 should not be comparable to lattice parameters, but rather to the separation between centers of polarization of the outermost atoms. His model calculations for a number of systems indeed show that d_0 is very close to 2 Å in all cases. Finally, Table 12.20 shows the resulting Hamaker coefficients obtained for a variety of particle materials in naphthalene.

The three-phase Hamaker coefficients, A_{pls}, are needed in order to predict adhesion (engulfment) or rejection of the particles in the experiments mentioned above. These coefficients, A_{pls}, were calculated using the two sets of A_{ij} data: A_{ij}^a and A_{ij}^c, as explained previously. This resulted in two types of coefficients: A_{pls}^a and A_{pls}^c, respectively, which are summarized in Table 12.21; they do not agree particularly well. Dividing A_{pls}^a by A_{pls}^c yields a correction factor, C, similar to that introduced in Equation 12.58. This factor, also given in Table 12.21, ranges from 1.49 to 2.0, well within the range of C values reported by Visser [96]. The average value is $C = 1.8$. It should be noted that the predictions of "attraction" and "repulsion" are identical for both calculations: the Hamaker coefficients for nylon-6,12, polystyrene, Teflon, and siliconized glass are positive, so that the theoretical prediction for these four systems is attraction (engulfment); for the other five materials, a negative Hamaker coefficient and hence repulsion is predicted.

Regarding the average value, $C = 1.8$, it has become apparent that [90] there is a fallacy in Visser's arguments. Generally speaking, the modifying factor, C, is intimately connected with the fact that the geometric mean combining rule, Equation 12.59, is only an approximation. As a result, the A_{pls} values obtained through the use of Equations 12.51 and 12.54 are always larger by a factor of 1.5–2.0 than those obtained using Equation 12.59. Therefore, the discrepancy between results from Lifshitz theory and Hamaker's microscopic theory (Equation 12.56) is solely due to the inadequacies of the geometric mean combining rule. If more refined work in the future should show that a discrepancy remains between these two theories, then

TABLE 12.19
Hamaker Coefficients A_{pp} and A_{ll} (in J) at Room Temperature

d_0 (Å)	Polystyrene $A_{pp} \times 10^{20}$	Teflon $A_{pp} \times 10^{20}$	Nylon-6,6 $A_{pp} \times 10^{20}$	PMMA $A_{pp} \times 10^{20}$	n-Decane $A_{ll} \times 10^{20}$	Polyethylene $A_{pp} \times 10^{20}$	PVA $A_{pp} \times 10^{20}$	PHFP $A_{pp} \times 10^{20}$
1.0	2.49	1.51	3.47	2.94	1.81	2.34	3.09	1.29
1.2	3.58	2.17	4.99	4.23	2.61	3.37	4.45	1.86
1.5	5.60	3.39	7.80	6.62	4.07	5.26	6.96	2.90
2.0	9.95	6.03	13.87	11.76	7.24	9.35	12.37	5.16
2.5	15.55	9.42	21.68	18.38	11.31	14.61	19.32	8.06
Literature values	7.6, 11.8, 5.6–6.4	4.8, 5.3, 5.6	12.4, 11.7	10.0, 10.2, 6.3	5.0, 4.6, 5.8	6.3, 10.0, 9.5	8.84	5.2
Average of literature values	8.47	5.23	12.05	8.83	5.13	8.43	8.84	5.2
Corresponding d_0 (Å)	1.85	1.86	1.86	1.73	1.68	1.90	1.69	2.01

TABLE 12.20

Hamaker Coefficients (in J) calculated from the energy of adhesion (A^a_{ij}) and from the energy of cohesion (A^c_{ij}) at 80°C for $d_0 = 1.82$ Å. Naphthalene: $A^a_{sl} = 7.32 \times 10^{-20}$, $A^c_{sl} = 7.35 \times 10^{-20}$, and $A^c_{ll} = 8.19 \times 10^{-20}$ J

	Acetal	Nylon-6	Nylon-6,6	Nylon-12	Nylon-6,10	Nylon-6,12	Polystyrene	Teflon	Siliconized glass
$A^a_{ps} \times 10^{20}$	8.10	8.07	8.02	7.84	7.63	7.24	6.74	4.97	4.23
$A^c_{ps} \times 10^{20}$	8.30	8.28	8.19	7.95	7.70	7.26	6.74	5.04	4.35
$A^a_{pl} \times 10^{20}$	9.19	9.15	9.07	8.83	8.57	8.09	7.49	5.44	4.64
$A^c_{pl} \times 10^{20}$	9.26	9.23	9.13	8.86	8.58	8.09	7.51	5.62	4.84

TABLE 12.21

Overall Hamaker Coefficients, A_{pls} (in J), of Particle Materials in Naphthalene at 80°C for $d_0 = 1.82$ Å and the Ratio $C \equiv A^a_{pls} / A^c_{pls}$

	Acetal	Nylon-6	Nylon-6,6	Nylon-12	Nylon-6,10	Nylon-6,12	Polystyrene	Teflon	Siliconized glass
$A^a_{pls} \times 10^{21}$	−2.10	−2.05	−1.85	−1.30	−0.75	0.20	1.26	4.25	5.23
$A^c_{pls} \times 10^{21}$	−1.20	−1.10	−1.00	−0.70	−0.40	0.10	0.70	2.60	3.50
C	1.75	1.85	1.86	1.86	1.88	2.00	1.80	1.63	1.49

TABLE 12.22

Theoretical Predictions and Experimental Observations of Particle Behavior at Solidification Fronts

		Naphthalene	Biphenyl	Thymol	o-Terphenyl	Salol	2-Phenyl-phenol	Pinacol
Acetal	ΔF^{Path}	+2.62	+1.99	+0.14	+1.12	+0.21	+0.01	+0.08
	A_{pls}	−3.27	−2.49	−0.25	−1.40	−0.25	−0.01	−0.10
	Observation	R	R	R	R	R	R	R
Nylon-6	ΔF^{Path}	+2.25	+1.77	+0.19	+1.00	+0.19	+0.01	+0.08
	A_{pls}	−2.81	−2.20	−0.24	−1.24	−0.22	−0.01	−0.10
	Observation	R	R	R	R	R	–	(E)
Nylon-6,6	ΔF^{Path}	+2.14	+1.68	+0.18	+0.92	+0.17	+0.01	+0.07
	A_{pls}	−2.67	−2.09	−0.22	−1.15	−0.23	−0.01	−0.09
	Observation	R	R	R	R	(E)	–	(E)
Nylon-12	ΔF^{Path}	+1.58	+1.29	+0.15	+0.63	+0.13	+0.01	+0.06
	A_{pls}	−1.97	−1.60	−0.19	−0.79	−0.15	−0.01	−0.07
	Observation	R	R	R	R	R	(E)	R
Nylon-6,10	ΔF^{Path}	+0.74	+0.72	+0.11	+0.26	+0.07	−0.01	+0.05
	A_{pls}	−0.92	−0.90	−0.14	−0.32	−0.07	+0.01	−0.06
	Observation	R	R	R	R	R	(E)	R
Nylon-6,12	ΔF^{Path}	−0.28	+0.02	+0.06	−0.24	−0.02	−0.02	+0.03
	A_{pls}	+0.35	−0.02	−0.06	+0.29	+0.02	+0.03	−0.04
	Observation	R	R	R	R	(E)	(E)	E

Polystyrene	ΔF^{Path}	-1.46	-0.77	+0.00	-0.80	-0.12	-0.03	+0.01
	A_{pls}	+1.82	+0.85	+0.00	+1.00	+0.14	+0.04	-0.01
	Observation	E	E	R	E	E	E	R
Teflon	ΔF^{Path}	-5.14	-3.24	-0.19	-2.52	-0.40	-0.06	-0.05
	A_{pls}	+6.43	+4.04	+0.24	+3.15	+0.50	+0.07	+0.06
	Observation	E	E	(E)	E	E	E	R
Siliconized glass	ΔF^{Path}	-6.41	-4.11	-0.23	-3.02	-0.46	-0.08	-0.06
	A_{pls}	+8.01	+5.14	+0.29	+3.77	+0.58	+0.10	+0.07
	Observation	E	E	E	E	E	E	E

a factor C as in Equation 12.58 might have to be reintroduced and interpreted as a difference between d_{Opls} and d_0 (see Equation 12.55).

The microscopic observations of Table 12.6 are repeated in Table 12.22, together with the results for ΔF^{adh} and the Hamaker coefficients, A_{pls}. It is to be noted that the predictions of engulfment and rejection have to be the same as in Table 12.6, except in those cases where ΔF^{adh} and ΔF^{eng} have different signs. Comparing Equation 12.20 with Equation 12.23, it is clear that $\Delta F^{eng} > 0$ and $\Delta F^{adh} < 0$ can indeed occur in some cases. In these cases, it is not clear a priori what should occur in the engulfing experiment. It is conceivable that the particle, after establishing an initial adhesive bond with the solidification front, becomes engulfed, although the later steps in the engulfing process are not favored thermodynamically, because the adhesive bond may be too strong to be broken. On the other hand, it is also conceivable that the initial adhesive bond, being fairly small, may be overcome by thermal motion or fluid flow (convection), possibly because of the unfavorable change in the surface free energy involved. In the case of nylon-6,12 in naphthalene, $\Delta F^{eng} > 0$ and $\Delta F^{adh} < 0$, which would support the second possibility, in view of the fact that the particles are rejected. In these borderline cases, the microscopic observations of "engulfment" and "rejection" might no longer reflect "attraction" and "repulsion" precisely.

Notwithstanding one special and explicable borderline case, it can be concluded that Hamaker coefficients derived from surface thermodynamics are consistent with those obtained by other means, as well as being self-consistent. There is general agreement between the sign of the Hamaker coefficients, A_{pls} (i.e., attraction and repulsion), and the microscopically observed behavior of the various types of particles at the solidification front of naphthalene. As seen from Table 12.22, A_{pls} can be positive as well as negative (the positive value implies engulfment and the negative value indicates repulsion). It is therefore clear that although van der Waals interactions between objects made of the same material are always attractive (in a medium or under vacuum), repulsive van der Waals interactions can exist between objects made of different materials (i.e., different surface free energies) separated in a medium, as in these experiments involving particles at the solidification fronts.

12.8.2 DETERMINATION OF REPULSIVE FORCES AND CRITICAL SEPARATION DISTANCES FOR PARTICLE ENGULFMENT

This section will discuss the relationship between the free energy of adhesion, ΔF^{adh}, of a particle at an advancing solidification front and the repulsive force, F_R, acting between the particle and the front in cases where the particle is pushed by the front. As shown in Section 12.4, a thermodynamic approach to calculate ΔF^{adh} using an equation of state for interfacial tensions has been employed to predict cases where particle pushing occurs [2,29,30,33]. There is good agreement between thermodynamic predictions and experimental observations based on pushing or engulfment criteria. A more rigorous examination, however, in cases where particle pushing occurs, is to see whether there is a correlation between ΔF^{adh} and F_R. This test becomes feasible as the direct measurement of F_R is possible using solidification experiments on an incline; that is, by bringing gravity into play.

Generally, the experiment to measure F_R involves the observation of particles being pushed up an incline by a solidification front. With increasing angle of tilt, an increasing fraction of the weight of the particle will oppose the repulsive force between particle and solidification front. If the particle material is denser than the liquid melt, there is a limiting or critical angle of inclination at which these two forces are equal, and the particles will be engulfed. From the angle of this incline and the density difference of the particle and the liquid melt, the value of F_R can be determined, as explained later.

The expectation of F_R to increase as ΔF^{adh} increases is based on the following theoretical expression [46,47]

$$F_R = 2\pi r \Delta F^{adh} \left(\frac{d_0}{h} \right)^2, \qquad (12.61)$$

where r is the radius of the particle, d_0 is the distance of adhesion; that is, approximately the atomic distance (2 Å) [90], and h is the particle-solid melt separation distance while the particle is being pushed. Equation 12.61 is obtained by combining Hamaker's equation [93] for $h \ll r$,

$$F_R = -\frac{A_{pls} r}{6h^2}, \qquad (12.62)$$

with the change in free-energy expression for bringing the particle into contact with the solid melt [46]

$$\Delta F^{adh} = -\frac{A_{pls}}{12\pi d_0^2}, \qquad (12.63)$$

where A_{pls} is the Hamaker coefficient for the interaction of the particle with the solid phase through the melt.

The new aspect of this investigation is the experimental determination of F_R in particle-melt systems where ΔF^{adh} is known. The ΔF^{adh} values are taken from previous work [2,29,30,33,47]. Once F_R is experimentally measured in systems where ΔF^{adh} is known, the relationship between these quantities is established and hence can be compared with the theoretical expectation (Equation 12.61). It will also be possible, from such a relationship, to calculate the separation distance, h, at the point of particle engulfment. The magnitude of this separation distance is new information and may help in understanding the behavior of a particle at a solidification front.

In a horizontal solidification cell, the forces affecting the particle are the repulsive force, F_R, between the particle and the solidification front, and the viscous drag force, F_D, of the liquid melt acting on the particle as it is pushed through this phase. Particle pushing will continue on the horizontal as long as $F_R > F_D$. However, if the solidification front velocity, V, and hence F_D increases to a point where $F_R < F_D$, then a particle that was pushed at lower solidification rates will become engulfed. At this critical velocity of engulfment, V_c, there is a balance between the forces favorable and those opposed to particle pushing. As the net force on the particle is zero at this point, the particle is no longer pushed and the solid melt grows around the particle, resulting in particle engulfment. However, this result does not allow for the determination of the magnitude of the repulsive force and other unknowns such as the separation distance between the particle and the solidification front when the force balance is achieved. In order to be able to determine F_R, it is desired to introduce a force whose influence can be easily manipulated, such as the force of gravity. This is done by performing solidification front experiments on an incline.

The solidification front experiment on an incline involves the observation, under a microscope, of a particle being pushed up an incline by the advancing solidification front. As the particle is pushed, three forces, F_g, F_D, and F_R, are acting on it. The force F_g is a component of the gravitational force:

$$F_g = v \Delta \rho \, g \sin w, \tag{12.64}$$

where v is the volume of the particle, $\Delta \rho$ is the density difference between the particle and the liquid melt, g is the acceleration due to gravity, and w is the angle of inclination. Force $F_D = F_D(V)$ is the viscous drag acting on the particle, which is a function of the velocity, V, at which the particle is pushed (see Section 12.2); F_R is the repulsive force between the particle and the solidification front.

The force balance on the particle is

$$F_R = F_D + F_g. \tag{12.65}$$

Particle engulfment occurs when $(F_D + F_g) \geq F_R$.

For a given angle of inclination, w, there is a critical velocity, V_c, which is the minimum solidification front velocity at which engulfment occurs; that is,

$$F_D(V_c) = F_R - F_g(w). \tag{12.66}$$

Increasing w will increase F_g. As F_R will be unchanged by changes in w, V_c, and thus F_D required to cause the transition from pushing to engulfment will decrease as the angle of inclination, w, increases. By increasing w to a critical angle of inclination, w_c, V_c will be zero and

$$F_R = F_g(w_c). \tag{12.67}$$

Thus, the magnitude of F_R can be obtained by determining $F_g(w_c)$.

Solidification front experiments on an incline were carried out to measure F_R in three particle matrix systems [47]. Acetal particles were used with benzophenone, biphenyl, and salol as melt materials. These particle-melt systems were selected because ΔF^{adh} was available from previous work [2,29,30,33,47] (see Table 12.23).

TABLE 12.23

The Free Energy of Adhesion, ΔF^{adh}, and the Density Differences, $\Delta\rho$, Between Acetal Particle and Three Different Melt Materials

Matrix Material	ΔF^{adh} (mJ/m²)	$\Delta\rho$ (kg/m³)
Salol	0.21	230
Benzophenone	0.66	290
Biphenyl	1.99	438

An apparatus suitable for determining the critical velocity of small particles in organic matrix materials, both on the horizontal and at positive and negative angles of inclination was designed to meet the requirements of this investigation [47]. The main elements of this apparatus are the same as described in Section 12.3; however, this apparatus is mounted on a tilting stage on which the microscope and the copper cell are placed. In the design of this stage, the frame on which the microscope is mounted is hung from a rotating axle. This axle is connected to a hand crank by means of a worm-and-spur gear. Rotating this axle changes the tilting angle of the microscope stage and thus that of the solidification cell. This arrangement provides angles of tilt from –90° to +90°.

Experiments were performed for a selection of inclination angles. For each angle there will be a different critical velocity, V_c, for particle engulfing. Theoretically, V_c should decrease as the angle of inclination, w, increases. If the density of the particle material is sufficiently large compared with that of the liquid melt, there will be a minimum angle, w_c, at which particle pushing will not occur; that is, $V_c = 0$. At this w_c, $F_R = F_g(w_c)$. Thus, the task of determining F_R is now to determine w_c and hence $F_g(w_c)$. Because experimentally it is not possible, with the present equipment, to measure solidification front velocities below 1 μm/s, determining the angle w_c at which $V_c = 0$ was not possible directly. Instead, V_c was determined for a selection of inclination angles less than w_c. The results of these experiments were extrapolated to find the angle w_c at which $V_c = 0$.

A typical experiment involved looking at a large number of acetal particles and their interaction with the solidification front. These particles, though not spherical, were sized according to their effective diameter. This diameter was computed from the average of the maximum and minimum cross-sectional dimensions of the particle. The solidification rate was varied over the course of an experiment, allowing the interaction of particles of various sizes and the solidification front to be observed at different velocities.

Plotting of the experimental observations—engulfment, pushing, and transition—for each solidification front velocity and particle diameter was used to determine the critical velocity, V_c, as discussed previously. As examples, plots of these data for acetal particles in benzophenone, at various inclinations, are shown in Figures 12.15 and 12.16. A positive inclination angle designates pushing the particle uphill.

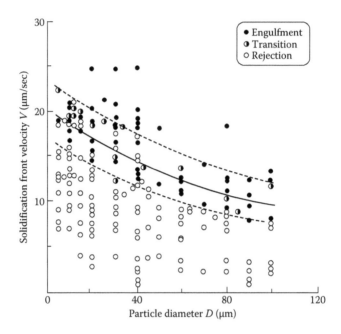

FIGURE 12.15 Behavior of acetal particles of diameter D at a solidification front advancing on an incline of $-5°$ at a velocity V in benzophenone.

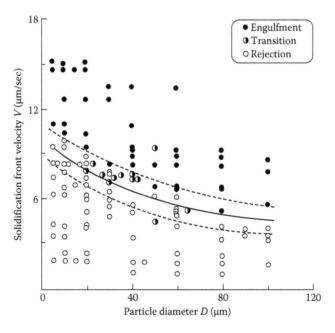

FIGURE 12.16 Behavior of acetal particles of diameter D at a solidification front advancing on an incline of $10°$ at a velocity V in benzophenone.

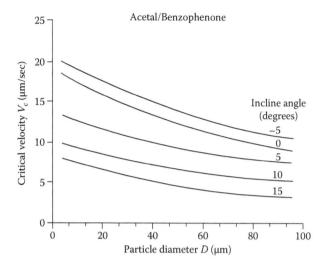

FIGURE 12.17 Critical velocity of acetal particles in benzophenone, at various angles of inclination, obtained from the mean of the band of the plots similar to those shown in Figures 12.15 and 12.16.

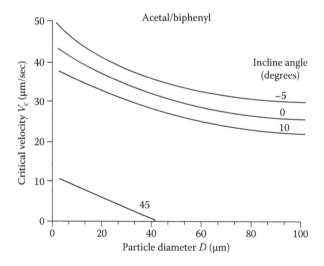

FIGURE 12.18 Critical velocity of acetal particles in biphenyl, at various angles of inclination, obtained from the mean of the band of the plots similar to those shown in Figures 12.15 and 12.16.

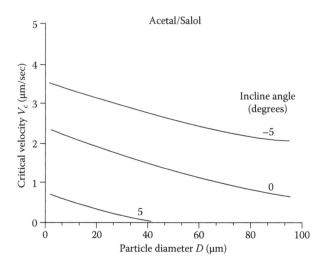

FIGURE 12.19   Critical velocity of acetal particles in salol, at various angles of inclination, obtained from the mean of the band of the plots similar to those shown in Figures 12.15 and 12.16.

TABLE 12.24

Parameters of the Straight Line fit, $V_c = A$ sin $w + B$, for a Particle Diameter of 20 μm and the Component of the Gravitational Force Per Unit Volume, $F_g(w_c)/v$, at the Critical Inclination Angle, w_c

Matrix Material	Slope A (μm/s)	Constant B (μm/s)	Correlation Coefficient	w_c (deg)	$F_g(w_c)/v$ (N/m³)
Salol	−16.2	1.71	0.992	6.1 ± 1.1	237 ± 43
Benzophenone	−34.6	14.7	0.979	25.3 ± 4.8	1216 ± 216
Biphenyl	−47.9	38.9	0.993	54.3 ± 6.8	3473 ± 294

Critical velocities for acetal particles in benzophenone, biphenyl, and salol for various inclination angles were obtained and are presented in Figures 12.17 through 12.19. For biphenyl and salol at high angles of inclination, only particles in the size range around 20 μm diameter were observed to be pushed. Thus, data for 20 μm particles were used for comparing the behavior of particles in the three different melts.

The V_c data for the 20 μm acetal particle as a function of the angle of inclination were fitted to a straight line using linear regression. Coefficients obtained for a curve of the form

$$V_c = A \sin w + B, \tag{12.68}$$

for each data set are given in Table 12.24. The correlation coefficient obtained was in every case 0.979 or better, signifying that the straight line is a good fit for the data.

From these curves, the critical angle of inclination, w_c, that is, that value of w for which V_c becomes zero, was computed. This angle for 20 μm acetal particles is given for all three matrix materials, as shown in Table 12.24. The error limits associated with this angle are 95% confidence limits, and the errors may result from the uncertainty in determining V_c. The error associated with V_c is taken as half the band width of the transition zone, on plots of the type given in Figures 12.15 and 12.16; that is, half the distance from the top line to the bottom line of these plots at a given particle size.

The different values of w_c for the three melt materials are a result of the different repulsive forces between the solidification front and the acetal particles in the three matrix materials. Values of ΔF^{adh} presented in Table 12.23 predict that the largest repulsive force will be between acetal and biphenyl while acetal and salol will have the smallest F_R. Taking w_c from Table 12.24 and $\Delta\rho$ from Table 12.23, values for the quantity $\Delta\rho g \sin w_c$ were calculated. This is the gravitational force per unit particle volume, F_g/v (see Equation 12.64). Since the acetal particles were irregularly shaped, their exact volume is not known. However, particles from the same source were used in all experiments and should thus be similar in volume. Therefore, F_g/v provides an unbiased comparison of the repulsive force in the three particle-melt systems. The values of F_g/v computed from w_c are also given in Table 12.24. The errors associated with the F_g/v values are 95% confidence limits and follow from the errors associated with w_c.

Relating ΔF^{adh} to $[F_g(w_c)]/v$ is achieved by plotting the data of Table 12.24, as shown in Figure 12.20. A line was fitted to the three data points using linear regression and the following result was obtained:

$$\Delta F^{adh} = 5.57 \times 10^{-4} \frac{F_g(w_c)}{v} + 0.039, \qquad (12.69)$$

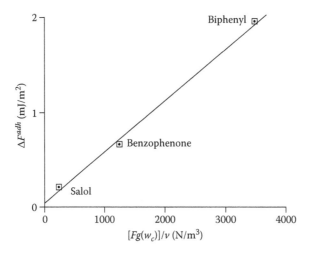

FIGURE 12.20 Free energy of adhesion, ΔF^{adh}, versus the forces of gravity, determined from solidification front experiments on an incline in benzophenone, bibenzyl, and salol.

TABLE 12.25

The Magnitude of the Repulsive Force and the Minimum Separation Distance Between Acetal Particles in Three Matrix Materials

Matrix Material	Particle Diameter (μm)	Repulsive Force F_R (nN)	Minimum Separation Distance h (nm)
Salol	20	0.0010	22
Benzophenone	20	0.0051	18
Biphenyl	20	0.0145	19

with a correlation coefficient of 0.999.

The expectation of a linear relationship between ΔF^{adh} and F_R assumes that the separation distance, h, at the point of engulfment is the same in all three particle-matrix systems (see Equation 12.61). This can be checked by rearranging Equation 12.61 and evaluating h for the known values of ΔF^{adh} and F_R:

$$h = \left(\frac{2\pi r \Delta F^{adh} d_0^2}{F_R} \right)^{1/2}.$$

(12.70)

In order to calculate F_R values from the results of $[F_g(w_c)]/v$, the particle volume, v, has to be estimated. It will be assumed that the particles are spheres of 20 μm diameter. As discussed above, this assumption will introduce a common error into the results for all three matrix materials. The F_R and h values thus obtained for acetal particles in salol, benzophenone, and biphenyl, are given in Table 12.25. Considering the influence of experimental errors, the separation distance, h, for acetal particles in all three matrix materials is essentially the same: approximately 20 nm. Such a constant h value for all these particle-matrix systems means a linear relationship between ΔF^{adh} and F_R, as expected.

The relationship given by Equation 12.61, an equation based on Hamaker's equation and thermodynamics, also predicts that ΔF^{adh} will be zero when F_R is zero. However, it may be noted that, from the empirical equation, Equation 12.69, $\Delta F^{adh} = 0.039$ mJ/m^2 when $F_R = 0$. This small deviation of the experimental results from the theory may be ignored if one realizes that Equation 12.69 contains: (1) errors from calculated ΔF^{adh} values, which include errors of contact angle measurements; and (2) errors from determination of w_c, which, as explained earlier, is obtained by extrapolation. Therefore, the agreement between the experimental results and the theory is acceptable.

In summary, the solidification front technique presented here is a direct force measurement method for the repulsive force between a particle and a solidification front. The relationship between ΔF^{adh} and F_R was found to be linear, in agreement with theory. Values for F_R were measured directly, as described in this section. The ΔF^{adh} values, taken from previous work, were calculated from temperature-dependent contact angle measurements using the equation of state for interfacial tensions. The fact that the theoretically expected relationship between ΔF^{adh} and F_R was achieved experimentally provides strong support for the thermodynamics approach and measurement techniques used to obtain these ΔF^{adh} values.

Good agreement with theory also provides confidence in the solidification front on an incline technique for studying the interaction forces in the particle/solidification front system. Although the experimental relationships obtained cannot be considered universal, based on only three sets of data, they do provide some insight into the interactions between a particle and a solidification front.

Results presented in Table 12.25 give estimates of the magnitude of the repulsive force, F_R, and the particle/solidification front separation distance, h, at the time of engulfment. Because the acetal particles used were not spherical, but irregular in shape, these results provide only estimates. Still, by considering these separation distance results, one sees that the processes that lead to engulfment begin at approximately 20 nm, a point when there is still liquid melt between the particle and the solid phase. This large separation distance at the point of engulfment would, according to theory [46], suggest that the interactions between particle and solid involve several molecular layers beyond the surface.

REFERENCES

1. S. N. Omenyi. *Attraction and Repulsion of Particles by Solidifying Melts*. PhD Thesis, University of Toronto, 1978.
2. S. N. Omenyi and A. W. Neumann. *Journal of Applied Physics* 47 (1976): 3956.
3. D. Shangguan, S. Ahuja, and D. M. Stefanescu. *Metallurgical Transactions A* 23 (1992): 669.
4. G. Kaptay. *Metallurgical and Materials Transactions A* 32 (2001): 993.
5. M. C. Flemings. *Solidification Processing*. McGraw-Hill, New York, 1974.
6. W. Kurz and D. J. Fisher. *Fundamentals of Solidification*. Trans Tech Publications, Aedermannsdorf, Switzerland, 1989.
7. S. S. L. Peppin, J. A. W. Elliott, and M. G. Worster. *Journal of Fluid Mechanics* 554 (2006): 147.
8. A. W. Neumann, C. J. van Oss, and J. Szekely. *Kolloid-Z, Z. Polymer.* 251 (1973): 415.
9. S. Torza and S. G. Mason. *Science* 162 (1969): 813.
10. C. Korber. *Quarterly Reviews of Biophysics* 21 (1988): 229.
11. J. W. Garvin, L. Mao, and H. S. Udaykumar. *American Society of Mechanical Engineers, Heat Transfer Division* 372 (2002): 219.
12. D. R. Uhlmann, B. Chalmers, and K. A. Jackson. *Journal of Applied Physics* 35 (1964): 2986.
13. V. H. S. Kuo and W. R. Wilcox. *Separation Science and Technology* 8 (1973): 375.
14. V. H. S. Kuo and W. R. Wilcox. *Industrial & Engineering Chemistry Process Design and Development* 12 (1973): 376.
15. V. L. Bronstein. *Journal of Crystal Growth* 52 (1981): 345.
16. W. E. Brower, M. J. Freund, M. D. Baudino, and C. Ringwald. *Cryobiology* 18 (1981): 277.
17. Ch. Körber and G. Rau. *Journal of Crystal Growth* 72 (1985): 649.
18. A. W. Rempel and M. G. Worster. *Journal of Crystal Growth* 205 (1999): 427.
19. F. Nota, R. Savino, and S. Fico. *Acta Astronautica* 59 (2006): 20.
20. M. S. Park, A. A. Golovin, and S. H. Davis. *Journal of Fluid Mechanics* 560 (2006): 415.

21. L. Hadji. *Physical Review E* 75 (2007): 04260.
22. A. E. Corte. *Journal of Geophysical Research* 67 (1962): 1085.
23. P. Hoekstra and R. D. Miller. *Journal of Colloid and Interface Science* 25 (1967): 166.
24. J. Cissé and G. F. Bolling. *Journal of Crystal Growth* 10 (1971): 67.
25. J. Cissé and G. F. Bolling. *Journal of Crystal Growth* 11 (1971): 25.
26. A. M. Zubko, V. G. Lobonov, and V. V. Nikonova. *Soviet Physics and Crystallography* 18 (1973): 239.
27. K. H. Chen and W. R. Wilcox. *Journal of Crystal Growth* 40 (1977): 214.
28. P. F. Aubourg. *Interaction of Second-Phase Particles with a Crystal Growing from the Melt*. PhD Thesis, Massachusetts Institute of Technology, 1978.
29. S. N. Omenyi, A. W. Neumann, and C. J. van Oss. *Journal of Applied Physics* 52 (1981): 789.
30. S. N. Omenyi, A. W. Neumann, W. W. Martin, G. M. Lespinard, and R. P. Smith. *Journal of Applied Physics* 52 (1981): 796.
31. A. W. Neumann, J. Szekely, and E. J. Rabenda. *Journal of Colloid and Interface Science* 43 (1973): 727.
32. R. Asthana and S. N. Tewari. *Journal of Materials Science* 28 (1993): 5414.
33. S. N. Omenyi, R. P. Smith, and A. W. Neumann. *Journal of Colloid and Interface Science* 75 (1980): 117.
34. M. A. Azouni and P. Casses. *Advances in Colloid and Interface Science* 75 (1998): 83.
35. Z. Wang, K. Mukai, and I. J. Lee. *ISIJ International* 39 (1999): 553.
36. G. F. Bolling and J. Cissé. *Journal of Crystal Growth* 10 (1971): 56.
37. A. A. Chernov and D. E. Temkin. In *1976 Crystal Growth and Materials*. Edited by E. Kaldis and H. J. Scheel, 3. North-Holland, Amsterdam, 1977.
38. A. A. Chernov, D. E. Temkin, and A. M. Mel'nikova. *Soviet Physics and Crystallography* 21 (1976): 369.
39. A. A. Chernov, D. E. Temkin, and A. M. Mel'nikova. *Soviet Physics and Crystallography* 22 (1977): 13.
40. A. A. Chernov, D. E. Temkin, and A. M. Mel'nikova. *Soviet Physics and Crystallography* 22 (1977): 656.
41. R. R. Gilpin. *Journal of Colloid and Interface Science* 74 (1980): 44.
42. D. M. Stefanescu, F. R. Juretzko, B. K. Dhindaw, A. Catalina, S. Sen, and P. A. Curreri. *Metallurgical and Materials Transactions A* 29 (1998): 1697.
43. A. V. Catalina, S. Mukherjee, and D. M. Stefanescu. *Metallurgical and Materials Transactions A* 31 (2000): 2559.
44. J. W. Garvin, Y. Yang, and H. S. Udaykumar. *International Journal of Heat Mass Transfer* 50 (2007): 2952.
45. J. W. Garvin, Y. Yang, and H. S. Udaykumar. *International Journal of Heat Mass Transfer* 50 (2007): 2969.
46. D. W. Francis. *Interfacial Tensions and van der Waals Interactions of Small Particles at Solid-Liquid Interfaces*. Ph.D. Thesis, University of Toronto, 1983.
47. R. P. Smith. *Applied Surface Thermodynamics for the Interaction of Small Particles with an Advancing Solidification Front*. Ph.D. Thesis, University of Toronto, 1984.
48. A. W. Neumann, R. J. Good, C. J. Hope, and M. Sejpal. *Journal of Colloid and Interface Science* 49 (1974): 291.
49. C. A. Ward and A. W. Neumann. *Journal of Colloid and Interface Science* 49 (1974): 286.
50. D. Li, J. Gaydos, and A. W. Neumann. *Langmuir* 5 (1989): 1133.
51. D. Li and A. W. Neumann. *Journal of Colloid and Interface Science* 137 (1990): 304.
52. D. Li and A. W. Neumann. *Journal of Colloid and Interface Science* 148 (1992): 190.

53. R. P. Smith, S. N. Omenyi, and A. W. Neumann. *Physiochemical Aspects of Polymer Surfaces.* Edited by K. L. Mittal, Vol. 1, 155–71. Plenum Press, New York, 1983.
54. A. W. Neumann, R. J. Good, P. Ehrlich, K. Basu, and G. J. Johnston. *Journal of Macromolecular Science-Physics* 37 (1973): 525.
55. A. W. Neumann. *Zeitschrift für Physikalische Chemie* 41 (1964): 339.
56. A. W. Neumann and W. Tanner. *Journal of Colloid and Interface Science* 34 (1970): 1.
57. S. K. Li, R. P. Smith, and A. W. Neumann. *Journal of Adhesion* 17 (1984): 105.
58. A. W. Neumann and W. Tanner. "Continuous Measurement of the Time Dependence of Contact Angles between Individual Fibres and Surfactant Solutions." 5th International Congress of Surface Activity, Vol. 2, 727. Barcelona, Spain, 1968.
59. A. W. Neumann and R. J. Good. In *Surface and Colloid Science.* Edited by R. J. Good and R. R. Stromberg, Vol. 11, 31. Plenum Press, New York, 1979.
60. A. W. Neumann, D. R. Absolom, W. Zingg, and C. J. van Oss. *Cell Biophysics* 1 (1979): 79.
61. S. K. Chang, O. S. Hum, M. A. Moscarello, A. W. Neumann, W. Zingg, M. Leutheusser, and B. Ruegsegger. *Medical Progress Through Technology* 5 (1977): 57.
62. W. Zingg, A. W. Neumann, A. B. Strong, O. S. Hum, and D. R. Absolom. *Biomaterials* 2 (1981): 156.
63. A. W. Neumann, D. R. Absolom, D. W. Francis, C. J. van Oss, and W. Zingg. *Cell Biophysics* 4 (1982): 285.
64. C. J. van Oss, C. F. Gillman, and A. W. Neumann. *Phagocytic Engulfment and Cell Adhesiveness as Surface Phenomena.* Marcel Dekker, New York, 1975.
65. D. F. Gerson. *Biochimica et Biophysica Acta* 602 (1980): 269.
66. J. F. Boyce, S. Schurch, and D. J. McIver. *Atherosclerosis* 37 (1980): 361.
67. C. J. van Oss. *Annual Review of Microbiology* 32 (1978): 19.
68. J. K. Spelt. *Surface Tension Measurements of Biological Cells Using the Freezing Front Technique.* M.A.Sc. Thesis, University of Toronto, 1980.
69. J. K. Spelt, D. R. Absolom, W. Zingg, C. J. van Oss, and A. W. Neumann. *Cell Biophysics* 4 (1982): 117.
70. S. N. Omenyi, R. S. Snyder, and C. J. van Oss. "Effects of Zero van der Waals and Zero Electrostatic Forces on Droplet Sedimentation." *54th Colloid and Surface Science Symposium.* ACS Lehigh, 1980, Abstract No. 163.
71. S. N. Omenyi, R. S. Snyder, C. J. van Oss, D. R. Absolom, and A. W. Neumann. *Journal of Colloid and Interface Science* 81 (1981): 402.
72. H. Shiffman. *A Preliminary Investigation of Principles Relating to Biological Cell Separation Using Solidifying Melts of Aqueous Solutions.* M.A.Sc. Thesis, University of Toronto, 1978.
73. S. N. Omenyi, A. W. Neumann, W. W. Martin, G. M. Lespinard, and R. P. Smith. *Journal of Applied Physics* 52 (1981): 796.
74. R. P. Smith. *The Development of a Technique to Determine Interfacial Tensions from Particle Behaviour at a Solid-Liquid Melt Interface.* M.A.Sc. Thesis, University of Toronto, 1981.
75. A. W. Neumann. *Advances in Colloid and Interface Science* 4 (1974): 105.
76. R. C. West, ed. *Handbook of Chemistry and Physics.* 51st ed., F5. Chemical Rubber Company, Ohio, 1970.
77. L. J. Cianciolo, R. J. Genco, M. R. Patters, J. McKenna, and C. J. van Oss. *Nature* 265 (1977): 445.
78. C. J. van Oss, J. M. Berstein, B. H. Park, L. J. Cianciolo, and R. J. Genco. *International Convocation on Immunology* 6 (1979): 311.
79. T. G. Thrasher, T. Yoshida, C. J. van Oss, S. Cohen and N. Rose. *Journal of Immunology* 110 (1973): 321.
80. G. Adam and C. Schumann. *Progress in Colloid and Polymer Science* 65 (1978): 200.

81. P. A. Albertsson. *Partition of Cells, Particles and Macromolecules*. Wiley Interscience, New York, 1971.

82. K. E. Magnusson, O. Stendahl, C. Tagesson, L. Edebo, and G. Johansson. *Acta Pathologica et Microbiologica Scandinavica* 85 (1977): 212.

83. O. Stendahl, K. E. Magnusson, C. Tagesson, R. Cunningham, and L. Edebo. *Infection and Immunity* 7 (1973): 573.

84. O. Stendahl, C. Tagesson, K. E. Magnusson, and L. Edebo. *Immunology* 32 (1977): 11.

85. D. F. Gerson and J. Akit. *Biochimica et Biophysica Acta* 602 (1980): 281.

86. M. R. Soulard, E. I. Vargha-Butler, H. A. Hamza, and A. W. Neumann. *Chemical Engineering Communications* 21 (1983): 329.

87. A. W. Neumann, E. I. Vargha-Butler, H. A. Hamza, and D. R. Absolom. *Colloids and Surfaces* 17 (1986): 131.

88. E. I. Vargha-Butler, D. R. Absolom, H. A. Hamza, and A. W. Neumann. In *Interfacial Phenomena in Coal Technology*. Vol. 32, 33. Edited by G. D. Botsaris and Y. M. Glazman, 33. Marcel Dekker, New York, 1988.

89. E. I. Vargha-Butler, T. K. Zubovits, R. P. Smith, I. H. L. Tsim, H. A. Hamza, and A. W. Neumann. *Colloids and Surfaces* 8 (1984): 231.

90. A. W. Neumann, S. N. Omenyi, and C. J. van Oss. *Colloid and Polymer Science* 257 (1979): 413.

91. A. W. Neumann, S. N. Omenyi, and C. J. van Oss. *Journal of Physical Chemistry* 86 (1982): 1267.

92. S. Nir. *Progress in Surface and Membrane Science* 8 (1976): 1.

93. H. C. Hamaker. *Physica* 4 (1937): 1058.

94. E. M. Lifshitz. *Soviet Physics–JETP* 2 (1956): 73.

95. I. E. Dzyaloshinskii, E. M. Lifshitz, and L. P. Pitaevskii. *Advanced Physics* 10 (1961): 165.

96. J. Visser. *Advances in Colloid and Interface Science* 3 (1972): 331.

97. F. M. Fowkes. *S.C.I. Monograph No. 25*. 3. Society of Chemical Industry, London, 1967.

98. J. N. Israelachvili. *Journal of the Chemical Society-Faraday Transactions* 269 (1973): 1929.

13 Line Tension and the Drop Size Dependence of Contact Angles

Robert David and A. Wilhelm Neumann

CONTENTS

13.1 INTRODUCTION

Line tension was introduced in Chapter 1 as part of the generalized theory of capillarity. Line tension is the work of formation (or excess energy) of a unit length of line phase, much as surface tension is the work of formation of a unit area of surface phase. The line phase considered is usually a three-phase line such as the circular edge around the base of an axisymmetric sessile liquid drop on a solid substrate. At this line the solid, liquid, and vapor phases meet.

The molecules near a three-phase line experience a different environment than those in the middle of a surface, or those in the interior of a bulk phase. The differing energetic state of the molecules near the three-phase line is the fundamental source of line tension. Line tension may be thought of as a correction to the surface tensions near a discontinuity (the edges of the surfaces), just as surface tension is a correction to bulk pressures at an interface (see Chapter 10).

For interfacial tension, a negative value would by definition cause spontaneous miscibility of the two adjacent phases and the disappearance of the interface. Such a negative tension is therefore impossible in equilibrium (see Chapter 9 for further discussion). However, a three-phase line is not a barrier separating two different phases, and so line tension can theoretically be either positive or negative.

A positive line tension signifies that energy is expended in creating line phase. Therefore, a system with a positive line tension tends to minimize the length of line phase; for a sessile drop, this corresponds to a constriction of the three-phase line and a consequent increase in the contact angle. A negative line tension signifies that energy is gained in creating line phase, and it therefore has the opposite influence.

The term "line tension" has also been used to describe tension between coexisting surface phases on a two-dimensional surface, especially domains in lipid membranes [1]. Another two-dimensional line tension is found around a drop that nearly wets a solid substrate, where a microscopically thin prewetting line advances ahead of the drop front and coexists with the dry surface; however, in this case the tension is usually called boundary tension [2]. The two-dimensional type of line tension must be positive to keep the domains separate. Despite sharing a name, the two line tension phenomena are fundamentally different, and this chapter will be restricted to line tension in three-dimensional systems.

With line phases of moderate curvature, the line energy in a thermodynamic system should scale as the total length of line phase present. As a system's dimensions shrink, the line energy becomes relatively more significant due to increasing line-to-surface and line-to-volume ratios. The relevant length scale is σ/γ, where σ is line tension and γ is surface tension (typically ~0.05 J/m²). Thus, a line tension of ~10^{-12} J/m would have a very minor effect at the micrometer level, and would become important only at an atomic length scale (at which line tension has no meaning anyway), whereas a line tension of ~10^{-6} J/m would be discernible at the millimeter level, and dominate at the micrometer level and below. Due to extremely wide-ranging results in the experimental literature, it is still debated which of these two cases is closer to reality [3]. Therefore, the breadth of practical applications in which line tension plays a role is as yet uncertain.

13.1.1 APPLICATIONS

The most obvious potential application for line tension is in microfluidic systems [4]. In these systems, tiny amounts of liquid are manipulated through submillimeter sized channels and reservoirs with the goal of miniaturizing and parallelizing biochemical analysis. Three-phase lines may occur, for example, between solid, liquid, and vapor, or between liquid and two different solid phases on a micropatterned surface. Line tension may become a consideration through its effect on wettability; control of wettability (e.g., via electrowetting or thermocapillary flow) is often used to actuate drops in microfluidic circuits [5].

As another example, the process of nucleation involves three-phase lines with very high curvature. For instance, line tension is a significant parameter in simulations of freezing [6]. A similar situation exists in heat transfer involving condensation or evaporation of small drops, as in heat pipes [7] or dropwise condensation [8,9]. In both cases, the efficiency of heat transfer is connected with the wettability of the condenser or wall surface, and so might be affected by line tension.

Another potential application of line tension is froth flotation in mineral processing, in which hydrophobic particles are removed from an aqueous solution by attachment to rising bubbles. Attachment requires the formation of a three-phase line between solid, liquid, and vapor. The expansion of such a line would be opposed by a positive line tension [10].

The opposite suggestion has been made for the case of ultrafine particles that penetrate cell membranes after being inhaled into the lungs. Here, other forces might induce expansion of the three-phase line beyond the equator of a spherical particle, at which point a positive line tension would favor complete immersion. This has been proposed to explain observations of greater immersion of small particles, compared to large particles, in the surfactant film on trachea walls [11].

A similar situation occurs when solid particles are used to stabilize emulsions and foams. In an emulsion, line tension may affect the ability of the particles to adsorb at the oil/water interface. In addition, the type of emulsion formed—oil in water or water in oil—depends on the contact angle formed at the three-phase line between oil, water, and solid, which may be affected by line tension [12]. Line tension has also been invoked at three-phase lines of oil, water, and air to explain the behavior of oil-based antifoams [13].

Finally, it has been suggested that line tension may affect cell adhesion, again due to the very small area of contact made by a filopodium of ~100 nm diameter [14]. In the analogous scenario of adhesion between very small areas of solids, the JKR theory [15] balances surface and elastic forces, but the possible influence of line tension appears to be unexplored as yet.

Thus, the value of line tension is of interest not only theoretically, but in a number of practical applications as well.

13.1.2 SIZE DEPENDENCE OF CONTACT ANGLES

In Chapter 1, the equilibrium condition at the contact line of a sessile drop, called the modified Young equation, was derived (Equation 1.65). It includes the effect of line tension σ as follows:

$$\gamma_{lv} \cos\theta = \gamma_{sv} - \gamma_{sl} - \sigma\kappa, \tag{13.1}$$

where γ_{lv} is the liquid–vapor interfacial tension, γ_{sv} is the solid–vapor interfacial tension, γ_{sl} is the solid–liquid interfacial tension, θ is the contact angle, and κ is the geodesic curvature of the line. In the usual situation of an axisymmetric drop, κ is the reciprocal of the drop radius r. Therefore, as expected from the line-to-surface ratio, the line tension term becomes insignificant as $r \rightarrow \infty$. If θ_∞ is defined as the contact angle in this limit, then

$$\cos\theta = \cos\theta_\infty - \frac{\sigma}{\gamma_{lv} r}. \tag{13.2}$$

Thus, line tension causes the contact angle to be a function of the drop size.

Equation 13.2 provides a method for measuring line tension. The contact angles of drops of different sizes are measured, and the line tension is obtained from the slope of a plot of $\cos\theta$ against $1/r$. However, this method is controversial because other factors may also cause size dependence of the contact angle, which has led some researchers to denote such measured values as pseudo-line tensions [16].

In Section 13.2, theoretical estimates for the value of line tension (based on its definition as the excess energy of a three-phase line) will be reviewed. Section 13.3 will cover experimental measurements of line tension, mostly based on the drop size dependence of contact angles. The uncertain relationship between theory and experiment for line tension will be discussed in Section 13.4.

13.2 THEORY

While the generalization of Gibbs's theory of capillarity provides a method for experimental measurement of line tension via the modified Young equation (Equation 13.1), it does not allow a theoretical estimate. Theoretical estimates are based on more detailed modeling of the region near the three-phase line, in some cases even at the molecular level. For simplicity, a two-dimensional model is often considered, in which a drop of liquid is replaced by a wedge with a straight contact line, with all quantities invariant along the contact line. Much of the theoretical work has focused on the value of line tension approaching wetting (i.e., as $\theta \rightarrow 0$).

In the following three sections, the three major approaches for modeling line tension will be described. They are: mean field or density functional theory (DFT) (Section 13.2.1), the more phenomenological interface displacement model (Section 13.2.2), and the computational method of molecular dynamics (Section 13.2.3).

13.2.1 DENSITY FUNCTIONAL THEORY

In DFT, the thermodynamic properties of each bulk phase are considered as functionals of the particle density (itself a function of space). For the study of line tension, the most important of these thermodynamic properties is the free energy. The modeling of the free energy in DFT as a smoothly varying field (as opposed to a sum of pairwise or higher order interactions between molecules) is called a mean field approximation.

DFT reduces the complete description of a phase to the determination of the correct density distribution. This approach greatly reduces the number of variables compared to a rigorous description, which would require molecule-by-molecule calculation (see Section 13.2.3). However, as a continuum theory, DFT must break down below a certain length scale. DFT is widely used in condensed matter physics beyond the particular application described here [17]. A one-dimensional DFT called gradient theory was used in Chapter 10 to estimate solid–liquid interfacial tensions.

In early work applying DFT to the estimation of line tension, Kerins and Widom [18] formulated three models for the contact line region, two of which will be mentioned here. In the first, crude model, they postulated the spatial variation of the free energy (not the particle density), and argued that line tension σ is of order γa, with γ an interfacial tension and a the width of an interface. This corresponded to σ of order 10^{-11} J/m. They also argued from this model that σ is more likely negative than positive.

In their second model, Kerins and Widom postulated the free energy Ω as a functional of the particle density ρ. The free energy contains contributions from the volume, surface, and lines phases (see Chapter 1). After subtracting the volume $(-PV)$ and surface (γA) contributions, the remainder (σL) gives the line tension. For a two-component system, the Ω functional had the form

$$\Omega(\rho_1,\rho_2) = F(\rho_1,\rho_2) + \sum_{i,j=1}^{2} m_{ij}(\nabla\rho_i \cdot \nabla\rho_j), \qquad (13.3)$$

where F achieved minima in the densities corresponding to each bulk phase. In this and subsequent papers, the weights m_{ij} have generally been taken to be zero for $i \neq j$ and one for $i = j$, thus serving to penalize rapid spatial changes in density. After assuming a general parameterized form for the density ρ, the authors numerically found the parameter values that minimized $\Omega(\rho)$, and calculated the resulting line tension σ to be of order 10^{-12} J/m.

Later work focused on the behavior of σ approaching wetting. Widom and Clarke [2] employed the first, simpler model of Kerins and Widom and found that σ approached zero in proportion to the contact angle θ near wetting. However, using a more complex version of the second model, Szleifer and Widom [19] found that σ changed sign from negative to positive as wetting was approached, with a possibly infinite value at wetting. The authors believed that the results in this work best reflected the normally observed first-order wetting transition. Their conclusions were confirmed later in more accurate calculations [20].

One limitation of the models of Widom and coworkers is that, due to assumed symmetry between the phases, they could not be applied to systems with a solid phase. However, Perković et al. [21] adapted the second model for just such a system. They also found the density ρ that minimized the free energy Ω without making any assumptions about the form of ρ. Their results were consistent with those of Koga and Widom [20].

A second limitation of the models of Widom and coworkers is that only local interactions are considered, as Ω depends only on ρ and its gradients (Equation 13.3). This has been addressed in the research of Dietrich's group. Compared to Widom and coworkers, Dietrich and his coworkers have used a simpler fluid density ρ and a more complex (physically motivated) free energy Ω.

This work culminated in an article by Bauer and Dietrich [22] that considered liquid drops or films on solid surfaces and corrected numerical problems present in earlier papers. The interactions of fluid molecules (both liquid and gas) with each other and with the solid substrate were modeled by Lennard-Jones potentials. The pairwise interactions were integrated over the density of the entire system to include nonlocal effects. The density distribution that minimized the free energy thus obtained was found numerically. Only sharp-kink density distributions were considered, meaning that the liquid and gaseous phases each had uniform density right up to their boundaries. The authors found line tension somewhat smaller than ε/d, with ε the well depth and d (often called σ) the molecular separation in the Lennard-Jones potential. This corresponds to line tension of order 10^{-12} J/m. Bauer and Dietrich also compared results from local and nonlocal calculations, and found only minor differences [22].

In the simplest DFT models, both the density distribution and the free energy functional are postulated and the line tension is calculated without any minimization. In early work, Tarazona and Navascués [23] took this approach and calculated σ of order -10^{-11} J/m using the properties of a number of dispersive, cryogenic liquids and dispersive solid substrates. (Dispersive molecules are nonpolar and interact only via London forces.) More recently, Qu and Li [24] estimated the line tension in several alkane–water–air systems. They assumed a linearized form for the density ρ in the three-phase region and calculated the resulting free energy using equations of state for liquids. They found positive line tensions of order 10^{-10} J/m.

Finally, Marmur [25] estimated line tension by considering only the excess liquid–solid interfacial energy near a contact line. (Liquid–liquid and solid–solid interactions, although just as significant, were neglected.) He calculated σ as the difference between the van der Waals interaction energy of a semi-infinite substrate with a drop shaped as a spherical cap; and a semi-infinite substract with a drop shaped as a semi-infinite cylinder with the same base radius as the spherical cap. He concluded that measured line tensions should not exceed 5×10^{-9} J/m in absolute value.

In Table 13.1, theoretical results for the magnitude of line tension from DFT (and other methods described below) are listed. The DFT approach to studying line tension is complex and all but the most simplified versions require lengthy numerical calculations. It is therefore notable that many of the essential results of DFT have also been demonstrated with a stripped-down approach called the interface displacement model.

TABLE 13.1
Theoretical Estimates of the Magnitude of Line Tension

Authors	Year(s)	Method	$|\sigma|$ (J/m)	Ref.
Widom and coworkers	1982–2007	density functional	10^{-12}	18–21
Bauer and Dietrich	1999	density functional	10^{-12}	22
Tarazona and Navascués	1981	density functional	10^{-11}	23
Qu and Li	1999	density functional	10^{-10}	24
Solomentsev and White	1999	interface displacement	10^{-11}	32
Dobbs	1999	interface displacement	10^{-12}	33
Checco et al.	2003	interface displacement	10^{-12}	34
Bresme and Quirke	1998–1999	molecular dynamics	10^{-11}	35,36
Werder et al.	2003	molecular dynamics	10^{-10}	37
Hirvi and Pakkanen	2006	molecular dynamics	10^{-12}	38
Schneemilch and Quirke	2007	molecular dynamics	10^{-11}	39

13.2.2 INTERFACE DISPLACEMENT MODEL

The interface displacement model focuses only on the shape of the liquid–vapor interface in the transition region between a macroscopic sessile drop on a solid substrate, and a surrounding thin film (Figure 13.1). The characterization of the liquid–vapor interface by only its thickness, with no density details preserved, is similar to the sharp-kink approximation used by Dietrich and coworkers with DFT (see the previous section). In keeping with the presence of the thin film surrounding the drop, small angle approximations are often used for the contact angle θ. While the early development of the theory was conducted by various researchers [26,27], we will follow the later presentation of Indekeu [28], who brought the interface displacement model into harmony with the results of DFT.

Suppose the profile of a two-dimensional drop in the transition region is $l(x)$, Figure 13.1. The free energy (i.e., the grand canonical potential, see Chapter 1) is given by a functional of the profile:

$$\Omega[l(x)] = \int_{-\infty}^{\infty} \left\{ V[l(x)] + \gamma_{lv} \left(\sqrt{1 + \left(\frac{dl}{dx}\right)^2} - 1 \right) \right\} dx - \Omega_0(x). \qquad (13.4)$$

The first term in the integral is the interface potential V, which gives the total surface energy of a system in which a horizontal liquid–vapor interface and a horizontal solid–liquid interface are separated by a distance l. The limiting values of V are γ_{sv} for the thin film $l = l_1$ (at $x = -\infty$) and $\gamma_{sl} + \gamma_{lv}$ for the macroscopic drop $l = \infty$ (at $x = \infty$). For convenience, V is often shifted by $-\gamma_{sv}$ so that the thin film is the zero energy reference. In between the $x = \pm\infty$ limits, V models the interaction between the two interfaces when they are close enough to affect each other. The interacting surfaces concept was historically introduced by Derjaguin as the disjoining pressure π [29], where

$$\pi(l) = -\frac{dV}{dl}. \qquad (13.5)$$

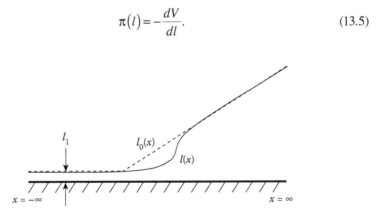

FIGURE 13.1 Liquid–vapor profile $l(x)$ in the transition region (solid line), and asymptotic profile $l_0(x)$, dotted line. The profiles do not change in the direction perpendicular to the page.

The second term in the integral in Equation 13.4 accounts for nonflatness of the liquid–vapor interface; the expression inside the parentheses is the difference between the arc length of the actual profile and its shorter arc length had it been piecewise horizontal. The square root is sometimes expanded to first order to obtain the so-called gradient-squared approximation. Finally, subtraction of the last term Ω_0 ensures that Ω is not infinite; Ω_0 is the free energy of the asymptotic profile $l_0(x)$ shown in Figure 13.1. This step eliminates surface energy from Ω, leaving just the excess line energy.

Gravitational energy can also be included in the free energy expression [30], but it has been omitted because it is negligible in the tiny dimensions of a typical transition region.

The liquid–vapor profile in the transition region will adopt the shape $l(x)$ that minimizes the free energy Ω; this minimum value of Ω is the line tension σ. The solution can be found using the calculus of variations [31]:

$$\sigma = \sqrt{2\gamma_{lv}} \int_{l_1}^{\infty} \left[\sqrt{V(l)} - \sqrt{-S} \right] dl, \qquad (13.6)$$

where S is the spreading coefficient, defined as $\gamma_{sv} - \gamma_{sl} - \gamma_{lv}$. Equation 13.6 allows calculation of the line tension from an assumed interface potential V. This is in contrast with the much more detailed input required to estimate line tension by DFT. An example of a form for V is shown in Figure 13.2; the shape of V depends on the assumed intermolecular forces.

Indekeu [28] showed that for first-order wetting with short-range forces, the interface displacement model predicted a negative value for σ that changed sign and became positive and finite at wetting. The result was the same for long-range (van der Waals) forces except that line tension was infinite at wetting. These findings matched DFT calculations by other researchers [20–22].

Quantitative calculations using the interface displacement model were carried out by Solomentsev and White, who considered a dispersive liquid on a solid substrate and estimated that σ was of order 10^{-11} J/m [32]. Dobbs [33] calculated the line tension for alkane lenses on water, finding σ of order 10^{-12} J/m. Checco et al. [34] also

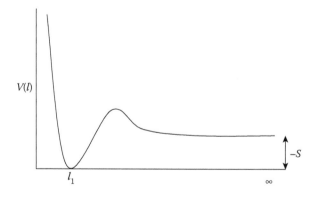

FIGURE 13.2 Example of an interface potential for partial wetting. The local minimum ($l = \infty$) is for a macroscopic drop and the global minimum ($l = l_1$) is for a thin film.

used the model to calculate line tension for dodecane on a methylated solid surface, obtaining a value of -2×10^{-12} J/m.

We now turn to direct, computationally intensive estimates of line tension.

13.2.3 MOLECULAR DYNAMICS

Computational estimates of line tension from first principles have been carried out using the method of molecular dynamics. In a molecular dynamics simulation, a large number of virtual particles, representing molecules of liquid, solid, and/or vapor, are confined in a three-dimensional virtual box. Interaction potentials between pairs of particles are defined and the system, beginning from some initial state, is allowed to relax to equilibrium by repeated application of Newton's Second Law in time increments of order 1 fs. The temperature and number of particles are selected such that the equilibrium state includes a three-phase line, eventually allowing evaluation of line tension (see below). With the computational power currently available, typical studies involve $\sim 10^4$ particles in boxes with side lengths of a few nanometers.

The system evolves with the number of particles, the volume, and the temperature held constant; that is, in the canonical (NVT) ensemble. This means that the Helmholtz free energy is minimized rather than the grand canonical free energy Ω. The final equilibrium is the same in either case; use of the Helmholtz potential only causes the search for equilibrium to include states with nonuniform chemical potential, even though none of these can ultimately be the equilibrium state (see chapter 1).

Early publications on line tension estimation via molecular dynamics were by Bresme and Quirke. They simulated a nanometer-sized solid particle at a liquid–vapor [35] and a liquid–liquid [36] interface. Because of uncertainty in measured contact angles, the authors chose not to calculate line tension from the modified Young equation (Equation 13.1), but to find it from the dependence of the system's free energy on the particle radius. Since only van der Waals forces were modeled, Bresme and Quirke interpreted their results for liquid argon, finding both positive and negative line tensions of order 10^{-12} or 10^{-11} J/m.

Werder et al. [37] simulated water drops on a graphite surface. Water–water interactions were modelled using point partial charges and water–graphite interactions using the Lennard-Jones potential. Positive line tension of order 10^{-10} J/m was found, with higher values in cases where the water–graphite interaction was strengthened.

Water drops were also studied by Hirvi and Pakkanen [38]. The drops were on simulated amorphous and crystalline polymer surfaces. In some cases the authors found no discernible line tension, and in other cases positive line tension up to order 10^{-11} J/m. Finally, Schneemilch and Quirke [39] simulated a similar system of water drops on hydrophobic PDMS (polydimethylsiloxane) surfaces. These authors also simulated partially oxidized PDMS surfaces, in which surface CH_3 groups were replaced by OH groups. Nonzero line tension was found only on the oxidized surfaces; it was of order $+ 10^{-11}$ J/m.

The results from this sample of molecular dynamics studies are included in Table 13.1, with the other theoretical results discussed above. The table shows that according to all three theoretical approaches, the magnitude of line tension is between 10^{-10} and 10^{-12} J/m. Dispersive systems (with weak intermolecular forces) as well as aqueous systems (with strong intermolecular forces) have been studied. In all cases, the magnitude of line tension is of the order of a surface tension multiplied by an interface thickness. This is unsurprising since line tension is modeled by considering the same intermolecular forces that give rise to interfacial tensions.

13.3 MEASUREMENT

Unlike liquid–fluid surface tension, line tension is not amenable to direct measurement as a force. Instead, it is usually measured from the drop size dependence of contact angles described by the modified Young equation (Equation 13.1). Various other methods have been used as well, although most of them amount to an indirect measurement of the contact angle, and are therefore also based on Equations 13.1 and 13.2.

Measurements can be divided into those in which a solid phase is present and those in which one is not. If a solid phase is present, a wider variety of experimental methods are feasible, but possible complications arise due to solid surface roughness and heterogeneity.

Both with and without solid phases, measurements have been made at different length scales, using different, suitable experimental techniques. The techniques used to observe contact angles can generally be grouped into three categories: (i) optical microscopy at the millimeter scale; (ii) interferometry at the scale of tens of micrometers; and (iii) atomic force microscopy (AFM) at the submicrometer scale. With optical microscopy, drops are usually viewed from the side and a single cross-section of the liquid–vapor interface is obtained. With the other two methods, drops are viewed from above, and the entire three-dimensional liquid–vapor interface can be mapped.

In the following sections, experimental results from the literature will be reviewed, starting with those that do not involve a solid surface (Section 13.3.1), followed by those that do (Section 13.3.2). Within each section, the results will be arranged by length scale. Section 13.3.3 will discuss a distinct method using patterned solid surfaces that involves intricate mathematical analysis, but gives a comparatively sensitive measurement of line tension. The interpretation of these data as a whole, and their connection to the theory of line tension discussed above, will be the subject of Section 13.4.

13.3.1 LIQUID–LIQUID SYSTEMS

Early measurements of line tension with no solid phase present included, among others, an analysis of the shapes of 3-singlets (merged droplets) in an emulsion [40], and a series of measurements by two groups using Newton black films [41–43]. These difficult measurements were plagued with uncertainty, and we will therefore focus on more recent results.

Recent line tension measurements with no solid phase have all used liquid lenses, that is, drops of one liquid floating on an immiscible second liquid. In this setting, Young's equation is replaced by the more general Neumann triangle relation, and hence the modified Young equation becomes a quadrilateral relation [44], depicted in Figure 13.3 (see also Chapter 2).

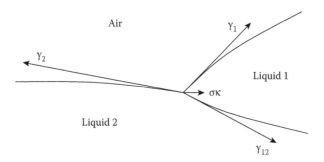

FIGURE 13.3 Quadrilateral relation, or vector balance, for a liquid lens. The line tension vector is drawn positive and greatly magnified compared to its typical length.

Lenses with diameters of a few millimeters have been studied in a series of experiments. It is possible to show [45] that with knowledge of all three interfacial tensions, plus two angles, the force balance in Figure 13.3 can be solved, allowing the length of the $\sigma\kappa$ vector to be found. Thus, a value of σ can be calculated for each lens, whereas in drop size dependence measurements using the modified Young equation (Equation 13.2), several drops of different diameters collectively produce one value of σ. Observing alkane lenses on water [45,46], Chen et al. measured line tensions of the order -10^{-6} J/m.

This experiment was later repeated for just dodecane lenses with updated technique [47]. Improvements included interfacial tension measurements with well-deformed drops (Chapter 3), and contact angle measurements using TIFA-AI (Chapter 5) rather than polynomial fitting. Negative line tension between 10^{-7} and 10^{-6} J/m was found. The results were statistically different from zero line tension based on the scatter in measured contact angles; as well, repeat experiments with purified dodecane showed that errors in interfacial tension values could not be responsible for the nonzero line tension result. Finally, a size dependence of the line tension was detected, with σ becoming more negative as lens diameter increased.

Three experiments have been reported that used lenses of ~0.1 mm diameter imaged from above via interferometry. Dussaud and Vignes-Adler measured line tension for octane lenses on saline [48]. Using the size dependence of contact angles, they obtained positive values of order 10^{-10} J/m. The octane lenses were near wetting and were created by the break-up of one much larger lens. The influence of the salt on the spreading of octane complicated the system and was not completely understood. It is unclear how the salt may have affected the value of the line tension.

In the second experiment using interferometry, Aveyard et al. [49] studied lenses of dodecane on water. Dodecanol was added to the dodecane in order to reduce the lens dihedral angle to a value allowing interferometric imaging. Line tension was found to be positive, of order 10^{-11} J/m, and independent of the dodecanol concentration.

In the third experiment, Takata et al. measured line tension using lenses of hexadecane on an aqueous surfactant solution [50]. The line tension was found as a function of the surfactant concentration, and was of order 10^{-11} J/m. Above a certain concentration, the line tension changed sign from positive to negative. This was believed to be associated with a transition of the hexadecane from partial to pseudo-partial wetting (i.e., from a lens to a lens in contact with a film).

Optical microscopy has also been used to measure line tension in liquid–liquid systems at a relatively small scale. Wallace and Schürch published two studies in which line tension was measured for drops of dibutyl phthalate on a liquid fluoro-carbon surface. The surrounding medium was not air, but saline, and a surfactant was spread on the interfaces between the saline and each of the other liquids. Due to the high density of the lower (fluorocarbon) phase, it remained nearly flat and Wallace and Schürch used the modified Young equation (Equation 13.1) to fit their data. In their first study, the authors found line tension of about $+10^{-8}$ J/m for drop sizes of order 0.1 mm [51]. In their second study, with drops about 10 times smaller, the line tension was smaller but still of the same sign and order of magnitude [52].

Finally, in the smallest scale liquid lens experiment, Stöckelhuber et al. used interferometry to measure line tension from the shape of a single water lens on a surface of dodecane [53]. They found $\sigma = +5.4 \times 10^{-9}$ J/m for a lens with a three-phase line diameter of 2.4 μm.

The line tension measurements with liquid lenses are summarized in Table 13.2. Unfortunately, four out of the seven experiments featured the presence of a surfactant or solute, possibly hindering meaningful comparison of the results. The reason for the differing signs of line tension is unclear (see Section 13.4.2). Regarding the magnitude of line tension, the majority of measurements produced results well out of the range of theoretical predictions.

The main purpose of studying liquid–liquid systems as opposed to liquid–solid systems is the elimination of the influence of surface imperfections. Indeed, contact angle hysteresis, commonly used to measure surface quality, was reported for some liquid lenses and was typically ≤ 1° [47,51], several times lower than in carefully prepared liquid–solid systems. Yet, as seen below, the general pattern of results in liquid–solid systems is identical to that in liquid–liquid systems.

The size dependence of line tension observed by David et al. [47] may empirically reconcile the discrepancy between the large magnitudes measured by Chen et al. and the smaller magnitudes measured by other researchers. This size dependence

TABLE 13.2
Measurements of Line Tension Using Liquid Lenses

Authors	Year(s)	Materials	Diameter (m)	σ (J/m)	Ref.
Chen et al.	1997–1998	alkanes/water/air	10^{-3}	-10^{-6}	45,46
David et al.	2009	dodecane/water/air	10^{-3}	-10^{-7}	47
Dussaud et al.	1997	octane/saline/air	10^{-4}	$+10^{-10}$	48
Aveyard et al.	1999	dodecane/water/air	10^{-4}	$+10^{-11}$	49
Takata et al.	2005	hexadecane/aqueous surfactant solution/air	10^{-4}	$±10^{-11}$	50
Wallace and Schürch	1988–1990	dibutyl phthalate/ fluorocarbon/saline	10^{-5}–10^{-4}	$+10^{-8}$	51,52
Stöckelhuber et al.	1999	water/dodecane/air	10^{-6}	$+10^{-8}$	53

is unexpected for a thermodynamic line tension and will be further discussed in Section 13.4. Finally, it is notable that the three smallest line tension magnitudes in Table 13.2 arose from the three experiments that examined systems near wetting. A decrease in line tension approaching wetting has also been observed in some liquid–solid systems, as seen below.

13.3.2 LIQUID–SOLID SYSTEMS

To date, a few dozen papers have been published reporting measurements of drop size dependence of contact angles on solids, with inferred values of line tension from the modified Young equation (Equation 13.2). In some cases, rough or heterogeneous substrates were used; these results will be discussed in Section 13.4.1. In this section, we will focus on recent results on smooth, homogeneous substrates, omitting older data (e.g., [15]) and some newer data for which solid surface quality was questionable. At the end of this section, solid surface methods other than drop size dependence will be discussed.

13.3.2.1 Millimeter Scale

At the millimeter scale, extensive data have been collected for the drop size dependence of contact angles measured by Axisymmetric Drop Shape Analysis (ADSA, Chapter 6). The first study, in 1987, was by Gaydos and Neumann [54], who used solid substrates of Teflon FEP. The surfaces were carefully prepared, with contact angle hysteresis of 2–6° for water and hexadecane. For a series of alkane liquids, as the drop diameters increased from about 2 to about 8 mm, the contact angle was observed to decrease by about 5°, resulting in calculated line tensions of approximately $+3 \times 10^{-6}$ J/m.

In similar experiments using different liquids and solids, and improved experimental technique, Li and Neumann [55] and Duncan et al. [56] found similar values for line tension. Amirfazli and coworkers then continued this line of research with measurements on self-assembled monolayers (SAMs) that presented methyl (CH_3) groups to the liquids. The measured values of line tension remained positive and of order 10^{-6} J/m [57].

A correlation between line tension and solid–liquid interfacial tension was noted by Duncan et al. [56]. Since in a liquid–solid–vapor capillary system there are two thermodynamic degrees of freedom (see Chapter 9), only two among the set $\{\gamma_{lv}, \gamma_{sv}, \gamma_{sl}, \sigma\}$ should be independent. Amirfazli and coworkers tested the empirical relationship between σ and γ_{sl} by making measurements in one system with very high interfacial tension and in another with low interfacial tension (i.e., a system near wetting). For the high energy system, the drop size dependence of contact angles of liquid tin on silica was measured at 900°C. An unmistakable contact angle variation of 20° was observed, corresponding to an extremely large line tension of $+1.55 \times 10^{-4}$ J/m [58]. For the low energy system, Amirfazli et al. measured contact angles of alkanes near wetting on methyl-terminated SAMs by imaging from above (ADSA-D, Chapter 6). In these systems, the line tension was as usual positive but smaller than the values that had been obtained farther from wetting, falling between 10^{-7} and 10^{-6} J/m [59]. Thus, the results of both experiments bore out the empirical correlation between σ and γ_{sl}.

Other researchers have used different approaches to measure line tension at the millimetre scale. Gu et al. [60] dipped fluorocarbon-coated solid cones into baths of liquid alkanes and measured the meniscus contact angles. As each cone became more submerged, the contact line radius increased (its curvature decreased), producing a varying contact angle according to Equation 13.1. In fact, in this geometry, line tension does not act in the same direction as solid–liquid interfacial tension, requiring a cosine correction in Equation 13.1. The remarkable consistency of the results of Gu et al. with those of Li and Neumann [55] and Duncan et al. [56] using very similar materials served to confirm the direction in which line tension acts.

In an experiment with the inverted geometry, Jensen and Li [61] allowed liquids to rise in the interiors of hollow conical solids. After correction of their results for gravity [62], good agreement was obtained with previous data [55,56,60], with line tension of order $+10^{-6}$ J/m.

Finally, in a liquid–solid–liquid system, Gu [63] used ADSA to measure contact angles of oil drops on a fluorocarbon-coated solid substrate immersed in water. He found line tension of $+8.2 \times 10^{-7}$ J/m.

The data reviewed in this section and those following are gathered in Table 13.3. Based on the results discussed so far, at the millimeter scale in systems with smooth, homogeneous solid substrates, and single component liquids with moderate interfacial tensions, line tension is large and positive with order of magnitude 10^{-6} J/m.

However, two exceptions, in which line tension in such systems was measured to be significantly smaller, have been reported. Line tensions of order $+ 10^{-7}$ J/m have recently been measured for a variety of liquids on films of Teflon AF 1600 and EGC-1700, two highly inert fluoropolymers [64]. The second exception is from earlier work by Drelich and will be covered in the following section.

13.3.2.2 Submillimeter Scale

In the 1990s, Drelich and coworkers including Whitesides and Good published a series of papers reporting line tension measurements principally at the 0.1 mm scale. Measuring the contact angles of air bubbles in water on polyethylene and oil drops in water on quartz [65], Drelich et al. found a nonlinear dependence of $\cos\theta$ on $1/r$; that is, a dependence not conforming to the modified Young equation (Equation 13.2). By fitting lines to sections of their data, Drelich and Miller calculated line tension of order 10^{-6} J/m at the millimeter level and no measurable line tension at the submillimeter level. They assigned negative signs to values for which the line tension vector pointed away from the aqueous phase; however, as the line tension pointed into the bubbles or oil drops it acted to decrease the length of the three-phase line and was therefore positive.

In subsequent research, Drelich et al. found similar nonlinear results for air bubbles in water or buffer on SAMs presenting methyl and carboxyl (COOH) groups [66]. However, in other experiments on like systems, Drelich et al. found no measurable line tension at any scale [67–69], and concluded that this was due to better surface quality [69].

Much of the data of Drelich et al. should be viewed with caution because in their dynamic bubble technique, the principal one used, intermediate contact angles

TABLE 13.3
Measurements of Line Tension from Size (or Curvature) Dependence of Contact Angles on Smooth, Homogeneous Solid Surfaces

Authors	Year(s)	Materials	Diameter (m)	σ (J/m)	Ref.
Gaydos and Neumann	1987	alkanes/Teflon/air	10^{-3}	$+10^{-6}$	54
Li and Neumann	1990	various/various/air	10^{-3}	$+10^{-6}$	55
Duncan et al.	1995	various/FC-721/air	10^{-3}	$+10^{-6}$	56
Amirfazli et al.	1998	various/CH_3/air	10^{-3}	$+10^{-6}$	57
Amirfazli et al.	1998	tin/silica/vacuum	10^{-3}	$+10^{-4}$	58
Amirfazli et al.	2003	alkanes/SAMs/air	10^{-3}	$+10^{-7}$	59
Gu et al.	1996	alkanes/FC-725/air	10^{-3}	$+10^{-6}$	60
Jensen et al.	1999–2003	alkanes/FC-721/air	10^{-3}	$+10^{-6}$	61,62
Gu	2001	silicone oil/FC-725/ water	10^{-3}	$+10^{-6}$	63
David et al.	2009	various/Teflon/air	10^{-3}	$+10^{-7}$	64
Drelich et al.	1993–1994	various	10^{-3}	$+10^{-6}$	65,66
			10^{-4}	$\leq10^{-7}$	
Drelich et al.	1994–1996	various	10^{-4}–10^{-3}	$\leq10^{-8}$	67,69
Sundberg et al.	2007	water/various/air	10^{-5}	$\pm10^{-8}$	70
Stöckelhuber et al.	1999	water/mica/air	10^{-5}	$+10^{-8}$	53
Wang et al.	2001	various/CH_3/air	10^{-5}	$\pm10^{-10}$	71
Aronov et al.	2007	water/various/vacuum	10^{-6}	$\pm10^{-9}$	72
Pompe and Herminghaus	2000–2002	various/various/air	10^{-6}	$\pm10^{-10}$	73,74
Mugele et al.	2002	hexaethylene glycol/ phenyl/air	10^{-3} 10^{-5}	$+10^{-6}$ $\leq10^{-10}$	75
Checco et al.	2003	alkanes/CH_3/air	10^{-6}	-10^{-10}	33
Seemann et al.	2001	polystyrene/various/air	10^{-6}	$\leq10^{-11}$	76
Rieutord and Salmeron	1998	sulphuric acid/mica/air	10^{-6}	-10^{-10}	77
Xu and Salmeron	1998	glycerol/mica/air	10^{-6}–10^{-5}	-10^{-11}	78
Yang et al.	2003	CO_2/CH_3/water	10^{-6}	-10^{-10}	79
Zhang and Ducker	2008	decane/CH_3/aqueous solution	10^{-7}	$+10^{-10}$	81

Note: Cases where no line tension was measurable are listed as $\leq x$, where x was the approximate resolution. The FC coatings were formerly products of 3M.

(between advancing and receding) were measured, while it is the advancing angle that should be used in Equation 13.2 (see Chapter 7).

For observation of drops with diameters in the tens of micrometers, other optical techniques have been employed. Sundberg et al. [70] used both interferometry and confocal microscopy to image water drops of this size. Their results conflicted, with

σ of order 10^{-7} J/m using the first method, and 10^{-8} J/m using the second. Volatile liquids at these volumes evaporate quickly, and it is unlikely that advancing angles were measured. The authors also found no measurable line tension at the millimeter scale using conventional imaging, but contact angle resolution was not stated, and was likely poor due to the use of ellipses to fit Laplacian drop shapes.

Stöckelhuber et al. [53] imaged similarly sized water drops on mica interferometrically, and measured positive line tension of 7.6×10^{-9} J/m. There appears to have been virtually no contact angle hysteresis.

Wang et al. [71] used the same imaging method to examine drops of octane and octene on methyl-terminated SAMs. They measured line tension as a function of temperature just below the wetting transition temperature of each liquid. In both cases, they found line tension of order 10^{-10} J/m, changing sign from negative to positive as wetting was approached, in qualitative agreement with theory (see Sections 13.2.1 and 13.2.2).

Experimental results at the submillimeter level for the magnitude of line tension are thus significantly lower than those at the millimeter level, but still often much higher than theoretical predictions.

13.3.2.3 Micrometer Scale

Different techniques are necessary to observe drops of a size comparable to the wavelength of visible light. Aronov et al. [72] studied drops of water on SiO_2 using environmental scanning electron microscopy. This technique allowed three-dimensional imaging, and the authors measured contact angles as a function of drop radius, and also as a function of contact line curvature for nonaxisymmetric drops. They found σ of order 10^{-9} J/m, although the relationship between $\cos\theta$ and $1/r$ was slightly nonlinear. The line tension was negative on a hydrophobized surface and positive on a hydrophilic surface, matching the sign change observed by Wang et al. [71] (and predicted by theory) on approach to wetting.

The remainder of studies at the micrometer level used AFM. AFM-based measurement of line tension was pioneered by Pompe and coworkers. Pompe and Herminghaus [73] measured three-dimensional drop shapes on microfabricated surfaces with stripes of alternating wettability. They calculated line tension from the relationship between the contact angle and the local curvature of the contact line (Equation 13.1), and also by inserting the measured microscopic drop profile into the interface displacement model (Section 13.2.2). Results from the two methods coincided, with line tensions of order 10^{-10} J/m. In another publication, Pompe [74] used the second method and found σ of the same order, changing sign from negative to positive on approach to wetting.

From the same group, Mugele et al. [75] measured line tension at both the millimeter and micrometer scales. While they measured $\sigma = +9 \times 10^{-7}$ J/m for millimeter-sized drops, the authors disavowed these results, citing uncertainties in the optical setup and fitting procedure. Their microscopic measurements produced only an upper bound for $|\sigma|$ of 10^{-10} J/m.

Checco et al. used noncontact AFM to measure line tension from the drop size dependence of contact angles of alkanes on methyl-terminated SAMs [33]. They found a nonlinear relationship between $\cos\theta$ and $1/r$ best fit by a negative line tension of order 10^{-10} J/m. They viewed this result as arising from surface heterogeneity.

Seemann et al. [76] formed micrometer-sized drops of polystyrene on coated silicon wafers and, using AFM, measured negative line tension of order 10^{-11} J/m. However, the contact angles were receding and the nonzero value for σ appears to rely mainly on a single data point.

Rieutord and Salmeron [77] used AFM to measure the contact angles of drops of sulphuric acid on mica. The authors employed an unconventional definition of the contact angle (measuring at the inflection point of the profile) and likely measured receding angles. Their data correspond to a negative line tension of order 10^{-10} J/m. In a second paper, Xu and Salmeron [78] studied microscopic glycerol drops of widely ranging sizes, with corresponding negative line tension about three times smaller. They modelled their results with an exponential interface potential.

Yang et al. [79] studied CO_2 bubbles formed at solid substrates in water, and measured $\sigma = -3 \times 10^{-10}$ J/m. However, Zhang et al. [80] argued that contact angles could not be measured with sufficient accuracy in such systems to calculate line tension. Both groups found much higher contact angles for the microscopic bubbles (measured through the liquid phase) than for macroscopic drops. In a later paper, Zhang and Ducker [81] used tapping-mode AFM to image decane drops on a methyl-terminated SAM immersed in diluted ethanol, and found $\sigma = +8 \times 10^{-11}$ J/m. However, the drops were deformed by the AFM tip and receding contact angles were likely measured. In this system, macroscopic and microscopic contact angles were similar.

In summary, at the micrometer scale, there is broad consensus in a number of drop size dependence studies using AFM that line tension, whether positive or negative, is of order 10^{-10} J/m. The results are reasonably close to theoretical predictions, but generally higher. In some cases, confirmation of specific theoretical predictions was possible: the sign change of line tension near wetting, and the shape of the liquid–vapor profile in the transition region. Thus, AFM studies have been much more successful at validating the theory of line tension than studies at larger length scales.

The disagreement between experiments at the macroscale and the microscale, as seen in Table 13.3, increases as the systems become larger, reaching, incredibly, a factor of 10^4 at the millimeter level.

13.3.2.4 Alternative Methods

Some line tension measurements that are based on alternative methods to drop size dependence of contact angles have been attempted.

Nguyen, Stechemesser, and coworkers measured line tension from the dynamics of expanding three-phase lines. They found σ of order $+10^{-6}$ J/m for submillimeter sized methylated glass spheres penetrating a water–air interface [82], and for expanding air bubbles on methylated quartz substrates [83]. The motion of the contact line was modeled by a molecular-kinetic theory. In the first study, the axisymmetry of the three-phase line was not verified. In the second, the surface tension of the water used was lower than the accepted value. Consequently, adsorption of impurities to the solid–gas interface during three-phase line expansion may have been a problem [83].

A small particle on a solid surface may be picked up (engulfed) by a liquid pendant drop that is lowered into contact with it. A positive line tension would oppose this, the more so for smaller particles. This idea was the basis of a measurement of line tension by Aveyard et al. [84]. The authors picked up fluorinated, submillimeter diameter particles with drops of dodecane and concluded that line tension was at most 10^{-7} J/m.

Vinke et al. [85] analyzed the adhesion of layers of particles to gas bubbles using a complicated model. In conjunction with experimental measurements, they calculated the dependence of the contact angles on the radii of the particles, obtaining very scattered positive line tensions of orders 10^{-12}, 10^{-11}, and 10^{-9} J/m for different particle coatings. The three-phase lines had diameters in the micrometer range.

Some researchers have attempted to measure line tension from heterogeneous nucleation data. Line tension strongly affects the condensation of extremely small liquid drops from supersaturated atmospheres. Alexandrov et al. [86] interpreted discrepancy between their data and classical nucleation theory in terms of a negative line tension of order 10^{-10} J/m. (The system they studied was actually liquid–liquid–vapor but is included here for convenience.) In liquid–solid–vapor systems, Hienola et al. [87] also found negative line tension of order 10^{-10} J/m.

Finally, in the most unusual approach, Teschke and de Souza [88] measured the force experienced by AFM tips on approach to millimeter-sized air bubbles in water on rough Teflon surfaces. They calculated the electric energy at the water–air interface and found a dependence on the solid surface roughness. They attributed this dependence to the effect of a positive three-phase line tension of order 10^{-8} J/m acting in contact line corrugations of ~10 nm radius of curvature.

In general, the alternative methods for line tension measurement are less direct than the drop size dependence of contact angles method and involve additional, sometimes complex theory, with attendant assumptions. While there is therefore more uncertainty associated with the results, the alternative methods are important because of the controversy surrounding the drop size dependence method. The results of the alternative methods are summarized in Table 13.4. They are relatively similar to the results reported for drop size dependence, with the largest magnitudes of σ found in the largest systems.

13.3.3 STRIPED SURFACES

A different strategy for measuring line tension, using striped heterogeneous surfaces, has been proposed. The measurement is made in a nonaxisymmetric system consisting of a liquid in contact with a chemically heterogeneous wall. Ideally, the surface of the solid wall is smooth, rigid, planar, and composed of two different materials that do not chemically react with the liquid. The two materials are arranged as alternating stripes (Figure 13.4). The liquid has a different contact angle on each material, resulting in a wavy, periodic contact line, as it tries to advance further on the more wetting stripes.

The shape of the contact line (mainly its amplitude) depends on the value of solid–liquid–vapor line tension. Relative to zero line tension, positive line tension reduces the amplitude of the waves due to the added energetic cost of forming the contact line. In the extreme case, infinite line tension would minimize the length of the contact line by flattening it to a straight line. Hence, it is in principle possible to determine the operative value of line tension in such a system by comparing the

TABLE 13.4
Measurements of Line Tension Using Alternative Methods

Authors	Year(s)	Method	Diameter (m)	σ (J/m)	Ref.
Gaydos	1992	striped surface	10^{-3}	$+10^{-6}$	94
Nguyen et al.	1997–1998	contact line dynamics	10^{-4}	$+10^{-6}$	82,83
Aveyard et al.	1996	particle pick-up	10^{-4}	$\leq 10^{-7}$	84
Vinke et al.	1991	particle-to-bubble adhesion	10^{-6}	$+10^{-10}$	85
Alexandrov et al.	1993	heterogeneous nucleation (liquid)	10^{-8}	-10^{-10}	86
Hienola et al.	2007	heterogeneous nucleation (solid)	10^{-8}	-10^{-10}	87
Teschke and de Souza	2007	electric field at interface	10^{-8}	$+10^{-8}$	88

Note: Cases where no line tension was measurable are listed as $\leq x$, where x was the approximate resolution.

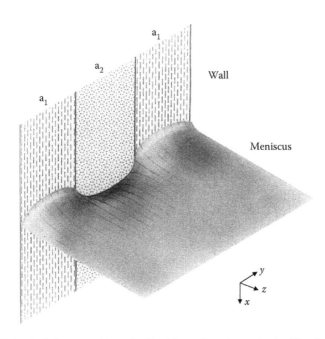

FIGURE 13.4 Artist's conception of a liquid meniscus in contact with a heterogeneous striped wall, and the coordinate system used in the analysis. (Reprinted from Boruvka, L., and Neumann, A. W., *Journal of Colloid and Interface Science*, 65, 315, 1978. With permission from Elsevier.)

experimental contact line with a series of theoretical contact lines, each of which is calculated assuming a different value of line tension.

Generating the theoretical contact lines for this approach involves a significant mathematical and computational effort [89–92]. Some preliminary considerations follow.

Two assumptions about the line tension are made. Firstly, in order to make the problem mathematically tractable, the line tensions on the two materials are assumed to be identical. The agreement (or lack thereof) between theoretical and experimental contact lines is expected to verify whether this assumption is good. Secondly, line tension is assumed positive. Experimentally, a negative line tension would cause the amplitude of the three-phase wave-like line to be in excess of that for the zero line tension case.

A further assumption is made about gravity. If the maximum stripe width is denoted by a, it is assumed that a is much less than the capillary length, defined as $(\gamma/\rho g)^{1/2}$, where γ is the liquid surface tension, ρ its density, and g is gravitational acceleration. This allows gravity to be neglected; the liquid–vapor interface at a distance of several a away from the wall may then be approximated by a planar surface called the Cassie plane. Nearer the wall, the liquid–vapor interface is described by the Laplace equation that, with no gravity, is

$$\frac{\Delta P}{\gamma} = \frac{1}{R_1} + \frac{1}{R_2} = 0, \tag{13.7}$$

where ΔP is the pressure difference across the liquid–vapor interface, and R_1 and R_2 are its principal radii of curvature. A surface for which the sum of the principal curvatures equals zero everywhere is known as a minimal surface. If the liquid–vapor interface is represented by the function $z(x, y)$, using the coordinate system shown in Figure 13.4, then Equation 13.7 is

$$\left(1+z_y^2\right) z_{xx} - 2z_x z_y z_{xy} + \left(1+z_x^2\right) z_{yy} = 0, \tag{13.8}$$

where subscripts denote partial derivatives. Whereas Equation 13.8 describes the liquid–vapor interface away from the wall, on the contact line the modified Young equation (Equation 13.2) holds.

The next two sections review the mathematics that have been used to generate the shape of the contact line for the cases of zero, small, and large line tension. For the zero line tension case, conformal mapping was used to integrate the Laplace equation analytically and obtain the exact theoretical shape of the liquid surface and the contact line. The small line tension case was solved numerically as a perturbation away from the zero line tension case. The large line tension case was solved numerically as a perturbation away from the infinite line tension case (flat contact line). Following the theory, the limited available experimental data will be reviewed.

13.3.3.1   Zero Line Tension

The method of characteristics was used to express the minimal surface equation (Equation 13.8) in the standard canonical form [89]

$$\frac{\partial^2 z}{\partial \alpha^2} + \frac{\partial^2 z}{\partial \beta^2} = 0, \tag{13.9}$$

using the coordinate transformation

$$\frac{\partial z}{\partial x} = \frac{\cos\alpha}{\sinh\beta}, \quad \frac{\partial z}{\partial y} = \frac{\sin\alpha}{\sinh\beta}. \tag{13.10}$$

The two new variables, α and β, which replace x and y, have a straightforward physical interpretation. A plane that is parallel to the wall and in front of it cuts the liquid surface in a periodic curve. The amplitude of this curve increases as the cutting plane approaches the wall. A tangent line to the periodic curve, drawn in the cutting plane, makes an angle with the horizontal that, measured counter-clockwise, is α. Thus, we may describe α as the local turning angle of the liquid surface. It varies between $\pm\pi/2$. The other new variable β is related to the local contact angle θ as follows:

$$\beta = -\ln\left(\tan\frac{\theta}{2}\right). \tag{13.11}$$

θ is defined at points away from the wall as the angle (measured through the liquid) between the tangent plane to the liquid–vapor interface and the wall. β achieves a minimum value on the contact line of the low energy stripe (the one with higher contact angle), and a maximum value on the contact line of the high energy stripe (the one with lower contact angle).

Figure 13.5a shows the α-β domain. The boundary of the rectangle represents the contact line in these coordinates. When the line tension is zero, the contact angle on each stripe is constant (see Equation 13.2); thus, β must vary only along the four-phase contact line formed by the two solids and the liquid–vapor meniscus (see Figure 13.5b).

Two consecutive mappings are applied in order to move the problem from the rectangular α-β domain where the solution is unknown to one of a standard number of domains in which a solution is easier to obtain. First, a Schwarz–Christoffel transformation (mapping polygons to half-planes or vice versa) maps the rectangular α-β domain into the complex upper ξ-half-plane. A subsequent bilinear transformation maps the upper ξ-half-plane onto the interior of a unit circle in the w-plane. The net result of the two mappings is that every point on the perimeter of the α-β rectangle is mapped to a corresponding point on the circumference of the circle, and every point in the interior of the rectangle is mapped into the interior of the circle. Once the solution is obtained in the w-plane the transformations can be undone via reverse mappings.

The solution in the α-β domain is shown in Figure 13.5a. The closed curves inside the rectangle correspond to points on the liquid surface away from the wall. The more interior curves correspond to greater distances from the wall, with the Cassie plane mapped onto the point at the center of the rectangle. Using the result in the α-β domain, Equation 13.10 was integrated numerically to return the solution

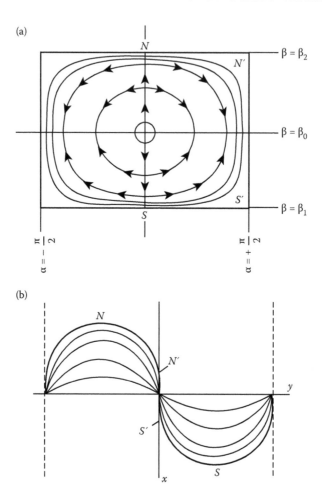

FIGURE 13.5 (a) Schematic of the position of the contact line after mapping to the α-β domain. The rectangular boundary is the three-phase contact line. The closed curves shown inside the rectangle correspond to points on the liquid–vapor interface at different fixed distances from the wall. (b) Schematic view of a liquid–vapor interface in contact with an equal width striped wall. The three-phase contact line is indicated by a heavy line labeled $SS'N'N$ while the thinner lines represent the liquid–vapor surface at various positions out from the wall ($z > 0$). (Reprinted from Boruvka, L., Gaydos, J., and Neumann, A. W., *Colloids and Surfaces*, 43, 307, 1990. With permission from Elsevier.)

to the original x-y domain (see Boruvka and Neumann [89] for details). Figure 13.5b shows the result in the x-y domain.

With the aid of Figures 13.5a and b the shapes of the liquid–vapor interface (at different distances away from the wall) in the α-β and x-y coordinates can be correlated. Beginning with the point S on the lower energy (higher contact angle) stripe where the turning angle α is zero, we proceed to the point S'. At S' the contact angle is still identical to that at S (as long as line tension is zero) so β remains unchanged, but α has changed from zero to $\pi/2$. Along the vertical section of the contact line that follows the solid–solid boundary, α remains fixed at $\pi/2$ while β changes from its lower constant value on one stripe to its higher constant value on the adjacent stripe. As the contact line leaves the solid–solid boundary at N', β now remains fixed while the turning angle α decreases in value from $\pi/2$ to zero at point N. Tracing the curve from N to smaller values of the coordinate y yields the remaining half of the rectangular boundary in the α-β domain.

It is notable that at points S' and N' of Figure 13.5b, the contact line experiences infinite curvature. Although the presence of these infinite curvature points does not violate either the classical Laplace or Young equations of capillarity, it does violate the assumption of moderate curvature upon which Gibbs' derivation of the theory of capillarity is based. To remove this difficulty, the effect of nonzero line tension was considered, beginning with line tension close to zero.

13.3.3.2 Nonzero Line Tension

In comparison with the zero line tension case, a positive line tension is expected to reduce the amplitude of the liquid surface waves, and shorten the portion of the contact line that follows the solid–solid boundary. Moreover, from Equation 13.2, the contact angle is no longer constant on each stripe, meaning that the contact line no longer follows a rectangular path in the α-β domain. The unknown shape of the contact line in the α-β domain prevents an analytical solution via conformal mapping.

Instead, a series of Fourier coefficients can be introduced to approximate the shape of the contact line numerically [90]. The mapping between the α-β domain and the unit circle in the w-plane is performed directly by a complex power series:

$$\alpha = \sum_{k=1}^{\infty} x_k \rho^k \sin(k\phi), \qquad \beta - \beta_0 = -\sum_{k=1}^{\infty} x_k \rho^k \cos(k\phi), \qquad (13.12)$$

where β_0 is the value of β for the Cassie plane, and (ρ, ϕ) are polar coordinates in the w-plane. The Fourier coefficients x_k need to be determined from three contact line boundary conditions, which in the w-plane lie along the perimeter of the unit circle (i.e., for $\rho = 1$). Two of these boundary conditions are the modified Young equation (Equation 13.2) for the contact line on each stripe, and the third is $\alpha = \pi/2$ on the four-phase boundary between the stripes. If the stripes have equal widths given by a, then it is found that the line tension σ enters the problem only in the form $\sigma/\gamma a$. More details are available elsewhere [90].

The solution for the contact line was obtained using the Newton-Raphson method with incremental loading. The initial set of Fourier coefficients, corresponding to the zero line tension case, was found by minimizing the difference between the numerical solution and the analytical solution (outlined above). These Fourier coefficients were then used as a springboard to other adjacent coefficients that represented contact line shapes with small, nonzero line tensions [90].

The case of large line tension was handled in a similar way [91]. In this case, the contact line crosses the solid–solid boundary at only one point. In the α-β domain, this means that the solution is completely detached from the rectangular boundary representing the zero line tension solution. The absence of the four-phase section of the contact line reduces the number of boundary conditions used for finding the Fourier coefficients from three to two. The solution is now found starting from the infinite line tension solution (flat contact line) in terms of small values of a parameter inversely proportional to the line tension σ. It was shown that for intermediate values of σ, the small and large line tension solutions were in close agreement [91].

Later work [92] addressed two technical problems still present in the numerical solutions for nonzero line tension: the program assumed that the contact angles on the respective stripes added to 180° (i.e., the Cassie plane was horizontal), and it did not converge when either contact angle was below 11°. The first issue was fixed, and the second circumvented by extrapolating solutions from those for contact angles above 11°. An example solution is shown in Figure 13.6.

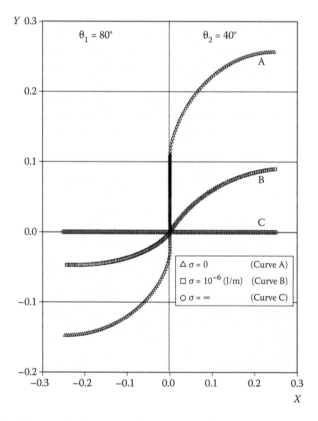

FIGURE 13.6 Computational solutions for the contact line on a striped wall with contact angles of 80° and 40°, for three different values of line tension. The stripes have equal widths, and both axes are normalized by twice this width. (Reprinted from Hoorfar, M., Amirfazli, A., Gaydos, J. A., and Neumann, A. W., *Advances in Colloid and Interface Science*, 114–115, 103, 2005. With permission from Elsevier.)

The solutions for contact lines on striped surfaces have been used to argue that positive line tension may reduce the magnitude of contact angle hysteresis on real surfaces, by minimizing corrugation of the contact line as it crosses high energy impurities on a low energy surface [91].

The problem of the shape of a contact line on a striped, heterogeneous surface has also been tackled using the finite element method [93]. Results were obtained for finite-sized drops, including the cases of gradual transition of wettability between the stripes, and varying line tension between the stripes. The qualitative features of the results were similar to those using the mapping method, but no quantitative comparison has been made.

Experimental measurements on striped surfaces have lagged behind theoretical and computational progress. One experiment has been performed with ethylene glycol on a surface of alternating stripes of silicon oxide and a CF_3-terminated SAM [94]. Positive line tension of order 10^{-6} J/m was found with stripes of order 1 mm width. More recent, preliminary observations have indicated a much smaller line tension for hexadecane on a heterogeneous CH_3/COOH-terminated SAM with 12.5 μm wide stripes [95].

13.4 DISCUSSION

Theory and experiment for line tension are summarized in Tables 13.1 through 13.4. Beyond the remarkably wide variation of experimental results, two observations may be made from these tables: (i) measured line tensions appear to be more or less proportional to the length of the contact line; and (ii) as noted earlier, it is the line tensions measured at the micrometer level that are generally in closer agreement with theory.

The first observation is made more explicit in Figure 13.7, which shows line tension data from the literature of 1987–2007. The dependence of line tension on

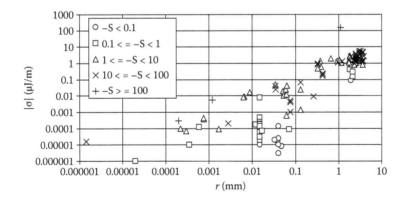

FIGURE 13.7 Line tension data from the literature, plotted against the typical size of the drops used in the measurement. Each symbol represents a single measurement, with sometimes several per paper. (Modified from David R., and Neumann, A. W., *Langmuir*, 23, 11999, 2007, American Chemical Society. With permission.)

the spreading coefficient $S = \gamma_{lv}(\cos\theta - 1)$ is also noticeable. Earlier, a correlation between σ and γ_{sl} was found (Section 13.3.2.1); both γ_{sl} and S vanish on approach to wetting.

Considering measurements of σ phenomenologically as $d(\cos\theta)/d(1/r)$, the trend in Figure 13.7 implies that $\cos\theta$ does not vary linearly with $1/r$ as in the modified Young equation (Equation 13.2), but rather

$$\cos\theta = -\frac{c(S)}{\gamma_{lv}} \ln\frac{r_0}{r}, \tag{13.13}$$

where c is found empirically to be roughly proportional to S (but independent of r), and r_0 depends on the liquid and solid [96]. Equation 13.13 predicts a nonlinear relationship between $\cos\theta$ and $1/r$ in individual experiments. This has in fact been observed by researchers who have measured contact angles over a wide enough range of drop radii in a single experiment [47,63,65,67,77]. Values of c (for moderate S) calculated from individual nonlinear plots of $\cos\theta$ versus $1/r$ were consistent with the value calculated from a plot similar to Figure 13.7 [96].

Figure 13.7, and Equation 13.13 that describes it, contradict the intuitive expectation that the energy of a contact line scales with the length of the line, producing a line tension independent of drop size. Using the interface displacement model, Indekeu [31] has explicitly shown that this expectation is true for intermolecular potentials decaying faster than d^{-5}, where d is intermolecular separation. Nonretarded van der Waals forces, which are the most common type of intermolecular force and that include dispersive forces, decay as d^{-6}. In many systems studied, these are assumed to be the only forces active. Therefore, the length dependence of measured line tensions contradicts not just intuition but also theory.

Assuming that line tension is in fact independent of drop size, according to Equation 13.2 the large values (~10^{-6} J/m) measured at the millimeter scale would preclude the formation of three-phase contact lines with diameters below about 10 μm. This, of course, is not the case. Thus, a line tension of 10^{-6} J/m cannot be a constant energy per unit length of line phase.

In sum, the quantity measured as line tension bears little resemblance to the theoretical concept of line tension. It has therefore been repeatedly suggested in the literature that the measurements above the micrometer level are flawed in some way.

Nevertheless, the supposition of experimental error is also not supported by Figure 13.7, in which a relatively consistent pattern is seen to emerge from a variety of experimental methods at different length scales, with different geometries and materials, performed in different labs. This suggests that the quantity measured is a real physical phenomenon. Still, it is clear that the physical basis of this phenomenon, even if it is associated with the three-phase line, cannot be the same intermolecular forces that are considered in theoretical studies of line tension.

One commonality of nearly all measurements of line tension is that they have been based on Equations 13.1 and 13.2. Thus, any deviation from Young's equation that can be appropriately linearized may be interpreted as an experimental measurement of line tension. Doubts about this interpretation, especially for large line tension measurements on solid surfaces, have led to these data being referred to as pseudo-line tension in the literature [16,65]. Pseudo-line tension has been attributed to solid surface heterogeneity (as opposed to excess energy at the three-phase line, the source of true line tension). Other proposed explanations for large line tension measurements, such as gravity, viscosity, solid surface deformation, and vapor adsorption, have been rejected [97,98].

In the next section, the evidence for solid surface heterogeneity as a cause of the drop size dependence of contact angles will be examined. Following that, in Section 13.4.2 the separate issue of the sign of line tension will be briefly discussed, and finally in Section 13.4.3 some conclusions will be drawn.

13.4.1 SOLID SURFACE HETEROGENEITY

To begin, this section will focus on experiments on solid surfaces at the millimeter scale, which have produced the largest magnitudes for line tension. Good and Koo [16] measured large, negative values of line tension on solid surfaces and proposed that they arose from corrugation of the three-phase line. They reasoned qualitatively that the shape of the liquid–vapor interface would be more affected by corrugations for small drops, resulting in smaller measured contact angles and a negative pseudo-line tension.

Simple mathematical models of drops with corrugated contact lines, based on the modified Young equation (Equation 13.1), were developed by Li et al. [99] and by Drelich and Miller [97]. Similar considerations apply whether the corrugation is caused by roughness or chemical heterogeneity. Li et al. calculated that a large drop size dependence of contact angles could be caused by heterogeneity. However, in later work, Amirfazli et al. [98] used the same model to make the opposite argument. Their calculations were based on model parameter values for high quality solid surfaces (SAMs on glass slides). Mugele et al. [75] also concluded that contact line corrugations on a high quality solid surface were too small to explain the large drop size dependence of contact angles that they observed.

On the experimental side, Drelich and coworkers manufactured homogeneous surfaces with low line tension, and showed that strong dependences of contact angles on drop size could be obtained for a variety of rough and heterogeneous surfaces, including partially methylated quartz [67], surfaces roughened with polishing paper [68], and partial monolayers of oleate [100]. Corrugated contact lines were observed by Drelich and coworkers on their heterogeneous surfaces, indicating heterogeneity at the submillimeter level. They concluded that line tension was near zero on high quality solid surfaces.

In contrast, Amirfazli and coworkers measured line tension on both homogeneous (CH_3-terminated) and heterogeneous (CH_3/COOH-terminated) SAMs and found similar, large values for line tension on each [57,98]. These surfaces were believed to be heterogeneous only at the molecular level, with the contact line observed to be smooth at the micrometer scale.

Values of contact angle hysteresis, an indicator of surface quality, were measured in many of these experiments. While some homogeneous surfaces of Drelich et al.

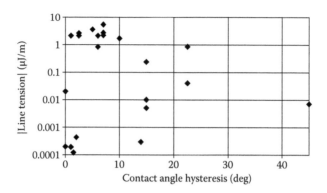

FIGURE 13.8   No correlation is seen between the magnitude of measured line tension and the contact angle hysteresis in literature data (From Checco, A., Guenoun, P., and Daillant, J., *Physical Review Letters*, 91, 186101, 2003; Chen, P., Susnar, S. S., Mak, C., Amirfazli, A., and Neumann, A. W., *Colloids and Surfaces A*, 45, 129–130, 1997; Wallace J. A., and Schürch. S., *Journal of Colloid and Interface Science,* 124, 452, 1988; Gaydos J., and Neumann. A. W., *Journal of Colloid and Interface Science* 120, 76, 1987; Li D., and Neumann. A. W., *Colloids and Surfaces* 43, 195, 1990; Duncan, D., Li, D., Gaydos, J., and Neumann. A. W., *Journal of Colloid and Interface Science* 169, 256, 1995; Amirfazli, A., Kwok, D. Y., Gaydos, J., and Neumann. A. W., *Journal of Colloid and Interface Science* 205, 1, 1998; Gu, Y., Li, D., and Cheng. P., *Journal of Colloid and Interface Science* 180, 212, 1996; Gu. Y., *Colloids and Surfaces A* 181, 215, 2001; Drelich, J., Miller, J. D., Kumar, A., and Whitesides. G. M., *Colloids and Surfaces A* 93, 1, 1994; Drelich, J., Wilbur, J. L., Miller, J. D., and Whitesides. G. M., *Langmuir* 12, 1913, 1996; Wang, J. Y., Betelu, S., and Law. B. M., *Physical Review E* 63, 031601, 2001; Pompe T., and Herminghaus. S., *Physical Review Letters* 85, 1930, 2000; Yang, J., Duan, J., Fornasiero, D., and Ralston. J., *Journal of Physical Chemistry B* 107, 6139, 2003; Amirfazli, A., Hänig, S., Müller, A., and Neumann. A. W., *Langmuir* 16, 2024, 2000; Checco, A., Schollmeyer, H., Daillant, J., Guenoun, P., and Boukherroub. R., *Langmuir* 22, 116, 2006). Each point represents a different solid or liquid substrate.

[70] had ~45° hysteresis and line tension of 10^{-8} J/m at the millimeter scale, a variety of surfaces have been produced by others with ~5° hysteresis and line tension of 10^{-6} J/m at the millimeter scale [54–56]. It is also notable that line tensions of order 10^{-6} J/m [45–47] have been measured on liquid surfaces with virtually zero hysteresis.

Thus, while there is convincing evidence that contact line corrugations on low quality (or intentionally rough or heterogeneous) solid surfaces can produce a large pseudo-line tension, similar measurements on high quality surfaces (and liquid surfaces) remain unexplained. Both large and relatively small values of line tension have been measured on high quality surfaces, and the line tension appears to vary independently of the contact angle hysteresis (Figure 13.8).

Finally, in nanoscale experiments, Checco et al. [102] have measured line tension of order 10^{-10} J/m, larger than the theoretical prediction for their system of 10^{-12} J/m, and have likewise attributed the difference to solid surface heterogeneity. They postulated that their drops, which were formed by condensation, had a tendency to nucleate on high energy sites, causing lower contact angles for the smaller drops. For drops of this size (10^{-6} m), even high quality solid surfaces may be heterogeneous at or near the drop length scale.

13.4.2 Sign of Line Tension

As mentioned earlier, there is no theoretical restriction on the sign of line tension. Theoretical predictions of a sign change for line tension near wetting have been discussed above (Section 13.2), as well as qualitatively similar experimental findings at the microscale (Section 13.3). Macroscale experiments have mostly produced positive line tensions, as seen in Tables 13.2 through 13.4. Of course, in light of the discussion in the preceding sections, the measured values at the microscale and the macroscale may or may not be the same physical quantity. As little as is understood about the magnitude of line tension, the sign of line tension has often taken a back seat to it and is thus even less clear.

It may be questioned whether a negative line tension would result in corrugation of a contact line becoming energetically favorable. For a sessile drop, the energy gained from an increased length of line phase would be offset by the energy cost of creating more liquid–vapor surface. Starting with the assumption that the equilibrium shape of a sessile drop three-phase line is in fact circular, a thermodynamic argument has been made that line tension must therefore be positive [102]. However, thermodynamic arguments do not apply below a certain length scale, and the most recent study of this issue that included consideration of the interface potential has shown that negative line tensions would not result in contact line corrugation, and are therefore plausible [103].

13.4.3 Conclusions

The subject of line tension and the drop size dependence of contact angles remains murky, with many questions still open. The theoretical, thermodynamic concept of line tension is well-defined and has a logical place in the theory of capillarity [104], but uncertainties appear as soon as it is applied to real systems.

Surface tensions of dispersive liquids are well understood in terms of van der Waals forces between molecules [105]. (For nondispersive liquids, the understanding is more qualitative.) Modeling line tension in a similar way inevitably results in a line tension magnitude approximately equal to a surface tension times an interfacial width; that is, about 10^{-11} J/m. This line tension cannot be the cause of any significant drop size dependence of contact angles above the micrometer scale.

Based on the large amount of data in the literature, and consistent results between different experimental methods, it can nevertheless be concluded that the large measured drop size dependence is a real effect. The wide variety of systems examined in earlier measurements at the millimeter scale suggested a universality of the effect, at least for low energy surfaces. However, the increasing number of measurements of smaller line tensions at this scale [64,67–69] now suggests a less fundamental source for the phenomenon. Efforts to identify this source as surface imperfections are unconvincing for high quality (or liquid) substrates, and yet no better alternative explanation is currently available. Even on intentionally rough and heterogeneous substrates, the drop size dependence of contact angles has never been quantitatively explained.

We are left with the possibility that three intrinsically different phenomena may induce corrections to the classical Young equation: a three-phase line excess energy, analogous to surface tension, called line tension; an irregular effect produced by grossly rough and heterogeneous substrates, called pseudo-line tension; and a more subtle effect present even in liquid–liquid systems that has no name.

REFERENCES

1. T. Baumgart, S. T. Hess, and W. W. Webb. *Nature* 425 (2003): 821.
2. B. Widom and A. S. Clarke. *Physica A* 168 (1990): 149.
3. A. Amirfazli and A. W. Neumann. *Advances in Colloid and Interface Science* 110 (2004): 121.
4. S. K. Sia and G. M. Whitesides. *Electrophoresis* 24 (2003): 3563.
5. A. A. Darhuber and S. M. Troian. *Annual Review of Fluid Mechanics* 37 (2005): 425.
6. E. Mendez-Villuendas and R. K. Bowles. *Physical Review Letters* 98 (2007): 185503.
7. V. Sartre, M. C. Zaghdoudi, and M. Lallemand. *International Journal of Thermal Sciences* 39 (2000): 498.
8. S. J. Gokhale, J. L. Plawsky, P. C. Wayner, Jr., and S. DasGupta 16 (2004): 1942.
9. S. Vemuri and K. J. Kim. *International Journal of Heat and Mass Transfer* 49 (2006): 649.
10. A. V. Nguyen, H. J. Schulze, and J. Ralston. *International Journal of Mineral Processing* 51 (1997): 183.
11. S. Schürch, M. Geiser, M. M. Lee, and P. Gehr. *Colloids and Surfaces B* 15 (1999): 339.
12. R. Aveyard, J. H. Clint, and T. S. Horozov. *Physical Chemistry Chemical Physics* 5 (2003): 2398.
13. P. R. Garrett, S. P. Wicks, and E. Fowles. *Colloids and Surfaces A* 282–283 (2006): 307.
14. D. C. Ellwood and B. A. Pethica. In *Microbial Adhesion to Surfaces*. 549. Ellis Horwood, Chichester, 1980.
15. K. L. Johnson, K. Kendall, and A. D. Roberts. *Proceedings of the Royal Society of London A* 324 (1971): 301.
16. R. J. Good and M. N. Koo. *Journal of Colloid and Interface Science* 71 (1979): 283.
17. W. Kohn, A. D. Becke, and R. G. Parr. *Journal of Physical Chemistry* 100 (1996): 12974.
18. J. Kerins and B. Widom. *Journal of Chemical Physics* 77 (1982): 2061.
19. I. Szleifer and B. Widom. *Molecular Physics* 75 (1992): 925.
20. K. Koga and B. Widom. *Journal of Chemical Physics* 127 (2007): 064704.
21. S. Perković, E. M. Blokhuis, and G. Han. *Journal of Chemical Physics* 102 (1995): 400.
22. C. Bauer and S. Dietrich. *European Physical Journal B* 10 (1999): 767.
23. P. Tarazona and G. Navascués. *Journal of Chemical Physics* 75 (1981): 3114.
24. W. Qu and D. Li. *Colloids and Surfaces A* 156 (1999): 123.
25. A. Marmur. *Journal of Colloid and Interface Science* 186 (1997): 462.
26. J. A. de Feijter and A. Vrij. *Journal of Electroanalytical Chemistry* 37 (1972): 9.
27. N. V. Churaev, V. M. Starov, and B. V. Derjaguin. *Journal of Colloid and Interface Science* 89 (1982): 16.

28. J. O. Indekeu. *International Journal of Modern Physics B* 8 (1994): 309.
29. B. V. Derjaguin, N. V. Churaev, and V. M. Muller. *Surface Forces.* Consultants Bureau, New York, 1987.
30. P. G. de Gennes. *Reviews of Modern Physics* 57 (1985): 827.
31. J. O. Indekeu. *Physica A* 183 (1992): 439.
32. Y. Solomentsev and L. R. White. *Journal of Colloid and Interface Science* 218 (1999): 122.
33. H. Dobbs. *Langmuir* 15 (1999): 2586.
34. A. Checco, P. Guenoun, and J. Daillant. *Physical Review Letters* 91 (2003): 186101.
35. F. Bresme and N. Quirke. *Physical Review Letters* 80 (1998): 3791.
36. F. Bresme and N. Quirke. *Physical Chemistry Chemical Physics* 1 (1999): 2149.
37. T. Werder, J. H. Walther, R. L. Jaffe, T. Halicioglu, and P. Koumoutsakos. *Journal of Physical Chemistry B* 107 (2003): 1345.
38. J. T. Hirvi and T. A. Pakkanen. *Journal of Chemical Physics* 125 (2006): 144712.
39. M. Schneemilch and N. Quirke. *Journal of Chemical Physics* 127 (2007): 114701.
40. S. Torza and S. G. Mason. *Kolloid-Z. und Z. Polymere* 246 (1971): 593.
41. D. Platikanov, M. Nedyalkov, and A. Scheludko. *Journal of Colloid and Interface Science* 75 (1980): 612.
42. D. Platikanov, M. Nedyalkov, and V. Nasteva. *Journal of Colloid and Interface Science* 75 (1980): 620.
43. P. A. Kralchevsky, A. D. Nikolov, and I. B. Ivanov. *Journal of Colloid and Interface Science* 112 (1986): 132.
44. P. Chen, J. Gaydos, and A. W. Neumann. *Langmuir* 12 (1996): 5956.
45. P. Chen, S. S. Susnar, A. Amirfazli, C. Mak, and A. W. Neumann. *Langmuir* 13 (1997): 3035.
46. P. Chen. *The Quadrilateral Relation and Line Tension Measurements in Liquid–Liquid–Fluid Systems.* Ph.D. Thesis, University of Toronto, 1998.
47. R. David, S. M. Dobson, Z. Tavassoli, M. G. Cabezas, and A. W. Neumann. *Colloids and Surfaces A* 333 (2009): 12.
48. A. Dussaud and M. Vignes-Adler. *Langmuir* 13 (1997): 581.
49. R. Aveyard, J. H. Clint, D. Nees, and V. Paunov. *Colloids and Surfaces A* 146 (1999): 95.
50. Y. Takata, H. Matsubara, Y. Kikuchi, N. Ikeda, T. Matsuda, T. Takiue, and M. Aratono. *Langmuir* 21 (2005): 8594.
51. J. A. Wallace and S. Schürch. *Journal of Colloid and Interface Science* 124 (1988): 452.
52. J. A. Wallace and S. Schürch. *Colloids and Surfaces* 43 (1990): 207.
53. K. W. Stöckelhuber, B. Radoev, and H. J. Schulze. *Colloids and Surfaces A* 156 (1999): 323.
54. J. Gaydos and A. W. Neumann. *Journal of Colloid and Interface Science* 120 (1987): 76.
55. D. Li and A. W. Neumann. *Colloids and Surfaces* 43 (1990): 195.
56. D. Duncan, D. Li, J. Gaydos, and A. W. Neumann. *Journal of Colloid and Interface Science* 169 (1995): 256.
57. A. Amirfazli, D. Y. Kwok, J. Gaydos, and A. W. Neumann. *Journal of Colloid and Interface Science* 205 (1998): 1.
58. A. Amirfazli, D. Chatain, and A. W. Neumann. *Colloids and Surfaces A* 142 (1998): 183.
59. A. Amirfazli, A. Keshavarz, L. Zhang, and A. W. Neumann. *Journal of Colloid and Interface Science* 265 (2003): 152.
60. Y. Gu, D. Li, and P. Cheng. *Journal of Colloid and Interface Science* 180 (1996): 212.
61. W. C. Jensen and D. Li. *Colloids and Surfaces A* 156 (1999): 519.
62. S. O. Asekomhe and J. A. W. Elliott. *Colloids and Surfaces A* 220 (2003): 271.
63. Y. Gu. *Colloids and Surfaces A* 181 (2001): 215.
64. R. David, M. K. Park, A. Kalantarian, and A. W. Neumann. *Colloid and Polymer Science* 287 (2009): 1167.
65. J. Drelich, J. D. Miller, and J. Hupka. *Journal of Colloid and Interface Science* 155 (1993): 379.

66. J. Drelich, J. D. Miller, A. Kumar, and G. M. Whitesides. *Colloids and Surfaces A* 93 (1994): 1.
67. J. Drelich and J. D. Miller. *Journal of Colloid and Interface Science* 164 (1994): 252.
68. J. Drelich, J. D. Miller, and R. J. Good. *Journal of Colloid and Interface Science* 179 (1996): 37.
69. J. Drelich, J. L. Wilbur, J. D. Miller, and G. M. Whitesides. *Langmuir* 12 (1996): 1913.
70. M. Sundberg, A. Månsson, and S. Tågerud. *Journal of Colloid and Interface Science* 313 (2007): 454.
71. J. Y. Wang, S. Betelu, and B. M. Law. *Physical Review E* 63 (2001): 031601.
72. D. Aronov, G. Rosenman, and Z. Barkay. *Journal of Applied Physics* 101 (2007): 084901.
73. T. Pompe and S. Herminghaus. *Physical Review Letters* 85 (2000): 1930.
74. T. Pompe. *Physical Review Letters* 89 (2002): 076102.
75. F. Mugele, T. Becker, R. Nikopoulos, M. Kohonen, and S. Herminghaus. *Journal of Adhesion Science and Technology* 16 (2002): 951.
76. R. Seemann, K. Jacobs, and R. Blossey. *Journal of Physics: Condensed Matter* 13 (2001): 4915.
77. F. Rieutord and M. Salmeron. *Journal of Physical Chemistry B* 102 (1998): 3941.
78. L. Xu and M. Salmeron. *Journal of Physical Chemistry B* 102 (1998): 7210.
79. J. Yang, J. Duan, D. Fornasiero, and J. Ralston. *Journal of Physical Chemistry B* 107 (2003): 6139.
80. X. H. Zhang, N. Maeda, and V. S. J. Craig. *Langmuir* 22 (2006): 5025.
81. X. H. Zhang and W. Ducker. *Langmuir* 24 (2008): 110.
82. A. V. Nguyen, H. Stechemesser, G. Zobel, and H. J. Schulze. *Journal of Colloid and Interface Science* 187 (1997): 547.
83. H. Stechemesser and A. V. Nguyen. *Colloids and Surfaces A* 142 (1998): 257.
84. R. Aveyard, B. D. Beake, and J. H. Clint. *Journal of the Chemical Society-Faraday Transactions* 92 (1996): 4271.
85. H. Vinke, G. Bierman, P. J. Hamersma, and J. M. H. Fortuin. *Chemical Engineering Science* 46 (1991): 2497.
86. A. D. Alexandrov, B. V. Toshev, and A. D. Scheludko. *Colloids and Surfaces A* 79 (1993): 43.
87. A. I. Hienola, P. M. Winkler, P. E. Wagner, H. Vehkamäki, A. Lauri, I. Napari, and M. Kulmala. *Journal of Chemical Physics* 126 (2007): 094705.
88. O. Teschke and E. F. de Souza. *Chemical Physics Letters* 447 (2007): 379.
89. L. Boruvka and A. W. Neumann. *Journal of Colloid and Interface Science* 65 (1978): 315.
90. L. Boruvka, J. Gaydos, and A. W. Neumann. *Colloids and Surfaces* 43 (1990): 307.
91. J. Gaydos and A. W. Neumann. *Advances in Colloid and Interface Science* 49 (1994): 197.
92. M. Hoorfar, A. Amirfazli, J. A. Gaydos, and A. W. Neumann. *Advances in Colloid and Interface Science* 114–115 (2005): 103.
93. J. Buehrle, S. Herminghaus, and F. Mugele. *Langmuir* 18 (2002): 9771.
94. J. Gaydos. *Implications of the Generalized Theory of Capillarity*. PhD Thesis, University of Toronto, 1992.
95. R. David, S. Meyyappan, A. Amirfazli, and A. W. Neumann. Unpublished data, University of Toronto, 2007.
96. R. David and A. W. Neumann. *Langmuir* 23 (2007): 11999.
97. J. Drelich and J. D. Miller. *Particulate Science and Technology* 10 (1992): 1.
98. A. Amirfazli, S. Hänig, A. Müller, and A. W. Neumann. *Langmuir* 16 (2000): 2024.
99. D. Li, F. Y. H. Lin, and A. W. Neumann. *Journal of Colloid and Interface Science* 142 (1991): 224.

100. J. Drelich, W.-H. Jang, and J. D. Miller. *Langmuir* 13 (1997): 1345.
101. A. Checco, H. Schollmeyer, J. Daillant, P. Guenoun, and R. Boukherroub. *Langmuir* 22 (2006): 116.
102. D. Li and D. J. Steigmann. *Colloids and Surfaces A* 116 (1996): 25.
103. S. Mechkov, G. Oshanin, M. Rauscher, M. Brinkmann, A. M. Cazabat, and S. Dietrich. *Europhysics Letters* 80 (2007): 66002.
104. L. Boruvka and A. W. Neumann. *Journal of Chemical Physics* 66 (1977): 5464.
105. J. N. Israelachvili. *Intermolecular and Surface Forces*. Academic Press, London, 1992.

Index